Human-Induced Climate Change
An Interdisciplinary Assessment

Bringing together many of the world's leading experts, this volume is a comprehensive, state-of-the-art review of climate change science, impacts, mitigation, adaptation, and policy. It provides an integrated assessment of research on the key topics that underlie current controversial policy questions.

The first part of the book addresses recent topics and findings related to the physical–biological earth system, including air pollution–climate interactions, climate interactions with the carbon cycle, and quantitative probability estimates of climate sensitivity and change. The next part of the book surveys estimates of the impacts of climate change for different sectors and regions, describes recent studies for individual sectors, and examines how this research might be used in the policy process. The third part examines current topics related to mitigation of greenhouse gases and explores the potential roles of various technological options that would limit greenhouse-gas emissions and enhance terrestrial carbon sinks. The last part focuses on policy design under uncertainty.

Dealing with the scientific, economic, and policy questions at the forefront of the climate change issue, this book will be invaluable for graduate students, researchers, and policymakers interested in all aspects of climate change and the issues that surround it.

Michael E. Schlesinger is Professor of Atmospheric Sciences in the Department of Atmospheric Sciences at the University of Illinois, Urbana-Champaign.

Haroon S. Kheshgi is Program Leader for Global Climate Change Science at ExxonMobil's Corporate Strategic Research.

Joel Smith is a Vice President with Stratus Consulting Inc. in Boulder, Colorado.

Francisco C. de la Chesnaye is Chief of the Climate Economics Branch for the United States Environmental Protection Agency.

John M. Reilly is Associate Director for Research in the Joint Program on the Science and Policy of Global Change, and a Senior Research Scientist in the Laboratory for Energy and Environment at the Massachusetts Institute of Technology.

Tom Wilson is a Technical Executive in the Climate Change Research Program at the Electric Power Research Institute (EPRI).

Charles Kolstad is a Professor in the Department of Economics at the Bren School of Environmental Science and Management, University of California, Santa Barbara.

Human-Induced Climate Change
An Interdisciplinary Assessment

Edited by

Michael E. Schlesinger
University of Illinois

Haroon S. Kheshgi
ExxonMobil Research and Engineering

Joel Smith
Stratus Consulting Inc.

Francisco C. de la Chesnaye
US Environmental Protection Agency

John M. Reilly
Massachusetts Institute of Technology

Tom Wilson
Electric Power Research Institute

Charles Kolstad
University of California, Santa Barbara

CAMBRIDGE UNIVERSITY PRESS
Cambridge, New York, Melbourne, Madrid, Cape Town, Singapore, São Paulo

Cambridge University Press
The Edinburgh Building, Cambridge CB2 8RU, UK

Published in the United States of America by Cambridge University Press, New York

www.cambridge.org
Information on this title: www.cambridge.org/9780521866033

© Cambridge University Press 2007

This publication is in copyright. Subject to statutory exception
and to the provisions of relevant collective licensing agreements,
no reproduction of any part may take place without
the written permission of Cambridge University Press.

First published 2007

Printed in the United Kingdom at the University Press, Cambridge

A catalog record for this publication is available from the British Library

ISBN 978-0-521-86603-3 hardback

Cambridge University Press has no responsibility for the persistence or
accuracy of URLs for external or third-party internet websites referred to in
this publication, and does not guarantee that any content on such
websites is, or will remain, accurate or appropriate.

Dedication of this book to Alan Manne

During the course of writing this book, the climate community lost one of its greatest treasures. On September 27, 2005 Alan Manne died while doing what he enjoyed most, horseback riding – an activity that he took up when he was 62 and passionately pursued for nearly two decades.

Alan was a gifted modeler who consistently produced ingenious solutions to important problems in both the private and public sectors. When asked to describe the goals of Operations Research, a discipline which he helped to found, he noted, "It is simple; it's doing the best you can with what you have." Adhering to this straightforward but powerful philosophy, he contributed enormously to our understanding of the challenges posed by global climate change, first in deepening our understanding of what would be required to slow the rate of global warming, and then in providing an elegant framework for thinking about benefits and costs.

Born in New York City on May 1, 1925, Alan received a bachelor's degree in economics from Harvard College at the age of 18. After serving in the Navy during the Second World War, he returned to Harvard, where he earned a doctorate in economics in 1950. He stayed on there as an instructor before accepting a position at the Rand Corporation as an economic analyst from 1952 to 1956. He subsequently served as a professor at Harvard, Yale and Stanford Universities, with a brief hiatus as an economist at the International Institute for Applied Systems Analysis in Vienna, Austria. He retired formally from teaching in 1992 when he became professor emeritus in Stanford's Department of Operations Research.

Over four decades, Alan applied economic models internationally, aiding India in developing its fertilizer industry, and developing industrial-resource and planning models for Mexico and Turkey. He also explored the economic consequences surrounding alternative fuels and energy conservation. More recently, he shifted his focus to addressing the risks posed by global climate change.

During his career he received many honors, including being named a fellow of the Econometric Society and the American Academy of Arts and Sciences, Member of the National Academy of Engineering, and recipient of the Larnder Memorial Prize of the Canadian Operations Research Society, Lancaster Prize of the Operations Research Society of America, and the Paul Frankel Award from the US Association of Energy Economists. He also received honorary degrees from Gutenberg University and from the University of Geneva.

Alan's contributions to our understanding of the impacts of human activity on climate extend well beyond his personal research. Over the years, he taught and mentored a generation of students who share his values and high standards. These include those who had the honor to learn from Alan directly in the classroom and through personal collaborations, and the broader community who benefited indirectly through meetings such as the Climate Change Impacts/Integrated Assessment meetings held annually in Snowmass, Colorado, for the past 11 years. As such, it is fitting that this volume be dedicated to Alan, and it is comforting that his work lives on through a new generation of researchers.

No tribute to Alan would be complete without acknowledging the others in the Manne family who provided the warm, gracious, and welcoming environment for so many of Alan's interactions with his colleagues, students, and friends. These include his dear wife Jacqueline, and three children, of whom he was so proud: Edward, Henry, and Elizabeth. His passion and devotion to his family – and theirs to him – were clear to all.

Richard Richels
January 8, 2006

Contents

List of contributors	page x
Preface	
FRANCISCO C. DE LA CHESNAYE, JAE EDMONDS, HAROON S. KHESHGI, CHARLES KOLSTAD, JOHN M. REILLY, RICHARD RICHELS, MICHAEL E. SCHLESINGER, JOEL SMITH AND TOM WILSON	xvii

Part I Climate system science

HAROON S. KHESHGI, COORDINATING EDITOR
Introduction — 1

1 The concept of climate sensitivity: history and development
NATALIA ANDRONOVA, MICHAEL E. SCHLESINGER, SURAJE DESSAI, MIKE HULME AND BIN LI — 5

2 Effect of black carbon on mid-troposphere and surface temperature trends
JOYCE E. PENNER, MINGHUAI WANG, AKSHAY KUMAR, LEON ROTSTAYN AND BEN SANTER — 18

3 Evaluating the impacts of carbonaceous aerosols on clouds and climate
SURABI MENON AND ANTHONY D. DEL GENIO — 34

4 Probabilistic estimates of climate change: methods, assumptions and examples
HAROON S. KHESHGI — 49

5 The potential response of historical terrestrial carbon storage to changes in land use, atmospheric CO_2, and climate
ATUL K. JAIN — 62

6 The albedo climate impacts of biomass and carbon plantations compared with the CO_2 impact
M. SCHAEFFER, B. EICKHOUT, M. HOOGWIJK, B. STRENGERS, D. VAN VUUREN, R. LEEMANS AND T. OPSTEEGH — 72

7 Overshoot pathways to CO_2 stabilization in a multi-gas context
T. M. L. WIGLEY, R. G. RICHELS AND J. A. EDMONDS — 84

8 Effects of air pollution control on climate: results from an integrated global system model
RONALD PRINN, JOHN M. REILLY, MARCUS SAROFIM, CHIEN WANG AND BENJAMIN FELZER — 93

Part II Impacts and adaptation

JOEL SMITH, COORDINATING EDITOR
Introduction — 103

9 Dynamic forecasts of the sectoral impacts of climate change
ROBERT MENDELSOHN AND LARRY WILLIAMS — 107

10 Assessing impacts and responses to global-mean sea-level rise
ROBERT J. NICHOLLS, RICHARD S. J. TOL AND JIM W. HALL — 119

11 Developments in health models for integrated assessments
KRISTIE L. EBI AND R. SARI KOVATS — 135

12	The impact of climate change on tourism and recreation JACQUELINE M. HAMILTON AND RICHARD S. J. TOL	147
13	Using adaptive capacity to gain access to the decision-intensive ministries GARY W. YOHE	156
14	The impacts of climate change on Africa ROBERT MENDELSOHN	161

Part III   Mitigation of greenhouse gases

	JOHN M. REILLY AND FRANCISCO C. DE LA CHESNAYE, COORDINATING EDITORS Introduction	167
15	Bottom-up modeling of energy and greenhouse gas emissions: approaches, results, and challenges to inclusion of end-use technologies JAYANT A. SATHAYE	171
16	Technology in an integrated assessment model: the potential regional deployment of carbon capture and storage in the context of global CO_2 stabilization J. A. EDMONDS, J. J. DOOLEY, S. K. KIM, S. J. FRIEDMAN AND M. A. WISE	181
17	Hydrogen for light-duty vehicles: opportunities and barriers in the United States JAMES L. SWEENEY	198
18	The role of expectations in modeling costs of climate change policies PAUL M. BERNSTEIN, ROBERT L. EARLE AND W. DAVID MONTGOMERY	216
19	A sensitivity analysis of forest carbon sequestration BRENT SOHNGEN AND ROBERT MENDELSOHN	227
20	Insights from EMF-associated agricultural and forestry greenhouse gas mitigation studies BRUCE A. MCCARL, BRIAN C. MURRAY, MAN-KEUN KIM, HENG-CHI LEE, RONALD D. SANDS AND UWE A. SCHNEIDER	238
21	Global agricultural land-use data for integrated assessment modeling NAVIN RAMANKUTTY, TOM HERTEL, HUEY-LIN LEE AND STEVEN K. ROSE	252
22	Past, present, and future of non-CO_2 gas mitigation analysis FRANCISCO C. DE LA CHESNAYE, CASEY DELHOTAL, BENJAMIN DEANGELO, DEBORAH OTTINGER-SCHAEFER AND DAVE GODWIN	266
23	How (and why) do climate policy costs differ among countries? SERGEY PALTSEV, JOHN M. REILLY, HENRY D. JACOBY AND KOK HOU TAY	282
24	Lessons for mitigation from the foundations of monetary policy in the United States GARY W. YOHE	294

Part IV   Policy design and decisionmaking under uncertainty

	CHARLES KOLSTAD AND TOM WILSON, COORDINATING EDITORS Introduction	303
25	Climate policy design under uncertainty WILLIAM PIZER	305
26	Climate policy assessment using the Asia–Pacific Integrated Model MIKIKO KAINUMA, YUZURU MATSUOKA, TOSHIHIKO MASUI, KIYOSHI TAKAHASHI, JUNICHI FIJINO AND YASUAKI HIJIOKA	314
27	Price, quantity, and technology strategies for climate change policy W. DAVID MONTGOMERY AND ANNE E. SMITH	328

Contents

28 What is the economic value of information about climate thresholds?
KLAUS KELLER, SEUNG-RAE KIM, JOHANNA BAEHR, DAVID F. BRADFORD AND
MICHAEL OPPENHEIMER   343

29 Boiled frogs and path dependency in climate policy decisions
MORT WEBSTER   355

30 Article 2 and long-term climate stabilization: methods and models for decisionmaking under uncertainty
FERENC L. TOTH   365

31 Whither integrated assessment? Reflections from the leading edge
HUGH M. PITCHER, GERALD M. STOKES AND ELIZABETH L. MALONE   377

32 Moving beyond concentrations: the challenge of limiting temperature change
RICHARD G. RICHELS, ALAN S. MANNE AND TOM M. L. WIGLEY   387

33 International climate policy: approaches to policies and measures, and international coordination and cooperation
BRIAN S. FISHER, A. L. MATYSEK, M. A. FORD AND K. WOFFENDEN   403

Colour plates appear between pages 302 and 303
Index   *414–426*

Contributors

Natalia Andronova
Department of Atmospheric
Oceanic and Space Science
University of Michigan
1541D Space Research Building
2455 Hayward Street Ann Arbor
MI 48109–2143
USA

Paul Bernstein
5265 Lawelawe Pl.
Honolulu
HI 96821
USA

David F. Bradford
(deceased)

Francisco C. de la Chesnaye
Climate Change Division
Office of Atmospheric Programs
US Environmental Protection Agency
1200 Pennsylvania Ave.
NW (6207J) Washington
DC 20460
USA

Benjamin J. DeAngelo
Climate Change Division
Office of Atmospheric Programs
US Environmental Protection Agency
1200 Pennsylvania Ave.
NW (6207J)
Washington
DC 20460
USA

Anthony Del Genio
NASA Goddard Institute for Space Studies
2880 Broadway New York
NY 10025
USA

Casey Delhotal
Climate Change Division
Office of Atmospheric Programs
US Environmental Protection Agency
1200 Pennsylvania Ave.; NW (6207J)
Washington
DC 20460
USA

Suraje Dessai
Tyndall Centre for Climate Change Research
School of Environmental Sciences
University of East Anglia
Norwich NR4 7TJ
UK

James J. Dooley
Pacific Northwest National Laboratory
Joint Global Change Research Institute at the University of Maryland
8400 Baltimore Avenue
Suite 201 College Park
MD 20740–2496
USA

Robert Earle
2125 E. Orange Grove Blvd. Pasadena
CA 91104
USA

Kristie L. Ebi
Exponent Health Group 1800 Diagonal Road
Suite 300 Alexandria
VA 22314
USA

Jae Edmonds
Pacific Northwest National Laboratory
Joint Global Change Research Institute at the University of Maryland
8400 Baltimore Avenue
Suite 201 College Park

MD 20740–2496
USA

Bas Eickhout
Global Sustainability and Climate Team
Netherlands Environmental Assessment Agency (MNP)
PO Box 303
3720 AH Bilthoven A. van Leeuwenhoeklaan 9
Bilthoven
The Netherlands

Benjamin Felzer
Ecosystems Center at the Marine Biological Laboratory
Starr Building
MBL Street
Woods Hole
MA 02543
USA

Brian S. Fisher
ABARE
Edmund Barton Building
Macquarie Street
Barton
ACT 2600 GOP Box 1563
Canberra
ACT 2600, Australia

M. A. Ford
ABARE
Edmund Barton Building
Macquarie Street
Barton
Act 2600 GOP Box 1563
Canberra
ACT 2600, Australia

Julio Friedmann
Lawrence Livermore National Laboratory
7000 East Ave Livermore
CA 94550–9234
USA

Junichi Fujino
National Institute for Environmental Studies
16–2 Onogawa
Tsukuba 305–8506
Japan

Dave Godwin
Stratospheric Protection Division
Office of Atmospheric Programs
US Environmental Protection Agency

1200 Pennsylvania Ave.
NW (6205J) Washington
DC 20460 USA

Jim W. Hall
School of Civil Engineering and Geosciences
University of Newcastle-upon-Tyne Newcastle
NE1 7RU
UK

Jacqueline M. Hamilton
Centre for Marine and Climate Research
Hamburg University
Leuschnerstr. 91
21031 Hamburg
Germany

Tom Hertel
Center for Global Trade Analysis
Purdue University
1145 Krannert Building 403 West State Street
West Lafayette
IN 47907–1145
USA

Monique Hoogwijk
Ecofys B.V.
PO Box 8408
NL-3503 RK
Utrecht
The Netherlands

Mike Hulme
Climatic Research Unit
University of East Anglia
Norwich
NR4 7TJ UK

Henry Jacoby
Joint Program on the Science and Policy of Global Change
Massachusetts Institute of Technology
E40–439 77 Massachusetts Ave.
Cambridge
MA 02139
USA

Atul K. Jain
Department of Atmospheric Sciences
University of Illinois at Urbana-Champaign
105 S. Gregory Street
Urbana
IL 61801
USA

Mikiko Kainuma
Social and Environmental Systems Division
National Institute for Environmental Studies
16–2 Onogawa
Tsukuba 305–8506
Japan

Klaus Keller
208 Deike Building
Department of Geosciences
The Pennsylvania State University
University Park
PA 16802
USA

Haroon S. Kheshgi
ExxonMobil Research and Engineering Company
Route 22 East Annandale
NJ 08801
USA

Seung-Rae Kim
Woodrow Wilson School and Department of Economics
Princeton University
Princeton
NJ 08544
USA

Charles Kolstad
Department of Economics
University of California
Santa Barbara
CA 93106–9210
USA

R. Sari Kovats
Centre on Global Change and Health
London School of Hygiene and Tropical Medicine
Keppel Street
London WC1E 7HT
UK

Akshay S. Kumar
Bowie New Town Center 4201 Northview Drive
Suite 404 Bowie
MD 20716
USA

Huey-Lin Lee
Center for Global Trade Analysis
Purdue University
1145 Krannert Building
403 West State Street
West Lafayette
IN 47907–1145
USA

Rik Leemans
Environmental Systems Analysis Group
Wageningen University and Research (WUR)
PO Box 47
6700 AA Wageningen
The Netherlands

Bin Li
Department of Atmospheric Science
University of Illinois at Urbana-Champaign
105 S. Gregory St.
Urbana
IL 61801
USA

Yun Li
Woodrow Wilson School
Robertson Hall
Princeton University
Princeton
NJ 08544
USA

Alan S. Manne
(deceased)

Toshihiko Masui
National Institute for Environmental Studies
16–2 Onogawa
Tsukuba 305–8506
Japan

Yuzuru Matsuoka
Kyoto University
Kyoto 606–8501
Japan

A. L. Matysek
ABARE
Edmund Barton Building
Macquarie Street
Barton
ACT 2600 GOP Box 1563
Canberra
ACT 2600, Australia

Bruce A. McCarl
Department of Agricultural Economics
Texas A & M University

College Station
TX 77843–2124
USA

Robert Mendelsohn
Yale FES
230 Prospect Street
New Haven
CT 06511
USA

Surabi Menon
Lawrence Berkeley National Laboratory
MS90KR109
1 Cyclotron Road Berkeley
CA 94720
USA

David Montgomery
1201 F St. NW Ste. 700
Washington
DC 20004
USA

Robert J. Nicholls
School of Civil Engineering and the Environment
University of Southampton
Southampton
SO17 1BJ
UK

Michael Oppenheimer
Department of Geosciences
Princeton University
Princeton
NJ 08544
USA

J. D. Opsteegh
KNMI
PO Box 201 De Bilt
3730 AE
The Netherlands

Deborah Ottinger-Schaefer
Climate Change Division
Office of Atmospheric Programs
US Environmental Protection Agency
1200 Pennsylvania Ave.
NW (6207J) Washington
DC 20460
USA

Sergey Paltsev
Joint Program on the Science and Policy of Global Change
Massachusetts Institute of Technology
E40–429
77 Massachusetts Ave. Cambridge
MA 02139
USA

Joyce Penner
1538 Space Research Building
Department of Atmospheric
Oceanic and Space Sciences
University of Michigan
2455 Hayward Street Ann Arbor
MI 48109–2143
USA

Hugh M. Pitcher
Joint Global Change Research Institute
8400 Baltimore Ave.
Suite 201 College Park
MD 20740
USA

Billy Pizer
Resources for the Future
1616 P Street
NW Washington
DC 20036
USA

Ronald Prinn
Department of Earth
Atmosphere
and Planetary Sciences
Massachusetts Institute of Technology
E54–1312 77 Massachusetts Ave.
Cambridge
MA 02139
USA

Navin Ramankutty
SAGE Nelson Institute for Environmental Studies
University of Wisconsin
1710 University Avenue
Madison
WI 53726
USA

John M. Reilly
Joint Program on the Science and Policy of Global Change
Massachusetts Institute of Technology
E40–433
77 Massachusetts Ave. Cambridge
MA 02139
USA

Richard Richels
Electric Power Research Institute
2000 L Street NW
Suite 805 Washington
DC 20036
USA

Steven K. Rose
Climate Change Division
Office of Atmospheric Programs
US Environmental Protection Agency
1200 Pennsylvania Avenue, NW (6207J)
Washington
DC 20460
USA

Leon D. Rotstayn
CSIRO Marine and Atmospheric Research
Aspendale
Vic. 3195 Australia

Benjamin J. Santer
Lawrence Livermore National Laboratory
7000 East Avenue Livermore
CA 94550
USA

Marcus Sarofim
Joint Program on the Science and Policy of Global Change
Massachusetts Institute of Technology
E40–411
77 Massachusetts Ave.
Cambridge
MA 02139
USA

Jayant A. Sathaye
MS 90–4000 Lawrence Berkeley National Laboratory
1 Cyclotron Road
Berkeley
CA 94720
USA

Michiel Schaeffer
Royal Netherlands Embassy
PO Box 20061
2500EB Den Haag
The Netherlands

Michael E. Schlesinger
Department of Atmospheric Science
University of Illinois at Urbana-Champaign
105 S. Gregory St. Urbana
IL 61801
USA

Uwe Schneider
Hamburg University
Centre of Marine and Atmospheric Sciences
Research unit Sustainability and Global Change
Bundesstrasse 55
D-20146 HAMBURG
Germany

Anne E. Smith
CRA International
1201 F Street NW
Suite 700 Washington
DC 20004–1204
USA

Joel Smith
Stratus Consulting Inc.
PO Box 4059
Boulder
CO 80306–4059
USA

Brent Sohngen
AED Economics
Ohio State University
2120 Fyffe Rd. Columbus
OH 43210–1067
USA

Gerald M. Stokes
Joint Global Change Research Institute
8400 Baltimore Ave.
Suite 201
College Park
MD 20740
USA

Bart J. Strengers
Global Sustainability and Climate Team
Netherlands Environmental Assessment Agency (MNP)
PO Box 303
3720 AH Bilthoven
A. van Leeuwenhoeklaan 9 Bilthoven
The Netherlands

James L. Sweeney
Stanford University Terman Engineering Center
Room 440
380 Panama Way
Stanford
CA 94305–4026
USA

Kiyoshi Takahashi
National Institute for Environmental Studies

List of contributors

16–2 Onogawa
Tsukuba 305–8506
Japan

Kok Hou Tay
Blk 21
St. George's Road #20–174 S(321021)
Singapore

Richard S. J. Tol
Hamburg University ZMK
Bundesstrasse 55
20146 Hamburg
Germany

Ferenc Toth
International Atomic Energy Agency
PO Box 100
Wagramer Strasse 5
A-1400 Vienna
Austria

Detlef van Vuuren
Global Sustainability and Climate Team
Netherlands Environmental Assessment Agency (MNP)
PO Box 303
3720 AH Bilthoven A. van Leeuwenhoeklaan 9
Bilthoven
The Netherlands

Chien Wang
Department of Earth
Atmosphere
and Planetary Sciences
Massachusetts Institute of Technology
E40–425 77 Massachusetts Ave.
Cambridge
MA 02139
USA

Minghuai Wang
1546 Space Research Building
Department of Atmospheric
Oceanic and Space Sciences
University of Michigan
2455 Hayward Street Ann Arbor
MI 48109–2143
USA

Mort Webster
217 Abernethy Hall

CB# 3435 Department of Public Policy
The University of North Carolina at Chapel Hill
Chapel Hill
NC 27599–3435
USA

Tom Wigley
National Center for Atmospheric Research
Boulder CO 80307–3000
USA

Larry J. Williams
Global Climate Research
Electric Power Research Institute
3420 Hillview Ave.
Palo Alto
CA 94304–1395
USA

Thomas Wilson
Electric Power Research Institute
3412 Hillview Ave.
PO Box 10412
Palo Alto
CA 94303
USA

Marshall Wise
Pacific Northwest National Laboratory
Joint Global Change Research Institute at the University of Maryland
8400 Baltimore Avenue
Suite 201 College Park
MD 20740–2496
USA

K. Woffenden
ABARE
Edmund Barton Building
Macquarie Street
Barton
ACT 2600 GOP Box 1563
Canberra
ACT 2600 Australia

Gary Yohe
Wesleyan University
238 Church Street
Middletown
CT 06459
USA

Preface

This volume of peer-reviewed chapters arose from a scientific meeting – the Stanford Energy Modeling Forum on Climate Change Impacts/Integrated Assessment (CCI/IA) – that has occurred annually now for 11 years during boreal summer in Snowmass, Colorado, under the leadership and direction of John Weyant. The concept for the CCI/IA meetings was developed by Richard Richels, Jae Edmonds, and Michael Schlesinger in October 1994 at the Third Japan–US Workshop on Global Change at the East–West Center, Honolulu, Hawaii. The objectives of these CCI/IA meetings were to improve: (1) the representation of the impacts of climate change in integrated assessment (IA) models, and (2) IA modeling of the climate-change problem by bringing together disciplinary experts from relevant scientific fields. A planning meeting was held in March 1995 at Dulles Airport. The first CCI/IA meeting was held in summer 1995, and the most recent meeting took place in summer 2005. The CCI/IA meetings have been sponsored by the Electric Power Research Institute, the US Department of Energy, the US Environmental Protection Agency, the US National Oceanographic and Atmospheric Administration, the US National Science Foundation, the Australian Bureau of Agricultural Resource Economics, the ExxonMobil Corporation, the National Institute for Environmental Studies of Japan, and the European Commission.

The initial meeting in 1995 was organized under what turned out to be a rather naïve assumption that the climate-change impact-modeling community would show up and hand off a set of damage functions to the integrated assessment modelers, and then the two groups could part and continue on their independent research paths. What resulted instead was a rich, sometimes heated, interchange among the two communities. From the perspective of an integrated assessment modeler, the IA community learned much about the complexities of representing climate impacts and saw the need to learn much more.

Eleven years and eleven meetings later we are far better informed, even if the representation of impacts of climate change in our integrated assessment models remains nascent. Models are always a simplification of the real-world processes they hope to represent. Understanding better the full complexity of systems, even if it is not possible to represent all this complexity in a model, can lead to more intelligent use of models and results because one is better able to describe the caveats, limits, and possible biases and omissions. Rather than a hand-off of results, the annual meeting has become instead an intensive set of tutorials for integrated assessment modelers, bringing together top scientists and economists who describe recent advances in their fields and in so doing challenge the integrated assessment community to represent these processes in their models. The very nature of the meeting and research is interdisciplinary.

Where does disciplinary research end and integrated assessment begin? Fortunately, the answer to this question has blurred as integrated assessment modeling has matured. In representing interactions between complex systems, these models provide new scientific research results and insights that cannot be demonstrated with component models run alone. They have become more than simple assessment tools. At the same time, these new findings have stretched the Earth system modeling community to more fully integrate human systems. The organization of this volume reflects the research approach of the community, juxtaposing results and chapters from some of the more complete integrated assessment models with results from more detailed models of one or a few components of the full Earth system.

Part I of the volume, **Climate system science**, addresses new topics and findings related to the physical–biological Earth system. Several chapters examine interactions between air pollution and climate including the roles of aerosols and tropospheric ozone. Another set focus on climate interactions with the carbon cycle, including chapters on land use, the biosphere, and emissions time paths in stabilization regimes. Finally, a recurring theme throughout the volume is uncertainty, represented in this section by chapters that have tried to establish quantitative probability estimates of climate sensitivity and change.

Part II of the volume, **Impacts and adaptation**, addresses the original focus of the Snowmass meetings. Chapters range from surveys that pull together estimates for different sectors and regions to provide a comprehensive evaluation of climate damages, to new studies and reviews for individual sectors ranging from water and agriculture, ecosystems and biodiversity, health, coastal effects of sea-level rise, and effects on tourism. Also included are chapters that address how this research might be used in the policy process, whether that relates to avoiding "dangerous interference" in the climate system, the key goal of the Framework Convention on Climate Change, or the more on-the-ground problem of working with planning ministries to improve the adaptive capacity of vulnerable people and activities.

Part III, **Mitigation of greenhouse gases**, addresses current topics related to mitigation of greenhouse gases. These chapters deal with the potential roles of various technological options that would limit greenhouse gas emissions, ranging from carbon dioxide capture and storage, to hydrogen systems to improvements in energy efficiency. Chapters cover options for reducing non-CO_2 greenhouse gases, and the potential for enhancing terrestrial carbon sinks. Also within this section are chapters that address the role of expectations and technical issues that arise in evaluating the costs of mitigation measures, and that draw on lessons from monetary policy on how to set policy under uncertainty.

Finally, Part IV **Policy design and decisionmaking under uncertainty**, is devoted to issues related to the design of mitigation policy. These chapters often draw on formal methods developed to deal with decisionmaking under uncertainty, including the acquisition of information to reduce uncertainty. They consider issues of communicating risk and uncertainty as well as issues of international and domestic policy design.

The volume as a whole provides a solid overview of research on key topics that underlay currently controversial policy questions, addressing these topics using different approaches. It succeeds in addressing topics of interest to the technical modeling community, providing new results of interest to the policy community, and in offering an overview of the rapidly developing field of integrated assessment modeling. It has been our pleasure to attend nearly all of the Snowmass meetings and have the opportunity to interact with the fantastic group of scientists that assemble each summer. We hope this volume conveys much of the excitement and vitality of the community to those who have, at least so far, not been able to attend this workshop, and in so doing encourages the next generation of researchers to join us in this endeavor.

Part I

Climate system science

Haroon S. Kheshgi, coordinating editor

Introduction

The Earth sciences form core disciplines contributing to the interdisciplinary assessment of human-induced climate change. Assessment exploits understanding gained from the huge, ongoing scientific endeavor to better understand the climate system as well as from interdisciplinary research to better understand how the climate system interacts with human activities. The behavior and response of the Earth system defines the links between human activities influencing the climate system, and climate system influences on society. Papers in this section provide examples of research and review of the analysis and modeling of the Earth system, and the application of such models to provide a framework to address questions associated with the following three sections of this book: impacts and adaptation, mitigation of Greenhouse gases, and policy design and decisionmaking under uncertainty.

This section examines key issues relevant to our ability to forecast future climate, construct and test models of the Earth system for use in integrated assessment, characterize uncertainty in forecasts, and analyze illustrative cases of interactions of human activities with the climate system. While the specific topics addressed in this section are hardly comprehensive, they do give examples of how interdisciplinary studies have not only drawn from fundamental understanding generated by the Earth sciences, but have also contributed to better understanding of the research needs to address societal questions and have begun to carry out this research in conjunction with specialists. Such interaction is leading to the continued and rapid advance of integrated assessment research.

From the earliest integrated assessments of climate change that sought to balance, to costs of mitigation with the benefits of avoiding climate change, climate sensitivity was seen as a key uncertainty in such analyses. Climate sensitivity is the ratio of change in global near-surface temperature to change in climate forcing, and is influenced by, for example, the uncertain response of clouds to climate forcing. Andronova and co-authors (Chapter 1) survey estimates of climate sensitivity. While the accurate determination of climate sensitivity has withstood continued efforts by climate scientists, there are a growing number of studies that document that climate sensitivity is, indeed, highly uncertain and seek to estimate its probability distribution. In particular, recent estimates include the possibility that climate sensitivity may be high, presenting challenging questions of how to plan for such an outcome should it prove true.

Observations of past climate change provide important information for testing models of the climate system, and potentially estimating model parameters such as climate sensitivity. The apparent difference in temperature trends of records of surface temperature with some records of tropospheric temperature has been a continuing controversy in climate science, given that climate models do not produce such a difference. However, existing model results considered did not include climate change driven by carbonaceous aerosols (from, for example, biomass burning); how might their inclusion affect this controversy? Penner and co-authors (Chapter 2) find that inclusion of carbonaceous aerosols actually has the opposite effect, making the discrepancy between models and *some* records even larger.

The forcing of climate by aerosols introduces a wide range of uncertainty in estimates of climate forcing. Menon and Del Genio (Chapter 3) review the range of model-based estimates of climate forcing of aerosols with a focus on carbonaceous aerosols. While the range of results of aerosol forcing is comparable to the absolute magnitude of forcing of greenhouse gases thus far, carbonaceous aerosols are simulated to have a much different, and more profound, effect on precipitation than on surface temperature. Carbonaceous aerosols present a challenge for integrated assessment models that rely on energy balance models of climate response.

Moving from records and models of past climate change, estimates of future climate change rely on simulations of climate models with their attendant uncertainties and assumptions. Kheshgi (Chapter 4) reviews methods that are being used to generate probabilistic estimates of climate change which are conditional on assumptions, with an emphasis on using past climate records for model calibration. Addressing assumptions provides a means of improving estimates of future climate change. The future acquisition of data will further constrain estimates and could, provided assumptions can be addressed, narrow the uncertainty of climate projections. Clearly, such assumptions should be considered if using generated probabilities in analyses of decisionmaking – the question remains how?

The long-term accumulation of carbon dioxide remains central to the concern for human-induced climate change. Models of the Earth system include models of global carbon cycle. Jain (Chapter 5) tests the sensitivity of a carbon cycle model for vegetation and soils to the effects of land-use change, climate change, and rising atmospheric CO_2. Offsetting effects leave large uncertainties in each effect despite constraints on the global carbon budget. Schaeffer and co-authors (Chapter 6) consider future scenarios for bioenergy and carbon sequestration in plants and soils as means to mitigate climate change. In their analysis, land-use limitations, effects on global carbon cycle, and changes in land-cover albedo each prove important in estimating the effectiveness of such options and their contribution to scenarios of the future.

The complex question of what actions are appropriate to manage climate risk has often been informed by illustrative analyses of CO_2 stabilization: what would be required to limit the concentration of CO_2 to some, as yet, undetermined level; and what would be the effects on climate? But how should other greenhouse gases be incorporated into such analyses? Wigley and co-authors (Chapter 7) consider the effects of accounting for non-CO_2 greenhouse gases (namely CH_4 and N_2O) for different effective CO_2 (i.e. radiative forcing from well-mixed GHGs) "targets" and time trajectories of

---

Human-induced Climate Change: An Interdisciplinary Assessment, ed. Michael Schlesinger, Haroon Kheshgi, Joel Smith, Francisco de la Chesnaye, John M. Reilly, Tom Wilson and Charles Kolstad. Published by Cambridge University Press. © Cambridge University Press 2007.

concentrations. They consider least-cost trajectories using an energy economics model and find that in such trajectories CO_2 concentrations often overshoot their ultimate specified concentrations. Furthermore, non-CO_2 greenhouse gas mitigation plays important roles, with CH_4 mitigation providing a sensitive means of mitigating the modeled pace of temperature change.

Actions to mitigate GHG emissions will interact with existing policies and efforts to control air pollution. Prinn and co-authors (Chapter 8) consider how limits on emissions of air pollutants (SO_x, NO_x, CO, and volatile organic compounds, VOCs) affect modeled temperature increase via interactions in their integrated global system model. Their model contains a rich set of interactions including atmospheric chemistry that affects the concentrations of the greenhouse gases, ozone and methane, and ozone's effects on the carbon cycle of plants and soils. Their results show that, overall, air pollution controls have weak, either positive or negative, effects on modeled global temperature; but they note that additional interactions such as the effects of air pollution policy on overall demand for fossil fuels have yet to be included.

These papers give examples of varied ways that climate system science is being treated in the interdisciplinary assessment of climate change. They range from the detailed modeling and analysis of data sets and processes, to the integration of the many factors that influence climate. The depth of climate system science, and its explicit consideration of uncertainty, forms a foundation for the integrated assessment of climate change, and is driving a trend towards more complicated models – models that are more directly coupled to the current state of understanding of climate system science. The more varied, and realistic, applications of integrated assessment are requiring a more comprehensive treatment of the climate system and are broadening the focus of climate system science research. Interdisciplinary research is leading to synergies that are adding value, and are contributing to the growth of integrated assessment research.

1

The concept of climate sensitivity: history and development

Natalia Andronova, Michael E. Schlesinger, Suraje Dessai, Mike Hulme and Bin Li

1.1 Introduction

The climate sensitivity concept (CSC) has more than a century of history. It is closely related to the concept of "climate forcing" or "radiative forcing," which was fully presented and discussed by successive IPCC Assessment Reports (e.g. see Chapter 6 Houghton *et al.*, 2001). According to CSC, a change in the equilibrium global near-surface air temperature (NST) of the Earth, ΔT, due to an external disturbance of the Earth's energy balance (radiative forcing), can be linearly related to a change in the net radiation at some level in the atmosphere, ΔF. Thus,

$$\Delta T = \lambda \Delta F, \quad (1.1)$$

where λ is the climate sensitivity, which characterizes the ability of the climate system to amplify or reduce the initial temperature change initiated by the external forcing. The climate sensitivity has been estimated using Eq. (1.1) most frequently from the NST change, ΔT_{2x}, resulting from the radiative forcing due to a doubling of atmospheric carbon dioxide concentration from pre-industrial levels, ΔF_{2x}:

$$\lambda = \frac{\Delta T_{2x}}{\Delta F_{2x}}. \quad (1.2)$$

Thus ΔT_{2x} has become a surrogate for λ and has played a central role throughout the history of IPCC in interpreting the output of numerical models, in evaluating future climate changes from various scenarios, and in attributing the causes of observed temperature changes.

Between the 1960s and 1980s various types of deterministic models were used to estimate climate sensitivity, leading to a wide range of results. However, it was a mixture of modeling results and expert assessment – the Charney report of 1979 (NAS, 1979) – that established the range of 1.5 °C to 4.5 °C that was later reported in all three IPCC Assessment Reports (IPCC, 1990, 1996, 2001).

Currently the primary reason for the substantial range in model-based estimates of climate sensitivity is widely believed to be differences in their treatment of feedbacks (Schlesinger, 1985, 1988, 1989; Cess *et al.*, 1996; Colman, 2003) – particularly cloud feedbacks. But systematic comparisons have not been made to confirm that this is true for the current generation of models. Within international climate modeling projects, the development of new models, together with both formal and informal model comparison exercises that are currently being conducted by various groups, suggests that a renewed focus on the reasons for different model-based estimates of climate sensitivity may be particularly useful at this time.

The probability density and cumulative distribution functions for ΔT_{2x} obtained recently using the instrumental temperatures and estimated forcing from the mid-nineteenth century to the present by four groups (Andronova and Schlesinger, 2001; Forest *et al.*, 2002; Gregory *et al.*, 2002; Knutti *et al.*, 2002), using different methods, indicate that the IPCC range for the climate sensitivity, $1.5\,°C \leq \Delta T_{2x} \leq 4.5\,°C$, is too narrow. This is consistent with the observation that experts routinely underestimate uncertainty (Kahneman *et al.*, 1982;

Human-induced Climate Change: An Interdisciplinary Assessment, ed. Michael Schlesinger, Haroon Kheshgi, Joel Smith, Francisco de la Chesnaye, John M. Reilly, Tom Wilson and Charles Kolstad. Published by Cambridge University Press. © Cambridge University Press 2007.

Shlyakhter, 1994). The four independent estimates indicate that there is no likelihood that $\Delta T_{2x} < 0.6\,°C$ (this is a particularly robust finding), but there is a 5% likelihood that $\Delta T_{2x} \geq 9\,°C$.

As we show here, some recent studies suggest that new insights into the likely range of climate sensitivity may be possible through comparisons of models and observational data – contemporary, historical and paleoclimatic. Other recent studies raise issues regarding the applicability of response/forcing relationships in the climate system – such as the degree of predictability of climate and the relevance of climate predictability for estimates of climate sensitivity, and the degree to which forcings such as those due to solar variability, well-mixed greenhouse gases, and aerosols may produce different responses.

In Section 1.2 we briefly review the history of the climate sensitivity concept. Section 1.3 presents some recent developments of this concept. Section 1.4 discusses some of its possible future developments. Concluding remarks are given in Section 1.5.

1.2 History of the climate sensitivity concept (CSC)

The concept of climate sensitivity and the development of this concept are directly related to empirical or model estimations that established a linear relationship between radiative forcing and near-surface air temperature. CSC originated from the concept of the greenhouse effect, introduced by Arrhenius at the end of the nineteenth century (Arrhenius, 1896). Arrhenius defined the greenhouse effect in terms of ΔT_{2x}. Almost a century later, Budyko (1972) and Sellers (1969) repeated Arrhenius's calculations using more comprehensive energy balance models (North, 1981), and strongly supported the concept of the greenhouse effect. As a result, the climate sensitivity concept was promulgated.

Until the third IPCC report (IPCC, 2001), the concept of climate sensitivity was based on calculations of the equilibrium NST change. However, the third IPCC report (Cubasch et al., 2001) defined three measures of climate sensitivity, with the differences between the measures arising from the different types of simulations performed with the climate model.

The equilibrium climate sensitivity λ_{eq} is given by Eq. (1.1) and is estimated from climate simulations in which the radiative forcing does not vary with time after an initial change, such as for an instantaneous CO_2 doubling. Most estimations were made for λ_{eq}, which we will review below.

The effective climate sensitivity (after Murphy, 1995) is given by

$$\lambda_{eff}(t) = \frac{\Delta T(t)}{F(t) - dH(t)/d(t)} \quad (1.3)$$

where $dH(t)/dt$ is the change in heat storage of the climate system – essentially the heat taken up or lost by the ocean, obtained from climate simulations with time-dependent forcing, $F(t)$. If the radiative forcing in Eq. (1.3) were made time-independent, then $dH(t)/dt$ would approach zero with increasing time as the climate system approaches equilibrium and $\lambda_{eff}(t)$ would approach λ_{eq}. Time series of the effective climate sensitivity demonstrate how feedbacks of the climate system evolve with time.

The transient climate sensitivity (after Murphy and Mitchell, 1995) is given by

$$\lambda_{trans} = \frac{\Delta T(t_{2x})}{F_{2x}} \quad (1.4)$$

where $\Delta T(t_{2x})$ is the change in NST when the CO_2 concentration increases to double the pre-industrial value in a transient climate simulation, particularly one in which the CO_2 concentration increases at 1% per year. Because the thermal inertia of the ocean $\Delta T(t_{2x}) < T_{2x}$, $\lambda_{trans} < \lambda_{eq}$. Calculations of the transient climate sensitivity are primarily used for comparison among coupled atmosphere/ocean general circulation models because these models take thousands of years to equilibrate. However, because the actual climate system is always facing external stresses from multiple simultaneous forcings, at any particular time the climate system is facing transient sensitivity. Under slowly evolving external forcings, the transient sensitivity may be approximated by λ_{eq}.

Model-based estimation of climate sensitivity varies considerably from model to model because of different parameterizations of physical processes that are not explicitly resolved in the respective models, such as clouds. For example, energy balance models (EBMs), typically one- or two-dimensional, generally predict only the NST and frequently only in terms of its globally averaged value. The value of ΔT_{2x} obtained by EBMs ranges from $0.24\,°C$ (Newell and Dopplick, 1979) to $9.6\,°C$ (Möller, 1963).

Radiative-convective models (RCMs) determine the vertical distribution of atmospheric temperature and the surface temperature, virtually always in terms of their globally averaged values. As their name indicates, radiative-convective models include the physical processes of radiative transfer and convection. The value of ΔT_{2x} obtained by RCMs ranges from $0.48\,°C$ (Somerville and Remer, 1984) to $4.2\,°C$ (Wang and Stone, 1980).

Atmospheric general circulation models (AGCMs) simulate the vertical and geographical distributions of temperature, surface pressure, wind velocity, water vapor, vertical velocity, geopotential height, ground temperature and moisture, and the geographical distribution of snow and ice on the ground. AGCMs have been coupled with different ocean and sea-ice models. These include a swamp ocean which has zero heat capacity but infinite water, a mixed-layer ocean which has a prescribed depth, and a full dynamic ocean GCM, which is the oceanic counterpart of the AGCM. Sea-ice models range from freezing a swamp ocean (where the temperature would otherwise drop below the freezing temperature of seawater) that has only thermodynamics and no transport by wind or currents, to a fully dynamic-thermodynamic sea-ice model. In all cases the atmosphere/ocean model has a finite horizontal and vertical resolution. The value of ΔT_{2x} obtained by AGCMs ranges from 1.3 °C (Washington and Meehl, 1983) to 5.2 °C (Wilson and Mitchell, 1987).

All of the above ranges of the climate sensitivity estimates are based on a set of individual model runs and should be treated in a statistical sense as equally plausible, because each model has different sets of feedbacks, induced by different model parameterizations, built to be consistent with various observational data obtained from monitoring of the different parts of the climate system.

The growing amount of empirical temperature data has resulted in numerous attempts to use these data to estimate the climate sensitivity inversely. Budyko (1972) made the earliest such estimate of the climate sensitivity using paleodata of northern hemisphere temperature and atmospheric carbon dioxide concentration for six paleoclimate epochs (a review of these earlier estimates can be found in Schlesinger [1985]).

The pre-instrumental proxy record consists of quantities that are sensitive to temperature, such as the thickness and isotopic composition of the annual growth rings of trees and corals, and the annual layers of glacial ice; the relative abundance and isotopic composition of planktonic (living near the sea surface) and benthic (living near the sea floor) foraminifera (shell-covered species) that, after dying, fall to the sea floor where they are covered up by the sedimentary material raining down from above; and the relative abundance of pollen in the annual growth layers of sediment at the bottom of lakes (Ruddiman, 2000). These proxy data are converted into temperatures using statistical relations that have been developed based on the present climate. The reconstructed temperature for any paleoclimate gives the temperature difference from the present, ΔT_s. This is used to estimate the climate sensitivity from

$$\Delta T_{2x} = \left(\frac{\Delta F_{2x}}{\Delta F}\right)\Delta T_s \qquad (1.5)$$

where ΔF is an estimate of the radiative forcing. The range using paleo-based methods is 1.4 °C (Hoffert and Covey, 1992), for the lowest estimation, to 6.0 °C (Barron, 1994), for the highest estimation. There are at least three factors that lead to uncertainty in the estimates of ΔT_{2x} by the paleo-calibration method. First, the proxy data for temperature are not global in extent, hence their global average is uncertain, and their conversion to temperature is also uncertain (Mann and Jones, 2003). Second, estimation of the radiative forcing for paleoclimates relative to the present climate is difficult and thus uncertain. Third, the sensitivity of paleoclimate temperature changes from the present climate may be different from the sensitivity of future human-induced temperature changes from the present because the active feedbacks in each period are different.

Tol and de Vos (1998) (TdV) made one of the first inverse estimations of the probability density function (pdf) of climate sensitivity. They used a simple statistical model and Bayesian updating in combination with expert opinion and observational constraints on the initial (prior) pdf of ΔT_{2x}. They found that large values of ΔT_{2x} cannot be excluded and that the posterior pdf is strongly dependent on the prior pdf.

Starting with the analysis by Andronova and Schlesinger (2001), referred to below as AS01, there have been a handful of estimates of the pdf and cumulative density function (cdf) for ΔT_{2x} using simplified climate models (SCM) to replicate the temperature changes observed since the middle of the nineteenth century. Andronova and Schlesinger (2001) used an SCM consisting of an EBM coupled to an upwelling – diffusion model of the ocean to simulate the change in hemispheric-mean temperatures from 1765 to the present for prescribed radiative forcing. Sixteen radiative forcing models (RFMs) were examined with all possible combinations of: (1) anthropogenic radiative forcing consisting of greenhouse-gas (GHG) forcing due to the increasing concentrations of CO_2, methane, N_2O, chlorofluorocarbons and tropospheric ozone, and the direct (clear air) plus indirect (cloudy air) radiative forcing by tropospheric sulfate aerosols (SO_4); (2) volcanic radiative forcing; and (3) solar radiative forcing. The values of ΔT_{2x} and the unknown sulfate forcing in reference year 1990, $\Delta F_{ASA}(1990)$, where ASA is anthropogenic sulfate aerosol, were estimated for each RFM by optimizing the fit of the simulated and observed global-mean temperatures (GMT) and the interhemispheric temperature differences (ITD), respectively. The difference between the observed and simulated GMT and ITD was bootstrapped to generate 5000 samples of the unforced noise to which was added the simulated temperature signal to create 5000 surrogate observational hemispheric temperature records. For each ensemble member thereof, the values of ΔT_{2x} and $\Delta F_{ASA}(1990)$ were estimated using the same procedure as used for the single real observational record. This method of estimating the climate sensitivity does not depend on priors for the estimated quantities, but rather on the combination of the three types of radiative forcing, anthropogenic, solar and volcanic, and the natural variability of the observed temperatures. The resulting (AS01) cdf for ΔT_{2x} has a mean value of 3.40 °C, a 90% confidence interval of 1.0 °C to 9.3 °C, and a 54% likelihood that ΔT_{2x} lies outside the IPCC range of 1.5 to 4.5 °C.

Gregory et al. (2002) (GEA02) estimated climate sensitivity from

$$\Delta T_{2x} = \frac{\Delta \overline{T}'}{\overline{Q}' - \overline{F}'}, \qquad (1.6)$$

where $\Delta \overline{T}'$ is the change in the observed global-mean near-surface temperature between 1861–1900 and 1957–1994, $\overline{Q}'$ is the change in the estimated radiative forcing between the two periods, and $\overline{F}'$ is the change in heat uptake between the two periods – calculated by an SCM for the earlier period and estimated from observations for the latter. Normal probability distributions are assumed for $\Delta \overline{T}'$, with 2σ between $0.302\,°C$ and $0.368\,°C$; $\overline{Q}'$, with 2σ between -0.3 and $+1.0\,W/m^2$; and $\overline{F}'$, with 2σ between 0.00 and $0.32\,W/m^2$. A singularity exists where $\overline{F}'$ approaches $\overline{Q}'$. When $\overline{F}'$ is larger than $\overline{Q}'$, this implies a negative sensitivity, which was rejected because it would make the climate system unstable to natural variations.

Forest et al. (2002) used the Massachusetts Institute of Technology two-dimensional (latitude–altitude) statistical-dynamical model to simulate climate change from 1860 to 1995, varying ΔT_{2x}, oceanic vertical heat diffusivity, K, and anthropogenic aerosol forcing for the 1980s decade relative to pre-1860. They used optimal fingerprint detection in latitude, altitude and time of the simulated climate change in the observed climate change for surface air temperature, ocean temperature (0 to 3000 m) and tropospheric temperature. Their optimal fingerprint detection algorithm was based on pattern matching for the patterns for surface, upper-air, and deep-ocean temperature changes. The comparison was made for four zonal bands (90° S–30° S, 30° S–0°, 0°–30° N, 30° N–90° N) for decadal-mean temperature for the 1946–95 period relative to the 1906–95 climatology using an observational data mask. A goodness-of-fit statistic was computed for each of the three quantities. In the optimal detection, noise estimates for each diagnostic were obtained from two atmosphere/ocean GCMs. The surface and upper-air diagnostics reject similar regions, namely low K and high ΔT_{2x}, while the ocean diagnostic rejects high K and high ΔT_{2x}. Rejection regions shift to higher ΔT_{2x} for increasing negative aerosol forcing. Bayes' theorem was used to update the pdfs of parameters from an assumed starting (prior) pdf. The resulting ΔT_{2x} cdfs for a uniform starting pdf is wider than for the expert prior pdf – the 90% confidence interval for the uniform prior is $1.4\,°C$ to $7.7\,°C$, and for the expert prior is $1.4\,°C$ to $4.1\,°C$. For further references, we will call the resulting cdfs (FUN02) and (FEX02), respectively.

Knutti et al. (2002) (KEA02) used a climate model of reduced complexity – a zonally averaged dynamical ocean model coupled to a zonally and vertically averaged energy- and moisture-balance model of the atmosphere – to make 25 000 Monte Carlo simulations of the changes in the surface temperature from 1900 to 2000 and the ocean heat uptake from 1955 to 2000. They used multiple values of ΔT_{2x}, distributed uniformly over $1\,°C$ to $10\,°C$, and indirect aerosol forcing in 2000, uniformly distributed over -2 to $0\,W/m^2$. They found that the climate sensitivity is only weakly constrained by the ocean heat uptake, which gave $\Delta T_{2x} = 5.7 \pm 3\,°C$ (one standard deviation). When they used the surface temperature as a constraint, it gave $\Delta T_{2x} = 4.6\,°C$, with a range of $1.8\,°C$ to $8.7\,°C$. As shown by KEA02, the 95% confidence interval for ΔT_{2x} is $2.2\,°C$ to $9.1\,°C$.

Knutti et al. (2003) used a neural-network-based climate-model substitute to produce a large number of ensemble simulations similar to Knutti et al. (2002). They assumed a uniform prior distribution of the climate sensitivity and found that the surface warming in 2100 exceeded the range projected by IPCC for almost half the ensemble members. They noted that reduction of the uncertainty in climate sensitivity requires a significant reduction in the uncertainties of the observational temperatures, as well as better constraints on the reconstructed radiative forcing.

Figure 1.1a summarizes all estimates of the climate sensitivity described in this section. In this figure, we sorted the estimates into two groups: the deterministic estimates, based on estimation of the single number, and the probabilistic estimates, based on constructing a probability density function (pdf) or cumulative density function (cdf). The left part of Figure 1.1a presents ranges of the climate sensitivity based on the deterministic estimates. The right side of this figure presents ranges of the probabilistic estimates based on the 90% confidence interval of the probabilistic estimates. Among the deterministic estimates we have included the IPCC range. Among the probabilistic estimates we have included an estimate given by the Charney report of 1979 (NAS, 1979), namely. "$3\,°C$ with a probable error of $\pm 1.5\,°C$," which we have interpreted as the 50% confidence interval, labeled as NRC79. The range cited by IPCC originated from NAS (1979). Also, in this figure we have included an expert elicitation of 16 climate "experts" performed by Morgan and Keith (Morgan and Keith, 1995) of Carnegie-Mellon University (CMU95) five years after the IPCC First Assessment Report was published. For this we combined the 16 experts' opinions in terms of their mean estimation and variance cited in Morgan and Keith (1995) into a single cdf, under the assumption that each of the 16 estimations is normally distributed. It is seen that the CMU95 cdf has a non-zero probability that $\Delta T_{2x} < 0$. This non-zero probability occurs because three of the "experts" had a non-zero probability that $\Delta T_{2x} < 0$.

The right vertical axis of Figure 1.1a shows the climate sensitivity estimates in terms of the climate system's total feedback, calculated as defined by Schlesinger (1985, 1988, 1989):

$$f = 1 - \frac{(\Delta T_{2x})_o}{\Delta T_{2x}}. \qquad (1.7)$$

Here $(\Delta T_{2x})_o = G_o \Delta F_{2x}$ is the change in global-mean near-surface air temperature without feedback due to a doubling of the pre-industrial CO_2 concentration with radiative forcing $\Delta F_{2x} = 3.71\,W/m^2$ (Myhre et al., 1998).

$$G_o = \frac{T_s}{(1-\alpha_p)S} = 0.30\,K/(W/m^2) \qquad (1.8)$$

is the gain of the climate system without feedback, with $T_s = 288\,K$ the present global-mean NST, $S = 1367\,W/m^2$ the present solar irradiance and $\alpha_p = 0.3$ the present planetary albedo. If $\Delta T_{2x} \geq (\Delta T_{2x})_o$, $f \geq 0$, and if $\Delta T_{2x} < (\Delta T_{2x})_o$, $f < 0$.

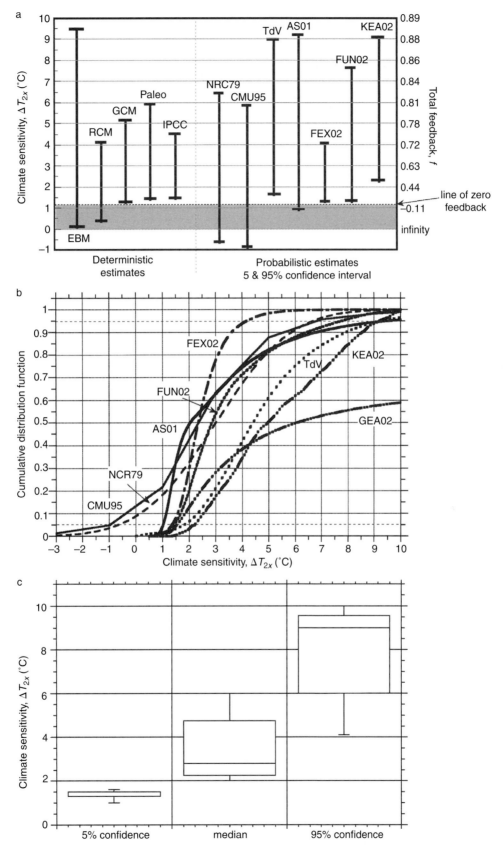

Figure 1.1 Estimations of the climate sensitivity (see explanations in the text).

As can be seen from Figure 1.1a, there are only a few estimates for which the net feedback of the climate system is negative.

Figure 1.1b presents cdfs of the probabilistic estimates of the climate sensitivity briefly described in this section. As can be seen, the lower end of the climate sensitivity estimates has a much smaller range of uncertainty than the upper end. For illustrative purposes Figure 1.1c summarizes the uncertainty range in terms of box plots for the 5th, 50th and 95th percentile values of the cdfs for ΔT_{2x}. This figure is based on six pdfs shown on Figure 1.1b: TdV, AS01, FEX02, FUN02, KEA02 and GEA02. There is disagreement about the median ΔT_{2x}, from 2.2 °C to 5.0 °C, and the 95th percentile, from 7.5 °C to 10.0 °C. These empirical studies, based on observations of the present climate, indicate that there is more than a 50% likelihood that ΔT_{2x} lies outside the canonical range of 1.5 °C to 4.5 °C, with disquietingly large values not being precluded.

1.3 Recent developments

The concept of equilibrium climate sensitivity served well for comparison of sophisticated climate models. In addition it helped to understand some of the models' feedbacks, which might work as well in the real climate system (Cess et al., 1996; Colman, 2003). However, researchers were always dissatisfied with the wide range of uncertainty in the estimates of climate sensitivity. This is mostly because: (1) uncertainties in the observational data do not allow reduction of the uncertainties in the magnitude of the models' parameters; (2) the model parameterizations, which reflect the models' inability to explicitly resolve all the physical processes in the climate system, may not represent actual feedbacks; and (3) it is not a trivial task to apply advanced mathematical methods to estimate climate sensitivity and, moreover, make an insightful interpretation of the results. This is why recently questions have been asked about the usefulness of the climate sensitivity concept, among which are these. (1) Is the climate sensitivity a robust characteristic of the climate system that is useful in climate economics and policy making? (2) If not, should we look for another important climate variable, or should we live with the uncertainties until they are resolved to some degree? (3) Should we look for another concept to characterize the climate system behavior? Some of those questions are nicely highlighted in the previous IPCC reports. The second IPCC report (IPCC, 1996, Chapter 2) introduced the concept of radiative forcing, stated how fast and slow feedbacks in the climate system relate to the climate sensitivity, and discussed the robustness of the linear relationship between forcing and response for different forcings. The third IPCC report (IPCC, 2001, Chapter 6) paid much more attention to the assessment of the concept of the radiative forcing as an important part of the concept of the climate sensitivity. Below we present some new developments.

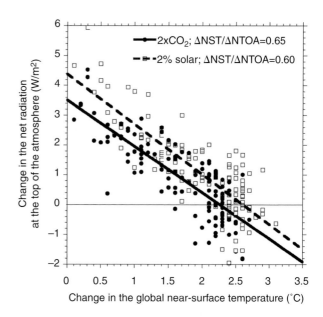

Figure 1.2 Linear regression between near-surface temperature, NST, and change in the net radiation at the top of the atmosphere, NTOA.

The procedure of calculating climate sensitivity from general circulation models, whether coupled to a non-dynamic model of the upper ocean (mixed-layer ocean model) or to a dynamic model of the full ocean, is complicated (Wetherald and Manabe, 1988). There are many uncertainties in how to perform the calculation, mostly related to where in the atmosphere and when the forcing and temperature changes should be sampled. Recently, Gregory et al. (2004) used a simple method of monitoring the ratio of the change in the global net radiation at the top of the atmosphere to the change in the global near-surface temperature, which approaches the equilibrium climate sensitivity as the model approaches its equilibrium climate change.

To illustrate this method, we applied it to compare the sensitivities of our 24-layer troposphere-stratosphere general circulation model (Yang et al., 2000; Yang and Schlesinger, 2002; Rozanov et al., 2002a, 2002b, 2004) coupled to a mixed-layer ocean model (24-L AGCM-ML) for two disturbances: a CO_2 doubling and a 2% increase in the amount of incoming solar radiation. Figure 1.2 presents a scatter diagram of the change in the monthly mean global net radiation at the top of the atmosphere (NTOA, at 1 hPa) against the change in monthly mean global NST for both experiments, together with respective linear regressions. In this figure, the change in NTOA initially is due to the perturbation and subsequently decreases as the change in NST increases and the climate system approaches its new equilibrium. The intercept of the regression line with the x-axis approximates the equilibrium change in NST due to the disturbance. The slope of the regression line, $\Delta NST/\Delta NTOA$, approximates the sensitivity of the model to the disturbances. Thus, if the regression lines for different disturbances are parallel to each other, then for these disturbances the model's sensitivities are the same. In the case presented in Figure 1.2, the 24-L AGCM-ML has approximately the same sensitivity to the CO_2 doubling and 2% increase in solar irradiance. Obviously, there are some shortcomings in this approach, which are discussed in Gregory et al. (2004). The simplicity of the presentation and interpretation, however, makes this method well suited for estimating the equilibrium climate sensitivity, without any additional calculations of the forcing.

Murphy et al. (2004) presented the Perturbed Physics Ensemble Method (PPEM) to estimate the pdf of climate sensitivity for an atmospheric general circulation model coupled to a mixed-layer ocean model. PPEM consists of running multiple realizations of the model with subjectively selected model parameters chosen randomly, one by one, from their subjectively prescribed range. In terms of computing costs, this method has become possible owing to the appearance of faster computers and massive parallelization of a model's code. One of the limitations of PPEM is that the set of the perturbed parameters is chosen by expert opinion and this may omit key model parameters. Another important limitation is that parameters are tested one by one and their synergetic effect on the climate sensitivity is not considered. And computer power still remains a major limiting factor for this kind of "massive" experiment. However, if it is possible to eliminate most of the shortcomings and use it for many GCMs, PPEM should give some very useful insights on GCMs and their climate sensitivities.

Figure 1.3 presents a comparison of the climate sensitivity values from Murphy et al. (2004) with other estimations. Figure 1.3a and 1.3b compares two cases from Murphy et al. (2004) with some of the estimates presented in Figure 1.1, namely the deterministic climate sensitivity range for existing GCMs, the expert elicitation CMU95, and FEX02 and AS01 which have the smallest and largest 90% confidence interval for climate sensitivity, respectively. The estimates of Murphy et al. (2004) have a 90% confidence interval for climate sensitivity of 1.8 °C to 5.2 °C for M1, where all model versions are assumed equally likely, and 2.4 °C to 5.2 °C for M2, which accounts for a reliability-based weighting of model versions according to the climate Prediction Index described in Murphy et al. (2004). It can be seen that both M1 and M2 are closest to FEX02, which uses an expert prior, and they are similar to the deterministic range for GCMs with a slightly higher minimum estimation. Recently, Stainforth et al. (2005), using the PPEM method, obtained the range of the climate sensitivity 1.5 °C to 11.5 °C by varying combinations of perturbations in six model parameters. This large range of uncertainties for a GCM brings up an old question of the validity of some GCM parameterizations, especially of cloud microphysics.

We note that the idea of computing the sensitivities of a model to its parameters is not new. Hall et al. (1982) applied the adjoint method (AM), formulated by Cacuci (1981) for use in numerical models, to calculate the sensitivities of a simple radiative-convective model and later the two-layer AGCM of Oregon State University (Hall, 1986) to a CO_2 doubling. The AM allowed the calculation of all linear sensitivities of each model parameter to the perturbation and to the initial conditions, all in one model simulation. However, at that time the application of the AM to larger models did not look feasible, because of the technical burden of re-writing the model code to include the model's adjoint equations and the mathematical problem of inverting large matrices. Thus, AM was not widely accepted by the "climate sensitivity community." But AM has been extensively applied by the "data assimilation community," who developed the Tangent and Adjoint Model Compiler (TAMC) to automatically generate adjoint model code. TAMC has been used to monitor and predict the linear tendencies of model variables. Further information is available at www.autodiff.com/tamc/.

Large numerical models calculate the climate sensitivity directly, but a similar value of the climate sensitivity may be obtained by different models even if they have different representations of the models' physical processes. Learning about new parameterizations for the subgrid-scale processes, inventing and applying new techniques for tuning model parameters, and systematic comparison of the models and their modules will definitely improve the models, if and only if the observational data have satisfactory temporal and spatial resolution and low observational errors. Thus most likely we will not be able to learn the "true" climate sensitivity from the large models soon, at least not from the IPCC Fourth Assessment Report (AR4) planned for 2007.

Inverse estimations of climate sensitivity depend heavily on the uncertainties in the observations. As an example, Figure 1.4 compares the observed historical hemispheric temperature departures of Jones and Moberg (2003), J2003, and Folland et al. (2001), F2001, and Figure 1.5 shows the influence of their differences on the estimate of climate sensitivity. Figures 1.4a and 1.4b present a comparison of these data for northern and southern hemispheres, respectively. It can be seen that the northern hemisphere data are very close to each other, while for the southern hemisphere the F2001 record is warmer in general than the J2003 SH record. Figures 1.4c and 1.4d shows the temporal behavior of the first two principal components of the temperature departure time evolution extracted using Singular Spectrum Analysis for the northern and southern hemispheres, respectively. Again, it can be recognized that in the southern hemisphere there are considerable differences in the temporal behavior of the two data sets. These differences make a major input into representation of the natural and forced variability of the observed temperature departure. Figure 1.5 shows the estimation of climate sensitivity using the inverse technique presented in Andronova and Schlesinger (2001), briefly described in the previous section. This technique is based on bootstrapping the residuals – the difference between the observed and simulated temperatures. Figure 1.5 shows the climate sensitivity estimated for four radiative forcing models used by Andronova and Schlesinger (2001): G – greenhouse gases only, GA – greenhouse gases and anthropogenic sulfate

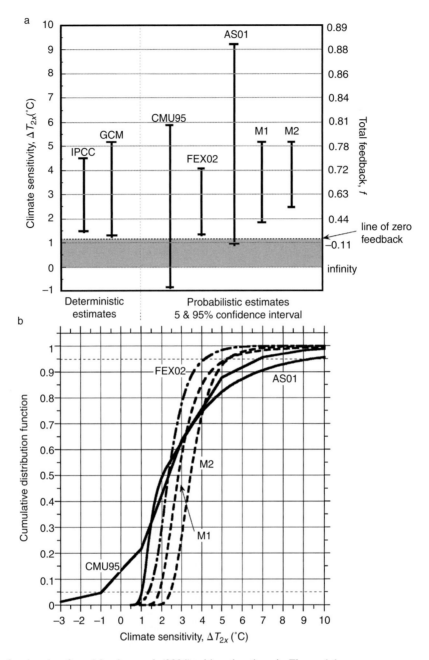

Figure 1.3 Comparison of estimation from Murphy *et al.* (2004) with estimations in Figure 1.1.

aerosol, GAS – GA plus the solar irradiance variability constructed by Lean and colleagues (Lean *et al.*, 1995; Fröhlich and Lean, 1998), and GTA – GA plus radiative forcing by tropospheric ozone. For 100 Monte Carlo estimates of the climate sensitivity, all three cases show a smaller median and range for the F2001 data than for the J2003 data. It is necessary to note that the difference between the two instrumental data sets arises from two different methods used to compile them: the optimal averaging and the area-weighted averaging (Rayner, personal communication) which coincide with increasing spatial coverage in both hemispheres. Until these data differences are reconciled, the uncertainty in climate sensitivity will remain large.

Retrospectively, we can look at how the range of the climate sensitivity evolves with time as new information on instrumental temperature departures (from the 1961–1990 average) is added. For this we estimate the climate sensitivity for different record lengths of the J2003 temperature departures starting in 1860 and ending in 1940 and then adding each 10 years of data thereto until the data reach 2000. Figure 1.6 illustrates the resulting evolution of the climate sensitivity for greenhouse gas, tropospheric ozone and aerosol

The concept of climate sensitivity

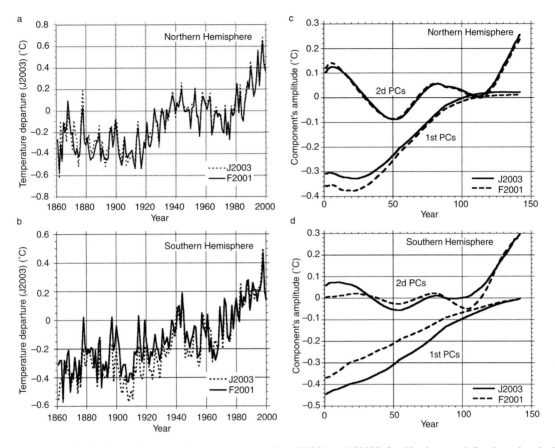

Figure 1.4 Comparison of the observed near-surface temperature data F2001 and J2003 for Northern and Southern hemispheres (a and b) and their two first principal components calculated using singular spectrum analysis (c and d).

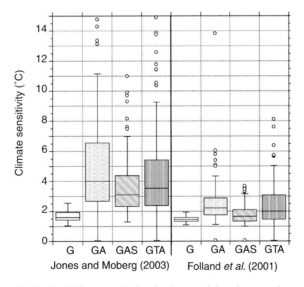

Figure 1.5 Estimation of the climate sensitivity for different radiative forcing models using two data sets. Each box encloses 50% of the data with the median value of the variable displayed as a line. The top (UQ) and bottom (LQ) of the box mark the limits of ±25% of the variable population. The lines extending from the top and bottom of each box mark the minimum and maximum values within the data set that fall within an acceptable range. Any value outside this range, that is greater than $UQ + 1.5(UQ - LQ)$ or less than $LQ - 1.5(UQ - LQ)$, called an outlier, is displayed as an individual point.

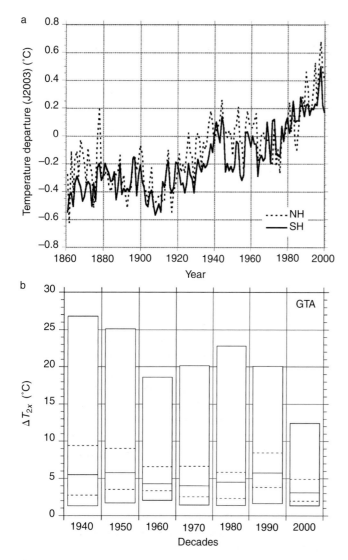

Figure 1.6 Observed temperature departure (Jones and Moberg, 2003) for northern and southern hemispheres (a) and evolution of the climate sensitivity uncertainties (b).

radiative forcing (GTA). Figure 1.6a presents the J2003 temperature departures for the northern and southern hemispheres for the entire period. Figure 1.6b presents the evolution of the climate sensitivity, in box-plot form, obtained by marching 10 years forward. It shows that starting in 1950 the cooling in the temperature record dramatically increases the uncertainty in the estimated climate sensitivity, which is based on the record from 1860 to 1950, because it becomes inconsistent with the increasing tendency in the greenhouse forcing. However, later on with the additional observational data, the uncertainty decreases because of the persistence of the GA signal and averaging out the natural climate variability. The main point of this illustration is that it takes a long time to reduce uncertainties in the estimation of climate sensitivity.

1.4 Future perspectives

The new IPCC process, started in preparation for the fourth scientific assessment (AR4) in 2007, began a new discussion about the concept of climate sensitivity. Most likely, in the new report the climate sensitivity concept will not be dismissed – it has proven to be useful. Thus, the answer to the question "Has any progress been made in the application and development of the concept of climate sensitivity for understanding climate change?" is "yes". The estimation of climate sensitivity has progressed from a single value estimate to a probability distribution to characterize uncertainty. In addition, many new methods for estimating climate sensitivity have been developed and understanding of climatic processes has definitely improved. However, because of the large uncertainty in the published estimates of climate sensitivity, which is difficult to apply in policy-relevant problems (see Dessai and Hulme (2004) for a review of the use of climate probabilities in decision making), the climate sensitivity concept has been questioned in a few aspects, some of which were discussed in the previous section. Also, reduction of the uncertainty in the estimates of climate sensitivity will probably not be possible in the time-frame of the AR4, hence policy makers will need to learn how to deal with this (Lempert and Schlesinger, 2000; Yohe et al., 2004). The range of uncertainty in the climate sensitivity, in comparison with just having a single value, requires policy makers to create new approaches in mitigation and abatement policy, particularly in rigorously counting all the possible outcomes for adaptation to, and coping with, climate change. The larger the change that would occur, the more difficult it would be for humans and natural systems to adapt.

From another viewpoint, because of regional heterogeneity of the temperature response to different forcings, there are some difficulties in "translating" the "global" climate sensitivity into regional climate changes. Regional climate sensitivities may be different from the global sensitivity because different feedbacks may be regionally but not globally important (Boer and Yu, 2003). Thus, further enrichment of the climate sensitivity concept requires the search for a set of climate variables that characterizes climate change in full on both regional and global scales. Understandably, the search for this set of variables will be limited to existing databases of observations and/or reconstructions, and will include all corresponding uncertainties.

Development of the methodological aspects of the climate sensitivity concept will mostly likely be related to understanding the climate feedbacks on the different spatial and temporal scales. Already there are a few intercomparison projects in progress (e.g., Working Group on Coupled Models; Global Analysis Interpretation and Modeling; Coupled Model Intercomparison Project) whose goals are to improve large numerical models (a brief description of some of them can be found at www.clivar.org/science/mips.htm). Furthermore, development and application of new mathematical methods for understanding climate feedback processes are also very important. Among the relatively new methods for the climate-change community is the neural network method (NNM) – a method that greatly speeds up model diagnostic calculations and which, like the adjoint method mentioned in the previous section, has not yet been used widely because of the burden of constructing the neural network for a particular model with thousands of variables. Definitely, the adjoint method will also be applied for climate feedback analysis, especially with the development of the Tangent and Adjoint Model Compiler.

1.5 Concluding remarks

The concept of climate sensitivity has served well during the past hundred years. As van der Sluijs *et al.* (1998) have argued, the concept of climate sensitivity "works as a 'boundary object' managing the interface between different social worlds (climate modeling, climate impacts research, climate policy making) and acts as an 'anchor' that fixes the scientific basis for the climate policy debate." Vast progress has been made, particularly with the process of science integration when climatology, mathematics, economics and policy come together to formulate demands for the empirical and model data to understand anthropogenic influences on climate and environment. This science integration taught climate scientists to present the uncertainties in estimates of climate sensitivity, and it taught policy makers not to ignore the climate-change problem. Demands from economists and policymakers stimulated climate scientists to pay attention to regional-scale problems, and particularly to study feedbacks in the climate system.

Thus, the climate sensitivity concept has not been dismissed. Rather, it is being enriched and further developed. The recent developments reported in this paper are likely to dispel the consensus-estimate range supported by the IPCC since its inception. As Pielke Jr (2001) has noted, efforts to reduce uncertainty via "consensus science" are misplaced and can only provide an illusion of certainty, which might constrain decision makers' options.

Acknowledgements

This material is based upon work supported by the National Science Foundation under Award No. ATM-0084270. Any opinions, findings, and conclusions or recommendations expressed in this publication are those of the authors and do not necessarily reflect the views of the National Science Foundation. S. D. is supported by a grant (SFRH/BD/4901/2001) from Fundação para a Ciência e a Tecnologia, in Portugal. We thank Chris Forest, Jonathan Gregory, Retto Knutti, James Murphy and Richard Tol for providing us their climate-sensitivity data. Special thanks to Chris Forest and James Sweeney for their thoughtful discussions of the paper and to Marty Hoffert for his thoughts about the possible range of climate sensitivity.

References

Andronova, N. G. and Schlesinger, M. E. (2001). Objective estimation of the probability density function for climate sensitivity. *Journal of Geophysical Research* **106** (D19), 22 605–22 612.

Arrhenius, S. (1896). On the influence of carbonic acid in the air upon the temperature of the ground. *The London, Edinburgh, and Dublin Philosophical Magazine and Journal of Science* **41**, 237–277.

Barron, E. J. (1994). Chill over the Cretaceous. *Nature* **370**, 415.

Boer, G. J. and Yu, B. (2003). Climate sensitivity and response. *Climate Dynamics* **20**, 415–429; doi: 10.1007/s00382-002-0283-3.

Budyko, M. I. (1972). *Human's Impact on Climate*. Leningrad: Gidrometeoizdat.

Budyko, M. I., Ronov, A. B. and Yanshin, A. L. (1987). *The History of the Earth's Atmosphere*. Berlin: Springer-Verlag.

Cacuci, D. G. (1981). Sensitivity theory for nonlinear systems. I. Nonlinear functional analysis approach. *Journal of Mathematical Physics* **22**, 2794–2802.

Cess, R. D., Zhang, M. H., Ingram, W. J. *et al.* (1996). Cloud feedback in atmospheric general circulation models: an update. *Journal of Geophysical Research* **101**, 12 791–12 794.

Colman, R. A. (2003). A comparison of climate feedbacks in general circulation models. *Climate Dynamics* **20**, 865–873.

Cubasch, U., Meehl, G. A., Boer, G. J. *et al.* (2001). Projections of future climate change. In *Climate Change 2001: The Scientific Basis. Contribution of Working Group I to the Third Assessment Report of the Intergovernmental Panel on Climate Change*, ed. J. T. Houghton, Y. Ding, D. J. Griggs, *et al.* Cambridge: Cambridge University Press, pp. 525–582.

Dessai, S. and Hulme, M. (2004). Does climate adaptation policy need probabilities? *Climate Policy* **4**, 107–128.

Folland, C. K., Rayner, N. A. Brown, S. J. *et al.* (2001), Global temperature change and its uncertainties since 1861. *Geophysical Research Letters* **28**, 2621–2624.

Forest, C. E., Stone, P. H., Sokolov, A. P., Allen, M. R. and Webster, M. D. (2002). Quantifying uncertainties in climate system properties with the use of recent climate observations. *Science* **295**, 113–117.

Fröhlich, C. and Lean, J. (1998). The Sun's total irradiance: cycles, trends and related climate change uncertainties since 1976. *Geophysical Research Letters* **25**, 4377–4380.

Gregory, J. M., Stouffer, R. J., Raper, S. C. B., Stott, P. A. and Rayner, N. A. (2002). An observationally based estimate of the climate sensitivity. *Journal of Climate* **15**, 3117–3121.

Gregory, J. M., Ingram, W. J., Palmer, M. A. *et al.* (2004). A new method for diagnosing radiative forcing and climate sensitivity, *Geophysical Research Letters* **31**, L03205.

Hall, M. C. G. (1986). Application of adjoint sensitivity theory to an atmospheric general circulation model. *Journal of Atmospheric Science* **43**, 2644–2651.

Hall, M. C. G., Cacuci, D. G. and Schlesinger, M. E. (1982). Sensitivity analysis of a radiative-convective model by the adjoint method. *Journal of Atmospheric Science* **39**, 2038–2050.

Hoffert, M. I. and Covey, C. (1992). Deriving global climate sensitivity from paleoclimate reconstructions. *Nature* **360**, 573–576.

IPCC (1990). *Scientific Assessment of Climate Change: Report of Working Group I*, ed. J. T. Houghton, G. J. Jenkins and J. J. Ephraums. Cambridge: Cambridge University Press. *Summary for Policymakers*, Bracknell: UK Meteorological Office, IPCC/WMO/UNEP.

IPCC (1996). *Climate Change 1995: The Science of Climate Change Contribution of Working Group I to the Second Assessment Report of the Intergovernmental Panel on Climate Change*, ed. J. T. Houghton, L. G. Meira Filho, B. A. Callander *et al.* Cambridge: Cambridge University Press.

IPCC (2001). *Climate Change 2001: The Scientific Basis. Contribution of Working Group I to the Third Assessment Report of the Intergovernmental Panel on Climate Change*, ed. J. T. Houghton, Y. Ding, D. J. Griggs, *et al.* Cambridge: Cambridge University Press.

Jones, P. D. and Moberg, A. (2003). Hemispheric and large-scale surface air temperature variations: an extensive revision and an update to 2001. *Journal of Climate* **16**, 206–223.

Kahneman, D., Slovic, P. and Tversky, A. (1982). *Judgment Under Uncertainty: Heuristics and Biases*. New York: Cambridge University Press.

Knutti, R., Stocker, T. F., Joos, F. and Plattner, G. K. (2002). Constraints on radiative forcing and future climate change from observations and climate model ensembles. *Nature* **416**, 719–723.

Knutti, R., Stocker, T. F., Joos, F. and Plattner, G.-K. (2003). Probabilistic climate change projections using neural networks. *Climate Dynamics* **21**, 257–272.

Lean, J., Beer, J. and Bradley, R. (1995). Reconstruction of solar irradiance since 1610: Implications for climate change. *Geophysical Research Letters* **22** (23), 3195–3198.

Lempert, R. J. and Schlesinger, M. E. (2000). Robust strategies for abating climate change. *Climatic Change* **45** (3–4), 387–401.

Mann, M. E. and Jones, P. D. (2003). Global surface temperature over the past two millennia. *Geophysical Research Letters* **30**, 1820; doi: 10.1029/2003GL017814.

Möller, F. (1963). On the influence of changes in CO_2 concentration in air on the radiative balance of the earth's surface and on the climate. *Journal of Geophysical Research* **68**, 3877–3886.

Morgan, M. G. and Keith, D. (1995). Subjective judgments by climate experts. *Environmental Science & Technology* **29**, 468A–476A.

Murphy, J. M. (1995). Transient response of the Hadley Centre coupled ocean-atmosphere model to increasing carbon dioxide. Part III: Analysis of global-mean response using simple models. *Journal of Climate* **8**, 496–514.

Murphy, J. M. and Mitchell J. F. B. (1995). Transient response of the Hadley Centre coupled ocean-atmosphere model to increasing carbon dioxide. Part II: Spatial and temporal structure of response. *Journal of Climate* **8**, 57–80.

Murphy, J. M., Sexton, D. M. H. Barnett, D. N. *et al.* (2004). Quantification of modelling uncertainties in a large ensemble of climate change simulations. *Nature* **430**, 768–772.

Myhre, G., Highwood, E. J., Shine, K. P. and Stordal, F. (1998). New estimates of radiative forcing due to well mixed greenhouse gases. *Geophysical Research Letters* **25**, 2715–2718.

NAS (1979). *Carbon Dioxide and Climate: A Scientific Assessment*. Washington, DC: US National Academy of Sciences.

Newell, R. E. and Dopplick, T. G. (1979). Questions concerning the possible influence of anthropogenic CO_2 on atmospheric temperature. *Journal of Applied Meteorology* **18**, 822–825.

North, G. (1981). Energy balance climate models. *Reviews of Geophysics and Space Physics* **19**, 91–121.

Pielke, R. A. Jr. (2001). Room for doubt. *Nature* **410**, 151.

Rozanov, E., Schlesinger, M. E., Andronova, N. G. *et al.* (2002a). Climate/chemistry effects of the Pinatubo volcanic eruption simulated by the UIUC stratosphere/troposphere GCM with interactive photochemistry. *Journal of Geophysical Research* **107** (D21), 4594, doi: 10.1029/2001JD000974.

Rozanov, E., Schlesinger, M. E. and Zubov, V. A. (2002b). The University of Illinois, Urbana-Champaign three-dimensional stratosphere-troposphere general circulation model with interactive ozone photochemistry: fifteen-year control run climatology. *Journal of Geophysical Research* **106** (D21), 27 233–27 254.

Rozanov, E., Schlesinger, M. E., Egorova, T. A. *et al.* (2004). Atmospheric response to the observed increase of solar UV radiation from solar minimum to solar maximum simulated by the UIUC climate-chemistry model. *Journal of Geophysical Research* **109**, D01110, doi: 10.1029/2003JD003796.

Ruddiman, W. F. (2000). *Earth's Climate, Past and Future*. New York: W. H. Freeman.

Schlesinger, M. E. (1985). Feedback analysis of results from energy balance and radiative-convective models. In *The Potential Climatic Effects of Increasing Carbon Dioxide*, ed. M. C. MacCracken and F. M. Luther. US Department of Energy, pp. 280–319.

Schlesinger, M. E. (1988). Quantitative analysis of feedbacks in climate model simulations of CO_2-induced warming. In *Greenhouse-Gas-Induced Climatic Change: A Critical Appraisal of Simulations and Observations*, ed. M. E. Schlesinger. Amsterdam: Elsevier, pp. 653–737.

Schlesinger, M. E. (1989). Quantitative analysis of feedbacks in climate model simulations. In *Understanding Climate Change*, ed. A. Berger, R. E. Dickinson and J. W. Kidson. Washington, DC: American Geophysical Union, pp. 177–187.

Sellers, W. D. (1969). A global climate model based on the energy balance of the earth–atmosphere system. *Journal of Applied Meteorology* **8**, 392–400.

Shlyakhter, A. I. (1994). Uncertainty estimates in scientific models: lessons from trends in physical measurements, population and energy projections. In *Uncertainty Modeling and Analysis: Theory and Applications*, ed. B. M. Ayyub and M. M. Gupta. North-Holland: Elsevier Scientific Publishers, pp. 477–496.

Somerville, R. C. J. and Remer, L. A. (1984). Cloud optical thickness feedbacks in the CO_2 climate problem. *Journal of Geophysical Research* **89**, 9668–9672.

Stainforth, D. A., Aina1, T. Christensen, C. et al. (2005). Uncertainty in predictions of the climate response to rising levels of greenhouse gases. *Nature* **433**, 403–405.

Tol, R. S. J. and de Vos, A. D. (1998). A Bayesian statistical analysis of the enhanced greenhouse effect. *Climatic Change* **38**, 87–112.

van der Sluijs, J., van Eijndhoven, J., Shackley, S. and Wynne, B. (1998). Anchoring devices in science for policy: the case of consensus around climate sensitivity. *Social Studies of Science* **28**, 291–323.

Wang, W.-C. and Stone, P. H. (1980). Effect of ice-albedo feedback on global sensitivity in a one-dimensional radiative-convective model. *Journal of Atmospheric Science* **37**, 545–552.

Washington, W. M. and Meehl, G. A. (1983). General circulation model experiments on the climatic effects due to a doubling and quadrupling of carbon dioxide concentration. *Journal of Geophysical Research* **88**, 6600–6610.

Wetherald, R. T. and Manabe, S. (1988). Cloud feedback processes in a general circulation model. *Journal of Atmospheric Science* **45**, 1397–1415.

Wilson, C. A., and Mitchell, J. F. B. (1987). A doubled CO_2 climate sensitivity experiment with a global climate model including a simple ocean. *Journal of Geophysical Research* **92**, 13 315–13 343.

Yang, F., and Schlesinger, M. E. (2002). On the surface and atmospheric temperature changes following the 1991 Pinatubo volcanic eruption: A GCM study. *Journal of Geophysical Research* **107** (D8); doi: 10.1029/2001JD000373.

Yang, F., Schlesinger, M. E. and Rozanov, E. (2000). Description and performance of the UIUC 24-layer stratosphere/troposphere general circulation model. *Journal of Geophysical Research* **105** (D14), 17 925–17 954.

Yohe, G., Andronova, N. and Schlesinger, M. (2004). To hedge or not to hedge against an uncertain climate future. *Science* **305**, 5695.

2

Effect of black carbon on mid-troposphere and surface temperature trends

Joyce E. Penner, Minghuai Wang, Akshay Kumar,
Leon Rotstayn and Ben Santer

2.1 Introduction

There is a continuing controversy over whether satellite-observed temperature trends in the mid and lower troposphere based on Microwave Sounding Unit (MSU) satellite data since 1979 are consistent with surface-observed trends. The satellite trends are as much as 0.14 °C/decade smaller than the surface-observed trends. However, the satellite-inferred temperatures must be corrected for drifts and calibration differences between different satellites, and different procedures for doing so among different groups have led to different mid-tropospheric trend estimates. Here, we examine whether model-predicted trends are consistent with the satellite-based trends from the University of Alabama in Huntsville (UAH), and from the Remote Sensing Systems (RSS) group. It is important to re-examine model results in light of new evidence that indicates that the inclusion of black carbon aerosols tends to cool the surface and heat the troposphere, whereas the satellite data imply the opposite. Unlike previous model studies, we include an estimate of the effects of direct forcing by fossil fuel organic matter and black carbon aerosols, and by biomass aerosols, on trend estimates, as well as direct and indirect sulfate aerosol forcing, stratospheric ozone forcing and long-lived greenhouse gas forcing. We use the quasi-steady state results from the Commonwealth Scientific and Industrial Research Organisation (CSIRO) global climate model with a q-flux ocean model to correct transient simulations from the National Center for Atmospheric Research (NCAR) parallel climate model and from the CSIRO model that did not include the effects of fossil fuel carbon and biomass burning aerosols.

Temperature trends for the period from 1979 to 1999 have been examined. The spatial signatures of the raw temperature trends for the MSU channel 2 temperatures from the transient simulations, and from the transient simulations adjusted for the effects of fossil fuel carbon and biomass burning aerosols, are consistent with the available data from both the RSS and UAH groups. However, the trend of the surface minus the MSU channel 2 temperatures is less variable than that of either the raw surface trends or raw MSU channel 2 trends. The trend in the difference between the surface and MSU channel 2 temperatures from the adjusted transient models are not consistent with the trend in the difference between the surface and MSU channel 2 temperatures inferred by UAH, but are consistent with the trend in the difference between the surface and MSU channel 2 temperatures inferred by RSS. This conclusion holds for the northern hemisphere and southern hemisphere as well as the global average.

The Microwave Sounding Unit on the NOAA (National Oceanic and Atmospheric Administration) polar-orbiting satellites has been measuring microwave emissions in several channels since 1979. MSU channel 2 (MSU2) measures temperature in a broad layer that extends from the surface to 50 hPa, with about 15% of its signal originating in the stratosphere above 200 hPa. Four separate groups have estimated the temperature trend associated with this channel. Christy and Norris (2004) (hereafter UAH) report a trend of 0.03 ± 0.05 °C/decade over the time period from 1979 to 2002, while Prabhakara et al. (1998), Mears et al. (2003) (hereafter, Remote Sensing Systems, or RSS), and Vinnikov and Grody (2003) reported trends of 0.11, 0.097 ± 0.02 and 0.24 ± 0.02 °C/decade over 1980 to 1996,

---

Human-induced Climate Change: An Interdisciplinary Assessment, ed. Michael Schlesinger, Haroon Kheshgi, Joel Smith, Francisco de la Chesnaye, John M. Reilly, Tom Wilson and Charles Kolstad. Published by Cambridge University Press. © Cambridge University Press 2007.

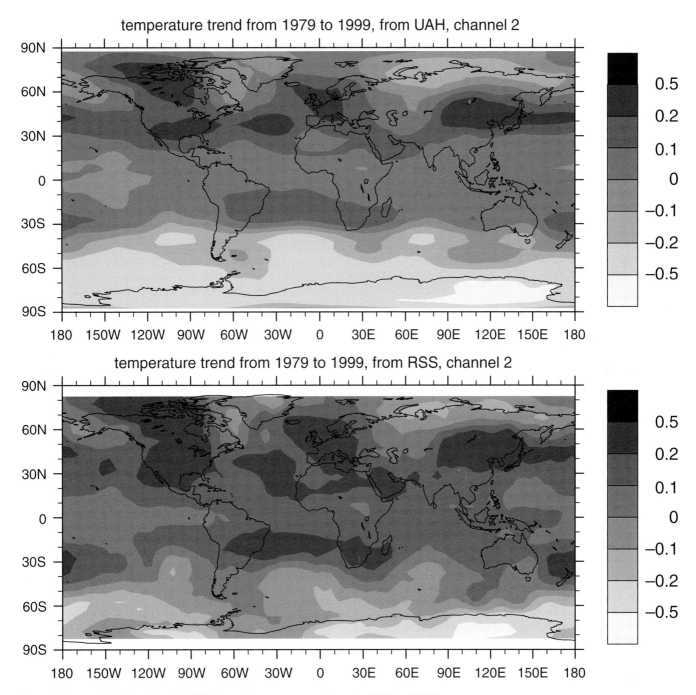

Figure 2.1 Trend (°C/decade) in MSU channel 2 temperatures from the UAH and RSS groups.

1979–2001, and 1978–2002, respectively. These estimates of the MSU2 trend have been compared with that measured at the surface. The surface warming from thermometer measurements has been estimated as $0.20 \pm 0.10\,°C$ per decade for 1979–98 (Santer et al., 2000a). The 95% confidence intervals quoted by Santer et al. account for temporal autocorrelation effects and are therefore nearly twice as large as estimates that do not account for this autocorrelation.

The fact that the surface temperature increase is larger than that estimated from the satellite measurements reported by three of the four groups has led to controversy regarding the adequacy of both the surface network of thermometer measurements and the satellite measurements themselves. Decreasing stratospheric ozone concentrations have decreased stratospheric temperatures over this time period, so this may explain some of the difference in trend estimates between the surface and the channel 2 measurements (Hansen et al., 1995). As shown in Figure 2.1, there is spatial coherence in the pattern of trends in the UAH MSU2 temperature data and the RSS MSU2 temperature data. Moreover, both data sets show a decreasing trend near Antarctica, where the decreases in stratospheric ozone associated with increasing halocarbon concentrations have been largest. Fu et al. (2004) attempted to remove the influence of stratospheric temperature trends within the MSU data by using a linear fit between the spatially averaged monthly temperature anomaly profiles from radiosonde-derived MSU channel 2 and channel 4 temperatures and the average radiosonde temperature between 300 and 850 hPa. The derived coefficients were then used to determine the 300 to 850 hPa temperatures from the MSU channel 2 and channel 4 temperatures. The 300 to 850 hPa temperature trends derived in this way are much closer to those determined from the surface temperature network.

UAH has developed a product that they call the 2LT temperature (Spencer and Christy, 1992) that is meant to decrease the influence of the stratosphere on mid-troposphere temperatures and therefore capture a temperature that is closer to the surface. But it may also have more biases and noise (Mears et al., 2003) than the direct channel 2 measurement. For these reasons, here we focus on a direct comparison of measured MSU2 temperature trends with those determined from model simulations. The model simulations that we discuss include the effect of stratospheric ozone losses on upper tropospheric and stratospheric temperatures and so may be used to compare modeled and observed temperature trends directly.

Previously, Santer et al. (2000a) compared model results from ECHAM (European Centre for Medium Range Forecasting, Hamburg) with UAH 2LT temperature trends, with surface temperature trends, and with the difference between surface and 2LT trends. The most realistic model simulation included forcing effects from changes in greenhouse gases, direct and indirect effects of sulfate aerosols tropospheric and stratospheric ozone changes, and Pinatubo volcanic aerosols. While the modeled global average trends from the surface and 2LT (through 1997) time series were consistent with the UAH and surface temperature data, those for the difference between the surface and 2LT temperatures (or the lapse rate) were not.

Hegerl and Wallace (2002) further examined the difference between the surface and 2LT temperatures and concluded that the variability associated with El Nino Southern Oscillation (ENSO) temperature changes in the tropics and with patterns associated with the relatively warmer continental surface temperatures and cooler ocean temperatures could not explain the observed lapse rate trends. As in Santer et al. (2000a) they concluded that the lapse rate trend was not consistent with the trends predicted by the ECHAM model when changes in long-lived greenhouse gases, sulfate direct and indirect forcing, and tropospheric ozone were included.

Following the above studies, Santer et al. (2003) examined the correspondence between the NCAR PCM transient model simulations and both channel 2 and channel 4 measured temperature trends. They considered the transient PCM runs that we consider here. These runs include a representation of the forcing by long-lived greenhouse gases, stratospheric and tropospheric ozone changes, the direct radiative effects of anthropogenic sulfate and volcanic aerosols, and solar irradiance changes. Indirect aerosol effects and the effects of trends in carbonaceous aerosols were not included. This study concluded that the model results were more consistent with the observed MSU2 temperatures from the RSS study, and also that the model fingerprint of temperature change was "detected" at the 5% significance level in both RSS and UAH cases if the natural variability from the PCM model was used to determine the "noise" in the model signature. This paper did not discuss the consistency between the observed and modeled surface and MSU2 temperature trends and the lapse rate (or difference between the surface and MSU2 temperature trends).

Finally, Douglass et al. (2004) revisited this problem, examining the surface trends (Jones et al., 1999), the satellite MSU 2LT trends (Christy et al., 2000), the radiosonde trends (Parker et al., 1997), and the NCAR/NCEP reanalysis data trends (Kistler et al., 2001) in comparison with the Hadley CM3 model (Tett et al., 2002), the NCAR PCM model (Meehl et al., 2003), and the GISS SI2000 model (Hansen et al., 2002). They stated that the global average trends from the models were not consistent with the MSU, radiosonde and the NCAR/NCEP reanalysis data but did not offer any statistical significance testing to demonstrate this. Moreover, they did not process the model results using the MSU 2LT weightings so that their conclusions about the differences between MSU data and the models are not quantitative in any case.

Since the above analyses were completed, new model simulations that include the effects of carbonaceous aerosols on surface and mid-troposphere temperatures have become available (Menon et al., 2002; Penner et al., 2003). These model results indicate that carbonaceous aerosols might cool the surface and warm the mid troposphere. Thus, there is clearly a need to revisit the issue of whether model simulations and observations are consistent. Is a trend in lapse rate expected when carbonaceous aerosols are included in transient climate simulations? Are differences between observations and the model-generated surface and MSU2 temperatures and lapse rates significant? These are the questions we wish to address.

In Section 2.2, we present a discussion of observed surface and mid-troposphere temperatures. Section 2.3 discusses the effects that different forcings may have on transient simulations, including the effects of carbonaceous aerosols. Because there are no transient simulations available that include carbonaceous aerosols, we briefly discuss historical data for the emissions of carbonaceous aerosols and use these data to derive a scaling factor for determining the effects of carbonaceous aerosols on the transient PCM simulations and a CSIRO transient model simulation. Section 2.4 presents the results of our model/data comparison.

2.2 Observed surface and mid-troposphere temperature trends

Figure 2.2 plots the time history of observed surface and MSU2 monthly average temperature anomalies for the globe, northern

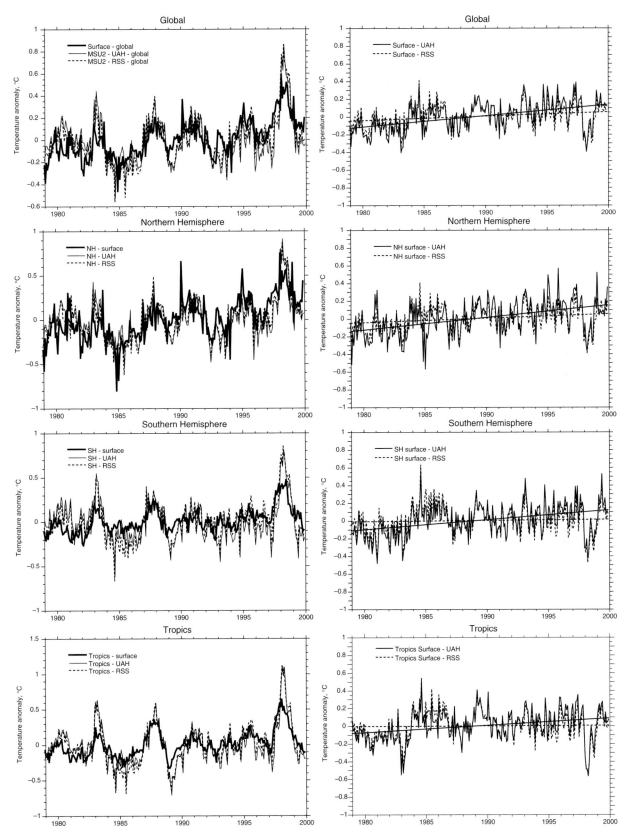

Figure 2.2 Time history of temperature anomaly in surface and MSU channel 2 temperatures (left) and the trend in the difference between the surface and MSU channel 2 temperatures (right).

hemisphere (NH, 0–90° N), southern hemisphere (SH, 0–90° S), and tropics (30° S to 30° N), as well as the difference in the surface and MSU2 temperatures. As has been noted previously, the surface temperature warms more rapidly than does the MSU2 temperature. Even though one might expect the trend in the difference between the surface and MSU2 temperatures to be larger in the southern hemisphere, where the MSU temperatures are affected by cooling from stratospheric ozone depletion, the trend in the difference between the surface and MSU2 temperatures is actually larger in the NH than in the SH. However, this appears to be due to the larger increases in the MSU temperature anomaly compared with that of the surface associated with the 1997–98 El Nino. This increase near the end of the record is especially evident in the southern hemisphere data, which then damps the overall trend of the difference in that hemisphere. We note that larger increases in the MSU channel 2 temperature anomalies compared with the surface were also recorded for the 1982–83 El Nino, but they do not stand out as clearly in the difference plots. This may partly be due to the near-simultaneous 1982 El Chichon volcanic eruption.

Trends were estimated for each time series and region in Figure 2.2 using a least squares fit to the monthly mean anomalies. In deriving these trends, we masked the data from channel 2 with the availability of the surface temperature data. Tables 2.1a and 2.1b summarize the trends for the surface and for the MSU2 temperatures, respectively, along with the 95% confidence intervals of the trends using an effective sample size adjusted for the lag-1 autocorrelation coefficient (Santer et al., 2000b). Table 2.1c summarizes the trends in the difference between the surface and MSU2 temperatures as well as the 95% confidence intervals for the difference trend. The trend in the difference between the surface and MSU2 temperatures can be estimated with much more confidence (lower confidence intervals) because of the strong correlation between the surface variability and the MSU2 variability.

One may note that the global RSS MSU2 trend is larger than the global UAH MSU2 trend by about 0.08 °C per decade over the time period from 1979 to 1999. Christy and Norris (2004) report that the corrections made by the UAH and RSS groups for the NOAA-9 sensor (which forms a critical link between the pre-1984 and post-1986 data) have a significant effect on the deduced trend. Therefore, we also considered the trend for the time period from 1987 to 1999, but the trend deduced by the two groups for this time period still differs by 0.06 °C/decade. The

Table 2.1a Trend in surface temperatures (°C/decade).
See Table 2.2 for a definition of acronyms.

	Global		NH		SH		Tropics	
Jones	0.18	±0.06	0.23	±0.07	0.11	±0.07	0.14	±0.15
PCM VGSSuO	0.15	±0.16	0.16	±0.12	0.12	±0.18	0.11	±0.24
PCM + FFC	0.15	±0.16	0.16	±0.12	0.12	±0.18	0.11	±0.24
PCM + FFC + BB	0.14	±0.16	0.16	±0.12	0.12	±0.18	0.10	±0.24
CSIRO GSSuO	0.14	±0.03	0.16	±0.04	0.11	±0.03	0.10	±0.05
CSIRO + FFC	0.14	±0.03	0.16	±0.04	0.11	±0.03	0.11	±0.05
CSIRO + FFC + BB	0.13	±0.03	0.15	±0.04	0.10	±0.03	0.10	±0.05
CSIRO + PCMV	0.11	±0.10	0.09	±0.09	0.12	±0.10	0.10	±0.21
CSIRO + V + FFC	0.11	±0.10	0.09	±0.09	0.12	±0.10	0.10	±0.21
CSIRO + V + FFC + BB	0.10	±0.10	0.09	±0.09	0.12	±0.10	0.09	±0.21

Table 2.1b Trend in MSU channel 2 temperatures (°C/decade).

	Global		NH		SH		Tropics	
UAH	0.05	±0.13	0.09	±0.12	0.00	±0.12	0.06	±0.21
RSS	0.13	±0.13	0.15	±0.12	0.10	±0.12	0.13	±0.22
PCM VGSSuO	0.10	±0.15	0.09	±0.13	0.10	±0.12	0.13	±0.17
PCM + FFC	0.10	±0.15	0.10	±0.13	0.10	±0.12	0.13	±0.17
PCM + FFC + BB	0.10	±0.15	0.10	±0.13	0.10	±0.12	0.13	±0.17
CSIRO GSSuO	0.13	±0.03	0.12	±0.03	0.14	±0.05	0.14	±0.05
CSIRO + FFC	0.14	±0.04	0.13	±0.03	0.14	±0.05	0.14	±0.05
CSIRO + FFC + BB	0.14	±0.04	0.14	±0.03	0.14	±0.05	0.14	±0.05
CSIRO + PCMV	0.10	±0.09	0.08	±0.08	0.13	±0.10	0.10	±0.13
CSIRO + V + FFC	0.11	±0.09	0.10	±0.08	0.13	±0.10	0.11	±0.13
CSIRO + V + FFC + BB	0.11	±0.09	0.10	±0.08	0.13	±0.10	0.10	±0.13

Monthly averaged temperatures have been masked to the availability of the surface temperature data.

Table 2.1c Trend in the difference between surface and MSU channel 2 temperatures (°C/decade).

	Global		NH		SH		Tropics	
Jones–UAH	0.13	±0.05	0.14	±0.05	0.11	±0.06	0.08	±0.07
Jones–RSS	0.05	±0.05	0.08	±0.05	0.01	±0.06	0.01	±0.07
PCM VGSSuO	0.05	±0.04	0.07	±0.04	0.02	±0.04	−0.02	±0.05
PCM + FFC	0.04	±0.04	0.06	±0.04	0.02	±0.04	−0.02	±0.05
PCM + FFC + BB	0.03	±0.04	0.05	±0.04	0.01	±0.04	−0.02	±0.05
CSIRO GSSuO	0.00	±0.03	0.04	±0.03	−0.03	±0.04	−0.03	±0.03
CSIRO + FFC	0.00	±0.03	0.03	±0.03	−0.03	±0.04	−0.03	±0.03
CSIRO + FFC + BB	−0.01	±0.03	0.02	±0.03	−0.04	±0.04	−0.04	±0.04
CSIRO + PCMV	0.00	±0.04	0.01	±0.05	−0.00	±0.05	−0.00	±0.05
CSIRO + V + FFC	0.00	±0.04	0.00	±0.05	−0.00	±0.05	−0.00	±0.05
CSIRO + V + FFC + BB	−0.01	±0.04	−0.01	±0.05	−0.01	±0.05	−0.01	±0.05

trend from each group is within the 95% confidence interval of the trend from the other group, and, from this perspective, each might be considered equally valid. Nevertheless, in the following, we examine the comparison of model results to both sets of data. We use the longer time period (from 1979 to 1999), since the 95% confidence interval of the trends is significantly reduced with the longer time period.

2.3 Modeled trends and the effects of carbonaceous aerosols

Two transient model simulations were available to us: the Parallel Climate Model (PCM) of the National Center for Atmospheric Research (NCAR) and Los Alamos National Laboratory (LANL) (Washington et al., 2000) and the R21 version of the Commonwealth Scientific and Industrial Organization (CSIRO) climate model. The version of the CSIRO model used here employed the ocean model described by Hirst et al. (2000), the AGCM described by Gordon et al. (2002), and the sulfur and cloud modifications described by Rotstayn and Lohmann (2002a, b). The PCM model includes a representation of volcanic forcing, greenhouse gas forcing, solar changes, sulfate aerosol direct forcing, and forcing from changes in stratospheric and tropospheric ozone (VGSSuO). The CSIRO transient run includes a representation of greenhouse gas forcing, solar changes, sulfate aerosol direct and first indirect forcing, and forcing from changes in stratospheric ozone (GSSuO). (Note that Table 2.2 summarizes the acronyms that we use to refer to different forcings in the models.)

Neither of the transient model runs includes all of the forcings that are thought to have occurred over the time period of the satellite observations. For example, the PCM model did not include sulfate aerosol indirect forcing. We may estimate the importance of this omission by using the forcing estimates in Ramaswamy et al. (2001). Ramaswamy et al. (2001) estimated that the sulfate direct forcing for the time period from 1991 to 1995 was −0.4 W/m² while that for 1981 to 1985 was −0.36 W/m². If we use the difference in these values to scale the indirect aerosol forcing (which was estimated to be

Table 2.2 Acronyms for forcings used in the model simulations in Table 2.1.

Acronym	Included forcing
VGSSuO	Volcanoes, greenhouse gases, solar, sulfate aerosols (direct forcing only for the PCM model or direct + indirect for CSIRO), ozone
FFC or FF OM + BC	Fossil fuel black carbon and organic matter direct forcing
BB	Biomass burning smoke direct forcing
PCMV or V	Volcanoes estimated from the PCM model

between 0 and −2 W/m² in 2000), we obtain a change in the indirect sulfate forcing of from 0 to −0.08 W/m² over this time period. If the larger of these two values is correct, then it may be important to consider the effects of indirect sulfate aerosol forcing over this time period.

The CSIRO model included indirect sulfate aerosol forcing, but did not include tropospheric ozone forcing or volcanic aerosol forcing. Ramaswamy et al. (2001) estimated the difference in forcing from volcanoes between the time period 1991 to 1995 and 1981 to 1985 as −0.53 W/m², while that from tropospheric ozone was 0.06 W/m². Thus, the omission of volcanic aerosol forcing in the CSIRO model might seriously affect their temperature trends over the time period of the satellite record. The omission of tropospheric O_3 may also be important. Below, we adjust the CSIRO modeled temperature trends using a volcano-only forcing run from the PCM model, but do not correct for its omission of tropospheric O_3 forcing.

In addition to the above influences, two prominent changes that have not been included in these model simulations are changes from fossil fuel black carbon (BC) and organic matter (OM) aerosols and changes from smoke aerosols produced during biomass burning. What are the expected effects of these different forcings on model simulated surface and MSU2 trends? We examine this question using a q-flux version of the CSIRO climate model. To calculate the direct effect of

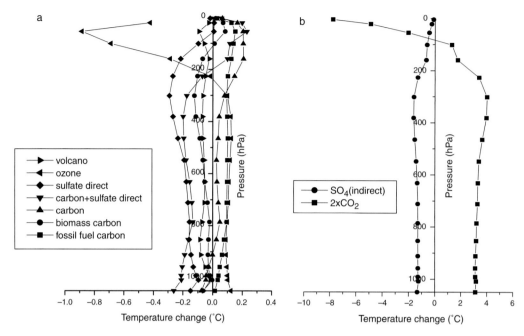

Figure 2.3 Difference in temperature for CSIRO q-flux runs for present day and pre-industrial conditions. The volcano and ozone profiles were derived from a transient PCM simulation (see text).

fossil fuel (FF) black carbon and organic matter aerosols (FF BC + OM or FFC) and biomass burning (BB) aerosols, we added the radiative transfer treatment for these aerosols described by Grant et al. (1999). The FF BC + OM aerosols were added assuming they were initially hydrophobic and became hydrophilic after a 1.15 day e-folding period. The BB aerosols were assumed to be hydrophilic. Following Penner et al. (1998), we assumed that the hydrophobic FF BC + OM aerosols did not take up water as the relative humidity changed, while the hydrophilic FF BC + OM aerosols and the BB aerosols followed the "Hardiman" distribution and did take up water. Dry and wet scavenging of hydrophilic aerosols followed the scheme for sulfate. Emissions for FF BC + OM and BB aerosols were those described by Liousse et al. (1996). Simulations for each q-flux run were carried out for 40 years, and the last 20 years were averaged to obtain the expected average temperature change.

Figure 2.3a shows the expected change in the global average vertical temperature profile from the direct effects of anthropogenic sulfate aerosols, the direct effects of fossil fuel (FF), black carbon (BC), and organic matter (OM) aerosols (labeled "fossil fuel carbon" in the figure or FF BC + OM or FFC in the tables), the direct effects of biomass burning (BB or "biomass carbon" in the figure), the combined direct effects of FF BC + OM and BB aerosols (labeled "carbon" in the figure) based on q-flux runs from the CSIRO model. In addition, Figure 2.3b shows the expected signature of the forcing from indirect sulfate aerosols and a $2 \times CO_2$ simulation. The top-of-atmosphere (TOA) radiative forcings used in these simulations are given in Table 2.3. The "instantaneous forcing" was calculated using

Table 2.3 Top-of-atmosphere forcing (i.e. W/m^2) for some of the simulations shown in Figure 2.3.

	NH	SH	Global
Instantaneous shortwave forcing			
FF BC + OM	0.657	0.023	0.340
BB	−0.015	−0.028	−0.021
Combined (FFC + BB)[a]	0.643	0.010	0.322
SO$_4$ only	−0.455	−0.068	−0.261
Relaxed shortwave (SW) forcing[b]			
FF BC + OM	0.753	0.078	0.415
BB	0.078	0.222	0.149
Combined (FFC + BB)[a]	0.718	0.242	0.479
SO$_4$ only	−0.448	0.152	−0.149
Relaxed longwave (LW) forcing[b]			
FF BC + OM	−0.465	−0.101	−0.283
BB	−0.134	−0.252	−0.193
Combined (FFC + BB)[a]	−0.469	−0.303	−0.386
SO$_4$ only	−0.043	−0.243	−0.143
Net LW and SW relaxed forcing[b]			
FF BC + OM	0.288	0.23	0.132
BB	0.132	−0.03	−0.044
Combined (FFC + BB)[a]	0.249	−0.61	0.093
SO$_4$ only	−0.491	−0.091	−0.292

[a] Combined (FFC + BB) refers to a simulation with FF BC + OM and BB aerosols.
[b] The relaxed forcing was calculated by fixing the sea surface temperature in the model and calculating the change in top-of-atmosphere fluxes (see Penner et al., 2003).

Table 2.4 Temperature change for the surface and MSU channel 2 for the simulations in Figure 2.3. All runs are for CSIRO except the volcano- and ozone-only runs.

	Global	NH	SH	Tropics
Surface temperature change (°C)				
FFC	0.01	0.03	0.00	0.03
BB	−0.06	−0.03	−0.10	−0.07
(FFC + BB)	−0.08	−0.05	−0.10	−0.07
$2 \times CO_2 - 1 \times CO_2$	3.12	3.07	3.18	2.81
SO_4 direct	−0.15	−0.23	−0.06	−0.14
SO_4 indirect	−1.34	−2.00	−0.67	−0.75
(FFC + BB) + SO_4 direct	−0.22	−0.29	−0.16	−0.20
PCM volcano only (94–99) − (74–79)	−0.06	−0.16	0.03	−0.03
PCM ozone only (1990–1999) − (1890–1899)	0.12	0.01	0.23	0.07
MSU2 temperature change (°C)				
FFC	0.10	0.19	0.02	0.04
BB	−0.07	−0.05	−0.08	−0.08
(FFC + BB)	0.08	0.16	0.00	0.00
$2 \times CO_2 - 1 \times CO_2$	2.95	2.88	3.03	3.44
SO_4 direct	−0.20	−0.26	−0.14	−0.22
SO_4 indirect	−1.27	−1.70	−0.85	−1.19
(FFC + BB) + SO_4 direct	−0.13	−0.11	−0.14	−0.25
PCM volcano only (94–99) − (74–79)	−0.06	−0.10	−0.02	−0.05
PCM ozone only (1990–99) − (1890–99)	−0.02	−0.04	−0.01	−0.06

FFC refers to a simulation with the direct effects of fossil fuel BC + OM while BB refers to a run with the direct effects of biomass burning aerosols.

two radiation calls at each time step of the GCM calculation to compute the top-of-atmosphere shortwave flux change, thereby guaranteeing that the only changes in the calculation of this forcing are those specified by the perturbed quantity. The "relaxed forcing" is estimated as the difference between two multi-year simulations with fixed sea surface temperature, so that the perturbation (i.e. the aerosols) is allowed to change the air and land temperatures, water vapor, and cloud fields through their direct radiative effects and through their indirect cloud microphysical effects. The relaxed forcing, therefore, includes some measure of the response of the system to these aerosols. The net (longwave, LW and shortwave, SW) relaxed forcing for sulfate aerosols is similar to that for the instantaneous SW forcing, indicating that the TOA instantaneous SW forcing is a good measure of the atmospheric response (Rotstayn and Penner, 2001). But the BC + OM net relaxed forcings all differ substantially from the instantaneous SW forcing and are generally smaller than the instantaneous forcing (Penner et al., 2003). This indicates that the surface temperature response to these forcings is probably smaller than indicated by the instantaneous SW forcings (see Table 2.5 below) and that the standard scaling between SW forcing and the climate model response (Rotstayn and Penner, 2001) must be modified. Moreover, the absorption by BC within the atmospheric column may change the MSU2 temperatures in ways that differ substantially compared with the surface response (see Penner et al., 2003, and Table 2.5).

Figure 2.3a also shows the effects of volcanic aerosols and ozone changes from PCM transient simulations on the vertical temperature change. The volcanic profile was derived from the difference in average temperatures for the volcanic-only transient simulation of the PCM model between the time periods 1994–99 and 1974–79 and the difference in average temperatures for an ozone-only transient simulation between the time periods 1990–99 and 1890–99. The effects on the surface and the MSU2 temperature change for these simulations are given in Table 2.4. As shown there, an increase in CO_2, sulfate direct forcing, and biomass aerosol direct forcing, as well as changes to tropospheric and stratospheric ozone, tends to warm the surface more than the MSU2 temperatures, while the combined forcing from FF BC + OM and BB aerosols most strongly tends to cool the surface relative to the MSU2 temperatures. The sulfate indirect forcing cools the surface more than the MSU2 temperatures on a global average basis, but this is dominated by the northern hemisphere signal since both the southern hemisphere and tropics contribute to a stronger warming at the surface relative to the MSU2 temperatures. The PCM volcanic effects are small on a global average basis, but tend to cool the surface more than the MSU2 weighted temperatures. This is also dominated by the northern hemisphere, because both in the tropics and southern hemisphere the surface warms more than the MSU2 temperature. We note that because almost 15% of the MSU2 temperature signal is due to the temperature in the stratosphere, the effects of cooling by

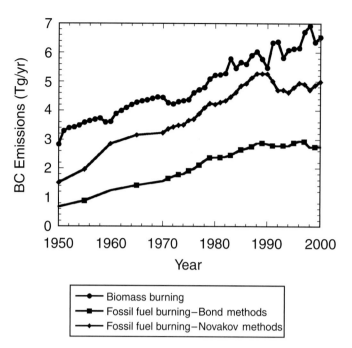

Figure 2.4 Emissions of black carbon (BC) from biomass burning from Ito and Penner (2005), and from fossil fuel burning using two methods: that of Bond *et al.* (2004) as applied in Ito and Penner (2005), and that of Novakov *et al.* (2003).

CO_2 in the stratosphere cause more cooling in the MSU2 temperatures than would be obtained from simply examining the difference between the surface and mid-tropospheric temperatures. Moreover, the effects of black carbon aerosols which tend to warm more in the stratosphere than at the surface, especially in the combined FF BC + OM and BB simulation, cause increases in the MSU2 temperatures that are particularly strong relative to the surface temperatures.

As is clear from Table 2.4, the trend expected in the MSU2 temperatures may be less than, greater than, or similar to the trend expected in the surface data, depending on the strength of the different forcings over any given time period. While reasonable estimates of the strength from different forcings during different time periods are available for most of the climate forcings of interest, those for the forcings associated with fossil fuel BC + OM aerosols and for biomass aerosols have not been available.

We have recently estimated historical fossil fuel BC and OM emissions, open biomass burning smoke emissions, and biofuel smoke emissions (Ito and Penner, 2005). Our estimated fossil fuel emissions used the emissions factors and methods summarized in the paper by Bond *et al.* (2004) except that the emission factors for on-road diesel fuels were adjusted to account for the decrease in emission factors associated with the introduction of improved technology. Although emission factors for other fossil fuel burning might also be expected to decrease as a result of technology changes, these changes were not included in the analysis. Nevertheless, the changes we define, which are based on the change in fossil fuel use, may represent the bulk of the emission changes from fossil fuels. However, there is clearly a need to better define the technology introduced in different countries as a function of time in order to obtain a better estimate of fossil fuel BC + OM emissions.

The emissions from open biomass burning in Ito and Penner (2005) were developed using a present-day inventory (Ito and Penner, 2004) that was regionally scaled by previous results from an inverse modeling study for CO for the year 2000 (Arellano *et al.*, 2004). Then, they used data from the TOMS satellite (Herman *et al.*, 1997; Torres *et al.*, 1998) to extend the inventory to the period from 1979 to 2000. The present-day biofuel inventory was developed using data sets provided by Yevich and Logan (2003) and data sets from FAOSTAT (2004) together with emission factors from Bond *et al.* (2004) to separately determine the sources of emissions from biofuel burning in developing and developed countries. Emissions in developed countries were derived from the country-specific FAOSTAT data for wood consumption for residential use from 1961 to 2000. In developing countries, residential biofuel emissions were extrapolated in time using the per capita household usage developed by Yevich and Logan (2003) and population statistics. The agro-industrial uses in each country together with crop production changes for 1961–2000 were used to estimate the agro-industrial combustion of these fuels.

These emissions estimates provide an opportunity to examine the possible effects of black carbon and organic matter emissions on the expected temperature trends during the period of the MSU satellite measurements. Figure 2.4 shows the global average emissions for black carbon from fossil fuel burning and from biomass burning (the sum of open burning and biofuel burning) based on the analysis of Ito and Penner (2005) over the time period 1950–2000. The trend in the global average fossil fuel emissions of BC for the time period 1979–2000 is very small, only 0.023 Tg BC/yr, while that from biomass burning is 0.07 Tg BC/yr. The biomass burning trend is made up of a trend for biofuels of 0.045 Tg BC/yr and a trend for open burning of 0.024 Tg BC/yr. The small trend estimated for fossil fuel emissions is made up of relatively larger positive trends in Asia (0.037 Tg BC/yr) and negative trends in Eastern Europe (−0.016 Tg BC/yr). During the time period from 1950 to 1979 the global estimated trend in fossil fuel BC emissions is much larger, about 0.05 Tg BC/yr. We may contrast these estimated trends in emissions from fossil fuel burning with those summarized by Novakov *et al.* (2003). The global average trend in submicron BC emissions (estimated as 0.85 times the total emissions) based on Novakov *et al.* (2003) for the time period 1979–1999 is 0.026 Tg BC/yr. Thus, the fossil fuel trend from Ito and Penner (2005) is similar to that from Novakov *et al.* (2003) although the emissions themselves are smaller.

The combined trend from open biomass burning BC emissions, biofuel BC emissions, and fossil fuel BC emissions over 1979–2000 may be estimated as 0.092 Tg BC/yr using the fossil fuel emissions based on Ito and Penner (2005) and as 0.095 Tg BC/yr using fossil fuel emissions based on Novakov *et al.* (2003). In the following, we use the estimated trends from the Novakov *et al.* (2003) fossil fuel estimates and the Ito and Penner (2005) biomass burning estimates to correct the predicted temperatures from the CSIRO and PCM transient climate model simulations for emissions changes due to both FF BC + OM emissions and the sum of FF BC + OM and BB emissions. For this correction, we fit a regression line to the estimated emissions for the period 1975–2000. This period was chosen to simulate the somewhat delayed response of the transient climate system to emissions prior to 1979. We then estimated the temperature for a given year, t, from:

$$T(x,y,z,t) = T_{\text{transient}}(x,y,z,t) \\ + \Delta T_{\text{q-flux}}(x,y,z) \times E_R(t)/E_{\text{q-flux}} \quad \text{(Eq. 2.1)}$$

where $T_{\text{transient}}$ is the temperature from the transient model simulation, $\Delta T_{\text{q-flux}}$ is the change in temperature calculated using the CSIRO q-flux model with present-day emissions and pre-industrial emissions, $E_R(t)$ are the BC emissions from the regression line (which increase by 0.05 Tg BC/yr and 0.13 Tg BC/yr for fossil fuel BC emissions and the sum of fossil fuel and biomass burning emissions, respectively) and $E_{\text{q-flux}}$ are the present-day BC emissions used in the q-flux runs (which were 6.6 and 12.3 Tg BC/yr for fossil fuel only and biomass plus fossil fuel emissions, respectively). We applied these corrections to both the PCM transient model run and the CSIRO transient model run. In determining the corrections, we considered both the possible trends from only fossil fuel BC emissions and the possible trends from the combined fossil fuel and biomass burning emissions. Trends were analyzed for the globe, the northern hemisphere, the southern hemisphere, and the tropics.

The use of Eq. (2.1) assumes that the temperature response for the carbonaceous aerosols would add linearly to the response from the transient model simulated forcings and that the trend in the temperature response may be estimated from the trend in emissions. Moreover, this approach also assumes that separate emissions trends for different regions do not significantly alter the global average temperature response or the temperature responses in the broad regions (NH, SH, and tropics) examined here. While several studies have examined the linearity of the response of the temperature change to different forcings (e.g. Rotstayn and Penner, 2001), the question of linearity between the temperature response for carbonaceous aerosols and other forcings has not been studied in a full general circulation model (but see Shine *et al.*, 2003). Here, we note that since the presence of black carbon affects the cloudiness in the model (Ackerman *et al.*, 2000; Penner *et al.*, 2003), it may interfere with the response of the climate model to other forcings, especially when the magnitude of the forcing depends on the amount of cloudiness. In particular, the response of the climate model to both sulfate aerosol direct and indirect forcing may change when black carbon is included. Therefore, the use of Eq. 2.1 in our analysis is only approximate.

Table 2.5 examines the adequacy of the linear assumption embodied in Eq. (2.1). Here we show the temperature change from a series of q-flux runs. The first, labeled "FFC only", considers only fossil fuel BC + OM aerosols, the second, "BB only", considers only biomass burning aerosols. The simulation labeled "combined FFC + BB" is the simulation that considered the combined direct forcing from fossil fuel BC + OM and biomass burning aerosols. The simulation labeled "SO_4 only" considered only sulfate direct aerosol effects, and the simulation labeled "all aerosols" considered the combined direct forcing from FFC, BB, and SO_4 aerosols. One might expect the direct sulfate forcing to decrease in areas where cloudiness increases as a result of carbonaceous aerosols and to increase in areas where cloudiness decreases as a result of carbonaceous aerosols. To examine whether the combined signal from the "all aerosols" case is a linear combination of the separate signals from the combined FFC + BB signal and the SO_4 signal we also formed the sum (FFC + BB) + SO_4. We also examined the linear combination of the separate FFC, BB, and SO_4 signals.

The table also shows the percentage difference between the (FFC + BB) + SO_4 signal (i.e. that from the combined FFC + BB simulation plus the SO_4 only simulation) and the "all aerosols" signal as well as the percentage difference between the FFC + BB + SO_4 signal (i.e. the sum of the individual FFC only, BB only, and SO_4 only simulations) and the "all aerosols" signal. For surface temperature, the difference between the sums of the different q-flux experiments and the "all aerosols" case is within 30% in the broad regions considered in our analysis. In addition, the "all aerosols" signal is more negative than the sum of the individual signals. Apparently a decrease in the overall cloudiness allows the sulfate

Table 2.5 Annual mean values for temperature change from q-flux runs.

		NH	SH	Tropics	Global
Surface temperature change (°C)					
	FFC only	0.03	0.00	0.03	0.01
	BB only	−0.03	−0.10	−0.07	−0.06
	Combined (FFC + BB)	−0.05	−0.10	−0.07	−0.08
	SO$_4$ only	−0.23	−0.06	−0.15	−0.15
	All aerosols together	−0.29	−0.23	−0.20	−0.26
	(FFC + BB) + SO$_4$	−0.28	−0.16	−0.21	−0.22
Percentage difference ((FFC + BB) + SO$_4$) − All aerosol		−2	−30	5	−14
	FFC + BB + SO$_4$	−0.23	−0.16	−0.19	−0.20
Percentage difference (FFC + BB + SO$_4$) − All aerosol		−19	−31	−6	−24
MSU2 temperature change (°C)					
	FFC only	0.19	0.02	0.04	0.10
	BB only	−0.05	−0.08	−0.08	−0.07
	Combined FFC + BB	0.16	0.00	0.00	0.08
	SO$_4$ only	−0.27	−0.14	−0.25	−0.20
	All aerosols together	−0.06	−0.16	−0.18	−0.11
	(FFC + BB) + SO$_4$	−0.11	−0.14	−0.25	−0.13
Percentage difference ((FFC + BB) + SO$_4$) − All aerosol		90	−12	39	15
	FFC + BB + SO$_4$	−0.13	−0.20	−0.29	−0.17
Percentage difference (FFC + BB + SO$_4$) − All aerosol		22	43	14	34
Surface − MSU2 tempertaure (°C)					
	FFC	−0.16	−0.01	−0.02	−0.09
	BB	0.02	−0.02	0.02	0.00
	Combined FFC + BB	−0.21	−0.09	−0.06	−0.15
	SO$_4$ only	0.04	0.08	0.10	0.06
	All aerosols together	−0.23	−0.07	−0.02	−0.15
	(FFC + BB) + SO$_4$	−0.17	−0.02	0.04	−0.10
Percentage difference ((FFC + BB) + SO$_4$) − All aerosol		−25	−74	−312	−35
	FFC + BB + SO$_4$	−0.10	0.05	0.10	−0.03
Percentage difference (FFC + BB + SO$_4$) − All aerosol		−56	−170	−692	−81

Note: FFC includes fossil fuel BC + OM while BB includes biomass burning aerosols
All Aerosols: Simulation with FFC, BB and SO$_4$ all in one single run
(FFC + BB) + SO$_4$: Addition of FFC + BB signal to the SO$_4$ signal
FFC + BB + SO$_4$: Addition of FFC signal, BB signal and SO$_4$ signal

aerosol to exert a more negative forcing on the climate system when all aerosols act together. Since this is probably due to the effect of BC on clouds, we might expect that the combined indirect sulfate aerosol forcing and FFC and BB forcing would be somewhat smaller (i.e. less negative) than the sum of the individual forcing experiments. The percentage differences between the sum of the individual cases and the "all aerosols" case are much larger for the diagnosed MSU2 temperatures, especially in the northern hemisphere, where the predicted temperature change in the "all aerosol" case is quite small, only 0.06 °C. As noted above, the predicted MSU2 temperature change is the result of both positive and negative changes throughout the atmosphere, especially for biomass burning aerosols (see Figure 2.3). As a result, the sum of individual cases is, in most cases, smaller (i.e. more negative) than the "all aerosols" case.

Table 2.5 also shows the difference between the surface temperature and MSU2 temperatures for the individual q-flux runs and the combined runs. Because the difference between the surface temperature and MSU2 temperature is very small, the percentage change between the sum of individual cases and the "all aerosols" case is, in some cases, quite large. However, the individual predicted temperature change in these cases is very small.

This examination demonstrates that the temperature change that results from summing the individual q-flux runs is similar to, but not exactly equal to, that from a combined run. In most cases, the sign of the temperature change from the sum of the individual cases is the same as that from the "all aerosols" case. The exception to this is in the tropics where the sign of the temperature change for the surface minus MSU2 for the (FFC + BB) + SO$_4$ case is positive, whereas that for "all aerosols" is negative. In addition, the temperature change for the surface minus MSU2 for the FFC + BB + SO$_4$ case in the southern hemisphere and in the tropics is positive, whereas that for the "all aerosols" case is negative. In both these cases the temperature changes are quite small. Nevertheless, we should not interpret differences between the observations and our constructed model results, which rely on Eq. (2.1), too literally, given the uncertainties in this technique. However, because of apparent uncertainties both in the effects of BC (Penner, 2003) and in BC emissions themselves (Bond et al., 2004), here we simply take Eq. (2.1) as given and consider whether the postulated model trends are or are not consistent with the observations.

2.4 Results and discussion

Table 2.1 summarizes the trends for the surface temperatures, MSU2 temperatures and the difference between the surface and the MSU2 temperatures for the transient model simulations and for the transient model simulations adjusted using Eq. (2.1). Because volcanic emissions during 1979–99 have generally acted to decrease temperatures, and because these emissions were not included in the CSIRO transient simulations, we include a set of trends which were derived by adding the temperature change from the PCM volcano-only simulation to that from the CSIRO transient temperatures adjusted for fossil fuel BC + OM and for the combined fossil fuel BC + OM and biomass aerosols (labeled CSIRO + V + FFC and CSIRO + V + FFC + BB, respectively). This procedure, while not very satisfying because the response of the CSIRO model to volcanic forcing may differ from that of the PCM model, at least lets us surmise how the addition of the volcanic signal would alter the calculated trends. Figure 2.5a shows the estimated global average temperature trends, while Figures 2.5b, c, and d show the northern hemisphere, southern hemisphere, and tropical trends. In addition, the 95% confidence intervals in the trends adjusted for temporal autocorrelation in each time series are included (Santer et al., 2000b).

As can be noted from Figure 2.5, the 95% confidence intervals for the PCM simulations are notably larger than those for the CSIRO model. This is because the former simulations represent the average trends for five different realizations. As such, the autocorrelation in this time series is much higher than that in the CSIRO time series resulting in fewer degrees of freedom and a much larger confidence interval than that in the single CSIRO time series. Of course the CSIRO simulation is but a single transient simulation – and a second or third transient may yield trends that are somewhat different from the trends reported here.

Comparing first the global trends which are shown in Figure 2.5a, we may note that all of the simulations and adjusted simulations are consistent with the global surface trend measured by Jones et al. (1999) and with both the UAH and RSS satellite-estimated MSU2 temperature trends. This agrees with the previous conclusions of Santer et al. (2000a) who compared satellite and surface trends with ECHAM model results. Nevertheless, when we examine the difference between the global average surface temperature trend and the MSU2 temperature trend, we see that none of the simulations that include an adjustment for the trends in fossil fuel BC + OM and biomass aerosols are consistent with the trend from the observations constructed from the difference between the Jones et al. (1999) surface temperatures and the UAH MSU2 temperatures, while all adjusted simulations are consistent (within the 95% confidence intervals) with the difference between the Jones et al. (1999) surface temperatures and the RSS MSU2 temperatures.

In the northern hemisphere and southern hemisphere, the 95% confidence intervals for the difference between surface and UAH observations do encompass the results for the PCM + FFC model, but do not encompass the results for the PCM + FFC + BB model. Moreover, they do not encompass any of the CSIRO model results. In the tropics, all of the models are consistent with the 95% confidence interval from the difference between the surface and UAH observations if they include the effects of volcanic aerosols. Thus, the adjustment of the CSIRO model transient simulation for fossil fuel BC + OM and biomass burning aerosols causes a significantly lower trend in the tropics compared to the (Jones − UAH) time series, but when the PCM volcano trend is added, the adjusted model is consistent with both the (Jones − UAH) time series and the (Jones − RSS) time series.

When we consider the difference between the surface and RSS observed trends, all of the model configurations are within the 95% confidence interval of the trend, regardless of whether or not fossil fuel BC + OM and biomass aerosols are included.

2.5 Conclusions

Our current results demonstrate that the difference in the surface and MSU channel 2 temperature trends from the models is consistent with the temperature trend constructed from the difference between the Jones and RSS time series irrespective of whether BC + OM from fossil fuel and biomass aerosols are included in the models. But the global average modeled trends, as well as the trends for the northern hemisphere and southern hemisphere, are not consistent with the difference between the Jones and UAH time series when fossil fuel BC + OM and biomass aerosols are included. We note that this conclusion is specific to the trends that we estimated and used for FF BC + OM and BB aerosols and that the

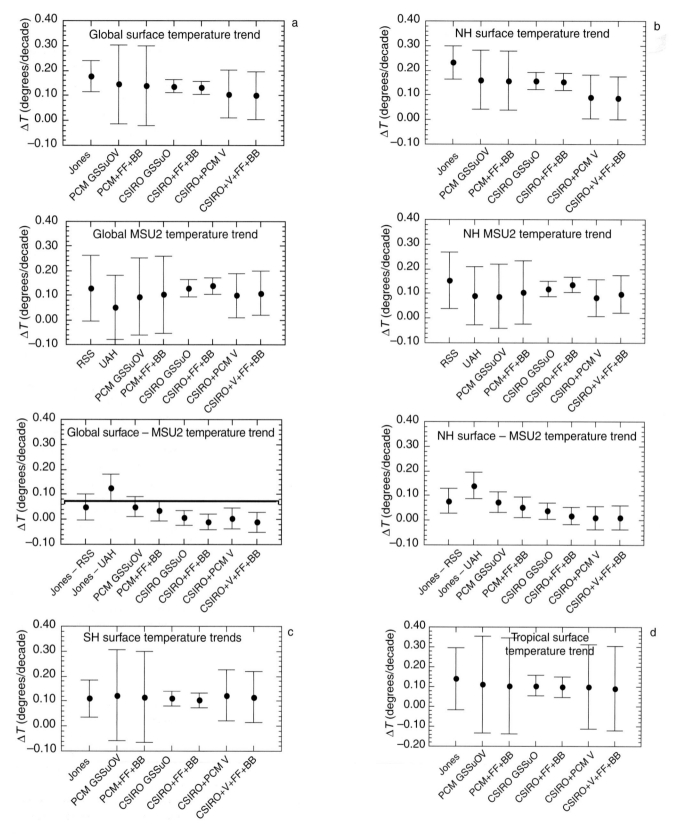

Figure 2.5 Trends and 95% confidence intervals for observations and models for (a) global average surface temperature (b) northern hemisphere, (c) southern hemisphere, and (d) the tropics.

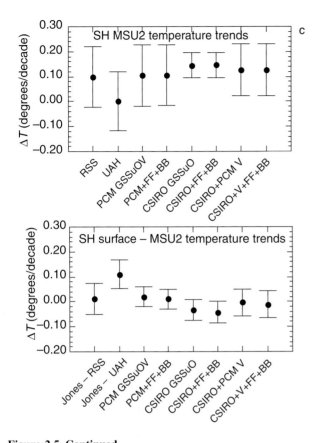
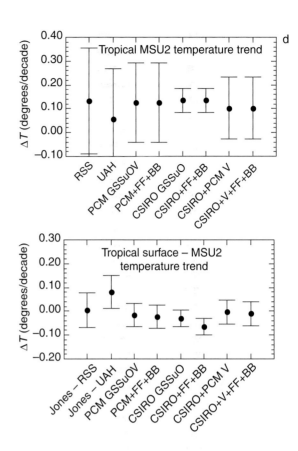

Figure 2.5 Continued

models may require improved historical emissions for both fossil fuel BC + OM and biomass aerosols. Thus, the disagreement between the Jones and UAH time series may point to deficiencies in both the models and the measurements. We also note that we were not able to include any reduction in emission factors associated with technology improvements in burning fossil fuels other than improvements in the emission factors for the use of diesel fuels in the transport sector. Adding improved emissions trends associated with changing technology for other fuels may lead to improved agreement between the model-simulated trends and the trends constructed from the surface minus satellite observations, because a decrease in FF BC emissions may tend to warm the surface more than the mid troposphere. Errors in our estimated biomass burning trends may not be so important, since the estimated difference in surface and MSU2 temperature trends from this source is small.

We have not been able to adjust for spatially specific changes in emissions. Thus, if emissions changes were much larger in Asia than they were in Africa, our adjustment procedure is unable to capture this change, because we used a single pattern of change and scaled the pattern to the global average change in emissions. Clearly, it will be important to re-examine the issue of the change in surface and MSU2 temperature trends with a set of transient simulations that include the full set of known forcings. Finally, we note that improvements should include the effect of BC absorption on ice and snow albedos. The latter might also significantly perturb atmospheric temperatures (Hansen et al., 2004).

Acknowledgements

We wish to thank Martin Dix for providing the CSIRO transient model simulation.

References

Ackerman, A. S., Toon, O. B., Stevens, D. E. et al. (2000). Reduction of tropical cloudiness by soot. *Science* **288**, 1042–1047.

Arellano, A. F. Jr., Kasibhatla, P. S., Giglio, L., van der Werf, G. R. and Randerson, J. T. (2004). Top-down estimates of global CO sources using MOPITT measurements. *Geophysical Research Letters* **31**, L01104; doi: 10.1029/2003GL018609.

Bond, T. C., Streets, D. G., Yarber, K. F. et al. (2004). A technology-based global inventory of black and organic carbon emissions from combustion. *Journal of Geophysical Research* **109**, D14203; doi: 10.1029/2003JD003697.

Christy, J. R. and Norris, W. B. (2004). What may we conclude about global tropospheric temperature trends? *Geophysical Research Letters* **31**, L06211; doi: 10.1029/2003GL019361.

Christy, J. R., Spencer, R. W. and Braswell, W. D. (2000). MSU tropospheric temperatures: dataset construction and radiosonde comparisons. *Journal of Atmospheric and Oceanic Technology* **17**, 1153–1170.

Douglass, D. H., Pearson, B. D. and Singer, S. F. (2004). Altitude dependence of atmospheric temperature trends: climate models versus observation. *Geophysical Research Letters* **31**, L13208; doi: 10.1029/2004GL020103.

FAOSTAT (2004). *Food and Agricultural Organization Statistical Database*. Rome, Italy. Available at http://apps.fao.org

Fu, Q., Johanson, C. M., Warren, S. G. and Seidel, D. (2004). Contribution of stratospheric cooling to satellite-inferred tropospheric temperature trends. *Nature* **429**, 55–58.

Gordon, H. B., Rotstayn, L. D., McGregor, J. L. et al. (2002). *The CSIRO Mk3 Climate System Model* [Electronic publication]. Aspendale: CSIRO Atmospheric Research (CSIRO Atmospheric Research technical paper No. 60). www.dar.csiro.au/publications/gordon_2002a.pdf.

Grant, K. E., Chuang, C. C., Grossman, A. S. and Penner, J. E. (1999). Modeling the spectral optical properties of ammonium sulfate and biomass burning aerosols; parameterization of relative humidity effects and model results. *Atmospheric Environment* **33**, 2603–2620.

Hansen, J. and Nazarenko, L. (2003). Soot climate forcing via snow and ice albedos. *Proceedings of the National Academy of Sciences USA* **101**, 423–428; doi: 10.1073/pnas.2237157100.

Hansen, J., Wilson, H. and Sato, M. (1995). Satellite and surface temperature data at odds? *Climatic Change* **30**, 103–117.

Hansen, J., Sato, M., Nazarenko, L. et al. (2002). Climate forcings in Goddard Institute for Space Studies SI2000 simulations. *Journal of Geophysical Research* **107**, (D18), 4347; doi: 10.1029/2001JD001143.

Hegerl, G. and Wallace, J. M. (2002). Influence of patterns of climate variability on the difference between satellite and surface temperature trends. *Journal of Climate* **15**, 2412–2428.

Herman, J. R., Bhartia, P. K., Torres, O. et al. (1997) Global distribution of UV-absorbing aerosols from Nimbus-7/TOMS data. *Journal of Geophysical Research* **102**, 16 911–16 922.

Hirst, A. C., O'Farrell, S. P. and Gordon, H. B. (2000). Comparison of a coupled ocean-atmosphere model with and without oceanic eddy-induced advection. Part I: Ocean spinup and control integrations. *Journal of Climate* **13**, 139–163.

Ito, A. and Penner, J. E. (2004). Global estimates of biomass burning emissions based on satellite imagery for the year 2000. *Journal of Geophysical Research* **109**, (D14)D14S05; doi: 10.1029/2003JD004423.

Ito, A. and Penner, J. E. (2005). Historical emissions of carbonaceous aerosols from biomass and fossil fuel burning for the period 1870–2000. *Global Biogeochemical Cycles* **19**, No. 2; GB2028; doi: 10.1029/2004GB002374.

Jones, P. D. and Moberg, A. (2003). Hemispheric and large-scale surface air temperature variations: an extensive revision and an update to 2001. *Journal of Climate* **16** (2), 206–223.

Jones, P. D., New, M., Parker, D. E., Martin, S. and Rigor, I. G. (1999). Surface air temperature and its changes over the past 150 years. *Reviews of Geophysics* **37**, 173–199.

Kistler, R., Collins, W., Saha, S. et al. (2001). The NCEP_NCAR 50-year reanalysis: monthly means. *Bulletin of the American Meterological Society* **82**, 247–267.

Liousse, C., Penner, J. E., Chuang C. et al. (1996). A three-dimensional model study of carbonaceous aerosols. *Journal of Geophysical Research* **101**, 19 411–19 432.

Mears, C. A., Schabel, M. C. and Wentz, F. J. (2003). A reanalysis of the MSU channel 2 tropospheric temperature record. *Journal of Climate* **16**, 3650–3664.

Meehl, G. A., Washington, W. M., Wigley, T. M. L., Arblaster, J. M. and Dai, A. (2003). Solar and greenhouse gas forcing and climate response in the twentieth century. *Journal of Climate* **16**, 426–444.

Menon, S., Hansen, J. Nazarenko, L. and Luo, Y. (2002). Climate effects of black carbon aerosols in China and India. *Science* **297**, 2250–2253.

Novakov, T., Ramanathan, V., Hansen, J. E. et al. (2003). Large historical changes of fossil-fuel black carbon aerosols. *Geophysical Research Letters* **30** (6), 1324; doi: 10.1029/2002GL016345.

Parker, D., Gordon, M., Cullum, D. P. N. et al. (1997). A new gridded radiosonde temperature data base and recent trends. *Geophysical Research Letters* **24**, 1499–1502.

Penner, J. E. (2003). Comments on "Control of fossil-fuel particulate black carbon and organic matter, possibly the most effective method of slowing global warming" by Jacobson. *Journal of Geophysical Research* **108** (D24), 4771; doi: 10.1029/2002JD003364.

Penner, J. E., Chuang, C. and Grant, K. (1998). Climate forcing by carbonaceous and sulfate aerosols. *Climate Dynamics* **14**, 839–851.

Penner, J. E., Zhang, S. Y. and Chuang, C. C. (2003). Soot and smoke aerosol may not warm climate. *Journal of Geophysical Research* **108** (D21), 4657; doi: 10.1029/2003JD003409.

Prabhakara, C., Iacovazzi, J. R., Yoo, J.-M. and Dalu, G. (1998). Global warming deduced from MSU. *Geophysical Research Letters* **25**, 1927–1930.

Ramaswamy, V., Boucher, O., Haigh, J. et al. (2001). Radiative forcing of climate change. In *Climate Change 2001: The Scientific Basis. Contribution of Working Group I to the Third Assessment Report of the Intergovernmental Panel on Climate Change*, ed. J. T. Houghton, Y. Ding, D. J. Griggs, et al. Cambridge: Cambridge University Press, pp. 349–416.

Rotstayn, L. D. and Lohmann, U. (2002a). Tropical rainfall trends and the indirect aerosol effect. *Journal of Climate* **15**, 2103–2116.

Rotstayn, L. D. and Lohmann, U. (2002b). Simulation of the tropospheric sulfur cycle in a global model with a physically based cloud scheme. *Journal of Geophysical Research* **107** (D21), 4592; doi: 10.1029/2002JD002128.

Rotstayn, L. D. and Penner, J. E. (2001). Forcing, quasi-forcing and climate response. *Journal of Climate* **14**, 2960–2975.

Santer, B. D., Wigley, T. M. L., Gaffen, D. J. *et al.* (2000a). Interpreting differential temperature trends at the surface and in the lower troposphere. *Science* **287**, 1227–1232.

Santer, B. D., Wigley, T. M. L., Boyle, J. S. *et al.* (2000b). Statistical significance of temperature trends. *Journal of Geophysical Research* **105**, 7337–7356.

Santer, B. D., Wigley, T. M. L., Meehl, G. A. *et al.* (2003). Influence of satellite data uncertainties on the detection of externally forced climate change. *Science* **300**, 1280–1284.

Shine, K. P., Cook, J., Highwood, E. J. and Joshi, M. M. (2003). An alternative to radiative forcing for estimating the relative importance of climate change mechanisms. *Geophysical Research Letters* **30**, No. 20, 2047; doi: 10.1029/2003GL018141.

Spencer, R. W. and Christy, J. R. (1992). Precision and radiosonde validation of satellite gridpoint temperature anomalies. Part II: A tropospheric retrieval and trends during 1979–1990. *Journal of Climate* **5**, 858–866.

Tett, S. F. B., Jones, G. S., Stott, P. A. *et al.* (2002). Estimation of natural and anthropogenic contributions to twentieth century temperature change. *Journal of Geophysical Research* **107** (D16), 4306; doi: 10.1029/2000JD000028.

Torres, O., Bhartia, P. K., Herman, J. R., Ahmad, Z. and Gleason, J. (1998). Derivation of aerosol properties from satellite measurements of backscattered ultraviolet radiation: theoretical basis. *Journal of Geophysical Research* **103**, 17 099–17 110.

Vinnikov, K. Y. and Grody, N. C. (2003). Global warming trend of mean tropospheric temperature observed by satellites. *Science* **302**, 269–272.

Washington, W. M., Weatherly, J. W., Meehl, G. A. *et al.* (2000). Parallel Climate Model (PCM) control and transient simulations. *Climate Dynamics* **16**, 755–774.

Yevich, R. and Logan, J. A. (2003). An assessment of biofuel use and burning of agricultural waste in the developing world. *Global Biogeochemical Cycles* **17**, (4), 1095; doi: 10.1029/2002GB001952.

3

Evaluating the impacts of carbonaceous aerosols on clouds and climate

Surabi Menon and Anthony D. Del Genio

3.1 Introduction

Any attempt to reconcile observed surface temperature changes within the past 150 years to changes simulated by climate models that include various atmospheric forcings is sensitive to the changes attributed to aerosols and aerosol–cloud–climate interactions, which are the main contributors that may well balance the positive forcings associated with greenhouse gases, absorbing aerosols, ozone related changes, etc. These aerosol effects on climate, from various modeling studies discussed in Menon (2004), range from $+0.8$ to $-2.4\,\text{W/m}^2$, with an implied value of $-1.0\,\text{W/m}^2$ (range from -0.5 to $-4.5\,\text{W/m}^2$) for the aerosol indirect effects. Quantifying the contribution of aerosols and aerosol–cloud interactions remains complicated for several reasons, some of which are related to aerosol distributions and some to the processes used to represent their effects on clouds. Aerosol effects on low-lying marine stratocumulus clouds that cover much of the Earth's surface (about 70%) have been the focus of most prior simulations of aerosol–cloud interaction effects. Since cumulus clouds (shallow and deep convective) are short-lived and cover about 15 to 20% of the Earth's surface, they are not usually considered as radiatively important. However, the large amount of latent heat released from convective towers, and corresponding changes in precipitation, especially in biomass regions owing to convective heating effects (Graf et al., 2004), suggest that these cloud systems, and aerosol effects on them, must be examined more closely. The radiative heating effects for mature deep convective systems can account for 10–30% of maximum latent heating effects and thus cannot be ignored (Jensen and Del Genio, 2003).

The first study that isolated the sensitivity of cumulus clouds to aerosols was from Nober et al. (2003) who found a reduction in precipitation in biomass burning regions and shifts in circulation patterns. Aerosol effects on convection have been included in other models as well (cf. Jacobson, 2002) but the relative impacts on convective and stratiform processes were not separated. Other changes to atmospheric stability and thermodynamical quantities due to aerosol absorption are also known to be important in modifying cloud macro/micro properties. Linkages between convection and boreal biomass burning can also affect the upper troposphere and lower stratosphere, radiation and cloud microphysical properties via transport of tropospheric aerosols to the lower stratosphere during extreme convection (Fromm and Servranckx, 2003). Relevant questions regarding the impact of biomass aerosols on convective cloud properties include the effects of vertical transport of aerosols, spatial and temporal distribution of rainfall, vertical shift in latent heat release, phase shift of precipitation, circulation and their impacts on radiation.

Over land surfaces, a decrease in surface shortwave radiation ($\sim$3–6 W/m^2 per decade) has been observed between 1960 to 1990, whereas increases of 0.4 K in land temperature that occurred during the same period have resulted in speculations that evaporation and precipitation should also have decreased (Wild et al., 2004). However, precipitation records for the period 1950–2000 over most land areas do not indicate any decrease, or for that matter any significant trend (Beck et al., 2005). Wild et al. (2004) speculate that the decrease in precipitation may be related to increased moisture advection from the oceans, which may well have some contributions from

---

Human-induced Climate Change: An Interdisciplinary Assessment, ed. Michael Schlesinger, Haroon Kheshgi, Joel Smith, Francisco de la Chesnaye, John M. Reilly, Tom Wilson and Charles Kolstad. Published by Cambridge University Press. © Cambridge University Press 2007.

aerosol–radiation–convection coupling that could modify circulation patterns and hence moisture advection in specific regions.

Other important aspects of aerosol effects, besides the direct, semi-direct, microphysical and thermodynamical impacts, include alteration of surface albedos, especially snow and ice covered surfaces, due to absorbing aerosols. These effects are uncertain (Jacobson, 2004) but may produce as much as $0.3 \, W/m^2$ forcing in the northern hemisphere that could contribute to melting of ice and permafrost and change in the length of the season, such as the early arrival of spring (Hansen and Nazarenko, 2004). Besides the impacts of aerosols on the surface albedos in the polar regions, and the thermodynamical impacts of Arctic haze (composed of water-soluble sulfates, nitrates, organic and black carbon (BC)), the dynamical response to Arctic haze (through the radiation–circulation feedbacks that cause changes in pressure patterns) is thought to have the potential to modify the mode and strength of large-scale teleconnection patterns such as the Barents Sea Oscillation that could affect other climate regimes, mainly in Europe (Rinke et al., 2004). Additionally, via the Asian monsoon, wind patterns over the eastern Mediterranean and lower stratospheric pollution at higher latitudes (Lelieveld et al., 2002) are thought to be linked to the pollutants found in Asia, indicating the distant climate impacts of aerosols.

Thus, it becomes difficult to quantify regional changes in precipitation or radiation in the context of a regional model, since long-range transport of aerosols and its dynamical impacts can modify distant climates and can affect boundary conditions that are imposed in regional simulations. On the other hand, the use of a global model to investigate regional changes may lead to problems associated with not resolving processes on coarse spatial scales. Here, we will examine global model simulations of the indirect effect for warm clouds, both stratiform and convective, in order to examine the influence of aerosols in modifying precipitation and radiation. Since stratus clouds cover large areas, spatial domain is not a constraint, whereas for convective clouds in biomass regions, the area covered by biomass burning may be large enough to examine regional influences. Section 3.2 describes the model and simulations that were performed in this study, Section 3.3 describes the results from the modeling study as well as results obtained for different regions, and finally in Section 3.4 we present the conclusions from this study.

3.2 Model description

We use the Goddard Institute for Space Studies (GISS) ModelE, the latest version of the GISS model II', which is described in more detail by Schmidt et al. (2005) and Hansen et al. (2005), coupled to the aerosol chemistry model of Koch et al. (1999, 2005). Cloud water is treated prognostically in the model (Del Genio et al., 1996), with representation of sources due to large-scale convergence and cumulus detrainment, and sinks due to autoconversion, accretion, evaporation, and cloud top entrainment, as well as precipitation enhancement due to the seeder-feeder effect. Stratiform cloud cover is a diagnostic function of relative humidity and stability allowing for subgrid vertical cloud fraction. Moist convection uses a quasi-equilibrium cloud base closure with entraining and non-entraining updrafts and a cumulus-scale downdraft. Main changes in the updated GCM of relevance to the indirect effect include a microphysics-based cumulus scheme described in Del Genio et al. (2005), and subgrid turbulence at all levels of the GCM (based on the second-order closure model of Cheng et al. 2003). Improvements to the cumulus scheme include a more physical partitioning of convective condensate between precipitation and detrainment into anvil clouds based on a microphysics scheme (using a Marshall–Palmer distribution for droplets, empirical relationships between droplet size and terminal velocity for liquid, graupel, and ice hydrometeors, and prescribed cumulus updraft speeds).

The aerosol emission data sets used in the model are derived from AEROCOM (an aerosol model intercomparison project; Frank Dentener, personal communication, 2004). Sulfate aerosols include fossil fuel sources that are country-based emissions for the year 2000 and both continuously erupting and explosive volcanoes (Andres and Kasgnoc, 1998; Halmer and Schmincke, 2003) as well as dimethyl sulfide (DMS) emissions. Carbonaceous aerosols (treatment described in Koch and Hansen, 2005) include fossil- and biofuel sources from Bond et al. (2004) and biomass sources. Biomass sources for sulfates and carbonaceous aerosols are from van der Werf et al. (2004). In addition, organic aerosols include a fraction of natural terpenes (15%) from secondary organics. Solubility for industrial carbonaceous aerosols is 100% for aged aerosols (e-folding time of 1 day). For the biomass components, solubilities of 80% and 60% are assumed for organic and BC, respectively. Sea-salt is also calculated based on model wind speeds (Koch et al., 2005) instead of through the source function used previously (Koch et al., 1999). All aerosols are treated as external mixtures, which may not accurately represent aerosol radiative properties for regions away from sources, where aerosols tend to form internal mixtures.

Improvements to the treatment of the aerosol indirect effect described in Menon et al. (2002a) (hereafter referred to as M02) include a semi-prognostic treatment of the cloud droplet number concentration (CDNC), addition of BC aerosol effects on clouds, and incorporation of subgrid vertical velocity effects on cloud droplet number. Several parameterizations have been used to determine CDNC, which is the main and critical link between the aerosols and cloud microphysics. The resulting values for the indirect effect vary by ~40%, depending on the assumptions or parameterizations used. Kiehl et al. (2000) find over a factor of 3 difference in their estimate of the indirect effect when using different CDNC schemes. Here, we mainly refer to simulations that use the parameterization from Gultepe and Isaac (1999), given as:

$$CDNC_{Land} = 298 \times \log_{10} N_{a,\,land} - 595$$
$$CDNC_{Ocean} = 162 \times \log_{10} N_{a,\,ocean} - 273 \quad (3.1)$$

where N_a is the aerosol concentration (cm^{-3}) calculated from the aerosol mass concentration and assumed sizes as used in Lohmann et al. (1999). The aerosol concentration N_a for land includes the effects of sulfates, organic and BC aerosols, whereas N_a for ocean includes, in addition, sea-salt aerosols. The values so derived for CDNC are representative of in-cloud CDNC values since measurements were for in-cloud values. Dust is assumed to be insoluble in these simulations and since the amount of dust roughly remains unchanged for present-day and pre-industrial simulations, its contribution to the indirect effect in terms of changes in CDNC is neglected. Future treatments will include dust formed on sulfate or sea-salt via heterogeneous chemical reactions (Bauer et al., 2005), which is not discussed in this work. Aerosol sources for present-day simulations include both anthropogenic (sulfates and carbonaceous aerosols from fossil fuel, biofuel and biomass sources) and natural sources from DMS, volcanic and biomass emissions (assumed to be 50% of the anthropogenic biomass source), organic aerosols from terpenes, sea-salt and dust. Aerosol sources for pre-industrial simulations include just the natural sources. Where applicable, we use "Δ" to denote differences between present-day and pre-industrial simulations. All simulations use fixed sea-surface temperatures derived for present-day climatologies. All results reported are based on 6 years of model simulations averaged over the past 5 years.

3.3 Aerosol indirect effect on warm clouds

Prior model estimates of the aerosol indirect effect (AIE) with the GISS GCM (M02) are based on simulations that distinguish between pre-industrial and present-day aerosol emissions. These estimates depend on the background concentration imposed for either aerosols or CDNC. As an example, in climate simulations for present-day versus pre-industrial aerosols, changing the minimum (background) CDNC from 10 cm^{-3} to 40 cm^{-3} produces almost a 50% change in the indirect effect estimate (Menon, 2004). Here, we use a value of 20 cm^{-3} for background CDNC for all our simulations based on surface observations that indicate values of ~10 to 20 cm^{-3} in pure marine background air with no wind (J.-L. Brenguier, personal communication). We perform several sets of simulations, shown in Table 3.1, to investigate the impacts of fossil- and biofuel BC aerosols, aerosol–convective cloud effects, and the semi-direct aerosol effects. Note that all simulations only include aerosol effects on warm cloud microphysics and that they include both the first indirect effect (changes in cloud reflectivity due to smaller droplet sizes;

Table 3.1 Model simulations and their description.

Simulation	Type
Exp A	Standard run with both types of aerosol indirect effects
Exp A_S	Like Exp A but with changes to the aerosol burden (2×anthropogenic sulfate for present-day aerosol burden and 10% of present-day biomass aerosols for pre-industrial aerosol burden)
Exp NBC	Like Exp A but without any fossil- and biofuel sources for BC
Exp 2BC	Like Exp A but with twice the fossil- and biofuel sources for BC
Exp CC1	Like Exp A but including aerosol effects on convective clouds
Exp CC2	Like Exp CC1 but with a different treatment for determining cloud droplet number concentration in convective clouds
Exp NIE	Like Exp A but with fixed cloud droplet number

Twomey, 1977) and the second indirect effect (changes in cloud lifetime and cloud cover from suppression of precipitation; Albrecht, 1989). Our indirect effect represents the net effects of aerosols on clouds since it includes implicit changes to cloud fields from other climate feedbacks.

Additional simulations similar to those listed in Table 3.1 were performed that include aerosol emissions for pre-industrial scenarios (year 1850). Present-day emissions (year 2000) include both natural and anthropogenic aerosols. Most of the treatment for the aerosol indirect effect follows that given in M02, except that we modulate the CDNC so that it changes with cloud cover and cloud water changes. Here, we use the formula of Beheng (1994) for autoconversion (the process that initiates the conversion of cloud water to precipitation), which is a function of cloud liquid water content and CDNC. The magnitude of the indirect effect, calculated as difference between present-day and pre-industrial simulations, for Exp A is about -1.36 W/m^2 obtained from differences between net radiation at the top of the atmosphere (TOA) and the direct aerosol effect.

Alternatively, the aerosol indirect effect can be obtained from the difference in net cloud radiative forcing, which then is -0.65 W/m^2. Differences between these methods of estimating the indirect effect, which are both approximate, are discussed in M02. They are dependent on the treatment used to separate radiation into that due to aerosols or that due to clouds. These values are significantly lower than our previous estimates which ranged from -1.55 to -4.4 W/m^2 for the indirect effect (change in net cloud forcing) and could be attributed to the minimum CDNC used here (as opposed to

Table 3.2 Globally averaged annual values of aerosol column burdens (mg/m^2) for the different simulations. Two values listed for M02 are for two sets of simulations of the first and second indirect effects that mainly differ in the treatment of the autoconversion scheme. The values separated by '/' are for present-day and pre-industrial aerosol burdens. Note that in M02 only total organic aerosol burden was available, which includes the biomass/terpene components.

Case	Sulfate total	OC (fossil- & biofuel)	OC (biomass & terpene)	BC (fossil- & biofuel)	BC (biomass)	Net cloud forcing (W/m^2)
M02	2.66/0.42	1.57/0.14		–	–	–4.36
	5.03/1.05	2.46/0.27		–	–	–2.41
Exp A	2.96/0.15	0.98/0.57	1.61/0.80	0.13/0.0	0.12/0.06	–0.65
Exp A_S	4.34/0.14	0.96/0.55	1.63/0.15	0.12/0.0	0.12/0.01	–1.03

10 cm^{-3} in M02); inclusion of dust and BC aerosols (which will increase the positive forcing); inclusion of a relative humidity factor that accounts for changes in extinction efficiencies due to aerosol growth (~36% increase in net cloud forcing); choice of autoconversion parameterization and other general improvements in Model E (e.g. improvements to the convective scheme resulting in increased coverage of marine stratus clouds along the western edges of the continent, which were previously deficient; Del Genio et al., 2005). Perhaps the more significant reason for the smaller values obtained here is the smaller anthropogenic aerosol burden and the larger assumed natural aerosol burden (increased biomass aerosols assumed for pre-industrial conditions compared with Koch et al., 1999) as shown in Table 3.2. In Exp A, we take anthropogenic aerosol column burdens to be 2.3 times the assumed natural aerosol burden; in M02 they were 4.6 to 6.5 times the natural burden. This smaller burden leads to the large reduction in the aerosol indirect effect magnitude.

As a sensitivity test, we performed an additional simulation, similar to Exp A but with increased anthropogenic sulfates (factor of 2 increase) for present-day conditions and one with reduced biomass aerosols (10% of present-day) for pre-industrial conditions (Exp A_S). The resulting anthropogenic column burden in Exp A_S is higher by a factor of 7 compared with the assumed natural burden. Compared with ΔExp A, the magnitude of the indirect effect increases by 60% for ΔExp A_S, mainly owing to the higher anthropogenic burden in Exp A_S. As shown in Table 3.2, in addition to the choice of the autoconversion parameterization (which leads to some of the differences in the two M02 simulations), aerosol burdens play a major role in determining changes in radiative forcing.

3.3.1 Black carbon aerosol effects on clouds

As shown in Table 3.3, the forcing efficiency, defined as the ratio of direct forcing to anthropogenic column burden, of fossil- and biofuel BC is quite high. Since the heating associated with these absorbing aerosols can lead to significant changes in surface energy budgets, precipitation, cloud cover and circulation (Ramanathan et al., 2001; Chung et al., 2002;

Table 3.3 Values of direct forcing and forcing efficiencies for the different species for Exp A.

Case	Sulfate total	OC (fossil- and bio-fuel, biomass)	BC (fossil- and biofuel)	BC (biomass)	Total
Direct forcing (W/m^2)	–0.29	–0.13	0.18	0.06	–0.18
Forcing efficiency (W/g)	–103	–106	1385	857	NA

Menon et al., 2002b), and since the sources for fossil- and biofuel BC may be controlled more readily (via controls on technology, e.g. emission filters for transportation, clean-coal burning technology, use of natural gas to replace biofuels, etc.) than that for biomass-derived aerosols, a closer inspection is warranted to understand the impacts of these absorbing aerosols on clouds and climate.

Present-day climate simulations (with twice the BC from fossil- and biofuel sources (Exp 2BC); with the standard amount (Exp A); and without any fossil- and biofuel BC (Exp NBC)) were compared with climate simulations for pre-industrial aerosol concentrations. For annual global changes, the major differences made by increasing amounts of BC were decreased net radiation (solar) at the surface, and decreased sensible heat flux, precipitation and soil evaporation. Increases in net TOA radiation, shortwave cloud forcing and water cloud optical depth were also observed for increasing BC. With respect to simulations for pre-industrial aerosol concentrations, these changes were within 2.5% except for changes to water cloud optical depths (>10%), net TOA radiation (>15%) and net heating (>15%) at the surface. Differences of 3% are also manifested in cloud top temperature (which decreases with increasing BC) for varying BC amounts. While global differences are not as high, regional changes are more significant and are discussed in more detail in Section 3.3.3. Although changes in optical depth are related to changes in cloud water paths and cloud droplet sizes, the increase in cloud optical depth is not related to changes in CDNC (~5% between Exp A, 2BC and NBC), but rather to changes due to the thermodynamical effects of heating (sensible and latent heat fluxes) that change the availability of water since precipitation is reduced.

Also of significance are changes to the zonally averaged vertical profiles of temperature at higher latitudes, which increase with increasing BC especially towards the poles and higher levels (indicative of upper tropospheric transport of warm air), and solar radiation heating rates that are much stronger in the northern hemisphere. This is not surprising since the impacts of BC-related forcing tend to be stronger in regions of high surface albedo (Hansen and Nazarenko, 2004). Increasing amounts of BC also lead to greater reduction in the extent of snow and ice cover (reductions seen in ΔExp 2BC and ΔExp A are a factor of 1.6 higher than those in ΔExp NBC), as was also indicated by Hansen and Nazarenko (2004), Wang (2004) and Roberts and Jones (2004); with stronger changes in the northern hemisphere and the Arctic than in the Antarctic. Also of significance is the change in sea-level pressure over all ocean regions and part of the NH continents (positive anomalies in ΔExp 2BC replace neutral changes found in ΔExp A and ΔExp NBC) indicative of changes in the circulation patterns that are stronger in ΔExp 2BC. Similar to the results of Wang (2004), we find changes in circulation and snow cover, as well as surface fluxes, with changing BC amounts. Analyses of the June to August precipitation fields for Exp A, NBC and 2BC do indicate a northward shift in the Intertropical Convergence Zone (ITCZ) in Exp 2BC and Exp A, as was also observed in simulations of BC climate effects by Wang (2004) and Roberts and Jones (2004). Without fossil- and biofuel BC this northward shift in the ITCZ is not evident.

Other uncertainties, besides vertical and horizontal distributions, are related to absorption properties of BC within clouds. In-cloud absorption by BC is thought to increase the absorption efficiency of clouds by 5% globally, with more significant regional changes (15–25%; Chuang et al., 2002). These results were for the first indirect effect only. Accounting for these BC absorption effects in cloud for both indirect effects, via a proxy enhancement factor (25%) that is analogous to increasing BC optical depth, results in a 12% increase in net cloud forcing (sum of shortwave and longwave) globally. Regionally, differences are much greater in regions with large BC emissions and co-located clouds. In terms of surface temperature response with BC in-cloud heating included, as shown in Figure 3.1 (see color plate section), global differences are negligible but regional differences are much greater in regions where BC emissions dominate, such as Asia, Europe and eastern United States. (Figure 3.2, in the colour plate section, shows the annual column burdens of carbonaceous aerosols from biomass and fossil- and biofuel sources.) Note that these responses are mainly a sensitivity test to indicate the lower bound of the change in temperature when BC in-cloud absorption is accounted for, since we use an atmosphere–only model with prescribed sea surface temperatures (Wang, 2004). A coupled ocean–atmosphere model would be more effective in simulating the full impact of observed surface temperature changes.

In an atmosphere model (Hadley Centre climate model) coupled to a slab-ocean model with four times as much fossil fuel BC as in Exp A, annual mean surface temperature change is ~0.436 K giving a climate sensitivity of 0.56 K/(W/m^2) (Roberts and Jones, 2004). However, effects of BC on cloud properties are not explicitly parameterized in these simulations. Within the same model the climate sensitivity to doubled CO_2 is about 0.91 K/(W/m^2). Another study on temperature response per unit of direct forcing (Jacobson, 2002), without the use of a slab-ocean model, reports a value of 1.4 K/(W/m^2) for BC and OC from fossil fuel. While a direct comparison to these values is not possible in the context of our simulations (since they include different methods on treatment of carbon optics and climate effects), we can compare the ratios of the relative magnitude of surface temperature changes to radiative flux values between ΔExp A, ΔExp NBC, and ΔExp 2BC. These simulations mainly differ in terms of the amounts of fossil- and biofuel BC. The values for ΔExp A, ΔExp NBC, and ΔExp 2BC are 0.12, 0.097, and 1.14 K/(W/m^2), respectively. The sensitivity of this version of the model coupled to a mixed ocean slab model to doubled CO_2 is ~0.66 K/(W/m^2) (Hansen et al., 2005). Although we cannot directly compare these BC and CO_2– related climate changes from the GISS GCM, because a coupled ocean–atmosphere model must be used to accurately estimate the climate sensitivity to BC, results from Roberts and Jones (2004) and Hansen et al. (2005) do suggest that climate sensitivity factors for BC may be quite comparable to that for CO_2.

3.3.2 Aerosol effects on convective clouds

Besides the radiative effects of fossil- and biofuel aerosols, aerosol effects from biomass burning are also important, especially near regions that are more active convectively. Here we examine aerosol effects on convective clouds, in addition to stratiform clouds. Since parameterizations for predicting CDNC from aerosols for convective clouds are not as well developed as those for stratiform clouds, we use the same droplet prediction scheme as for stratiform clouds (Exp CC1), following the approach of Nober et al. (2003). Using empirical data from Lahav and Rosenfeld (2000), we modify the precipitation rate as a function of cloud base CDNC and temperature, such that for temperatures below 263 K and for low CDNC values (<750 cm^{-3}), precipitation remains unchanged; and with increasing CDNC and temperatures above 263 K precipitation reduces to 25% (750 cm^{-3} < CDNC < 1000 cm^{-3}) and 0% (CDNC > 1000 cm^{-3}), respectively. These simulations only account for aerosol effects on warm convective clouds.

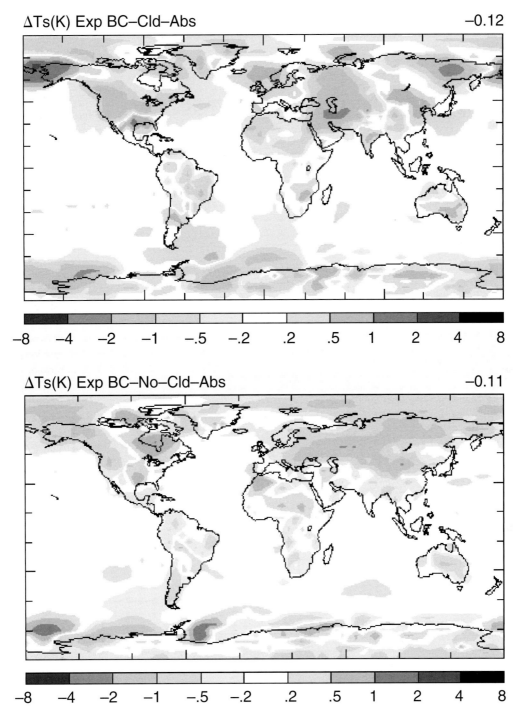

Figure 3.1 Model simulated annual surface temperature change (K) for year 2000 – Year 1850 for simulations that account for BC absorption in-cloud (top panel) and that do not account for BC (bottom panel).

The suppression of rain in warm convective clouds should lead to increased water contents and therefore should also have an effect on freezing rates for cold clouds and precipitation changes (Andreae et al., 2004) via the Bergeron–Findeisen process, which facilitates ice-phase precipitation processes at the expense of warm-phase processes.

In addition to Exp CC1, we conducted a second simulation (Exp CC2) with CDNC based on observations from the Cirrus

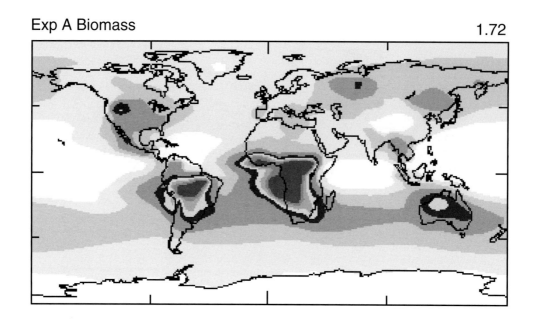

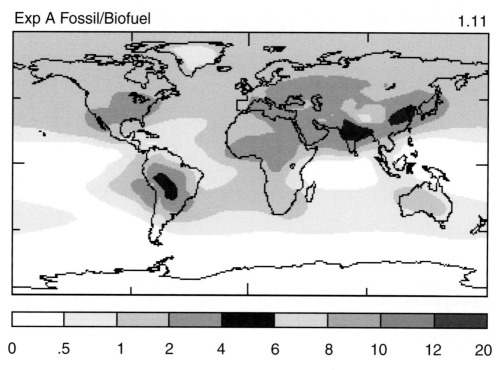

Figure 3.2 Annual values of carbonaceous aerosol column burden distribution (mg/m^2) from biomass (top panel) and fossil- and biofuel sources (bottom panel). Global mean values are on the right-hand side of the figure.

Regional Study of Tropical Anvils – Florida Area Cirrus Experiment (CRYSTAL FACE) data for warm cumulus clouds (Conant *et al.*, 2004). The relationship we use is given as:

$$\text{CDNC} = 10^{[0.433 + 0.815 \log N_a + 0.280 \log(w)]} \quad (3.2)$$

Here, w is the updraft velocity in m/s. Since model diagnostics do not account for cloud-scale updraft velocities we use the following notation

$$w = \bar{w} + \sqrt{0.66 \times \text{TKE}} \quad (3.3)$$

where $\bar{w}$ is the grid average updraft velocity and TKE is the turbulent kinetic energy (Ye Cheng, personal communication). Most of the climate diagnostics for ΔExp CC2 are similar to those for ΔExp CC1 except for larger differences in the net radiation at TOA, cloud water optical depths, and net heating at the surface, leading to differences in soil evaporation, sensible heat flux, and precipitation. Larger differences are seen regionally and are discussed in Section 3.3.3.

Changes with the addition of aerosol-convective cloud effects (Exp CC1/CC2 versus Exp A) include increases in the liquid water path (40%), water and ice cloud optical depths (increases 20%, and over a factor of 2, respectively), net TOA radiation and net ground heating (70% decrease), shortwave (16%) and longwave (7%) cloud forcing. Increase in cloud cover (water or ice, convective or total) is less than 5% and reduction in total precipitation is between 2 and 10%. With the increase in cloud water (almost a factor of 2), effective water cloud particle sizes (product of cloud effective radii and optical depth) are ~50% larger. More importantly, with aerosol-convective cloud effects the level of precipitation formation may be shifted higher since most of the cloud water increase is at higher levels. Shifts from warm to cold precipitation processes should increase the height at which latent heat is released in clouds with similar rainfall amounts as well as increasing the water available for ice processes, lightning and more intense convective storms (Andreae et al., 2004).

Changes in aerosol concentration for present-day versus pre-industrial simulations (higher by a factor of 2.7) result in values of 0 and $0.22\,\text{W/m}^2$ for the indirect effect from changes in net cloud radiative forcing for ΔExp CC1 and ΔExp CC2, respectively, and $-0.43\,\text{W/m}^2$ for both ΔExp CC1 and ΔExp CC2 when accounting for the difference in net TOA radiation and the direct effect. Despite similar increases in aerosol burdens, between ΔExp CC1, ΔExp CC2, and ΔExp A, this reduction in the net cloud radiative forcing obtained when including aerosol–convective cloud effects may arise from changes in the longwave component that are somewhat comparable to the shortwave. Some differences may also arise from changes in aerosol burden as indicated from a sensitivity test conducted to understand changes in climate diagnostics for differences in biomass aerosols. At present, we assume that pre-industrial biomass aerosols are ~50% of present-day biomass emissions. These estimates are highly uncertain because of the uncertainties in separating natural and anthropogenic contributions to biomass burning over the past 150 years (Dorothy Koch, personal communication). Thus, we perform similar simulations to Exp CC2 but with 10% of biomass for pre-industrial simulations. As compared with ΔExp CC2, the AIE is now slightly negative ($-0.07\,\text{W/m}^2$).

Nober et al. (2003) find no significant signal in the radiation budgets in their simulations, although it is not clear if they refer to changes between simulations with and without aerosol–convective cloud effects or changes with respect to changing aerosol concentrations. In our case, we do find increased cloud radiative forcings (more negative) for Exp CC1 and CC2 compared with Exp A, and smaller cloud radiative forcings for ΔExp CC1 and ΔExp CC2 compared with ΔExp A. Although aerosol–convective cloud effects may appear to be radiatively unimportant in their simulations, Nober et al. (2003) do find a weakening of the Walker circulation during June to August caused by the negative precipitation anomaly of the summer monsoon over the Indian subcontinent and surrounding regions including the southwest Pacific, which are correlated to changes in the velocity potential fields. In our simulations, zonally averaged profiles of vertical velocity and vertical transport of latent heat for June to August indicate increased vertical velocities and increased vertical transport of latent heat in the Southern Tropics for Exp CC1/CC2 versus Exp A, along with increased water contents at higher atmospheric levels. Comparison of observed June–July–August precipitation fields (e.g. from the global precipitation climatology project, http://precip.gsfc.nasa.gov/gifs/sg_v2.0259.gif) with present-day simulations for Exp A, CC1 and CC2 (see Figure 3.3a in colour plate section) indicates that Exp CC2 may exaggerate the reduction in precipitation in the Amazonia, but Exp CC1 does simulate the precipitation amounts over the tropical convective regions and mid-eastern Amazonia slightly more realistically than Exp A. This can be seen in Figure 3.3b in the colour plate section, which shows the difference in precipitation between Exp CC1 and Exp A (the decrease and increase in precipitation over the biomass burning regions are in the same direction as observations).

3.3.3 Regional impacts of aerosols on clouds and climate

Changes in global annual values may obscure significant regional as well as seasonal changes. Thus, it becomes necessary to focus on regional changes within a global model since although aerosol sources vary regionally, transport and teleconnections may play a role in modifying climate at a distance. Such changes have been found in several other studies that investigated the role of Asian and European/Russian aerosols in affecting the distant Arctic climate (Koch and Hansen, 2005). Increases in aerosol concentrations in North America via transport of dust plumes (that include entrained pollutants) from Mongolia and China (DeBell et al., 2004) have also been observed. Seasonality can also play a strong role in the types of aerosols produced (fossil or biomass based production) and thus may change values of the resulting radiative fluxes (Menon, 2004). Here, we focus on those regions that have the largest changes in either biomass or fossil- and biofuel based aerosols. The seasons are chosen based on the availability of observations from large field campaigns so that model simulations of certain variables may be compared to observed values.

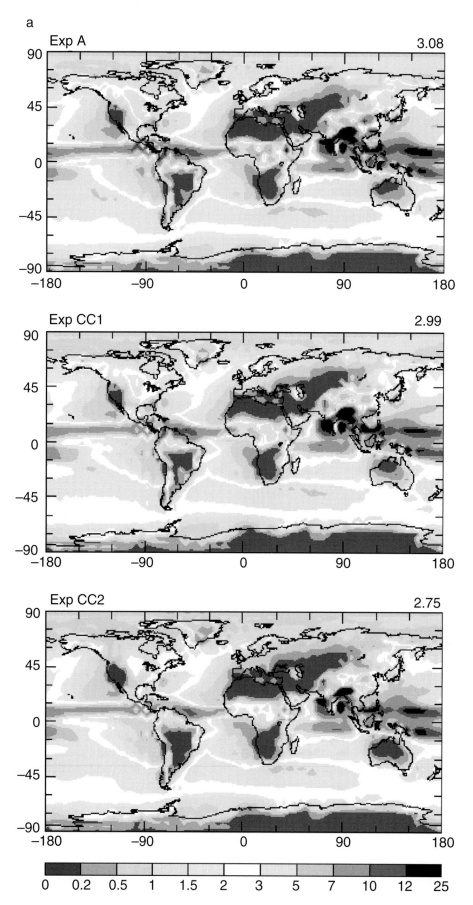

Figure 3.3 June–July–August precipitation (mm/day) fields for the year 2000 from Exp A, Exp CC1 and Exp CC2 (a), and change in precipitation between Exp CC1 and Exp A (b). Global mean values are indicated on the right-hand side.

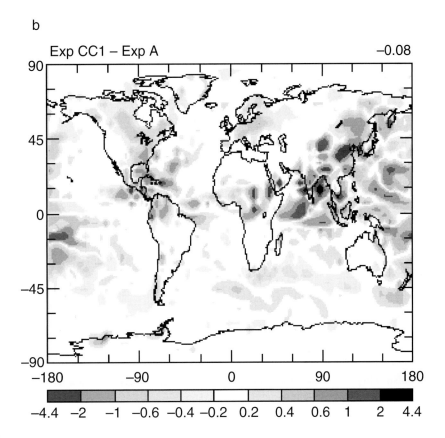

Figure 3.3 Continued

Black carbon aerosol effects on regional climate

Aerosol effects over China are thought to be partly responsible for the anomalous changes in the precipitation field observed over the past 30 years, with the north being more prone to droughts, while the south is subject to large-scale flooding (Xu, 2001; Menon et al., 2002b). Several other studies have investigated changes in sunshine duration, diurnal cycle of temperature, and the anomalies of the northwestern Pacific subtropical high (Kaiser and Qian, 2002; Gong et al., 2004), which are correlated to these anomalous precipitation fields. Similarly, over the Indian subcontinent and surrounding regions, Ramanathan et al. (2001) and Chung et al. (2002) find large spatial changes in the distribution of precipitation related to changes in surface forcing, from absorbing aerosols.

Here, we show values of radiative fluxes at the surface, TOA, and the atmospheric column along with precipitation for select regions over India and China in Tables 3.4a and b, respectively. The region over China was based on prior results from Menon et al. (2002b) and Menon (2004) to highlight changes in the northern/southern regions based on changes due to BC. Over India, we chose a region based on field data that were available from the Indian Ocean Experiment (INDOEX) as given in Ramanathan et al. (2001). During January to March 1999, radiative forcings at the surface, TOA and atmosphere were −20, −2, 18 W/m^2 for the direct aerosol effect and −6, −5, 1 W/m^2 when including the first aerosol indirect effect

Table 3.4a Simulated changes in the top of the atmosphere (TOA), surface, and atmospheric net radiation budgets (W/m^2), and precipitation over India/Indian Ocean region.

Indian Ocean (Jan–Mar) 0–20° N, 40–100° E	ΔTOA (W/m^2)	ΔSfc. (W/m^2)	ΔAtmos. (W/m^2)	ΔPrec. (mm/d)
Exp A	−2.97	−7.33	4.36	0.35
Exp NBC	−2.07	−3.52	1.45	−0.08
Exp 2BC	−2.06	−5.71	3.65	0.01

Table 3.4b Similar to Table 3.4a but for differences over northern and southern China.

China (Jun–Aug) 90–120° E		ΔTOA (W/m^2)	ΔSfc. (W/m^2)	ΔAtmos. (W/m^2)	ΔPrec. (mm/d)
Exp A	34–42° N	−6.46	−5.38	−1.08	−0.39
	18–30° N	−5.69	−7.77	2.08	0.05
Exp NBC	34–42° N	−7.81	−5.17	−2.64	−0.10
	18–30° N	−6.56	−6.90	0.34	0.24
Exp 2BC	34–42° N	−3.87	−6.48	2.61	−0.77
	18–30° N	−5.41	−9.28	3.87	−0.13

(Ramanathan et al., 2001). Values as high as 33 W/m^2 have been reported based on observed radiative fluxes and aerosol properties at a site in India during the dry season (January–April) (Pandithurai et al., 2004) for the aerosol direct effect. In our simulations, the forcings computed for the Indian sub-continent and Ocean region for the January–March period, shown in Table 3.4a, are for differences between present-day and pre-industrial aerosol emissions, similar to those from Ramanathan et al. (2001). As shown in Table 3.4a, our TOA and atmospheric fluxes are generally higher than those obtained from INDOEX. In both regions, precipitation appears to increase with increasing atmospheric fluxes (reduced surface fluxes), which may be partly due to the response of convection to the atmospheric heating, although it is not that easily demonstrated without separating it from the feedbacks that exist when linking the dynamical response of convection to heating or stability conditions, as shown in Chung et al. (2002). Over China, precipitation decreases more in the north than in the south, consistent with changes in the atmospheric fluxes for each of the simulations, although there is no direct relationship between them across the north and south domain. This suggests that along with the atmospheric aerosol content, surface and meteorological conditions also play an important role in modifying the response of precipitation to atmospheric heating.

Other climate aspects of absorbing aerosols are the direct heating effects on cloud cover – also called the "semi-direct effect" (Hansen et al., 1997). Although the semi-direct effect (calculated as the difference between TOA net irradiance for simulations with and without aerosols, due to changes in cloud fields) (Johnson et al., 2004) may be associated with positive forcing (due to changes in cloud cover that can increase the absorbed shortwave), recent studies have suggested that the sign may be dependent on the vertical aerosol distribution. In Tables 3.5a and 3.5b we show results

Table 3.5a Local radiative forcings for different vertical aerosol distributions.

Condition	Semi-direct effect (W/m^2)	Direct effect (W/m^2)
Absorbing aerosol in boundary layer	22.8	20.5
Absorbing aerosol in and above boundary layer	10.2	15.4
Absorbing aerosol above boundary layer	−9.5	0.7
Scattering aerosol above boundary layer	−0.1	−7.4

Source: Johnson et al. (2004)

Table 3.5b Local radiative forcings for different vertical distribution of aerosols over China (18–50° N, 90–130° E).

Condition	Semi-direct effect (W/m^2)	Direct effect (W/m^2)
Standard aerosol distribution	−5.39	5.82
All aerosols in first layer (0.42 km)	0.97	−0.06
All aerosols in fourth layer (3.8 km)	−5.74	8.37
All aerosols in seventh layer (10.5 km)	−19.8	15.3

Source: Menon, (2004).

from a large eddy scale model (Johnson et al., 2004), and a GCM based study (Menon, 2004) that investigated regional forcings associated with different vertical distributions of absorbing aerosols. As shown in Table 3.5a, with absorbing aerosols above the boundary layer, the semi-direct effect can be negative, mainly owing to the increases in liquid water path (LWP) fields and entrainment effects that led to a shallower moister boundary layer with higher LWP (Johnson et al., 2004). In the GCM study, the effects of vertical velocity, which got stronger as aerosols were distributed higher, led to increased cloud formation and LWP, which in turn changed the semi-direct effect from positive to negative (Menon, 2004). Penner et al. (2003) also find negative semi-direct effects for biomass aerosols injected in the mid troposphere, mainly by accounting for the longwave forcing (longwave forcing decreases when high cloud cover decreases).

On a global average, the semi-direct effect, as indicated in a simulation comparable to Exp A but without the indirect aerosol effects (Exp NIE-CDNC is kept a constant 174 cm^{-3} for land and ~60 cm^{-3} for ocean), is approximately −0.08 W/m^2. Regions of negative values are co-located with high biomass burning regions, and changes to both longwave cloud radiative forcing and high cloud cover are negative compared to ΔExp A. Thus, these negative longwave effects outweigh the positive forcings due to cloud reduction, which then give us a semi-direct effect that is negative in regions of strong biomass burning. However, Koren et al. (2004) find that for heavy smoke conditions in the Amazonia, the cumulus cloud cover decreases by 38% as compared to clean conditions, with an overall instantaneous forcing of 8 W/m^2. Another study in the Middle East during the Kuwait oil fires found that depending on the stability of the atmosphere and humidity profiles, large deep convective clouds can form at the top of the smoke layer (Rudich et al., 2003). Thus, vertical profiles of temperature

Carbonaceous aerosols and climate

(heating), (and thus of aerosols) and humidity may play a leading role in changing the sign of the semi-direct aerosol effects due to changes in cloud cover. These various results are difficult to interpret since observations of vertical aerosol distributions remain sparse.

Effects of biomass aerosols over Amazonia

During the active fire season large amounts of forest are burned in Amazonia, and changes in carbon fluxes, as well as cloud properties, are significantly different from the areas not affected by burning as shown by satellite records (Kaufman and

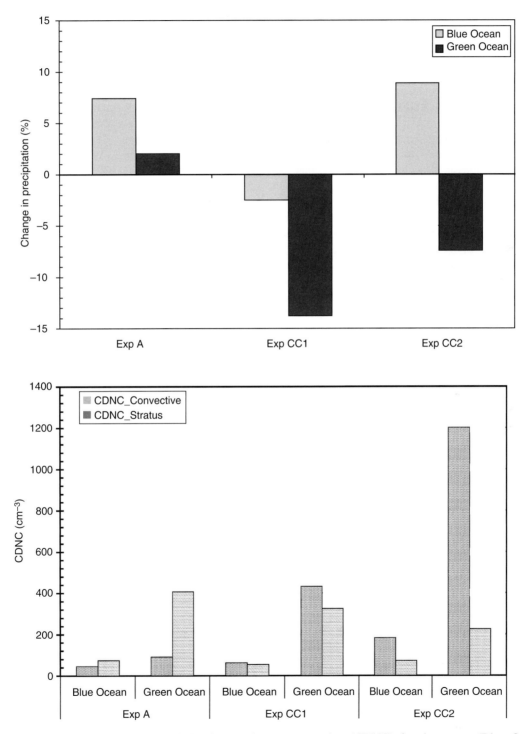

Figure 3.4 Change in precipitation and cloud droplet number concentration (CDNC) for the ocean (Blue Ocean) and the biomass source (Green Ocean) regions for the different model simulations.

Fraser, 1997). Recent results from a large field campaign conducted over Amazonia during the fire season indicate changes in deep convective cloud properties and precipitation when Amazonia was separated into a green ocean (biomass source region), a blue ocean (ocean) and a polluted smoky region (Andreae et al., 2004). Separating Amazonia similarly, on the basics of aerosol–concentrations, we compare model computations of precipitation and CDNC for simulations with and without aerosol convective cloud effects in Figure 3.4. Since we use seasonal means for August to October, specific days of active burning are not accounted for and thus, we mainly contrast values for the ocean (5° N–5° S, 30–40° W), "blue ocean", versus regions with large biomass emissions (5–15° S, 50–75° W), "green ocean". We compare our CDNC values with those from the SMOCC campaign (Andreae et al., 2004) of 200, 600 and 2200 cm^{-3} for the blue ocean, green ocean and smoky clouds, respectively. The CDNC values for Exp CC1 and CC2 are comparable to those measured within uncertainties that are bound to exist when comparing single point/time values with averages from the model over larger domains.

As shown in Figure 3.4, precipitation changes from the ocean to the biomass regions are stronger for Exp CC1 and CC2, mainly because of the substantial increase in CDNC obtained for convective clouds. Without aerosol–convective cloud effects, no decrease in precipitation is evident. In addition, zonally averaged vertical profiles of particle size indicate that the size of particles needed to participate in coalescence processes (12–14 μm) is at ~900 hPa in Exp A, but between 600 and about 900 hPa in both Exp CC1 and CC2, indicative of a shift in onset of precipitation. Andreae et al. find a shift in the onset of precipitation under smoky conditions that tend to favor the formation of large ice hydrometeors, although the change in the amount or areal distribution of precipitation was not evident, because the increased intensity of storms compensates for the effects of the suppression of precipitation initiation (time and level for cloud droplet sizes to reach the coalescence size). As shown in Figure 3.4, both ΔExp CC1 and CC2 indicate a larger decrease in precipitation from the ocean region to the biomass regions compared with that seen in ΔExp A. However, verification of such changes with observations is not possible and thus these values are mainly indicative of the sensitivity of precipitation to changing CDNC in convective clouds.

3.4 Conclusion

In the newly updated version of the GISS GCM, the aerosol indirect effect was found to be −0.65 W/m^2 (or alternatively −1.36 W/m^2), much lower than previous values obtained in Menon et al. (2002a) that ranged from −2.4 to −4.4 W/m^2. These lower values were mainly due to smaller differences in the anthropogenic aerosol burdens currently assumed, changes in the aerosol extinction efficiency due to aerosol effects on relative humidity, (36%), and addition of BC and dust aerosols, amongst other general improvements in the updated model. Including aerosol effects on convective clouds shifts the cloud water content to higher levels, as well as changing the initiation of precipitation formation to higher levels. The percentage change in precipitation for the clean ocean regions versus the biomass regions was found to vary by a factor of 2 to 4 mainly owing to the increase in cloud droplets in convective clouds. The aerosol indirect effect with aerosol–convective cloud effects included ranged from +0.22 to −0.43 W/m^2 and was much lower than estimates without aerosol–convective cloud effects, some of which may be related to changes in the longwave cloud forcing, which was of similar magnitude to the shortwave cloud forcing. The global magnitude of the aerosol semi-direct effect was found to be close to 0 (−0.08 W/m^2) since longwave effects of biomass aerosols (lofted higher) cancel out positive forcings from a reduction in cloud cover. Changes in the vertical profiles of temperature and humidity as well as the liquid water path can influence the sign of the semi-direct effect through changes in the cloud cover.

Several simulations were conducted to examine the effects of BC aerosols from fossil- and biofuel and carbonaceous aerosols from biomass burning. For twice the amount of fossil- and biofuel BC emissions, with all indirect effects included, the ratio of surface temperature to TOA radiative fluxes (for present-day versus pre-industrial aerosol emissions) increased by more than a factor of 10 (changes from 0.097 to 1.14 K W^{-1} m^2). While the forcing associated with BC may be somewhat linear, the temperature response to the forcing is non-linear since it includes changes in cloud fields and other feedbacks. Climate sensitivity to doubled CO$_2$ in a similar version of the model coupled to a slab-ocean model is ~0.66 K W^{-1} m^2 (Hansen et al., 2005). These results and those from Roberts and Jones (2004) (climate sensitivity of 0.56 and 0.91 K/(W/m^2) for four times the fossil fuel BC amount and twice the CO$_2$), as well as results based on the forcing efficiency of BC (Table 3.3) and Jacobson (2002), indicate that despite small atmospheric burdens (~0.13 mg/m^2) and short life times (~3–7 days), BC aerosol emissions have a significant influence on climate. Regional changes in precipitation, examined over India and China, were found to be related to the amount of atmospheric heating, with higher atmospheric fluxes corresponding to larger changes (positive) in precipitation, though we do not discount the influence of surface and meteorological conditions that may also lead to similar changes. The changes described in this paper are for aerosol effects on warm clouds only, and to fully understand the net aerosol effects on climate, various aerosol effects on cold cloud microphysics must be included in future simulations.

Acknowledgements

We acknowledge support from the NASA Climate Modeling and Analysis Program managed by Don Anderson and Tsengdar Lee; the NASA GWEC project and the DOE ARM project. Additionally,

SM acknowledges support from LBNL's LDRD project (supported by the Director, Office of Science, Office of Basic Energy Science, of the US Department of Energy under Contract No. DE–AC02–05CH11231). The aerosol model used in this work was developed by Dorothy Koch who provided expertise and guidance in its use as well as providing useful comments to this work. We gratefully acknowledge all her contributions to this work and also acknowledge comments and guidance on model simulations by Larissa Nazarenko.

References

Andreae, M. O., Rosenfeld, D., Artaxo, P. et al. (2004). Smoking rain clouds over the Amazon. *Science* **303**, 1337–1342.

Andres, R. J. and Kasgnoc, A. D. (1998). A time-averaged inventory of subaerial volcanic sulfur emissions. *Journal of Geophysical Research* **103** (D19), 25 251–25 261.

Beck, C., Grieser, J. and Rudolf, B. (2005). A new monthly precipitation climatology for the global land areas for the period 1951 to 2000. *Climate Status Report 2004*, Offenbach: German Weather Service.

Beheng, K. D. (1994). A parameterization of warm cloud microphysical conversion processes. *Atmospheric Research* **33**, 193–206.

Bond, T. C., Streets, D. G., Yarber, K. F. et al. (2004). A technology-based global inventory of black and organic carbon emissions from combustion. *Journal of Geophysical Research* **109** (D14), D14203.

Cheng. Y., Canuto, V. M. and Howard, A. M. (2003). An improved model for the turbulent PBL. *Journal of Atmospheric Science* **60**, 3043–3046.

Chuang, C. C., Penner, J. E., Prospero, J. M. et al. (2002). Cloud susceptibility and the first aerosol indirect forcing: sensitivity to BC and aerosol concentrations. *Journal of Geophysical Research* **107** (D21), 4564, doi: 10.1029/2000JD000215.

Chung, C. E., Ramanathan, V. and Kiehl, J. T. (2002). Effects of South Asian absorbing haze on the northeast monsoon and surface-heat exchange. *Journal of Climate* **15**, 2462–76.

Conant, W. C. VanReker, T. M., Rissman, T. A. et al. (2004). Aerosol-cloud drop concentration closure in warm cumulus. *Journal of Geophysical Research* **109**, D13204, doi:10.1029/2003J004324.

DeBell, L. J., Vozzella, M., Talbot, R. W. and Dibb, J. E. (2004). Asian dust storm events of spring 2001 and associated pollutants observed in New England by the Atmospheric Investigation, Regional Modeling, Analysis and Prediction (AIRMAP) monitoring network. *Journal of Geophysical Research* **109**, doi: 10.1029/2003JD003733.

Del Genio, A. D., Yao, M.-S. and Lo, K.-W. (1996). A prognostic cloud water parameterization for global climate models. *Journal of Climate* **9**, 270–304.

Del Genio, A. D., Kovari, W., Yao, M.-S. and Jonas, J. (2005). Cumulus microphysics and climate sensitivity. *Journal of Climate* **18**, 2376–2387.

Fromm, M. D. and Servranckx, R. (2003). Transport of forest fire smoke above the tropopause by supercell convection. *Geophysical Research Letters* **30**, 1542, doi: 10.1029/2002GL016820.

Gong, D.-Y., Pan, Y.-Z. and Wang, J.-A. (2004). Changes in extreme daily mean temperatures in summer in eastern China during 1955–2000. *Theoretical and Applied Climatology* **77**, 25–37.

Gultepe, I. and Isaac, G. A. (1999). Scale effects on averaging of cloud droplet and aerosol number concentrations: observations and models. *Journal of Climate* **12**, 1268–1279.

Halmer, M. M. and Schmincke, H. U. (2003). The impact of moderate-scale explosive eruptions on stratospheric gas injections. *Bulletin of Volcanology* **65** (6), 433–440.

Hansen, J. E. and Nazarenko, L. (2004). Soot climate forcing via snow and ice albedos. *Proceedings of the National Academy of Sciences USA* **101**, 423–428.

Hansen, J., Sato, M. and Ruedy, R. (1997). Radiative forcing and climate response. *Journal of Geophysical Research* **102**, 6831–6864.

Hansen, J. E. Sato, M., Ruedy, R. et al. (2005). Efficacy of climate forcings. *Journal of Geophysical Research* **110**, D18104, doi: 10.1029/2005JD005776.

Jacobson, M. Z. (2002). Control of fossil-fuel particulate black carbon and organic matter, possibly the most effective method of slowing global warming. *Journal of Geophysical Research* **107** (D19), 4410, doi: 10.1029/2001JD001376.

Jacobson, M. Z. (2004). Climate response of fossil fuel and biofuel soot, accounting for soot's feedback to snow and sea ice albedo and emissivity. *Journal of Geophysical Research* **109**, D21201, doi: 10.1029/2004JD004945.

Jensen, M. and Del Genio, A. D. (2003). Radiative and microphysical characteristics of deep convective systems in the tropical western Pacific. *Journal of Applied Meteorology* **42**, 1234–1254.

Johnson, B. T., Shine, K. P. and Forster, P. M. (2004). The semi-direct aerosol effect: impact of absorbing aerosols on marine stratocumulus. *Quarterly Journal of the Royal Meteorological Society* **130**, 1407–22.

Kaiser, D. P. and Qian, Y. (2002). Decreasing trends in sunshine duration over China for 1954 –1998: indication of increased haze pollution? *Geophysical Research Letters* **29**, 2042, doi: 10.1029/ 2002GL016057.

Kaufman, Y. J. and Fraser, R. S. (1997). The effect of smoke particles on clouds and climate forcing. *Science* **277**, 1636–1639.

Kiehl, J. T., Schneider, T. L., Rasch, P. J., Barth, M. C. and Wong, J. (2000). Radiative forcing due to sulfate aerosols from simulations with the NCAR Community climate model (CCM3). *Journal of Geophysical Research* **105**, 1441–1457.

Koch, D. and Hansen, J. (2005). Distant origins of Arctic soot: a Goddard Institute for Space Studies ModelE experiment. *Journal of Geophysical Research* **110**, D04204, doi: 10.1029/2004JD005296.

Koch, D., Jacob, D., Tegen, I., Rind, D. and Chin, M. (1999). Tropospheric sulfur simulation and sulfate direct radiative forcing in the Goddard Institute for Space Studies general circulation model. *Journal of Geophysical Research* **104**, 13 791–13 823.

Koch, D., Schmidt, G. and Field, C. (2005). Sulfur, sea salt and radionuclide aerosols in GISS ModelE GCM. *Journal of Geophysical Research* **110**, D04204, doi: 10.1029/2004JD005296.

Koren, I., Kaufmann, Y. J., Remer, L. A. and Martins, J. V. (2004). Measurement of the effect of Amazon smoke on inhibition of cloud formation. *Science* **303**, 1342–1345.

Lahav, R. and Rosenfeld, D. (2000). Microphysical characterization of the Israel clouds from aircraft and satellites. *Proceedings of 13th International Conference on Clouds and Precipitation* **2**, 732.

Lelieveld, J., Berresheim, H., Borrmann, S. *et al.* (2002). Global air pollution crossroads over the Mediterranean. *Science* **298**, 794–799.

Lohmann, U., Feichter, J., Chuang, C. C. and Penner, J. E. (1999). Prediction of the number of cloud droplets in the ECHAM GCM. *Journal of Geophysical Research* **104** (D8), 9169–9198.

Menon, S. (2004). Current uncertainties in assessing aerosol effects on climate. *Annual Review of Environment and Resources* **29**, 1–31.

Menon, S., Del Genio, A. D., Koch, D. and Tselioudis, G. (2002a). GCM simulations of the aerosol indirect effect: sensitivity to cloud parameterization and aerosol burden. *Journal of Atmospheric Science* **59**, 692–713.

Menon, S., Hansen, J., Nazarenko, L. and Luo, Y. (2002b). Climate effects of black carbon aerosols in China and India. *Science* **297**, 2250–2253.

Nober, F. J., Graf, H.-F. and Rosenfeld, D. (2003). Sensitivity of the global circulation to the suppression of precipitation by anthropogenic aerosols. *Global Planetary Change* **37**, 57–80.

Pandithurai, G., Pinker, R. T., Takamura, T. and Devara, P. C. S. (2004). Aerosol radiative forcing over a tropical urban site in India. *Geophysical Research Letters* **31**, L12107, doi: 10.1029/2004GL019702.

Penner, J. E., Zhang, S. Y. and Chuang, C. C. (2003). Soot and smoke aerosol may not warm climate. *Journal of Geophysical Research* **108**, 4657, (D21) doi: 10.1029/2003JD003409.

Ramanathan, V., Crutzen, P. J., Lelieveld, J. *et al.* (2001). Indian Ocean Experiment: An integrated analysis of the climate forcing and effects of the great Indo-Asian haze. *Journal of Geophysical Research* **106** (D22), 28 371–28 398.

Rinke, A., Dethloff, K. and Fortmann, M. (2004). Regional climate effects of Arctic haze. *Geophysical Research Letters* **31**, L16202; doi: 10.1029/2004GL020318.

Roberts, D. L. and Jones, A. (2004). Climate sensitivity to black carbon aerosols from fossil fuel combustion. *Journal of Geophysical Research* **109**, D16202; doi: 10.1029/2004JD004676.

Rudich, Y., Sagi, A. and Rosenfeld, D. (2003). Influence of the Kuwait oil fire plumes (1991) on the microphysical development of clouds. *Journal of Geophysical Research* **108**, (D15), doi: 10.1029/2003JD003472.

Schmidt, G. A., Ruedy, R., Hansen, J. E. *et al.* (2005). Present day atmospheric simulations using GISS ModelE: comparison to in-situ, satellite and reanalysis data. *Journal of Climate* **19**, 153–192.

Twomey, S. (1977). The influence of pollution on the shortwave albedo of clouds. *Journal of Atmospheric Science* **34**, 1149–1152.

van der Werf, G. R., Randerson, J. T., Collatz, G. J. *et al.* (2004). Continental-scale partitioning of fire emissions during the 1997 to 2001 El Nino/La Nina period. *Science* **303**, 73–76.

Wang, C. (2004). A modeling study on the climate impacts of black carbon aerosols. *Journal of Geophysical Research* **109**, D03106, doi: 10.1029/2003JD004084.

Wild, M., Ohmura, A., Gilgen, H. and Rosenfeld, D. (2004). On the consistency of trends in radiation and temperature records and implications for the global hydrological cycle. *Geophysical Research Letters* **31**, L1201, doi: 10.1029/2003GL019188.

Xu, Q. (2001). Abrupt change of the mid-summer climate in central east China by the influence of atmospheric pollution. *Atmospheric Environment* **35**, 5029–5040.

4

Probabilistic estimates of climate change: methods, assumptions and examples

Haroon S. Kheshgi

4.1 Introduction to approaches to estimating future climate change

An important approach to assessment of the risks of climate change relies on *estimates* of future climate based on a variety of methods including: simulation of the climate system; analysis of the sensitivity of climate system simulations to model parameters, parameterizations and models of climate system; and model-based statistical estimation constrained by a variety of historical data. The results of each of these methods are contingent on assumptions. Increasingly sophisticated approaches are being applied, and the implications, or even enumeration, of assumptions is becoming increasingly complex. This chapter gives an overview of the progression of methods used to estimate change in the future climate system and the climate sensitivity parameter. Model-based statistical estimation has the potential of synthesizing information from emerging climate data with models of the climate system to arrive at probabilistic estimates, provided all important uncertain factors can be addressed. A catalog of uncertain factors is proposed including consideration of their importance in affecting climate change estimates. Addressing and accounting for factors in the catalog, beginning with the most important and tractable, is suggested as an orderly way of improving estimates of future climate change.

A variety of methods have been used to generate projections or estimates. The simulations of state-of-the-art models discussed in Section 4.2 are often used as scenarios (plausible representations) of how climate might change in the future (Mearns *et al.*, 2001). Consideration of simulations of one or more models with a range of model parameters is another general approach that is often used to give some indication of the level of uncertainty in future projections (see for example Schneider, 2001; Kheshgi and Jain, 2003), and is discussed in Section 4.3. Observation-based measures of climate clearly provide information that has increasingly been synthesized with our understanding of climate through statistical estimation methods. However, the ideal of synthesizing all knowledge via statistical estimation is subject to numerous approximations that must somehow be accounted for if confidence in such estimates is to be assessed. Model-based statistical estimation using historical climate data as constraints is undergoing rapid development, and is discussed in detail in Section 4.4, the primary focus of this chapter.

Figure 4.1 shows a variety of information and results that illustrate different types of analysis that have been used to infer future climate. The logarithmic dependence of radiative forcing of climate on CO_2 concentration is well established (Ramaswamy *et al.*, 2001). If climate sensitivity was constant, it would define an equilibrium temperature change that was linearly dependent on radiative forcing as shown in Figure 4.1 for the benchmark climate sensitivities of 1.5, 2.5, and 4.5 °C for a doubling of CO_2 concentration. Climate sensitivity might be inferred from the results of state-of-the-art models of climate or from paleo-analogs (see Section 4.3). The curves in Figure 4.1 indicate a range of model projections of annual-averaged global mean temperature (GMT) and CO_2 for the IPCC SRES A1B scenario (IPCC, 2000) by combining the parameterized range of responses of ocean and terrestrial carbon cycle models with an energy balance model (EBM) with a climate sensitivity range of 1.5 to 4.5 °C (Kheshgi and Jain, 2003). In this example parametric sensitivity based on

Human-induced Climate Change: An Interdisciplinary Assessment, ed. Michael Schlesinger, Haroon Kheshgi, Joel Smith, Francisco de la Chesnaye, John M. Reilly, Tom Wilson and Charles Kolstad. Published by Cambridge University Press. © Cambridge University Press 2007.

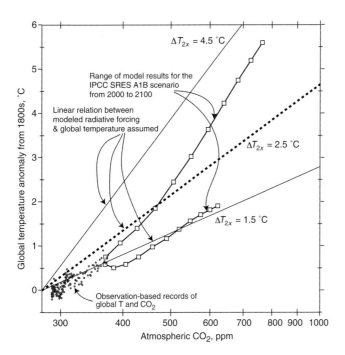

Figure 4.1 Annual global mean temperature and atmospheric CO_2 concentration. Lines show the relation of equilibrium GMT and CO_2 assuming constant climate sensitivity of 1.5, 2.5 and 4.5 °C as an anomaly from Holocene conditions (c. CO_2 of 275 ppm). Circles show the annual GMT from 1865 to 2002 (Jones *et al.*, 2004) taken as an anomaly from the 1865–1899 average vs. CO_2 from instrumental (Keeling and Whorf, 2004) and ice core records (Enting *et al.*, 1994). Squares show the range of model projections of GMT and CO_2 for the time period 2000 to 2100 for the IPCC SRES A1B scenario (IPCC, 2000) by combining the parameterized range of responses of ocean and terrestrial carbon cycle models with an EBM with a climate sensitivity range of 1.5 to 4.5°C (Kheshgi and Jain, 2003).

the simulation of the results of multiple models (see Section 4.3) is used to generate a future temperature range.

Observation-based records of GMT and atmospheric CO_2 concentration are also shown in Figure 4.1. Such data have been used in many analyses to estimate, for example, climate model parameters such as climate sensitivity using statistical estimation methods (e.g. regression; see Section 4.4). Of course, many additional – and less certain – factors such as climate system dynamics (e.g. ocean heat uptake and climate variability), and forcing factors in addition to CO_2 (aerosols, non-CO_2 greenhouse gases, solar and land cover) must be accounted for in such an estimation. In studies that begin to address these factors as well as including additional observational constraints, the probability distribution of climate sensitivity is both broader and higher than would be suggested by the comparison in Figure 4.1. Yet different studies give different results: see, for example, estimates of climate sensitivity summarized by Andronova *et al.* in Chapter 1 of this volume. This leaves open the question of what confidence should be assigned to such quantitative results (cf. Manning *et al.*, 2004).

In light of this gap, the robust lessons to be learnt from climate estimation problems are not necessarily the quantitative results. Analysis (e.g. sensitivity to additional uncertain parameters, model structure, and other types of assumptions) of estimation problems can, for example, indicate key data constraints or process uncertainties (see Kheshgi *et al.*, 1999). Analysis can indicate potential modes of uncertainty and the associated research needs.

This chapter considers the assumptions implicit in model-based statistical estimation of future climate, with the intent to provide some context with which to assess confidence and guide future study. It does not consider, however, the contribution of uncertainty in future human activities. The use of statistical estimation (e.g. climate model parameter estimation), which is undergoing rapid development and holds perhaps the greatest promise for quantifying probabilities in climate forecasting, is the primary focus, and is discussed in greater detail in Section 4.4. In Section 4.4, I propose a catalog of the many factors – ranging from climate model parameters to radiative forcing factors to uncertainty of the observations that could be used as constraints – that may be important in the statistical estimation of future climate. Examining how these factors are accounted for in existing studies provides some context to judge the completeness of the study, and the reliability of their results. Such a catalog could potentially be used to guide further improvement of models by accounting for the most important and tractable elements of the catalog.

4.2 State-of-the-art climate models

Application of state-of-the-art global climate models (e.g. coupled atmosphere–ocean general circulation models, GCMs) to simulate future climate has dominated discourse on climate change. Simulations have been used as plausible futures, or scenarios of climate change. Such simulations, however, do not provide a sufficient basis for the assessment of risk. Complete probabilistic information is not provided. And no model validation on the scales of interest is yet possible.

State-of-the-art climate models are subject to well-known uncertainties and gaps in our understanding of the climate system that prevent confident estimates of future climate based on their a-priori results. Uncertain processes include those such as cloud feedbacks, which are well established, have been characterized parametrically, and present large and, so far, unrelenting uncertainties in climate simulations (see Cess *et al.*, 1990). Models also must resolve the response of systems, such as ocean circulation, that are known to have exhibited complex behavior. The complex dynamics of paleoclimate is evident in records; yet it remains unclear what exactly are the mechanisms for such behavior, how one might validate a model for such behavior, and what are the limits of its predictability (NRC, 2002).

State-of-the-art climate models do generate climate variability. The generation of multiple realizations of a model – an ensemble – is becoming an increasingly common practice in weather prediction and now in climate modeling. Ensembles can be used to characterize the statistics of model variability when generated by a chaotic or random process. Ensembles can also be used to separate random variability from modeled climate signals, and it has been recognized that they have a clear (but limited) role in climate attribution studies using climate models whose simulations contain both a climate signal and climate variability (i.e., noise) (Barnett et al., 2000). While an ensemble of climate projections can be used to generate a probabilistic climate forecast, it includes only a partial estimate of uncertainty – only uncertainty due to model variability is included, and the well-known uncertainties inherent in climate models and their forcing are omitted. Such ensemble results should not be confused with more complete probabilistic estimates of future climate. Furthermore, care must be applied in how they are used in climate attribution studies since they do not represent the full range of signal and noise uncertainty. The ensemble approach can, however, be extended (as discussed in Section 4.3) to include realizations of different models and parameter values to begin to form a more complete probabilistic estimate.

To forecast the real world, models must also account for all important drivers and feedbacks that may occur in the future. This is a daunting task given that we have not yet adequately included even the climate forcing agents that we know have been important in the past. For example, the range of aerosol forcing agents and their interactions with the climate system have not been included in climate change attribution studies (Mitchell et al., 2001), even though they are recognized as being the most important uncertainty in climate forcing (Ramaswamy et al., 2001). Extending from state-of-the-art global climate models to models of the entire climate system introduces additional uncertainties. These include the additional model elements needed to forecast atmospheric composition (e.g. carbon cycle and atmospheric chemistry) and effects on radiative transfer (e.g. aerosol effects on clouds and future solar insolation change). To study how these additional factors, or scenarios of human activity, affect the climate system, the sensitivities of climate system models – often highly aggregated "simple" models (Harvey et al., 1997) – are commonly tested.

4.3 Sensitivity to parameters, parameterizations and models

Parametric sensitivity can be used to assess uncertainty *if* model structure (including parameterization) is accurate and provided there is information about the true values of model parameters. If such were the case, a model can be used to propagate the probability distribution of parameters to yield a probability distribution for future climate.

A great deal of attention has been paid to the equilibrium climate sensitivity. As used in simple energy balance models (EBMs), it is the primary parameter taken to be the source of uncertainty in climate forecasts (for a given scenario of human activities). Methods to estimate climate sensitivity have relied on rough theory (Lapenis, 1998), expert elicitation (Morgan and Keith, 1995), regression (Andronova and Schlesinger, 2001; Forest et al., 2002; Gregory et al., 2002), ranges of GCM results (Cubasch et al., 2001), and the reconstruction of paleoclimates (Hoffert and Covey, 1992).

It may be that the empirical estimation from paleoclimate gives the most confident current estimate of climate sensitivity as some believe (Hansen, 2004). Inferring climate sensitivity from paleoclimate reconstructions has the advantages of large amplitude and long duration possibly reducing effects of variability and ocean heat uptake. However, reconstruction of the climate system – including past temperature, greenhouse gases, ice sheet positions, aerosols, and solar insulation – adds uncertainty. Using paleoclimate to infer climate response also relies on it being an appropriate analog for the future; for example, climate sensitivity is assumed to be constant even though feedbacks such as those from sea ice may have acted in different ways than at present. The fact that CO_2 concentration lagged changes in temperature during the last glaciation (Prentice et al., 2001) gives pause to those using changes in paleoclimate as a simple indicator of future climate change. And what caused glacial/interglacial changes (from about 200 to 300 ppm) in atmospheric pCO_2 remains a central unresolved question (Archer et al., 2000a). Paleoclimate interpretation is complicated further by evidence for abrupt climate change, where climate changes more rapidly than radiative forcing (NRC, 2002). The potential for complex behavior (e.g. multiple equilibrium states) has led experts to question the existence of a constant equilibrium climate sensitivity (Morgan and Keith, 1995). Yet, if these issues could be addressed, then paleoclimate might be used as a constraint on climate sensitivity, and could be synthesized with the statistical estimation (Section 4.4) using, for example, Bayesian statistics.

Recently, Stainforth et al. (2005) reported the results of a large ensemble of GCM results using one model with a range of model parameters. This test of model sensitivity to parameters – a huge computational task – resulted in a wide range of climate sensitivity (from less than 2 °C to more than 11 °C). An obvious extension of this approach would be to test its sensitivity to different climate models or parameterizations. And questions to be addressed include the selection of parameters and ranges, and the possible invalidation of some results evaluated by comparison to current climate.

A range of GCM model responses, for example, was used to generate the range of model response presented by the IPCC (Cubasch et al., 2001). This was done by calibrating an EBM to each response of a set of seven coupled GCMs by adjusting EBM parameters such as climate sensitivity and ocean heat diffusivity. While the IPCC did not assign a probability to any of these models, or the range of temperature change reported, the assumption that each of the seven sets of parameters had equal probability has been used to generate probability distributions for future temperature change (see Wigley and Raper, 2001).

Consideration of multiple models can provide some information about plausible model structures as well as providing information about uncertainty in projections. A coupled GCM result used in model intercomparisons, for example, is generally based on the modeling group's best judgement. Therefore, a range of models represents the range of best estimates, and not a full estimate of uncertainty. Furthermore, models may make similar assumptions, and may possibly all make the wrong assumptions. Since models give differing results, we know that all models cannot be correct; however, all models could be incorrect. Therefore, if model results cannot be rejected based on, for instance, comparisons to data, then the envelope of model results must be contained within the uncertainty range, thereby forming a lower bound on the uncertainty range (cf. Kheshgi and Jain, 2003).

Producing ensembles from multiple GCMs will allow incorporation of both modeled variability and model differences. However, without some estimate of the likelihood that a model is true, such an approach does not provide probabilistic information. Constraining model results by statistical comparison to observed changes in climate holds some promise in overcoming this barrier (cf. Murphy *et al.*, 2004).

4.4 Statistical estimation using observational constraints

Historical climate data can be used to constrain the behavior of climate models. This approach has the advantage of incorporating emerging climate data, thereby improving results with the acquisition of new data. Furthermore, probabilistic results can be generated subject to a range of assumptions. Statistical estimation studies commonly rely on a least-squares approach. They range from the simple extrapolation of past trends (e.g. Tsonis and Elsner, 1989), to model-based extrapolation via Bayesian parameter estimation (e.g., Tol and De Vos, 1998), to likelihood estimation involving sampling of parameter space of models of intermediate complexity (e.g. Forest *et al.*, 2002) and of GCMs (e.g. Murphy *et al.*, 2004).

4.4.1 Introduction to components of an estimation problem

Ideally, statistical estimation would apply all observation-based constraints to all degrees of freedom (i.e. uncertainties) in modeled climate. However, in practice, idealized climate models and data modeling (approximating measurements with a mathematical representation) are used to simplify the estimation problem. The way in which climate and observations are modeled remains critical to calculated results, and so requires careful consideration in order to interpret the results and the confidence in findings that may be justified.

In the idealization of a climate estimation problem, it is common to break up the problem into the following components (addressed in detail in the indicated sections):

1. Modeled climate (Section 4.4.2)
   a. Modeled climate response to forcing
   b. Climate forcing: observations and modeling
   c. Modeled climate variability
2. Modeled observations (Section 4.4.3)

Each of these components of the estimation problem contains uncertain factors to be defined (prior information) and/or estimated. I propose Table 4.1 as a preliminary catalog that defines various uncertain factors influencing the estimation of future climate. The factors are described in the context of EBMs, where the three components above (i.e. 1. a–c) of a climate system model are more easily separable. Some factors should be expected to have a *high* contribution to the uncertainty of a statistical estimation result, some may have a *low* contribution, and some may be highly *speculative*. In Table 4.1, I give my judgement (discussed in the following subsections) of their relative importance. The importance of the various factors should be a key consideration in the choices made in designing a statistical climate estimation problem. Such choices are always made, either implicitly or explicitly, in every estimation study. Such a catalog may prove useful in judging the implications of idealized (e.g. that do not account for some uncertain factors) estimation studies and prioritizing what factors most need work.

Given a well-posed estimation problem with models for climate and observations defined, estimation methods can be applied. In Section 4.4.4, the results of a variety of estimation studies are considered.

4.4.2 Modeled climate

In statistical estimation problems, which apply constraints from past observed climate change to estimate future climate, EBMs are commonly used. EBMs generally separate the modeling of radiative forcing, climate response to radiative forcing, and the modeling of climate variability. Whether or not the real climate system is well approximated with this modeling approach depends on the validity of such models. There are many tests for partial validation of climate models (Schneider, 1989), yet they are incomplete since they do not necessarily apply for the model calibration periods or future conditions of interest. Notwithstanding, in this section modeling of each of these three aspects of the climate system is considered individually. In studies using GCMs, the three components discussed here can all be contained within the same model, yet the discussion that follows

Table 4.1 Uncertain factors in the statistical estimation of future climate.

Uncertain factors	Magnitude of contribution to uncertainty in statistical estimates
Climate response to forcing	
Amplitude of climate response (e.g. climate sensitivity)	High
Ocean heat uptake over a 30-year timescale	High: timescale of past GHG increase
Ocean heat uptake over a 100-year timescale	High: timescale of climate forecasts
Non-linear ocean transport (e.g. stratification and thermohaline circulation change)	Speculative: unclear role over past century
Variability in ocean transport	Speculative: indications from ocean heat content observations
Non-linear feedbacks (e.g. clouds)	Speculative: hypothesized mechanisms have not been confirmed
Radiative forcing	
Well-mixed GHGs	Low for past: based on observations
	High for future: uncertainty in carbon cycle and atmospheric chemistry
Solar (temporal pattern)	High: uncertain prior to satellite measurement, and in the future; potential for amplification to be confirmed
Land cover	Low: limited magnitude on global scale, but important locally
Aerosols: (i) amplitude	High: largest uncertainty in past radiative forcing
(ii) temporal pattern	High: timing of various aerosol sources uncertain
(iii) spatial pattern	High: relative importance of different aerosol types uncertain
(iv) interactions with climate system	Speculative: importance of interactions on global scale not well established
Tropospheric ozone	Moderate: limited magnitude over the past
Climate variability	
5 yr	Low: shorter than the timescale for past climate trends
30 yr	High: variability on timescale of GHG trend not strongly constrained
100 yr	Moderate: indications from pre-instrumental record of changes on century timescale
Spatial pattern	High: relative to spatial pattern signal
Uncertainty in observation-based records	
Instrumental record of GMT	Low: uncertainty smaller than observed increase
Instrumental record of hemispheric MT difference	High: differences between records are large
Tropospheric temperature	Speculative: models and data have yet to be reconciled
Ocean heat content	High: sparse coverage
Pre-instrumental temperature records	High: differences between records are large

and the factors in Table 4.1 can largely be translated to apply to GCM-based studies as well.

Criteria for models used in estimation problems differ from criteria for the development of state-of-the-art climate model simulations. For model-based estimation it is important not to neglect uncertainties and gaps in understanding in model representations; in fact, it is critical to include the least certain factors. However, for state-of-the-art climate models used for simulation, common assumptions are often used and facilitate benchmarking, with speculative or uncertain mechanisms not included. Factors such as sulfate aerosol direct forcing are included; they are essential to provide consistency with data in model evaluation. The implicit requirement that models used in statistical estimation mimic such state-of-the-art models used for simulation gives a possibly false perception of confidence in such estimation models. An outcome is a bias towards current state-of-the-art model consensus, and a neglect of modeled climate forcing, variability, and response uncertainties, some of which have measurable effects (e.g. effect of carbonaceous aerosols on radiative transfer; see Ramanathan et al., 2001) and some of which remain speculative (e.g. forcing from ice condensation nuclei; see Penner et al., 2001).

Modeled climate response to forcing

The response of climate to radiative forcing can be modeled with increasingly complicated models for atmosphere and ocean general circulation. Over large spatial and temporal scales, the temperature results of such models have been well approximated by EBMs whose behavior is that of a *linear system with constant coefficients* (Wigley and Raper, 1990). EBMs have gained wide use in scenario analyses (Harvey *et al.*, 1997). Alternatively, climate model results have been directly modeled by their empirical orthogonal functions (EOFs) to simplify their use in detection/attribution studies. Furthermore, EOF patterns have been used in combination with EBMs to model the evolution of climate patterns (Hooss *et al.*, 2001). Each of these successive modeling steps can be a good approximation of detailed model results over aggregate spatial and temporal scales.

Conservation of energy is the fundamental principle of energy balance models. What is not apparent, however, is whether an EBM must be linear or have constant coefficients. The response of a linear system with constant coefficients can be summarized by a Green's function or transfer function. Such a function for temperature is commonly parameterized by few coefficients such as climate sensitivity and some measure associated with ocean heat uptake. Model simulation of past temperature change is highly affected by choice of climate sensitivity and ocean heat uptake over a 30-year timescale (similar to the timescale for GHG increase). Climate sensitivity will strongly affect estimates of future climate as well. And since ocean heat uptake over a 30-year timescale affects estimation of climate sensitivity, it will strongly affect estimates of future temperature (even if it does not directly affect the future energy balance to a large extent). For these reasons the uncertainty of climate response to radiative forcing is often approximated by two uncertain parameters – climate sensitivity, and one for ocean heat uptake (e.g. Tol and De Vos, 1998; Andronova and Schlesinger, 2001; Forest *et al.*, 2002).

Considering a mixing time for ocean heat of $O(1000\,\text{years})$ compared with a time period of decades associated with calibration to past observed climate change, a period of interest of many decades for future climate projections allows for some greater uncertainty (flexibility in the shape of the Green's function) than can be accommodated by one parameter. In Table 4.1 it is proposed that ocean heat uptake over a 100-year timescale can also have a high effect on future estimates of temperature change. In EBMs, for example, different schemes for ocean transport (e.g. a 1D upwelling–diffusion versus a diffusion model) result in Green's functions of different shape. And schematic models for ocean transport and GCMs have differing behavior caused by the model for transport and the numerical schemes applied (Archer *et al.*, 2000b).

Is climate well approximated as a linear system with constant coefficients? This question must also be faced when considering ocean carbon cycle, which is related to ocean heat uptake because each shares a common ocean transport field (although their response functions need not be the same because the boundary conditions and mixing properties are different). Analysis of ocean carbon uptake has relied heavily on the use of tracers such as radiocarbon to constrain models for ocean transport. It is recognized, however, that such an approach could lead to spuriously accurate estimates of ocean uptake if it is assumed that ocean transport is time-invariant. This has led to estimation problems that allow ocean transport parameters to vary with time (Enting and Pearman, 1987), or include uncertain parameters which have the effect of decoupling the long-time-averaged constraint of natural radiocarbon from tracers for contemporary ocean behavior (Kheshgi *et al.*, 1999). Observational study of ocean carbon cycle has progressed to examining tracers that could be indicating change in ocean transport (Joos *et al.*, 2003). Furthermore, the variation of ocean transport of carbon from one hemisphere to another has long been an issue in the estimation of carbon sources and sinks inferred from gradients in the atmospheric field (see for example Tans *et al.*, 1990) and could also lead to corresponding variations in ocean heat transport. Since variation in ocean transport is a key question for global carbon cycle, is this a key mode of uncertainty in climate modeling due to variation in the response of ocean heat uptake? One way to begin to address this question is to include modeled variation in ocean heat uptake and constrain this with records of ocean temperature (Levitus *et al.*, 2000; Levitus *et al.*, 2005) or other tracers (and accounting for the errors in such observation-based records).

How should uncertainty related to unsteady or non-linear climate response be included in estimation problems? In Table 4.1, three different ways are included. Perhaps the best established is non-linear ocean transport due to stratification and shifts in thermohaline circulation. Another factor may be variability in ocean transport as appears evident in some aggregations of ocean temperature data (Gregory *et al.*, 2004; Levitus *et al.*, 2005). Yet another class of potential non-linear effects has to do with feedbacks from, for example, clouds that may have a stabilizing or destabilizing effect (Stocker *et al.*, 2001). While speculative, each of these may be important. Yet it is not obvious how to model or include them in a statistical estimation problem. Their inclusion may currently be considered intractable and, therefore, their potential existence is a caveat on estimation results.

Climate forcing: observations and modeling
The attribution of climate change, as well as the calibration of climate models, depends on estimates of climate forcing factors such as the well-mixed greenhouse gases (GHGs), short-lived GHGs such as tropospheric ozone, anthropogenic and natural (mineral dust, volcanoes, sea salt, dimethyl sulfide ...) aerosols, solar effects, and changes in land cover.

The history of radiative forcing of well-mixed GHGs is tied to observations of their historical atmospheric concentrations and is well established relative to other forcing agents (Ramaswamy *et al.*, 2001); its uncertainty is accordingly rated *low* in Table 4.1.

Aerosols are recognized as the largest source of uncertainty in anthropogenic radiative forcing and are assessed to "*mostly produce negative radiative forcing*" of climate (IPCC, 2001). The cooling effect of aerosols via their effects on clouds have been assessed as a large contributor to the uncertainty in overall radiative forcing (Schwartz and Andreae, 1996). Aerosols, however, can also cause warming effects. Black carbon aerosols from combustion have been estimated to have varying effects on local, regional and global climate (Jacobson, 2002; Penner *et al.*, 2003; Wang, 2004). Furthermore, the darkening of snow by soot can have an additional warming effect (Hansen *et al.*, 2004; Hansen and Nazarenko, 2004). And the indirect effect of aerosols in nucleating ice clouds can also have a potentially strong warming effect (Penner *et al.*, 2001). Often aerosols are assumed, in attribution (e.g. Mitchell *et al.*, 2001) or model calibration (e.g. Forest *et al.*, 2002) studies, to have a known pattern in time (history) and space with only the amplitude unknown. The pattern (even its sign) of aerosol effects, however, contains more than one degree of freedom and might be better represented as outlined in Table 4.1. Furthermore, interactions between aerosols and the climate system might not be well represented by radiative forcing of climate.

Solar forcing has received great attention by earth scientists owing to its apparent role in the 100-kyr cycle of the ice ages via orbital cycles (Imbrie and Imbrie, 1980), and the Little Ice Age (Eddy, 1976; Grove, 1988). However, prior to measurement of solar insulation via satellite beginning in the 1970s, the basis for quantitative estimates of solar forcing is currently weak (Sofia, 2004). Furthermore, there are hypotheses for climate effects of solar activity being amplified through changes in clouds and atmospheric chemistry. And as pointed out by Andronova and Schlesinger (2001), estimates of climate sensitivity based on the instrumental record of climate change are highly sensitive to uncertainty in solar radiative forcing. While solar forcing might not lead to an appreciable change in future climate, it does have a significant impact on future estimates of climate that rely on calibration with historical climate records.

Modeled climate variability
Climate variability will inevitably contribute to future climate change, but, perhaps more importantly for the purpose of making climate forecasts, has contributed to past climate change in ways that are only partially understood. Understanding of climate variability is based on observation-based records, and the modeling of phenomena that drive the variability. Given an understanding of climate variability, statistical modeling can be used to represent variability.

Variability of most interest in either the study of the detection and attribution of climate change, or the calibration of climate model parameters, is on the timescale of the modeled past anthropogenic signal – about 30 years (Kheshgi and White, 1993) – and it is variability on this timescale that is deemed to have *high* importance for climate estimation in Table 4.1. Empirical evidence of temperature variability exists in both the instrumental and proxy records of climate. For a stationary noise process, a long temperature record is needed to constrain a statistical model of temperature variability. If the variability in aggregate measures of surface temperature is indeed small as has been estimated in some reconstructions of northern hemisphere temperatures (Mann *et al.*, 1998; Mann and Jones, 2003), then natural variability may well have a limited contribution to estimation of climate model parameters. Reconstruction of past variability, however, has been shown to be dependent on the calibration approach applied to tree rings (Esper *et al.*, 2004). Additionally, the aggregation of proxy data that has been scaled (calibrated) to match the instrumental record has been shown with synthetic data to underrepresent the magnitude of variability (von Storch *et al.*, 2004).

Climate variability generated from GCMs is increasingly being assumed as a surrogate for true climate variability in studies of the detection and attribution of climate change (Zwiers and Zhang, 2003). GCM variability is seen to be of roughly the right amplitude or perhaps an underestimate of actual variability (Risbey and Kandlikar, 2002). This has the practical advantage that it provides patterns of variability at the resolution of climate models, and makes possible the studies of the detection and attribution based on patterns, but with the underlying assumption that the GCM patterns of variability are correct (Mitchell *et al.*, 2001). Use of patterns of climate variability in climate model calibration does enter into estimation problems in different ways through statistical models of variability based on climate models (Forest *et al.*, 2002) or on the instrumental record (Andronova and Schlesinger, 2001), and variability based directly on realizations of GCMs (Murphy *et al.*, 2004).

4.4.3 *Modeled observations*

Estimation of future climate relies heavily on observations of recent climate. Temperature, for example, is commonly measured and contains some measurement error and bias; however, the temperature records that are used average, correct and fill in the individual measurements to arrive at gross averages of temperature anomalies (Jones and Wigley, 1990). Alternate records of, for example, GMT anomaly (the absolute GMT cannot be estimated with comparable accuracy) do contain quantitative differences of tenths of a degree (Folland *et al.*, 2001). Estimation of uncertainty of such records does present challenges (NRC, 2000); nevertheless, uncertainty estimates of tenths of a degree have been assessed (Folland *et al.*, 2001). This magnitude of uncertainty is small compared with the change in GMT of roughly $0.6\,°C$ over the past century, and has commonly been neglected in both studies of the detection and attribution of climate change, and the calibration of climate models (it is given *low* importance in Table 4.1).

In moving beyond GMT to, for example, hemispheric contrast in temperature, however, an uncertainty of tenths of a degree becomes larger relative to the size of temperature contrast anomalies. And for observation-based measures such as changes in ocean heat content (Levitus *et al.*, 2000; Forest *et al.*, 2002; Levitus *et al.*, 2005) the uncertainty of this measure is also of importance. In cases where these types of data are used, neglecting their uncertainty in statistical estimation could lead to spurious overconstraint of future estimated climate: see Table 4.1. Of course, if their uncertainty is too large, then they should not be expected to form effective constraints and this could be a justification for not applying such observation-based records as constraints.

Reconciling differences in tropospheric and surface temperature changes has been identified as a near-term priority (CCSP, 2004). Understanding the uncertainty of differences in tropospheric and surface temperature records is critical in this activity. Given sound estimates including their uncertainty, it may be that tropospheric temperature will provide a validation test for climate models (cf. Santer *et al.*, 2000) rather than be used, at least initially, as a means of model calibration (Forest *et al.*, 2002).

Estimates of the uncertainty in proxy temperature indirectly could affect climate estimation though the modeling of variability for high time-resolution records for, for example, the past millennium (see Section 4.4.2) and for estimation of *equilibrium* conditions for past climate epochs (Covey *et al.*, 1996; Kheshgi and Lapenis, 1996).

Recognition of the importance of additional observation-based constraints, of course, should be expected in the future. For example, records of shortwave radiation reaching the Earth's surface could be useful for the attribution of climate change (Liepert *et al.*, 2004) as well as for statistical estimation. However, inconsistencies remain among data sets, and between data sets and the behavior of GCMs (Charlson *et al.*, 2005). Applications of data sets, which constrain specific mechanisms in the climate system, form one path to improving forecasts of climate.

4.4.4 Statistical estimation: methods, assumptions and examples

This section considers methods, assumptions and examples from the progression of studies that use observation-based records to constrain estimates of climate models. Many studies have performed parameter estimation to address attribution questions, estimate parameters such as climate sensitivity, and compute estimates of future climate.

Climate change attribution studies often perform linear regression of spatial and/or temporal patterns to estimate amplitudes of climate patterns modeled to be due to, for example, greenhouse and aerosol components. In fact, Mitchell *et al.* (2001) defines *optimal detection* as the solution of a linear regression problem.[1] This problem is defined as:

$$\mathbf{y} = \mathbf{Ga} + \mathbf{u} \quad (4.1)$$

where $\mathbf{y}$ is a vector of "observations" that is modeled to be a linear combination of the matrix of signal patterns $\mathbf{G}$ with the vector of signal amplitudes $\mathbf{a}$ which make up the parameters to be estimated, and noise $\mathbf{u}$. The calculated uncertainty of the amplitude estimate is given by

$$\mathbf{C_{aa}} = (\mathbf{G}^T \mathbf{C_{uu}}^{-1} \mathbf{G})^{-1} \quad (4.2)$$

where $\mathbf{C_{uu}}$ and $\mathbf{C_{aa}}$ are the covariance matrices of noise $\mathbf{u}$ and amplitude estimate $\mathbf{a}$.

Kheshgi and White (1993) considered this regression problem for the simple application where the time-series of annual average global mean temperature (GMT) was used to estimate a parameter that is non-linearly related to the traditional climate sensitivity parameter. Note that in (4.2) the uncertainty of the estimated parameter is independent of the temperature data that form a constraint in the case where $\mathbf{C_{uu}}$ is independent of $\mathbf{a}$. Furthermore, if a scenario for future climate signals (radiative forcing) exists, then the uncertainty can be estimated for the future. In this way Kheshgi and White (1993) gave future estimates of learning, showing that for this idealized problem climate sensitivity would be better known within decades, with the quantitative result dependent on factors such as the covariance of the noise model (C), the assumed future signal pattern (G) and the amount and type of data constraints chosen (determining rank and content of both G and C): see Figure 4.2. It should be expected that analyses that specify climate variability as an independent process, as in (4.1), will calculate a monotonic decay of uncertainty even with minor non-linearities such as the relation between the amplitude parameter in (4.1) and the climate sensitivity parameter as commonly used in EBMs. However, if $\mathbf{C_{uu}}$ were dependent on $\mathbf{y}$ as it would be, for example, if it were derived from goodness of fit (see for example Andronova and Schlesinger, 2001), then the uncertainty cannot be calculated without future data.

Calculated uncertainty may, however, be higher if the specific analyzed model system (4.1) were poor; if, for example, the degrees of freedom were found not to represent

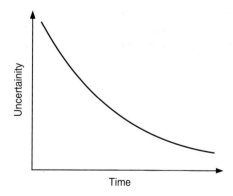

Figure 4.2 Monotonic decrease in uncertainty based on a linear estimation problem with acquisition of additional data. Rate of learning and amplitude of uncertainty dependent on modeled signal and noise according to Eq. (4.2). Based on quantitative example given in Kheshgi and White (1993).

---

[1] Climate change detection might be better classified as a hypothesis testing problem (Kheshgi and White, 2001) rather than a linear regression.

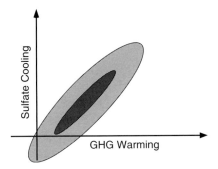

Figure 4.3 Probability density contours of the amplitude of sulfate cooling and greenhouse warming of GMT estimated to have occurred in the recent past as estimated in climate change attribution studies (cf. Mitchell *et al.*, 2001).

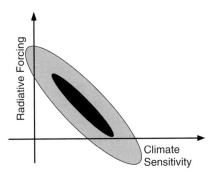

Figure 4.4 Probability density contours of sulfate radiative forcing and climate sensitivity for a *linearized* parameter estimation problem.

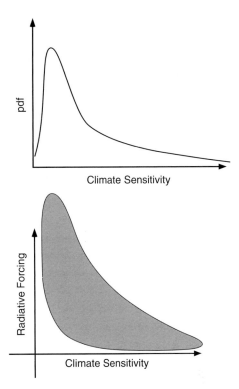

Figure 4.5 The probability density function (pdf) of climate sensitivity (top) and the associated probability density contour of sulfate radiative forcing and climate sensitivity for a non-linear parameter estimation problem.

those of the real system, the noise model **u** were found to be a poor approximation of the real system, or the observation-based records were found to have consequential uncertainties. For example, including the amplitude of a specified sulfate aerosol forcing history increases parameter uncertainty and leads to covariance between this parameter and estimated climate sensitivity. In attribution studies (see Mitchell *et al.*, 2001), this is often illustrated by an ellipse representing a probability density contour in parameter space: see Figure 4.3. The tendency for the ellipse to lie along the diagonal axis is a result of the high covariance of sulfate cooling and greenhouse warming patterns. For a *linearized* parameter estimation problem with two degrees of freedom – for example climate sensitivity and aerosol cooling – the axis of the ellipse would be reflected as shown in Figure 4.4. Including additional degrees of freedom in the estimation problem (4.1–4.2) can significantly increase the calculated uncertainty of the estimate. For example, the uncertainty estimate of climate sensitivity made by Kheshgi and White (1993) would be higher if the amplitude of sulfate radiative forcing were assumed to be another degree of freedom. On the other hand, the estimate could be lower if additional data records were applied as additional constraints.

The probability distributions of signal amplitudes often assumed to be Gaussian in attribution studies need not be so. For example, the climate sensitivity parameter that would result in a Gaussian distribution of current greenhouse-gas-induced temperature anomaly need not be Gaussian; it could have a tail at the high sensitivity end that could extend to higher values as illustrated in Figure 4.5. This result can occur as a consequence of non-linear dependence on parameters of the estimation problem along with temporal covariation of aerosol and GHG forcing (Andronova and Schlesinger, 2001; Forest *et al.*, 2002). In a Bayesian estimation problem, it could also be a consequence of the specification of a prior distribution of, for example, aerosol forcing, which can also lead to a skewed distribution of climate sensitivity as demonstrated by Tol and De Vos (1998).

The consequence of a skewed probability distribution of climate sensitivity, all other things being equal, is a skewed estimate of future temperature anomaly as shown in Figure 4.6. For example, over the calibration period, the uncertainty of observation-based data constraints and noise from climate variability can result in a virtually symmetric distribution of simulated past GMT (made symmetric by the correlation of skewed climate sensitivity with estimates of other parameters such as sulfate radiative forcing amplitude), while resulting in a skewed estimate of future climate change (as seen in Forest *et al.*, 2002).

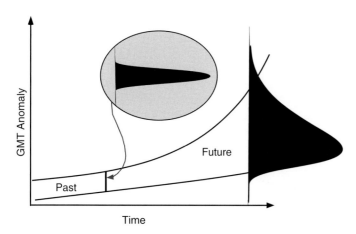

Figure 4.6 An illustrative example of the probability distribution of estimated past and future GMT anomaly based on a climate model calibration resulting in a skewed estimate of climate sensitivity as in Figure 4.5.

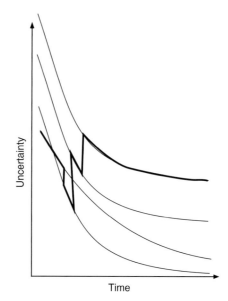

Figure 4.7 The evolution of estimated uncertainty with the acquisition of additional (future) data. Each curve represents a learning curve like that illustrated in Figure 4.2 where future acquisition of data further constrains uncertainty for a defined estimation problem. Alternate curves exist for alternate estimation problems where inclusion of additional sources of uncertainty (uncertain parameters, e.g., shown in Table 4.1) leads to higher estimated uncertainty, and inclusion of additional types of data constraints (e.g. ocean heat content) leads to lower estimated uncertainty with different potential learning rates. Adopted uncertainty estimates may switch between the results of alternate estimation problems leading to either decreasing or increasing estimates of uncertainty.

Estimates of climate sensitivity by reconstruction of the climate system of paleo-epochs such as the last glacial maximum (LGM: about 20 kyr before present), have not so far resulted in skewed estimates of climate sensitivity (Hoffert and Covey, 1992; Covey *et al.*, 1996), and do not appear prone to the aerosol–GHG covariation problem that is seen in the instrumental record. Indeed, there is evidence (Anderson and Charlson, 1990) that during the LGM there was a much greater quantity of (mineral) aerosols in the atmosphere (thought to be a cooling agent), and lower GHG concentrations, than in the Holocene (the last 10 kyr prior to significant human effects). Recognizing all the caveats at play in applying the paleocalibration approach to future climate change, paleocalibration may be seen as limiting the nearly unbounded estimates (on the high end) of climate sensitivity based solely on climate model calibration with the instrumental record (Andronova and Schlesinger, 2001).

Recently, methods for assessing model uncertainty have been combined by producing multiple realizations of multiple GCMs, sampling a prior (expert opinion) distribution of model parameters, and estimating the likelihood of outcomes by comparison to climate records. For example, Murphy *et al.* (2004) consider 29 parameters in a set of GCMs, and present telling results as to which GCM parameters are of importance in calculating climate sensitivity. Such approaches have, in principle, the potential of addressing many of the assumptions present in studies with simpler climate models. However, each of the factors, for example, in Table 4.1 would need to be addressed, including those related to radiative forcing uncertainty (e.g. aerosols), particularly of the past if estimation of likelihood relies on comparison to past data. The use of multiple models highlights the limited ability to validate climate models for the purpose of long-term climate forecasting. Although climate models may be (or can be made to be) consistent with climate records, they still might not contain all the key phenomena that have determined past climate change.

In this way, the use of multiple models can underestimate uncertainty (see Kheshgi and Jain, 2003).

Statistical estimation can generate probability distributions, and estimates of future learning, of parameters such as climate sensitivity, which has been shown to be an important factor in theoretical constructs of decision-making under uncertainty (Nordhaus, 1992; Webster, 2003; Richels *et al.*, 2006). Statistical estimation problems continue to evolve, with additional observation-based constraints being applied, climate models being refined, and additional modes of uncertainty being included. As we move into the future, uncertainty estimates may not decline monotonically as is shown in Figure 4.2, but may change sporadically as illustrated in Figure 4.7, as additional sources of uncertainty are recognized and adopted (leading to step increases in estimated uncertainty) and additional types of data constraints are applied (leading to step decreases in estimated uncertainty). We may learn that we cannot predict climate change as accurately as we had previously thought, or we may reduce estimated uncertainty by application of future data or by independent reduction of sources of uncertainty. The challenge of drawing robust conclusions from the integrated assessment of climate change must somehow account for the incomplete nature of climate forecasting.

4.5 Conclusions

Statistical estimation methods provide powerful means of drawing inferences from large and diverse assemblages of information, particularly with *well-defined* data and models. However, without a means of validating the models used in statistical estimation, to differentiate what we know from what we do not know, estimation results will unavoidably be contingent upon assumptions. Applications of statistical estimation of future climate are subject to the following assumptions.

Uncertainty in climate model response: models used for the response of climate to climate forcing are generally assumed to be exact, except for a limited number of parameters (for example two, one associated with climate sensitivity and the other associated with ocean heat uptake, sometimes constrained by prior information). Inclusion of additional uncertain factors (cf. Table 4.1) presents a path to improving estimation studies. However, how to fully include model structure uncertainty (e.g. in GCM-based studies) remains an open question.

Uncertainty in climate variability: climate variability is often assumed to be independent of climate forcing (or gives this approximate response in climate models), and sometimes independent of climate response to forcing, and represented by a random stationary noise process. Uncertainty in noise statistics is not generally included. For spatial patterns, climate variability statistics often rely on GCM simulations.

Uncertainty in data constraints: uncertainty in applied data constraints is often neglected since it can be small compared with climate variability. While this may be a valid approximation for the trend in GMT, it is not valid for some patterns (e.g. hemispheric temperature difference) or additional data constraints (e.g. ocean heat content). Construction of observation-based records largely omits the quantification of its uncertainty including the covariance that would be needed for inclusion in estimation studies.

Forcing factors: current analyses tend to consider only the best described forcing factors (e.g. greenhouse and sulfate aerosols). Less certain, yet known, factors are often omitted (e.g. biomass burning aerosols). Factors are assumed to be exact (e.g. greenhouse gas forcing) or to contain uncertainty only in their amplitude (e.g. aerosol forcing) with spatial and temporal patterns assumed exactly known.

Selection of scales and data constraints: the selection of time and space scales over which to aggregate analysis is often driven by approaches to simplify analysis, the implications of idealized model-based statistical analyses (e.g. selection of most detectable EOF patterns), data quality considerations, and judgments about robust features of climate system models (e.g. scales at which model structure uncertainty is not evident in model intercomparisons).

Assumptions for each of these components of the climate estimation problem must be examined to test the robustness of estimation results. Given a simplifying set of assumptions, many remaining assumptions can be tested via, for example, sensitivity analysis. Important, established sources of uncertainty (rated *high* in Table 4.1) could be incorporated in estimation problems with less-well-established factors (rated *speculative* in Table 4.1) neglected and recognized as assumptions awaiting better characterization before their inclusion (should they prove important). Results can be used in the scientific process to gather and/or apply additional data to constrain key remaining uncertainties. Unless confidence in climate forecasts can be scientifically justified, integrated assessments and policy analyses that are sensitive to climate forecasts will be suspect. Statistical estimation provides a path forward to incorporate emerging data in the analysis of decision making under uncertainty, but clearly work remains in understanding the implications of assumptions made in such analyses.

References

Anderson, T. L. and Charlson, R. J. (1990). Ice-age dust and sea salt. *Nature* **345**, 393.

Andronova, N. G. and Schlesinger, M. E. (2001). Objective estimation of the probability density function for climate sensitivity. *Journal of Geophysical Research* **106**, 22 605–22 611.

Archer, D. E., Winguth, A., Lea, D. and Mahowald, N. (2000a). What caused the glacial/interglacial atmospheric pCO_2 cycles? *Reviews of Geophysics* **38**, 159–189.

Archer, D. E., Eshel, G., Winguth, A. *et al.* (2000b). Atmospheric pCO_2 sensitivity to the biological pump in the ocean. *Global Biogeochemical Cycles* **14**, 1219.

Barnett, T. P., Hegerl, G., Knutson, T. and Tett, S. (2000). Uncertainty levels in predicted patterns of anthropogenic climate change. *Journal of Geophysical Research* **105**, 15 525–15 542.

CCSP (2004). *Strategic Plan for the US Climate Change Science Program*. Washington, DC: Climate Change Science Program, US Government.

Cess, R. D., Potter, G. L., Blanchet, J. P. *et al.* (1990). Intercomparison and interpretation of cloud-climate feedback processes in nineteen atmosphere general circulation models. *Journal of Geophysical Research* **95**, 16 601–16 615.

Charlson, R. J., Valero, F. P. J. and Seinfeld, J. H. (2005). In search of balance. *Science* **308**, 806–807.

Covey, C., Sloan, L. C. and Hoffert, M. I. (1996). Paleoclimate data constraints on climate sensitivity: the paleocalibration method. *Climatic Change* **32**, 165–184.

Cubasch, U., Meehl, G. A., Boer, G. J. *et al.* (2001). Projections of future climate change. In *Climate Change 2001: The Scientific Basis. Contribution of Working Group I to the Third Assessment Report of the Intergovernmental Panel on Climate Change*, ed. J. T. Houghton, Y. Ding, D. J. Griggs *et al.* New York: Cambridge University Press, pp. 525–582.

Eddy, J. A. (1976). The Maunder Minimum. *Science* **192**, 1189–202.

Enting, I. G. and Pearman, G. I. (1987). Description of a one-dimensional carbon cycle model calibrated by the techniques of constrained inversion. *Tellus* **39B**, 459–476.

Enting, I. G., Wigley, T. M. L. and Heimann, M. (1994). *Future Emissions and Concentrations of Carbon Dioxide: Key Ocean/Atmosphere/Land Analyses*. Australia: CSIRO.

Esper, J., Frank, D. C. and Wilson, R. J. S. (2004). Climate reconstructions: low-frequency ambition and high-frequency ratification. *Eos* **85**, 113, 120.

Folland, C. K., Karl, T. R., Christy, J. R. et al. (2001). Observed climate variability and change. In *Climate Change 2001: The Scientific Basis. Contribution of Working Group I to the Third Assessment Report of the Intergovernmental Panel on Climate Change*, ed. J. T. Houghton, Y. Ding, D. J. Griggs et al. New York: Cambridge University Press, pp. 99–181.

Forest, C. E., Stone, P. H., Sokolov, A. P., Allen, M. R. and Webster, M. D. (2002). Quantifying uncertainties in climate system properties with the use of recent climate observations. *Science* **295**, 113–117.

Gregory, J. M., Ingram, W. J., Palmer, M. A. et al. (2002). An observationally based estimate of climate sensitivity. *Journal of Climate* **15**, 3117–3121.

Gregory, J. M., Banks, H. T., Stott, P. A., Lowe, J. A. and Palmer, M. D. (2004). Simulated and observed decadal variability in ocean heat content. *Geophysical Research Letters* L15312; doi: 10.1029/2004GL020258.

Grove, J. M. (1988). *The Little Ice Age*. New York: Methuen.

Hansen, J. E. (2004). Defusing the global warming time bomb. *Scientific American* **290**, 69–77.

Hansen, J. E. and Nazarenko, L. (2004). Soot climate forcing via snow and ice albedos. *Proceedings of the National Academy of Sciences USA* **101**, 423–428.

Hansen, J., Bond, T., Cairns, B. et al. (2004). Carbonaceous aerosols in the industrial era. *EOS* **85**, 241, 244.

Harvey, L. D. D., Gregory, J., Hoffert, M. et al. (1997). *An Introduction to Simple Climate Models Used in the IPCC Second Assessment Report. IPCC Technical Paper II*. Bracknell, UK: Intergovernmental Panel on Climate Change.

Hoffert, M. I. and Covey, C. (1992). Deriving global climate sensitivity from paleoclimate reconstructions. *Nature* **360**, 573–576.

Hooss, G., Voss, R., Hasselmann, K., Maier-Reimer, E. and Joos, F. (2001). A nonlinear impulse response model of the coupled carbon-climate system (NICCS). *Climate Dynamics* **18**, 189–202.

Imbrie, J. and Imbrie, J. Z. (1980). Modeling the climatic response to orbital variations. *Science* **207**, 943–953.

IPCC (2000). *Emissions Scenarios: A Special Report of Working Group III of the Intergovernmental Panel on Climate Change*, ed. N. Nakicenovic and R. Swart. Cambridge: Cambridge University Press.

IPCC (2001). *Climate Change 2001: The Scientific Basis. Contribution of Working Group I to the Third Assessment Report of the Intergovernmental Panel on Climate Change*, ed. J. T. Houghton, Y. Ding, D. J. Griggs et al. New York: Cambridge University Press.

Jacobson, M. Z. (2002). Control of fossil-fuel particulate black carbon and organic matter, possibly the most effective method of slowing global warming. *Journal of Geophysical Research* **107** (D19), 4410; doi: 10.1029/2001JD001376.

Jones, P. D. and Wigley, T. M. L. (1990). Global warming trends. *Scientific American* **263**, 84–91.

Jones, P. D., Parker, D. E., Osborn, T. J. and Briffa, K. R. (2004). Global and hemispheric temperature anomalies – land and marine instrumental records. In *Trends: A Compendium of Data on Global Change*. Oak Ridge, Tennessee, USA: Oak Ridge National Laboratory.

Joos, F., Plattner, G.-K., Stocker, T. F., Körtzinger, A. and Wallace, D. W. R. (2003). Trends in marine dissolved oxygen: implications for ocean circulation changes and the carbon budget. *EOS* **84**, 197, 201.

Keeling, C. D. and Whorf, T. P. (2004). Atmospheric CO_2 records from sites in the SIO air sampling network. In *Trends: A Compendium of Data on Global Change*. Oak Ridge, Tennessee, USA: Carbon Dioxide Information Analysis Center, Oak Ridge National Laboratory.

Kheshgi, H. S. and Jain, A. K. (2003). Projecting future climate change: implications of carbon cycle model intercomparisons. *Global Biogeochemical Cycles* **17**, 1047; doi: 10.29/2001GB001842.

Kheshgi, H. S. and Lapenis, A. G. (1996). Estimating the uncertainty of zonal paleotemperature averages. *Palaeogeography, Palaeoclimatology, Palaeoecology* **121**, 221–237.

Kheshgi, H. S. and White, B. S. (1993). Does recent global warming suggest an enhanced greenhouse effect? *Climatic Change* **23**, 121–139.

Kheshgi, H. S. and White, B. S. (2001). Testing distributed parameter hypotheses for the detection of climate change. *Journal of Climate* **14**, 3464–3481.

Kheshgi, H. S., Jain, A. K. and Wuebbles, D. J. (1999). Model-based estimation of the global carbon budget and its uncertainty from carbon dioxide and carbon isotope records. *Journal of Geophysical Research* **104**, 31 127–31 144.

Lapenis, A. G. (1998). Arrhenius and the Intergovernmental Panel on Climate Change. *EOS* **79**, 271.

Levitus, S., Antonov, J., Boyer, T. P. and Stephens, C. (2000). Warming of the world ocean. *Science* **287**, 2225–2229.

Levitus, S., Antonov, J. and Boyer, T. (2005). Warming of the world ocean, 1955–2003. *Geophysical Research Letters* **32**; doi: 10.1029/2004GL021592.

Liepert, B. G., Feichter, J., Lohmann, U. and Roeckner, E. (2004). Can aerosols spin down the water cycle in a warmer and moister world? *Geophysical Research Letters* **31**; L06207; doi: 10.1029/2003GL019060.

Mann, M. E. and Jones, P. D. (2003). Global surface temperatures over the past two millennia. *Geophysical Research Letters* **30**, 1820; doi: 10.029/2003GL017814.

Mann, M. E., Bradley, R. S. and Hughes, M. K. (1998). Global-scale temperature patterns and climate forcing over the past six centuries. *Nature* **392**, 779–787.

Manning, M., Petit, M., Easterling, D. et al. (eds) (2004). *IPCC Workshop on Describing Scientific Uncertainties in Climate Change to Support Analysis of Risk and Options*. Boulder, Colorado: IPCC.

Mearns, L. O., Hulme, M., Carter, T. R. et al. (2001). Climate scenario development. In *Climate Change 2001: The Scientific*

Basis: *Contribution of Working Group I to the Third Assessment Report of the Intergovernmental Panel on Climate Change*, ed. J. T. Houghton, Y. Ding, D. J. Griggs *et al.* New York: Cambridge University Press, pp. 739–768.

Mitchell, J. F. B., Karoly, D. J., Hegerl, G. C. *et al.* (2001). Detection of climate change and attribution of causes. In *Climate Change 2001: The Scientific Basis. Contribution of Working Group I to the Third Assessment Report of the Intergovernmental Panel on Climate Change*, ed. J. T. Houghton, Y. Ding, D. J. Griggs *et al.* New York: Cambridge University Press, pp. 695–738.

Morgan, M. G. and Keith, D. W. (1995). Subjective judgements by climate experts. *Environmental Science and Technology* **29**, 468–476A.

Murphy, J. M., Sexton, D. M. H., Barnett, D. N. *et al.* (2004). Quantification of modelling uncertainties in a large ensemble of climate change simulations. *Nature* **430**, 768–772.

Nordhaus, W. D. (1992). An optimal transition path for controlling greenhouse gases. *Science* **258**, 1315.

NRC (2000). *Reconciling Observations of Global Temperature Change*. Washington, DC: National Academy Press.

NRC (2002). *Abrupt Climate Change: Inevitable Surprises*. Washington, DC: National Academy Press.

Penner, J. E., Andreae, M., Annegarn, H. *et al.* (2001). Aerosols, their distribution and indirect effects. In *Climate Change 2001: The Scientific Basis. Contribution of Working Group 1 to the Third Assessment Report of the IPCC*, ed. J. T. Houghton, Y. Ding, D. J. Griggs *et al.* New York: Cambridge University Press, pp. 289–348.

Penner, J. E., Zhang, S. Y. and Chuang, C. C. (2003). Soot and smoke aerosol may not warm climate. *Journal of Geophysical Research* **108**(D21); doi: 10.1029/2003JD003409.

Prentice, C., Farquhar, G., Fasham, M. *et al.* (2001). The carbon cycle and atmospheric CO_2. In *Climate Change 2001: The Scientific Basis. Contribution of Working Group I to the Third Assessment Report of the Intergovernmental Panel on Climate Change*, ed. J. T. Houghton, Y. Ding, D. J. Griggs *et al.* New York: Cambridge University Press, pp. 183–237.

Ramanathan, V., Crutzen, P. J., Kiehl, J. T. and Rosenfeld, D. (2001). Aerosols, climate, and the hydrological cycle. *Science* **294**, 2119–2124.

Ramaswamy, V., Boucher, O., Haigh, J. *et al.* (2001). Radiative forcing of climate change. In *Climate Change 2001: The Scientific Basis. Contribution of Working Group I to the Third Assessment Report of the Intergovernmental Panel on Climate Change*, ed. J. T. Houghton, Y. Ding, D. J. Griggs *et al.* New York: Cambridge University Press, pp. 349–416.

Richels, R., Manne, A. S. and Wigley, T. M. L. (2006). Moving beyond concentrations: the challenge of limiting temperature change, ed. M. E. Schlesinger, F. C. de la Chesnaye, J. Edmonds *et al.* Cambridge:Cambridge University Press.

Risbey, J. S. and Kandlikar, M. (2002). Expert assessment of uncertainties in detection and attribution of climate change. *Bulletin of the American Meteorological Society* **83** 1317–1326.

Santer, B. D., Wigley, T. M. L., Gaffen, D. J., *et al.* (2000). Interpreting differential temperature trends at the surface and in the lower troposphere. *Science* **287**, 1227–1231.

Schneider, S. H. (1989). *Global Warming*. San Francisco: Sierra Club Books.

Schneider, S. H. (2001). What is "dangerous" climate change? *Nature* **411**, 17–19.

Schwartz, S. E. and Andreae, M. O. (1996). Uncertainty in climate change caused by aerosols. *Science* **272**, 1121–1122.

Sofia, S. (2004). Variations of total solar irradiance produced by structural changes of the solar interior. *EOS* **85**, 217.

Stainforth, D. A., Aina, T., Christensen, C., *et al* (2005). Uncertainty in predictions of the climate response to rising levels of greenhouse gases. *Nature* **433**, 403–406.

Stocker, T. F., Clarke, G. K. C., Le Treut, H. *et al.* (2001). Physical climate processes and feedbacks. In *Climate Change 2001: The Scientific Basis. Contribution of Working Group I to the Third Assessment Report of the Intergovernmental Panel on Climate Change*, ed. J. T. Houghton, Y. Ding, D. J. Griggs *et al.* New York: Cambridge University Press, pp. 417–470.

Tans, P. P., Fung, I. Y. and Takahashi, T. (1990). Observational constraints on the global atmospheric CO_2 budget. *Science* **247**, 1431–1438.

Tol, R. S. J. and De Vos, A. F. (1998). A Bayesian statistical analysis of the enhanced greenhouse effect. *Climatic Change* **38**, 87–112.

Tsonis, A. A. and Elsner, J. B. (1989). Testing the global warming hypothesis. *Geophysical Research Letters* **16**, 795–797.

von Storch, H., Zorita, E., Jones, J. M. *et al.* (2004). Reconstructing past climate from noisy data. *Science* **306**, 679–682.

Wang, C. (2004). A modeling study on the climate impacts of black carbon aerosols. *Journal of Geophysical Research* **109**(D03106); doi: 10.1029/2003JD004084.

Webster, M. (2003). The curious role of "learning" in climate policy: should we wait for more data? *The Energy Journal* **23**, 97–119.

Wigley, T. M. L. and Raper, S. C. B. (1990). Natural variability of the climate system and detection of the greenhouse effect. *Nature* **344**, 324–327.

Wigley, T. M. L. and Raper, S. C. B. (2001). Interpretation of high projections for global-mean warming. *Science* **293**, 451–454.

Zwiers, F. W. and Zhang, X. (2003). Towards regional climate change detection. *Journal of Climate* **16**, 793–797.

5

The potential response of historical terrestrial carbon storage to changes in land use, atmospheric CO_2, and climate

Atul K. Jain

5.1   Introduction

Modeling and measurement studies indicate that ocean and land ecosystems are currently absorbing slightly more than 50% of the human CO_2 fossil emissions (Prentice et al., 2001). However, a significant question remains regarding the sources and sinks of carbon over land governed by changes in land covers and physiological processes that determine the magnitude of the carbon exchanges between the atmosphere and terrestrial ecosystems. Most of these processes are sensitive to climate factors, in particular temperature and available soil water (Post et al., 1997). It is also likely that these processes are sensitive to changes in atmospheric CO_2. Moreover, the climate variation is not uniformly distributed throughout the Earth's surface or within ecosystem types. Therefore, simulations of terrestrial carbon storage must take into account the spatial variations in climate as well as non-climate factors that influence carbon storage, such as land-cover type and soil water holding capacity, that interact with climate. Estimates should also account for land-cover changes with time. Because the changes in land cover, mainly from forest to croplands or forest to pasturelands, shorten the turnover of carbon above and below ground, they act to reduce the sink capacity of the biosphere.

Historical changes in biospheric carbon storage and exchange with the atmosphere are commonly simulated with globally aggregated biospheric models (Jain et al., 1996; Kheshgi and Jain, 2003), mostly in response to changes in atmospheric CO_2 and climate, or changes in land-cover types (Houghton and Hackler, 2001; Houghton, 2003). However, simulations of terrestrial fluxes with geographically distributed terrestrial biosphere models in response to combined changes in climate, CO_2, and land cover are less common (IMAGE Team, 2001; McGuire et al., 2001; Prentice et al., 2001).

Accordingly, here we use a global terrestrial carbon dynamics model that describes the land ecosystem, CO_2, and climate changes with a compartment model of carbon in vegetation and soil. We investigate how the historical land-use, climate, and CO_2 changes over the period 1900–1990 may have affected the net land–atmosphere fluxes of CO_2. In particular, we use two different historical data sets for land-cover change to estimate the uncertainties in the terrestrial fluxes. We then compare the model-based estimated range of terrestrial fluxes with the data-based estimated range of flux values (Prentice et al., 2001) for the 1980s to evaluate the model performance.

5.2   Methods

5.2.1   The model

In this study we employed the terrestrial component of our Integrated Science Assessment Model (ISAM-2, Jain and Yang, 2005). The parameterization, performance and validation of ISAM-2 have been previously discussed (Jain and Yang, 2005). Here I briefly describe its main features. The ISAM-2 model simulates the carbon fluxes to and from different compartments of the terrestrial biosphere with 0.5-by-0.5 degree spatial resolution. Each grid cell is completely occupied by at least one

Human-induced Climate Change: An Interdisciplinary Assessment, ed. Michael Schlesinger, Haroon Kheshgi, Joel Smith, Francisco de la Chesnaye, John M. Reilly, Tom Wilson and Charles Kolstad. Published by Cambridge University Press. © Cambridge University Press 2007.

Historical terrestrial carbon storage

Table 5.1 Net primary productivity, and vegetation, soil, and total carbon storage by land cover type in 1765 (Jain and Yang, 2005).

Land cover type	Net primary productivity (NPP)		Vegetation carbon (VC)		Soil organic carbon (SOC)		Total carbon (TC)	
	kgC m^2 yr	GtC/yr	kgC m^2	GtC	kgC m^2	GtC	kgC m^2	GtC
Tropical evergreen	1.07	13.7	20.87	268	10.65	133	31.53	402
Tropical deciduous	0.77	6.0	16.54	128	5.96	50	22.50	168
Temperate evergreen	1.02	5.1	21.50	109	14.03	73	35.53	190
Temperate deciduous	0.98	5.0	21.27	108	11.00	35	32.27	146
Boreal	0.32	6.4	7.07	142	26.26	496	33.34	667
Savanna	0.81	9.7	2.01	24	19.22	206	21.23	278
Grassland	0.15	4.0	0.33	9	4.67	131	5.00	114
Shrubland	0.22	3.2	0.82	12	8.78	110	9.61	127
Tundra	0.13	1.2	0.73	7	21.74	241	22.47	203
Desert	0.09	0.7	0.60	5	5.60	34	6.20	<1
Polar ice/desert	0.00	0.0	0.00	0	11.64	5	11.64	4
Cropland	0.38	3.3	1.62	14	1.60	39	3.21	50
Pastureland	0.12	1.0	1.46	13	2.86	40	4.32	33
Total		59.3		838		1594		2419

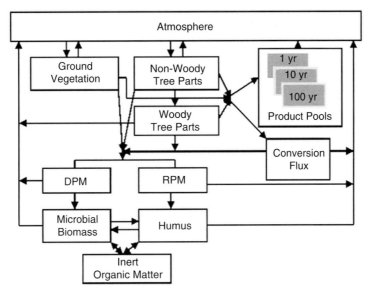

Figure 5.1 Schematic diagram of the terrestrial ecosystem component of the Integrated Science Assessment Model (ISAM). Ground vegetation, non-woody, and woody tree parts represent vegetation carbon pools. DPM and RPM are, respectively, easily decomposable plant material and resistant plant material, in above and below ground litter pools, respectively. The three soil pools are microbial, humus, and inert organic matter. The conversion carbon fluxes are associated with the land changes representing slash carbon left on the ground that is consecutively transferred to litter pools, and carbon released from burning of plant material. The part of the biomass harvested is transferred to three product pools with different turnover times.

of the 12 natural land coverage classifications and/or croplands (Table 5.1). Each grid is also assigned one of the 105 soil types based on the FAO-UNESCO Soil Map of the World (Zobler, 1986, 1999). Within each grid cell, the carbon dynamics of each land-coverage classification is described by a compartment model of carbon in vegetation and soil, which is depicted in Figure 5.1. The vegetation carbon partitioned into three vegetation reservoirs (ground vegetation, GV; non-woody tree part, NWT; woody tree part, WT) is based on the global ratios of the contents of these reservoirs (Jain et al., 1996). Jain and Yang (2005) provide the initial carbon contents for various land cover types.

In the model net primary productivity, NPP (kg C m^2/yr, the photosynthetically fixed carbon minus the autotrophic respiration) by ground vegetation and trees varies with vegetation carbon according to plant growth equations (King et al., 1995). The increase in the rate of photosynthesis by terrestrial biota, relative to pre-industrial times (1765), is modeled using the function of King *et al.* (1995) and a biotic growth factor (β) of Polglase and Wang (1992), which was derived based on the biochemical model of Farquhar and von Caemmerer (1982). According to Polglase and Wang (1992), β is a decreasing function of atmospheric CO_2 and increasing function of temperature. While NPP may increase with increasing CO_2, the rate of increase per increase in atmospheric CO_2 will decline if the temperature is constant but can increase with increasing temperature. The vegetation component of the model contains temperature feedback through the respiration and photosynthesis rates to and from the vegetation reservoirs, which follow the Q_{10} formulation described in Kheshgi *et al.* (1996). In the model, increases in either temperature or CO_2 lead to increases in NPP.

The litter and soil carbon dynamics are estimated based on the Rothamsted soil turnover model (Jenkinson, 1990). This model uses monthly plant material as an input. The incoming plant material enters the litter reservoirs (DPM is easily decomposable plant material, and RPM is resistant plant material) and undergoes decomposition and releases CO_2. The decomposed material is then distributed into the atmosphere, BIO, and HUM reservoirs. The soil is also assumed to contain a small amount of inert organic matter (IOM) in a separate reservoir. The exchange rates of carbon are modified by environmental factors including temperature, soil moisture deficit, soil temperature, and the plant protection factors (Jain and Yang, 2005). The decomposition rates are temperature-dependent according to Post *et al.* (1997). We calculate actual soil water (mm) and soil water pressure (kPa) for each grid cell with the monthly climatic water budget model of Thornthwaite and Mather (1957) as implemented by Pastor and Post (1985). The soil hydraulic characteristics for the Rothamsted soil moisture function and the Thornthwaite and Mather (1957) water balance calculations are derived from soil depth and texture information for each FAO soil type (Zobler, 1986, 1999), rooting depth estimates (Webb *et al.*, 1991), and relationships between soil texture and water content at the critical pressure (Rawls *et al.*, 1982).

The land-use emissions due to changes in land cover are calculated using the same bookkeeping approach as Houghton *et al.* (1983) for modeling ecosystems affected by land-use changes. In their model, annual changes in vegetation and soil following the land cover changes were prescribed using the synoptic response curves for different ecosystems. However, in this study the changes in carbon stocks following the land cover changes are affected by the changes in the NPP and soil respiration and the effects of changing environmental conditions on these fluxes. We consider two types of activities leading to land cover changes: clearing of natural ecosystems for croplands and pastureland, and recovery of abundant croplands and pasturelands to pre-conversion natural vegetation.

Within a grid cell, cleared natural vegetation is replaced by croplands/pasturelands. With changes in natural vegetation, a specified amount of carbon is released from the three vegetation carbon pools (GV, NW, and WTP) based on the relative proportions of the carbon contained in these reservoirs. A fraction of the released carbon is transferred to litter reservoirs as slash left on the ground. The rest is either released to the atmosphere by the burning of plant material to help clear the land for agriculture/pasturelands (Conversion Flux in Figure 5.1) or transferred to wood and/or fuel product reservoirs (Product Pools in Figure 5.1). This study does not account for the carbon transport between grid cells or regions. The carbon added to the litter reservoir decays according to the decomposition rates of the litter reservoir as discussed earlier, and then carbon is released to the atmosphere. Carbon stored in the product reservoirs is released to the atmosphere at a variety of rates from three general reservoirs with turnover times of 1 year (agriculture and agriculture products), 10 years (paper and paper products) and 100 years (lumber and long-lived products) (Houghton and Hackler, 2001). We use the fractions of total cleared vegetation assigned to each product pool, and the vegetation amount burned and/or left as slash, from Houghton and Hackler (2001), which varies with land cover type and region. We also assume that carbon is transferred from soil reservoirs to the litter reservoir.

In the case of croplands and pasturelands abandonment, the area of abandoned land is returned to the area of the pre-conversion natural vegetation type. Natural vegetation is then allowed to regrow from the extant state of the grid cell at the time of abandonment.

5.2.2 The data

The monthly temperature and precipitation data used in this study are the CRU TS 2.0 observation data set of the Tyndall Center (Mitchell *et al.*, 2004). These climate data are available for the period 1900–2000 and the resolution of this data set is 0.5 degree. For initialization of the ISAM model back over the period between 1755 and 1899, we generated the climate data over this time period by randomly selecting yearly climate data between the period 1900 and 1920.

Estimates of atmospheric CO_2 concentration from ice cores (Neftel *et al.*, 1985; Friedli *et al.*, 1986) and direct measurements given by Keeling *et al.* (1982) are specified from 1765 through 1958. The average of annual concentrations from the Mauna Loa (Hawaii) and South Pole Observatories (Keeling and Whorf, 2000) is specified for the period from 1959 through 1990.

The global distributions for natural vegetation in 1765 are estimated by superimposing the 1765 cropland data of Ramankutty and Foley (1999), and pastureland data of Klein Goldewijk (2001), over the potential vegetation data set of Ramankutty and Foley (1999). Ramankutty and Foley (1999) derived the potential vegetation data primarily from Loveland and Belward (1997) and Haxeltine and Prentice (1996) vegetation data sets. For the land cover changes starting in 1765, we superimpose the historical cropland/pasturelands data sets over the initial natural vegetation data set.

Over the period 1765–1990, we calculated historical land-use and net terrestrial biospheric carbon fluxes due to changes

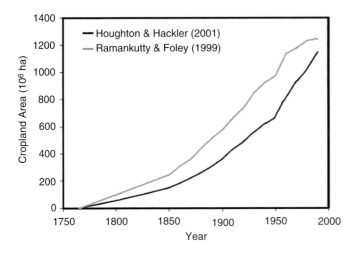

Figure 5.2 Global mean cropland areas based on Houghton and Hackler (2001) and Ramankutty and Foley (1999) data sets.

in land cover for croplands and abandonment based on data sets by Houghton (1999, 2000, 2003) (HH hereafter) and Ramankutty and Foley (1998, 1999) (RF hereafter) data sets (Figure 5.2). Since HH data used by Houghton also provide the information for pastureland changes, we also calculated terrestrial biospheric fluxes for changes in pastureland area in addition to croplands. Accordingly, we combined the HH-based pastureland changes data with HH and RF cropland data sets.

5.2.3 Model simulation experiments

After running the model to equilibrium in 1765, we performed transient experiments through 1990. To estimate the marginal effects of increasing CO_2, climate change, and land-cover changes for croplands on the terrestrial carbon cycle during the historical time period, we performed three experiments using our ISAM model. In the first experiment $E1$, atmospheric CO_2 and climate were varied over the historical time period. In experiment $E2$, atmospheric CO_2 and land cover changes for cropland and pasturelands were varied. In experiment $E3$, atmospheric CO_2, climate, and land cover changes for cropland and pasturelands were varied. The land-use emissions due to land cover changes for croplands and pasturelands were estimated by subtracting $E1$ from $E3$, and the effect of climate change was determined by subtracting $E2$ from $E3$. The marginal effect of increasing CO_2 was determined by subtracting the land-use emission and climate effects from experiment $E3$.

5.3 Results

This section presents the ISAM estimated global and annual mean changes in the NPP and terrestrial ecosystem carbon from 1900 to 1990 using the observed historical atmospheric CO_2 data and historical climate change data (precipitation and temperature data) based on reconstructed climate data from 1765 to 1899 and based on observation data from 1900 to 1990. I also used HH and RF croplands data sets, which are compared below, to simulate the effects of land cover changes for croplands between 1900 and 1990. In the following, first we compare ISAM-estimated terrestrial biospheric fluxes associated with changes in CO_2, climate, and land-use.

5.3.1 Net land–atmosphere carbon flux

The net land–atmospheric flux of CO_2 is the sum of land–atmosphere caused by changes in climate and CO_2 fertilization, and CO_2 flux associated with land cover changes for croplands as discussed above. Overall, ISAM-estimated cumulative net land–atmosphere flux based on the RF data set between 1765 and 1990 was –0.6 GtC/yr as compared to estimated flux based on the HH data of –0.2 GtC/yr. Figure 5.3a compares ISAM-estimated 10-year running mean net land–atmosphere fluxes of CO_2 based on the ISAM-HH and ISAM-RF cases between 1900 and 1990. From around 1900 to 1990, there were substantial differences in the net land–atmosphere fluxes calculated based on the two data sets.

The differences are particularly pronounced in the 1980s, when fluxes based on both data sets begin declining (Figure 5.3a). However, the decline based on RF data was much steeper than that based on HH data. Most importantly, model results for the 1980s derived from the RF data set demonstrate the terrestrial biosphere acting as a sink of atmospheric CO_2 (–0.57 GtC/yr) whereas those derived from the HH data set demonstrate the biosphere acting as a source (0.04 GtC/yr) (Table 5.2). The model results indicate that these large differences are mainly due to the differences in the estimated net biospheric CO_2 uptake in the tropics based on the two data sets during the 1980s (Figure 5.4a). In general, the model in the tropics simulates net sink activity based on RF data and source activities based on HH data (Table 5.2). The results based on both data sets indicate that the terrestrial biosphere was a sink of atmospheric CO_2 in the subtropics (Figure 5.4a). However, in spite of much smaller land-use fluxes in northern high latitudes (Figure 5.4d), particularly between 45°N and 70°N, the model simulates net source activities based on both data sets. This is because the estimated climate-related carbon released to the atmosphere is much higher in these latitudes (Figure 5.4b) than the CO_2 fertilization-related carbon sink (Figure 5.4c). Moreover, since the climate- and CO_2-related terrestrial fluxes during the 1980s are approximately the same based on two data sets (Figure 5.3b and 5.3c); the differences in net fluxes are mainly due to differences in the estimated net land-use sources (Figure 5.3d), which are discussed below.

5.3.2 Climate and CO_2 fertilization feedbacks

Figures 5.3b and 5.3c show that ISAM-estimated (GtC/yr) 10-year running mean fluxes of climate and CO_2 fertilization feedbacks based on HH and RF data sets from 1900 and 1990 are almost identical, although there are some minor differences

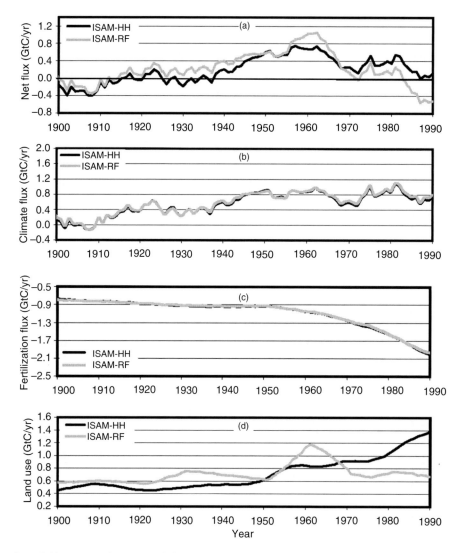

Figure 5.3 (a) The yearly and 10-year running mean of ISAM-estimated global and annual historical land–atmosphere carbon flux (GtC/yr) between 1900 and 1990. This can be partitioned into the flux attributed to (b) climate change, (c) increase in CO_2 concentrations, and (d) cropland expansion and abandonment. Positive values represent net carbon release to the atmosphere and negative values represent net carbon storage in terrestrial biosphere. The ISAM-HH and ISAM-RF results are based on data sets by Houghton (Houghton, 1999, 2000, 2003), and Ramankutty and Foley (1998, 1999), for cropland changes.

in yearly fluxes (not shown here). Over the period 1900–1990, the integrated climate and CO_2 fertilization feedback fluxes based on RF were only 1% higher than for HH.

Figure 5.3b shows that, prior to 1910, climate feedback had negligible effect on CO_2 uptake by the terrestrial ecosystems. However, there was a net release of CO_2 from the biosphere to the atmosphere from 1910 to 1990, as a consequence of increased rates of soil decomposition of current vegetation primarily to the north of the tropics due to temperature and precipitation changes (Figure 5.4b). From these results, it is clear that the temperature-enhanced productivity of plants has a smaller effect in the model than the temperature-enhanced autotrophic and heterotrophic respirations. Figure 5.3c, on the other hand, shows a continuous increase of CO_2 storage by the terrestrial ecosystems due to the fertilization effect over

the period 1900–1990, primarily due to enhancement of plant productivity. Our model results suggest that the storage of CO_2 due to CO_2 fertilization feedback is significantly higher (2.1 GtC/yr by 1990) than the CO_2 release due to climate feedback (0.8 GtC by 1990).

Our model results show that during the 1980s, the effects of climate tend to promote carbon storage between $20°$ S and $20°$ N, whereas climate change promotes CO_2 release in the subtropics and high latitudes, particularly in the northern hemisphere (Figure 5.4b). On the other hand, our model results show that CO_2 fertilization feedback enhances carbon storage all over the globe (Figure 5.4c). In particular, the effect is relatively higher in the tropics and in the north of the tropics (Figure 5.4c). With climate change alone, global biospheric carbon storage declines by 0.7 GtC/yr during the

Table 5.2 ISAM-estimated land-use emissions and terrestrial net land–atmosphere carbon flux over the periods 1765–1990 and 1980–1989 based on HH (Houghton and Hackler, 1999, 2001; Houghton, 1999, 2000, 2003) and RF (Ramankutty and Foley, 1998, 1999) land-cover change for cropland data sets. Model also incorporates the effects of climate and increasing CO_2 concentrations for simulations based on both data sets.

	Land-use emission (GtC)						Net land–atmosphere carbon emission (GtC)					
	1765–1990			1980–1989			1765–1990			1980–1989		
Regions	ISAM-HH GtC	ISAM-RF GtC	%change[a]	ISAM-HH GtC	ISAM-RF GtC	%change[a]	ISAM-HH GtC	ISAM-RF GtC	%change[a]	ISAM-HH GtC	ISAM-RF GtC	%change[a]
Tropical America	13.6	14.9	10	2.2	2.4	9	−19.5	−16.5	−15	−1.0	−0.5	−50
Tropical Africa	16.0	4.8	−70	4.8	0.8	−83	−0.9	−13.0	44	0.7	−3.3	−[b]
Tropical Asia	32	34.7	8	5.8	3.4	−41	20.2	23.2	15	3.9	1.4	−64
Total tropics	61.5	54.3	−12	12.8	6.7	−48	−0.3	−6.3	21x	3.6	−2.4	−[b]
North America	16.6	22.3	34	0.5	0.1	−80	5.6	13.3	137	0.4	0.2	−50
Europe	4.5	6.2	38	0.1	0.0	−[b]	1.3	3.5	169	−0.5	−0.5	0
N. Africa & Middle East	1.4	1.7	21	−0.3	0.1	−133	−2.1	−1.9	−9	−0.7	−0.4	−43
Former Soviet Union	7.7	16.1	109	−0.5	0.0	−[b]	−9.4	−1.1	−94	−0.1	0.8	−[b]
China	14.5	11.5	−20	0.2	−0.3	−[b]	6.3	2.9	−88	−0.8	−1.1	37
Pacific developing	2.5	1.3	−48	0.3	0.1	−66	−7.6	−8.3	9	−1.5	−1.7	13
Total non-tropical regions	47.1	59.1	25	0.5	0.0	−[b]	−5.8	8.4	−[b]	−3.2	−3.1	−[b]
Global	108.6	113.4	4	13.3	6.7	−50	−6.1	2.1	−[b]	0.4	−5.7	−[b]

[a] Relative to ISAM-HH.
[b] Either RF value is zero or HH and RF have different signs (+/−). In this situation it is not appropriate to calculate the percentage difference.

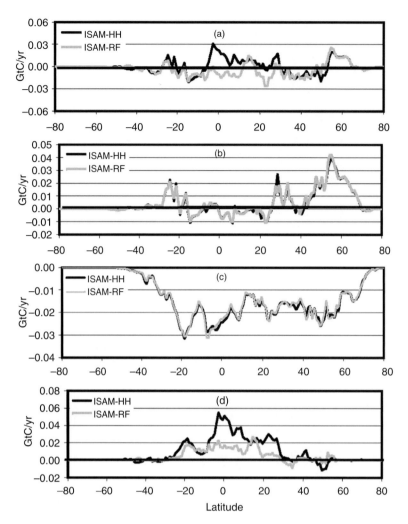

Figure 5.4 Partitioning of the ISAM-estimated annual mean latitudinal distribution of (a) net ecosystem carbon (GtC/yr) attributed to changes in (b) climate, (c) increase in CO_2 concentrations, (d) cropland expansion and abandonment. The results are compared for HH (Houghton and Hackler, 2001; Houghton, 1999, 2000, 2003), and RF (Ramankutty and Foley, 1998, 1999) cropland data cases for the 1980s. Positive values represent net carbon release to the atmosphere and negative values represent net carbon storage in terrestrial biosphere.

1980s, whereas CO_2 fertilization promotes carbon storage by about 2.0 GtC/yr.

5.3.3 Land use emissions

The carbon emissions (GtC/yr) estimated from ISAM for the period 1900–1990 derived from HH and RF data sets of cropland changes are shown in Figure 5.3d. Both HH and RF data sets show a generally increasing rate of change of area for cropland until about 1960. Thereafter, both data sets reveal different trends until 1990. The RF data set show a sharp decrease in the rate of change in area for croplands between 1960 and 1990. In contrast, data based on HH show first a decreasing trend between 1950 and 1970, then an increasing trend through the 1980s, even though HH-based cropland changes stabilize or decrease during the late 1980s. This is because emission rates do not immediately follow the rates of changes for croplands; rather, they depend on the amounts and turnover rates of the product pools (forest products have slower turnover rates relative to agriculture and paper products, so forest products release emissions over longer timescales).

The simulation based on both data sets indicates that the combined land-use component for non-tropical regions, and for Europe and the Former Soviet Union, was approximately neutral during the 1980s (Table 5.2), because the effects of carbon storage associated with forest regrowth are approximately balanced by releases associated with the decomposition of agriculture, paper, and wood products. The model results also show large differences in the regional land-use emissions based on the two data sets. For the period 1980–89, the emissions results based on the RF data set in some regions were appreciably lower than HH-based emissions (see column 7 of Table 5.2). In absolute terms, the HH-based land-use emissions for tropical regions were substantially higher than the RF-based emissions (Table 5.2). In terms of global results, the ISAM-estimated land-use emissions for 1980–89 based on RF data (6.6 GtC/yr) are about 50% lower than HH-based estimates (13.3 GtC/yr) (Figure 5.4d and Table 5.2).

5.4 Discussion

Over the period 1900–1950, our terrestrial ISAM model results for the land-use flux, based on two different sets of land-use data for cropland changes (HH and RF), exhibited similar trends (Figure 5.3d). However, the results were substantially different thereafter, particularly after around 1960 when the fluxes based on HH data show a constant increasing trend until 1990, whereas the fluxes based on RF data show a sharp decreasing trend. The increasing net land uptake of atmospheric CO_2 based on the RF data set during the 1980s is mainly due to the decline in the net land-use source during the 1980s (Figure 5.3d), which does not occur in the HH data owing to higher deforestation rates.

The differences in the land-use emissions based on HH and RF data are reflected in the differences in the net land–atmosphere biospheric fluxes of CO_2 for these two data cases. The higher the land-use emission (as in case of ISAM-HH), the lower the net terrestrial carbon storage and vice versa. Therefore, model results for the 1980s derived from the HH data set show the terrestrial biosphere acting as a source of atmospheric CO_2 (0.04 GtC/yr), while those derived from the RF data set demonstrate the biosphere acting as a sink (–0.57 GtC/yr) (Table 5.2).

While changes in area for croplands have been the dominant source of land-use emissions for CO_2 since pre-industrial time, there are other changes in land-use, such as pasturelands (Houghton, 2003), that may well lead to higher land-use emissions. According to Houghton (2003), pasturelands activities were responsible for about 15% of the total land-use emissions in 1980s. Since the land cover changes for pasturelands are the second dominant source (after croplands) of land-use emissions (Houghton, 2001, 2003), and Houghton (2001) also provides the pastureland change activities for the period 1765–1990, we extended our model simulations for changes in area to pasturelands in addition to croplands. Because RF data does not provide the information for the historical pastureland changes, we incorporate the historical pastureland changes based on Houghton (2001) not only into HH (defined here as ISAM-HH(C + P) case) but also into RF cropland data sets (defined here as ISAM-RF (C + P) case).

As expected, the ISAM-estimated combined land emissions scaled upward approximately the same amount in both cases (Figure 5.5b), whereas the estimated range of values for the net terrestrial uptake (sink) became smaller than the cropland-only case (Figure 5.5a). The ISAM-estimated range of values for the pastureland changes in addition to croplands (lower and higher range of values are based on ISAM-RF(C + P) and ISAM-HH(C + P), respectively) for land-use emissions in the 1980s were 1.50 to 2.06 GtC/yr (1.83 GtC/yr was middle of the

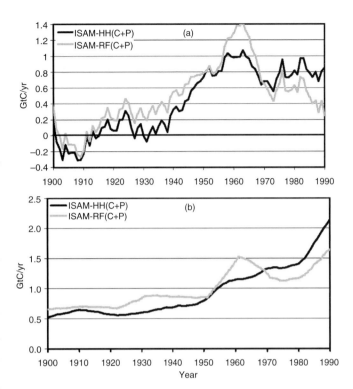

Figure 5.5 ISAM-estimated (a) net land–atmosphere flux, and (b) land-use emission between 1900 and 2000 associated with climate, CO_2 increase, and cropland and pastureland changes. For the ISAM-HH (C + P) case the data on croplands and pasturelands changes are taken from Houghton and Hackler (2001), and Houghton (1999, 2000, 2003). For the ISAM-RF(C + P) case the cropland data are based on Ramankutty and Foley (1998, 1999), and pastureland data are based on Houghton and Hackler (2001), and Houghton (1999, 2000, 2003).

range value), whereas the estimated range of values for the net terrestrial sink were 0.43 to 0.83 GtC/yr (0.63 GtC/yr was middle of the range value). Most importantly, the terrestrial biosphere no longer acts as a sink for atmospheric CO_2 for RF cropland data during the 1980s.

If the land-use emissions are indeed higher, such as in the case of cropland plus pastureland data, the terrestrial ecosystems may not act as a sink for atmospheric CO_2 during the 1980s, in contrast to suggestions by recent studies based on CO_2 and O_2 data (Prentice et al., 2001). Based on the CO_2 and O_2 data, the estimated net land–atmosphere CO_2 flux for the 1980s was –0.2 ± 0.7 GtC/yr. However, if the terrestrial biosphere was indeed acting as a sink for atmospheric CO_2, then the magnitude of our model estimated net terrestrial sink may be underestimated or land-use emissions may be overestimated.

A number of processes that might enhance the ISAM net terrestrial sink processes and lower the net land–atmosphere flux values include: the contribution of nitrogen deposition (Prentice et al., 2001 and references therein); fire suppression leading to woody encroachment (Houghton et al., 1999); recovery from past natural disturbances, sedimentation and spatial redistribution of carbon in products (House et al.,

2003); uptake of carbon during weathering processes on land, and transport of carbon from land areas to the ocean via rivers (Prentice et al., 2001). Consideration of these processes is beyond the scope of the present analysis but should be considered in the development of more detailed models.

In regard to the data on land cover change used in this study, we believe that large uncertainty in these two sets of data, which provides extreme high and low estimates of land clearing for croplands, merits comprehensive investigation. The causes for the differences will require further analysis using the full range of environmental monitoring at all scales, including integration of the recent land cover changes in forest cover with high resolution satellite measurements. Some of these efforts are already under way. For example, two recent studies of tropical deforestation based on satellite data (Achard et al., 2002; DeFries et al., 2002) suggest that the FAO-based rates of deforestation in tropical forests might be overestimated. Achard et al. (2002) found rates 23% lower than the FAO for the 1990s, while DeFries et al. (2002) estimated 54% lower than those reported by the FAO. However, in his most recent paper, Houghton (2003) questioned the inconsistency between these two studies. Houghton (2003) pointed out that the estimates of rates of change based on these two studies are as different from each other as they are from those of the FAO. He also questioned the reliability of the satellite-based data on percentage tree cover of DeFries et al. (2002) for tropical Africa, where the greatest differences are found (Houghton, 2003). DeFries et al. (2002) have also noted that tropical Africa is the most uncertain region because of difficulties in detecting patchy clearings, and spurious data sources.

In conclusion, our results leave open the possibility that the discrepancy in the magnitudes of the modeled and the data-based net land–atmosphere fluxes may be due to the limitations of terrestrial ecosystem models or the overestimation of the land-use sources. Finding the missing sink in the terrestrial biosphere will require continued refinements of both terrestrial biospheric sink capacity and/or the global land-use emission estimates from changes in forest covers, or refinement of O_2 and CO_2 based estimates.

Acknowledgements

This research was supported in part by the Office of Science (BER), US Department of Energy under Award No. DOE-DE-FG02–01ER63463.

References

Achard, F., Eva, H.D. Stibig, H.J. et al. (2002). Determination of deforestation rates of the world's humid tropical forests. *Science* **297**, 999–1002.

DeFries, R.S., Houghton, R.A., Hansen, M.C. et al. (2002). Carbon emissions from tropical deforestation and regrowth based on satellite observations for the 1980s and 1990s. *Proceedings of the National Academy of Sciences* **99**(22), 14 256–14 261.

Farquhar, G.D. and von Caemmerer, S. (1982). Modelling of photosynthetic response to environmental conditions. In *Physiological Plant Ecology II. Water Relations and Carbon Assimilation*. Vol. 12B of *Encyclopedia of Plant Physiology. New Series*, ed. O.L., Lange P.S. Nobel, C.B.Osmond, and H. Ziegler. Berlin: Springer-Verlag, pp. 549–588.

Friedli, H., Lötscher H., Oeschger H., Siegenthaler, U. and Stauffer, B. (1986). Ice core record of $^{13}C/^{12}C$ ratio of atmospheric CO_2 in the past two centuries. *Nature* **324**, 237–238.

Haxeltine, A. and Prentice, I.C. (1996). BIOME3: An equilibrium terrestrial biosphere model based on ecophysiological constraints, resource availability, and competition among plant functional types. *Global Biogeochemical Cycles* **10**, 693–709.

Houghton, R.A. (1999). The annual net flux of carbon to the atmosphere from changes in land-use 1850–1990. *Tellus* **51B**, 298–313.

Houghton, R.A. (2000). A new estimate of global sources and sinks of carbon from land-use change. *EOS* **81**, supplement 281.

Houghton, R.A. (2003). Revised estimates of the annual net flux of carbon to the atmosphere from changes in land-use 1850–2000. *Tellus* **55B**, 378–390.

Houghton, R.A. and Hackler, J.L. (2001) *Carbon Flux to the Atmosphere from Land-Use Changes: 1850 to 1990*. ORNL/CDIAC-131, NDP-050/R1 (http://cdiac.esd.ornl.gov/ndps/ndp050.html), Oak Ridge, Tennessee: Carbon Dioxide Information Analysis Center, US Department of Energy, Oak Ridge National Laboratory.

Houghton, R.A., Hobbie, J.E., Melillo, J.M. et al. (1983). Changes in the carbon content of terrestrial biota and soils between 1860 and 1980 – a net release of CO_2 to the atmosphere. *Ecological Monographs* **53**, 235–262.

Houghton, R.A., Hackler, J.L. and Lawrence, K.T. (1999). The US carbon budget: contributions from land-use change. *Science* **285**, 547–578.

House, J.I., Prentice, I.C. Ramankutty, N., Houghton, R.A. Heimann, M. (2003). Reconciling apparent inconsistencies in estimates of terrestrial CO_2 sources and sinks. *Tellus*, **55B** 345–363.

IMAGE team (2001). *The IMAGE 2.2 Implementation of the SRES Scenarios: Climate Change Scenarios Resulting from Runs with Several GCMs*. RIVM CD-ROM Publication 481508019, Bilthoven: National Institute of Public Health and the Environment, The Netherlands.

Jain, A.K. and Yang, X. (2005). Modeling the effects of two different land cover change data sets on the carbon stocks of plants and soils in concert with CO_2 and climate change. *Global Biogeochemical Cycles* **19** GB 2015; doi: 10.1029/2004GB002349.

Jain, A.K., Kheshgi, H.S. and Wuebbles, D.J. (1996). A globally aggregated reconstruction of cycles of carbon and its isotopes. *Tellus* **48B**, 583–600.

Jenkinson, D.S. (1990). The turnover of organic carbon and nitrogen in soil. *Philosophical Transactions of the Royal Society of London B* **329**, 361–368.

Keeling, C.D. and Whorf, T.P. (2000). Atmospheric CO_2 records from sites in the SIO air sampling network. In *Trends: A Compendium of Data on Global Change*. Oak Ridge, Tennessee: Carbon Dioxide Information Analysis Center, Oak Ridge National Laboratory.

Keeling, C. D., Bacastow, R. B. and Whorf, T. P. (1982). Measurements of the concentration of carbon dioxide at Mauna Loa Observatory, Hawaii. In *Carbon Dioxide Review: 1982*, ed. W. C. Clarke. New York: Oxford University Press.

Kheshgi, H. S. and Jain A. K. (2003). Projecting future climate change: implications of carbon cycle model intercomparison. *Global Biogeochemical Cycles* **17**(2), 1047; doi: 10.1029/2001GB001842.

Kheshgi, H. S., Jain, A. K. and Wuebbles, D. J. (1996). Accounting for the missing carbon sink with the CO_2 fertilization effect. *Climatic Change* **33**, 31–62.

King, A. W., Emanuel, W. R. Wullschleger, S. D. and Post, W. M. (1995). In search of the missing carbon sink: a model of terrestrial biospheric response to land-use change and atmospheric CO_2. *Tellus* **47B**, 501–519.

Klein Goldewijk, C. G. M. (2001). Estimating global land-use change over the past 300 years: the HYDE 2.0 database. *Global Biogeochemical Cycles* **15**, 417–433.

Loveland, T. R. and Belward, A. S. (1997). The IGBP-DIS global 1 km land cover data set, DISCover: first results. *International Journal of Remote Sensing* **18**, 3291–3295.

McGuire, A. D., Sitch, S. Clein, J. S. *et al.* (2001). Carbon balance of the terrestrial biosphere in the twentieth century: analyses of CO_2, climate and land-use effects with four process-based ecosystem models. *Global Biogeochemical Cycles*, **15**(1), 183–206.

Mitchell, T. D., Carter, T. R., Jones, P. D., Hulme, M. and New, M. (2004). *A Comprehensive Set of High-resolution Grids of Monthly Climate for Europe and the Globe: the Observed Record (1901–2000) and 16 Scenarios (2001–2100)*. Tyndall Centre Working Paper No. 55. Norwich: Tyndall Centre for Climate Research, University of East Anglia.

Neftel, A., Moor, E., Oeschger, H. and Stauffer, B. (1985). Evidence from polar ice cores for the increase in atmospheric CO_2 in the past two centuries. *Nature* **315**: 45–47.

Pastor, J. and Post, W. M. (1985). *Development of a Linked Forest Productivity-Soil Process Model*. Technical Report ORNL/TM-9519, Oak Ridge, Tennessee: Oak Ridge National Laboratory.

Polglase, P. J. and Wang, Y. P. (1992). Potential CO_2-enhanced carbon storage by the terrestrial biosphere. *Australian Journal Botany* **40**, 641–656.

Post, W. M., King, A. W. and Wullschleger, S. D. (1997). Historical variations in terrestrial biospheric carbon storage. *Global Biogeochemical Cycles* **11**, 99–110.

Prentice, C., Farquhar, G. H Fasham, M. *et al.* (2001). The carbon cycle and atmospheric CO_2. In *Climate Change 2001: The Scientific Basis. Contribution of Working Group I to the Third Assessment Report of the Intergovernmental Panel on Climate Change*, ed. J. T. Houghton, Y. Ding, D. J. Griggs *et al.* New York: Cambridge University Press, pp. 183–237.

Ramankutty, N. and Foley, J. (1998). Characterizing patterns of global land-use: an analysis of global croplands data. *Global Biogeochemical Cycles*, **12**, 667–685.

Ramankutty, N. and Foley, J. A. (1999). Estimating historical changes in global land cover: croplands from 1700 to 1992. *Global Biogeochemical Cycles* **13**, 997–1027.

Rawls, W. J., Brakensiek, D. L. and Saxton, K. E. (1982). Estimation of soil water properties. *Transactions of the ASAE* **25**, 1316–1328.

Thornthwaite, C. W. and Mather, J. R. (1957). Instructions and tables for computing potential evapotranspiration and the water balance. *Publ. Climatol.* **10**, 183–311.

Webb, R. S., Rosenzweig, C. E., and Levine, E. R. (1991). *A Global Data Set of Soil Particle Size Properties*. NASA Technical Memorandum 4286. New York: NASA, Goddard Space Flight Center, Institute for Space Studies.

Zobler, L. (1986). *A World Soil File for Global Climate Modelling*. NASA Technical Memorandum 87802. New York: NASA Goddard Institute for Space Studies.

Zobler, L. (1999) *Global Soil Types, 1-Degree Grid (Zobler). Data Set*. Available on-line from Oak Ridge National Laboratory Distributed Active Archive Center at www.daac.ornl.gov.

6

The albedo climate impacts of biomass and carbon plantations compared with the CO_2 impact

M. Schaeffer, B. Eickhout, M. Hoogwijk, B. Strengers,
D. van Vuuren, R. Leemans and T. Opsteegh

6.1 Introduction

Changes in land use and the consequent changes in land-cover properties modify the interactions between the land surface and the atmosphere locally and regionally (Kabat et al., 2004). Important factors in these interactions are the biochemical fluxes of CO_2 and other trace gases, and the biophysical fluxes of energy and water vapor. Modeling studies, well validated with detailed observations, show that changing land-use in the past centuries influenced local, regional, and probably also global climate patterns (e.g. IPCC, 2001). Historically, land-use mediated climate change appears to be an important factor (Brovkin et al., 1999). In mid to high latitudes, for example, land-use changes influence surface-air temperature because of the large difference in surface albedo between different land covers, such as cropland and forest in snow-covered conditions (Robinson and Kukla, 1985; Bonan et al., 1995; Harding and Pomeroy, 1996; Sharrat, 1998). Emission scenarios, required to estimate future climate change, nowadays often include detailed changes in land-use patterns and the consequent changes in sources and sinks of trace gases (e.g. Strengers et al., 2004). The biophysical consequences on the climate systems are, however, often neglected. It is therefore important to examine the role of land-use changes in determining future climates (Pielke Sr et al., 2002).

Future land-use change does not only include deforestation and afforestation as a consequence of expanding or contracting agriculture. Other land uses, such as plantations for carbon sequestration or energy production (to substitute fossil fuels), are likely to become more important. We define a "carbon plantation", or C-plantation, as the land cover resulting from the planting, managing, and harvesting of trees on formerly non-forested lands with the specific aim of achieving a maximum net uptake of CO_2 from the atmosphere into vegetation, litter, or soil. A "biomass plantation" refers to the land cover resulting from the activities aimed at providing biomass as a primary energy carrier.

Currently, integrated assessment modelers are beginning to include such land uses in studies of mitigation strategies, but only the impact on CO_2 fluxes is included (Leemans et al., 1996). However, if applied at a large scale, biomass and C-plantations might not only influence climate change by reducing global greenhouse gas (GHG) concentrations, but also have an impact on the energy fluxes between the land surface and the atmosphere. This was initially brought forward for C-plantations by Betts (2000), who estimated the effect of coniferous C-plantations in the northern hemisphere on planetary albedo. His calculations suggest that the global warming by albedo changes associated with high-latitude C-plantations would outweigh global cooling by carbon sequestration.

Here we will compare the importance of the biophysical climate impact of future land-use changes for two major mitigation options: production of biofuels and permanent C-plantations. We will largely focus on the impact of surface albedo changes as the obvious non-CO_2 impact of land-use change on climate in the extra-tropics. Our analysis is crucial to decide if such biophysical effects also need to be included in integrated assessment models for an adequate evaluation of the effectiveness of mitigation options that modify the land surface.

Human-induced Climate Change: An Interdisciplinary Assessment, ed. Michael Schlesinger, Haroon Kheshgi, Joel Smith, Francisco de la Chesnaye, John M. Reilly, Tom Wilson and Charles Kolstad. Published by Cambridge University Press. © Cambridge University Press 2007.

Albedo impacts of biomass and carbon plantations

Our analysis starts by applying our integrated assessment model IMAGE 2 (Alcamo *et al.*, 1998) to develop a scenario projection of future abandoned agricultural land. This provides a more realistic estimate of the geographical potential for carbon and biomass plantations than that of Betts (2000). IMAGE 2.2 has been used for a wide range of scientific and policy-advice studies, including a major contribution to the IPCC Special Report on Emissions Scenarios (IPCC, 2000; IMAGE team, 2001a), determining the importance of different feedback processes (Leemans *et al.*, 2002), exploring the geographical carbon sequestration potential of C-plantations (J. G. van Minnen, B. J. Strengers, B. Eikhout, R. Leemans and R. Swart, forthcoming) and estimating the geographical potential for modern biomass (Hoogwijk *et al.*, 2005). This paper combines the isolated results of these studies and provides a more comprehensive analysis. IMAGE 2 has been combined with ECBilt-CLIO, an atmosphere/ocean general circulation model of intermediate complexity, which was used extensively in studies of climate variability and change (Opsteegh *et al.*, 1998; Schaeffer *et al.*, 2002). This model will complete the evaluation by allowing assessment of the climate impacts of changes in albedo in addition to the CO_2 changes.

In Section 6.2, we will build on the previous IMAGE-2.2 work as mentioned above to develop a set of scenarios for the future large-scale application of biomass and C-plantations. In Section 6.3, we will describe relevant features of the models and outline the experiments, which are analyzed in Section 6.4 (impact on carbon fluxes, surface albedo, and climate). A short discussion in Section 6.5 concludes the paper.

6.2 Scenarios and assumptions

6.2.1 Scenario development

In this study we use the IMAGE implementation of the scenarios in the IPCC's Special Report on Emission Scenarios (SRES; IPCC, 2000; IMAGE team, 2001a). The IPCC SRES scenarios were developed by several modeling teams to explore the different possible pathways regarding greenhouse-gas emissions in the twenty-first century. In contrast to earlier IPCC scenarios, the SRES scenarios describe not only plausible trajectories for anthropogenic emissions of climate-relevant substances, but also consistent trajectories of human activities (the so-called "drivers"). Uncertainties obviously play a major role over such a time period, and the SRES modeling teams explicitly paid attention to this by developing a set of multiple, storyline-based scenarios in an attempt to map out some of the plausible futures.

The new scenarios are grouped into four scenario families which differ in their outlook along two major axes: (1) strong globalization and international governance versus more regional emphasis (A1 and B1 vs. A2 and B2); and (2) strong focus on economic development policies versus a development process more focused on social and environmental objectives (A1 and A2 vs. B1 and B2). Most of the modeling teams of the SRES focused mainly on the energy system and related emissions on the basis of the four storylines. Two teams (IMAGE and AIM) were able to elaborate the SRES storylines also in terms of geographically explicit land-use

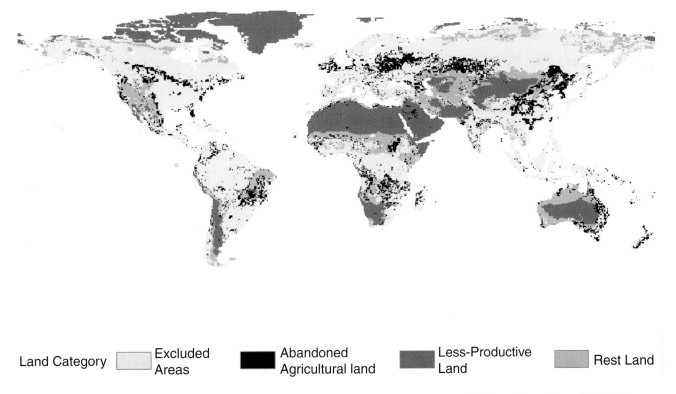

Land Category — Excluded Areas — Abandoned Agricultural land — Less-Productive Land — Rest Land

Figure 6.1 Estimate of abandoned cropland and pasture in 2050 for the IPCC SRES A1B scenario (from Hoogwijk *et al.*, 2005b).

scenarios (Strengers *et al.*, 2004). This work showed that for two out of the four scenario families (A1 and B1), large areas of agricultural land might be abandoned in the course of the twenty-first century as a result of further increases in agricultural yields – and stabilizing and even declining global population. In this study the biomass and carbon plantation experiments are based on the A1b scenario.

6.2.2 Geographic potential for biomass and carbon plantations

In literature, the use of biomass and C-plantations is considered both on more degraded, or marginal lands, and on productive (often abandoned agricultural) land. As the first uses less-productive lands, production will mostly be extensive. Here, we concentrate on the productive grounds. In agreement with most studies, we also assume that biomass or C-plantation will mainly take place on areas not used for the production of food, fodder and forestry products. Moreover, we assume that neither biomass nor C-plantations will take place on areas that are currently forests (for example because of conservation policies) (van Minnen *et al.*, forthcoming).

The assumptions above imply that in temperate zones excess agricultural areas could provide the most interesting locations to use for either biomass or carbon plantations. In Figure 6.1, we show results from Hoogwijk *et al.* (2005), indicating for the IPCC SRES A1b scenario the locations that would have become available by the year 2050 after abandoning agricultural land, because of a surplus of agricultural area, or a shift towards more productive sites (driven by climate change). The time-dependent version of the map in Figure 6.1 defines the geographical locations for the potential future application of biomass and C-plantations as assessed in the present study.

Please note that the scenario experiments in this paper are illustrative and assume extreme situations in order to investigate the impact of the surface-albedo changes. These sensitivity experiments differ from other long-term land-use scenarios using the IMAGE-2.2 model, which have matched demand for land and resources of land, resulting in a more plausible future trend.

6.3 Description of models and further specification of scenario experiments

6.3.1 IMAGE-2.2 model and experiment set-up

The general objective of IMAGE 2.2 is to explore the long-term dynamics of global environmental change. The model consists of several linked modules (Alcamo *et al.*, 1998; IMAGE team, 2001a). The main driving forces are economic and demographic trends at a regional level. Energy system drivers (production and consumption flows) are simulated in TIMER (Targets IMage Energy Regional model; de Vries *et al.*, 2001) together with related emissions of GHG and regional air pollutants. Ecosystem, crop, and land-use modules are used to compute land use on the basis of regional consumption, production and trading of food, animal feed, fodder, grass, and timber, and local climatic and soil properties. The exchange of CO_2 between terrestrial ecosystems and the atmosphere is simulated (Leemans *et al.*, 2002; Strengers *et al.*, 2004). The atmospheric and ocean models calculate changes in atmospheric composition by employing the emissions and by taking oceanic CO_2 uptake and atmospheric chemistry into consideration. Subsequently, changes in climatic properties are computed (Eickhout *et al.*, 2004). Changes in climate and CO_2 concentration feed back on the land-cover projections. Of special interest for the present paper are the characteristics of TIMER and the terrestrial modules, the latter functioning on a $0.5° \times 0.5°$ grid.

6.3.2 The IMAGE energy model TIMER

In this paper, we use the TIMER 1.0 energy model to assess the potential impact of increased use of biofuel on fossil-fuel greenhouse gas emissions (for description see de Vries *et al.* [2001] and van Vuuren *et al.* [2004]). TIMER describes the energy consumption and production for 10 primary energy carriers for which market shares are determined on the basis of assumed consumer preferences and costs. In the model, biofuels mainly compete in the transport sector (with oil-based alternatives) and in the electric power sector (offsetting natural gas and coal use). Under the IMAGE-A1B scenario, biofuels reach a production level of about 250 exajoule (1 EJ = 10^{18} J) worldwide in 2100, i.e. around 15% of global primary energy consumption. For the purpose of this analysis, however, we have defined the use of biofuels exogenously. A set of simple assumptions was made to determine how biofuels would be used:

- In the "no-biofuel" variant, the production of biofuels in northern hemisphere (NH) regions has been set to zero. The supply of biofuels in all regions is reduced by the impact of this reduction on global biofuel supply. As a result, the market shares for biofuels are replaced by fossil-fuel based alternatives, mostly oil and natural gas.
- In the "exogenous-biofuel" variant, the supply of biofuel in NH regions is set exogenously to the maximum potential as identified by Hoogwijk *et al.* (2005).[1] For simplification, it is assumed that the increased biomass supply will equally increase biomass use in all regions globally – and across all sectors. In transport, the additional biomass is assumed to substitute oil use. In the electricity sector, it is assumed that

---

[1] TIMER uses an efficiency of 40–60% to convert primary energy carriers such as biomass to electricity. In the transport sector, TIMER only uses secondary biomass fuels. As Hoogwijk *et al.* (2005) give total primary energy numbers for biofuels, we had to partly convert these numbers. Based on the assumption that 50% of the biofuels will be used in transport using a conversion efficiency from primary biomass to transport fuel, the primary energy production numbers have been multiplied by 0.7.

biomass will substitute for natural gas use (a more conservative estimate than substituting biomass for coal; moreover, natural gas has the largest market share in new electric power capacity). Indirectly, some further changes may occur as the additional biomass affects the prices of fossil fuels.

In both cases, the impact on emissions is calculated – with noticeable changes for greenhouse gas emissions and sulfur emissions.

6.3.3 The IMAGE terrestrial models

The land-cover model allocates the agricultural demand grid cell by grid cell within each region, giving preference to cells with the highest probability for satisfying this demand, on the basis of a set of heuristic land-use rules (Alcamo et al., 1998). Crops are assigned to agricultural cells according to their potential crop productivity, which is based on climate circumstances, soil characteristics, and the CO_2 concentration (Leemans and van den Born, 1994). A central assumption is that agricultural high-yield areas and areas close to established infrastructure are preferable locations to retain food- and fodder-growing purposes.

In the normal mode, abandoned agricultural land is transformed to natural vegetation as determined by the natural vegetation model (a modified version of the BIOME-model; Prentice et al., 1992) on the basis of climate- and soil conditions, and changes in atmospheric CO_2 concentrations. Calibration has led to an improved parameter setting for the different land-cover types (Leemans et al., 2002).

When abandoned agricultural land is covered completely by biomass plantations (in our case "woody biofuels" like eucalyptus, willow and poplar), the carbon uptake is simulated consistently with the crop productivity that was used as input for the energy potential (see above and Hoogwijk et al. [2005]). The high carbon uptake of biomass plantations is partly offset by higher soil respiration fluxes.

In the case of C-plantations, the most suitable tree type out of six representative types is chosen by the model on the basis of climate constraints and the highest calculated average net primary productivity (NPP) of the different types under climate and soil conditions of the grid cell in question. The additional carbon sequestration of C-plantations, compared with natural vegetation types, is implemented in the carbon model on the basis of a literature survey (Schaeffer et al., 2006; van Minnen et al., forthcoming). In this study it is assumed C-plantations are used for CO_2 sequestration only and will not be harvested for timber or fuel.

6.3.4 The three land-use change experiments with IMAGE

Three modified versions of the IMAGE-2.2 A1B scenario for the time period 2000–2100 will be explored. The first experiment assumes no demand for biofuels in the NH extra-tropical regions in IMAGE 2.2 (IM-nat). All former agricultural land returns to vegetation types given by the natural vegetation model. This forms the non-mitigation "baseline scenario" used as a benchmark to compare the effectiveness of the two plantation mitigation options.

In the second experiment, IM-C, all abandoned cropland will be used for permanent C-plantations instantaneously and the land–atmosphere CO_2 fluxes are modified accordingly. For the energy sector there will be no difference from the IM-nat experiment. In IM-C the abandoned land is only transformed into C-plantations when the NPP of these plantations is substantially higher than the natural regrowing vegetation (somewhat limiting the total area of abandoned agricultural land available).

In the third experiment, IM-bio, all abandoned cropland will be used for biomass plantations. For this scenario, the exogenous-biofuel variant of IMAGE-TIMER (see above) is used on the basis of the technical potential of this abandoned area (in accordance with Hoogwijk et al., 2005). Since we assume short rotations of 5 years for the biomass plantations, the resulting reduction in emissions in the energy sector is offset by an additional amount of carbon emitted through burning of this above-ground biomass harvest.

In Figure 6.2 (see colour plate section) the different land-use change scenarios are visualized. Because we assume biomass crops are planted on all abandoned cropland, the total area of biomass plantations is larger than of C-plantations in the IM-C experiment with the additional "NPP constraint". In Figure 6.3, the areas of plantations over time are given for the selected regions, showing that in all regions the largest area increase occurs in the first half of the twenty-first century, except in China.

6.3.5 ECBilt-CLIO model and experiment set-up

For exploring the climate impacts of the IMAGE-2.2 experiments, and thus the effectiveness of the two climate-change mitigation options, we will use the coupled atmosphere/ocean/cryosphere model ECBilt-CLIO. The atmospheric component ECBilt (Opsteegh et al., 1998) was developed for research on the relative importance of the physical feedbacks in the extra-tropics of the climate system on decadal and longer timescales. Since the main application of this model is in the extra-tropics, a quasi-geostrophic approach for the dynamical core of the model was adopted. The neglected geostrophic terms in the vorticity and thermodynamic equations are included as a time- and space-varying forcing. In the extra-tropics, the description of atmospheric motion of the model is comparable to that of atmospheric general circulation models (GCMs). The resolution of ECBilt is about 5.6° lat/lon (T21), with three vertical levels. The ocean model CLIO is a GCM with a dynamic sea-ice component and a relatively sophisticated parameterization of vertical mixing (Goosse and Fichefet, 1999). The horizontal resolution of CLIO is 3° in

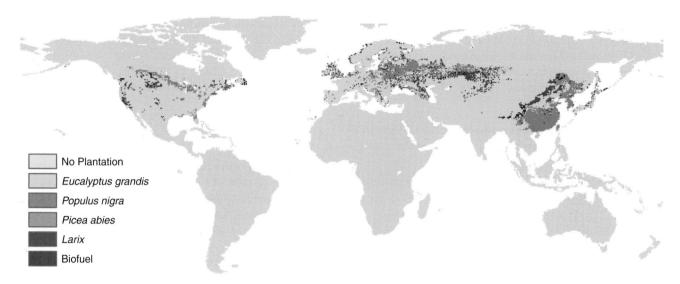

Figure 6.2 Carbon-plantation tree types for the year 2100 in the IM-C experiment. Because of the extra surplus-NPP constraint on C-plantations and bioclimatic limits, the total area of these is smaller than that of biomass plantations. The additional area of biomass plantations in the IM-bio experiment is indicated in red. Land-cover changes for regions other than the northern hemisphere regions selected for the sensitivity experiments in this paper are not shown here.

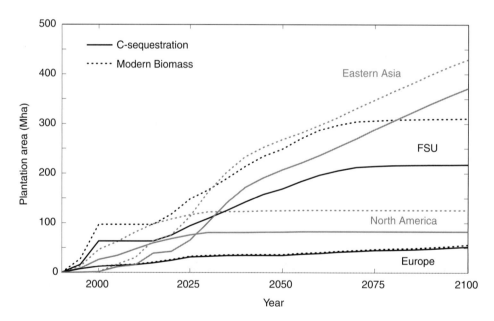

Figure 6.3 Areas of plantations for the selected regions for both the carbon-sequestration (IM-C) and modern biomass (IM-bio) scenarios.

latitude and longitude, and there are 20 unevenly spaced vertical levels in the ocean. The coupled model (ECBilt-CLIO) was recently used to study the influence of mid- to high-latitude atmosphere/ocean/sea-ice interactions on climate variability and change (Goosse et al., 2001; Renssen et al., 2001, Goosse et al., 2002; Schaeffer et al., 2002, 2004, 2005). The present-day climatology is explored in more detail by Goosse et al. (2001). The estimated global climate sensitivity is 1.7 °C for a doubling of CO_2, which is at the low end of the estimated range (Kattenberg et al., 1996; Andronova and Schlesinger, 2001).

Physical parameterizations in ECBilt were kept as simple as possible. A bucket model of uniform depth (15 cm) represents soil moisture. Probably the most important surface/atmosphere energy feedback in the mid to high latitudes is the albedo–snow feedback. In the updated snow model in ECBilt (Schaeffer et al., 2006), the snow-layer depth is used to determine the albedo by interpolation between snow-free albedo and completely snow-covered albedo. In addition, the snow albedo decreases through snow aging. Constant values for these no-snow, fresh-snow and old-snow albedo types are assigned for each vegetation type (see tables in Schaeffer et al. [2006]).

In the scenario experiments, the albedo of the grid cells of ECBilt-CLIO is determined by using land-cover classes from the higher-resolution IMAGE-2.2 maps and area-weighting the albedo types. In this modified ECBilt version, the simulated present-day albedo generally compares well with observations (Schaeffer et al., 2006).

All ECBilt-CLIO experiments start by using initial conditions from a single model spin-up forced by historical GHG concentrations from the years 1850 to 1970. Before this spin-up, the model was brought into equilibrium by running 1000 years with constant 1850 GHG concentrations. For each of the three modified IMAGE-2.2 A1b experiments, ECBilt-CLIO was run twice: with GHG changes only, and with both GHG and land-cover changes. Since we expected that the difference in climate impact between the scenarios is small, all experiments were performed in ensemble mode, by applying tiny random distortions to the atmospheric initial conditions. Each of 20 individual simulations within one ensemble represent different evolutions of the climate system for the same external forcing, with equal probability of occurrence. The spread in the results among the individual ensemble members thus provides an indication of the influence of internal, or natural, climate variability on climate change projections. Compared with a single climate model run, the ensemble mean provides a better estimate of the mean climate response by reducing the sampling error.

6.4 Impacts of plantations on CO_2, albedo and climate

6.4.1 Impacts on CO_2

In the no-biofuels variants (IM-nat and IM-C), the use of modern biomass in 2100 is reduced from 250 EJ to 150 EJ per year compared with the original A1b scenario (i.e. only from the tropical northern hemisphere and the southern hemisphere), increasing CO_2 emissions with about 1.6 GtC per year. In the IM-bio case, the worldwide use of modern biomass is increased to 440 EJ/yr (of which 290 EJ/yr from NH). As a result, oil use is reduced considerably (by 140 EJ/yr in 2100). Smaller reductions occur for natural gas (65 EJ/yr), coal and nuclear/solar/wind.

Carbon dioxide emissions from fossil fuels are reduced from 17 GtC/yr in 2100 for the IM-nat and IM-C cases to 12 GtC/yr for IM-bio (see Figure 6.4). Because of harvesting and burning of biofuels from biomass plantations, the gross emission flux from biofuels in IM-bio reaches a level of 7 GtC per year in 2100 (see Figure 6.4). This emission flux more than offsets the gain of using biofuels in the energy sector (5 GtC per year in 2100). This result is in line with the assumption that the burning of biofuels occurs less efficiently than the burning of gas or oil.

Figure 6.4 also shows the terrestrial uptake fluxes of the three different experiments. The terrestrial uptake is highest in the IM-bio case, since in this case the plantations are harvested each 5 years, implying a steep carbon uptake curve at the beginning of each new rotation period for these fast-growing crops. Logically, the IM-nat case returns the lowest terrestrial uptake of carbon, since no management is applied to the abandoned agricultural land and the vegetation types are not primarily selected on the basis of optimal carbon sequestration. In the case of IM-C, the terrestrial uptake is higher than IM-nat, benefiting from management of C-plantations, but lower than IM-bio, since we assumed no harvest of the C-plantations.

The consequences for the atmospheric CO_2 concentrations are depicted in Figure 6.5. The differences between IM-C and IM-bio for the complete carbon cycle can be neglected in

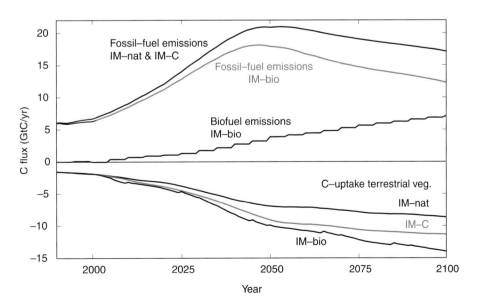

Figure 6.4 Anthropogenic and terrestrial carbon fluxes for the three land-use experiments.

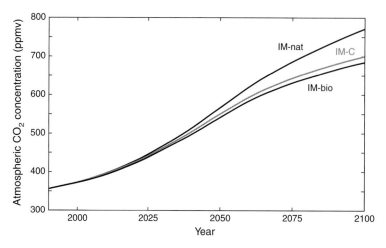

Figure 6.5 Atmospheric CO_2 concentration in the three land-use experiments.

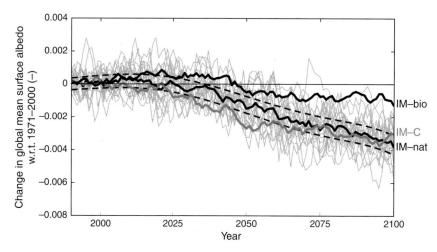

Figure 6.6 Global and annual mean surface albedo in ECBilt-CLIO with respect to 1971–2000. Thin gray lines are individual ensemble members of the no-plantation baseline experiment. Thick lines are ensemble means of the no-plantation baseline (IM-nat), carbon-plantation (IM-C) and biomass-plantation (IM-bio) experiments. The dashed lines indicate the distance to the baseline case beyond which ensemble means differ from the baseline with 95% significance.

light of the uncertainties involved in the carbon budget calculations. As mitigating options, the two mitigation experiments result in a similar reduction of 70–80 ppmv CO_2 by the year 2100 compared with the non-mitigation experiment IM-nat.

6.4.2 Impacts on albedo

The short-rotation crops of the biomass plantations effectively mean that current agricultural land will remain in production, whereas this is replaced by dense forest in the carbon plantation, and natural vegetation in the no-plantation case. Forested areas have a lower albedo than agricultural areas, especially at mid to high latitudes in winter owing to snow-masking (Robinson and Kukla, 1985; Bonan *et al.*, 1995; Harding and Pomeroy, 1996; Sharrat, 1998). Figure 6.6 shows that the no-plantation and carbon-plantation experiments result in a lower global-mean surface albedo than the biomass-plantation case. In all experiments, global warming gradually reduces the snow cover, so that a gradual decline in surface albedo is superimposed on the trends related to land-use change alone. Thus, albedo does also decrease in the biomass plantation experiment, as well as in the CO_2-only experiments (not shown).

6.4.3 Impacts on climate

The increase in CO_2 concentration to 770 ppmv in the IM-nat case of IPCC SRES A1b leads to a rise in global-mean surface-air temperature (SAT) of 0.8 °C by the year 2100 with respect to the 1971–2000 average (Figure 6.7). This is relatively modest compared to other climate models, because of the low climate sensitivity (see Section 6.3) and because only changes in GHGs are included in the ECBilt-CLIO experiments, not the expected decrease in sulfur emissions that

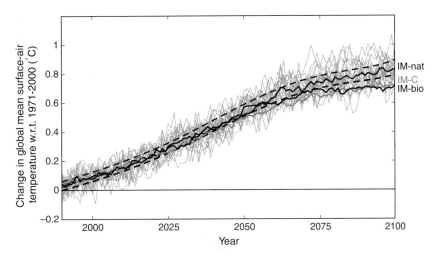

Figure 6.7 Global and annual mean surface-air temperature in ECBilt-CLIO with respect to 1971–2000 for the CO_2-only experiments, i.e. without taking into account differences in albedo. Legend as in Figure 6.6.

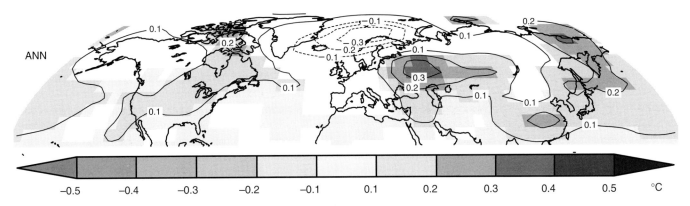

Figure 6.8 Difference in annual-mean surface-air temperature in 2071–2100 (°C) of carbon-plantation with respect to biomass-plantation ensemble mean, including albedo effects. Contours are plotted for all model grid cells. The grid cells for which the difference between the two ensemble means is significant above the 95% level (2-tailed t-test) are colored.

would lead to further global warming. By the year 2050, the reduction in CO_2 concentrations in the IM-C and IM-bio experiments cause annual-mean SAT to diverge from the IM-nat baseline significantly at the 95% level (1-tailed t-test). Note that this significance level is given by internal climate variability alone. Obviously, the carbon-budget calculations in IMAGE 2.2 introduce additional uncertainties of a different category, some of which were studied by Leemans et al. (2002). Figure 6.7 shows that the 70–80 ppmv reduction in CO_2 concentrations in the two mitigation scenarios moderates global-mean temperature increase by about 0.1 °C in 2100, thus mitigating global warming over the twenty-first century by more than 10%.

Although the CO_2 pathways of the two mitigation scenarios are comparable, the albedo difference causes different climate impacts (see Figure 6.8, colour plates section). In the regions of major land-use changes (see Figure 6.2) surface-air temperature is higher in the carbon plantation scenario. The difference reaches 0.3 °C in some regions, which is about 25% of the full climate change signal over the twenty-first century in those regions in the baseline case and up to several times the local temperature difference between the mitigation scenarios and the baseline caused by changes in CO_2 concentration only. In central and western Eurasia, heating in spring increases evaporation, thereby reducing the soil moisture content. The less intensive hydrological cycle in summer then reduces latent heat fluxes, amplifying the albedo-induced summer heating and extending this heating far outside the area of land-use change itself (Schaeffer et al., 2006).

Globally, the difference between the two mitigation scenarios increases rapidly in the first half of the twenty-first century (Figure 6.9, gray line) when large areas of agricultural land are taken out of production in areas with significant snow cover. The less-productive agricultural land in these areas is first in line to be selected for abandonment. The global temperature difference after the 2050s stays more or less constant, because additional areas in Russia and China have less snow cover and because the snowline slowly retreats northward

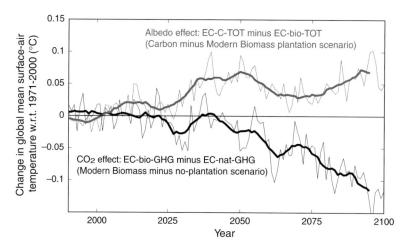

Figure 6.9 Difference between ensemble averages in global and annual mean surface-air temperature with respect to 1971–2000. The black line indicates the global cooling resulting from lower net CO_2 emissions in the two mitigation scenarios (shown only for biomass plantation CO_2-only experiment) compared with the no-plantation baseline. The gray line shows the relative global warming in the C-plantation scenario compared with the biomass plantations scenario, caused by the difference in the pathway of surface albedo.

owing to global warming, thus lowering the albedo difference between cropland and forests near the retreating snowline.

In contrast to the albedo impact, it is well after the year 2050 that any significant effect of the gradually reducing CO_2 concentration can be detected in the two mitigation options compared with the baseline (Figure 6.9, black line). The size of the two effects becomes comparable in the last decades of the twenty-first-century model simulation. Thus, although the timing is different, the climate impact of the albedo difference between the two mitigation scenarios is comparable in size to the impact of the reduced CO_2 concentration in the two mitigation options compared to the baseline. As illustrated for soil moisture in the above, (regional) climate feedbacks complicate this global picture. However, these reinforce the conclusion that other climate effects than just the impact of changes in CO_2 concentrations need to be taken into account when assessing mitigation options that involve land-use changes.

6.5 Discussion and conclusions

Our analysis shows that the large-scale application of carbon and biomass plantations in the northern extra-tropics causes changes in surface albedo and CO_2 concentration. The lower albedo of permanent carbon plantations results in significantly higher regional and global-mean surface-air temperatures than in the biomass-plantation scenario. The impact on CO_2 concentrations of these two mitigation options is comparable.

The changes in climate due to albedo changes accelerate rapidly on the short-to-medium term, while the CO_2 concentrations gradually increase up to the end of the twenty-first century. By that time, the CO_2 effect on climate has become somewhat larger than that of albedo change. Relative to the impact of changes in CO_2, albedo change thus gradually becomes less important, but is the dominant forcing in these scenarios over the larger part of the twenty-first century.

Carbon plantations are often brought forward as an option to "buy time" for the medium term, allowing a longer delay of transformations in the energy sector. Since we found that on the short-to-medium term the albedo effect is strongly unfavorable for the carbon plantations case, it is questionable whether carbon plantations in the northern hemisphere could actually fulfill such an effective role in medium-term mitigation efforts. On the centennial timescale over which the albedo effect becomes relatively less important, the permanency of carbon plantations is also questionable (e.g. Barford et al., 2001).

The IPCC A1B land-use scenario, which forms the basis for our sensitivity experiments, is subject to some uncertainty typical for geographical patterns of climate change. We have used IMAGE 2.2 earlier with climate change patterns from a variety of coupled general circulation models (IMAGE team, 2001b). The principal conclusion is that even at the end of the twenty-first century the sensitivity of land-cover changes to climate-change patterns is low on the spatial scale of IMAGE-2.2 regions (sub-continental scale) and the uncertainty is largely driven by socio-economics. Moreover, the difference is necessarily much smaller between the baseline and the two mitigation scenarios, because the difference in CO_2 concentration is "only" 70–80 ppmv, so the difference between climate change patterns of different CGCMs is also smaller. A remaining issue forms a basic shortcoming in an exercise using a chain of models. In our analyses a feedback of albedo-induced climate change in the climate model to the land-cover model is absent. For example, the relative regional warming in the carbon-plantation scenario by albedo differences might affect CO_2 uptake and thus the global CO_2 concentration. However, we consider this of secondary importance.

The climate sensitivity of our climate model is on the lower side of the range spanned by CGCMs (IPCC, 2001; Raper et al., 2002), as well as the range found plausible in dedicated climate-sensitivity estimates (Andronova and Schlesinger, 2001;

Forest et al., 2002; Harvey and Kaufmann, 2002; Knutti et al., 2002). This is an important aspect to consider, since we are concerned with the relative impact of changes in greenhouse-gas concentration and land-surface albedo, and these two drivers act on climate through different mechanisms. Drijfhout et al. (1999) argue that an earlier, but on essential points comparable, version of ECBilt was in general a relatively insensitive climate model. Not only was the climate sensitivity roughly half of the mean of CGCMs, the amplitude of mid-latitude (internal) climate variability and the response to a change in (external) solar forcing was also about half of that found in CGCMs. The relative contribution of these components to the total climate variations was, however, comparable. Recently, we have compared the response of ECBilt with five other Earth System Models of Intermediate Complexity (EMICs) to a time series of historical (1700–2000) land-use changes (Brovkin et al., 2005). Compared with these models, the response of ECBilt is again on the lower side of the mean. These considerations lead us to believe that the relative impact of the changes in greenhouse-gas concentrations and land-surface albedo that we have presented here is reasonably robust.

In this paper, we have concentrated on the albedo impact of land-use changes, because this is the main driver of land-use related climate change in the extra-tropics. However, the processes are more complicated in the tropics (Pielke Sr et al., 2002). Here the hydrological cycle plays a crucial role, and albedo differences are not the single most important driver (Niyogi et al., 2002; Kabat et al., 2004). In addition, teleconnections between the tropics and the extra-tropics give rise to a complicated picture of land-use changes in the tropics causing changes in climate over large distances (Avissar, 1995). Our current analysis explicitly excluded the tropics because the climate model that we used is not viable for assessing the climate impacts of land-use changes in the tropics. The relative importance of these biophysical climate impacts of land-use changes might be much lower compared with CO_2 changes, since plantations are potentially much more productive in the tropics (van Minnen et al., forthcoming). Thus, our main conclusion that biophysical climate impacts are highly unfavorable to carbon plantations probably does not apply to the tropics.

In the regions we have analyzed in this paper, there might be other important considerations than climate impacts to take into account when comparing carbon and biomass plantations. For example, we have not dealt with the economic aspects. Hoogwijk et al. (2005) indicate that part of the geographic potential for biomass use might be available at relatively low costs. The larger share of the potential might in the medium future be available at production costs up to 0.05 $/kWh to generate electric power. This implies an emission-reduction cost of up to 100 US$/tC. For transport, biofuel costs could come down to around 10 $/GJ in the next decades, compared with 7 US$/GJ for gasoline, implying emission-reduction costs of up to 150 US$/tC. Beyond 70% of the potential, however, production costs tend to increase rapidly. Cost estimates of carbon sinks vary significantly. Strengers et al. (2005), for example, estimate that these costs would be of the order of 10–150 US$/tC – but at the same time, they questioned whether it would be possible to implement the full technical potential given all kinds of implementation barriers, most of which could be represented as additional costs.

At the moment, we have assumed that the patterns of food production would not shift in response to additional biofuel production. Alternatively, one could imagine that some food-producing areas (e.g. in Eastern Europe) might be attractive as biomass growing areas (for Eastern Europe particularly in response to European policies towards a higher energy self-sufficiency), thus implying additional agricultural land for food production in other areas of the world. On top of such considerations of food production and energy supply, other services of land-use forms and ecosystems might be considered, such as the influence on land degradation and surface run-off. Finally, one could imagine that cultural and regional preferences could play a role in decision making on land-use forms such as biomass and carbon plantations, like concerns for nature and biodiversity conservation, or values related to preserving country life in developed countries. The resulting changes in land-use patterns could also partly change the results of the albedo calculations, but analysis of these is far beyond the scope of this paper.

Summarizing, although many more considerations will be taken into account in "real-world" decisionmaking on land use, the extra-tropical albedo effect assessed here is strong enough to justify the idea that any climate-change mitigation analysis including carbon and biomass plantations is at best "incomplete" without including this effect. In our analysis, without such impacts the large-scale use of carbon plantations or biofuels in the northern hemisphere has similar benefits in avoiding climate change. Including such impacts, however, almost completely erodes the climate-change mitigation impact that carbon plantations might have via the carbon cycle.

References

Alcamo, J., Leemans, R. and Kreileman, E. E. (1998). *Global Change Scenarios in the 21st Century: Results from the IMAGE 2.1 Model.* London: Elsevier Science.

Andronova, N. and Schlesinger, M. E. (2001). Objective estimation of the probability distribution for climate sensitivity. *Journal of Geophysical Research* **106** (D19), 22 605–22 612.

Avissar, R. (1995). Recent advances in the representation of land-atmosphere interactions in general circulation models. *Reviews of Geophysics* **33**, 1005–1010.

Barford, C. C., Wofsy, S. C., Goulden, M. L. et al. (2001). Factors controlling long- and short-term sequestration of atmospheric CO_2 in a mid-latitude forest. *Science* **294**, 1688–1691.

Betts, R. A. (2000). Offset of the potential carbon sink from boreal forestation by decreases in surface albedo. *Nature* **408**, 187–190.

Bonan, G. B., Chapin F. S. III, Thompson, S. L. *et al.* (1995). Boreal forest and tundra ecosystems as components of the climate system. *Climatic Change* **29**, 145–167.

Brovkin, V., Ganopolski, A., Claussen, M. *et al.* (1999). Modelling climate response to historical land cover change. *Global Ecology and Biogeography* **8**(6), 509–517.

Brovkin, V., Claussen, M., Driesschaert, T. *et al.* (2005). Biogeophysical effects of historical land cover changes simulated by six Earth system models of intermediate complexity. *Climate Dynamics* **26**(6), 587–600.

de Vries, H. J. M., van Vuuren, D. P., den Elzen, M. G. J. *et al.* (2001). *The TARGETS IMage Energy Regional (TIMER) Model*. Bilthoven: National Institute for Public Health and the Environment (RIVM).

Drijfhout, S. S., Haarsma, R. J., Opsteegh, J. D. *et al.* (1999). Solar-induced versus internal climate variability in a coupled climate model. *Geophysical Research Letters* **26**, 205–208.

Eickhout, B., den Elzen, M. G. J., Kreileman, G. J. J. *et al.* (2004). *The Atmosphere–Ocean System of IMAGE 2.2: A Global Model Approach for Atmospheric Concentrations, and Climate and Sea Level Projections*. Bilthoven: National Institute for Public Health and the Environment (RIVM).

Forest, C. E., Stone, P. H., Sokolov, A. P. *et al.* (2002). Quantifying uncertainties in climate system properties with the use of recent climate observations. *Science* **295**, 113–117.

Goosse, H. and Fichefet, T. (1999). Importance of ice-ocean interactions for the global ocean circulation: a model study. *Journal of Geophysical Research* **104**, 23 337–23 355.

Goosse, H., Selten, F. M., Haarsma, R. J. and Opsteegh, J. D. (2001). Decadal variability in high Northern latitudes as simulated by an intermediate-complexity climate model. *Annals of Glaciology* **33**, 525–532.

Goosse, H., Selten, F. M., Haarsma, R. J. and Opsteegh, J. D. (2002). A mechanism of decadal variability of the sea-ice volume in the Northern Hemisphere. *Climate Dynamics* **19**, 61–83.

Harding, R. J. and Pomeroy, J. W. (1996). The energy balance of the winter Boreal landscape. *Journal of Climate* **9**, 2778–2787.

Harvey, L. D. D. and Kaufmann, R. K. (2002). Simultaneously constraining climate sensitivity and aerosol radiative forcing. *Journal of Climate* **15**(20), 2837–2861.

Hoogwijk, M., Faaij, A., de Vries, B. and Turkenburg, W. (2005a). Potential of biomass energy out to 2100, for four IPCC SRES land-use scenarios. *Biomass and Bioenergy* **29**(4), 225–227.

Hoogwijk, M., Faaij, A., Eickhout, B. *et al.* (2005b). Potential of biomass energy out to 2100, for four IPCC SRES land-use scenarios. *Biomass and Bioenergy* (accepted).

IMAGE team (2001a). The IMAGE 2.2 implementation of the SRES scenarios. A comprehensive analysis of emissions, climate change and impacts in the 21st century. Main disc. Bilthoven: National Institute for Public Health and the Environment (RIVM).

IMAGE team (2001b). The IMAGE 2.2 implementation of the SRES scenarios. Climate change scenarios resulting from runs with several GCMs. Supplementary disc. Bilthoven: National Institute for Public Health and the Environment (RIVM).

IPCC (2000). *Emissions Scenarios. A Special Report of Working Group III of the Intergovernmental Panel on Climate Change*, ed. N. Nakicenovic and R. Swart. Cambridge: Cambridge University Press.

IPCC (2001). *Climate Change 2001: The Scientific Basis. Contribution of Working Group I to the Third Assessment Report of the Intergovernmental Panel on Climate Change*, ed. J. T. Houghton, Y. Ding, D. J. Griggs *et al.*, Cambridge: Cambridge University Press.

Kabat, P., Claussen, M., Dirmeyer, P. A. *et al.* (2004). *Vegetation, Water, Humans and the Climate. A New Perspective on an Interactive System*. Berlin: Springer Verlag.

Kattenberg, A., Giorgi, F., Grassl, H. *et al.* (1996). Climate models – Projections of future climate. In *Climate Change 1995: The Science of Climate Change. Contribution of Working Group I to the Second Assessment Report of the International Panel on Climate Change*, ed. J. T. Houghton, L. G. Meira Filho, B. A. Callander *et al.* Cambridge: Cambridge University Press, p. 572.

Knutti, R., Stocker, T. F., Joos, F. and Plattner, G.-K. (2002). Constraints on radiative forcing and future climate change from observations and climate model ensembles. *Nature* **416**, 719–723.

Leemans, L. and van den Born, G.-J. (1994). Determining the potential global distribution of natural vegetation, crops and agricultural productivity. *Water, Air and Soil Pollution* **76**, 133–161.

Leemans, R., van Amstel, A., Battjes, C. *et al.* (1996). The land cover and carbon cycle consequences of large-scale utilizations of biomass as an energy source. *Global Environmental Change* **6**(4), 335–357.

Leemans, R., Eickhout, B., Strengers, B. *et al.* (2002). The consequences of uncertainties in land use, climate and vegetation responses on the terrestrial carbon. *Science in China*, **45**, 126–142.

Niyogi, D. S., Xue, Y.-K. and Raman, S. (2002). Hydrological feedback in a land-atmosphere coupling: comparison of a tropical and a midlatitudinal regime. *Journal of Hydrometeorology* **3**, 39–56.

Opsteegh, J. D., Haarsma, R. J., Selten, F. M. and Kattenberg, A. (1998). ECBILT: a dynamic alternative to mixed boundary conditions in ocean models. *Tellus* **50A**, 348–367.

Pielke, R. A. Sr, Marland, G., Betts, R. A. *et al.* (2002). The influence of land-use change and landscape dynamics on the climate system: relevance to climate-change policy beyond the radiative effect of greenhouse gases. *Philosophical Transactions of the Royal Society of London* **360**, 1705–1719.

Prentice, I. C., Cramer, W., Harrison, S. P. *et al.* (1992). A global biome model based on plant physiology and dominance, soil properties and climate. *Journal of Biogeography* **19**, 117–134.

Raper, S. C. B., Gregory, J. M. and Stouffer, R. J. (2002). The role of climate sensitivity and ocean heat uptake on AOGCM transient temperature response. *Journal of Climate* **15**, 124–130.

Renssen, H., Goosse, H., Fichefet, T. and Campin, J.-M. (2001). The 8.2 kyr BP event simulated by a global atmosphere–sea-ice–ocean model. *Geophysical Research Letters* **28**(8), 1567–1570.

Robinson, D. A. and Kukla, G. (1985). Maximum surface albedo of seasonally snow-covered lands in the northern hemisphere. *Journal of Climate and Applied Meteorology* **24**, 402–411.

Schaeffer, M., Selten, F. M., Opsteegh, J. D. and Goosse, H. (2002). Intrinsic limits to predictability of abrupt regional climate change in IPCC SRES scenarios. *Geophysical Research Letters* **29**(16), 14:1–14:4.

Schaeffer, M., Selten, F. M., Opsteegh, J. D. and Goosse, H. (2004). The influence of ocean convection patterns on high-latitude climate projections. *Journal of Climate* **17**(22), 4316–4329.

Schaeffer, M., Selten, F. M. and Opsteegh, J. D. (2005). Shifts of means are not a proxy for changes in extreme winter temperatures in climate projections. *Climate Dynamics* **25**(1), 51–63.

Schaeffer, M., Eickhout, B., Hoogwijk, M. *et al.* (2006). CO_2 and albedo climate impact of extratropical carbon and biomass plantations. *Global Biogeochemical Cycles* **20**, doi: 10.1029/2005 GB 002 581.

Sharrat, B. S. (1998). Radiative exchange, near-surface temperature and soil water of forest and cropland in interior Alaska. *Agricultural and Forest Meteorology* **89**, 269–280.

Strengers, B., Leemans, R., Eickhout, B. J. *et al.* (2004). The land-use projections and resulting emissions in the IPCC SRES scenarios as simulated by the IMAGE 2.2 model. *Geo Journal* **61**, 381–393.

Strengers, B., van Minnen, J. G. and Eickhout, B. J. (2005). *The Role of Carbon Plantations in Mitigating Climate Change: Potentials and Costs*. Bilthoven: Netherlands Environmental Assessment Agency, National Institute for Public Health and the Environment (RIVM).

van Vuuren, D. P., Eickhout, B., Lucas, P. L. and den Elzen, M. G. J. (2004). Long-term multi-gas scenarios to stabilise radiative forcing: exploring costs and benefits within an integrated assessment framework. *Energy Policy* (in press).

7

Overshoot pathways to CO_2 stabilization in a multi-gas context

T. M. L. Wigley, R. G. Richels and J. A. Edmonds

7.1 Introduction

Stabilization of the climate system requires stabilization of greenhouse-gas concentrations. Most work to date has considered only stabilization of CO_2, where there are choices regarding both the concentration stabilization target and the pathway towards that target. Here we consider the effects of accounting for non-CO_2 gases (CH_4 and N_2O), for different CO_2 targets and different pathways. As primary cases for CO_2 we use the standard "WRE" pathways to stabilization at 450 ppm or 550 ppm. We also consider a new "overshoot" concentration profile for CO_2 in which concentrations initially exceed and then decline towards a final stabilization level of 450 ppm, as might occur if an initial target choice were later found to be too high.

Emissions reductions for CH_4 and N_2O are optimized for the different pathways using an energy-economics model (MERGE). The optimization procedure minimizes the total cost of emissions reductions. The CH_4 and N_2O emissions reductions lead to substantially reduced future warming and future sea-level rise relative to stabilization cases where likely emissions reductions for these gases are ignored. For central climate and sea level model parameter values the reductions are 0.3–0.4 °C and 2–3 cm in 2100 and 0.9–1.0 °C and about 14 cm in 2400. Reduced CH_4 and N_2O emissions also allow larger CO_2 emissions by reducing the magnitude of climate feedbacks on the carbon cycle.

For the overshoot case, the initial choice leads to a significant overshoot of the final temperature stabilization level and a noticeable warming rate commitment, both of which are reduced through reductions in CH_4 emissions. We note that concentration stabilization does not lead immediately to climate stabilization. The "lag" effect of oceanic thermal inertia means that global-mean temperature continues to rise (albeit slowly) for centuries. This effect is much more pronounced for sea level.

Article 2 of the UNFCCC challenges us to stabilize the climate system through stabilizing the concentrations of greenhouse gases. Concentration pathways (or "profiles") for CO_2 have been proposed that achieve stabilization at a range of levels between 350 and 1000 ppm (Enting *et al*., 1994; Wigley *et al*., 1996; Wigley, 2000). The Wigley–Richels–Edmonds (WRE) profiles described in Wigley *et al*. (1996) take account of economic factors, and have been shown in the IPCC Third Assessment Report to be close to optimum in terms of the costs of emissions reductions (Hourcade and Shukla, 2001, Section 8.4). In estimating the emissions requirements and their climate consequences for the WRE pathways, however, no account has been taken of the possible effect of stabilizing other greenhouse gases. As noted above, in this paper we expand the WRE analyses into more realistic territory by considering the effects of optimized, cost-effective reductions in CH_4 and N_2O emissions. We also explore an alternative type of pathway where CO_2 concentrations overshoot the chosen target before declining. Such a pathway might occur if we initially adopted too high a concentration target and found ourselves having to make rapid adjustments in the future, a mid-course correction; or may be chosen as a deliberate strategy for economic, technological or political reasons. We assess the consequences for CO_2 emissions and for climate of following such an overshoot pathway.

Human-induced Climate Change: An Interdisciplinary Assessment, ed. Michael Schlesinger, Haroon Kheshgi, Joel Smith, Francisco de la Chesnaye, John M. Reilly, Tom Wilson and Charles Kolstad. Published by Cambridge University Press. © Cambridge University Press 2007.

The WRE concentration profiles assume a specific baseline no-climate-policy scenario. To account for inertia in the energy production system and the large costs that would be associated with the premature retirement of existing fossil-fuel-based capital stock, they further assume that initial departures from the baseline are slow.

The profiles have been defined for a range of stabilization targets. At present there is little agreement as to what constitutes an appropriate concentration target for CO_2 (Mastrandrea and Schneider, 2004; Wigley, 2004a). At best, we can adopt a near-term hedging strategy balancing the risks of acting too aggressively (stabilization target too low) and incurring unnecessarily high mitigation costs, against the risks of acting too tentatively (stabilization target too high) and incurring environmental damages that could otherwise have been avoided. It is the latter case that is considered here. To illustrate this possibility we consider the 450 ppm and 550 ppm CO_2 stabilization targets and define an overshoot profile for stabilization at 450 ppm. For the overshoot case we assume that concentrations initially follow the WRE550 profile, and that a mid-course correction is made in 2020. This is an idealized situation, chosen to illustrate what a change in stabilization target might mean in terms of more stringent emissions controls and reduced climate damages.

This is not the first time that overshoot profiles have been considered. In the original stabilization profiles developed for the IPCC (Enting et al., 1994) and the WRE modifications of these, the concentration pathways for stabilization at 350 ppm rise above the stabilization target before declining back to 350 ppm. Other overshoot examples are given in Swart et al. (2002) and O'Neill and Oppenheimer (2004). The present work is distinguished from these studies in two ways: (1) the overshoot profile is defined a priori with emissions calculated by an inverse procedure, and (2) climate feedbacks on the carbon cycle are accounted for. In Wigley (2004b), where the same standard WRE profiles are used, an overshoot pathway initially following the WRE650 profile and then changing course to stabilize at 550 ppm is considered. The methods used in that paper are the same as used here, but the present analysis considers the emissions and climate implications in far more detail.

7.2 Future CO_2, CH_4 and N_2O concentrations

Carbon dioxide concentrations are specified to follow updated WRE concentration stabilization profiles. Concentrations for CH_4 and N_2O are based on a multi-gas emissions reduction strategy, using cost-optimized emissions estimates that are consistent with the CO_2 profiles.

A number of recent scientific advances require the original WRE profiles to be updated: to ensure consistency with models used in the IPCC Third Assessment Report (Joos et al., 2001; Prentice, 2001; Khesghi and Jain, 2003), to include climate feedbacks on the carbon cycle (see Tables 7.1 and 7.2); and to account for improved baseline (no-climate-policy) emissions scenarios as published in the Special Report on Emissions Scenarios (SRES; IPCC: Nakićenović and Swart, 2000). (For details on the MAGICC carbon cycle model, see Wigley [1993, 2000]; and Wigley and Raper [2001].) The updated WRE concentration profiles presented here (and previously in Wigley [2004b]) have been designed to minimize differences from the original profiles, so the main consequences of these updates are for the emissions requirements (lower emissions because climate feedbacks on the carbon cycle are now accounted for). For the baseline, except for CH_4 and N_2O (see below), we now use the median of the set of 35 complete SRES scenarios out to 2100 and assume constant emissions thereafter (the "P50" scenario).

In order to stabilize the climate system in a cost-effective manner, a balanced portfolio of emissions reductions across the various greenhouse gases must be used. For gases potentially controlled under the Kyoto Protocol we consider only CO_2, CH_4, and N_2O, since the other gases have relatively minor effects on climate. For climatically active gases not considered by the Kyoto Protocol (sulfate and other aerosols, tropospheric and stratospheric ozone) we assume that these will be controlled by pollution rather than climate policies (Smith et al., 2000, 2005).

For CO_2 we prescribe the concentration profiles. For CH_4 and N_2O, however, we prescribe emissions. These emissions depend on an estimate of the warming target for the adopted CO_2 concentration profile – a lower warming target makes it more cost-effective to reduce the emissions of non-CO_2 gases early (Manne and Richels, 2001). Since we consider only one value for the climate sensitivity (see below) the warming target is directly linked to the CO_2 concentration stabilization target.

For the WRE450 and WRE550 profiles, we calculate cost-effective CH_4 and N_2O emissions reductions as in Manne and Richels (2001), except that we have re-run the Manne and Richels energy-economics model (MERGE) for the specified concentration stabilization cases using the P50 baseline for non-CO_2 gases. For the mid-course correction ("overshoot") case, which has the same CO_2 concentration target as WRE450, we scale the CH_4 and N_2O emissions reductions

Table 7.1 Comparison of climate feedbacks on the carbon cycle for different carbon cycle models used in the IPCC TAR. Results shown (ppm) are start-year 2100 concentrations for the IS92a emissions scenario (Leggett et al., 1992) using central climate model parameters (see Prentice, 2001).

MODEL	No climate feedbacks	With climate feedbacks	Increase due to climate feedbacks
Bern[a]	651	703	52
ISAM[b]	682	723	41
MAGICC	673	714	41

[a] Joos et al. (2001)
[b] Kheshgi and Jain (2003)

Table 7.2 Comparison of start-year concentration projections (ppm) for the SRES illustrative emissions scenarios (IPCC: Nakićenović and Swart, (2000) for the carbon cycle models used in the IPCC Third Assessment Report (Bern: Joos et al., 2001; ISAM: Kheshgi and Jain, 2003). The MAGICC model is used in the present analyses.

Scenario	2050			2100		
	Bern	ISAM	MAGICC	Bern	ISAM	MAGICC
A1B	522	532	530	703	717	711
A1T	496	501	498	575	582	574
A1FI	555	567	565	958	970	983
A2	522	532	529	836	856	855
B1	482	488	485	540	549	537
B2	473	478	475	611	621	616
IS92a	499	508	504	703	723	714

progressively from the WRE550 case initially to the WRE450 case in proportion to the overshoot concentration transition. These three sets of emissions reductions are applied to the P50 baseline emissions scenario. Note that the resulting absolute emissions for CH_4 and N_2O differ from those in Manne and Richels (2001), not least because the no-policy baseline used here is different.

Figure 7.1 (see colour plates) shows the concentration profiles used here. Figure 7.1a shows the revised monotonic (WRE) CO_2 stabilization profiles and the new overshoot profile compared with the baseline (P50) no-climate-policy case (shown only out to 2100). Figure 7.1b shows the corresponding concentrations for methane. For the WRE450 case it is cost-effective to reduce methane emissions substantially after 2020 and to continue to reduce these emissions almost linearly to 2150, the end of the multi-gas analysis period. At this point virtually all anthropogenic CH_4 emissions have been eliminated. Concentrations stabilize soon after this at close to the pre-industrial level. For the WRE550 case it is not cost-effective to reduce CH_4 emissions significantly until after 2050. Eventually, however, almost all anthropogenic CH_4 emissions are eliminated and concentrations again stabilize at close to the pre-industrial level. The implication that all anthropogenic emissions of CH_4 may be eliminated cost-effectively is subject to some uncertainty, and may be an artifact of the relatively simple gas-cycle models used in MERGE. Nevertheless, the qualitative result that large CH_4 emissions reductions can be made cost-effectively is undoubtedly realistic.

For N_2O, emissions reductions are cost-effective almost immediately in the WRE450 case and after 2020 in the WRE550 case. The effect of the pre-2020 emissions reductions in the WRE450 case is small. In both the WRE450 and WRE550 cases, N_2O emissions peak around 2050 and then decline steadily over the next 100 years to a level about 10% below the current level. This leads (slowly owing to the long lifetime of N_2O) to concentration stabilization at a level that is still substantially above the present level (Figure 7.1c).

7.3 Implications for CO_2 emissions

Figure 7.2 (see colour plates) shows the implied fossil-fuel CO_2 emissions, cumulative fossil emissions, and the rate of change of fossil emissions. Calculating emissions is a more complex process than in the original WRE process (Wigley et al., 1996). With climate feedbacks on the carbon cycle now included, the emissions results depend on projected changes in temperature, in turn dependent on climate model parameters (most importantly on the climate sensitivity) and on the emissions of non-CO_2 gases. For climate model parameters, we use the median values adopted in Wigley and Raper (2001). The median climate sensitivity is an equilibrium warming for a CO_2 doubling of 2.6 °C. For non-CO_2 gases, the interplay between emissions reductions in CO_2 and those for CH_4 and N_2O is accounted for in the cost-optimization procedure.

Having defined emissions for the non-CO_2 gases, it is a simple matter to calculate the CO_2 emissions to meet any of the three concentration profiles considered here. We do this by running the MAGICC coupled gas-cycle/climate model (Raper et al., 1996; Wigley and Raper, 2001, 2002) in inverse mode. We consider the CO_2 emissions results first, then the implications for global-mean temperature and sea level and for the rate of change of temperature. In all cases we discuss results both ignoring and including the effects of CH_4 and N_2O emissions reductions.

For CO_2 emissions (Figure 7.2a), the change in concentration stabilization target from 550 ppm to 450 ppm requires a rapid change in the emissions trajectory at the mid-course decision point, even though the concentration transition has been chosen to be very gradual. Emissions at this point are, however, much larger than they would be if the WRE450 pathway had been followed. Beyond this rapid transition, the emissions differential between the original WRE550 trajectory and the lower-target, mid-course correction trajectory remains quite small for about 100 years. It is not until the twenty-second and twenty-third centuries that emissions in the mid-course correction (overshoot) case drop to levels substantially below WRE550

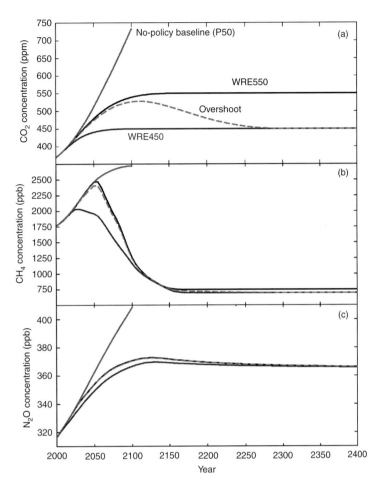

Figure 7.1 (a) Revised WRE and a new overshoot concentration stabilization profile for CO_2 compared with the baseline (P50) no-climate-policy scenario. (b) Methane concentrations based on cost-effective emissions reductions (Manne and Richels, 2001) corresponding to the WRE450, WRE550, and overshoot profiles for CO_2. The baseline (P50) no-climate-policy scenario result is shown for comparison. (c) Nitrous oxide concentrations based on cost-effective emissions reductions (Manne and Richels, 2001) corresponding to the WRE450, WRE550, and overshoot profiles for CO_2. The baseline (P50) no-climate-policy scenario result is shown for comparison.

emissions. Emissions for this case drop to below those for WRE450 after about 2130.

The effects of including CH_4 and N_2O emissions reductions can be seen by comparing the bold and dashed lines in Figure 7.2a. Reduced emissions of these gases lead to lower warming and, hence, a smaller climate feedback on the carbon cycle. This in turn allows the CO_2 emissions required to match any given concentration profile to be higher, an indirect benefit of the multi-gas emissions reduction strategy. The CO_2 emissions difference rises from near zero in 2050 to 0.7–0.8 GtC/yr in 2200.

For cumulative emissions (Figure 7.2b) the asymptotic values are only weakly dependent on the pathway to stabilization. This asymptotic limit occurs only after many centuries, however, so it is misleading, as is sometimes done, simply to identify a concentration target with a particular value of cumulative emissions. Substantially higher cumulative emissions occur in the mid-course correction case compared with WRE450 for times out to 2300, even though both have the same concentration target. Cumulative emissions are an indicator of mitigation costs, since larger cumulative emissions imply a slower transition to non-fossil energy sources.

Comparing the WRE550 and the 550-to-450 overshoot case, the transition from WRE550 to the overshoot case will undoubtedly incur additional mitigation costs. Following WRE550 implies a particular technological pathway, and, as noted in the original WRE paper, additional costs must be incurred by any decision to deviate from a previously chosen pathway. However, the additional costs may be relatively small, especially if the possibility of a change in concentration target is anticipated. Both pathways require a continual transition from fossil to non-fossil energy production, and the magnitude of the technological challenge this represents is at least partly related to the rate of reduction of CO_2 emissions. Since there is little change in the maximum rate of emissions reduction as one goes from the WRE550 to the 550-to-450

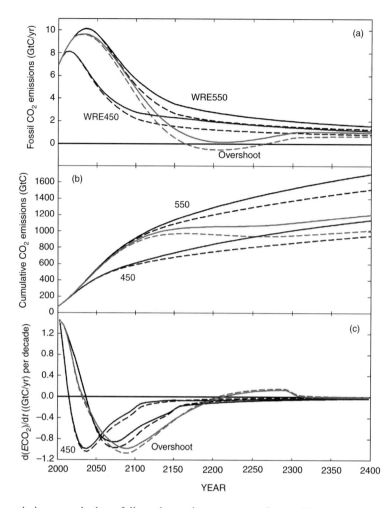

Figure 7.2 (a) Fossil CO_2 emissions required to follow the various concentration stabilization profiles shown in Figure 7.1a. The emissions account for climate feedbacks on the carbon cycle (see Table 7.1), and were derived using an inverse version of the MAGICC coupled gas-cycle/climate model. Two sets of results are shown: assuming median SRES emissions for non-CO_2 gases to 2100 and constant emissions for these gases after 2100 (dashed lines); and employing cost-effective reductions in CH_4 and N_2O emissions (full lines). (b) Cumulative fossil CO_2 emissions for the various stabilization profiles. (c) Rates of change of fossil CO_2 emissions $dECO_2/dt$ required to follow the various stabilization profiles.

case, the additional challenge may be small. Balanced against this, one must compare the overshoot case with WRE450. From an emissions reduction point of view, following the overshoot trajectory to 450 ppm would be less costly than following WRE450. The overshoot case, however, is likely to incur larger climate damages – see below.

Figure 7.2c shows the rates of change of fossil CO_2 emissions. Emissions begin to decline between 2015 and 2035. The rate of decline is an indicator of the rate of transition from fossil to non-fossil energy sources, and so provides an indication of the technological challenges (e.g. Hoffert *et al.*, 2002) that are demanded by this transition. A number of points should be noted. First, the maximum rates of emissions decline do not occur when concentrations are declining, but earlier, when the rate of concentration increase is declining rapidly. Second, with the chosen concentration profiles, the maximum rates of emissions decline are similar in all cases

(around 1 GtC/yr per decade). The maximum decline-rate point, however, occurs much later in the overshoot case than in the corresponding monotonic (WRE450) case allowing more time to develop the required non-fossil energy and/or sequestration technologies. Reductions in CH_4 and N_2O emissions reduce the maximum rate of decline in CO_2 emissions, but only slightly.

7.4 Temperature and sea-level implications

In Figure 7.3 (see colour plates) we compare global-mean temperature and sea-level changes for the three stabilization profiles with and without the effects of CH_4 and N_2O emissions reductions. We show results for only one set of climate/sea-level model and radiative forcing parameters, the central values used in the IPCC Third Assessment Report (Church and Gregory, 2001; Cubasch and Meehl, 2001;

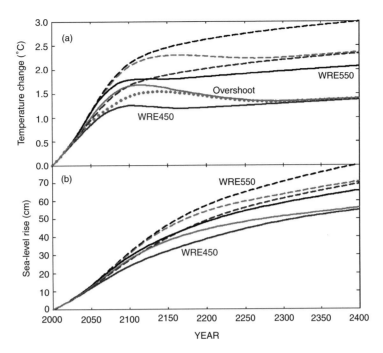

Figure 7.3 Global-mean warming for the various stabilization profiles using central estimates for all climate model parameters. Two sets of results are shown: assuming median SRES emissions for non-CO_2 gases to 2100 and constant emissions for these gases after 2100 (dashed lines); and employing cost-effective reductions in CH_4 and N_2O reductions (full lines). The dotted magenta temperature line for the 550 to 450 (overshoot) case shows the result of doubling the reductions in CH_4 emissions. (b) Global-mean sea-level rise for the various stabilization profiles using central estimates for all climate and sea-level model parameters. Two sets of results are shown: assuming median SRES emissions for non-CO_2 gases to 2100 and constant emissions for these gases after 2100 (dashed lines); and employing cost-effective reductions in CH_4 and N_2O reductions (full lines).

Wigley and Raper, 2001). Figure 7.3a shows that, because of oceanic thermal inertia, concentration stabilization does not lead immediately to climate stabilization – global-mean temperature continues to rise (albeit slowly) for centuries in the monotonic (WRE450 and WRE550) cases. The effect is much more pronounced for sea level (see below). For all three CO_2 pathways, the effects of CH_4 and N_2O emissions reductions are substantial – a temperature reduction around 0.4 °C in 2100 rising to almost 1 °C in 2400. When CH_4 and N_2O emissions reductions are included, global-mean temperature effectively stabilizes (residual warming rate of less than 0.1 °C per century) at about 1.4 °C above present for the WRE450 and overshoot cases. Global-mean temperature reaches about 2 °C above present for the WRE550 case in 2100 with a residual (thermal inertia) warming trend of about 0.1 °C per century.

A comparison of the WRE550 and overshoot case quantifies the effects of a mid-course correction where an initial 550 ppm target was changed to a 450 ppm target because of perceived climate risks. The key question here is whether such a mid-course correction might avert these risks. The correction clearly reduces the magnitude of future temperature changes, with the difference increasing from about 0.1 °C in 2100 to more than 0.6 °C by 2400. This would clearly lead to reduced climate impacts, but only on very long timescales. A more important benefit of the mid-course correction may be in reducing the risk of exceeding some threshold that might lead to a non-linear response of the climate system with greatly exacerbated damages.

Comparing the WRE450 and the overshoot case quantifies the effects of deliberately choosing the overshoot pathway. This must lead to greater changes in climate for any given stabilization target. Figure 7.3a shows that temperatures in the overshoot case (full magenta line in Figure 7.3a) are substantially higher (up to 0.6 °C) than for the corresponding monotonic WRE450 pathway (full blue line), with significant differences as early as 2050. Furthermore, the initial choice of a higher target leads to a large overshoot of the final temperature stabilization level (note the decline in temperature in the overshoot case – full magenta line – from around 2100 onwards). This is particularly so when CH_4 and N_2O emissions reductions are included, where the temperature overshoot is almost 0.4 °C. The overshoot magnitude can be reduced by implementing more stringent reductions in CH_4 emissions. The magenta dotted temperature line in Figure 7.3a shows the result of doubling the reductions in CH_4 emissions (in which case, anthropogenic emissions drop to zero by 2100, after which they are assumed to remain at zero), a strategy that approximately halves the amount of overshoot.

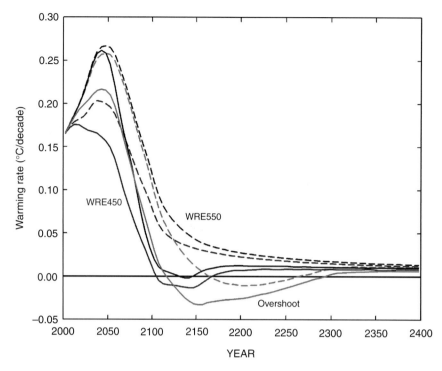

Figure 7.4 Rates of change of global-mean temperature (°C/decade) for the temperature projections shown in Figure 7.3a.

Sea-level rise considers the sum of oceanic thermal expansion plus ice melt and other factors considered in the IPCC TAR (Church and Gregory, 2001). (The melt models are as used in the TAR except for melt from glaciers and small ice sheets – GSICs – where an improved model is used; Wigley and Raper, 2005). Results are shown in Figure 7.3b. As is well known, atmospheric composition stabilization does not lead to sea-level stabilization (Wigley, 1995, 2005). Even in 2400, sea level is continuing to rise at 5–7 cm per century. Compared with WRE450, sea-level rise in the overshoot case is about 6 cm higher over 2100–2200, with the differential decreasing slowly thereafter. Reductions in CH_4 and N_2O emissions have a marked effect, leading to sea levels that are 2–3 cm lower in 2100 and around 14 cm lower in 2400.

As an additional indicator of potential differential impacts we consider the rate of global-mean warming (Figure 7.4, in colour plates section). Rate of warming has been suggested as an important factor in determining how the thermohaline circulation may change (Cubasch and Meehl, 2001), and faster warming would clearly affect the response of natural ecosystems and the ability of human systems to adapt to a changing climate (see also O'Neill and Oppenheimer, 2004). Figure 7.4 shows that, when cost-effective reductions in CH_4 and N_2O emissions are included, a mid-course correction to a lower target reduces the maximum warming rate from 0.26 °C/decade for WRE550 to 0.22 °C/decade for the mid-course correction case (compared with about 0.18 °C/decade for the WRE450 case). If CH_4 and N_2O emissions reductions are not included, the maximum warming rates in the WRE550 and mid-course correction cases are virtually the same. Thus, non-CO_2 emissions reductions can provide considerable leverage in determining the maximum warming rate; but there is still a warming *rate* commitment from the initial choice of stabilization target. Warming rate may be an important factor in determining thresholds for dangerous interference with the climate system, and a consideration of this may lead to the requirement of larger non-CO_2 emissions reductions (Manne and Richels, 2001).

7.5 Conclusions

The case where CO_2 concentrations overshoot the eventual stabilization target has been presented here either as an example of a mid-course correction to a lower CO_2 stabilization target (in order to avoid a perceived increase in the risk of high climate damages), or to simulate a deliberate choice to follow an overshoot pathway for economic, technological or political reasons. For the 450 ppm target case considered here, the overshoot pathway allows much greater cumulative CO_2 emissions out to beyond 2200. In addition, while there is little difference between the maximum rates of emission reduction in the overshoot and monotonic cases, the maximum occurs more than 50 years later in the overshoot case, allowing more time for developing and implementing non-fossil technologies. Thus, for a given concentration stabilization target, the mitigation costs of an overshoot pathway could be substantially less than for a monotonic pathway.

The overshoot pathway, however, has larger absolute warming and a larger maximum warming rate, so climate damages and the risk of passing a threshold for non-linear climate response will be increased. Assessing the costs of the increment in climate damages is beyond our current ability. The issue is whether or not a global-mean temperature difference of around 0.6 °C or sea-level rise difference of 6 cm (the maximum differences in Figure 7.3) or a larger warming rate would lead to sufficiently larger damages (or a sufficient increase in the risk of non-linear responses in climate or climate impacts) to offset the reduced mitigation costs.

For the mid-course correction case, where an initial 550 ppm target is changed to a 450 ppm target, cumulative emissions must be reduced substantially. However, the additional technological challenge may be minor, since there is little change in the maximum rate of emissions reduction. Temperatures necessarily stabilize at a lower level after the mid-course correction, but the two trajectories diverge only slowly. The maximum warming rate, however, is significantly reduced by the mid-course correction, reducing the risk of non-linear responses.

Acting too tentatively and choosing too high a stabilization target initially clearly amplifies the risk of dangerous interference with the climate system. The initial choice of stabilization target is an important factor in determining the peak warming and warming rate. On the other hand, deliberately choosing an overshoot pathway to stabilization (i.e., choosing a higher initial stabilization target) may have advantages in terms of the costs of reducing emissions and the time available to develop non-fossil technologies, but the associated greater climate risks may be unacceptable. Further work is required to better quantify the climate damages side of the cost-benefit equation.

The advantage of a multi-gas climate stabilization strategy based on economic principles is clearly illustrated in our work. Substantial reductions in warming and sea-level rise occur when cost-effective reductions in CH_4 and N_2O emissions are included in the analysis, and CH_4 emissions reductions provide considerable leverage in reducing future warming and warming rates.

References

Church, J. A. and Gregory, J. M. (co-ordinating lead authors) (2001). Changes in sea level. In *Climate Change 2001: The Scientific Basis. Contribution of Working Group I to the Third Assessment Report of the Intergovernmental Panel on Climate Change*, ed. J. T. Houghton, Y. Ding, D. J. Griggs *et al.* Cambridge: Cambridge University Press, pp. 639–693.

Cubasch, U. and Meehl, G. A. (co-ordinating lead authors) (2001). Projections of future climate change. In *Climate Change 2001: The Scientific Basis. Contribution of Working Group I to the Third Assessment Report of the Intergovernmental Panel on Climate Change*, ed. J. T. Houghton, Y. Ding, D. J. Griggs *et al.* Cambridge: Cambridge University Press, pp. 525–582.

Enting, I. G., Wigley, T. M. L. and Heimann, M. (1994). *Future Emissions and Concentrations of Carbon Dioxide: Key Ocean/Atmosphere/Land Analyses*. CSIRO Division of Atmospheric Research Technical Paper No. 31.

Hoffert, M. L., Caldeira, K., Benford, G. *et al.* (2002). Advanced technology paths to global climate stability: energy for a greenhouse planet. *Science* **298**, 981–987.

Hourcade, J.-C. and Shukla, P. (co-ordinating lead authors) (2001). Global, regional and national costs and ancillary benefits of mitigation. In *Climate Change 2001: Mitigation, Contribution of Working Group 3 to the Third Assessment Report of the Intergovernmental Panel on Climate Change* ed. B. Metz, A. Davidson, R. Swart and J. Pan). Cambridge: Cambridge University Press, pp. 544–552.

IPCC (2000). *Emissions Scenarios: A Special Report of Working Group III of the Intergovernmental Panel on Climate Change*, ed. N. Nakićenović and R. Swart. Cambridge: Cambridge University Press.

Joos, F., Prentice, I. C., Sitch, S. *et al.* (2001). Global warming feedbacks on terrestrial carbon uptake under the Intergovernmental Panel on Climate Change (IPCC) emissions scenarios. *Global Biogeochemical Cycles* **15**, 891–908, doi: 10.1029/2000GB001375.

Kheshgi, H. S. and Jain, A. K. (2003). Projecting future climate change: implications of carbon cycle model inter-comparisons. *Global Biogeochemical Cycles* **17**, 1047; doi: 10.1029/2001GB001842 (see also http//:frodo.atmos.uiuc.edu/isam).

Leggett, J., Pepper, W. J. and Swart, R. J. (1992). Emissions scenarios for the IPCC: an update. In *Climate Change 1992: The Supplementary Report to the IPCC Scientific Assessment*, ed. J. T. Houghton *et al.* Cambridge: Cambridge University Press, pp. 71–95.

Manne, A. S. and Richels, R. G. (2001). An alternative approach to establishing trade-offs among greenhouse gases. *Nature* **410**, 675–677.

Mastrandrea, M. D. and Schneider, S. H. (2004). Probabilistic integrated assessment of 'dangerous' climate change. *Science* **304**, 571–575.

O'Neill, B. C. and Oppenheimer, M. (2004). Climate change impacts sensitive to path to stabilization. *Proceedings of the National Academy of Sciences* **101**, 16 411–16 416.

Prentice, I. C. (co-ordinating lead author) (2001). The carbon cycle and atmospheric carbon dioxide. In *Climate Change 2001: The Scientific Basis. Contribution of Working Group I to the Third Assessment Report of the Intergovernmental Panel on Climate Change*. ed. J.T. Houghton, Y. Ding, D.J Griggs *et al.* Cambridge: Cambridge University Press, pp. 183–237.

Raper, S. C. B., Wigley, T. M. L. and Warrick, R. A. (1996). Global sea level rise: past and future. In *Sea-Level Rise and Coastal Subsidence: Causes, Consequences and Strategies*, ed. J. Milliman and B. U. Haq. Dordrecht: Kluwer Academic Publishers, pp. 11–45.

Smith, S. J., Wigley, T. M. L. and Edmonds, J. A. (2000). A new route toward limiting climate change? *Science* **290**, 1109–1110.

Smith, S. J., Pitcher, H. and Wigley, T. M. L. (2005). Future sulfur dioxide emissions. *Climatic Change* **73**, 267–318.

Swart, R., Mitchell, J. F. B., Morita, T. and Raper, S. C. B. (2002). Stabilisation scenarios for climate impact assessment. *Global Environmental Change* **12**, 155–165.

Wigley, T. M. L. (1993). Balancing the carbon budget. Implications for projections of future carbon dioxide concentration changes. *Tellus* **45B**, 409–425.

Wigley, T. M. L. (1995). Global-mean temperature and sea level consequences of greenhouse gas concentration stabilization. *Geophysical Research Letters* **22**, 45–48.

Wigley, T. M. L. (2000). Stabilization of CO_2 concentration levels. In *The Carbon Cycle*, ed. T. M. L. Wigley and D. S. Schimel. Cambridge: Cambridge University Press, pp. 258–276.

Wigley, T. M. L. (2004a). Choosing a stabilization target for CO_2. *Climatic Change* **67**, 1–11.

Wigley, T. M. L. (2004b). Modeling climate change under no-policy and policy emissions pathways. In *The Benefits of Climate Change Policies: Analytical and Framework Issues*. Paris: OECD Publications, pp. 221–248.

Wigley, T. M. L. (2005). The climate change commitment. *Science* **307**, 1766–1769.

Wigley, T. M. L. and Raper, S. C. B. (2001). Interpretation of high projections for global-mean warming. *Science* **293**, 451–454.

Wigley, T. M. L. and Raper, S. C. B. (2002). Reasons for larger warming projections in the IPCC Third Assessment Report. *Journal of Climate* **15**, 2945–2952.

Wigley, T. M. L. and Raper, S. C. B. (2005). Extended scenarios for glacier melt due to anthropogenic forcing. *Geophysical Research Letters* **32**, L05704, doi: 10.1029/2004GL021238.

Wigley, T. M. L., Richels, R. and Edmonds, J. A. (1996). Economic and environmental choices in the stabilization of atmospheric CO_2 concentrations. *Nature* **379**, 240–243.

8

Effects of air pollution control on climate: results from an integrated global system model

Ronald Prinn, John M. Reilly, Marcus Sarofim, Chien Wang and Benjamin Felzer

8.1 Introduction

Urban air pollution has a significant impact on the chemistry of the atmosphere and thus potentially on regional and global climate. Already, air pollution is a major issue in an increasing number of megacities around the world, and new policies to address urban air pollution are likely to be enacted in many developing countries irrespective of the participation of these countries in any explicit future climate policies. The emissions of gases and microscopic particles (aerosols) that are important in air pollution and climate are often highly correlated because of shared generating processes. Most important among these processes is combustion of fossil fuels and biomass which produces carbon dioxide (CO_2), carbon monoxide (CO), nitrogen oxides (NO_x), volatile organic compounds (VOCs), black carbon (BC) aerosols, and sulfur oxides (SO_x, consisting of some sulfate aerosols, but mostly SO_2 gas which subsequently forms white sulfate aerosols). In addition, the atmospheric lifecycles of common air pollutants such as CO, NO_x, and VOCs, and of the climatically important methane (CH_4) and sulfate aerosols, both involve the fast photochemistry of the hydroxyl free radical (OH). Hydroxyl radicals are the dominant "cleansing" chemical in the atmosphere, annually removing about 3.7 gigatonnes (1 GT = 10^{15} g) of reactive trace gases from the atmosphere; while the environmental impact of each gas is different, this amount is similar to the total mass of carbon removed annually from the atmosphere by the land and ocean combined (Ehhalt, 1999; Prinn, 2003).

In this paper, we report exploratory calculations designed to show some of the major effects of specific global air pollutant emission caps on climate. In other words, could future air pollution policies help to mitigate future climate change or exacerbate it? For this purpose, we will need to consider carefully the connections between the chemistry of the atmosphere and climate. These connections are complex and their non-linearity is exemplified by the fact that concentrations of ozone in urban areas for a given level of VOC emission tend to increase with increasing NO_x emissions until a critical CO-dependent or VOC-dependent NO_x emission level is reached. Above that critical level, ozone concentrations actually decrease with increasing NO_x emissions, emphasizing the need for air pollution policies to consider CO, VOC, and NO_x emission reductions jointly rather than independently. In contrast, concentrations of ozone in the global middle and upper troposphere, where this gas exerts its major tropospheric radiative forcing of climate, are determined significantly by long-range transport from urban areas below and the stratosphere above, and ozone production by contrast is limited by low ambient NO_x levels.

We emphasize that this paper is not intended as a review of the extensive literature on various aspects of air pollution and climate, but rather as an opportunity to highlight some new results from one specific integrated approach to this subject. In order to interpret the results of our calculations presented later, however, it is necessary to understand some of the reasons for the above complexity and non-linearity in air chemistry. Hence, the next section provides a review of the key issues,

Human-induced Climate Change: An Interdisciplinary Assessment, ed. Michael Schlesinger, Haroon Kheshgi, Joel Smith, Francisco de la Chesnaye, John M. Reilly, Tom Wilson and Charles Kolstad. Published by Cambridge University Press. © Cambridge University Press 2007.

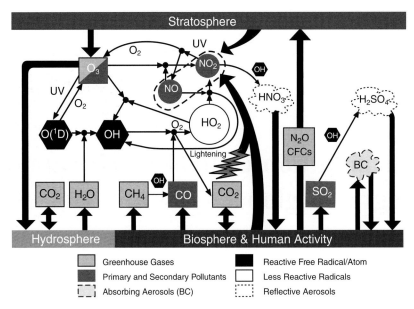

Figure 8.1 Summary of the chemistry in the troposphere important in the linkage between urban air pollution and climate (after Prinn, 1994, 2003). Volatile organic compounds as VOCs (not shown) are similar to CH_4 in their reactions with OH, but they form acids, aldehydes, and ketones in addition to CO.

aimed especially at the non-expert. In two sections following that, we introduce the global model that we use for our calculations, and present and interpret the results. We end with a summary and concluding remarks.

8.2 A chemistry primer

The ability of the lower atmosphere (troposphere) to remove most air pollutants depends on complex chemistry driven by the relatively small amount of the Sun's ultraviolet light that penetrates through the upper atmospheric (stratospheric) ozone layer (Ehhalt, 1999; Prinn, 2003). This chemistry is also driven by emissions of NO_x, CO, CH_4, and VOCs and leads to the production of O_3 and OH. Figure 8.1 reviews, with much simplification, the chemical reactions involved (Prinn, 1994). This chemistry is important to climate change because it involves both climate-forcing greenhouse gases (H_2O, CH_4, O_3) and air pollutants (CO, NO, NO_2, O_3). It also involves aerosols (H_2SO_4, HNO_3, BC) that influence climate through reflecting sunlight (cooling the Earth) or absorbing sunlight (warming the atmosphere). Aerosols and gases also affect the productivity of ecosystems (negatively, through their exposure to O_3, and to H_2SO_4 and HNO_3 in acid rain; and positively, through deposition of nitrogen as nitrate or ammonium), and negatively affect human health (through inhalation). Also important are short-lived free radicals and atoms in two forms: very reactive species such as $O(^1D)$ and OH, and less reactive ones such as HO_2, $O(^3P)$, NO, and NO_2.

Referring to Figure 8.1, when OH reacts with CH_4 the CH_4 is converted mostly to CO in steps that consume OH and also produce HO_2. The OH in turn converts CO to CO_2, NO_2 to HNO_3, and SO_2 to H_2SO_4. The primary OH production pathway occurs when H_2O reacts with the $O(^1D)$ atoms that come from dissociation of O_3 by ultraviolet (UV) light. Within about a second of its formation, on average, OH reacts with other gases, either by donating its O atom (e.g. to CO to form CO_2 and H) or by removing H (e.g. from CH_4 to form CH_3 and H_2O). The H and CH_3 formed in these ways attach rapidly to O_2 to form hydroperoxy (HO_2) or methylperoxy (CH_3O_2) free radicals which are reactive but much less so than OH. If there is no way to rapidly recycle HO_2 back to OH, then levels of OH are kept relatively low. The addition of NO_x emissions into the mix significantly changes the chemistry. Specifically, a second pathway is created in which NO reacts with HO_2 to form NO_2 and to reform OH. Ultraviolet light then decomposes NO_2 to produce O atoms (which attach to O_2 to form O_3) and reform NO. Hence NO_x (the sum of NO and NO_2) is a catalyst which is not consumed in these reactions. The key net ozone-producing reactions being catalyzed by NO_x are $CO + 2O_2 \rightarrow CO_2 + O_3$ and related reactions involving CH_4, and other VOCs. The production rate of OH by the above secondary path in polluted air is about five times as fast as the above primary pathway involving $O(^1D)$ and H_2O (Ehhalt, 1999). The reaction of NO with HO_2 does not act as a sink for HO_x (the sum of OH and HO_2) but instead determines the ratio of OH to HO_2. Calculations for polluted air suggest that HO_2 concentrations are about 40 times as great as OH concentrations (Ehhalt, 1999). This is due mainly to the much greater reactivity of OH compared with HO_2. The ultimate sinks for the HO_x and NO_x free radicals involve formation of H_2O_2, HNO_3, and other water-soluble gases which can be removed by rainfall and surface deposition. In addition to CO, the oxidation of VOCs produces water-soluble gases (e.g. aldehydes, ketones, organic acids) as well as organic aerosols, all of which can also be removed by wet and dry deposition.

If emissions of air pollutants that react with OH, such as CO, VOCs, CH_4, and SO_2, are increasing, then keeping all else constant, OH levels should decrease. This would increase the lifetime and hence concentrations of CH_4. However, increasing NO_x emissions should increase tropospheric O_3 (and hence the primary source of OH), as well as increasing the recycling rate of HO_2 to OH (the second source of OH). This OH increase should lower CH_4 concentrations. Thus an increase (decrease) in OH causes a decrease (increase) in CH_4 and other greenhouse gases and thus a decrease (increase) in radiative forcing of climate. Climate change will also influence OH. Higher ocean temperatures should increase H_2O in the lower troposphere and thus increase OH production through its primary pathway. Higher atmospheric temperatures also increase the rate of reaction of OH with CH_4, decreasing the concentrations of both. Greater cloud cover will reflect more solar ultraviolet light, thus increasing OH above the clouds and decreasing it below, and vice versa.

Added to these interactions involving gases are those involving aerosols. For example, increasing SO_2 emissions and/or OH concentrations should lead to greater concentrations of sulfate aerosols which are a cooling influence. Accounting for all of these interactions, and other related ones (see e.g. Prinn [2003]), requires that a detailed interactive atmospheric chemistry and climate model be used to assess the effects of air pollution reductions on climate.

8.3 Integrated Global System Model

For our calculations, we use the MIT Integrated Global System Model (IGSM). The IGSM consists of a set of coupled submodels of economic development and its associated emissions, natural biogeochemical cycles, climate, air pollution, and natural ecosystems (Prinn et al., 1999; Reilly et al., 1999; Webster et al., 2002, 2003). It is specifically designed to address key questions in the natural and social sciences that are amenable to quantitative analysis and are relevant to environmental policy. To account for the very long timescales of oceanic circulation, the carbon cycle, and land ecosystems, the IGSM is initialized ("spun up") by integrating the model from 1765–1977 or 1860–1977 using historical forcings. Production runs then proceed from 1977 through 2100 (Prinn et al., 1999; Webster et al., 2003). The current structure of the IGSM is shown in Figure 8.2.

Chemically and radiatively important trace gases and aerosols are emitted as a result of human activity. The Emissions Prediction and Policy Analysis (EPPA) submodel incorporates the major relevant demographic, economic, trade, and technical issues involved in these emissions at 5-year intervals at the national and global levels. Natural emissions of these gases are also important and are computed monthly in the Natural Emissions Model (NEM) which is driven by IGSM predictions of climate and ecosystem states around the world.

The coupled atmospheric chemistry and climate submodel is in turn driven by the combination of these anthropogenic and natural emissions. This submodel, with 20-minute time steps, includes atmospheric and oceanic chemistry and circulation, and land hydrological processes. The atmospheric chemistry component has sufficient detail to include its sensitivity to climate and different mixes of emissions (including the feedback between CH_4 emissions and OH), to agree reasonably well with ozone trace gas observations, and to address the effects on climate of policies proposed for control of air pollution and vice versa (Wang et al., 1998; Mayer et al., 2000). Of particular importance to the calculations presented here, the urban air pollution (UAP) submodel is based upon, and designed to simulate, the detailed non-linear chemical and dynamical processes in current 3D urban air chemistry models (Mayer et al., 2000). For this purpose, the emissions calculated in the EPPA submodel are divided into two parts: urban emissions which are processed by the UAP submodel before entering the global chemistry/climate submodel, and non-urban emissions which are input directly into the large-scale model. Also for this purpose, three categories of polluted urban areas are defined based on three threshold levels of NO_x emissions per unit area to best capture the chemical non-linearities (Mayer et al., 2000). Also, the fraction of the global total emissions that occur in urban areas generally increases with time (see Section 8.4). The UAP allows simultaneous consideration of control policies applied to local air pollution and global climate. It also provides the capability to assess the effects of air pollution on ecosystems, and to predict levels of irritants important to human health in the growing number of megacities around the world. The atmospheric and oceanic circulation components in the IGSM are simplified compared with the most complex models available, but they capture the major processes and, with appropriate parameter choices, can mimic quite well the zonal-average behavior of the complex models (Sokolov and Stone, 1998; Sokolov et al., 2003). We use the version of the IGSM with 2D atmospheric and 2D oceanic submodels here, although the latest version has a 3D ocean to capture better the deep ocean circulations that serve as heat and CO_2 sinks (Kamenkovich et al., 2002, 2003). The 2D/2D version we use here resolves separately the land and ocean (LO) processes at each latitude and so is referred to as the 2D-LO-2D version.

The outputs from the coupled atmospheric chemistry and climate model then drive a Terrestrial Ecosystems Model (TEM; Xiao et al., 1998) which calculates key vegetation properties monthly including production of vegetation mass, land–atmosphere CO_2 exchanges, and soil nutrient contents in 18 globally distributed ecosystems. TEM then feeds back its

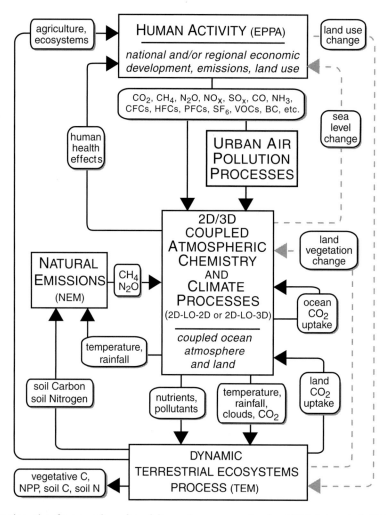

Figure 8.2 Schematic illustrating the framework, submodels, and processes in the MIT Integrated Global System Model (IGSM). Feedbacks between the component models that are currently included, or proposed for inclusion in later versions, are shown as solid or dashed lines respectively (adapted from Prinn *et al.*, 1999).

computed CO_2 fluxes to the climate/atmospheric chemistry submodel, and its soil nutrient contents to NEM, to complete the IGSM interactions. The current IGSM does not include treatment of black carbon aerosols (see Figure 8.1). Detailed studies with a global 3D chemistry and climate model indicate multiple, regionally variable, and partially offsetting effects of BC on absorption and reflection of sunlight, reflectivity of clouds, and the strength of lower tropospheric convection (Wang, 2004). These detailed studies also suggest important BC-induced changes in the geographic pattern of precipitation, not surprisingly since aerosols have important and complex effects on cloud formation, and on whether clouds will even produce precipitation. Methods to capture these effects in the IGSM are currently being explored. In light of the difficulty in simulating these and other regional effects, the numerical results presented here are limited to temperature and sea-level effects, primarily at the global and hemispheric level.

8.4 Numerical experiments

To investigate, at least qualitatively, some of the important potential impacts of controls of air pollutants on temperature, we have carried out runs of the IGSM in which emissions are held constant from 2005 to 2100 for individual pollutants, or combinations of these pollutants from all anthropogenic activities including agriculture. These are compared with a reference run (denoted "ref") in which there is no explicit policy to reduce greenhouse gas emissions and no specific new policies to reduce air pollution (see Reilly *et al.*, 1999; Webster *et al.*, 2002). In the reference, urban emissions account for 16, 28, and 27% of global

Effects of air pollution control on climate

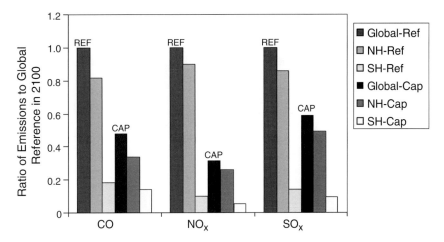

Figure 8.3 Global, northern hemispheric (NH), and southern hemispheric (SH) emissions in the year 2100 of CO/VOC, NO_x, and SO_x, when they are capped at 2005 levels (CAP), are shown as ratios to emissions in the reference (REF) case (no caps).

total CO, NO_x, and SO_x emissions in 2000 and 63, 75, and 64% in 2100, respectively.

Specifically, in five runs of the IGSM, we consider caps at the reference 2005 levels of emissions of the following air pollutants:

1. NO_x only (denoted "NO_x cap"), with urban CO, NO_x, and SO_x emissions being 30, 31, and 26% of their 2100 global totals;
2. CO plus VOCs only (denoted "CO/VOC cap"), with urban CO, NO_x, and SO_x emissions being 42, 75, and 64% of their 2100 totals;
3. SO_x only (denoted "SO_x cap"), with urban CO, NO_x, and SO_x emissions being 63, 75, and 58% of their 2100 totals;
4. Cases (1) and (2) combined (denoted "3 cap"), with urban CO, NO_x, and SO_x emissions being 18, 31, and 26% of their 2100 totals;
5. Cases (1), (2), and (3) combined (denoted "all cap"), with urban CO, NO_x, and SO_x emissions being 18, 31, and 29% of their 2100 totals.

Cases (1) and (2) are designed to show the individual effects of controls on NO_x and reactive carbon gases (CO, VOC), although such individual actions are very unlikely. Case (3) addresses further controls on emissions of sulfur oxides from combustion of fossil fuels and biomass, and from industrial processes. Cases (4) and (5) address combinations more likely to be representative of a real comprehensive approach to air pollution control. Note that, relative to the reference cases (2) and (3), the caps on NO_x emissions that occur in cases (1), (4), and (5) lead to smaller total number, and a shift from the higher to the lower intensity categories, of polluted urban areas (see urban emission percentages above, and Section 8.3).

One important caveat in interpreting our results is that we are neglecting the effects of air pollutant controls on (a) the overall demand for fossil fuels (e.g. leading to greater efficiencies in energy usage and/or greater demand for non-fossil energy sources), and (b), the relative mix of fossil fuels used in the energy sector (i.e. coal versus oil versus gas). Consideration of these effects, which may be very important, will require calculation in the EPPA model of the impacts of NO_x, CO, VOC, and SO_x emission reductions on the cost of using coal, oil, and gas. Such calculations have not yet been included in the current global economic models (including EPPA) used to address the climate issue. Such inclusion requires relating results from existing very detailed studies of costs of meeting near-term air pollution control to the more aggregated structure, and longer time horizon, of models used to examine climate policy.

In Figure 8.3 we show the ratios of the emissions of NO_x, CO/VOC, and SO_x in the year 2100 to the reference case in 2100 when their emissions are capped at 2005 levels. Because these chemicals are short-lived (hours to several days for NO_x, VOCs, and SO_x, few months for CO), the effects of their emissions are largely restricted to the hemispheres in which they are emitted (and for the shortest-lived pollutants restricted to their source regions). Figure 8.3 therefore shows hemispheric as well as global emission ratios. For calibration, the reference emissions in 2100 of CO_2, CH_4, SO_x, NO_x, CO, and VOCs are 24 GtC/yr, 864 megatonne (Mt) CH_4/yr, 235 Mt SO_x/yr, 397 Mt NO_x/yr, 2.6 Gt CO/yr, and 530 Mt VOC/yr respectively (1 Mt = 10^{12} g). The reference global emissions of NO_x, CO/VOC, and SO_x in 2100 are about 4, 2.5, and 1.5 times their 2000 levels. Note that while our reference emissions have some similarities to the central IPCC IS92a scenarios (IPCC, 2000), they differ specifically in having substantially increasing SO_x, NO_x, CO, and VOC emissions through 2100 since

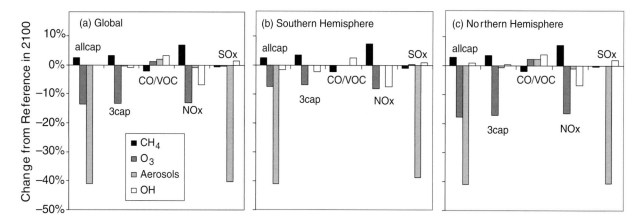

Figure 8.4 Concentrations of climatically and chemically important species (CH_4, O_3, aerosols, OH) in the five cases with capped emissions are shown as percentage changes from their relevant global or hemispheric average values in the reference case for the year 2100 (a) Global-average; (b) southern hemispheric; and (c) northern hemispheric concentrations.

they assume no specific air pollution policies beyond those already existing (Webster *et al.*, 2002).

8.4.1 Effects on concentrations

In Figure 8.4, the global and hemispheric average lower tropospheric concentrations of CH_4, O_3, sulfate aerosols, and OH in each of the above five capping cases are shown as percentage changes from the relevant global or hemispheric reference. Note that the global production of O_3 and destruction of CO and CH_4 are both dominated by non-urban area chemistry, whereas the transformations of the much shorter-lived SO_x, NO_x, and VOCs also have a significant contribution from urban area chemistry. From Figure 8.4a, the major global effects of capping SO_x are to decrease sulfate aerosols and slightly increase OH (owing to lower SO_2 which is an OH sink). Capping of NO_x leads to decreases in O_3 and OH and an increase in CH_4 (caused by the lower OH which is a CH_4 sink). The CO and VOC cap increases OH and thus increases sulfate (formed by OH and SO_2) and decreases CH_4. Note that the CO and VOC cap without a NO_x cap has only a small effect on O_3. Combining NO_x, CO, and VOC caps leads to a substantial O_3 decrease (driven largely by the NO_x decrease) and a slight increase in CH_4 (the enhancement due to the NO_x caps being partially offset by the opposing CO/VOC caps). Finally, capping all emissions causes substantial lowering of sulfate aerosols and O_3 and a small increase in CH_4.

The two hemispheres generally respond somewhat differently to these caps because of the short air pollutant lifetimes and dominance of northern over southern hemispheric emissions (Figures 8.4b and 8.4c). The northern hemisphere contributes the most to the global averages and therefore the globe responds to the caps similarly to the northern hemisphere (compare Figures 8.4a and 8.4c). The southern hemisphere shows very similar decreases in sulfate aerosol from its

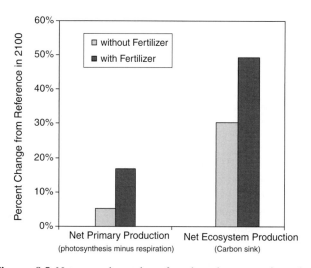

Figure 8.5 Net annual uptake of carbon by vegetation alone (net primary production) and vegetation plus soils (net ecosystem production, the land carbon sink) for the NO_x/SO_x/CO plus VOC capped (allcap) case is shown for the year 2100 as a percentage change from the reference case. The results show the effects with optimal nitrogen use through fertilization on cropland (with fertilizer) or with levels of nitrogen in croplands assumed to be the same as those in equivalent natural ecosystems (without fertilizer).

reference when compared to the northern hemisphere when either SO_x or all emissions are capped (compare Figures 8.4b and 8.4c).

When compared to the southern hemisphere, the northern hemispheric ozone levels decrease by much larger percentages below their northern hemisphere reference when either NO_x, NO_x/CO/VOC, or all emissions are capped. Capping NO_x emissions leads to significant decreases in non-urban OH and thus increases in methane in both hemispheres (Figures 8.4b and 8.4c). Because methane has a long lifetime (about 9 years; Prinn *et al.*, 2001) relative to the interhemispheric mixing time

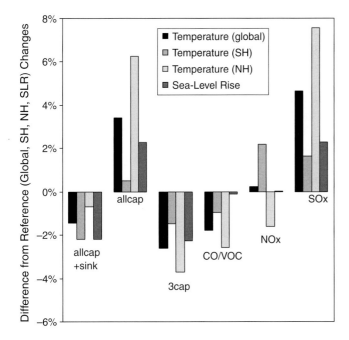

Figure 8.6 Impacts of the five types of air pollution caps on the global, northern hemispheric and southern hemispheric average temperature increases, and the global sea-level rise (SLR), between 2000 and 2100 are shown as changes from their average values (global or hemispheric) in the reference case. The equivalent results are also shown for the case where the enhanced sink due to the ozone decrease is included along with the caps on all pollutants. For this case, we assume the average of the fertilized and non-fertilized sink enhancements from Figure 8.5.

(about 1 to 2 years), its global concentrations are influenced by OH changes in either hemisphere alone, or in both. Hence CH_4 also increases in both hemispheres when NO_x/CO/VOC or all emissions are capped even though the OH decreases only occur in the southern hemisphere in these two cases (see Figures 8.4b and 8.4c).

8.4.2 Effects on ecosystems

Effects of air pollution controls on the land ecosystem sink for carbon can be significant owing to reductions in ozone-induced plant damage (Figure 8.5; see also Felzer et al. [2004, 2005]). Net primary production (NPP, the difference between plant photosynthesis and plant respiration), as well as net ecosystem production (NEP, which is the difference between NPP and soil respiration plus decay, and represents the net land sink), both increase when ozone decreases. This is evident in the case illustrated in Figure 8.5 where all pollutants are capped and ozone decreased by about 13% globally (Figure 8.4a). The effect is even greater when we assume that croplands receive optimal levels of nitrogen fertilizer ("with Fertilizer" case; Felzer et al., 2004, 2005). The land sink (NEP) is increased by 30 to 49% or 0.6 to 0.9 Gt of carbon (in CO_2) in 2100 through the illustrated pollution caps (Figure 8.5).

These ecosystem calculations do not include the additional positive effects on NPP and NEP of decreased acid deposition, and decreased exposure to SO_2 and NO_2 gas, that would result from the pollution caps considered. They also do not include the negative effects on NPP and NEP of decreasing nutrient nitrate and possibly sulfate deposition that also arise from these caps.

8.4.3 Economic effects

If we could confidently value damages associated with climate change, we could estimate the avoided damages in dollar terms resulting from reductions in temperature due to the lowered level of atmospheric CO_2 caused by the above increases in the land carbon sink achieved with the ozone caps. We could similarly value the temperature changes due to caps in other pollutants besides ozone. But monetary damage estimates suffer from numerous shortcomings (e.g. Jacoby, 2004). Felzer et al. (2004, 2005) valued increases in carbon storage in ecosystems due to decreased ozone exposure in terms of the avoided costs of fossil-fuel CO_2 reductions needed to achieve an atmospheric stabilization target. The particular target they examined was 550 ppm CO_2. The above extra annual carbon uptake (due to avoided ozone damage) of 0.6 to 0.9 Gt of carbon is only 2 to 4% of year 2100 reference projections of anthropogenic fossil CO_2 emissions (which reach nearly 25 Gt C/yr in 2100 according to Felzer et al. [2005]). However, as these authors point out, this small level of additional uptake can have a surprisingly large effect on the cost of achieving a climate policy goal. Here we conduct a similar analysis using a 5% discount rate, and adopting the policy costs associated with 550 ppm CO_2 stabilization, to estimate the policy cost savings that would result from the increased carbon uptake through 2100 in the "allcap" compared with the reference scenarios shown in Figure 8.5. The Net Present Value (NPV) cost savings through 2100 discounted to the present (2005) are $2.5 trillion ("without Fertilizer") to $4.7 trillion ("with Fertilizer") (1997 dollars; trillion = 10^{12}). These implied savings are 12 to 22% of the total NPV cost of the 550 ppm stabilization policy.

The disproportionately large economic value of the additional carbon uptake has two reasons. One reason is that the fossil carbon reduction savings are cumulative; the total additional 2000–2100 carbon uptake is 36 Gt (without Fertilizer) and 75 Gt (with Fertilizer), or about 6 to 13 years of fossil carbon emissions at current annual rates. A second reason is that the additional uptake avoids the highest marginal cost options. This assumes that the implemented policies would be cost-effective in the sense that the least costly carbon reduction options would be used first, and more costly options would be used later only if needed. An important caveat here is that, as shown in Felzer et al. (2004, 2005), a carbon emissions reduction policy also reduces ozone precursors so that an additional cap on these precursors associated with air pollution policy results in a smaller additional reduction, and less avoided ecosystem damage. A pollution cap as examined here, assuming there was also a 550 ppm carbon policy in place, leads to only a 0.1 to 0.8 Gt increase in the land sink in 2100 (compare 0.6 to 0.9 Gt in Figure 8.5) and a cumulative 2000–2100 increase of carbon uptake of 13 to 40 Gts of carbon, which is about one-half of the above increased cumulative uptake when the pollution cap occurs assuming there is no climate policy.

8.4.4 Effects on temperature and sea level

The impact of these various pollutant caps on global and hemispheric mean surface temperature and sea-level changes from 2000 to 2100 are shown in Figure 8.6 as percentages relative to the global-average reference case changes of 2.7 °C and 0.4 meters respectively.

In the reference run, CO_2 contributes about 80% of the total 6.6 W/m^2 increase in radiative forcing of climate change from 1990 to 2100. Smaller or larger contributions by CO_2 will lead to larger or smaller percentage changes than those shown in Figure 8.6. Also, these illustrated changes are very significant when compared with the changes induced by the non-CO_2 forcings, including ozone and aerosols, alone (which contribute only 20% of the reference increase in radiative forcing from 1990 to 2100). The largest increases in temperature and sea level occur when SO_x alone is capped, owing to the removal of reflecting (cooling) sulfate aerosols. Because most SO_x emissions are in the northern hemisphere, the temperature increases are greatest there. For the NO_x caps, temperature increases in the southern hemisphere (driven by the CH_4 increases), but decreases in the northern hemisphere (because the cooling effects of the O_3 decreases exceed the warming driven by the CH_4 increases). This result confirms the earlier conclusion of Mayer *et al.* (2000) that lowering NO_x alone has little impact on global mean temperatures because the effects of the CH_4 increases offset those of the O_3 decreases. For CO, and VOC reductions, there are small decreases in temperature driven by the accompanying aerosol increases and CH_4 reductions, with the greatest effects being in the northern hemisphere where most of the CO and VOC emissions (and aerosol production) occur.

When NO_x, CO, and VOCs are all capped, the non-linearity in the system is evidenced by the fact that the combined effects are not simple sums of the effects from the individual caps. Ozone decreases and aerosol increases (offset only slightly by CH_4 increases) lead to even less warming and sea-level rise than obtained by adding the CO/VOC and NO_x capping cases. Finally, the capping of all emissions yields temperature and sea-level rises that are smaller but qualitatively similar to the case where only SO_x is capped, but the rises are greater than expected from simple addition of the SO_x-capped and CO/VOC/NO_x-capped cases. Nevertheless, the capping of CO, VOC, and NO_x serves to reduce the warming induced by the capping of SO_x.

Note that these climate calculations in Figure 8.6 omit the cooling effects of the CO_2 reductions caused by the lessening of the inhibition of the land sink by ozone (Figure 8.5). This omission is valid if we presume that anthropogenic CO_2 emissions, otherwise restricted by a climate policy, are allowed to increase to compensate for these reductions. This was the basis for our economic analysis in the previous section. To illustrate the lowering of climate impacts if we allowed the sink-related CO_2 reductions to occur, we show a sixth case in Figure 8.6 ("allcap + sink") which combines the capping of all air pollutant emissions with the enhanced carbon sink from Figure 8.5. Now we see that the sign of the warming and sea-level rise seen in the "allcap" case is reversed in the "allcap + sink" case. If we could value this lowering of climate impacts, it would provide an alternative to the economic analysis in Section 8.4.3.

8.5 Summary and conclusions

To illustrate some of the impacts of air pollution policy on climate change, we examined five highly idealized but informative scenarios for placing caps on emissions of SO_x, NO_x, CO plus VOCs, NO_x plus CO plus VOCs, and all of these pollutants combined. These caps kept global emissions at 2005 levels through 2100 and their effects on climate were compared with a reference run with no caps applied. Our purpose was not to claim that these scenarios are in any way realistic or likely, but rather that they served to illustrate quite well the complex interactions between air pollutant emissions and changes in temperature and sea level.

In general, placing caps on NO_x alone, or on NO_x, CO and VOCs together, leads to lower ozone levels, and thus less radiative forcing of climate change by this gas, and to less inhibition by ozone of carbon uptake by ecosystems which also leads to less radiative forcing (this time by CO_2). Less radiative forcing by these combined effects means less warming and less sea-level rise.

Placing caps on NO_x alone also leads to decreases in OH and thus increases in CH_4. These OH decreases and CH_4 increases are lessened (but not reversed) when there are simultaneous NO_x, CO, and VOC caps. Increases in CH_4 lead to greater radiative forcing. In this respect, the comprehensive modeling study of CH_4 by Kheshgi *et al.* (1999) concluded that the most effective means of reducing CH_4 may be to raise OH by maintaining NO_x emissions while lowering CO and VOC emissions. Our results qualitatively confirm this, but the CH_4 reductions are modest in our case (Figure 8.4, CO/VOC caps). Placing caps on SO_x leads to lower sulfate aerosols, less reflection of sunlight back to space by these aerosols (direct effect) and by clouds seeded with these aerosols (indirect effect), and thus to greater radiative forcing of climate change due to solar radiation. Enhanced radiative forcing by these aerosol and CH_4 changes combined leads to more warming and sea-level rise. Hence these impacts on climate of the pollutant caps partially cancel each other.

Specifically, depending on the capping case, the 2000–2100 reference global average climate changes are altered only by +4.8 to −2.6% (temperature) and +2.2 to −2.2% (sea level). Except for the case of the NO_x cap alone, the alterations of temperature are of the same sign but significantly greater in the northern hemisphere (where most of the emissions and emission reductions occur) than in the southern hemisphere. Note that for the NO_x alone caps, the temperature decrease caused by ozone reductions is greater than the temperature increase driven by methane increases in the northern hemisphere, whereas the opposite is true in the southern hemisphere (Figure 8.6).

We have previously emphasized (and estimated) the significant uncertainties in IGSM projections of pollutant emissions and concentrations, and resultant climatic changes (Webster et al., 2002, 2003). These uncertainties apply to the six specific IGSM runs used in this study, but by expressing our results as percentage changes from the reference, the effects of many of the highly correlated uncertainties among the runs (e.g. total fossil fuel usage, radiative forcing by aerosols, sensitivity of climate to radiative forcing, heat and carbon uptake by the ocean) are minimized. While we explicitly include urban atmospheric chemistry, our use of a 2D model for the non-urban atmospheric chemistry, and for climate, needs to be tested using a 3D model (although, again, some of the uncertainties induced by use of the 2D model may be effectively minimized through our statement of results as percentages).

It is well established that urban air pollution control policies are beneficial for human health and downwind ecosystems. As far as ancillary benefits are concerned, our calculations suggest that air pollution policies may have only a small influence, either positive or negative, on mitigation of global-scale climate change. However, even small contributions to climate change mitigation can be disproportionately important in economic terms. This occurs because, as we show in the case of increased carbon uptake, these effects mean that the highest cost measures for climate change mitigation, those occurring at the margin, can be avoided. To further check on the validity of our conclusions, future work should include:

1. the effects of air pollution policy on overall demand for fossil fuels and individual demands for coal, oil, and gas;
2. the effects of caps on black carbon (as a regulated air pollutant) on climate;
3. the effects on ecosystems of changes in deposition rates of acids, nitrates, and sulfates, and levels of exposure to SO_2 and NO_2 resulting from air pollution reductions.

Acknowledgements

This research was supported by the US Department of Energy, US National Science Foundation, and the Industry Sponsors of the MIT Joint Program on the Science and Policy of Global Change: Alstom Power (France), American Electric Power (USA), BP p.l.c.(UK/USA), ChevronTexaco Corporation (USA), CONCAWE (Belgium), DaimlerChrysler AG (Germany), Duke Energy (USA), J-Power (Electric Power Development Co., Ltd) (Japan), Electric Power Research Institute (USA), Electricité de France, Enel Corporation (Italy), ExxonMobil Corporation (USA), Ford Motor Company (USA), General Motors (USA), Mirant (USA), Murphy Oil Corporation (USA), Oglethorpe Power Corporation (USA), RWE/Rheinbraun (Germany), Shell International Petroleum (Netherlands/UK), Southern Company (USA), Statoil (Norway), Tennessee Valley Authority (USA), Tokyo Electric Power Company (Japan), TotalFinaElf (France), Vetlesen Foundation (USA).

References

Ehhalt, D. H. (1999). Gas phase chemistry of the troposphere. *Topics in Physical Chemistry* **6**, 21–109.

Felzer, B., Kicklighter, D., Melillo, J. et al. (2004). Effects of ozone on net primary production and carbon sequestration in the conterminous United States using a biogeochemistry model. *Tellus* **56B**, 230–248.

Felzer, B., Reilly, J., Melillo, J. et al. (2005). Effects of ozone on carbon sequestration and climate policy using a biogeochemical model. *Climatic Change* **73**, 345–373.

IPCC (2000). *Emissions Scenarios: A Special Report of Working Group III of the Intergovernmental Panel on Climate Change*, ed. N. Nakicenovic and R. Swart. Cambridge: Cambridge University Press.

Jacoby, H. D. (2004). Informing climate policy given incommensurable benefits estimates. *Global Environmental Change Part A* **14** (3), 287–297; MIT JPSPGC Reprint 2004–7.

Kamenkovich, I. V., Sokolov, A. P. and Stone, P. (2002). An efficient climate model with a 3D ocean and statistical-dynamical atmosphere. *Climate Dynamics* **1**, 585–598.

Kamenkovich, I. V., Sokolov, A. P. and Stone, P. (2003). Feedbacks affecting the response of the thermohaline circulation to increasing CO_2: a study with a model of intermediate complexity. *Climate Dynamics* **21**, 119–130.

Kheshgi, H. S., Jain, A. K., Kotamarthi, V. R. and Wuebbles, D. J. (1999). Future atmospheric methane concentrations in the context of the stabilization of greenhouse gas concentrations. *Journal of Geophysical Research* **104**, 19 183–19 190.

Mayer, M., Wang, C., Webster, M. and Prinn, R. G. (2000). Linking local air pollution to global chemistry and climate. *Journal of Geophysical Research* **105**, 20 869–20 896.

Prinn, R. G. (1994). The interactive atmosphere: global atmospheric–biospheric chemistry. *Ambio* **23**, 50–61.

Prinn, R. G., (2003). The cleansing capacity of the atmosphere. *Annual Reviews Environment and Resources* **28**, 29–57.

Prinn, R. G., Jacoby, H., Sokolov, A. et al. (1999). Integrated Global System Model for climate policy assessment: feedbacks and sensitivity studies. *Climatic Change* **41**, 469–546.

Prinn, R. G., Huang, J., Weiss, R. *et al.* (2001). Evidence for substantial variations of atmospheric hydroxyl radicals in the past two decades. *Science* **292**, 1882–1888.

Reilly, J., Prinn, R., Harnisch, J. *et al.* (1999). Multi-gas assessment of the Kyoto Protocol. *Nature* **401**, 549–555.

Sokolov, A. and Stone, P. (1998). A flexible climate model for use in integrated assessments. *Climate Dynamics* **14**, 291–303.

Sokolov, A., Forest, C. E. and Stone, P. (2003). Comparing oceanic heat uptake in AOGCM transient climate change experiments. *Journal of Climate* **16**, 1573–1582.

Wang, C. (2004). A modeling study on the climate impacts of black carbon aerosols. *Journal of Geophysical Research* **109**, D03106, doi: 10.1029/2003JD004084.

Wang, C., Prinn, R. and Sokolov, A. (1998). A global interactive chemistry and climate model: formulation and testing. *Journal of Geophysical Research* **103**, 3399–3417.

Webster, M. D., Babiker, M., Mayer, M. *et al.* (2002). Uncertainty in emissions projections for climate models. *Atmospheric Environment* **36**, 3659–3670.

Webster, M. D., Forest, C. E., Reilly, J. M. *et al.* (2003). Uncertainty analysis of climate change and policy response. *Climatic Change* **61**, 295–320.

Xiao, X., Melillo, J., Kicklighter, D. *et al.* (1998). Transient climate change and net ecosystem production of the terrestrial biosphere. *Global Biogeochemical Cycles* **12**, 345–360.

Part II

Impacts and adaptation

Joel Smith, coordinating editor

Introduction

Estimation of impacts of climate change has long been a challenge for integrated assessment models (IAMs). Uncertainty about regional climate change and how it would translate into impacts is a key reason why it is difficult to estimate impacts of climate change. Yet, perhaps the greatest challenge to characterizing climate change impacts is in applying a single metric to represent impacts. Use of such a single metric allows for a single expression of impacts correlating to emissions or changes in climate. Other key elements of IAMs can be readily expressed in a single metric. Greenhouse gas emissions can be expressed as tonnes of CO_2 equivalent. Change in the atmosphere can be expressed through metrics such as radiative forcing (corresponding to the combined forcings of greenhouse gases and aerosols) or increase in global mean temperature. Costs of controlling net greenhouse gas emissions can be expressed in (appropriately) monetary terms.

Boiling the impacts of climate change down to a single metric is challenging for several reasons. First and foremost, many impacts of climate change are not in market sectors, but are on natural ecosystems. Expressing the value of such impacts in monetary terms can be difficult. A key impact of climate change on natural ecosystems is global or regional extinction of species. The value of such species is not typically traded in the market place, so ascribing values can be challenging.[1] Many other impacts of climate change are on societal systems whose values are not entirely captured in market transactions. Human health is a good example. Climate change could lead to increased loss of life from spread of infectious disease, increased extreme weather events, increased heat stress, and other causes. Value of lost wages and medical costs can capture the market value of lives, but studies of willingness-to-pay (WTP) to avoid mortality show that there is much greater value ascribed through a WTP process than forgone wages and medical costs.

In contrast to relying on one metric to measure impacts of climate change, Schneider *et al.* (2000) proposed five "numeraires" for quantifying climate change impacts. The numeraires are:

- monetary loss
- loss of life
- quality of life
- species or biodiversity loss
- distribution of impacts

What these suggest are a number of ways of measuring impacts. These can better capture a variety of impacts, but can make aggregating to single metric difficult. Some, such as quality of life, can be difficult to measure and aggregate in a broadly acceptable manner. Jacoby (2004) also discusses the difficulty of measuring all impacts in a single aggregate metric. He proposes examining geophysical impacts at a global scale, regional impacts using natural units, and examining aggregate impacts studies to assess the relationship between change in climate and impacts.

Is the climate change impacts literature moving in a direction that would better enable integrated assessment models to incorporate impacts? The papers in this section address recent developments on some of these topics. Some are overviews of impacts, while others explore new ground.

In Chapter 9, Robert Mendelsohn and Larry Williams estimate the dynamic monetary impacts of climate change on five sectors: agriculture, energy, forestry, water, and coastal resources. The paper examines the relative impacts of climate change, how the sectors are affected as global mean temperature increases, and the distribution. This is the kind of information that can feed into IAMs. Having results expressed in a single metric, in this case monetary impacts, can be particularly useful. Yet, it is important to remember that results are limited to market sectors. As noted above, estimating monetary impacts in non-market sectors is more challenging.

Nicholls *et al.* (Chapter 10) explore analysis of sea-level rise and impacts on coastal resources at two policy-relevant scales: (1) regional to global scales serving the mitigation debate; and (2) sub-national to national scales serving adaptation policy. This includes consideration of a range of impact and economic methods and appropriate metrics such as number of people at risk of flooding, loss of wetlands, loss of rice production, and monetary impacts (reviewing the literature on studies which aggregate monetary impacts). They do not attempt to aggregate the results of their studies. Directions for future development are provided.

Kristie Ebi and Sari Kovats (Chapter 11) take on the important issue of how IAMs can incorporate human health impacts. Human health is a critical issue for climate change impacts, indeed it is one of the five numeraires of Schneider *et al.* (2000). It has been a topic that has been difficult to quantify given uncertainties about baseline conditions (including the association between weather conditions and health outcomes and estimates of the adaptation baseline) and how the burden of climate-sensitive diseases might change under different scenarios. Ebi and Kovats review three trends in recent models: models that project the potential future health impacts of climate change (such as the prevalence of malaria), models that project the implications for economic activity of changes in the intensity and range of diseases, and models that begin to describe how population health may evolve over time. Finally, the authors discuss issues involved with integrating estimates of health impacts of climate change into IAMs.

One of the challenges in including climate change impacts in IAMs has been the lack of comprehensiveness of damage estimates. As noted above, it has been less of challenge for IAMs to estimate damages to market sectors than estimating

---

[1] To be sure, it is not an insurmountable task. Natural resource economists have long applied techniques for estimating values of ecosystems, habitats, and species.

Human-induced Climate Change: An Interdisciplinary Assessment, ed. Michael Schlesinger, Haroon Kheshgi, Joel Smith, Francisco de la Chesnaye, John M. Reilly, Tom Wilson and Charles Kolstad. Published by Cambridge University Press. © Cambridge University Press 2007.

damages to non-market sectors. Even within market sectors, there is the matter of whether the published literature has comprehensively assessed market impacts. Indeed, it has not. While there are a number of studies on such important market sectors as agriculture and coastal resources, some potentially large sectors, such as energy (that is the impacts of climate change on energy production) and tourism have received little attention in the literature. In Chapter 12, Jacqueline Hamilton and Richard S. J. Tol review the literature on climate change and tourism. Tourism is a rapidly growing sector of the global economy and is a key sector for many countries that are vulnerable to climate change. This paper reviews literature on the importance of climate for tourism, e.g., the importance of a desirable climate for selection of vacation locations. Hamilton and Tol review the literature on how climate change may affect the demand for and supply of tourism. They identify important gaps in the literature that should be addressed, in order, among other things, to improve the quality of estimates of climate change impacts on tourism so those estimates can be incorporated into IAMs.

A difficulty in estimating impacts of climate change is estimating how societies will adapt to climate change. Often, relatively simple assumptions are made about adaptation, e.g., on the one extreme, that those affected by climate change will do little to change behavior, or, on the other extreme, that those affected by climate change will make the most efficient adaptation that is possible (see for example Adams *et al.* [1999]). Clearly, each case is an exaggeration of how societies will respond to climate change. But these differences in how adaptation is addressed can result in substantial differences in estimation of climate change impacts. A further concern is that the studies upon which IAMs rely for estimates of climate change impacts have tended to ignore a very important actor in adaptation, namely governments. Government policies and decision on natural resource management will be critical for determining vulnerability to climate change as well as efficiency and effectiveness of adaptation responses. In Chapter 13, Gary Yohe discusses how adaptation to climate change can be considered by appropriate government entities. The paper reviews the state of knowledge on climate change adaptation, and in particular, insights on the process of adaptation that can be gained from the economics literature. Finally, turning to a topic that has received very little attention, Yohe addresses how government ministries which are responsible for decisions on development can incorporate adaptation into their decisionmaking.

In his second contribution to this volume (Chapter 14), Mendelsohn examines the potential impacts of climate change on agriculture in Africa. In many studies of total damages from climate change, agriculture often is estimated to comprise a large or even the largest share of damages. There have been a number of studies on global agriculture, e.g., Rosenzweig and Parry (1994), Darwin *et al.* (1995), and more recently, Fischer *et al.* (2002). All suggest that agricultural production could shift to higher latitudes. This could be particularly harmful to Africa, where agriculture makes up a large share of the region's economy. A detailed study of Africa, as Mendelsohn has shown, sheds more light not only on potential impacts of climate change on this important sector, but also how one particularly vulnerable region could be affected. Geographically detailed studies of the impacts of climate change can be important to understand the distributional consequences of global changes. Although the detailed analyses may be more than the IAMs need to model, the results are important for IAMs to capture.

The literature on climate change impacts is quite wide, but for the most part not particularly deep. This is not the fault of the many thoughtful authors who have worked for many years to improve our understanding of potential impacts of climate change. Rather it is the result of the breadth and complexity of climate change impacts. The papers in this section offer some new and interesting insight either into the state of the climate change impacts literature or by pushing the frontiers of research into important, yet under-studied areas. All of this is critical to provide more robust and meaningful estimates of climate change impacts to be used in IAMs. Beyond IAMs, it is important that our understanding of climate change impacts be improved to lead to better policymaking on control of greenhouse gas emissions and adaptation to its effects.

References

Adams, R. M., Hurd, B. H. and Reilly, J. (1999). *Agriculture and Global Climate Change*. Arlington, Virginia: Pew Center on Global Climate Change.

Darwin, R., Tsigas, M., Lewandrowski, J. and Raneses, A. (1995). *World Agriculture and Climate Change: Economic Adaptations*. Agricultural Economic Report Number 703. Washington, DC: US Department of Agriculture.

Fischer, G., Shah, M. and van Velthuizen, H. (2002). *Climate Change and Agricultural Vulnerability*. Vienna: International Institute of Applied Systems Analysis.

Jacoby, H. (2004). Informing climate policy given incommensurable benefits estimates. In *The Benefits of Climate Change Policies: Analytical and Framework Issues*, ed. J. Corfee-Morlot and S. Agrawala. Paris: OECD.

Rosenzweig, C. and Parry, M. L. (1994). Potential impact of climate change on world food supply. *Nature* **367**, 133–138.

Schneider, S. H., Kuntz-Duriseti, K. and Azar, C. (2000). Costing nonlinearities, surprises and irreversible events. *Pacific and Asian Journal of Energy* **10**(1), 81–106.

9

Dynamic forecasts of the sectoral impacts of climate change

Robert Mendelsohn and Larry Williams

9.1 Introduction

It is well documented that the increasing levels of carbon dioxide and other greenhouse gases are likely to change future climates across the planet (IPCC, 2001a). The Third Assessment Report (TAR) of the Intergovernmental Panel on Climate Change (IPCC) has synthesized many studies describing the qualitative market and non-market impacts of climate change (IPCC, 2001b). However, quantitative estimates of impacts are rare. Most economic impact analyses of climate have focused on comparative equilibrium analyses. They have explored the difference between current conditions and what would occur if greenhouse gases doubled (see Pearce et al., 1996). Economists have rarely tackled the more difficult task of forecasting how impacts might unfold over the century. In this paper, we combine the power of the most sophisticated climate models with recent economic research to predict the path of climate impacts over time.

The modeling begins with forecasts of greenhouse gas emissions in the absence of mitigation. These emissions, in turn, lead to projections of increasing greenhouse gas concentrations in the atmosphere (IPCC, 2001a). We then use the climate predictions of six dynamic atmosphere–ocean general circulation models (AOGCMs) based on these concentrations. We combine these forecasts with two climate response functions: a relatively pessimistic experimental and a relatively optimistic cross-sectional model (Mendelsohn and Schlesinger, 1999; Mendelsohn and Neumann 1999; Mendelsohn, 2001). These response functions reflect the range of quantitative impact results found in other studies as well (Pearce et al., 1996; Tol 2002a, b). Combined with background information about each country in the world, these response functions generate 12 forecasts of climate impacts each decade for each sector for each country.

Understanding the path of global warming impacts is very important. First, the appropriate level of mitigation over time depends on the path of damages. Abatement costs need to be balanced against the stream of aggregate damages that emissions would otherwise cause (Nordhaus, 1991). How much should be spent on mitigation, and how that should change over time, therefore depends on the size and timing of the damages. Second, countries may be interested in learning which sectors are vulnerable to climate damage and when in order to determine what adaptations are worthwhile. Third, countries might want to know what the overall impacts of climate change are in order to seek compensation from the international community. Fourth, countries might want to know when impacts will be large enough to detect and where one should look for them. The predictions would help design effective monitoring programs.

There are a number of caveats that should be kept in mind. Empirical studies of climate impacts are not readily available in all places and for all effects. (1) Most of the empirical studies have been done in the temperate zone; few studies have been done in either the tropics or the high latitudes (IPCC, 2001b). (2) Impacts to many non-market sectors have not been quantified, although they are likely to occur. For example, although scientists know that climate will affect biodiversity, health, water quality, and air pollution, the value of these impacts has not been measured. (3) The results are uncertain. We have tried to represent uncertainty by showing a range of effects, which is why there are 12 scenarios.

Human-induced Climate Change: An Interdisciplinary Assessment, ed. Michael Schlesinger, Haroon Kheshgi, Joel Smith, Francisco de la Chesnaye, John M. Reilly, Tom Wilson and Charles Kolstad. Published by Cambridge University Press. © Cambridge University Press 2007.

Table 9.1 Temperature (T) and precipitation (P) change predictions for climate models in 2100.

Region	PCM		CSIRO		CCSR		HAD2		HAD3		CGCM 1	
	T(C°)	P(%)	T(C°)	P(%)	T(C°)	P(%)	T(C°)	P(%)	T(C°)	P(%)	T(C°)	P(%)
L. America	2.0	6.5	3.2	1.6	4.2	−0.7	3.6	−16.9	4.7	−22.0	4.9	−4.1
Africa	2.3	11.9	3.8	1.9	4.7	13.1	3.9	11.6	4.4	3.2	6.2	−10.3
Asia	2.5	19.1	3.5	1.4	3.8	−11.6	3.7	13.6	4.4	9.1	5.2	−6.4
Oceania	2.0	6.8	3.1	1.0	3.3	−15.1	2.1	−2.8	3.6	4.5	4.4	−18.5
N. America	2.5	7.2	4.6	2.8	3.1	29.9	5.8	−12.3	5.6	14.6	5.7	3.3
W. Europe	2.1	7.1	3.7	14.9	3.4	1.4	5.0	−3.6	4.5	−7.1	3.7	−4.0
USSR & EE	3.1	12.9	4.8	12.5	3.3	10.6	5.8	1.0	5.6	−4.8	5.6	7.8
Globe	2.5	16.5	3.7	2.6	3.8	−4.5	4.0	7.7	4.6	4.1	5.2	−5.6

All climate measures are weighted by population not area.

(4) Extreme events are not modeled. The analysis does not include such phenomena as the shutdown of the thermohaline circulation or the collapse of the West Antarctic Ice Sheet.

9.2 Climate models

The origin of global warming lies in the emissions of greenhouse gases (IPCC, 2001a). The burning of fossil fuels for energy and a host of other man-made activities results in sizeable and growing emissions of carbon dioxide, nitrogen oxide, methane, and other miscellaneous greenhouse gases. Greenhouse gases are accumulating in the atmosphere (IPCC, 2001a). In this study, we assume that they will continue to accumulate for the next century at a steady rate. One set of model runs assumes that atmospheric levels will increase at 1% a year and the other set assumes that they will follow the IPCC IS92a predicted levels. Both sets of predictions assume that there will be no mitigation.

The greenhouse gases will cause increased amounts of heat to be trapped in the Earth's lower atmosphere. We rely on six AOGCMs that have dynamic atmosphere–ocean capabilities to predict dynamic climate paths. The six models, ranked in ascending order on the basis of their 2100 population-weighted temperature change predictions, are the Parallel Climate Model (PCM) (Washington *et al.*, 2000), CSIRO (Gordon and O'Farrell, 1997), Hadley Meteorological Centre (HAD2) (Johns, 1996), CCSR (Emori *et al.*, 1999), Hadley Meteorological Centre (HAD3) (Gordon *et al.*, 2000), and the Canadian Climate Centre Model (CGCM1) (Boer *et al.*, 2000). HAD2 and CGCM1 assume that CO_2 concentrations will reach 808 ppm by 2100 while the other four models assume that CO_2 concentrations will only reach 685 ppm by 2100. All six scenarios also account for other greenhouse gases and for sulfates, which tend to cool the atmosphere.

Table 9.1 presents the population-weighted temperature and precipitation changes for each continent and the globe according to each model for 2100. Climate scientists routinely present warming climatologies as area-weighted temperatures and precipitations (IPCC, 2001a). We rely on population-weighted measures for each country because market impacts occur where people live (Williams *et al.*, 1998). Starting with cubic spline interpolations, to 1/2 degree grids, the predictions of each model are weighted by population and then added up to obtain country averages.

The AOGCM models predict a range of climate change over time and space. The global population-weighted change in temperature varies from 2.5 °C to 5.2 °C across these six models with PCM the lowest and CGCM1 the highest. Warming is expected to increase with latitude (IPCC, 2001a) and PCM, CSIRO, CCSR, and HAD3 follow this accepted pattern. In contrast, HAD2 predicts that warming in the tropics (except for Oceania) will exceed warming in the rest of the planet and CGCM1 predicts that Africa will get especially hot. The predicted precipitation also varies across the models. For example, PCM predicts higher precipitation in every continent with Asia at the top of the list. In contrast, HAD2 predicts large losses in Latin America and North America, and CCSR predicts losses in Asia.

Although all the models predict global temperature increasing, the decadal changes by region follow random steps. For example, HAD2 starts with the highest prediction of warming by 2010 of 1.2 °C even though its century-long prediction lies in the middle of the models. In contrast, the two models with the highest century-long predictions, HAD3 and CGCM1, predict only a mild warming by 2010 of 0.8 °C. The decade of greatest warming varies from model to model. CSIRO predicts a large temperature increase in 2020–2030 of 0.51 °C, CCSR predicts a 0.82 °C warming in 2060–2070, and CGCM1 predicts a 0.73 °C warming in 2080–2090. Precipitation changes for countries are also random. For example, HAD2 predicts a sharp drop in precipitation by 2010 concentrated in Asia that disappears by 2020. These sharp climate changes, in turn, affect the path of predicted impacts over time.

We do not use the AOGCMs to predict sea-level rise. Instead we assigned sea-level rise scenarios to climate scenarios to mirror the range of possibilities highlighted by the IPCC report (IPCC, 2001a). The low end of the range, a sea-level rise of

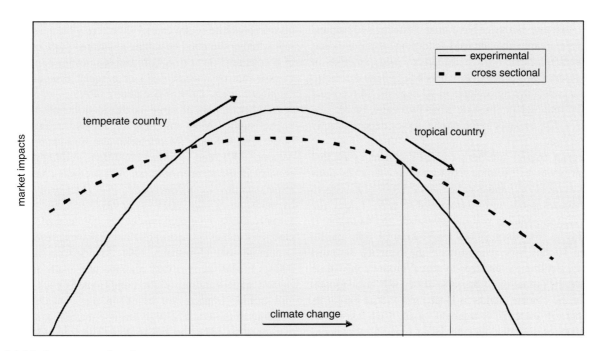

Figure 9.1 Market response functions.

0.3 m by 2100, was assigned to the PCM scenario. The high end of the range, an increase of 0.9 m by 2100, was assigned to the CGCM1 scenario. The remaining climate scenarios were assigned the expected sea-level rise scenario of 0.5 m by 2100. In all the scenarios, sea-level rise is expected to increase monotonically over the century.

9.3 Impact model

There are two sources of evidence to measure the sensitivity of market impacts to climate change: cross-sectional studies and experimental studies (Mendelsohn and Neumann, 1999; IPCC, 2001b; Mendelsohn, 2001). The experimental approach quantifies the links between climate and final impacts in each sector using controlled experiments. For example, with agriculture, the experimental approach would measure how each crop grows in different climate conditions (in laboratories or experimental fields), what affect this would have on farmers' yields, how this would change farmers' decisions to plant the crop, how all these choices would affect supply, and then how prices would change (see Adams et al., 1999). In contrast, the cross-sectional approach compares how outcomes in a sector change depending upon which climate zone the firm or household is in. For example, a cross-sectional study of agriculture would compare the profits or farm values of farmers in different climate settings to quantify the role of climate (Mendelsohn et al., 1994).

Both the experimental and cross-sectional approaches have complementary strengths and weaknesses. An important strength of the cross-sectional approach is that it automatically includes adaptation as part of impacts. This is a weakness of the experimental approach. Experimental-simulation models only include adaptation to the extent that the analyst models it.

The cross-sectional approach also tends to be representative as large areas are studied by design. The high cost of experiments limits the number of sites that can be tested with the experimental approach. The experimental approach, however, very carefully controls settings and isolates the climate and carbon fertilization effects. These are weaknesses of the cross-section approach, which cannot measure carbon fertilization impacts and struggles to control confounding influences. The weakness of each method is the strength of the other.

Another reason to include both experimental and cross-sectional results in this study is that the two types of studies tend to capture the range of results found throughout the literature. The experimental studies predict large impacts in the market sectors (high climate sensitivity) whereas the cross-sectional studies predict small impacts (low climate sensitivity) for a given climate change. By using both approaches, we reflect the range of climate sensitivity results found in the impact literature.

There are three major features of climate change that cause impacts: temperature, precipitation, and carbon dioxide. The effect of temperature change depends on the initial level of temperature. Many recent studies find a hill-shaped relationship between the welfare in each economic sector and temperature (Mendelsohn and Schlesinger, 1999; Mendelsohn and Neumann, 1999; Mendelsohn, 2001). A hypothetical example of the hill shape is shown in Figure 9.1. The exact shape of the hill is different for each sector. For example, the temperature that leads to the lowest energy costs is relatively cool whereas the temperature in which trees grow fastest is relatively warm. Because the response function of most sectors is hill-shaped, the impact of temperature change depends on initial temperatures. High-latitude countries tend to benefit from warming, temperate countries have mixed effects, and tropical countries are more quickly damaged. Additional precipitation is generally beneficial unless there are already large amounts of rain in a season. Carbon dioxide is beneficial because it helps forestry and especially agriculture in all regions. The coastal sector is the only sector immune from the major forces discussed above. The impacts to the coastal sector depend strictly on sea-level rise.

Most impact studies are partial equilibrium analyses that assume that prices will remain fixed (two notable exceptions are Adams *et al.* [1999] and Sohngen *et al.* [2002]). If climate change decreases goods and services and so raises prices, the partial equilibrium studies will underestimate the damages. If climate change increases goods and services, the partial equilibrium studies will overestimate the benefits. In most cases, we expect that the changes to global production will be small so that the partial equilibrium analyses will be adequate. However, if the climate change is large, that will no longer be true. Further, if there is a regional good, such as water, regional changes in price can be large even if net global changes are not. This is one reason why many watershed studies are general equilibrium analyses (Hurd *et al.*, 1999; Jenkins *et al.*, 2001); they can capture changes in the marginal value of water in the watershed as either the urban or agricultural demand or the water supply changes.

The next generation of impact studies will need to rely on general equilibrium models for each sector (see, for example, Sohngen *et al.*, 2002). This is a demanding task, because if goods are traded, price is determined by global markets. National general equilibrium models of traded goods are not adequate because their price effects are largely based on assumptions about what will happen in the rest of the world (outside the model).

This analysis is one of the few studies that measure all impacts over time (see also Tol, 2002a,b). However, there are several dynamic studies of the coastal (Yohe *et al.*, 1999) and timber (Sohngen *et al.*, 2002) sectors. These sectors contain large capital stocks of forests or structures that are slow and difficult to adjust. The more rapidly climate changes, the larger the damages in these sectors. For example, rapid change may cause some of the trees to die before they can be harvested, and coastal properties may be inundated while they still have many decades of useful life. Dynamic adaptation is also important to model. For example, building sea walls too early will increase the cost of protecting coasts tremendously. Cutting trees down before they mature gives much lower returns. Adaptations must be timed in response to the speed of climate change.

Dynamic studies require an interest rate to make intertemporal decisions. The interest rate is the value of time and it is used to compare costs in the present with costs in the far future. Although some analysts argue that a zero interest rate should be used for climate change, it is not prudent for society to use a different value of time for climate change relative to other public investment decisions (Pearce, 2003). All investments into social capital should rely on the same interest rate whether the good be a dam, a national park, or the defense of a coastline. As a result, impact studies tend to use either a market rate of interest (5%) or a social discount rate (3%) (Sohngen *et al.*, 1999; Yohe *et al.*, 1999).

Another important issue is the location of the economic impact studies. Most empirical impact studies have been done in the United States (Mendelsohn and Neumann, 1999; Mendelsohn, 2001) and so reflect conditions in a highly developed country in the temperate zone. A few studies have measured impacts in developing countries (IPCC: McCarthy *et al.*, 2001; Mendelsohn *et al.*, 2001). Very few economic studies have actually measured global impacts (Sohngen *et al.*, 2002). Most global estimates are extrapolations from measurements in the temperate zone (Nordhaus, 1991; Mendelsohn and Williams, 2004). In this study, every attempt was made to reflect our current understanding of the conditions in each region. The mid-latitude impacts are based on our considerable understanding of the US. The high-latitude impacts are based on an extrapolation from the US studies. The low-latitude impacts are based on a small sample of developing country or global studies. The uncertainty surrounding the impact estimates in the low and high latitudes is high.

The final critical element in the country predictions is capturing other pertinent features of each country. The model includes numerous characteristics from each country such as arable land, length of coasts, population, and GDP. Some of these variables are permanent but, for example, the latter two variables clearly change over time. The model includes a prediction of how these variables are expected to change over time (IPCC, 1994). GDP is expected to continue to grow more rapidly in developing countries and more slowly in the developed world. Population is expected to increase but at an ever slower rate until it stabilizes in the second half of the next century. These world forecasts suggest world GDP would reach 200 trillion (year 2000 USD) by 2100 and world population would reach 9.5 billion. These projections are exogenous to this model but are a source of additional uncertainty over both the level of effects and the dynamics.

9.4 Results

We calculate impacts each decade for each of the six climate scenarios using the cross-sectional and experimental climate impact sensitivities for a total of twelve dynamic scenarios. The agricultural impacts dominate the impacts in the other four market sectors. Agriculture accounts for all the global net benefits predicted by PCM and over 50% of the damages in HAD3 and CGCM1 by 2100 in the experimental scenarios.

The dynamic path of agricultural impacts in the 12 scenarios is shown in Figure 9.2. With the sensitive experimental response function, impacts are expected to be quite small at first but then they grow rapidly as climate warms towards the end of the century. There are two exceptions to this general

Dynamic forecasts 111

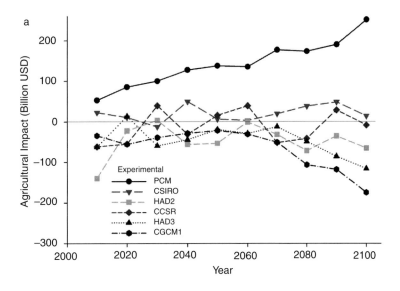

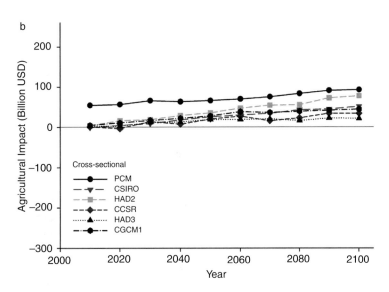

Figure 9.2 Agricultural impacts over time.

rule. With the PCM scenario, agricultural impacts are beneficial even as early as 2010 as the benefits of CO_2 fertilization and increased precipitation combine with a very mild warming to increase production. The other exception is with HAD2, which predicts very large though temporary agricultural damages in 2010 fueled by very large reductions in precipitation in southern Asia (-20%) coupled with very high temperature increases ($1.5\,°C$) in the low latitudes. With PCM, these early benefits simply increase linearly with time, but with HAD2, the damages virtually disappear by 2020 and only resurface near the end of the century. With the cross-sectional response function, the net impacts in agriculture are slightly beneficial for the first 70 years of the century. After that, the climate models vary with some predicting small benefits and some small losses. The only exception to this rule is the PCM climate scenario, which predicts linearly increasing benefits throughout the century.

The results in Figure 9.2, however, hide some important regional differences that are specific to agriculture. As can be seen in Table 9.2, all the scenarios predict that global warming is likely to increase agricultural production in the mid-to-high latitudes. The temperatures in the mid-to-high latitudes are currently either slightly too cool or near the optimum for agriculture. Warming will either improve conditions or at least make only a slight difference since the sector is at the flat part of the top of the response hill (see Figure 9.1). Carbon fertilization from higher CO_2 levels will then cause large increases in production. In contrast, the low-latitude countries are currently too warm, and warming is likely to result in sharp drops in production. With the experimental response function, the

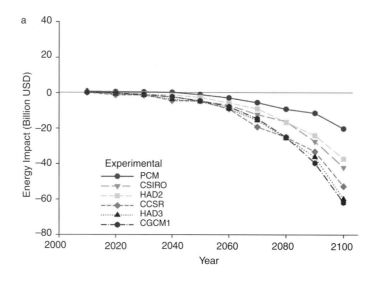

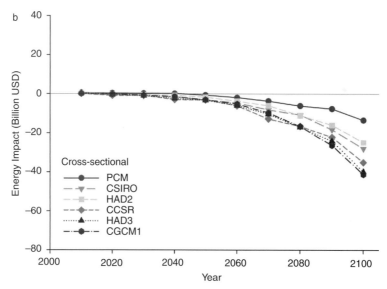

Figure 9.3 Energy impacts over time.

Table 9.2 Agricultural impacts by region in 2100.

Climate model	Climate response function			
	Experimental		Cross-sectional	
	Low latitude	High latitude	Low latitude	High latitude
CGCM1	−321	188	−9	54
HAD3	−297	218	−9	31
HAD2	−273	285	0	78
CSIRO	−211	239	−3	35
CCSR	−175	167	−1	36
PCM	−7	258	12	81

Impacts are in billions of dollars. Positive values are benefits and negative values are damages. The low latitudes include Latin America, Africa, South Asia, and the Pacific, and the high latitudes are the rest of the world.

climate models predict large damages for agriculture in the low latitudes in all scenarios except the mild PCM. The cross-sectional results are similar but much smaller. According to the cross-sectional models, agriculture is not very sensitive to climate change.[1]

As shown in Figure 9.3, the impacts in the energy sector follow a very different pattern than agriculture. Impacts start very small and they simply become more harmful over time. Impacts increase with warming but they are also sensitive to the growth of the economy. Impacts are predicted to be slightly larger with the experimental than the cross-sectional

---

[1] The cross-sectional models cannot measure carbon fertilization effects. We attribute small fertilization effects to the cross-sectional response function to contrast with our assumption in the experimental results.

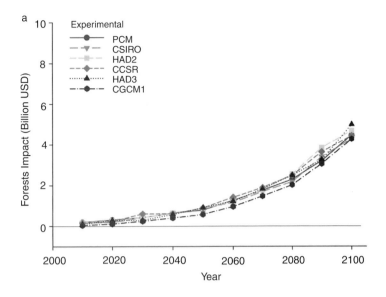

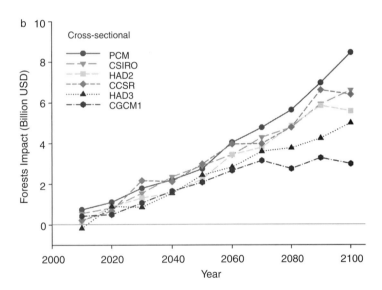

Figure 9.4 Forestry impacts over time.

approach but the range of impacts in energy is smaller than with agriculture.

The timber sector, in contrast, is expected to do well with warming, as shown in Figure 9.4. A warmer, wetter, CO_2-enriched world will result in forests expanding and growing faster. That is, the model suggests that the world is slightly cooler than the optimal growing conditions for timber (subtropics). In all 12 scenarios, warming will cause timber benefits to rise over time. However, in the cross-sectional model, the high temperature increases of CGCM1 suggest there is a limit to these benefits and that eventually warming will become harmful to this sector as well. However, in all cases, the benefits in this sector are small because the sector itself is small relative to GDP.

The impacts in water, Figure 9.5, follow the same general pattern seen in energy (Figure 9.3). As warming proceeds, the damages are expected to increase over time. Water becomes very valuable with warming because agricultural and urban demands for water increase while the supply of available water generally declines. Precipitation varies greatly across regions and climate models. However, with greater evapotranspiration and more rapid melting of winter snows, watershed systems are predicted to have less available water even with small precipitation increases. Systems can adapt to these changes by storing more water and allocating the water efficiently across users, but these public adaptations will require effective governmental coordination and funding (Mendelsohn, 2000).

The coastal impacts, as shown in Figure 9.6, simply depend on sea-level rise. If sea-level rise is very modest, the damages will be small as coastline development will adapt without much change. In scenarios where the sea is expected to rise more quickly,

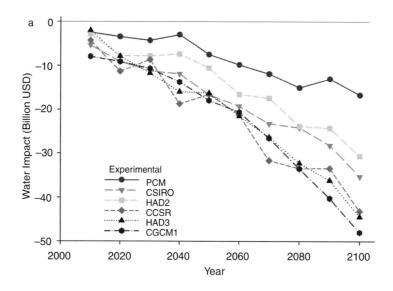

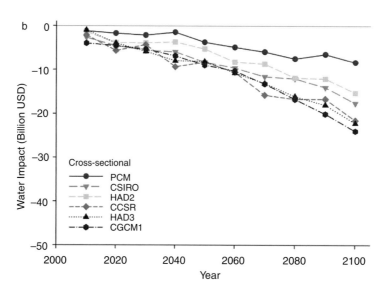

Figure 9.5 Water impacts over time.

damages will increase especially near the end of the century. If the seas rise rapidly, coastal buildings will have to be protected or lost to inundation. The analysis assumes that protection will proceed efficiently with the government timing the size of the protection to the rate of sea-level rise. Except in the most dramatic scenario, the damages from sea-level rise are expected to be small as the cost of protection is spread out across the century.

The aggregate net global market impacts in each decade are displayed in Figure 9.7. They reflect a change in welfare from the baseline that would have occurred if climate remained the same. With the experimental sensitivity model, the net impacts generally linger near zero through 2060 for all the climate scenarios except the moderate PCM scenario. After 2060, damages start to increase slowly for the CCSR and HAD2 scenarios and more rapidly for the HAD3 and CGCM1 scenarios. Net damages are not apparent in the CSIRO scenario until 2100. The net impacts in the PCM scenario are beneficial and increase linearly with time. These results can largely be explained by the global temperature predictions of each climate model. The mild climate forecasts with PCM and the small climate changes at the beginning of the century lead to net benefits or no effects. However, the larger climate changes at the end of the century with the more severe climate scenarios often lead to large damages. With the cross-section model, the net impacts for the globe are near zero for the century for the CSIRO, CCSR, CGCM1, and HAD3 climate scenarios. With the HAD2 scenario and especially the PCM scenario, the impacts are beneficial. The cross-sectional impacts are smaller than the experimental impacts as expected because it is less climate sensitive.

As a fraction of GDP, these net damages are relatively small. They are small at first, because they hardly differ from

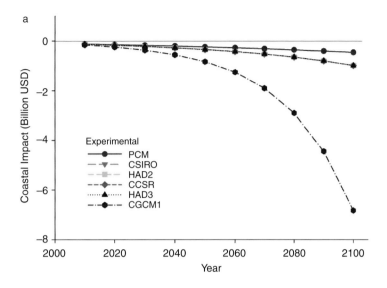

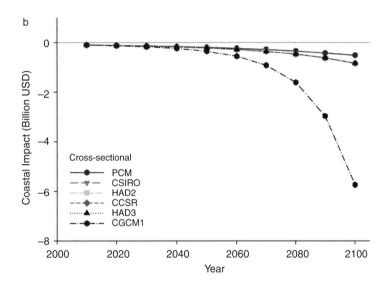

Figure 9.6 Coastal impacts over time.

zero. Later in the century, they remain small because GDP has grown so much faster than the damages have grown. For example, world GDP is forecasted to be $200 trillion by 2100 whereas even in the worst scenario, the net climate damages are only $300 billion.

Although net global impacts may be relatively small, they hide much larger changes that are evident in each region as can be seen in Figure 9.8 (see colour plates). The map illustrates the impacts in every country in 2100 for the PCM and HAD3 climate scenarios. Only the results of the experimental response function are shown. The impacts for the cross-section response function are similar but far more muted. Although most of the world benefits in the mild PCM climate scenario, there are damages in Africa, India, Central America, and the South Pacific. In contrast to what is happening to the low latitudes, the high latitudes are predicted to enjoy large benefits.

The damages in the HAD3 climate scenario are both more pronounced and more widespread. There are extensive damages especially to Africa but also to Latin America and Southern Asia. Virtually all countries in the low latitudes are at risk. These damages even begin to extend to the mid latitudes by 2100. Nonetheless, high-latitude countries are predicted to benefit in this scenario. These regional effects are driven primarily by what is happening to the agriculture sector.

9.5 Conclusion

This study combines natural science and economics to predict the future impact of climate change. The 12 forecasts provide a detailed description of how impacts may unfold decade to decade for each country of the world. Net market impacts for the entire globe are small at first and only after 2060 do the

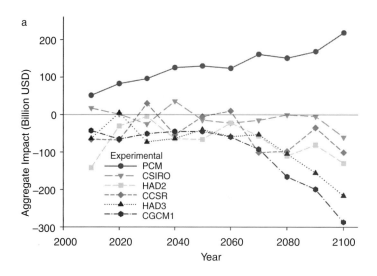

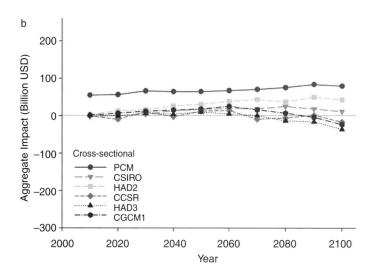

Figure 9.7 Aggregate market impacts over time.

more severe scenarios predict large enough temperature increases to generate sizeable net negative global impacts. Net global effects are predicted to be modest primarily because large damages in the low latitudes are offset by benefits in the temperate and high latitudes.

Although net global impacts may remain small into the end of the century, impacts in the tropical and subtropical regions can be harmful by 2020. Warming in the low latitudes is especially damaging to agriculture and these damages increase as warming becomes more severe over time. Asia is particularly vulnerable in large part owing to the size of the Asian agriculture sector. As a fraction of GDP, however, Africa is the most vulnerable region since a large fraction of the economy in sub-Saharan Africa is dependent on agriculture. The model also predicts that agriculture in the former USSR, Eastern Europe, and Western Europe will benefit from warming in most scenarios. North America will also benefit except for climate scenarios that predict a reduction in precipitation in the region.

The study suggests that the first market impacts to become identifiable will be in the agricultural sector in the high and low latitudes. The study predicts benefits in the high latitudes and damages in the low latitudes. Ironically, very few empirical impact studies have been done in these regions, so these forecasts of future impacts remain uncertain. In the mid latitudes, the first evidence of climate impacts may not occur until the end of the century.

The study, however, does not measure all effects. For example, the study omits impacts to health and ecosystems because of the absence of reliable welfare measures and the inclusion of adaptation. The study also does not examine extreme events or climate variance. In this case, it is the absence of a link to near-term emissions of greenhouse gases that keep these phenomena from being included. With

Dynamic forecasts

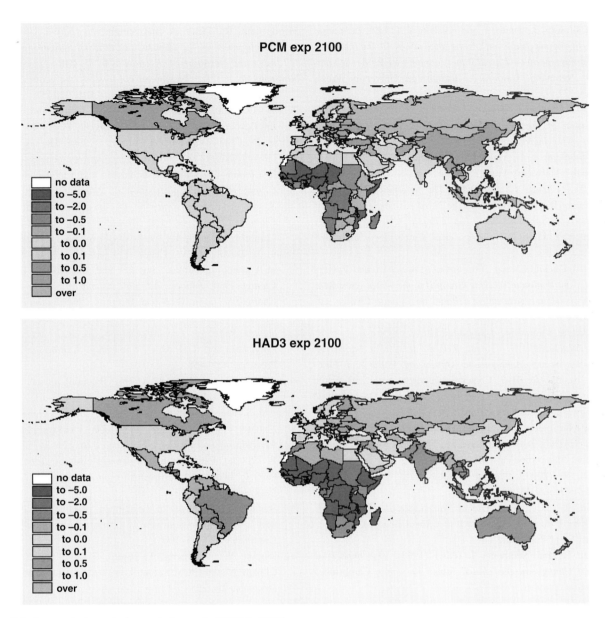

Figure 9.8 Aggregate impacts (percent change in GDP) in 2100

additional research, many of these effects can be included in the analysis and they may well change the results.

Acknowledgements

Funding was provided by the Electric Power Research Institute. The views are the authors' alone.

References

Adams, R., McCarl, B., Segerson, K. *et al.* (1999). The economic effects of climate change on US agriculture. In *The Impact of Climate Change on the United States Economy*, ed. R. Mendelsohn and J. Neumann. Cambridge: Cambridge University Press, pp. 18–54.

Boer, G., Flato, G. and Ramsden, D. (2000). A transient climate change simulation with greenhouse gas and aerosol forcing: projected climate for the 21st century. *Climate Dynamics* **16**, 427–450.

Emori, S., Nozawa, T., Abe-Ouchi, A., Namaguti, A. and Kimoto, M. (1999). Coupled ocean–atmospheric model experiments of future climate change with an explicit representation of sulfate aerosol scattering. *Journal of the Meteorological Society Japan* **77**, 1299–1307.

Gordon, C., Senior, C., Banks, H. *et al.* (2000). The simulation of SST, sea ice extents, and ocean heat transports in a version of the Hadley Centre coupled model without flux adjustments. *Climate Dynamics* **16**, 147–168.

Gordon, H. and O'Farrell, S. (1997). Transient climate change in the CSIRO coupled model with dynamic sea ice. *Monthly Weather Research* **125**, 875–907.

Hurd, B., Callaway, J., Smith, J. and Kirshen, P. (1999). Economic effects of climate change on US water resources. In *The Impact of Climate Change on the United States Economy*, ed. R. Mendelsohn and J. Neumann. Cambridge: Cambridge University Press, pp. 133–177.

IPCC (1994). *Radiative Forcing of Climate Change and an Evaluation of the IPCC IS92 Emissions Scenarios. Special Report of the Intergovernmental Panel on Climate Change*, ed. J.T. Houghton, L.G. Meira Filho, J. Bruce et al. Cambridge: Cambridge University Press.

IPCC (2001a). *Climate Change 2001: The Scientific Basis. Contribution of Working Group I to the Third Assessment Report of the Intergovernmental Panel on Climate Change*, ed. J.T. Houghton, Y. Ding, D.J. Griggs, et al. Cambridge: Cambridge University Press.

IPCC (2001b). *Climate Change 2001: Impacts, Adaption, and Vulnerability. Contribution of Working Group II to the Third Assessment Report of the Intergovernmental Panel on Climate Change*, ed. J.J. McCarthy, O.F. Canziani, N.A. Leary, D.J. Dokken and K.S. White. Cambridge: Cambridge University Press.

Jenkins, M.W., Lund, J.R., Howitt, R.E. et al. (2001). *Improving California Water Management: Optimizing Value and Flexibility*, Report No. 01–1, Center for Environmental and Water Resources Engineering, University of California, Davis; http://cee.engr.ucdavis.edu/faculty/lund/CALVIN/

Johns, T. (1996). *A Description of the Second Hadley Centre Coupled Model (HADCM2)*, Climate Research Technical Note 71, United Kingdom Meteorological Office.

Mendelsohn, R. (2000). Efficient adaptation to climate change. *Climatic Change* **45**, 583–600.

Mendelsohn, R. (ed.) (2001). *Global Warming and the American Economy: A Regional Assessment of Climate Change*. England: Edward Elgar Publishing.

Mendelsohn, R. and Neumann, J. (eds.) (1999). *The Impact of Climate Change on the United States Economy*. Cambridge: Cambridge University Press.

Mendelsohn, R. and Schlesinger, M. (1999). Climate response functions. *Ambio* **28**, 362–366.

Mendelsohn, R. and Williams, L. (2004). Comparing forecasts of the global impacts of climate change. *Mitigation and Adaptation Strategies for Global Change* **9**, 315–333.

Mendelsohn, R., Nordhaus, W. and Shaw, D. (1994). The impact of global warming on agriculture: a Ricardian analysis. *American Economic Review* **84**, 753–771.

Mendelsohn, R., Dinar, A. and Sanghi, A. (2001). The effect of development on the climate sensitivity of agriculture. *Environment and Development Economics* **6**, 85–101.

Nordhaus, W. (1991). To slow or not to slow: the economics of the greenhouse effect. *Economic Journal* **101**, 920–37.

Pearce, D. (2003). The social cost of carbon and its policy implications. *Oxford Review Of Economic Policy* **19**, 362–384.

Pearce, D., Cline, W., Achanta, A. et al. (1996). The social cost of climate change: greenhouse damage and the benefits of control. In *Climate Change 1995: Economic and Social Dimensions of Climate Change. Contribution of Working Group III to the Second Assessment Report of the Intergovernmental Panel on Climate Change*, ed. J.P. Bruce, H. Lee and E.F. Haites. Cambridge: Cambridge University Press, pp. 179–224.

Reilly, J., Baethgen, W., Chege, F.E. et al. (1996). Agriculture in a changing climate: impacts and adaptations. In *Climate Change 1995: Impacts, Adaptations and Mitigation of Climate Change: Scientific-Technical Analyses. Contribution of Working Group II to the Second Assessment Report of the Intergovernmental Panel on Climate Change*, ed. R.T. Watson, M.C. Zinyowera and R.H. Moss. Cambridge: Cambridge University Press, pp. 427–468.

Rosenzweig, C. and Parry, M. (1994). Potential impact of climate change on world food supply. *Nature* **367**, 133–138.

Sohngen, B., Mendelsohn, R. and Sedjo, R. (2002). A global model of climate change impacts on timber markets. *Journal of Agricultural and Resource Economics* **26**, 326–343.

Tol, R. (2002a). Estimates of the damage costs of climate change. Part 1: benchmark estimates. *Environmental and Resource Economics* **21**, 47–73.

Tol, R. (2002b). Estimates of the damage costs of climate change. Part 2: dynamic estimates. *Environmental and Resource Economics* **21**, 135–160.

Washington, W., Weatherly, J., Meehl, G. et al. (2000). Parallel Climate Model (PCM): control and transient scenarios. *Climate Dynamics* **16**, 755–774.

Williams, L., Shaw, D. and Mendelsohn, R. (1998). Evaluating GCM output with impact models. *Climatic Change* **39**, 111–133.

Yohe, G., Neumann, J. and Marshall, P. (1999). The economic damage induced by sea level rise in the United States. In *The Impact of Climate Change on the United States Economy*, ed. R. Mendelsohn and J. Neumann. Cambridge: Cambridge University Press, pp. 178–208.

10

Assessing impacts and responses to global-mean sea-level rise

Robert J. Nicholls, Richard S. J. Tol and Jim W. Hall

10.1 Introduction

One of the more certain impacts of human-induced climate change is a rise in global-mean sea level (Nicholls and Lowe, 2004). While the impacts of this sea-level rise are confined to coastal areas, these include the most densely populated land areas on Earth and they support important and productive ecosystems that are sensitive to sea-level change. Further, coasts are also experiencing significant human-induced modification, so sea-level rise and climate change are an *additional* stress, which amplifies their impacts (Bijlsma, 1996; Kremer et al., 2005).

During the twenty-first century, global-mean sea-level rise will likely be less than 1 metre (Church and Gregory, 2001), but still potentially directly affecting at least 200 million people based on 1990 population (Hoozemans et al., 1993; Mimura, 2000). Over the longer term (many centuries), a much larger sea-level rise exceeding 10 m is possible under some emission pathways owing to ablation of the Greenland Ice Sheet and collapse of the West Antarctic Ice Sheet, among other changes (Oppenheimer and Alley, 2004; Nicholls and Lowe, 2005). Further, the high human exposure to sea-level rise is increasing rapidly because of global population growth and coastward migration. Therefore, any global assessment of the climate change issue must include the coastal implications.

A fundamental result that has long been recognized by climate scientists, but less considered by policy, is that irrespective of future greenhouse gas emissions, there is a "commitment to sea-level rise" (Nicholls and Lowe, 2004; 2005). This is due to the slow response of the oceans and large continental ice sheets to global warming, and as a result, global-mean sea levels will rise far into the future even if anthropogenic greenhouse gas emissions were immediately reduced to zero. Therefore, the policy response to sea-level rise needs to combine mitigation and adaptation, as the two policies will be most effective when combined. Adaptation is required for the committed sea-level rise, and mitigation will reduce the future rate of change and the ultimate commitment. To support the response to climate change in coastal areas, a range of scales of assessment are required from local to global, with different questions and decisions being informed as we move to bigger scales. This paper reviews methods and future direction of development at the two broadest scales of assessment:

- Regional to global scale assessments to assess the benefits of different mitigation policies, and mixtures of mitigation and adaptation;
- Sub-national to national assessments to support strategic coastal planning and the development of adaptation strategies within the context of wider coastal development and management goals.

Both of these scales of assessment have significant challenges in terms of representing the impacts of sea-level rise when compared with more detailed case studies, as detailed process-based modeling is not feasible, and the relevant data are often difficult to find, heterogeneous, and of poor quality. These problems condition how the impacts are described: generally quite simple spatially aggregated impact and adaptation algorithms are used. It is important that the methods are unbiased and hence errors are reduced by aggregation

Human-induced Climate Change: An Interdisciplinary Assessment, ed. Michael Schlesinger, Haroon Kheshgi, Joel Smith, Francisco de la Chesnaye, John M. Reilly, Tom Wilson and Charles Kolstad. Published by Cambridge University Press. © Cambridge University Press 2007.

Table 10.1 The main natural system effects of relative sea-level rise, including interacting climatic and non-climatic factors. Some interacting factors (e.g., sediment supply) appear twice as they can be influenced both by climate and non-climate factors.

Natural system effect		Interacting factors	
		Climate	Non-climate
Inundation, increased flood and storm damage	Surge (sea)	Wave and storm climate, erosion, sediment supply	Sediment supply, flood management, erosion, land claim
	Backwater effect (river)	Run-off	Catchment management and land use
Wetland loss (and change)		CO_2 fertilization, sediment supply	Sediment supply, migration space, direct destruction
Erosion (direct and indirect morphological change)		Sediment supply, wave and storm climate	Sediment supply
Saltwater intrusion	Surface waters	Run-off	Catchment management and land use
	Groundwater	Rainfall	Land use, aquifer use
Rising water tables/impeded drainage		Rainfall, run-off	Land use, aquifer use, catchment management

Source: Adapted from Nicholls (2002).

(e.g. Hoozemans *et al.*, 1993). Also, the methods and data requirements must generally become simpler as we consider larger areas, as discussed for UK flood assessment by Hall *et al.* (2003a).

In the paper, following a brief overview of impacts and responses to sea-level rise, both scales of assessment are considered in more detail. Global studies have generally explored potential impacts of sea-level rise (e.g., number of people affected, areas of wetland lost), or the economic implications of sea-level rise (e.g., impact and adaptation costs) with relatively simple aggregated methods. Here, both these approaches are examined. Sub-national to national assessments are at a variable state of development and studies which genuinely influence coastal management policy are in their infancy. Here we focus mainly on recent experience within the United Kingdom, as the issue of adapting to sea-level rise and climate change and its wider implications are being explored in significant detail.

10.2 Sea-level rise, impacts and responses

This section outlines the nature of sea-level change and its consequences in both physical and human terms, and presents the main scenarios that are used.

Relative sea-level rise occurs when the sea rises relative to the land and can be produced by a rise in ocean level, a fall in land elevation (subsidence), or a combination of the two effects. While the media focuses on projections of global-mean sea-level rise, any impacts are manifest through changes in relative sea level and these changes will vary from place-to-place. Through the twentieth century, global-mean sea level is estimated to have risen 10 to 20 cm (Church and Gregory, 2001). In some limited locations, land uplift was sufficient to counteract this effect and relative sea levels fell (e.g., northern Baltic), while subsidence added to the global rise in many other areas (e.g. deltaic settings). Thus, relative sea levels have risen widely around the globe. From 1990 to 2100, global-mean sea level may rise 9 to 88 cm (Church and Gregory, 2001) with a greater rise being possible, especially if Antarctica becomes a positive contribution. Any acceleration in global-mean sea-level rise will increase the rate of relative rise and increase the potential for impacts.

Relative sea-level rise has a wide range of natural system effects (Table 10.1). In addition to raising mean sea level, all the coastal processes that operate around sea level are raised. Therefore, the immediate effect is submergence and increased flooding of coastal land, as well as saltwater intrusion of surface waters. Longer-term effects also occur as the coast adjusts to the new environmental conditions, including increased erosion, wetland losses and change, and saltwater intrusion into groundwater. These lagged changes interact with the immediate effects of sea-level rise and often exacerbate them. While there are quantitative approaches for analysis of each impact, some of these effects remain poorly understood, and further research is required on the long-term physical response to sea-level rise, such as lagoon and estuary interaction with the open coast (Stive, 2004).

Table 10.2 links natural system effects to their most important direct socio-economic impacts by sector. Indirect impacts of sea-level rise are not shown as they are more difficult to analyse, but they have the potential to be important in many sectors, such as human health. Examples of possible indirect triggers of health impacts include the nutritional impacts of loss of agricultural production in coastal

Table 10.2 The more significant *direct* socio-economic impacts of relative sea-level rise on different sectors in coastal zones, including uncertain cases.

Sector	Natural system effect (from Table 10.1)						
	Inundation, flood and storm damage				Saltwater intrusion		
	Surge	Backwater effect	Wetland loss	Erosion	Surface	Ground	Rising water tables
Water Resources	•				•	•	
Agriculture	•	•			•	•	•
Human health	•	•			•	•	
Fisheries	?	?	•		•		
Tourism	•		•	•	•		
Human settlements	•	•			•	•	
Coastal biodiversity	?	?	•	•	•	•	•

Source: Adapted from Nicholls (2002).

areas, the release of toxic materials from eroded landfills, and changes to disease vectors caused by waterlogging and rising water tables. Other indirect effects include changes in agricultural production (as land is lost or ground- and surface waters become salinized) and the capital market (as investment is diverted to coastal protection). Thus, sea-level rise can produce a cascade of socio-economic impacts, although their magnitude will depend on our ability to adapt.

Climate change and sea-level rise is affecting an evolving world, and any analysis should consider sets of internally consistent climate and socio-economic scenarios (Klein and Nicholls, 1999). This paper follows the available literature, and two main sets of consistent climate and socio-economic scenarios are considered:

- The unmitigated IS92a scenario (greenhouse gas concentrations grow at 1%/year) and related stabilization scenarios that lead to stabilization at 750 ppm CO_2 and stabilization at 550 ppm CO_2 (termed the S750 and S550 emission scenarios, respectively) (see Nicholls and Lowe, 2004). Here these different climate change scenarios are imposed on a single socio-economic pathway (Hulme *et al.*, 1999; Arnell *et al.*, 2002).
- The Special Report on Emission Scenarios (SRES) (IPCC, 2000) combines a range of emission scenarios with a range of distinct socio-economic pathways (see also Arnell *et al.* [2004]; Nicholls [2004]).

There is not a unique correspondence between the emissions and socio-economic scenarios as the climate response to a known level of emissions varies widely because of the uncertainties in climate sensitivity. However, to simplify the analyses an explicit link between these two factors is often assumed. For example, in the national study by Evans *et al.*

Table 10.3 An example of linking climate and socio-economic scenarios: UKCIP02 climate scenarios and Foresight socio-economic futures for the UK Foresight Flood and Coastal Defence Assessment.

SRES[a]	UKCIP02[b]	Global average sea-level rise (1990 to 2080s)[c]	Foresight futures[d]
B1	Low emissions	23 (9–48)	Global Sustainability
B2	Medium–low emissions	26 (11–54)	Local Stewardship
A2	Medium–high emissions	30 (13–59)	National Enterprise
A1F1	High emissions	36 (16–69)	World Markets

Source: Hall *et al.*, 2003b; Evans *et al.*, 2004a.
[a]Special Report on Emissions Scenarios (IPCC, 2000).
[b]UK Climate Impacts Programme 2002 scenarios (Hulme *et al.*, 2002).
[c]UK Climate Impacts Programme 2002 scenarios (Hulme *et al.*, 2002). Figures in brackets are IPCC ranges associated with same SRES emissions scenarios.
[d]Department of Trade and Industry (2003).

(2004a), national climate scenarios and socio-economic scenarios were linked as shown in Table 10.3.

10.3 Regional to global assessments

Regional to global assessments face significant difficulties in terms of both availability of data and appropriate methods and algorithms. Hence, the questions that are posed need to be realistic to address at this scale. To date, one of the most influential and widely used regional to global assessment of coastal zones has been the Global Vulnerability Assessment (or GVA) (Hoozemans *et al.*, 1993). Many subsequent impact and economic analyses have used the GVA base data with appropriate scenarios, including studies discussed here (e.g., Nicholls

2004; Tol, 2004). More recent regional studies have been developed independently, such as the Asia-Pacific Vulnerability Analysis (Mimura, 2000).[1] Examples of regional to global impact and economic analyses are now discussed in turn.

10.3.1 Impact analyses

After considering the limitations already outlined, the impact assessments evaluated fairly simple parameters. For example, the GVA assessed potential impacts on three distinct elements of the coastal zone and one possible adaptive response (Hoozemans et al., 1993):

- *population risk* – population at risk of surge flooding (and potential upgrade costs for dikes, as discussed under economic analyses);
- *ecosystem loss* – coastal wetlands (of international importance) at loss;
- *agricultural impacts* – rice production at change (for south, southeast and east Asia only).

The results were only considered meaningful when aggregated to 20 regions, and globally. A series of subsequent impact studies (e.g., Nicholls, 2004) refined the methods and results for coastal flooding and wetland losses, but the underlying data are the same. These two impacts are now considered in turn.

Coastal flooding

Increased flooding and storm damage are a major impact of sea-level rise (Table 10.1). Aggregated national data on surge elevations, land elevations, population density, and protection standards were developed and applied at the scale of 192 polygons (essentially the coastal countries existing in 1990). Available analyses were designed to explore two simple questions which are appropriate at this scale (e.g., Nicholls, 2004; Nicholls and Lowe, 2004):

1. Is global-mean sea-level rise a serious issue, if unmitigated?
2. What are the benefits of climate mitigation in terms of impacts avoided?

The exposure to flooding, and the likelihood of experiencing flooding can be estimated by examining the competing influence of:

- relative sea-level rise (due to global rise and local subsidence, if appropriate);
- coastal population change; and
- possible upgrades of dike standards, indexed to GDP/capita.[2]

In all the scenarios, there are large increases in coastal population, especially if coastward migration continues. The effect of sea-level rise on extreme water levels (i.e., surge height) is an explicit part of the analysis: relative sea-level rise simply displaces these extreme water levels upwards. Exposure is measured as a function of the population living in the flood plain. Risk is measured using the average number of people experiencing flooding per year, taking account of water levels, dike standards and the size of the exposed population (also termed people at risk).

For 1990, Hoozemans et al. (1993) estimated that about 200 million people lived in the coastal flood plain (below the estimated 1 in 1000 year surge-flood elevation), or about 4% of the world's population. Nicholls et al. (1999) estimated that on average 10 million people per year experienced flooding in 1990, rising to 140 million people per year with a 1-m rise scenario and no other change. These results have been validated against independent national estimates (see also Nicholls, 2004), and indicate that sea-level rise might have significant impacts, if there is no adaptation.

Nicholls and Lowe (2004) explored the impacts of global-mean sea-level rise under both unmitigated (IS92a) and stabilization scenarios (S750 and S550) from 1990 to the 2140s. This allows examination of broad system sensitivities to sea-level rise over timescales which allow the commitment to sea-level rise to be explored. For each emission scenario, three climate sensitivities are considered given a wide range of global-mean rise from 20 to 137 cm rise over the period 1990 to the 2140s. Improving dike standards are considered, but only responding to existing climate variability (i.e. surge height in 1990), and the analysis is deliberately considering an artificial world that is completely ignoring adaptation to human-induced sea-level rise.[3] The number of people at risk is illustrated in Figure 10.1, showing a number of important points:

- Without global sea-level rise (Figure 10.1a), the number of people experiencing flooding per year will increase significantly, owing to increasing coastal populations (i.e., exposure) to the 2050s, and then diminish as improvements in dike standards (due to rising GDP/capita) become the most important factor controlling flood risk.
- While sea-level rise is a slow onset hazard, subsequent impacts are significant, potentially affecting many millions or even hundreds of millions of additional people per year. The possible range of impacts increases with time, with actual impacts being controlled by climate sensitivity, among other factors.

---

[1] Covering 30 °E to 165 °W and 90 °N to 60 °S.

[2] To simplify the analysis, all adaptation for coastal flooding is treated as dike construction.

[3] This is termed "in-phase evolving protection."

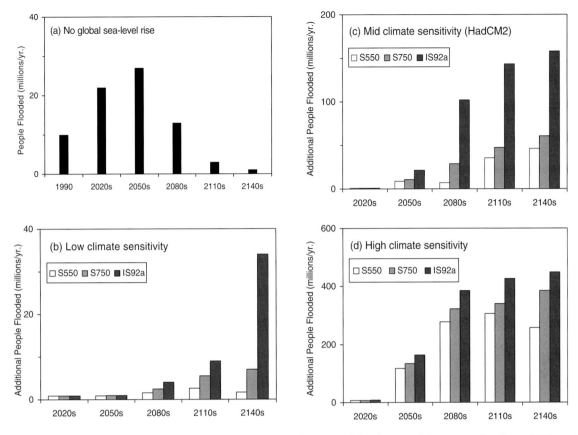

Figure 10.1 Coastal flooding under the IPCC 'S' stabilization experiments and in-phase evolving protection from 1990 to the 2140s, which compares unmitigated (IS92a) impacts with those under the S750 and S550 stabilization scenarios. (a) People flooded per year without any global sea-level rise; (b) additional people flooded per year through sea-level rise assuming low climate sensitivity; (c) as (b) for mid climate sensitivity (HadCM2); (d) as (b) for high climate sensitivity. Note the varying scale of the y-axis. Reprinted from Nicholls and Lowe (2004): *Global Environmental Change* 229–244, Copyright (2004), with permission from Elsevier.

- While climate mitigation has benefits (i.e. avoided impacts) in all cases, flood impacts increase with time in nearly all cases, even under stabilization scenarios. Thus, it is not immediately clear what impacts are avoided and what impacts are simply delayed because of mitigation. Of course, delaying impacts is a benefit in itself.

Nicholls (2004) found that the different SRES socio-economic scenarios are equally important to the different SRES climate scenarios: the A2 world is more vulnerable to coastal flooding than the other socio-economic scenarios with or without sea-level rise, even though sea-level rise is higher under the A1FI scenario. Also, factors such as the interaction of global-mean sea-level rise and human-induced subsidence require more analysis, as realistic scenarios of subsidence could greatly exacerbate flood impacts.

Flood risk is highly sensitive to assumptions about dike height. Extending the existing analyses, Table 10.4 shows the number of people flooded in the 2080s under the SRES scenarios given a range of different protection scenarios. If protection standards are constant over time, the potential impacts are immense. However, protection standards are dynamic.

Under "In-phase evolving protection", flood risk is reduced at least fivefold, and is at or below 1990 levels for all scenarios, except for the A2 scenario (see above). Allowing for upgraded adaptation which allows for sea-level rise, flood risk is further reduced to well below 1990 levels in all cases.

Thus, sea-level rise has major implications for coastal flood risk, with some regions having particular problems (especially the small island regions of the Caribbean, Indian Ocean, and Pacific Ocean, and the continental regions of the southern Mediterranean, West Africa, East Africa, South Asia, and Southeast Asia). Mitigation alone cannot avoid all these impacts. Raising dikes (and by implication other flood management strategies) appears a viable response strategy, but the wider feasibility and consequences of adaptation have not been considered.

Coastal wetlands
Wetland loss and change are already widespread problems around the world's coast, and this will be exacerbated by sea-level rise (Table 10.1) (Nicholls, 2004). Thus it illustrates an issue where multiple stresses are important, and further the response to sea-level rise is non-linear. Lastly, flood defences

Table 10.4 Global flood risk in the 2080s for the SRES climate scenarios[a] and a range of different protection scenarios, expressed as average number of people flooded per year.[b]

Scenario	Global-mean sea-level rise (1990 to 2080s) (cm)	People flooded by protection scenario (millions of people/year)		
		Constant protection[c]	In-phase evolving protection[d]	Upgraded protection[e]
A1FI	34	127	10	1
A2	28	169	34	4
B1	22	79	3	<1
B2	25	127	6	1

[a]SRES scenarios are described by Nicholls (2004).
[b]Population scenarios assume that coastal populations grow at twice national trends.
[c]Constant dike heights based on 1990 standards.
[d]Increasing dike heights in phase with increasing GDP/capita, but with no allowance for sea-level rise.
[e]As In-phase evolving protection, with additional upgrade of the dike standard by one class for sea-level rise (protection classes are < 1:1, 1:10, 1:100 and 1:1000, respectively).

hinder the response of coastal wetlands to sea-level rise and promote greater losses, so wetland loss is also influenced by dike construction.

In the analyses considered here, coastal wetlands comprise the saltmarshes, mangroves, and associated unvegetated intertidal areas (and exclude coral reefs) contained in a database of internationally important wetlands (Hoozemans et al., 1993). Nicholls (2004) presents the models of coastal wetland response to sea-level rise as used here. The impact indicator is the percentage loss of wetland by area, relative to the 1990 stock. The assessment considers the potential for both (1) vertical accretion and (2) horizontal wetland migration. Vertical accretion is based on the tidally weighted *rate* of relative sea-level rise: wetland loss occurs when the rate of sea-level rise exceeds a defined threshold. Given the uncertainty concerning this threshold, a range is used. Horizontal response assesses migratory potential based on both natural and human barriers (due to dikes).

Figure 10.2 shows the possible losses of coastal wetlands due to sea-level rise under the IS92a, S750 and S550 scenarios used in the flood analysis. (Other loss factors are not considered.) Wetland losses are predicted in all cases, but the uncertainties are significant. For the unmitigated scenarios, the losses could be as high as 42% of the world's coastal wetlands, and the losses are still increasing in the 2140s. Mitigation both reduces and stabilizes the rate of sea-level rise and hence wetland losses are reduced, and losses of coastal wetlands are approaching a maximum under stabilization by the 2140s. Therefore, it appears that stabilization has important long-term benefits for coastal wetlands, and unlike coastal flooding, additional impacts due to sea-level rise after the 2140s might be minimal.

10.3.2 Economic analyses

Economic analysis of sea-level rise represent impacts in monetary terms and can be divided into three groups of studies concerning:

1. direct costs of sea-level rise;
2. total economic costs of sea-level rise; and
3. adaptation to sea-level rise.

As with the impact analyses, these studies concern, in principle, all the effects listed in Table 10.1 and Table 10.2 although, in practice, attention has focused on a subset of effects, especially land loss (due to erosion and inundation) and wetland loss. In most of the studies that follow, the authors report results from the national level upwards, although given the underlying data, individual country results should be treated with caution (e.g., Hoozemans et al., 1993).

Direct cost estimates

Direct cost estimates are common across climate change impact literature as they are simple to estimate and easy to explain. The direct cost of change is the product of price and impact, including adaptation and residual damage costs. For sea-level rise, the simple direct cost is the product of land loss and land value for unprotected areas, plus the product of length of dikes and dike costs for protected areas.

The GVA included one of the first, global direct cost estimates: the costs of dike building for a 1 m rise in sea level, including allowances for surge characteristics (Hoozemans et al., 1993). Based on the experience of Delft Hydraulics, it was estimated that the unit costs of dike building do not differ much from place to place, with unit cost differences being mainly determined by the experience in building dikes. Hence, the unit costs of dike building decrease with experience and may be higher in a developing country than in a developed country.

The GVA, as well as later studies, show that the costs of coastal protection are small compared to the gross domestic product in most countries, which suggests that protection will be widespread. However, in countries that have a long coastline relative to land area and that are relatively poor, coastal protection may be very expensive, perhaps prohibitively so. This is particularly the case for small and poor

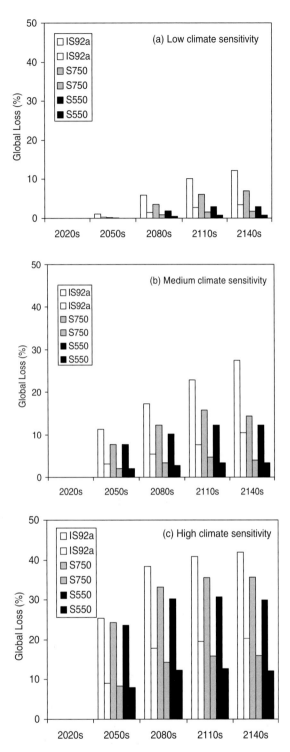

Figure 10.2 Wetland losses (relative to 1990) due to sea-level rise under the IPCC 'S' Stabilization experiments from 1990 to the 2140s, which compares unmitigated (IS92a) impacts with those under the S750 and S550 stabilization scenarios. The results compare (a) low climate sensitivity, (b) medium climate sensitivity, and (c) high climate sensitivity. No other loss factors are considered. Reprinted from Nicholls and Lowe (2004), with permission from Elsevier.

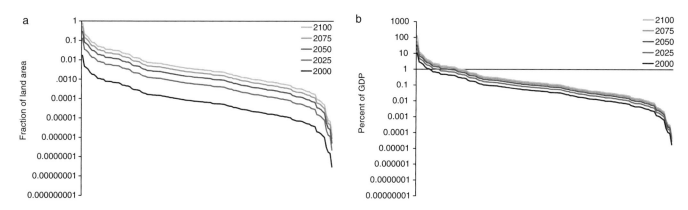

Figure 10.3 Loss of dryland (fraction of total area in 2000; panel(a)) and its value (percentage of GDP; panel(b)) without protection. Countries are ranked as to their values in 2100.

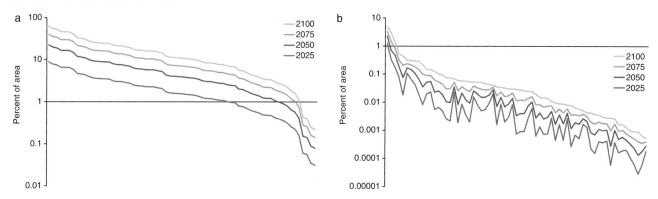

Figure 10.4 Loss of wetland (fraction of total area in 2000; panel(a)) and its value (percentage of GDP; panel(b)) without protection (left panels). Countries are ranked as to their values in 2100.

islands, such as the vulnerable island regions identified by the impact analyses.

Fankhauser (1994) extends, in a fairly crude way, the direct cost estimates of the GVA made in 1993 to include the loss of wetlands and drylands. For drylands, Fankhauser uses the average OECD value of croplands,[4] linearly scaled using per capita income to other regions. Although beach front land is generally more expensive than land further away from the coast, the use of average values is justified because land loss redistributes the beach front land rather than destroying it (Yohe, 1990; Penning-Rowsell *et al*., 2003). Again, the assumption is that protection will be widespread, and the costs of dike building dominate the costs of land loss. For wetlands, Fankhauser also used the average OECD value, linearly extrapolated using per capita income to other regions. The value is derived from monetization studies, rather than market studies. Again, the conclusion is that protection will be widespread. While the data are uncertain, and the extrapolation is simplistic, no one has yet provided a superior alternative. Tol (1995) added the direct costs of migration to Fankhauser (1994) – owing to the widespread application of protection, forced migration is much smaller than suggested in many studies of environmental refugees (of order thousands/year). This demonstrates the importance of considering protection options.

Figures 10.3 to 10.5 (see colour plates section) illustrate some national direct cost estimates based on the results of Tol (2004) for the SRES A1B scenario (IPCC, 2000). They draw on essentially the same assumptions as the studies mentioned above, but better data sets on various factors such as potential land loss provided in Bijlsma (1996) are included. Figure 10.3 shows dryland loss, as a fraction of the area in 2000, per country in 2000, 2025, 2050, 2075, and 2100. Coasts are assumed to be unprotected and factors such as tides and surges are ignored. Exposure differs by seven orders of magnitude. The most exposed countries are the Maldives (77% land loss in 2100), Micronesia (21%), Macau (18%), Vietnam (15%) and Bangladesh (10%), reflecting the vulnerable small island and deltaic locations noted in the impact analysis. Exposure increases over time. In the case of the Maldives, the projected land loss is 13% in 2025, 29% in 2050, 51% in 2075 and 77% in 2100.

Figure 10.3 also shows the costs of dryland loss, as a percentage of GDP. As dryland value is assumed to be linear in income density, the rank order is the same. The most exposed countries are the Maldives (122% of GDP in 2100), Micronesia (12%), Macau (10%), Vietnam (8%) and Bangladesh

---

[4] Fankhauser assumes that urban areas will be protected.

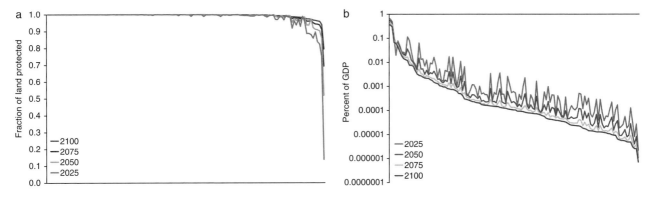

Figure 10.5 Protection level (fraction of coast protected; left panel) and the costs of protection (percent of GDP; right panel). Countries are ranked as to their protection level.

(5%). The economic exposure grows less rapidly than the physical exposure, as for the scenarios considered the rate of the sea-level rise is faster than the rate of economic growth. In the case of the Maldives, the costs of land loss are 19% of GDP in 2025, 32% in 2050, 55% in 2075, and 122% in 2100.

Figure 10.4 shows wetland loss, as a fraction of the 2000 area, per country in 2000, 2025, 2050, 2075, and 2100. The wetland loss model is different from that used to derive Figure 10.2: losses are linear in sea-level rise and coastal protection with the functions being derived from the GVA. Coasts are again assumed to be unprotected, so coastal protection is not a loss driver here. Exposure differs by three orders of magnitude. The most exposed countries are the Benin (62% wetland loss in 2100), Congo (Kinshasa) (55%), Cameroon (52%), Bangladesh (46%), and Pakistan (46%). Exposure increases over time, but at a decelerating rate as wetland stocks decline. In the case of Benin, the projected wetland loss is 9% in 2025, 22% in 2050, 40% in 2075, and 62% in 2100.

Figure 10.4 also shows the costs of wetland loss, as a percentage of GDP. Exposure differs by six orders of magnitude. As wetland value is non-linear, the rank order is different from the rank order of wetland area. The most exposed countries are the Bahamas (4.8% of GDP in 2100), Papua New Guinea (3.0%), Belize (1.4%), Malaysia (0.5%), and Senegal (0.4%). The economic exposure grows faster than the physical exposure, as people grow richer, while wetlands get scarcer: both of these trends increase the value of wetlands. For the Bahamas, the value of wetland loss amounts to 1.3% of GDP in 2025, 2.3% in 2050, 3.3% in 2075, and 4.8% in 2100.

Economy-wide impact estimates

Direct cost estimates are only rough estimates of the true economic impact, as they ignore the fact that land prices may rise if land is lost, food prices may rise if agricultural land gets scarcer, etc. The appropriate way to estimate these additional effects is to use a computable general equilibrium model (CGE). CGEs consider markets for all goods and services simultaneously, taking international trade and investment into account.

Darwin and Tol (1998) use the FARM CGE to estimate the economy-wide implications of sea-level rise for the IS92a scenario, using the data from the studies of Hoozemans *et al.* (1993), Fankhauser (1994) and Tol (1995). There are two important results. First, the total economic impact is some 20% larger than the direct cost estimate. The fact that the total economic cost is larger than the direct cost estimate is not surprising: resource loss implies an overall deflation of the economy. Note that, given the uncertainties, 20% is not a very large difference. Second, the spatial pattern of total economic costs is quite different from the spatial pattern of direct costs. The reason is competitive advantage. Countries with relatively limited land area are hit by both land loss and loss of competitive advantage, while countries with relatively plentiful land gain in competitive advantage. For Australia, Darwin and Tol (2001) report net economic *gains* as this country takes over food production from other parts of the world where prime agricultural land is lost due to sea-level rise.

Bosello *et al.* (2007) also investigate the impacts of sea-level rise using a CGE, using essentially the same impact data as Darwin and Tol (2001), and produce similar findings. Bosello *et al.* (2007) use a superior treatment of capital investments in coastal protection. They find a spatial disconnect between where the dikes are built and where the economic pain is felt. The economic pain is largest in those countries where capital is least productive; the forced investment in coastal protection further reduces the availability of capital.

Adaptation analysis

A range of adaptation models have been used as outlined below. The adaptation model in the GVA is simple: the design frequency of the coastal protection is determined by per capita income (Table 10.5) and applied uniformly across an entire nation. The class boundaries are based on expert knowledge at

Table 10.5 Protection classes as applied by Hoozemans *et al.* (1993) and Nicholls (2004).

Protection class (and status)	Hoozemans *et al.* (1993)		Nicholls (2004)		
			GDP/capita (US$)		
	GDP/capita (US$)	Design frequency	If deltaic coast	If non-deltaic coast	Design frequency
low	< 600	1/1 to 1/10	< 2400	< 600	1/10
medium	600 to 2400	1/10 to 1/100	2400 to 5000	600 to 2400	1/100
high	> 2400	1/100 to 1/1000	> 5000	> 2400	1/1000

Delft Hydraulics. The adaptation model of Nicholls (2004) is essentially the same as the GVA, but distinguishes adaptation in deltaic and non-deltaic countries, and considers the time required to adapt.

Fankhauser (1994) sets the level of adaptation based on an analysis of the costs of coastal protection (including additional wetland loss) and its benefits, that is, the avoided costs of land loss. The model consists of two components, the height of the dike and the length of the dike. Dike building is the only form of coastal protection considered. Ignoring the variability of tides and storms, the optimal dike height equals sea-level rise, as building a higher dike would not provide additional benefits, while a lower dike would be ineffective. Fankhauser (1994) assumes that income and capital goods are equally distributed across the population of a country. He also assumes that there is a linear gradient in population density varying from twice the average to zero. This implies that the benefits of coastal protection are linear in population density. Fankhauser uses dynamic programming (with an analytical solution) to determine at which population density land is worth protecting; and hence the optimal fraction of coast protected is determined.

The national adaptation analysis of Yohe *et al.* (1995; 1996) is also based on cost-benefit analysis, but it is based on a sample of individual sub-national coastal segments, rather than national coverage. They use dynamic programming (with a numerical solution) to determine whether to protect for each segment. As numerical optimization is computationally expensive, Yohe *et al.* (1996) use the sampling to upscale their results from a small number of coastal segments to the entire US coast. While these numerical techniques allow for non-linear relationships and more realistic behavior, they can only be applied at a fine resolution and are currently too computationally expensive to apply at a global scale. A combination of detailed case studies and upscaling is more feasible, but has yet to be done.

Tol (2004) combines the adaptation model of Fankhauser (1994) with Tol's vulnerability model, illustrated in Figures 10.3 and 10.4. The scenario is SRES A1B (IPCC, 2000). Figure 10.5 (colour plates section) shows the optimal protection level for coastal countries based on land loss. Using this method, almost all countries are predicted to have full protection as soon as 2025. There are a few rank reversals through time, as some countries grow faster than others. The least protected countries in 2100 are Kiribati (74% of the *vulnerable* coastline protected), New Caledonia (94%), Fiji (96%), New Zealand (96%) and Sweden (96%). Figure 10.5 also shows the costs of protection, expressed as a percentage of GDP. Relative protection costs differ by seven orders of magnitude. The countries which pay most for coastal protection are Micronesia (0.36% of GDP in 2100), Palau (0.30%), Tuvalu (0.07%), Kiribati (0.06%), and the Marshall Islands (0.04%).[5] Relative costs fall over time because under this scenario, economic growth exceeds the growth in protection costs. In the case of Micronesia, protection costs are 0.63% in 2025, 0.54% in 2050, 0.44% in 2075, and 0.36% in 2100. A cursory comparison of protection costs and the costs of dryland loss indicates why protection levels are so high: protection costs are about 1% of the dryland loss costs.

If the assumptions in Tol's (2004) model and scenario are broadly correct and if countries roughly follow an economic calculus when responding to sea-level rise, then the main impacts of sea-level rise will not be increased erosion and flood risks, but rather increased investment in coastal protection. Wetlands, of course, would be doubly hit, first directly by sea-level rise and second by coastal squeeze due to dike building, as discussed in earlier sections.

10.4 Sub-national to national assessments

While there are many existing national assessments of the impact of sea-level rise (e.g., Nicholls and Mimura, 1998; www.survas.mdx.ac.uk), most of these treat sea-level rise as the main coastal driver. The goal of this section is to consider those studies that have been linked to coastal management and policy development in a more holistic manner.

Coastal management decisions, which include adaptation to sea-level rise, take place at national and sub-national scales. Thus whilst global assessment of the impacts of sea-level rise provides evidence for global climate policy negotiations, more localized methods are required to inform coastal management in practice. Trans-national coastal issues do exist, particularly in relation to the origin and fate of fine-grained sediments

---

[5] Earlier, we state that the costs of coastal protection may be prohibitive for small and poor islands. Under the assumptions of the SRES A1B scenario (IPCC, 2000), these islands are still small but no longer poor at the end of the century.

which may be transported over very long distances, but they are fairly limited. The case for trans-national cooperation in coastal management in terms of sharing good practice is recognized in the (faltering) progress towards European guidance on Integrated Coastal Zone Management (Commission of the European Communities, 2000) and in European data collection initiatives such as the Eurosion project (RIKZ et al., 2004). However, in practice coastal management decisions have been localized phenomena and all too often excessively parochial in their outlook.

In the UK and elsewhere, a consequence of short-term parochial approaches to coastal management has been a chronic reduction in supply of sediments to the coast over the course of the twentieth century due to direct and indirect causes and a coastline that is in an increasingly over-stressed state that is far from equilibrium. A combination of harmful coastal practices (which can hardly be referred to as "coastal management" as the effects have more often than not been unintentional) and twentieth-century sea-level rise means that the form of the coast is visibly changing at most inhabited coastal sites. Wide beaches and mud-flats backed by natural environments have been squeezed between rising sea levels and engineered structures on the back-shore. Coastal steepening has also been observed (Taylor et al., 2004).

Awareness of the potential impacts of climate change on the coast has added further justification to the shift towards more strategic coastal management that progressively gained acceptance during the last two or three decades of the twentieth century. The strategic objectives could not until quite recently be backed up with the quantified strategic modeling tools required to appraise options and make resource allocation decisions. Modeling and analysis methods that had been dominated by the needs of site-specific engineering design could not be upscaled to deal with the requirements of strategic coastal management, even less so the more recent requirements of strategic climate impacts assessment on the coast, which require extended timescales (decades and longer) of assessment (Hall et al., 2005).

The emphasis here is on the methods that have been developed in the UK over the past five years or so, which have been motivated by the need to provide quantified national-scale assessments of the financial risk of all flooding (river and coastal) as evidence to the UK government's periodic spending reviews (Hall et al., 2003a). Allocation of government resources to flood and coastal management is increasingly based upon these quantified risk assessments. When the UK government's Office of Science and Technology launched a project to look at the risks of flooding and coastal erosion over a timescale of 2030–2100 (Evans et al., 2004a) the national-scale method could be adapted to deal with modified climate and socio-economic vulnerability. At the same time, more detailed methods have been under development, aimed at supporting strategic coastal management. These strategic methods for assessing the reliability of flood defence systems have much in common with dike system reliability analysis that has been applied in the Netherlands in recent years (Lassing et al., 2003; Buijs et al., 2004). In the following subsections, we examine first the national-scale analysis and then provide some brief insights into sub-national-scale analysis.

10.4.1 National-scale flood risk analysis

Operating at a national rather than global scale of assessment provides the opportunity to develop assessment methodologies that are specifically adapted to nationally available data sets. Aerial and satellite remote sensing campaigns are providing new topographic and land-use data. Commercial organizations are generating and marketing increasingly sophisticated data sets of the location and nature of people and buildings. In 2002 the Environment Agency, the organization responsible for operation of flood defences in England and Wales, introduced a National Flood and Coastal Defence Database (NFCDD), which for the first time provided in a digital database an inventory of flood defence structures and their overall condition. Together, these new data sets now enable flood and coastal risk assessment that incorporates probabilistic analysis of flood defence structures and systems (Hall et al., 2003c). Once the necessary data sets are held in a Geographical Information System (GIS) they can then be manipulated in order to explore the impact of future flood management policy and scenarios of climate change. Details of the approach can be found in Hall et al. (2003a, 2005). The main results are presented here.

Comparison of the extent of the Indicative Floodplain with residential, commercial and land-use databases revealed that in England and Wales there are 2.5 million people in coastal floodplains, which compares with approximately 2 million in fluvial floodplains, even though the latter occupy a greater area (Halcrow et al., 2001). These data were combined with relationships between flood depth and economic damage that have been developed from empirical analysis of past flooding events (Penning-Rowsell et al., 2003). The national-scale risk analysis yielded an estimated Expected Annual Damage due to fluvial and coastal flooding of £1.0 billion, with an uncertainty range between £0.6 billion and £2.1 billion. Of this £0.5 billion is attributable to coastal floodplains, mostly along the south, east and northwest coasts of England.

A scenarios approach was adopted for the analysis of future coastal flood risk, which involved combined use of climate and socio-economic scenarios (Table 10.3). Future coastal flood risk is greatly influenced by coastal management policy and practice, perhaps more so than it is by changes outside the control of the coastal manager, such as climate change or economic growth. However, in the analysis described here current coastal defence alignment and form, as well as the levels of investment in maintenance and renewal, were in the first instance kept the same across all scenarios. Clearly coastal management policy will change in the future and will tend to reflect the nature and public expectations of future society; that is, coastal management is scenario-dependent.

However, the aim of the Foresight study was to inform present-day policymakers and in order to do so, the present-day coastal management policy was subjected to particular scrutiny, by analyzing its effectiveness in a range of scenarios. Changing scenarios were superimposed on this fixed coastal management policy (including the current pattern of expenditure and technical approach), in order to assess the capacity of the current policy to cope with long-term changes.

The results of the scenarios analysis are summarized in Table 10.6. No discounting is applied to economic risks. Risk is estimated at time points in the future using today's prices. Large increases in the number of people occupying the coastal floodplain in the UK are envisaged in the relatively loosely regulated World Markets and National Enterprise scenarios. Most of this increase is predicted to occur by the 2050s, representing predictions of very rapid growth in the first half of this century which is envisaged to approach a limit associated with a fairly stable population and spatial constraints. Coastal floodplain occupancy is kept stable in the Global Sustainability and Local Stewardship scenarios. However, increasing flood frequency, primarily caused by sea-level rise, means that even with stable numbers of people in the coastal floodplain, the number of people at risk from flooding more frequently than 1:75 years will increase in all scenarios, assuming that current coastal management policies and practices are continued into the future.

In all scenarios other than the low growth, environmentally/socially conscious Local Stewardship scenario, annual economic flood damage from both coastal and fluvial sources is expected to increase considerably over the next century assuming the current flood defence policies are continued in future. A roughly 20-fold increase by the 2080s is predicted in the World Markets scenario, which is attributable to a combination of much increased economic vulnerability (higher floodplain occupancy, increased value of household/industrial contents, increasing infrastructure vulnerability) together with increasing flood frequency. Coastal flooding makes an increasing contribution to national flood risk, increasing from 26% in 2002 to 46% in the 2080s in the World Markets scenario and 63% in the Global Sustainability scenario.

Environmental effects could be significant with the decline in coastal grazing marsh being of most concern (Nicholls and Wilson, 2001; Evans *et al.*, 2004a).[6] Climate mitigation was explored by combining a low emissions climate scenario with a World Markets socio-economic scenario. While this reduces national flood risk by about 25% in the 2080s, significant adaptation is still required, supporting earlier global analysis. Evans *et al.* (2004b) explored a wide range of soft and hard adaptation responses in different combinations. This suggests that flood risk can be reduced while sustaining the environment if a diverse portfolio of adaptation responses are adopted. The research also examined the feasibility of implementing alternative coastal management approaches within the context of different socio-economic scenarios, highlighting how the success of many approaches depend upon the socio-economic context (Evans *et al.*, 2004b; Hall *et al.*, 2006).

These projections are of course subject to very considerable uncertainties. Whilst there is a commitment to sea-level rise, as already discussed, there is considerable disagreement about changes in other critical climate variables, for example the duration, frequency, and intensity of storms, and the resulting extreme water levels which are strongly influenced by the storm-driven surge climate. Future changes in coastal morphology were based entirely upon expert judgement. Detailed quantified morphological systems analysis is for the time being only feasible on a sub-national scale. Notwithstanding these uncertainties, the analysis described above has been a high profile stimulus to the UK government (*The Guardian*, April 22; Brown, 2004). In particular, it has highlighted the growing contribution that coastal flood risk is projected to make to total flood risk in the UK during the twenty-first century (Hall *et al.*, 2006).

10.4.2 Sub-national-scale analysis

The analysis outline above forms part of a tiered set of methods that are based on the same systems reliability concepts, but are based on data sets of increasingly high resolution and methods of increasing accuracy. The use of higher-resolution methods enables processes that were represented only in approximate terms in the national-scale analysis to be based upon more quantified modeling:

- Joint probability analysis of waves and water levels represent random loading on coastal defences.
- Models of long-term morphological change provide quantified information on the lowering of beach levels in front of coastal defences under different scenarios of climate change and coastal management (Walkden and Hall, 2005).
- Surveys of the coastal defences allow more detailed reliability analysis.
- Remotely sensed topographic data, together with tidal information, allows two-dimensional modeling of floodplain inundation. This is particularly important in large complex coastal floodplains, where features such as highway embankments can modify flood inundation.

Flood risk analysis studies of this type are being conducted in the Netherlands (Lassing *et al.*, 2003; Buijs *et al.*, 2004) and the UK (Dawson *et al.*, 2005a) and are now being modified to represent the effects of climate and socio-economic change (Dawson *et al.*, 2005b).

---

[6] Coastal grazing marsh is a largely artificial coastal habitat created by land claim of former saltmarsh over many decades or even centuries. While they are artificial, they have environmental value and many are designated.

Table 10.6 Summary of UK scenarios analysis (no adaptation).

	2002		World Markets 2050s		World Markets 2080s		National Enterprise 2080s		Local Stewardship 2080s		Global Sustainability 2080s	
	Coastal	Coastal and fluvial	Coastal	Coastal and fluvial	Coastal	Coastal and fluvial	Coastal	Coastal and fluvial	Coastal	Coastal and fluvial	Coastal	Coastal and fluvial
Number of people within the indicative floodplain (millions)	2.5	4.5	3.1	6.2	3.4	6.9	3.2	6.3	2.5	4.5	2.6	4.6
Number of people exposed to flooding (depth > 0 m) with a frequency > 1:75 years (millions)	0.9	1.6	1.6	3.3	1.8	3.5	1.8	3.6	1.3	2.3	1.4	2.4
Expected annual economic damage (residential and commercial properties) (£billions)	0.5	1.0	10.6	14.5	13.5	20.5	10.1	15.0	1.0	1.5	3.1	4.9
Annual economic damage relative to Gross Domestic Product per capita		0.09%				0.14%		0.31%		0.05%		0.06%
Expected annual economic damage (agricultural production) (£millions)	2.2	5.9	28.6	41.6	20.7	34.4	74.0	135.7	35.8	63.5	18.9	43.9

Source: After Hall *et al.* (2005).

A rather different cross-sectoral approach was adopted in the RegIS analysis, applied to the northwest and Anglian regions of the UK. This linked together impacts of coastal change and river flooding to land use, water and biodiversity (Nicholls and Wilson, 2001; Holman et al., 2005a,b). This also demonstrated the high sensitivity of coastal flooding to sea-level rise found in Evans et al. (2004a), while coastal ecosystems appear more sensitive to changes in coastal management policy than sea-level rise, especially the widespread adoption of managed realignment which will lead to profound changes, such as major declines in coastal grazing marsh. A fully coupled series of meta-models for the two regions is being developed within the RegIS-2 Project, and the method could be applied nationally.

10.5 Discussion/conclusion

The potential for coastal impacts of sea-level rise is significant and requires careful consideration within the broader issue of human-induced climate change. As the preceding analyses agree, a range of large impacts are possible, although those impacts could vary from impact damage to the required adaptation investment costs.

Comparing results at different scales of assessment, Tol (2004) suggests that the rational economic response to land loss due to sea-level rise is widespread investment in coastal protection, while Evans et al. (2004b) argues for a portfolio of responses to improve flood management including a mixture of protection and retreat which can achieve multiple goals, including avoiding coastal squeeze and sustaining coastal ecosystems. While these two studies are assessing different impacts, one can argue that Tol's "protect rather than abandon response" at national and greater scales should be interpreted as a "planned adaptation rather than abandon response" at sub-national scales. Different nations will follow different development pathways depending on national priorities. In the UK and the wider European Union, protection of human coastal use and sustaining coastal ecosystems are both policy goals, demanding a portfolio of responses. Discussions with Chinese colleagues suggests that they might follow a quite different adaptation approach with a strong emphasis on protection (e.g., Lau, 2003; Li et al., 2004). More detailed assessments of the sub-national response are clearly required where a full range of soft and hard adaptation responses can be explored in a range of combinations.

While the studies reviewed here show that mitigation will reduce impacts, and also that there are feasible and economically viable adaptation options, much more work is required to understand the choices implied by this result. Hence, there is a need to continue to explore the implications of sea-level rise (and wider climate and other changes) which require the further development of data sets, methods and their results across the range of scales to support the range of different decisions that are required (from different mixtures of adaptation and mitigation to strategic coastal management and planning). Also, the effects of emission reduction on coastal protection require more detailed exploration; such effects include not only reduced economic growth and hence reduced demand for safety from coastal hazards, and reduced availability of capital for coastal protection, but also an increased importance of agriculture (if biomass is used for energy) and tidal energy. To promote consistent thinking from global climate policy, down to coastal management, it is important that these analyses use nested approaches and consistent assumptions (e.g., Hall et al., 2003a).

At the regional to global scales, the impact studies and the economic studies presented here provide complementary insights about the system response to sea-level rise. Combining them will lead to more consistent and comparable results. The DINAS-COAST Project is considering all the impacts shown in Table 10.1, except waterlogging, within an integrated model. These are linked to a costing and adaptation module which will control a range of possible responses: (1) raising flood dikes, (2) beach nourishment, and (3) wetland nourishment. Hence the coupled behavior of the natural and human systems can be explored in more detail, giving more realistic estimates of impacts, costs, and adaptation for the world's coasts for a range of unmitigated and mitigated sea-level-rise scenarios.

Future developments in national-scale analysis are set in opposite directions. On the one hand, methods that are at present used for strategic assessment at a sub-national scale will be aggregated so that national-scale assessments can benefit from improved process representation. For example, national-scale coastal assessments would benefit from process-based representation of coastal morphodynamics and sediment budgets. National-scale assessment tools will become more complex, following trends in GCMs. Improved national-scale assessment tools will allow simulation and optimization of subtle portfolios of options. On the other hand, reduced complexity versions are required for uncertainty analysis, which requires multiple model realizations, and for exploratory policy analysis, where more or less instant results are desirable. These reduced-form versions can also be made available to broader stakeholder groups to allow education and participation. Thus, whilst at present national-scale appraisal is based on laborious analysis using a limited number of models that are the preserve of a small number of specialists, in future we expect to see an expanding family of models, of varying complexity tailored for the different needs of stakeholders.

Given the slow response of sea-level rise to the mitigation of climate change the issues raised in this paper need to be assessed far into the future (for many decades and longer). To provide better context for these assessments, it is important to move from climate change assessment to more comprehensive global change analysis of coastal zones (Kremer et al., 2005). This is essential for strategic coastal management, and even at the regional to global scale, it provides context to deliberations on climate change. Many of the results already presented are beginning to address this issue, but this needs to be developed further.

Acknowledgements

Nicholls and Tol have been supported by the DINAS-COAST Project which received funding from the EU under contract no. EVK2–2000–22024. Hall and Nicholls have been supported by the Tyndall Centre for Climate Change Research. Tol has been supported by the Michael Otto Foundation for Environmental Protection.

References

Arnell, N. W., Cannell, M. G. R., Hulme, M. *et al.* (2002). The consequences of CO_2 stabilisation for the impacts of climate change? *Climatic Change* **53**, 413–446.

Arnell, N. W., Livermore, M. J. L., Kovats, S. *et al.* (2004). Climate and socio-economic scenarios for climate change impacts assessments: characterising the SRES storylines. *Global Environmental Change* **14**, 3–20.

Bijlsma, L. (1996). Coastal zones and small islands. In *Impacts, Adaptations and Mitigation of Climate Change: Scientific-Technical Analyses*, ed. R. T. Watson, M. C. Zinyowera and R. H. Moss. Cambridge: Cambridge University Press, pp. 289–324.

Bosello, F., Lazzarin, M., Roson, R. and Tol, R. S. J. (2007). Economy-wide estimates of the implications of climate change: sea level rise. *Environmental and Resource Economics*, in press.

Brown, P. (2004). Global warming floods threaten 4m in UK. *The Guardian*, April 22; http://education.guardian.co.uk/higher/news/story/0,9830,1200347,00.html.

Buijs, F. A. van Gelder, P. H. A. J. M. and Hall, J. W. (2004). Application of Dutch reliability methods for flood defences in the UK. *Heron* **49** (1) 33–50.

Church, J. A. and Gregory, J. M. (2001). Changes in sea level. In *Climate Change 2001: The Scientific Basis. Contribution of Working Group I to the Third Assessment Report of the Intergovernmental Panel on Climate Change*, ed. Houghton, J. T., Ding, Y., Griggs, D. J. *et al.* Cambridge: Cambridge University Press, pp. 639–693.

Commission of the European Communities (2000). *Communication from the Commission to the Council and the European Parliament on Integrated Coastal Zone Management: a Strategy for Europe.* COM(2000) 547 final, Brussels, 27.09.2000.

Darwin, R. F. and Tol, R. S. J. (2001). 'Estimates of the economic effects of sea level rise', *Environmental and Resource Economics* **19** (2), 113–129.

Dawson, R. J., Hall, J. W., Sayers, P. B., Bates, P. D. and Rosu, C. (2005a). Sampling-based flood risk analysis for fluvial dike systems. *Stochastic Environmental Research and Risk Analysis* **19**(6), 388–402.

Dawson, R. J., Hall, J. W., Bates, P. D. and Nicholls, R. J. (2005b). Quantified analysis of the probability of flooding in the Thames Estuary under imaginable worst case sea-level rise scenarios, *International Journal of Water Resources Development: Special Edition on Wates and Disasters* **21** (4), 577–591.

Department of Trade and Industry (2003). *Foresight Futures 2020: Revised Scenarios and Guidance*. London: Department of Trade and Industry.

Evans, E. P., Ashley, R., Hall, J. W. *et al.* (2004a). *Foresight Flood and Coastal Defence Project: Scientific Summary Volume I: Future Risks and Their Drivers*. London: Office of Science and Technology.

Evans, E., Ashley, R., Hall, J. *et al.* (2004b). *Foresight, Future Flooding. Scientific Summary Volume II: Managing Future Risks*. London: Office of Science and Technology.

Fankhauser, S. (1994). Protection vs. retreat: the economic costs of sea level rise. *Environment and Planning A* **27**, 299–319.

Halcrow, HR Wallingford and John Chatterton Associates (2001). *National Appraisal of Assets at Risk from Flooding and Coastal Erosion, Including the Potential Impact of Climate Change*. London: Department of Environment, Food and Rural Affairs.

Hall, J. W., Dawson, R. J., Sayers, P. B. *et al.* (2003a). Methodology for national-scale flood risk assessment. *Water and Maritime Engineering*, ICE **156** (3), 235–247.

Hall, J. W., Evans, E. P, Penning-Rowsell, E. C. *et al.* (2003b). Quantified scenarios analysis of drivers and impacts of changing flood risk in England and Wales: 2030–2100. *Global Environmental Change Part B: Environmental Hazards* **5**(3–4), 51–65.

Hall, J. W., Meadowcroft, I. C., Sayers, P. B. and Bramley, M. E. (2003c). Integrated flood risk management in England and Wales. *Natural Hazards Review*, ASCE **4** (3), 126–135.

Hall, J. W., Sayers, P. B and Dawson, R. J. (2005). National-scale assessment of current and future flood risk in England and Wales. *Natural Hazards* **36** (1–2), 147–164.

Hall, J. W., Sayers, P. B., Walkden, M. J. A. and Panzeri, M. (2006). Impacts of climate change on coastal flood risk in England and Wales: 2030–2100. *Philosophical Transactions of the Royal Society* **364**, 1027–1049.

Holman, I. P., Rounsevell, M. D. A., Shackley, S. *et al.*, (2005a). A regional, multi-sectoral and integrated assessment of the impacts of climate and socio-economic change in the UK: Part I. Methodology. *Climatic Change* **71** (1–2), 9–41.

Holman, I. P., Nicholls, R. J., Berry, P. M. *et al.* (2005b). A regional, multi-sectoral and integrated assessment of the impacts of climate and socio-economic change in the UK: Part II Results. *Climatic Change* **71** (1–2), 43–73.

Hoozemans, F. M. J., Marchand, M. and Pennekamp, H. A. (1993). *A Global Vulnerability Analysis: Vulnerability Assessment for Population, Coastal Wetlands and Rice Production on a Global Scale*, 2nd ed. The Netherlands: Delft Hydraulics.

Hulme, M., Mitchell, J., Ingram, W. *et al.* (1999). Climate change scenarios for global impact studies. *Global Environmental Change* **9**, S3–S19.

Hulme, M., Jenkins, G. J., Lu, X. *et al.* (2002). *Climate Change Scenarios for the United Kingdom: The UKCIP02 Scientific Report*. Norwich: Tyndall Centre for Climate Change Research, School of Environmental Sciences, University of East Anglia.

IPCC (2000). *Special Report on Emissions Scenarios: A Special Report of Working Group III of the Intergovernmental Panel on Climate Change*. Cambridge: Cambridge University Press.

Klein, R. J. T. and Nicholls, R. J. (1999). Assessment of coastal vulnerability to sea-level rise. *Ambio* **28**, 182–187.

Kremer, H. H., Le Tisser, M. D. A. *et al.* (eds) (2005). *Land-Ocean Interactions in the Coastal Zone: Science Plan and Implementation Strategy*. IGBP Report 51/IHDP Report 18. Stockholm: IGBP Secretariat.

Lassing, B. L., Vrouwenvelder, A. C. W. M. and Waarts, P. H. (2003). Reliability analysis of flood defence systems in the Netherlands. In *Safety and Reliability*, Proc. ESREL 2003, Maastricht, 15–18 June 2003, ed. T. Bedford and P. H. A. J. M van Gelder. Lisse: Swets and Zietlinger, pp. 1005–1014.

Lau, M. (2003). *Integrated Coastal Zone Management in the People's Republic of China: An Assessment of Structural Impacts on Decision-making Processes*. Research Unit Sustainability and Global Change **FNU-28**. Centre for Marine and Climate Research, Hamburg University, Hamburg.

Li, C., Fan, D., Deng, B and Korotaev, V. (2004). The coasts of China and issues of sea-level rise. *Journal of Coastal Research*, Special Issue No. 43, 36–49.

Mimura, N. (2000). Distribution of vulnerability and adaptation in the Asia and Pacific Region. In *Global Change and Asia-Pacific Coasts*, ed. N. Mimura and H. Yokoki, Proceedings of APN/SURVAS/LOICZ Joint Conference on Coastal Impacts of Climate Change and Adaptation in the Asia-Pacific Region, Kobe, Japan, November 14–16, 2000. Asian-Pacific Network for Global Change Research, Kobe, and the Center for Water Environment Studies, Ibaraki University, pp. 21–25.

Nicholls, R. J. (2002). Rising sea levels: potential impacts and responses. In *Global Environmental Change*. ed. R. Hester and R. M. Harrison. Issues in Environmental Science and Technology, No. 17. Cambridge: Royal Society of Chemistry, pp. 83–107.

Nicholls, R. J. (2004). Coastal flooding and wetland loss in the 21st century: changes under the SRES climate and socio-economic scenarios. *Global Environmental Change* **14**, 69–86.

Nicholls, R. J. and Lowe, J. A. (2004). Benefits of mitigation of climate change for coastal areas. *Global Environmental Change* **14**, 229–244.

Nicholls, R. J. and Lowe, J. A. (2005). Climate stabilisation and impacts of sea-level rise. In *Avoiding Dangerous Climate Change*, ed. H. J. Schnellnhubes, W. Crames, N. Nahicenovic, T. Wigley, and G. Yohe. Cambridge: Cambridge University Press, pp. 195–202.

Nicholls, R. J. and Mimura, N. (1998). Regional issues raised by sea-level rise and their policy implications. *Climate Research*, **11**, 5–18.

Nicholls, R. J. and Wilson, T. (2001). Integrated impacts on coastal areas and river flooding. In *Regional Climate Change Impact and Response Studies in East Anglia and North West England (RegIS)*, ed. I. P. Holman and P. J. Loveland. Final Report of MAFF project no. CC0337, pp. 54–101.

Nicholls, R. J., Hoozemans, F. M. J. and Marchand, M. (1999). Increasing flood risk and wetland losses due to global sea-level rise: regional and global analyses. *Global Environmental Change* **9**, S69–S87.

Oppenheimer, M. and Alley, R. B. (2004). The West Antarctic Ice Sheet and long term climate policy. *Climatic Change* **64**, 1–10.

Penning-Rowsell, E. C., Johnson, C., Tunstall, S. M. *et al.* (2003). *The Benefits of Flood and Coastal Defence: Techniques and Data for 2003*. Flood Hazard Research Centre, Middlesex University, London.

RIKZ, EUCC, IGN, UAB, BRGM, IFEN and EADS (2004). *Living with Coastal Erosion in Europe: Sediment and Space for Sustainability, Part 1: Major Findings and Policy Recommendations of the EUROSION Project*. www.eurosion.org/reports-online/part1.pdf

Stive, M. J. F. (2004). How important is global warming for coastal erosion? *Climatic Change* **64**, 27–39.

Taylor, J. A., Murdock, A. P. and Pontee, N. I. (2004). A macroscale analysis of coastal steepening around the coast of England and Wales. *Geographical Journal* **170** (3) 179–188.

Tol, R. S. J. (1995). The damage costs of climate change: toward more comprehensive calculations. *Environmental and Resource Economics* **5**, 353–374.

Tol, R. S. J. (2004). *The Double Trade-Off between Adaptation and Mitigation for Sea Level Rise: An Application of FUND*. Research Unit Sustainability and Global Change **FNU-48**. Hamburg University and Centre for Marine and Atmospheric Science, Hamburg.

Walkden, M. J. A. and Hall, J. W. (2005). A predictive mesoscale model of soft shore erosion and profile development. *Coastal Engineering* **52**(6), 535–563.

Yohe, G. W. (1990). The cost of not holding back the sea: toward a national sample of economic vulnerability. *Coastal Management* **18**, 403–431.

Yohe, G. W., Neumann, J. E. and Ameden, H. (1995). Assessing the economic cost of greenhouse-induced sea level rise: methods and applications in support of a national survey. *Journal of Environmental Economics and Management* **29**, S-78–S-97.

Yohe, G. W., Neumann, J. E., Marshall, P. and Ameden, H. (1996). The economics costs of sea level rise on US coastal properties. *Climatic Change* **32**, 387–410.

11

Developments in health models for integrated assessments

Kristie L. Ebi and R. Sari Kovats

11.1 Introduction

The Energy Modeling Forum (EMF) has facilitated the development of integrated assessment models of economic activity and the resulting anthropogenic emissions, the relevant Earth system responses to the forcings from these emissions, and the costs of policies to mitigate anthropogenic emissions. Initially these integrated assessment models focused solely on the market impacts of climate change. EMF has been a leader in developing cross-disciplinary interactions between integrated assessment modelers and health researchers, with the result that some of the recent developments in population health models arose from the EMF workshops held in Snowmass. The appropriate and robust incorporation of health impacts into integrated assessment models will help policymakers visualize the dimensions of the possible health impacts of climate change and will inform the design of climate and adaptation policy responses.

The current burden of disease due to environmental factors is considerable (WHO, 2002). If trends continue, climate change is very likely to increase the burden of climate-sensitive diseases, including morbidity and mortality due to temperature extremes, extreme weather events (i.e., floods and droughts), air pollution, food- and water-borne diseases, and vector- and rodent-borne diseases; some of these relationships are better understood than others (McMichael *et al.*, 2001). Climate change is one of many factors that will affect the burden of these health outcomes. Significant changes are expected to continue to occur in major factors determining population health, such as per capita income, provision of medical care, obesity, and access to adequate nutrition, clean water, and sanitation. These non-climate changes will have larger population health impacts than the changes in climate that are projected to occur by the end of the century.

Interest in quantifying the possible future health impacts of climate variability and change has led to the development of a range of models. Models are evolving from simple descriptions of associations between a weather variable and a disease outcome, to complex, multi-determinant models that integrate exposures and processes that influence the occurrence of disease (Ebi and Gamble, 2005). Because of inherent uncertainties, a particular challenge is creating not-implausible assumptions of how population health and adaptive capacity may evolve over time. There is a range of approaches to model how population health may change in the future; one is the health transition paradigm that is based on an understanding of how population health has evolved since the mid-nineteenth century.

This chapter reviews some recent models that project (1) future health impacts of climate change; (2) the health benefits of controlling greenhouse gas emissions; and (3) the economic costs of the health impacts of climate change. Health transitions are then discussed as a basis for the development of models of how population health may evolve over time. We conclude with a discussion of future directions for the development of health models for incorporation into integrated assessment models.

11.2 Projecting the health impacts of climate change

The future health impacts of climate change can be projected for an individual disease, such as the model by van Lieshout *et al.* (2003) that projects the populations at risk of *Falciparum*

Human-induced Climate Change: An Interdisciplinary Assessment, ed. Michael Schlesinger, Haroon Kheshgi, Joel Smith, Francisco de la Chesnaye, John M. Reilly, Tom Wilson and Charles Kolstad. Published by Cambridge University Press. © Cambridge University Press 2007.

malaria, or projected for a set of climate-sensitive diseases, such as the study of the Global Burden of Disease due to climate change by McMichael *et al.* (2004).

11.2.1 Individual disease models

There is considerable ongoing research aimed at better understanding the associations between weather/climate and health outcomes. Approaches used to identify and quantify exposure–response relationships between weather variables and climate-sensitive diseases include ecological studies and time series methods (Ebi and Patz, 2002). In ecological studies, group-level relationships are explored through spatial or temporal variations in exposure and outcome. These studies take advantage of routinely reported, large aggregated databases of health outcomes, such as deaths or hospital visits. As with other epidemiological methods, potential confounders must be identified and their effects removed from the analysis. The advent of GIS (geographic information system) and geo-referencing of disease and exposure data have facilitated the investigation of the spatial distribution of many vector-borne diseases and/or their vectors. For example, several research groups have independently modeled the relationships between climate factors and the distribution of *Falciparum* malaria at global, regional, and national levels because climate is one determinant of malaria transmission and because the current burden of disease is high.

MARA/ARMA is a regional model of the climate determinants of malaria transmission that has been validated for sub-Saharan Africa (Craig *et al.*, 1999). This is a biological model based on the minimum and mean temperature constraints on the development of the *Plasmodium falciparum* parasite and the *Anopheles* vector, and on the precipitation constraints on the survival and breeding capacity of the mosquito. This model is the basis for several projections of the possible impacts of climate change on malaria transmission (e.g. Ebi *et al.*, 2005; McMichael *et al.*, 2004; Tanser *et al.*, 2003).

A global biological model of malaria transmission developed by Martens (Martens, 1998; Martens *et al.*, 1999) was used by van Lieshout *et al.* (2004) to estimate the potential impacts of climate change on the length of seasonal transmission and populations at risk for a range of climate and population scenarios. The population at risk was calculated as the population living in areas where climatic conditions were suitable for malaria occurrence at least one month per year. The transmission potential was used as a comparative index for estimating the impact of changes in environmental temperature and precipitation patterns on the risk of malaria. The transmission potential was derived from the basic reproduction rate (defined as the average number of secondary infections produced when a single infected individual is introduced into a potential host population in which each member is susceptible).

Simulations were driven by the SRES climate scenarios (A1FI, A2, B1, and B2). The population scenarios linked to the SRES emission scenarios were used, but the GDP scenarios were not thought to be either useful or appropriate.

Using expert judgement, countries were classified according to their current vulnerability and malaria control status. This vulnerability incorporated both socio-economic status, as a measure for adaptive capacity, and climate because malaria at the fringes of its climate-determined distribution is easier to control than malaria in endemic regions. Thus, current malaria control status was used as an indicator of adaptive capacity.

For those countries that currently have a limited capacity to control the disease, the model projected that by the 2080s the additional populations at risk would be in the range of 90 million (A1FI) to 200 million (B2b). Population growth was a more important driver in the projection of absolute numbers of future populations at risk than changes in climate. The greatest impact was under the B2 scenario, reflecting population growth in risk areas in Eurasia and Africa. The model projected climate-induced changes in the potential distribution of malaria in many poor and vulnerable regions of the world. However, climate change was not projected to affect malaria transmission in many least developed countries in sub-Saharan Africa; the climate in these countries is already suitable for year-round transmission. Thus, the impact of climate change on malaria is likely to be confined to the highland and desert fringes of the current malaria distribution. However, in order to provide information for local or national planning purposes, models should be driven and validated with local data.

11.2.2 Applying a quantitative relationship between socio-economic development and malaria

Challenges to applying a quantitative relationship between socio-economic development and malaria incidence include:

- The lack of a suitable indicator for socio-economic development for global statistical analysis. Social capital, an indicator of equity in income distribution within countries, is often a more important indicator than GDP. For example, a study by Halstead *et al.* (1985) reported that certain countries and regions have a standard of health far higher than might be expected for their GDP (e.g. Costa Rica and Cuba outperform Brazil on most health indicators);
- Political instability can undermine the influence of development (Russia);
- Economic development can increase transmission temporarily (as has been observed in relation to deforestation, population movement, and water development projects); and
- Many control programs depend on external/donor funding (e.g. Vietnam).

National aggregate indicators such as GDP and GNP are economic averages that have traditionally been used to estimate the level of national economic growth. By their nature, they average over many important and relevant factors, and have several disadvantages. Even when expressed on a per capita basis, they:

- Fail to include non-market (and therefore non-priced) subsistence production, including much of women's

work (this is particularly important in the context of low- and middle-income countries that are most affected by vector-borne diseases such as malaria);
- Do not reflect the disparities that may exist within countries between the rich and the poor, between urban and rural areas, across regions, between different ethnic groups, and between men and women; and
- Do not take proper account of the complexity of the relationship between marginal changes in income and health and how this varies between countries. For example, the relationship between malaria and income in Europe is not the same as the relationship within Africa.

Further, the relationship between economic development and malaria is two-way. Poor economic development is an effect of malaria as well as a cause (Sachs, 2001). In addition to the direct costs of treating and preventing malaria and of lost productivity, malaria has been shown to slow economic growth in low-income African countries, thus creating an ever-widening gap in prosperity between malaria-endemic and malaria-free countries. The reduced growth in countries with endemic malaria was estimated to be over 1% of GDP per year (Sachs, 2001). The cumulative effect of this "growth penalty" is severe and restrains the economic growth of the entire region.

11.2.3 Global Burden of Disease study

The Global Burden of Disease study began in 1992 with the objective of quantifying the burden of disease and injury in human populations (Murray and Lopez, 1996a). The burden of disease refers to the total amount of premature death and morbidity within a population. The goals of the study were to produce the best possible evidence-based description of population health, the causes of lost health, and likely future trends in health in order to inform policymaking. The WHO Global Burden of Disease 2000 project (GBD) updated the 1990 study (Murray et al., 2003). It drew on a wide variety of data sources to develop internally consistent estimates of incidence, health state prevalence, severity and duration, and mortality for over 130 major health outcomes, for the year 2000 and beyond. To the extent possible, the GBD synthesized all relevant epidemiologic evidence on population health within a consistent and comprehensive framework, the comparative risk assessment. Twenty-six risk factors were assessed, including major environmental, occupational, behavioral, and lifestyle risk factors. Climate change was one of the environmental risk factors assessed (McMichael et al., 2004).

The GBD used a summary measure of population health, the Disability Adjusted Life Years lost (DALYs). The attributable burden of DALYs for a specific risk factor was determined by estimation of the burden of specific diseases related to the risk factor; estimation of the increase in risk for each disease per unit increase in exposure to the risk factor; and estimation of the current population distribution of exposure, or future distribution as estimated by modeling exposure scenarios. Counterfactual or alternative exposure scenarios to the current distribution of risk factors were created to explore distributional transitions towards a theoretical minimum level of exposure (e.g. for exposure to carcinogens, the theoretical minimum level of exposure would be no exposure). Risk factors were chosen to be among the leading causes of the world burden of disease that had a high likelihood of being causally associated with disease, had reasonably complete data, and were potentially modifiable.

For climate change, the questions addressed were what will be the total health impact caused by climate change between 2000 and 2030 and how much of this burden could be avoided by stabilizing greenhouse gas emissions (WHO, 2002; McMichael et al., 2004). The alternative exposure scenarios defined were unmitigated emission trends (i.e., approximately following the IPCC IS92a scenario), emissions reductions resulting in stabilization at 750 ppm CO_2 equivalent by 2210 (S750), and emissions reductions resulting in stabilization at 550 ppm CO_2 equivalent by 2170 (S550). Climate change projections were generated by the HadCM2 global climate model (Johns et al., 2001). The health outcomes included in the analysis were chosen based on sensitivity to climate variation, predicted future importance, and availability of quantitative global models (or feasibility of constructing them). The health outcomes selected were the direct impacts of heat and cold, episodes of diarrheal disease, cases of *Falciparum* malaria, fatal unintentional injuries in coastal floods and inland floods/landslides, and non-availability of recommended daily calorie intake (as an indicator for the prevalence of malnutrition). Global and WHO specific region estimates were generated.

In the year 2000, the mortality attributable to climate change was 154 000 (0.3%) deaths, and the attributable burden was 5.5 million (0.4%) DALYs, with approximately 50% of the burden due to malnutrition (McMichael et al., 2004). Note that these estimates are for a year when the amount of climate change since baseline (1990) was near zero; therefore, future disease burdens would be expected to increase with increasing climate change, unless effective adaptation measures were implemented. About 46% of the DALYs attributable to climate change were estimated to have occurred in the WHO Southeast Asia region, 23% in countries in the Africa region with high child mortality and very high adult male mortality, and 14% in countries in the Eastern Mediterranean region with high child and adult male mortality.

For each health outcome, ranges of estimates were projected for relative risks attributable to climate change in 2030 under the alternative exposure scenarios (McMichael et al., 2004). For example, Table 11.1 provides relative risk estimates for malaria; these were based on the MARA/ARMA model. Sources of uncertainty in the relative risk estimates include uncertainties in the climate projections, the degree to which the MARA/ARMA model applies to other regions, the relationship between the increase in the population at risk and disease incidence in each region, and the effectiveness of

Table 11.1 Range of estimates for the relative risks of malaria attributable to climate change in 2030, under the alternative exposure scenarios.

Region	Relative risks		
	Unmitigated emissions	S750	S550
Africa region	1.00–1.17	1.00–1.11	1.00–1.09
Eastern Mediterranean region	1.00–1.43	1.00–1.27	1.00–1.09
Latin America and Caribbean region	1.00–1.28	1.00–1.18	1.00–1.15
Southeast Asia region	1.00–1.02	1.00–1.01	1.00–1.01
Western Pacific region[a]	1.00–1.83	1.00–1.53	1.00–1.43
Developed countries[b]	1.00–1.27	1.00–1.33	1.00–1.52

[a] Without developed countries.
[b] And Cuba.
Source: Campbell-Lendrum et al. (2003).

malaria control programs. Therefore, the lower uncertainty estimate assumed no effect of climate change on the risk of malaria, and the upper estimate was a doubling of the mid-range estimate. The upper estimate of the projected potential increase in malaria risk was large in some regions (e.g. 1.83 for the Western Pacific region) and small in others (e.g. 1.02 for the Southeast Asia region).

The study concluded that if the understanding of broad relationships between climate and disease was realistic, then unmitigated climate change was likely to cause significant health impacts through at least 2030 (Campbell-Lendrum et al., 2003).

11.3 Projecting the health benefits of controlling greenhouse gas emissions

The main focus of integrated assessment modeling has been on understanding the complex interactions between climate change and economic growth. In addition to market impacts, climate change will affect non-market sectors, with both positive and negative impacts. Controlling the emissions of greenhouse gases will have health benefits if projected changes decrease the ambient concentrations of air pollutants.

Yang et al. developed a method for integrating health effects from exposure to criteria air pollutants in the United States to impacts on the US economy, using the Emissions Prediction and Policy Analysis (EPPA) model (Yang et al., 2004). Although this model did not consider climate change, it holds considerable promise for projecting the possible health consequences of future changes in concentrations of air pollutants under different assumptions of economic growth. There were three steps in model development (Yang et al., 2004). The first was to identify the pollutants to be included, the amount of exposure that would be experienced, and the exposure–response relationship for health endpoints included. The second was to identify how and where to introduce these effects into the CGE (computable general equilibrium) Social Accounting Matrix (SAM) and to expand the matrix to cover non-market components. The third was to estimate the physical losses and the value of those losses in terms consistent with the expanded SAM. Extending the model to include health effects involved valuation of non-wage (leisure) time, not only of the existing labor force but also of children and the elderly, and inclusion of a household production sector to capture the economic effects of morbidity and mortality from acute exposures to air pollutants.

The EPPA model projected emissions of a range of air pollutants from energy, industry, agriculture, and biomass burning. The health endpoints associated with exposure to $PM_{10/2.5}$ (particulate matter less than 10 or less than 2.5 μm), NO_2, SO_2, O_3, CO, nitrates, and sulfates were taken from a study that had quantified impacts of air pollution on health (the European Union study *ExternE: Externalities of Energy*, Vol. 1, which included endpoints associated with both acute and chronic exposures, including mortality, asthma, bronchitis, restricted activity days, cough, and heart disease) (Holland et al., 1998). The CGE SAM assumed that air-pollution-related morbidity and mortality lowered productivity, consumption, and labor, which then lowered income.

It was assumed that a change in pollution concentration changed the cumulative (lifetime) exposure. The relationship between health effects (including death) and chronic exposures was modeled as a stock-flow problem because cancer and cardio-pulmonary events associated with air pollutants only occur after years of chronic exposure. Thus, the benefits of lower pollution today (or further losses due to higher levels of exposure) are only fully realized by those exposed to these levels over their lifetime.

To test this model, it was applied to the US for the period 1970 to 2000 and the resulting estimates of economic damage from air pollution were compared with estimates from a USEPA study (USEPA, 1989, 1999). The estimated benefits were somewhat lower than the EPA estimates, primarily because the EPPA model estimated that the benefits from reduced chronic exposure to air pollution will be only gradually realized.

The model focused on health and economic impacts in the United States, where economic impacts would be expected to be relatively large compared with the health impacts. In developing countries, the opposite would be expected, with large health impacts from current and projected future concentrations of air pollutants.

11.4 Projecting the economic costs of the health impacts of climate change

The Framework for Uncertainty, Negotiation and Distribution (FUND) integrated assessment model incorporates a range of damage functions to estimate the "costs" of climate change on the economy and the population (including health, agriculture, and coastal flooding) (Tol 2002a, b). The impacts of climate change on the scenarios of economic and population growth were calculated as dead-weight losses to the economy that reduced long-term economic growth. Changes in population growth were linked to projected mortality from selected climate-sensitive diseases. To deal with differences between countries, the value of a statistical life was set at 200 times the per capita income. Issues related to this approach include valuing human life differently in rich versus poor countries. The value of a statistical life did not vary by the age at death; thus, the death of a child and the death of an aged person within a country were assigned the same cost.

Tol used FUND to explore interactions between climate change, economic growth, and mortality from malaria (Tol, in press). Global malaria mortality was estimated from three malaria models, then scaled by the corresponding increase in the global mean temperature and averaged (Martens et al., 1995, 1997; Martin and Lefebvre, 1995; Morita et al., 1995). The yearly, regional burden of malaria mortality was taken from Murray and Lopez (1996a, b), expressed as the fraction of total population. It was assumed that changes in malaria incidence were proportional to changes in potential malaria distribution. The model assumed that malaria mortality was zero at a per capita income of $3100 (range 2100–4100).

The model found that climate change-induced malaria in sub-Saharan Africa slowed growth perceptibly when the parameter denoting the per capita income at which vector-borne mortality becomes zero was five times as large as its best guess value (Tol, in press). An analytically tractable approximation of the model suggested that the effect of malaria on economic growth had to be at least 30 times as large as the best guess in order to reverse growth. For example, without the effect of malaria, incomes in sub-Saharan Africa were projected to grow from about $500 to approximately $17 000 per person per year by 2100. When the effect of malaria was included in the model with all parameters set at their best guesses, this value fell by about $700. A higher sensitivity of malaria to climate change decreased income by about $1600 (from $17 000). The largest effect was due to climate sensitivity; increasing sensitivity to 4.5 °C for a doubling of CO_2 lead to an income reduction of about $3800.

Based on the results of a series of sensitivity analyses, Tol concluded that climate change may reverse economic growth through an increase in malaria only if climate change is rapid, economic growth is slow, and the effect of ill-health on growth is large, such as in parts of sub-Saharan Africa. However, slowing economic growth could have significant implications for development.

The model is structured so that per capita income is a key determinant of the burden of malaria. However, malaria is also a determinant of economic growth in low- and middle-income countries (Sachs, 2001). In addition, because the world's poorest countries tend to be in high-risk tropical and subtropical regions, national rates of malaria incidence would be expected to have a high correlation with per capita income (McMichael and Martens, 2002). In reality, environmental (including geography and prevailing climate), economic, and social factors determine the range and intensity of malaria transmission. Malaria and other disease models are needed that are designed to link with economic models to further our understanding of how development may alter the burden of climate-sensitive diseases.

The results from this model contrast with the population health model developed by Pitcher et al., (in press), described in the next section, which found that although per capita income was a significant predictor of life expectancy, other factors were more important.

11.5 Health transitions

Because most data sets used to study the relationships between weather and climate-sensitive diseases cover relatively recent and short time periods, the relationships described relate to recent conditions, including climatic, economic, social, etc. There is often an underlying assumption that these relationships will remain relatively constant under future climatic conditions. However, exposure–response relationships change over temporal and spatial scales for a variety of reasons. For example, Table 11.2 shows changes in deaths over the twentieth century in the UK (Carson et al., 2006). The periods were chosen to avoid wars and other periods where events such as the 1918/19 influenza epidemic would have affected mortality rates. Childhood mortality decreased dramatically from 38.5% of deaths in 1900–10 to 4.9% in 1954–64, while mortality among those 65 years of age and older increased from 29.4% of deaths to 79.7% by 1986–96. Clearly, using rates from one period to project rates in a subsequent period would not have been accurate.

Over the nineteenth and twentieth centuries, fertility and mortality rates in developed countries transitioned from high to low, and morbidity and mortality patterns changed from predominantly infectious to chronic diseases. These and related changes are referred to as the health transition. Public health

Table 11.2 Change in population health and deaths over the twentieth century, UK.

Percentage of deaths by age and cause	Period			
	1900–1910	1927–37	1954–64	1986–96
0–14 years	38.5%	13.3%	4.9%	1.5%
15–64 years	32.0%	40.5%	31.4%	18.8%
65+ years	29.4%	46.1%	63.7%	79.7%
Cardiovascular	12.1%	27.9%	33.3%	42.3%
Respiratory	18.9%	20.0%	14.1%	14.0%
Other	69.0%	52.1%	52.6%	43.7%
% of deaths attributable to cold	12.5	11.2	8.7	5.4
(95% CI)	(10.1, 14.9)	(8.4, 14.0)	(5.9, 11.5)	(4.1, 6.7)

has been, and continues to be, a key determinant of population health. The life expectancy of a child born in the United States in 1900 was 47 years (CDC, 1999). Today, life expectancy is about 80 years; 25 years of that increase has been attributed to improvements in public health such as vaccine development, improvement in motor vehicle safety, safer workplaces, control of infectious diseases, decline in morbidity and mortality from cardiovascular diseases and stroke, safer and healthier foods, healthier mothers and babies, family planning, fluoridation of water, and recognition of the hazards of tobacco (Bunker et al., 1994; CDC, 1999). Plausible population health models need to consider how population health in developed and developing countries is likely to change over the coming century; this can be informed by the changes experienced by developed countries as they developed.

Traditionally, the health transition is described in three stages (Martens and Hilderink, 2002). The first stage is the age of pestilence and famine, characterized by infectious diseases as the dominant causes of death and by inadequate provision of food and fresh water, resulting in low levels of life expectancy. Many developing countries continue to experience high rates of infectious diseases. The second stage is the age of receding pandemics, characterized by access to clean water and adequate nutrition leading to a reduction in the prevalence of infectious disease, thus reducing mortality rates and increasing life expectancy. Developed countries entered this stage in the middle of the nineteenth century, and many countries with economies in transition are currently in this stage. The third stage is the age of chronic diseases, characterized by chronic diseases as the dominant causes of death, with further increases in life expectancy. In this stage, countries experience improvements in medical care and social determinants that influence health, and literacy rates are high and fertility rates are low.

Although these stages are descriptive of the process of development for developed countries, this pattern may have limited applicability to currently developing countries because many of these countries experience triple health burdens, with high rates of infectious and chronic diseases, and with problems with the health care delivery systems. Public health infrastructures and medical care communities are trying to reduce outbreaks of infectious diseases, while at the same time trying to reduce the health effects of high levels of air pollution and other environmental exposures. Increasing economic growth in many developing countries is proving difficult because health is a critical input to economic growth. The reality of countries that have remained poor and unhealthy despite international interventions demonstrates how little is known about how to stimulate economic development. In addition, population health and economic growth are being increasingly affected by conditions that transcend national borders and political jurisdictions, such as macroeconomic policies associated with international financial institutions, global trade agreements, water shortages, pollution that crosses borders, etc. (Labonte and Spiegel, 2003). These global issues could affect the health transition in complex and non-linear ways.

What, then, could be some pathways for population health to evolve over the coming century? Possibilities include an age of emerging infectious diseases characterized by the re-emergence of currently controlled infectious diseases and the emergence of new infectious diseases (Martens and Hilderink, 2002). Factors that could influence this include travel and trade, microbiological resistance, human behavior, breakdowns in health systems, and increased pressure on the environment (Barrett et al., 1998). The consequences could include increasing rates of infectious diseases in developed and developing countries, with falling life expectancy and lower levels of economic activity. A number of worldwide events and trends support this possibility, including malaria parasites becoming increasingly drug resistant, HIV/AIDS lowering life expectancy in several African countries, West Nile virus moving into the North American continent, and severe acute respiratory syndrome (SARS) moving quickly, via travelers, across multiple continents.

Another possible health transition is changed health risks owing to changes in lifestyle and the environment. This could be envisioned as increasing life expectancy through improvements in medical technology (Martens and Hilderink, 2002), or as decreasing life expectancy owing to an increased burden of lifestyle-related diseases. In support of the latter, the United Nations Population Division's 2002 Revision anticipates a more serious and prolonged impact of the HIV/AIDS epidemic in the most affected countries than previous revisions (United Nations, 2003). In the 2002 Revision, the dynamics of the epidemic were assumed to remain unchanged until 2010, after which prevalence rates were assumed to decline. Under the medium population growth variant, the 2002 Revision projected a lower population in 2050 than did the 2000 Revision by 0.4 billion people (8.9 billion instead of 9.3 billion). About half of the 0.4 billion difference resulted from an increase in the number of projected deaths, the majority from the higher projected levels of HIV/AIDS, and the remainder from a reduction in the number of projected births, mostly due to lower expected future fertility levels. Over the current decade, the number of excess deaths because of AIDS among the 53 most affected countries was estimated at 46 million, with the total number of excess deaths projected to increase to 278 million by 2050.

Obesity, with its attendant increased burden of diabetes and other chronic diseases, is another disease that could decrease life expectancy. Obesity has become a worldwide epidemic. Figure 11.1 shows the prevalence of obesity among adults in the United States in 1985, 1990 and 2001 (Mokdad et al., 1999, 2001). Globally, obesity levels range from below 5% in China, Japan, and certain African nations, to over 75% in urban Samoa, with at least 300 million adults clinically obese worldwide (www.who.int/dietphysicalactivity/publications/facts/obesity/en, accessed 2 April, 2005). The increased mortality associated with obesity could decrease life expectancy despite advances in medical technology. These examples raise the question of how to more accurately capture mixed trends

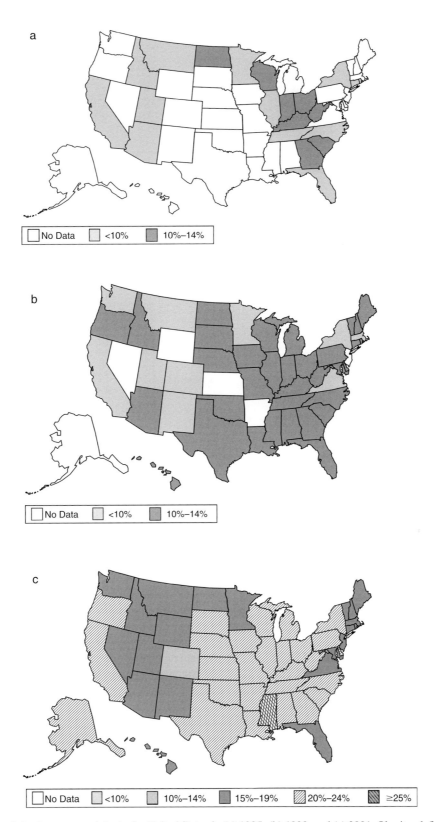

Figure 11.1 Prevalence of obesity among adults in the United States in (a) 1985; (b) 1990; and (c) 2001. Obesity, defined as a body mass index of 30 or greater (or approximately 30 pounds overweight for a woman of height 5' 4") increased in the United States from <10% in most states in 1985 to 20–24% in most states in 2001 (Mokdad et al., 1999, 2001).

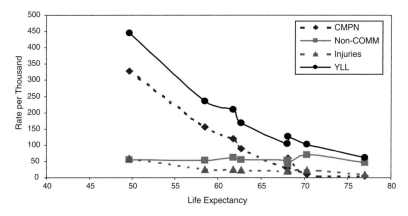

Figure 11.2 Rates of communicable (CMPN) and non-communicable diseases (non-COMM) and injuries by years of life lost (YLL) (Pitcher *et al.*, in press).

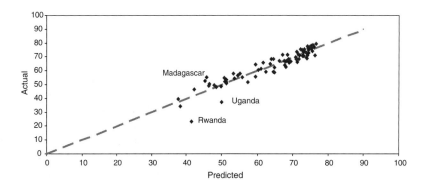

Figure 11.3 Comparison of actual and predicted life expectancy for final model (Pitcher *et al.*, in press).

of increasing and decreasing health burdens, and their implications for life expectancy, into models.

Another possible health transition is into an age of sustained health, with investments in social and medical services leading to a reduction in lifestyle-related diseases, and the elimination of most environment-related infectious diseases (Martens and Hilderink, 2002). Although desirable, this transition does not currently appear plausible.

11.5.1 Population health model

Pitcher *et al.* (in press) constructed a top-down model, using life expectancy as the indicator of population health because life expectancy explains much of the pattern in age-specific death rates. Country-level health data were categorized into communicable diseases (including maternal, perinatal and nutritional deficiencies), non-communicable diseases, and injuries. Rates of non-communicable diseases and injuries vary little across life expectancies; communicable diseases are the primary source of variability in life expectancy, as shown in Figure 11.2. Non-communicable diseases dominate disability and death in Established Market Economies, where life expectancies are higher.

Variations in life expectancy across 91 countries were described using a population-based statistical model based on cross-sectional data. Life expectancies across the countries varied from about 50 to 80 years. Factors reported or suspected to be associated with life expectancy were explored. Variables associated in the model with life expectancy included climate (climate within latitudinal bands was used as an indicator), per capita income in purchasing power parity, education (adult literacy), nutrition (caloric intake), access to simple medical care (an index constructed of access to immunization, oral rehydration therapy, and medical care at birth), water (access to clean water and sanitation), energy (electricity use per capita), transportation (passenger cars per capita), and medical expenditures in purchasing power parity. Income inequality as measured by the Gini coefficient was only weakly associated with life expectancy. A regression model that included all variables had an R^2 of 0.91. A parsimonious model was developed that included only literacy, access to clean water and sanitation, simple medical attention, and an indictor variable for sub-Saharan Africa; the R^2 value was 0.90. Income was not a significant addition to this model. Because income is included in global change scenarios, the log of income was added to the final model; however, it is difficult to account properly for the exogenous impacts of income when variables are highly correlated. Visual inspection of Figure 11.3 shows that the fit between life expectancy, as predicted by this model, and actual life expectancy was rather good, except for a few countries with low

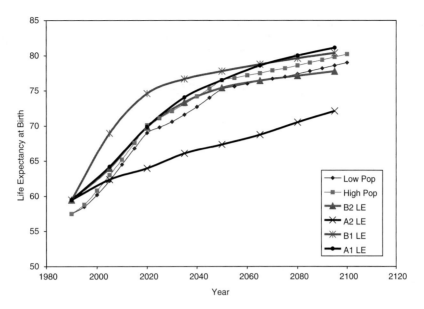

Figure 11.4 Comparison of life expectancy in India under four SRES storylines (Pitcher *et al.*, in press).

life expectancy; Rwanda, Uganda, and Madagascar accounted for one half of the unexplained variation in the model.

The model was then used to develop trajectories of life expectancy in India based on the four families of IPCC SRES storylines (IPCC, 2000). Figure 11.4 shows the results along with the low and high longevity assumptions underlying the SRES. Life expectancy in A1 and B1 are projected to rise much more rapidly than in either A2 or B2. Life expectancy in A2 is projected to be much lower than in the other storylines, with the difference remaining large up to 2100, when life expectancy is projected to be nearly 10 years lower than in the other trajectories.

Future population health will be determined by a range of factors, including how the current epidemics of HIV/AIDS, tuberculosis, malaria, and other diseases will be addressed, the pathways of economic development, the development of medical and other technology, how adaptive capacity will evolve, and others. Within this context, the population health model of Pitcher *et al.* suggests that investments designed to promote population health may make a large difference in life expectancy.

11.6 Future directions in the development of health impact models

Increased interest in the design of early warning systems and of models to project disease burdens under different possible future conditions is encouraging the health sector to undertake more model development. Challenges to the development of health models include incomplete understanding of disease mechanisms, including how key health determinants influence risks and vulnerabilities; how to incorporate critical thresholds or non-linearities into models; appropriate methods for aggregation of different diseases with different relationships with weather/climate, including valid approaches for extrapolation of the exposure–response relationship derived from a different population; how to characterize critical uncertainties; how to incorporate, on appropriate scales, the processes of adaptation and adaptive capacity; and model validation (Ebi and Gamble, 2005).

Effectively incorporating scale interactions into models can be a challenge. Population health is affected by a multitude of factors that operate at different scales, from community (e.g. whether there is an effective early warning system for heat waves) to individual levels (e.g. whether an individual spends adequate time in cooled environments during a heat wave) (Ebi and Gamble, 2005). Climate change is a global phenomenon whose impacts will be felt at local scales. Local actions, such as land-use changes that contribute to heating or cooling of the environment, can amplify or ameliorate larger-scale climate forces. Methods need to be developed for incorporating effects across multiple scales to develop models that are more accurate.

Another scale issue is that few studies of the relationships between weather and climate-sensitive diseases have been conducted across a wide range of geographic regions. Thus, regional and global models are based on the extrapolation of relationships from a limited number of areas, which raises questions of the validity and applicability of the model projections for other regions.

Modelers of climate-sensitive diseases should evaluate approaches to incorporating critical thresholds or non-linearities into models. For example, mortality typically exhibits a curvilinear relationship with ambient temperature, with mortality rates increasing with increasing ambient temperature (above the temperature at which mortality is at a minimum). In some cities and regions, there is a sharp increase in mortality at very high temperatures. However, most models fit a simple linear term above the temperature at which mortality is at a minimum. This is an example of how models can mask regional and local differences that may be important for understanding future vulnerability. The existence of thresholds and how thresholds may change over temporal and spatial scales must be considered when extrapolating an exposure–response relationship derived from a different population.

Model uncertainty needs to be dealt with explicitly. There are multiple sources of uncertainty in developing models, including whether the appropriate model structure was chosen, whether the key processes incorporated into the models are well understood and described, how underlying variables will change over time (i.e., the rate, speed, and regional extent of climate change; changes in economic development, technology, etc.), how populations in different regions will respond to climate change, and the effectiveness of mitigation and adaptation strategies, policies, and measures (IPCC, 2001; Kovats et al., 2003).

Another challenge to model development arises because the sensitivity and adaptive capacity of exposed populations vary considerably depending on factors such as population density, level of economic and technological development, local environmental conditions, pre-existing health status, and the quality and availability of health care and public health infrastructure (Woodward et al., 1998). The extent of these changes is inherently uncertain. A better understanding is needed of how to model the measures currently in place that affect vulnerability, and the policies and measures that could be implemented (along with an estimate of their effectiveness). Different assumptions about the effectiveness of current and future adaptation measures lead to dramatically different estimates of climate change impacts. Overconfidence (or underconfidence) in the effectiveness of the adaptation process can lead to underestimation (or overestimation) of the future burden of disease due to climate change and of the assessment of co-benefits of mitigation policies.

The larger and more complex a model, the more difficult validation may become, particularly when data are limited. For example, although a number of groups have developed models to project how malaria might spread under particular assumptions about changes in temperature and precipitation models, limited data are available for validation. Approaches need to be developed to provide policymakers with confidence that decisions based on model results will be robust.

Increased interest by both health and integrated assessment modelers in projecting non-market impacts of climate change will undoubtedly further model development. Because integrated assessment models can be useful to answer questions from policymakers on possible climate change impacts under different policy responses, and because the range of questions that will be raised is wide, it is unlikely that one modeling approach will be appropriate for all questions. Therefore, the best way forward is to develop a range of approaches and models (that are explicit about underlying assumptions and the evidence base for those assumptions) through collaboration among health and integrated assessment modelers.

Health modelers need to take more systems-based approaches that integrate micro- to macro-level exposures and processes, and they need to work with integrated assessment modelers to appropriately incorporate variables for exploring the possible consequences of policies responses to climate change. Integrated assessment modelers need to work with the health sector (including health economists) to ensure that key health determinants are appropriately included in their models. Together, these communities can develop models to explore, for example, the costs and benefits of spending on mitigation now in lieu of spending on development and health infrastructure, and how each pathway could influence current and future health burdens; and the benefits to health of climate policies, modeled at appropriate scales (i.e., what would be the public health impact for a particular country of implementing a particular policy) (Ebi and Gamble, 2005).

11.7   Conclusions

Growing interest in projecting the potential health-related impacts of climate change has resulted in the development of a range of models taking diverse approaches to integrate the multi-scale exposures and processes that influence the range and intensity of disease occurrence. Currently, most integrated assessment models that include health do so as a module linked with projected changes in temperature and other weather factors through exposure–response relationships. The complexity of health models has begun to increase as models incorporate the interactions among health and the demographic structure of a population and its economic activities, providing feedback to policy options that influence greenhouse gas emissions. Developments in integrating health into integrated assessment models are likely to lead to more valid modeling of complex interactions among climate, ecology, and human elements in ways that more accurately portray the actual dynamics of the entire system.

References

Barrett, R., Kuzawa, C. W., McDade, T. and Armelagos, G. J. (1998). Emerging and re-emerging infectious diseases: the third epidemiologic transition. *Annual Review of Anthropology* **27**, 247–271.

Bunker, J. P., Frazier, H. S. and Mosteller, F. (1994). Improving health: measuring effects of medical care. *Milbank Quarterly* **72**, 225–58.

Campbell-Lendrum, D. H., Corvalan, C. F. and Pruss-Ustun, A. (2003). How much disease could climate change cause? In *Climate Change and Human Health: Risks and Responses*, ed. A. J. McMichael, D. Campbell-Lendrum, C. F. Corvalan et al. Geneva: WHO/WMO/UNEP.

Carson, C., Hajat, S., Armstrong, B. and Wilkindan, P. (2006). Declining vulnerability to temperature-related mortality in London over the 20th Century. *American Journal of Epidemiology* **164**, 77–84.

Centers for Disease Control and Prevention (CDC). (1999). Ten great public health achievements: United States, 1900–1999. *Morbidity and Mortality Weekly Review* **48**, 241–243.

Craig, M. H., Snow, R. W. and le Sueur, D. (1999). A climate-based distribution model of malaria transmission in sub-Saharan Africa. *Parasitology Today* **15**, 105–111.

Ebi, K. L. and Gamble, J. L. (2005) Summary of a workshop on health scenarios development: strategies for the future. *Environmental Health Perspectives* **113**, 335–338.

Ebi, K. L. and Patz, J. A. (2002). Epidemiologic and impacts assessment methods. In *Environmental Change, Climate and Health: Issues and Research Methods*, ed. P. Martens and A. J. McMichael. Cambridge: Cambridge University Press.

Ebi, K. L., Hartman, J., McConnell, J. K., Chan, N. and Weyant, J. (2005). Climate suitability for stable malaria transmission in Zimbabwe under different climate change scenarios. *Climatic Change* **73**, 375–393.

Halstead, S. B., Walsh, J. A. and Warren, K. (eds.) (1985). *Good Health at Low Cost: Proceedings of a Conference held at the Bellagio Conference Center, Bellagio, Italy, April 29–May 3, 1985*. New York: Rockefeller Foundation.

Holland, M., Berry, J. and Forster, D. (eds.). (1998). *ExternE, Externalities of Energy*, Vol. 7: *Methodology, 1998 Update*. Brussels: European Commission, Directorate-General XII, Science Research and Development.

IPCC (2000). *Emissions Scenarios: A Special Report of Working Group III of the Intergovernmental Panel on Climate Change*, ed. N. Nakicenovic and R. Swart. Cambridge: Cambridge University Press.

IPCC (2001). *Climate Change 2001: Impacts, Adaptation, and Vulnerability. Contribution of Working Group II to the Third Assessment Report of the Intergovernmental Panel on Climate Change*, ed. J. J. McCarthy, O. F. Canziani, and N. A. Leary, D. J. Dokken and K. S. White. Cambridge: Cambridge University Press.

Johns, T. C., Gregory, J. M., Stott, P. A. and Mitchell, J. F. B. (2001). Correlations between patterns of 19th and 20th century surface temperature change and HadCM2 climate model ensembles. *Geophysical Research Letters* **28** (6), 1007–1010.

Kovats, R. S., Ebi, K. L. and Menne, B. (2003). *Methods for Assessing Human Health Vulnerability and Public Health Adaptation to Climate Change*. Copenhagen: WHO/Health Canada/UNEP.

Labonte, R. and Spiegel, J. (2003). Setting global health research priorities: burden of disease and inherently global health issues should both be considered. *British Medical Journal* **326**, 722–723.

Martens, P. and Hilderink, H. (2002). Human health in transition: towards more disease or sustained health? In *Transitions in a Globalising World*, ed. P. Martens and J. Rotmans. Exton: Swets and Zeitlinger, pp. 61–84.

Martens, P., Kovats, R. S., Nijhof, S. *et al.* (1999). Climate change and future populations at risk from malaria. *Global Environmental Change* **9**, S89–107.

Martens, W. J. (1998). *Health and Climate Change: Modeling the Impacts of Global Warming and Ozone Depletion*. London: Earthscan.

Martens, W. J. M., Jetten, T. H., Rotmans, J. and Niessen, L. W. (1995). Climate change and vector-borne diseases: a global modeling perspective. *Global Environmental Change* **5** (3), 195–209.

Martens, W. J. M., Jetten, T. H. and Focks, D. A. (1997). Sensitivity of malaria, schistosomiasis and dengue to global warming. *Climatic Change* **35**, 145–156.

Martin, P. H. and Lefebvre, M. G. (1995). Malaria and climate: sensitivity of malaria potential transmission to climate. *Ambio* **24** (4), 200–207.

McMichael, A. J. and Martens, P. (2002). Global environmental changes: anticipating and assessing risk to health. In *Environmental Change, Climate and Health: Issues and Research Methods*, ed. P. Martens and A. J. McMichael. Cambridge: Cambridge University Press, pp. 1–17.

McMichael, A., Githeko, A., Akhtar, R. *et al.* (2001). Human health. In *Climate Change 2001: Impacts, Adaptation, and Vulnerability Contribution of Working Group II to the Third Assessment Report of the Intergovernmental Panel on Climate Change*, ed. J. J. McCarthy, O. F. Canziani, N. A. Leary, D. J. Dokken and K. S. White. Cambridge: Cambridge University Press, pp. 451–85.

McMichael, A. J., Campbell-Lendrum, D., Kovats, S. *et al.* (2004). Global climate change. In *Comparative Quantification of Health Risks: Global and Regional Burden of Disease due to Selected Major Risk Factors*, ed. M. Ezzati, A. Lopez, A. Rodgers and C. Murray, Geneva: World Health Organization, pp. 1543–1649.

Mokdad, A. H., Serdula, M. K., Dietz, W. H. *et al.* (1999). The spread of the obesity epidemic in the United States, 1991–1998. *Journal of the American Medical Association* **282**, 1519–1522.

Mokdad, A. H., Bowman, B. A., Ford, E. S. *et al.* (2001). The continuing epidemics of obesity and diabetes in the United States. *Journal of the American Medical Association* **286**, 1195–1200.

Morita, T., Kainuma, M., Harasawa, H., Kai, K. and Matsuoka, Y. (1995). *An Estimation of Climatic Change Effects on Malaria, Working Paper*. Tsukuba: National Institute for Environmental Studies.

Murray, C. J. L. and Lopez, A. D. (eds.) (1996a). *The Global Burden of Disease: A Comprehensive Assessment of Mortality and Disability from Diseases, Injuries, and Risk Factors in 1990 and Projected to 2020*. Cambridge: Harvard University Press.

Murray, C. J. L. and Lopez, A. D. (1996b). *Global Health Statistics*. Cambridge: Harvard School of Public Health.

Murray, C. J. L., Ezzati, M., Lopez, A. D., Rodgers, A. and Vander Hoorn, S. (2003). Comparative quantification of health risks: conceptual framework and methodological issues. *Population Health Metrics* at http://www.pophealthmetrics.com/content/1/1/1.

Pitcher, H., Ebi, K. L. and Brenkert, A. (in press). Population health model for integrated assessment models. *Climatic Change*.

Sachs, J. (2001). *Macroeconomics and Health: Investing in Health for Economic Development*. Report of the Commission on Macroeconomics and Health. Geneva: World Health Organization.

Tanser, F. C., Sharp, B. and le Sueur, D. (2003). Potential effect of climate change on malaria transmission in Africa. *Lancet* **362**, 1792–1798.

Tol, R. S. J. (2002a). Estimates of the damage costs of climate change. Part I: benchmark estimates. *Environmental and Resource Economics* **21**, 47–73.

Tol, R. S. J. (2002b). Estimates of the damage costs of climate change. Part II: dynamic estimates. *Environmental and Resource Economics* **21**, 135–160.

Tol, R. S. J. (in press). Climate, development and malaria: an application of FUND. *Climatic change*: Population Division, United Nations. Available at: www.un.org/esa/population/publications/wpp2002/WPP2002-HIGHLIGHTSrev1.PDF; Also *Climatic Change*, in press.

United Nations. (2003). *World Population Prospects: The 2002 Revision, Highlights*. Population Division, United Nations; http://www.un.org/esa/population/publications/wpp2002/WPP2002-HIGHLIGHTSrev1.PDF.

US Environmental Protection Agency (USEPA) (1989). *The Benefits and Costs of the Clean Air Act 1970 to 1990*. Washington, DC: Office of Air and Radiation, Office of Policy.

US Environmental Protection Agency (USEPA) (1999). *The Benefits and Costs of the Clean Air Act 1990 to 2010*. Washington, DC: Office of Air and Radiation, Office of Policy.

van Lieshout, M., Kovats, R. S., Livermore, M. T. J. and Martens, P. (2004). Climate change and malaria: analysis of the SRES climate and socio-economic scenarios. *Global Environmental Change* **14**, 87–99.

Woodward, A., Hales, S. and Weinstein, P. (1998). Climate change and human health in the Asia Pacific region: who will be the most vulnerable? *Climate Research* **11**, 31–38.

World Health Organization (WHO) (2002). *World Health Report 2002: Reducing Risks, Promoting Healthy Life*. Geneva: WHO.

Yang, T., Reilly, J. and Paltsev, S. (2004). Air pollution and health effects: toward an integrated assessment. In *Coupling Climate and Economic Dynamics*, ed. A. Haurie and L. Viguier. Dordrecht: Kluwer Publishers.

12

The impact of climate change on tourism and recreation

Jacqueline M. Hamilton and Richard S. J. Tol

12.1 Introduction

Tourism is one of the largest and fastest-growing economic sectors. Tourism is obviously related to climate, as the majority of tourists prefer spending time outdoors and travel to enjoy the sun or landscape. It is therefore surprising that the tourism literature pays little attention to climate and climatic change (e.g., Witt and Witt, 1995), perhaps because climate is deemed constant and beyond control. It is equally surprising that the literature on climate change impact pays little attention to tourism compared with the coverage of other important sectors (Smith et al., 2001), but this can perhaps be explained by the fact that most climate change impact studies are done by field experts (generic climate change impact experts are rare).

The situation is now slowly changing (e.g. Nicholls, 2004). Five branches of literature have started to grow. First, there are studies that examine the impact of climate change on tourism in a qualitative way (e.g. Viner and Agnew, 1999). Second, there are a few studies (e.g. Breiling and Charamza, 1999) that relate the fates of particular tourist destinations to climate change. Third, there are studies (e.g. Scott and McBoyle, 2001) that use indicators of the attractiveness of certain weather conditions to tourists to examine the impact of climate change. Fourth, there are a few studies (e.g. Maddison, 2001) that build statistical models of the behavior of certain groups of tourists as a function of weather and climate, and there are similar studies on recreational behavior.[1] Finally, there are a few studies (e.g. Hamilton et al., 2005a) that use simulation models of the tourism sector. This paper reviews these studies. Before looking at the impact studies, we review the studies in which the relationship between climate and tourism and recreation demand is examined. Section 12.3 provides an overview of climate change impact studies. Section 12.4 concludes the review and focuses on future avenues of inquiry.

Note that there is no separate section on adaptation. Most of the "impacts" discussed below are in fact "adaptations," that is, tourists deciding to travel elsewhere or tourist operators relying on artificial rather than natural snow.

This paper does not review the environmental consequences of tourism and recreation, including the emission of greenhouse gases (e.g. Gössling, 2002), let alone the implications of climate policy for tourism and recreation (e.g., Piga, 2003).

12.2 The importance of climate and weather for tourism and recreation

Whether in the process of deciding on the destination and the right time for a holiday or in the daily choices made about recreation activities whilst on holiday, climate and weather play an important role (Hamilton and Lau, 2005). One would suspect that, "last minute" holidays and short breaks apart, tourist destination choice is affected by the expected weather (climate) rather than the actual weather. For daily recreation choices, actual weather is the decisive factor in decision making. The importance of climate for tourism has been classified by de Freitas (2003) according to its aesthetic, physical, and thermal aspects. There is growing evidence,

---

[1] The difference between tourism and recreation is that the former includes at least one overnight stay away from home.

Human-induced Climate Change: An Interdisciplinary Assessment, ed. Michael Schlesinger, Haroon Kheshgi, Joel Smith, Francisco de la Chesnaye, John M. Reilly, Tom Wilson and Charles Kolstad. Published by Cambridge University Press. © Cambridge University Press 2007.

however, that climate has significant neurological and psychological effects (Parker, 2001), which may also have some influence on the choice of holiday destination. In the literature, there are two broad types of study where the importance of climate and weather for tourism and recreation have been examined: attitudinal studies and behavioral studies.

12.2.1 Attitudinal studies

Two kinds of attitudinal studies were found, those that examine the subjective rating of climate compared with the ratings from indices of weather data and those that examine the significance of climate in the image and the attractiveness of particular destinations. Thermal comfort indices have been developed in order to capture the complexity of the thermal aspect of climate, which is argued to be a composite of temperature, wind, humidity, and radiation. Special modifications of such indices have been used to assess the suitability of certain climates for tourism (e.g. Amelung and Viner, in press). The basis of these indices, however, is subjective and arbitrary according to de Freitas (2003). In a case study, carried out at a beach in Queensland, Australia, on 24 days spread over a single year, de Freitas (1990) finds that the relationship between HEBIDEX, an index of body – atmosphere energy budget, and the subjective rating of the weather by beach users is highly correlated. Furthermore, he finds that the optimal thermal conditions for beach users are not at the minimum heat stress level but at a point of mild heat stress. Using the thermal comfort index, predicted mean vote, Thorson et al. (2004) find a positive relationship between thermal comfort and urban park use for recreational activities in Göteborg, Sweden. They also find, however, that there is a discrepancy in the subjective rating of the weather and the rating of the weather according to the index. The majority of those surveyed said that the weather was "acceptable" when it was "warm" or "hot" according to the calculated index. These levels are associated with heat stress.

In spite of the popularity of studies of destination image in the tourism literature, only one of the 142 destination image papers that are reviewed by Pike (2002) specifically deals with weather.[2] This was a study by Lohmann and Kaim (1999), who note that there is a lack of empirical evidence on the importance of weather/climate on destination choice. Using a representative survey of German citizens, the importance of certain destination characteristics was assessed. Landscape was found to be the most important aspect even before price considerations. Weather and bioclimate were ranked third and eighth respectively for all destinations. Moreover, they found that although weather is an important factor, destinations are also chosen in spite of the likely bad weather. Measuring the importance of destination characteristics is also the focus of a study by Hu and Ritchie (1993). They review several studies from the 1970s and find that "natural beauty and climate" are of universal importance in defining the attractiveness of destinations. In their own study, they examine the image of Hawaii, Australia, Greece, France, and China using a survey of Canadian citizens. They find that climate is the second most important characteristic for the group of tourists on a "recreational" holiday. For the group of tourists taking an "educational" holiday, climate ranks twelfth. When the images of the countries are compared, Hawaii is found to have the most attractive climate. Climate and access to the sea are ranked the most important characteristics of destinations in a survey of German tourists who were departing from the Hamburg area to various destinations abroad in the summer of 2004 (Hamilton and Lau, 2005).

12.2.2 Behavioral studies

Some behavioral studies examine the patterns of daily recreational use at particular sites in terms of weather data. For example, Dwyer (1988) has estimated a daily site use model, for an urban forest in Chicago, USA, using data on noon temperature, percentage sunshine, percentage rain, and snow depth. In this study, data on wind-chill are not useful for estimating the use levels. Demand is highest on the sunniest days and on days that are exceptionally warm, especially when these conditions occur in late spring or in early summer. High temperatures in July decrease demand. Brandenburg and Arnberger (2001) predict daily use levels of the Danube Flood Plains National Park in Austria. They find that using standard climate data does not produce any satisfactory results. Instead they use the Physiological Equivalent Temperature (PET), the occurrence of precipitation and cloud cover, to estimate the number of visitors per day in total and for four groups: cyclists, hikers, joggers, and dog walkers. The PET value is very important in determining the use levels, particularly for cyclists and hikers.

Other studies examine the statistical relationship between tourism demand and weather. For example, Agnew and Palutikof (2001) model domestic tourism and international inbound and outbound tourism using a time series of UK tourism and weather data. The results show that temperature is the strongest indicator of domestic demand. In contrast, wetter weather increases the demand for trips abroad in the current period and in the year following. Snow dependent activities are the focus of a survey of US college students carried out in 1997 and 1998 by Englin and Moeltner (2004). Using data on price, weekly conditions at ski resorts in California and Nevada, and the participant's income they find that although demand increases as snow amount increases, trip demand is more responsive to changes in price. As said before, tourism (as opposed to recreation) is likely to be affected by the expected weather (climate) rather than the actual weather.

Another set of studies uses climate data to capture the role of expected weather in destination choice and consequently

---

[2] These papers were published in the period 1973 to 2000.

demand. Lise and Tol (2002) study the holiday travel patterns of tourists from a range of OECD countries. The data and method are crude, but the results suggest that people from different climates have the same climate preferences for their holidays: The climate of Southern France and California is preferred by everyone, regardless of the home climate. (Of course, this does not imply that all tourists travel to these places; climate is not the only factor in tourist destination choice.) Bigano et al. (2004) confirm this result, using less crude econometrics for a much wider range of countries including African and Asian ones. However, Bigano et al. also find that people from hotter places tend to have sharper preferences. That is, while Southern France is preferred by people from both hot and cold places, people from hot places would feel much worse about going elsewhere than Southern France than would people from cold places (see Figure 12.1.)

Also using climate data, the Pooled Travel Cost Model (PTCM) has been applied to the demand of tourists from the UK, the Netherlands, and Germany for a range of countries (Maddison, 2001; Lise and Tol, 2002; Hamilton, 2003). Hamilton (2003) includes the possibility of taking a vacation in the origin country, whereas the other PTCM studies only examine outbound tourism. The studies include temperature and temperature squared in their estimation of demand. The estimated coefficients on these studies allow the optimal temperature to be calculated, that is, where demand is highest. Demand for a country by tourists from the UK is maximized when the quarterly maximum daytime temperature is 31 °C; for tourists from Germany, the demand peaks where the mean monthly temperature is 24 °C. In the Dutch study, the coefficients were not significant.

12.3 The impact of climate change on tourism and recreation

Climate change impact studies can be categorized as qualitative and quantitative. Qualitative impact studies provide information about vulnerabilities and the likely direction of change; nevertheless, they do not provide estimates of changes in demand. Quantitative impact studies can be categorized further into four groups: first, studies that predict changes to the supply of tourism services; second, studies that use tourism climate indices to predict the change in climatic attractiveness; third, studies that use the statistical relationship between demand and weather or climate to estimate the changes in demand; and finally, simulation models of tourism flows.

12.3.1 Qualitative impact studies

Qualitative studies rely on experts' opinions on the likely impact of climate change. For example, Perry (2000) discusses the impact that climate change will have on tourism at the Mediterranean. The main impact caused by an increase in temperature will be a "doughnut"-shaped pattern of demand: in the shoulder season there will be more visits than in the summer season. In addition, he expects that there will be an increase in the demand for long winter holidays particularly from the older generations. The indirect effect of beach erosion caused by sea-level rise will reduce demand and increase the need for planning restrictions in the coastal zone. According to Gable (1997), Caribbean coastal areas will also experience a drop in demand through beach loss. Viner and Agnew (1999) describe the current climate and market situation for the most popular tourist destinations of the British. In addition, the consequences for demand for these destinations under a changed climate in the 2020s and 2050s are discussed. Currently warm resorts such as those in the Eastern Mediterranean are expected to become less attractive as temperature and humidity increase. As summer weather becomes more favorable and reliable in temperate countries, tourism is expected to increase. All of these studies rely on a synthesis of existing work on the physical impacts of climate change or on the expert opinion of the authors; a direct link with demand (or supply) of tourism is not made. Nevertheless, these studies highlight the range of impacts that climate change will have. One qualitative study, however, has involved the tourism industry. This a study by Krupp (1997), where tourism experts from the local tourist industry discussed the impact of climate change on tourism in the West coast of Schleswig-Holstein in Germany. Krupp finds that an increase in summer precipitation will reduce the willingness of the tourism industry to make new investments in tourism infrastructure and facilities.

12.3.2 Impact on the supply of tourism services

Predicting changes in the supply of tourism services has been applied to the winter sports industry. Breiling and Charamza (1999) analyze the impact of a 2 °C change in temperature on seasonal snow-cover depth for all districts in Austria. They estimate that these changes will reduce the length of the ski season and the usability of ski facilities. Warming will have strong impact on low-altitude resorts, which the authors expect will disappear first, and the remaining resorts will become more expensive. Similar studies have been carried out for winter sports tourism in Scotland (Harrison et al., 1999), Switzerland (Abegg, 1996; Elsasser and Messerli, 2001), Finland (Kuoppamaeki, 1996) and Canada (Scott et al., 2001). These studies find a general decline in natural skiing conditions, although this will be less of a problem at high-altitude sites. Moreover, the use of snow-making machines is also temperature dependent, which restricts the adaptation options available.

12.3.3 Impact on climatic attractiveness

The index approach, discussed in Section 12.2, has been used to examine the impact of climate change on the climatic attractiveness of tourist destinations. Scott and McBoyle (2001) apply the tourism index approach to the impact of climate change on city tourism in several North American cities. Cities

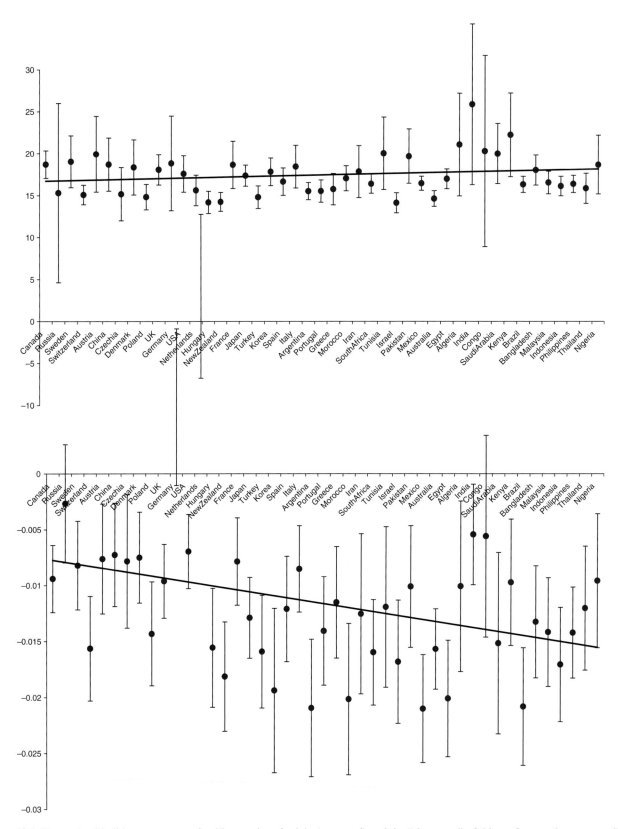

Figure 12.1 The optimal holiday temperature for 45 countries of origin (top panel) and the "sharpness" of this preference (bottom panel); the more negative the sharpness coefficient, the steeper the preference, and the faster the decline in attractiveness of non-optimal destinations. Climate attractiveness is given by $\alpha T + \beta T2$ where T is temperature and $\alpha > 0$ and $\beta < 0$ are parameters. The optimal temperature is $1/2\, \alpha/\beta$; β is the sharpness coefficient for illustration, a linear trend is also shown. The countries of origin are ranked according to their temperature. Source: Bigano *et al.* (2004).

are ranked according to their climatic appropriateness for tourism and the relationship between tourist accommodation expenditures is examined. Then this ranking is recalculated using data from scenarios of (the lower and middle bounds of) climate change for the 2050s and the 2080s. The attractiveness of the cities improves in both time slices. Improvements in spring are the largest of all the seasons. The authors predict an increase in revenue from tourist accommodation for Canadian cities. Amelung and Viner (in press) also use a tourism index to analyze tourism potential for Europe. The attractiveness of a location for tourists depends on temperature, precipitation, humidity, and wind in a very non-linear way. They calculate the index values for spatial resolution of $0.5° \times 0.5°$ using monthly data for the 2020s, 2050s, and 2080s. The results show that climate change would shift tourists towards higher latitudes and altitudes. In addition, there would be a shift from summer to spring and autumn in some destinations, and from spring and autumn to winter in other destinations.

12.3.4 The impact on demand

There are studies that examine the impact of climate change on recreation demand. For example, the impact of climate change in the United States on eight recreation activities is examined by Loomis and Crespi (1999). They estimate demand equations relating the number of activity days to temperature and precipitation. Under a scenario of a $+2.5\,°C$ change in temperature and a 7% reduction in precipitation in 2060, they predict sharp reductions in the number of skiing days (−52%) and increases in the number of days spent playing golf (14%), at the beach (14%) and at reservoirs (9%). Mendelsohn and Markowski (1999) estimate the impact of climate change on a range of recreation activities using the same climate change scenario and time-frame as Loomis and Crespi (1999). The aggregate impact is estimated in terms of welfare and ranges from a reduction of 0.8 billion 1991\$ to an increase of 26.5 billion 1991\$. Using the contingent visitation approach, Richardson and Loomis (2004) find that temperature is a positive determinant of demand for visits to the Rocky Mountain National Park in Colorado. Moreover, depending on the climate scenario, they estimate an increase in recreational visits from 9.9% to 13.6% in 2020. These three studies examine domestic demand within the United States either at site level or for each state, and although they refer to recreation, tourist trips (involving an overnight stay) have been included in their data.

The following studies examine national demand for domestic and international tourism. In the study by Agnew and Palutikof (2001), a $1\,°C$ increase in the summer temperature leads to a 1–5% increase in UK domestic demand. Maddison (2001) finds that the number of visits and the consumer surplus per year increase as the temperature increases although only until the optimal maximum daily temperature. More specifically, he finds that for a climate change scenario for the 2030s a temperature increase of $2\,°C$ and a reduction in precipitation of 15% in summer leads to a 1.3% reduction in trips from the UK to Greece and a 2.2% increase in trips to Spain. For the other seasons, there are increases in the number of trips and in the consumer surplus. Hamilton (2003) uses climate data and the coefficients on the climate variables from the statistical estimation of the demand by German tourists to construct a climate index. For a selection of European countries, the climate index is calculated for the period 1961–91, and for an arbitrary scenario for August, new values of the climate index were calculated. For the observed data, Spain, Portugal, and Greece have the most attractive climate in August. Under the scenario of climate change (a $2\,°C$ temperature increase, a 15% decrease in precipitation, and a 10% decrease in the number of wet days per month), however, the value of the climate index actually falls for these countries: it has become too hot. Germany's climate in August increases in attractiveness. Thus, we can expect a shift away from Mediterranean holidays in the summer months to domestic holidays.

12.3.5 Impact on global tourism flows

In the impact studies reviewed so far, the following gaps are evident. First, the possibility of substitution between destinations has been neglected in all studies. Second, the studies have focused on particular areas or particular origin nationalities; the global picture has yet to be filled in. Third, climate as a "push" factor has also been overlooked. By this, we mean that the climate of the tourists' home country motivates them to take a holiday to somewhere with a climate that they prefer. This may be to escape the heat of summer months or to get away from a cold and wet winter (or even summer). This is important as climate change may reduce (or increase, when countries get too hot) the need to go elsewhere to spend time in a suitable climate. Hamilton et al. (2005a, b) seek to fill these gaps with the Hamburg Tourism Model (HTM), which is a global model of flows from and to 207 countries that includes the climate of countries as a factor in both the estimation of demand to travel and the demand for a particular destination.

The model is calibrated for 1995, using data for total international departures and arrivals. Bilateral tourism flows are generated by the model, independent of data (e.g. WTO, 2003). The simulations are driven by five variables: general attractiveness, distance, population, income, and temperature. General attractiveness is a calibration parameter. It is kept constant. It represents all factors influencing tourist destination choice that are mentioned above, but are not explicitly included in the model. Distance is assumed constant and has no effect on the results. It is relevant to construct the 1995 tourism pattern, however. The effect of population growth is (assumed to be) simple: more people implies more tourists. The effect of per capita income is twofold. First, richer people travel more frequently. Second, tourists avoid poor countries. A world that grows ever richer – HTM runs on the SRES scenarios (IPCC, 2000) – thus sees more tourists, and many more tourists from developing countries. In addition, developing countries become more attractive as tourist destinations.

The annual temperature is the index for climate. There are two quadratic relationships. First, cool destinations become more attractive as they get warmer, and warm destinations become less attractive. Second, cool countries generate fewer international tourists as they get warmer, and warm countries generate more. Put together, these two effects generate an interesting pattern. Climate change shifts international tourists towards the poles and up the mountains. However, climate change also reduces the total number of tourists, because international tourism is dominated by the Germans and the British, who would prefer to take their holidays in their home countries (after climate change has made Germany and the UK more pleasant) (see Figure 12.2.) The reduction in international tourism because of climate change is, however, dwarfed by the growth due to population and economic growth.

Hamilton *et al.* (2005b) slightly change the simulation model, initially allowing tourism to grow more rapidly with economic growth but then assuming saturation of demand. This does not greatly change the results, although tourists from hot and poor countries gain in importance; these tourists would increasingly seek to escape to cooler places during their holiday. Recent model developments include the explicit modeling of domestic tourism, and an increased spatial resolution. Next steps are the inclusion of seasons and more climate variables.

Few of the climate change and tourism studies reviewed above extend into economics. Those that do (e.g. Mendelsohn and Markowski, 1999; Maddison, 2001), offer a straightforward welfare analysis limited to an estimate of the direct costs and benefits. The sole exception is the study by Berritella *et al.* (2004). Using the Hamburg Tourism Model of Hamilton *et al.* (2003) as an input, Berritella *et al.* use the GTAP5 computable general equilibrium model to analyze the economy-wide implications. Climate change impacts on tourism are represented as two additive shocks. First, there is a transfer of income from the countries that receive fewer tourists to those that receive more; this is because the GTAP data are based on the gross *domestic* product. This, of course, partially cancels out in the regional aggregation. Second, there is a shift in demand as people consume different things whilst on holiday (e.g., tourists buy fewer household appliances and more entertainment).

The results show that the global impact is negligible. There is, however, substantial redistribution. Countries in Western Europe, the subtropics and the tropics are negatively affected. North America, Eastern Europe and the former Soviet Union, and Australasia are positively affected. The negative impacts may amount to –0.3% of GDP by 2050, the positive impacts to 0.5% of GDP. These numbers are large compared with other monetized impacts of climate change (e.g., Smith *et al.*, 2001).

12.4 Discussion and conclusion

The review of the literature reveals a rather scattered field. There are a number of case studies on climate change impacts, and there are a number of related studies that could be re-interpreted as climate change impact studies, but each of these studies is unique. The studies use widely different methods and resolutions, looking at different periods and different places. A comprehensive, quantitative message does not emerge from this diversity. One clear, qualitative message does arise, however: climate change could well have substantial effects on tourism and recreation. As a corollary, because tourism and recreation is an important and fast-growing sector, the economic ramifications may be substantial too.

Qualitatively, there is more consensus. Climate change would reduce the number of sites suitable for winter sports holidays, particularly skiing. Summer tourism would shift towards the poles, towards higher latitudes, and towards spring and autumn. For tourists and travel operators, the impacts of climate change may not matter much, even if large. For site-specific tourist operators, impacts may be large as well as important. This is primarily a matter of redistribution, with winners as well as losers.

But the impact of climate change should be seen in its context. The tourism and recreation industry is growing rapidly, and it is very adaptive. Climate change may accelerate or decelerate growth, but is unlikely to change growth into decline (or vice versa). Competition is high, new attractions, destinations, and niches continuously emerge, technological progress is rapid, and the system is used to coping with natural disasters, epidemics, and political events. Gradual climate change does not pose a particular threat to such a versatile sector, but may well threaten particular locales. In some places, tourists come for one thing only, and would stop coming if that one thing – be it snow or a beach – disappears.

Our literature review reveals a number of serious gaps. We have already mentioned the diversity of the research to date. More coordinated research would be welcome. Although it is clear that climate is an important consideration in destination choice, it is not clear what aspects of climate tourists pay particular attention to. Is the Mediterranean popular because the weather is nice, or because the weather is predictable? The UK and Germany will get warmer, but weather variability is not likely to decrease. Most reviewed studies assume that tourists would go elsewhere; however, tourists may also take their holidays at different times of the year. The relative importance of spatial and temporal substitution and the extent of institutional restrictions on seasonality, such as school holidays, are unknown. Exceptions are ski and island resorts; the former may lose their snow, and the latter may disappear altogether. Here, the scarcity effect has yet to be studied. The final remaining ski resort with real snow, and the final remaining atoll, would be able to extract a considerable monopoly rent. Scenarios for tourism and recreation have not been developed. Tastes, technologies, and relative prices all

Impact on tourism and recreation

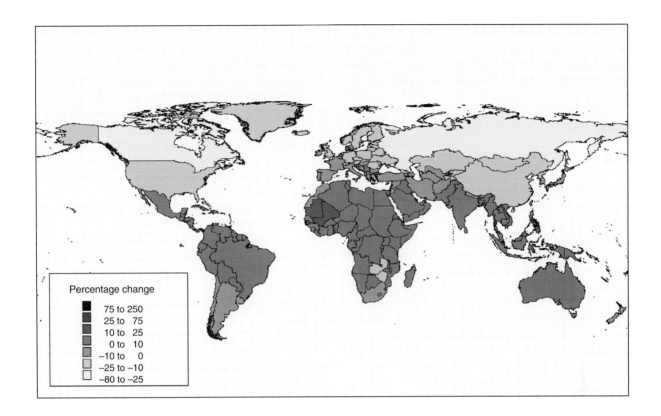

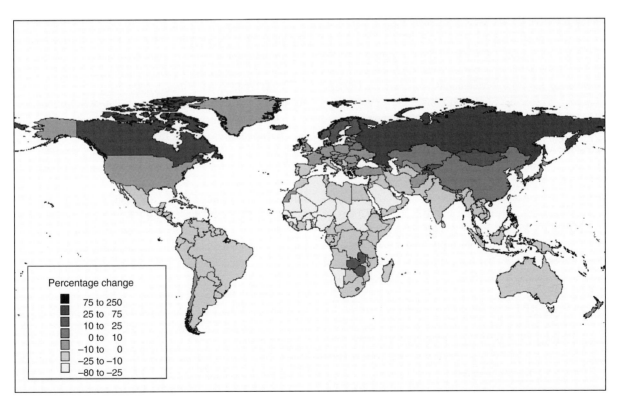

Figure 12.2 The change in departures (top) and arrivals (bottom) as a result of a 1 °C global warming in 2025.
Source: Hamilton *et al.* (2005a).

change, and would make the impact of climate change greater or smaller, but it is not known by how much. These and other questions will, we hope, be the subject of future research. Results so far indicate that such research would be fruitful and worthwhile.

Acknowledgements

The CEC DG Research through the DINAS-Coast project (EVK2–2000–22024), the US National Science Foundation through the Center for Integrated Study of the Human Dimensions of Global Change (SBR-9521914) and the Michael Otto Foundation for Environmental Protection provided welcome financial support. All errors and opinions are ours. The comments of two anonymous referees helped improve the paper. Parts of this paper were presented at various occasions, notably the Tenth EMF Workshop on Climate Change Impacts and Integrated Assessment in Snowmass. Attendees had many comments, some more relevant than others. John Weyant and Susan Sweeney deserve our deepest gratitude for a decade of excellent work. Financial support by the CEC DG Research through the DINAS-Coast project (EVK2–2000–22024) and by the Michael Otto Foundation for Environmental Protection is gratefully acknowledged.

References

Abegg, B. (1996). *Klimaänderung und Tourismus: Klimafolgenforschung am Beispiel des Wintertourismus in den Schweizer Alpen*. Zurich: vdf Hochschuleverlag an der ETH.

Agnew, M. D. and Palutikof, J. P. (2001). Climate impacts on the demand for tourism. In *International Society of Biometeorology, Proceedings of the First International Workshop on Climate, Tourism and Recreation*, ed. A. Matzarakis and C. de Freitas. Retrieved from www.mif.uni-freiburg.de/isb/ws/report.htm.

Amelung, B. and Viner, D. (in press). The vulnerability to climate change of the Mediterranean as a tourist destination. In *Climate Change and Tourism: Assessment and Coping Strategies*, ed. B. Amelung, K. Blazejczyk, A. Matzarakis and D. Viner. Dordrecht: Kluwer Academic Publishers.

Berritella, M., Bigano, A., Roson, R. and Tol, R. S. J. (2004). *A General Equilibrium Analysis of Climate Change Impacts on Tourism*. Research Unit Sustainability and Global Change Working Paper **FNU-49**. Hamburg: Hamburg University and Centre for Marine and Atmospheric Science; *Tourism Management*, in press.

Bigano, A., Hamilton, J. M. and Tol, R. S. J. (2004). *The Impact of Climate on Holiday Destination Choice*. Research Unit Sustainability and Global Change Working Paper **FNU-55**. Hamburg: Hamburg University and Centre for Marine and Atmospheric Science; *Climatic Change*, in press.

Brandenburg, C. and Arnberger, A. (2001). The influence of the weather upon recreation activities. In *International Society of Biometeorology Proceedings of the First International Workshop on Climate, Tourism and Recreation*, ed. A. Matzarakis and C. de Freitas. Retrieved from www.mif.uni-freiburg.de/isb/ws/report.htm.

Breiling, M. and Charamza, P. (1999). The impact of global warming on winter tourism and skiing: a regionalised model for Austrian snow conditions. *Regional Environmental Change* **1** (1), 4–14.

de Freitas, C. R. (1990). Recreation climate assessment. *International Journal of Climatology* **10**, 89–103.

de Freitas, C. R. (2003). Tourism climatology: evaluating environmental information for decision making and business planning in the recreation and tourism sector. *International Journal of Biometeorology* **48**, 45–54.

Dwyer, J. F. (1988). Predicting daily use of urban forest recreation sites. *Landscape and Urban Planning* **15** (1–2), 127–138.

Elsasser, H. and Messerli, P. (2001). The vulnerability of the snow industry in the Swiss Alps. *Journal of Mountain Research and Development* **21** (49), 335–339.

Englin, J. and Moeltner, K. (2004). The value of snowfall to skiers and boarders. *Environmental and Resource Economics* **29** (1), 123–136.

Gable, F. J. (1997). Climate change impacts on Caribbean coastal areas and tourism. *Journal of Coastal Research* **27**, 49–70.

Gössling, S. (2002). Global environmental consequences of tourism. *Global Environmental Change* **12**, 283–302.

Hamilton, J. M. (2003). *Climate and the Destination Choice of German Tourists*. Research Unit Sustainability and Global Change Working Paper **FNU-15** (revised). Hamburg: Centre for Marine and Climate Research, Hamburg University.

Hamilton, J. M. and Lau, M. A. (2005). The role of climate information in tourist destination choice decision making. In *Tourism and Global Environmental Change*, ed. S. Gössling and C. M. Hall. London: Routledge.

Hamilton, J. M., Maddison, D. J. and Tol, R. S. J. (2005a). Climate change and international tourism: a simulation study. *Global Environmental Change* **15** (3), 253–266.

Hamilton, J. M., Maddison, D. J. and Tol, R. S. J. (2005b): The effects of climate change on international tourism. *Climate Research* **29**, 255–268.

Harrison, S. J., Winterbottom, S. J. and Shephard, C. (1999). The potential effects of climate change on the Scottish tourist industry. *Tourism Management* **20**, 203–211.

Hu, Y. and Ritchie, J. R. B. (1993). Measuring destination attractiveness: a contextual approach. *Journal of Travel Research* **32** (2), 25–34.

IPCC (2000). *Emissions Scenarios: A Special Report on Working Group III of the Intergovernmental Panel on Climate Change*, ed. N. Nakicenovic and R. Swart. Cambridge University Press, Cambridge.

Krupp, C. (1997). *Klimaänderungen und die Folgen: eine exemplarische Fallstudie über die Möglichkeiten und Grenzen einer interdisziplinären Klimafolgenforschung*. Berlin: edition Sigma.

Kuoppamaeki, P. (1996). *Impacts of Climate Change from a Small Nordic Open Economy Perspective: The Potential Impacts of Climate Change on the Finnish Economy*. Helsinki: ETLA.

Lise, W. and Tol, R. S. J. (2002). Impact of climate on tourism demand. *Climatic Change* **55** (4), 429–449.

Lohmann, M. and Kaim, E. (1999). Weather and holiday destination preferences, image attitude and experience. *The Tourist Review* **2**, 54–64.

Loomis, J. B. and Crespi, J. (1999). Estimated effects of climate change on selected outdoor recreation activities in the United States. In *The Impact of Climate Change on the United States Economy*, ed. R. O. Mendelsohn and J. E. Neumann. Cambridge: Cambridge University Press, pp. 289–314.

Maddison, D. J. (2001). In search of warmer climates? The impact of climate change on flows of British tourists. *Climatic Change* **49**, 103–208.

Mendelsohn, R. and Markowski, M. (1999). The impact of climate change on outdoor recreation. In *The Impact of Climate Change on the United States Economy*, ed. R. O. Mendelsohn and J. E. Neumann. Cambridge: Cambridge University Press, pp. 267–288.

Nicholls, S. (2004). Climate change and tourism. *Annals of Tourism Research* **31** (1), 238–240.

Parker, P. (2001). *Physioeconomics: The Basis for Long-Run Economic Growth*. Cambridge: The MIT Press.

Perry, A. (2000). Impacts of climate change on tourism in the Mediterranean: adaptive responses. *Nota di Lavoro* **35.2000**. Milan: Fondazione Eni Enrico Mattei.

Piga, C. A. G. (2003). Pigouvian taxation in tourism. *Environmental and Resource Economics* **26**, 343–359.

Pike, S. (2002). Destination image analysis: a review of 142 papers from 1973 to 2000. *Tourism Management* **23**, 541–549.

Richardson, R. B. and Loomis, J. B. (2004). Adaptive recreation planning and climate change: a contingent visitation approach. *Ecological Economics* **50** (1–2), 83–99.

Scott, D. and McBoyle, G. (2001). Using a "Tourism Climate Index" to examine the implications of climate change for Climate as a Tourism Resource. In *International Society of Biometeorology Proceedings of the First International Workshop on Climate, Tourism and Recreation*, ed. A. Matzarakis and C. de Freitas. Retrieved from www.mif.uni-freiburg.de/isb/ws/report.htm.

Scott, D., McBoyle, G., Mills, B. and Wall, G. (2001). Assessing the vulnerability of the alpine skiing industry in Lakelands Tourism Region of Ontario, Canada to climate variability and change. In *International Society of Biometeorology Proceedings of the First International Workshop on Climate, Tourism and Recreation*, ed. A. Matzarakis and C. de Freitas. Retrieved from www.mif.uni-freiburg.de/isb/ws/report.htm.

Smith, J. B., Schellnhuber, H.-J., Mirza, M. M. Q. *et al.* (2001). Vulnerability to climate change and reasons for concern: a synthesis. In *Climate Change 2001: Impacts, Adaptation, and Vulnerability*, ed. J. J. McCarthy, O. F. Canziani, N. A. Leary, D. J. Dokken and K. S. White. Cambridge: Cambridge University Press, pp. 913–967.

Thorsson, S., Lindqvist, M. and Lindqvist, S. (2004). Thermal bioclimatic conditions and patterns of behaviour in an urban park in Göteborg, Sweden. *International Journal of Biometeorology* **48**, 149–156.

Viner, D. and Agnew, M. (1999). *Climate Change and its Impacts on Tourism*. Godalming: WWF-UK.

Witt, S. F. and Witt, C. A. (1995). Forecasting tourism demand: a review of empirical research, *International Journal of Forecasting* **11**, 447–475.

WTO (2003). *Yearbook of Tourism Statistics*. Madrid: World Tourism Organization.

13

Using adaptive capacity to gain access to the decision-intensive ministries

Gary W. Yohe

13.1 Introduction

The Conference of the Parties (COP) of the Framework Convention on Climate Change (UNFCCC) has begun to focus its attention on establishing mechanisms by which the incremental costs of adaptation to long-term climate change by developing countries might be supported by developed countries through their contributions to one or more adaptation funds. Meanwhile, the Intergovernmental Panel on Climate Change (IPCC), for the Fourth Assessment Report (AR4), has continued its interest in the interface between adaptation to climate change and climate variability, on the one hand, and sustainable development, on the other. Notwithstanding significant effort to promote cross-fertilization across these interests, it is not obvious that either process will, without some guidance, allow climate issues to gain access to the deliberations that are conducted behind the closed and sometimes locked doors of what might be termed the "decision-intensive ministries" – the ministries within which development planning is conducted and by which development policies are implemented. This short paper offers some thoughts (hypotheses, really) about how the climate community might use the emerging links between researchers' understandings of how adaptations work and practitioners' understandings of how attractive development might be promoted.

Of course, it is not the case of that decisionmakers concerned with development issues do not recognize the possibility that climate change may cause harm over the long term. The doors are closed primarily because the decisionmakers already have full plates. They worry about how to promote economic growth and productivity gains. They worry about the implications of lowering trade barriers. They are concerned about equity issues and the distribution of resources across their societies. They focus on better provision of health services and education. They try to take sustainability into account, but their primary objectives involve promoting as much near-term progress as possible. It is no wonder that the long-term implications of climate change seldom appear high on the list of the multiple stresses to which they must respond. Nonetheless, planners ignore climate at their own risk, especially when long-term investment decisions can lock their economies into specific development trajectories from which it could be expensive to deviate.

Section 13.2 begins with a brief review of the conclusions about adaptation offered in the Third Assessment Report (TAR) of the IPCC and the degree to which its emphasis on site-specific and path-dependent factors continues to hold. The fundamental lesson for the research community is that the answer to a question like "Will this adaptation work here or there?" is an empirical issue – sometimes it will and sometimes it will not. An equally brief review of some significant contributions to the recent economics literature, described in Section 13.3 because they focus on issues of interest in the decision-intensive ministries, suggests that this sort of "it depends" conclusion is nothing new to them. Moreover, Section 13.3 notes that the factors that determine their questions about whether a particular policy or program will promote economic growth or reduce poverty (e.g.) are the same factors upon which the relative efficacy of adaptation depends. This convergence of experience across researchers and practitioners is then used, in Section 13.4, to suggest

Human-induced Climate Change: An Interdisciplinary Assessment, ed. Michael Schlesinger, Haroon Kheshgi, Joel Smith, Francisco de la Chesnaye, John M. Reilly, Tom Wilson and Charles Kolstad. Published by Cambridge University Press. © Cambridge University Press 2007.

strategies for opening the doors through which climate issues must pass if they are to become part of mainstream development planning. Some concluding remarks try to provide some context to the overall discussion.

13.2 The state of knowledge about adaptation in 2004

The authors of Chapter 18, in their contribution of Working Group II to the Third Assessment Report of the IPCC (2001), included four fundamental insights among the points that they wanted to emphasize.

1. The vulnerability of any system to an external stress (or collection of stresses) is a function of exposure, sensitivity, and adaptive capacity.
2. Human and natural systems tend to adapt autonomously to gradual change and to change in variability.
3. Human systems can also plan and implement adaptation strategies in an effort to reduce potential vulnerability or exploit emerging opportunities even further.
4. The economic cost of vulnerability to an external stress is the sum of the incremental cost of adaptation plus any residual damages that cannot be avoided.

The Chapter 18 writing team was also careful to note that even systems which might face similar climate-induced stresses would, by virtue of their location and their level of development, confront the future manifestations of climate change from extraordinarily dissimilar socio-economic circumstances.

In addition to this now obvious diversity in socio-economic context, the TAR also recognized that any system's environment varies *idiosyncratically* from day to day, month to month, year to year, and decade to decade (see Mearns *et al.* [1997] or Karl and Knight [1998]). It follows that changes in the mean conditions that define those environments could actually be experienced most noticeably through changes in the nature and/or frequency of variable conditions that materialize across short timescales, and that adaptation necessarily involves reaction to this sort of variability. This is the fundamental point in Hewitt and Burton (1971), Yohe *et al.* (1996), Downing (1996), and Yohe and Schlesinger (1998). Some researchers, such as Smithers and Smit (1997), Smit *et al.* (1999, 2000) and Downing *et al.* (1997), have used the concept of "hazard" to capture these sorts of stimuli, and have claimed that adaptation is warranted whenever either changes in mean conditions or changes in variability have significant consequences. For most systems, though, changes in mean conditions over short periods of time fall within a "coping range" – a range of circumstances within which, by virtue of the underlying resilience of the system, significant consequences are not observed (see Downing *et al.* [1997] or Pittock and Jones [2000]). There are limits to resilience for even the most robust of systems, of course. It is therefore critically important to understand the boundaries of systems' resilience; how, exactly, are the thresholds determined beyond which the consequences of experienced conditions become significant?

Some of these critical boundaries are determined by physical properties, of course; but others are determined by socio-economic context and social preferences. Even across this bifurcation, the first TAR conclusion listed above has become a strong foundation from which to approach vulnerability analyses across a multitude of contexts. More specifically, adopting a slightly different emphasis provides the insight that any system's vulnerability to climate change and climate variability will be determined not only by *its exposure to the impacts of climate and its baseline sensitivity to those impacts*, but also by *its ability to cope with new sources of stress – i.e., its adaptive capacity*. This evolving approach exploits its recognition that all three of these factors, but perhaps most fundamentally the role of adaptive capacity in defining socio-economic thresholds of tolerance to climate-related stress, clearly depend on path-dependent and site-specific circumstances. To sort through the implications of this insight, Yohe and Tol (2002) have suggested that the determinants of adaptive capacity include:

1. the range of available technological options for adaptation;
2. the availability of resources and their distribution across the population;
3. the structure of critical institutions, the derivative allocation of decisionmaking authority, and the decision criteria that would be employed;
4. the stock of human capital including education and personal security;
5. the stock of social capital including the definition of property rights;
6. the system's access to risk-spreading processes;
7. the ability of decisionmakers to manage information, the processes by which these decisionmakers determine which information is credible, and the credibility of the decisionmakers themselves; and
8. the public's perceived attribution of the source of stress and the significance of exposure to its local manifestations.

This second-tier list of critical factors identifies some of the fundamental sources of diversity across paths and locations, and so it reinforces perhaps the most important conclusion from Chapter 18 of TAR:

Current knowledge of adaptation and adaptive capacity is insufficient for reliable prediction of adaptations; it also is insufficient for rigorous evaluation of planned adaptation options, measures and policies of governments. [...] Given the scope and variety of specific adaptation options across sectors, individuals, communities and locations, as well as the variety of participants – public and private – involved in most adaptation initiatives, it is probably infeasible to systematically evaluate lists of adaptation measures; improving and applying knowledge on the constraints and opportunities for enhancing adaptive capacity is necessary to reduce vulnerabilities associated with climate change. IPCC (2001), p. 880

This is not to say that all is lost. The take-home message is simply that research has a long way to go if it is to come to grips with the diversity of the socio-political-economic environments that produce wide ranges of sensitivities and imply enormous variances in adaptive capacity.

Adger and Vincent (2004), in a contribution to the IPCC Expert Meeting on Uncertainty held in Maynooth, Ireland, in the spring of 2004, take this warning to the next step. They observed that uncertainty is pervasive and that adaptive capacity essentially describes the adaptation space within which decisionmakers might find feasible adaptation options. They continued to argue that diversity in context makes it easier to anticipate change in generic adaptive capacity than adaptation, per se, so that linking the determinants of adaptive capacity to drivers and therefore perhaps to the available policy levers can help explain why the "magic" works sometimes in some places, but not at other times in other places. Their argument conforms well with a "weakest link" hypothesis authored earlier by Yohe and Tol (2002): the overall capacity of a system to adapt to an external stress (be it climate-related or not), is a function of the weakest of the underlying determinants of adaptive capacity.

It follows that the question of whether adaptation X will work in place Y at time T is largely an empirical one; and the determinants of adaptive capacity provide researchers and decisionmakers alike with a list of factors to consider in their analyses of exactly why. Local decisionmakers will have the best information about what will or will not work, to be sure; but the research community can, by using the underlying determinants of adaptive capacity to organize their thoughts, find common lessons across a wide range of locations and contexts. They can even discover attractive links between the determinants of success in coping with climate change and climate variability, on the one hand, and success in achieving other policy goals (such as promoting sustainable development), on the other.

13.3 Some insights from the economics literature

To illustrate this point more explicitly, consider the literature examining the link between economic policy levers (such as opening an economy more completely to international trade) and domestic planning and policy objectives (such as increased productivity growth, improved general welfare in the short and long term, and reduced poverty). This literature has shown repeatedly that the answer to the question of "What works where?" in an economic development arena is also essentially empirical. It also suggests strongly that the determinants of success or failure in these areas map well onto the determinants of adaptive capacity recorded above. Finally, many studies which examine the relative efficacy of various economic interventions have confirmed, in entirely different contexts, strong variants of the "weakest link" hypothesis.

Lucas (1988), for example, argued in a widely cited paper that human capital externalities are large enough to explain differences between the long-run growth rates of poor and rich countries. Moretti (2004) built on the work of Lucas, as well as the contributions of others who struggled with some significant statistical problems, to concentrate attention on the productivity spillovers that can be expected from human capital. He hypothesized that these spillovers, if they existed at all, would make manufacturing plants located in cities with higher levels of human capital more productive in the sense of producing greater output from the same inputs. His hypothesis was confirmed empirically when he showed that plants located in US cities where the fraction of college graduates grew faster experienced larger increases in productivity and correspondingly larger increases in wages.

Guiso et al. (2004) expanded the scope of analysis when they explored the role of social capital in supporting the successful application of financial structures. Conducting empirical analyses on data compiled in Italy, they found strong evidence that social capital matters most when education levels are low and law enforcement is weak. Recognizing that their results were site-specific and path-dependent, they also wondered whether or not their results would apply to developing countries that were plagued by both low levels of human capital and diminished stocks of social capital. When they focused their attention on interaction effects, they noted that trust (the component of social capital that they could quantify) was much less important in regions where the court system was more efficient or when people were more educated. Since they argue that neither characteristic prevails across much of the developing world, they conclude that social capital is "to be very important in explaining the success (or lack thereof) of developing countries" (p. 553).

Meanwhile, Rozelle and Swinnen (2004) looked across transition experiences of central European countries from the former Soviet Union. They observed that countries which grew steadily a decade or more after implementing their economic reforms had supported the reforms by creating macroeconomic stability, reforming property rights, hardening budget constraints, *and* creating institutions that facilitate exchange and develop an environment within which contracts can be enforced and new firms can enter. Order and timing did not matter, but success depended on meeting all of these underlying objectives at some point in the transition – a clear manifestation of what could be deemed a variant of the previously described "weakest link" hypothesis.

Finally, Winters et al. (2004) reviewed a long literature from the past three decades that explores the likelihood that trade liberalization can reduce poverty. This literature is littered with contradictory conclusions and statistical problems, but these authors concluded that a positive effect depends critically upon the existence and stability of markets, the ability of actors to handle changes in risk, access to technology, access to resources, competent and honest government, *and* policies that promote conflict resolution and promote human capital accumulation. The match between this list of characteristics required for success in promoting long-term growth, site-specific productivity gains, and improved equity (all concerns of the denizens of the decision-intensive ministries) and the determinants of adaptive capacity inspired by the TAR is strong. Both include references to strong and skilled governance, appropriate distributions of resources and access to resources, strong stocks of human capital, and overall stability. Just as in the climate arena, whether or not the links between an economic intervention (or an adaptation) and its desired outcomes are strong, weak, or actually run in a direction that is opposite to that predicted by theory or process analysis was found to be essentially an empirical question in nearly every instance.

13.4 Opening the doors to the decision-intensive ministries

A number of possible keys to gain access to the decision-intensive ministries can now be identified even though climate change may not be a fundamental concern in their deliberations (recall that these are the ministries within which development planning is conducted and by which development policies are implemented). These keys do not rely on the elevation of climate change in the list of stresses to which these ministries must respond. They depend, instead, on a commonality of underlying determinants for success – success in promoting the minister's objectives, to be sure, but also success in promoting the ability to cope with climate change and climate variability.

First of all, the precursors of sustained support of economic growth and improved well-being match the determinants of adaptive capacity quite well. The decision-intensive ministries are already familiar with these precursors, and they are already concerned with seeking ways of strengthened the "weakest links" that support the connections between policy implementation and success. The first key to bringing climate into their agendas is simply to convince decisionmakers in the development ministries that they are already working on these problems. Indeed, recognizing climate could provide them more ammunition when they negotiate for claims to scarce economic resources.

Second, the complexities of trying to predict what will work and what will not is an empirical issue in both contexts, but the critical ministries already have experience in coping with this complexity. Preparing and planning for adaptation by strengthening the determinants of adaptive capacity can simultaneously work as a hedge against climate impacts and as a means of improving prospects for sustainable development by supporting (for example) productivity growth (or at least adding to the insulation that protects productivity initiatives from external stress). Cast as a risk-reducing tool, improving adaptive capacity can also be seen as a tool to complement mitigation. This improves stability, and that improves productivity growth by making investment more attractive.

Finally, preparing for negotiations within the COP about adaptation and accessing the adaptation funds will require a thorough understanding of the state of the art of adaptation and the sensitivity of outcomes to the underlying determinants. The carrot of international support for adaptation efforts that will also promote growth is an incentive to be proactive in understanding how the empirical analyses will play out "in country" and what they mean for negotiations conducted at the highest levels of government.

13.5 Concluding remarks

None of the keys noted in Section 13.4 has been explored completely, at this point; but the evidence is certainly there to support a more thorough investigation of the associations on which they rely. Uncertainty in our ability to predict what will work, where, and when in our response to the climate problem is an empirical issue for which a significant number of case studies scattered across locations and sectors that span the variance of critical drivers will be required. This complication should not, however, discourage the attention of the decision-intensive ministries. They already know that the effectiveness of the policies that they contemplate all the time, such as opening trade or imposing environmental restrictions on industrial activity, may or may not work to increase productivity, improve general welfare (including equity considerations), or reduce poverty in a specific sector or across a specific region. Determining which of these policies will work is the equivalent empirical question with which they have some familiarity. Indeed, noting that the determinants of these more mainstream "adaptations" to other external stresses are the same as those for the capacity to adapt to climate stress suggests that they are already confronting exactly the same empirical question. Climate, therefore, is not a new issue to be added to an already clogged agenda. It is, instead, an additional incentive for the careful examination of how and why policies designed to promote productivity in an interdependent world might function.

Carefully designing the criteria by which applications to the various adaptation funds will be evaluated can open the doors to the critical ministries more quickly by offering another source of support for their initiatives. It is here, by suggesting ways of "mainstreaming" responses to climate risks into the development process, that the coincidence of underlying determinants can suggest how the global community might respond to calls by Ian Burton (2004) and others to close the "adaptation gap."

Acknowledgements

The author, while laying claim to all remaining errors, gratefully acknowledges the contributions of Joel Smith, two anonymous referees, and all of the participants in the 2004 Snowmass sessions. Their comments added focus to the arguments presented here so that a relatively simple story could be told without being cluttered by extraneous comments.

References

Adger, N. and Vincent, K. (2004). Uncertainty in adaptive capacity. In the Proceedings of an IPCC Expert Meeting on *Describing Scientific Uncertainty in Climate Change to Support Analysis of Risk and of Options* (Maynooth, Ireland). Boulder, CO: IPCC Working Group I Technical Support Unit, pp. 57–59.

Burton, I. (2004). Climate change and the adaptation deficit. In *Building the Adaptive Capacity*, ed. A. Fenech, D. MacIver, H. Auld, R. Bing Rong and Y. Yin. Toronto: Environment Canada, pp. 25–33.

Downing, T. E. (ed.) (1996). *Climate Change and World Food Security*. Berlin: Springer.

Downing, T. E., Ringius, L., Hulme, M. and Waughray, D. (1997). Adapting to climate change in Africa. *Mitigation and Adaptation Strategies for Global Change* **2**, 19–44.

Guiso, L., Sapienza, P. and Zingales, L. (2004). The role of social capital in financial development. *American Economic Review* **94**, 526–556.

Hewitt, J. and Burton, I. (1971). *The Hazardousness of a Place: A Regional Ecology of Damaging Events*. University of Toronto.

IPCC (2001). *Climate Change 2001: Impacts, Adaptation, and Vulnerability: Contribution of Working Group II to the Third Assessment Report of the Intergovernmental Panel on Climate Change*, ed. J. J. McCarthy, O. F. Canziani, N. A. Leary, D. J. Dokken and K. S. White. Cambridge: Cambridge University Press.

Karl, T. R. and Knight, R. W. (1998). Secular trends of precipitation amount, frequency and intensity in the United States. *Bulletin of the American Meteorological Society* **79**, 231–241.

Lucas, R. E. (1988). On the mechanics of economic development. *Journal of Monetary Economics* **22**, 3–42.

Mearns, L. O., Rosenzweig, C. and Goldberg, R. (1997). Mean and variance change in climate scenarios: methods, agricultural applications and measures of uncertainty. *Climatic Change* **34**, 367–396.

Moretti, E. (2004). Workers' education, spillovers, and productivity: evidence from plant-level production functions. *American Economic Review* **94**, 656–690.

Pittock, B. and Jones, R. N. (2000). Adaptation to what and why? *Environmental Monitoring and Assessment* **61**, 9–35.

Rozelle, S. and Swinnen, J. F. M. (2004). Success and failure of reform: insights from the transition of agriculture. *Journal of Economic Literature* **42**, 433–458.

Smit, B., Burton, I., Klein, R. J. T. and Street, R. (1999). The science of adaptation: a framework for assessment. *Mitigation and Adaptation Strategies for Global Change* **4**, 199–213.

Smit, B., Burton, I., Klein, R. J. T. and Wandel, J. (2000). An anatomy of adaptation to climate change and variability. *Climatic Change* **45**, 223–251.

Smithers, J. and Smit, B. (1997). Human adaptation to climatic variability and change. *Global Environmental Change* **7**, 129–146.

Winters, L. A., McCulloch, N. and McKay, A. (2004). Trade liberalization and poverty: the evidence so far. *Journal of Economic Literature* **42**, 72–115.

Yohe, G. and Schlesinger, M. E. (1998). Sea-level change: the expected economic cost of protection or abandonment in the United States. *Climatic Change* **38**, 437–472.

Yohe, G. and Tol, R. (2002). Indicators for social and economic coping capacity: moving toward a working definition of adaptive capacity. *Global Environmental Change* **12**, 25–40.

Yohe, G., Neumann, J., Marshall, P. and Amaden, H. (1996). The economic cost of greenhouse-induced sea-level rise for developed property in the United States. *Climatic Change* **32**, 387–410.

14

The impacts of climate change on Africa

Robert Mendelsohn

14.1 Background

Many scientists, economists, and policymakers agree that the world is facing a threat from climate warming (IPCC, 2001a). The degree of the impact and its distribution is still debated. The current evidence suggests that countries in temperate and polar locations may benefit from small economic advantages because additional warming will benefit their agricultural sectors (Mendelsohn et al., 2000; Mendelsohn and Williams, 2004). Many countries in tropical and subtropical regions are expected to be more vulnerable because additional warming will affect their marginal water balance and harm their agricultural sectors (IPCC, 2001b; Tol, 2002; Mendelsohn and Williams, 2004). However, little empirical research has been done on tropical countries, so that little is known about the extent of these damages. The problem is expected to be most severe in Africa, and especially sub-Saharan Africa, where the people are poor, temperatures are high, precipitation is low, technological change has been slow, and agriculture dominates domestic economies. African farmers have adapted to their current climates, but climate change may well force large regions of marginal agriculture out of production in Africa.

The agriculture sector is a major contributor to the current economy of most African countries. Across the continent, agriculture averages 21% of gross domestic product (GDP) but this value ranges from 10% to 70% of the GDP of individual countries (Mendelsohn et al., 2000). Future development is likely to reduce agriculture's share of GDP. With an optimistic forecast of future development, agriculture's share of GDP in 2100 could shrink to as little as 3% across the continent and to 8% across sub-Saharan countries. However, if Africa's future development path is not so bright, agriculture will continue to play a very large role. Because agriculture is one of the most vulnerable sectors to climate change, this suggests that the African economy will continue to be vulnerable to warming.

Even without climate change, there are serious concerns about agriculture in Africa because of water supply variability, soil degradation, and recurring drought events. A number of countries in Africa currently face semi-arid conditions that make agriculture challenging. Further, efforts to increase agricultural productivity in Africa have been particularly difficult to sustain. Over the past 40 years, sub-Saharan African agriculture has the slowest record of productivity increase in the world (Evenson and Gollin, 2003).

Agronomic experts are concerned that the agriculture sector in Africa will be especially sensitive to future climate change and any increase in climate variability (Rosenzweig and Parry, 1994). The current climate is already marginal with respect to precipitation in many parts of Africa. Further warming in these semi-arid locations is likely to be devastating to agriculture there. Even in the moist tropics, increased heat is expected to reduce crop yields. Agronomic studies suggest that yields could fall dramatically in the absence of costly adaptation measures. Given that the current farming technology is basic, and that incomes are low, farmers will have few options to adapt. At present, public infrastructure such as roads, long-term weather forecasts, and agricultural research and extension are inadequate to secure appropriate adaptation. Unfortunately, none of the empirical studies of climate impacts in Africa have explored what adaptations would be efficient for either African farmers or African governments.

Human-induced Climate Change: An Interdisciplinary Assessment, ed. Michael Schlesinger, Haroon Kheshgi, Joel Smith, Francisco de la Chesnaye, John M. Reilly, Tom Wilson and Charles Kolstad. Published by Cambridge University Press. © Cambridge University Press 2007.

Table 14.1 Climate predictions by model for 2100.

Region	Observed		PCM		HAD3		CCC	
	Temp (°C)	Precipitation (cm/mo)	ΔTemp (°C)	% Prec	ΔTemp (°C)	% Prec	ΔTemp (°C)	% Prec
World	18.5	9.5	+2.46	+15.5	+4.60	+3.2	+7.11	+4.8
Africa	22.8	8.1	+2.26	+11.9	+4.41	+4.1	+8.07	−15.7

Although there are well-established concerns about climate change effects in Africa, there is little quantitative information concerning how serious these effects will be. Existing studies cover only a small fraction of Africa, and few of the African studies include data of actual farmer behavior, specifically how farmers might adapt by changing planting dates, harvest dates, use of fertilizer, and crop choice. Existing studies mostly examine how individual crops, usually grain crops, behave in controlled experiments.

14.2 The analytical framework

In order to better understand the magnitude of the potential effects of climate change on African agriculture, we developed a simulation, using an IPCC forecast of future atmospheric carbon dioxide levels by 2100. Three alternative climate models are used to predict the climates in each country: the Parallel Climate Model (PCM) (Washington *et al.*, 2000), Hadley (HAD3) (Gordon *et al.*, 2000), and Canadian Climate Model (CCC) (Boer *et al.*, 2000). These specific scenarios were chosen because they provide a wide range of predicted climate changes. Table 14.1 presents the global and African predictions for 2100 of each model for temperature and precipitation. PCM has the mildest predictions of climate change whereas CCC has the most severe.

The climate changes were then evaluated with an impact model, GIM (Global Impact Model, V.2.5) (Mendelsohn *et al.*, 2000; Mendelsohn and Williams, 2004). GIM begins with the climate forecasts and then calculates impacts in each market sector using climate response functions (Mendelsohn and Schlesinger, 1999). There are two climate response functions that roughly correspond to experimental versus cross-sectional data analysis. In the case of agriculture, experimental data has been built from crop research in laboratories and fields. This in turn has led to crop simulation models that predict how individual crops will behave in different climate conditions. The most sophisticated of these models, in turn, have an economic component that describes how farmers will change their behavior as conditions change. The cross-sectional analysis, in contrast, examines actual outcomes across climate conditions. The actual outcomes already reflect adjustments or adaptations that farmers have made to different climate conditions.

The experimental response function tends to be more climate-sensitive than the cross-sectional response function, because it has historically underestimated climate adaptation. The response function consequently predicts higher damages as climate warms. Of course, it is also possible that the cross-sectional response function is in error as well because of unobserved variables not taken into account in the analysis. We present the results of both climate response functions in order to bracket the magnitude of possible impacts.

GIM is a global model that examines the impact on all market sectors of the economy. The literature has identified five market sectors that are sensitive to climate: agriculture, forestry, energy, water, and coasts (IPCC, 2001b; Mendelsohn *et al.*, 2004; Pearce *et al.*, 1996; Tol, 2002). In this paper, we focus on Africa alone. Using available data on cropland and other relevant features of each sector in each country, we predict how the tentative change in climate will affect each country. Non-market effects are not measured.

The analysis does not look at global price changes. In principle, climate could alter the global supply of food or timber sufficiently to change prices. For example, global warming is expected to have a beneficial effect on forests which would translate into lower timber prices (Sohngen *et al.*, 2002). However, a parallel impact on food is not expected, as the reduction in supply in some regions is offset by increases in other regions (Reilly *et al.*, 1996). For simplicity, we consequently assume prices are unchanged.

14.3 Results

In Table 14.2 we examine the predicted effects for Africa for each sector given each of the climate scenarios for 2100. The impacts are sensitive to both temperature and precipitation. However, in general, the impacts become more severe as the temperature increases. Further, the high sensitivity estimates lead to much larger impacts than the low sensitivity estimates. In fact, for the mild PCM scenario, the low sensitivity climate impacts are mildly beneficial to Africa. That is, the precipitation gains under PCM are sufficient to outweigh the warming according to the cross-sectional analysis. With the experimental response function, even the mild PCM scenario is predicted to generate damages of 17.2 billion USD for all of Africa.

As temperatures warm, however, damages rise quickly. With the high sensitivity climate response function, a large fraction of Africa's agricultural GDP is lost under the warmer scenarios. With CCC, total damages are predicted to rise to 187 billion USD for Africa, of which 179 billion USD is in agriculture. The low sensitivity response function yields much lower damage estimates, but it still predicts damages as high

The impacts of climate change on Africa

Table 14.2 African climate impacts by economic sector for alternative climate scenarios and response functions in 2100 (billion USD/yr).

Sector	PCM		HAD3		CCC	
	High response	Low response	High response	Low response	High response	Low response
Total	−17.2	−0.2	−100.7	−8.4	−187.1	−21.4
	(−0.8%)	(−0.0%)	(−1.0%)	(−0.1%)	(−1.9%)	(−0.2%)
Agriculture	−14.9	1.8	−95.4	−5.9	−179.0	−11.5
	(−4.5%)	(−0.1%)	(−28.6%)	(−1.8%)	(−53.7)	(3.4%)
Forestry	0.3	−0.2	0.3	−1.0	+0.2	−2.8
	(−0.9%)	(−0.7%)	(−0.9%)	(−3.6%)	(−0.7%)	(−7.0%)
Energy	−2.3	−1.5	−4.6	−3.1	−7.0	−6.3
	(−2.4%)	(−1.6%)	(−4.8%)	(−3.2%)	(−7.1%)	(−6.4%)
Water	−0.4	−0.2	−0.9	−0.5	−1.3	−0.8
	(−15.8%)	(−7.9%)	(−33.7%)	(−18.7%)	(−48.8%)	(−30.0%)
Coastal	−0.0	−0.0	−0.0	−0.0	−0.01	−0.0

Values in parentheses are percentage changes in that sector. No percentage values are available for the coast. Coastal impacts are projected to be less than $100 million per year because of expected widespread use of retreat and the loss of only low-value land. GDP growth rate is assumed to be 3% per year.

as 21 billion USD for CCC for Africa. Although these are very large numbers, if Africa enjoys a 3% growth rate over the next century, the impacts will still be a small percentage of GDP. With growth, the worst case scenario mentioned above would be just 2% of GDP.

With the mild scenario, the harmful effects of higher temperatures are balanced by the beneficial effects of higher CO_2, yielding small impacts in agriculture. However, with the warmer scenarios, damages to agriculture begin to dominate. This is especially true for the high response scenarios. Because the impacts to agriculture explain so much about what will happen to Africa, we focus on the agricultural impacts in the remainder of this paper.

One of the important results from the simulation is that effects are likely to be different across the African continent. The initial climate conditions are quite different as precipitation varies a great deal across sub-regions. Central Africa is generally very moist, as are parts of West Africa, whereas the rest of Africa is mostly semi-arid to arid. Initial temperatures also vary. West Africa, the Sahara, and East Africa (except for high elevation areas in Kenya and Ethiopia) are the warmest regions. In contrast, the Mediterranean coast of North Africa, the high elevation region in Kenya and Ethiopia, and non-desert areas in Southern Africa are more temperate. These temperate areas are particularly important because they are the most productive agricultural areas in Africa. Finally, economic conditions vary significantly across Africa. Currently in northern and southern Africa, agriculture represents only 16% and 8%, respectively, of GDP, whereas in the other African regions, agriculture counts for 30 to 43% of GDP.

The impacts to each region depend on the climate scenario, as shown in Table 14.3. Examining the high response results across scenarios reveals that with the relatively mild and wet PCM scenario, agriculture in the temperate southern region is barely affected and agriculture in the eastern region actually benefits from climate change. Agriculture in the Saharan and western regions of Africa endures the largest losses both in absolute terms and as a percentage of GDP. With the more severe Hadley scenario, agriculture in every region suffers losses of about one-quarter of agricultural GDP except for the Saharan region which loses three-quarters of its agricultural GDP. Finally, with the very severe Canadian forecast, the losses in every region are larger. The Saharan region loses all of its agriculture. The north and south lose less than other regions because of their initial moderate temperatures. Overall over half of African agriculture is lost.

The impacts are not at all as severe with the low response function. For the mild scenarios, there are benefits across Africa except in the western and Saharan regions. Farming in the south, in particular, does very well with the PCM scenario. With the Hadley scenario, impacts to all regions except the south are harmful. As a percentage of agricultural GDP, however, these harmful impacts remain below 3%. Only with the CCC scenario does one begin to see noticeable damage according to the low response model. With the hot and dry CCC scenario, agriculture in the Sahara, south, and central regions is hard hit and farming in the whole continent falls by 3.5%.

The overall impact to GDP or income depends not only on the sectors that are affected by climate but also on the size of the non-agricultural sector. Table 14.4 presents two economic development scenarios. In one scenario, Africa stays the way it is now. There is no growth in GDP. Under the growth scenario, overall GDP is projected to increase at 3% a year and agricultural GDP is projected to grow at 1.5% a year through 2100. Although these rates of growth may seem low for a developing country, sub-Saharan Africa has seen effectively no growth over the past 20 years. The two scenarios are intended to provide a sense of how the impact of climate change depends on the growth of the non-agricultural economy. If the economy grows quickly, the climate impacts will

Table 14.3 Impact of climate change on agriculture in African regions in 2100 (billion USD/yr).

Regions	PCM		HAD3		CCC	
	High response	Low response	High response	Low response	High response	Low response
West	−8.9	−0.2	−31.9	−1.1	−72.2	−2.1
	(−7.9%)	(−0.2%)	(−28.4%)	(−1.0%)	(−71.3%)	(−1.9%)
East	+5.6	+0.1	−14.7	−0.7	−33.0	−1.0
	(+7.6%)	(+0.1%)	(−19.6%)	(−0.9%)	(−44.2%)	(−1.4%)
South	−0.7	+1.0	−11.4	+0.2	−16.2	−2.3
	(−1.5%)	(+2.3%)	(−25.4%)	(+0.5%)	(−36.2%)	(−5.2%)
West-Central	−2.0	0.4	−11.8	−0.6	−25.9	−2.0
	(−4.2%)	(+0.8%)	(−24.9%)	(−1.3%)	(−54.4%)	(−4.1%)
North	−3.0	0.7	−10.4	−1.2	−12.0	−3.8
	(−7.6%)	(+1.3%)	(−26.2%)	(−2.9%)	(−30.0%)	(−9.6%)
Sahara	−6.0	−0.1	−15.1	−0.1	−19.8	−0.3
	(−30.3%)	(−0.9%)	(−76.2%)	(−0.1%)	(−100.0%)	(−1.9%)
Total	−14.7	+1.0	−95.4	−2.6	−179.0	−11.5
	(−4.4%)	(+0.6%)	(−28.6%)	(−1.2%)	(−53.7%)	(−3.5%)

Impacts as percentage of agricultural GDP in brackets. Agricultural GDP growth rate is assumed to be 1.5% per year.

Table 14.4 Effect of GDP growth rates on total climate impact per GDP across African regions.

Regions	Growth rate	PCM	HAD3	CCC
West	3%	−0.6%	−2.2%	−5.0%
	0%	−13.5%	−31.2%	−35.9%
East	3%	+0.7%	−1.9%	−4.2%
	0%	+17.7%	−29.4%	−41.7%
South	3%	−0.1%	−0.5%	−0.7%
	0%	−0.7%	−7.8%	−8.3%
Central	3%	−0.3%	−1.5%	−3.2%
	0%	−5.9%	−23.5%	−26.0%
North	3%	−0.1%	−0.3%	−0.4%
	0%	−2.2%	−6.8%	−7.8%
Sahara	3%	−1.2%	−3.1%	−4.1%
	0%	−22.1%	−37.2%	−37.9%
Total	3%	−0.2%	−1.0%	−1.9%
	0%	−3.2%	−15.6%	−18.0%

Impacts measured in terms of %GDP.

be a small fraction of overall GDP. Even with the CCC scenarios and high response function, impacts are only 2% for Africa and no more than 5% for any region with growth. However, if the African economy stays the way it is now, the impacts will be 18% for all of Africa, and the west, east, and Sahara regions will see losses equal to over one-third of their economies.

The impacts on individual countries vary over an even wider range (Table 14.5). Table 14.5 presents the impacts in each country in each climate scenario with the high climate response function and 3% growth scenario. The values of agriculture and overall GDP in 2100 are also shown given the growth rate assumptions in this scenario. One of the striking results in Table 14.5 is that some countries will lose their agricultural sectors entirely. Under the HAD3 scenario, Botswana, Sudan, and Zambia lose their agricultural sectors. Of course, some isolated pockets of agriculture would probably survive in these countries, but the model is projecting that there would be very little GDP left. With the CCC scenario, the list of countries with no agricultural GDP expands to Botswana, Burkina Faso, Central African Republic, Chad, Gambia, Guinea-Bissau, Niger, Nigeria, Sudan, Togo, and Zambia.

14.4 Conclusion

Across the different climate projections, different response functions, and different growth rates, there is a wide range of projections of what might happen to Africa. Although Africa could mildly benefit from warming in the best of circumstances, most of the projections imply that African agriculture will be vulnerable to climate change. Future outcomes may not be as severe as the most pessimistic projection, but it is likely that climate change will be relatively harmful to Africa. The modeling results presented here indicate that the potential damages may be large both in absolute terms and as a fraction of agricultural GDP. Even with the low climate sensitivity model, warming is predicted to lead to damages in all but the mildest climate change scenarios.

These effects have been presented here in the aggregate for the continent, regions, or individual countries. However, it should be remembered that the vast majority of African citizens are farmers. A severe reduction in their livelihood could be devastating on a household level even if agricultural GDP persists in their country.

Table 14.5 Country-specific estimates of GDP and climate impacts by model (billion USD).

Country	GDP 2000	GDP 2100	AGRGDP 2100	Climate impacts		
				PCM	HAD3	CCC
Algeria	55.5	1347.2	12.7	−1.2	−3.5	−3.9
Angola	6.3	152.4	10.1	−0.5	−3.5	−3.9
Benin	1.8	44.5	4.3	−0.3	−1.3	−2.4
Botswana	1.2	28.0	0.2	−0.0	−0.2*	−0.2*
Burkina Faso	2.8	68.9	4.7	−1.0	−2.7	−4.8*
Burundi	1.2	29.1	3.5	0.1	−0.5	−0.9
Cameroon	12.2	295.8	17.1	−1.5	−3.9	−10.8
Cape Verde	0.3	7.1	0.2	−0	−0	−0.01
Central African Rep.	1.2	29.0	2.6	0.0	−1.2	−2.6*
Chad	1.1	26.3	2.0	−1.4	−1.9	−2.0*
Comoros	0.2	5.3	0.4	0.1	0.0	−0.1
Congo	2.1	51.9	1.5	−0.1	−0.3	−0.3
Cote d'Ivoire	9.7	236.0	23.3	−0.4	−2.8	−4.7
Djibouti	0.3	8.4	0.7	−0.0	−0.0	−0.0
Egypt	33.9	824.3	12.5	−0.9	−1.4	−2.3
Equatorial Guinea	0.2	3.8	0.5	−0.1	−0.1	−0.2
Ethiopia	6.2	150.9	14.0	3.6	−3.5	−12.1
Gabon	3.2	77.6	1.7	−0.1	−0.5	−0.6
Gambia	0.2	5.0	0.4	−0.1	−0.2	−0.4*
Ghana	5.7	139.6	15.0	−0.5	−2.9	−4.9
Guinea	2.5	60.2	4.7	−0.2	−0.6	−1.0
Guinea-Bissau	0.2	4.4	0.4	−0.2	−0.3	−0.4*
Kenya	9.2	222.8	14.7	2.0	−0.3	−2.5
Lesotho	0.9	20.7	1.1	0.0	−0.0	−0.0
Liberia	1.1	26.7	2.1	−0.1	−0.2	−0.5
Libya	24.0	582.7	6.1	−0.7	−1.1	−1.7
Madagascar	2.7	64.5	4.3	−0.2	−0.8	−1.8
Malawi	1.5	37.4	2.8	−0.1	−1.0	−1.5
Mali	2.2	53.5	5.7	−0.7	−2.4	−4.5
Mauritania	1.0	24.2	1.9	−0.1	−0.3	−0.4
Mauritius	2.2	52.5	1.4	−0.0	0.1	−0.1
Morocco	23.0	559.7	6.2	−0.7	−4.5	−4.7
Mozambique	1.2	30.3	4.2	−0.5	−2.6	−3.7
Namibia	1.6	39.1	1.2	−0.1	−0.7	−0.8
Niger	2.3	55.7	4.3	−1.3	−2.1	−4.4*
Nigeria	29.6	718.1	47.3	−4.4	−17.4	−47.6*
Rwanda	2.3	54.7	4.4	0.1	−0.4	−0.8
Sao Tome/Principe	0.0	1.1	0.1	0	−0	−0
Senegal	4.9	119.6	5.6	−1.5	−2.4	−3.9
Seychelles	0.3	7.4	0.6	−0	−0	−0
Sierra Leone	0.8	20.6	2.0	−0.1	−0.2	−0.5
Somalia	1.1	26.3	3.6	−0.3	−0.7	−1.0
South Africa	89.8	2181.9	27.1	−0.5	−6.1	−9.8
Sudan	13.8	335.4	8.8	−2.6	−8.8*	−8.8*
Swaziland	0.7	17.3	0.9	0.0	−0.1	−0.2
Togo	1.4	34.6	2.5	−0.3	−1.6	−2.5*
Tunisia	10.5	255.9	2.52	−0.8	−1.8	−1.9
Uganda	4.4	107.9	16.8	0.5	−2.8	−5.8
United Rep. Tanzania	3.2	78.1	11.0	0.5	−1.0	−2.7
Zaire	9.2	224.2	14.2	−0.8	−5.5	−10.3
Zambia	3.2	77.6	2.4	0.1	−2.4*	−2.4*
Zimbabwe	6.3	154.1	4.2	−0.1	−2.4	−3.5
Total	402	9780	333	−17.2	−100.7	−187.1

The scenario shown is with high climate response and 3% growth of GDP. Impacts are in billions of USD/yr. Asterisk signifies agriculture GDP goes to zero.

How much climate change will affect future African countries also depends upon the magnitude of their non-farming economy. If the non-agricultural component of the economy grows quickly, the magnitude of the impact as a fraction of GDP will be small in most African countries. However, if development stalls, and economies do not grow, climate warming could lead to substantial reductions in GDP per capita. Development is an important adaptation strategy for Africa.

It is important to mention a serious pitfall of this study. The results in this study come from extrapolations of empirical work in other parts of the world. Economic studies of climate change impacts remain rare in Africa. There remains an urgent need to conduct empirical work in Africa itself. This research is under way in an 11 country project being led by the Center for Environmental Economics and Policy in Africa (CEEPA) of South Africa. It will be interesting to compare what this new project predicts with the results in this paper.

References

Boer, G., Flato, G. and Ramsden, D. (2000). A transient climate change simulation with greenhouse gas and aerosol forcing: projected climate for the 21st century. *Climate Dynamics* **16**, 427–450.

Evenson, R. and Gollin, D. (2003). Assessing the impact of the green revolution, 1960–2000. *Science* **300**, 758–762.

Gordon, C., Senior, C., Banks, H. *et al.* (2000). The simulation of SST, sea ice extents, and ocean heat transports in a version of the Hadley Centre coupled model without flux adjustments. *Climate Dynamics* **16**, 147–168.

IPCC (2001a). *Climate Change 2001: The Scientific Basis. Contribution of Working Group I to the Third Assessment Report of the Intergovernmental Panel on Climate Change*, ed. J.T. Houghton, Y. Ding, D.J. Griggs *et al.* Cambridge: Cambridge University Press.

IPCC (2001b). *Climate Change 2001: Impacts, Adaptation, and Vulnerability. Contribution of Working Group II to the Third Assessment Report of the Intergovernmental Panel on Climate Change*, ed. J.J. McCarthy, O.F Canziani, N.A. Leary, D.J. Dokken and K.S. White. Cambridge: Cambridge University Press.

Mendelsohn, R. and Schlesinger, M.E. (1999). Climate-response functions, *Ambio* **28**, 362–366.

Mendelsohn, R. and Williams, L. (2004). Comparing forecasts of the global impacts of climate change. *Mitigation and Adaptation Strategies for Global Change* **9**, 315–333.

Mendelsohn, R., Morrison, W., Schlesinger, M. and Andronova, N. (2000). Country specific market impacts from climate change. *Climatic Change* **45**, 553–569.

Pearce, D., Cline, W., Achanta, A. *et al.* (1996). The social cost of climate change: greenhouse damage and the benefits of control. *Climate Change 1995: Economic and Social Dimensions of Climate Change. Contribution of Working Group III to the Second Assessment Report of the Intergovernmental Panel on Climate Change*, ed. J.P Bruce, H. Lee and E.F. Haites. Cambridge: Cambridge University Press, pp. 179–224.

Reilly, J., Baethgen, W., Chege, F.E. *et al.* (1996). Agriculture in a changing climate: impacts and adaptations. In *Climate Change 1995: Impacts, Adaptations and Mitigation of Climate Change: Scientific-Technical Analyses. Contribution of Working Group II to the Second Assessment Report of the Intergovernmental Panel on Climate Change*, ed. R.T. Watson, M.C. Zinyowera and R.H. Moss. Cambridge: Cambridge University Press.

Rosenzweig, C. and Parry, M. (1994). Potential impact of climate change on world food supply. *Nature* **367**, 133–138.

Sohngen, B., Mendelsohn, R. and Sedjo, R. (2002). A global model of climate change impacts on timber markets. *Journal of Agricultural and Resource Economics* **26**, 326–343.

Tol, R. (2002). Estimates of the damage costs of climate change. Part 1: benchmark estimates. *Environmental and Resource Economics* **21**, 47–73.

Washington, W., Weatherly, J., Meehl, G. *et al.* (2000). Parallel Climate Model (PCM): control and transient scenarios. *Climate Dynamics* **16**, 755–774.

Part III

Mitigation of greenhouse gases

John M. Reilly and Francisco C. de la Chesnaye, coordinating editors

Introduction

An important component of integrated assessment models is an evaluation of the costs and technology implications of mitigating greenhouse gases. Many existing Integrated Assessment (IA) models started as global energy models that were first turned towards examination of the cost and technology implications of reducing CO_2 emissions. Further representations of Earth system processes were later added to describe the impacts of policies on atmospheric concentrations of greenhouse gases and on changes in climate and its economic effects. For those IA models with an energy model heritage, the implications of greenhouse gas mitigation for the economy, energy use, and energy technology remains a relative strength. Developments in modeling energy–economy interactions quite apart from the broader goals of IA have continued. The demands of IA have meant that the energy model components have had to expand to cover additional anthropogenic activities that contribute to climate change and the mitigation opportunities associated with them. Papers in this section address recent advances in this regard.

The topics addressed here can be broadly grouped in the following areas: first, better characterization of the technological opportunities for reducing greenhouse gases; second, broadening the scope of the models to consider non-CO_2 greenhouse gases and also to consider land-use change and carbon sequestration in vegetation and soils; and third, better characterization of the impact of climate policy on the economy. A cross-cutting theme of this section is the ongoing exchange of knowledge and insight gained from detailed, sectoral models and from IA models (including the economic components within them). The former are able to address rich details of a sector or a technology, while the latter focus on the effects of mitigation on the broader economy but often must greatly simplify their representation of specific sectors and technology options. The exchange has been fruitful and, as the papers in this section demonstrate, there remain many areas where IA models can further benefit from the detailed models and analysis efforts. The specific topics addressed are hardly a comprehensive review in this regard, but the reader will certainly get a feel for the lively interchange that has led to rapid advance in IA modeling.

With regard to characterization of technological opportunities, Edmonds and co-authors (Chapter 16) address the issue of carbon capture and sequestration, particularly focusing on sequestration. They show that sequestration opportunities have typically been modeled as essentially unconstrained and large at the global level, but constrained at the regional level. This reflects a recurring theme in mitigation analysis: a mitigation solution that, at first look, appears to be an unlimited backstop technology turns out to be constrained or subject to increasing costs on closer inspection. Sweeney (Chapter 17) considers hydrogen in fuel cell vehicles, the latest hope in GHG mitigation technology. He finds that there are a number of daunting technological challenges that must be solved if the cost of this option is to be brought within reasonable bounds. Two key challenges are producing hydrogen without emitting CO_2, which may depend on relatively inexpensive carbon capture and sequestration, and reducing the cost of fuel cells so that hydrogen can fulfill its promise as a competitive fuel for vehicles. Sathaye (Chapter 15) takes on the issue of modeling improvements in end-use efficiency using a detailed "bottom-up" model. More aggregated energy models that are part of IAMs have been unable to advance far beyond the Autonomous Energy Efficiency Improvement (AEEI) parameter which is no more than an extrapolation of past experience. Sathaye's effort is another in a long line of studies that would hope to improve this component of IAMs, or at least better inform choices for AEEI values.

A major advance over the past several years has been the integration of non-CO_2 greenhouse gas abatement in IAMs. Non-CO_2 gases, more potent than CO_2, include methane, nitrous oxide, and a group of fluorinated compounds. De la Chesnaye and co-authors (Chapter 22) report on further efforts to investigate abatement opportunities through detailed assessments of technological options to provide an improved understanding of how these opportunities vary between a range of economic sectors and among countries. Their work has been instrumental for IAMs and has been broadly incorporated in climate economic models.

McCarl and co-authors (Chapter 20) consider the agriculture sector, which is responsible for methane and nitrous oxide emissions and can be either a source or sink of CO_2. While their modeling approach is not an IAM as defined for the purposes of this volume, it does focus on integration by recognizing the strong interactions among mitigation options within the agriculture sector. They argue that in this sector, simple abatement curves separately identified for different options are likely to be misleading because there are often strong interactions, and mitigation options compete for the same land. They see the need for IAMs to vastly improve their representation of sink mitigation options, and towards that end, they identify several lessons and insights they have gained in their efforts to model these activities in the United States. Sohngen and Mendelsohn (Chapter 19) focus on global forestry and sequestration, to show how a forest sector model can be integrated, or at least loosely coupled, with an IAM that seeks to develop an optimal path of emissions and abatement of greenhouse gases. At least one message is common to these two papers: greenhouse gas mitigation in forests and agriculture cannot be seen as an activity independent of the use of forests for timber or the land for food, fiber, or energy production. Timber demand and the supply of land for forestry will affect carbon emissions from forests, and carbon sequestration will in turn affect optimal forest harvest and rotations. At the same time, the use of land for greenhouse gas mitigation will affect agricultural markets.

Economics has strongly framed the issues and economic tradeoffs in climate change policy. The final set of papers address key recent contributions in this regard. Paltsev and co-authors (Chapter 23) explain why the much-used "carbon price" may not be a good indicator of the real cost of climate policy in an economy. They use a computable general equilibrium model to illustrate that the carbon price can be a poor indicator of economic burden among countries. Perhaps more revolutionarily, they show that the cherished rule of economists – equating the carbon price across sectors and economies – does not necessarily lead to lower economic costs, and emissions trading that equates carbon prices can actually increase rather than reduce the economic cost of climate policy. They trace these findings to interactions of climate policy with existing policies, another key element of "integration." Bernstein and co-authors (Chapter 18) are also keenly interested in better characterizing the economic impact of climate policies on the economy. Their focus is on how expectations about climate policy in the distant future affect the current economy – and how a new policy proposal may affect those expectations. Among the more important conclusions of their far-ranging analysis is that policy uncertainty, while probably inevitable, in itself means that studies that consider a policy scenario as if it were certain (virtually all to date) are likely to underestimate the cost of climate policy.

Finally, Yohe (Chapter 24) also addresses the issue of setting policy given that there is uncertainty about climate change. He recognizes that uncertainty is not unique to climate policy and draws from a long history of economic analysis of monetary policy, where decisions by the Federal Reserve, for example, must necessarily be made in the absence of complete information on future economic conditions. He shows how currently available methods can provide the information and insight needed by policymakers to cope with the uncertainties presented by potential climate change impacts. He concludes that once the uncertainty about the effect of a policy is recognized, doing nothing in the near term is as much of a policy decision as doing something.

These are a rich set of papers, each tackling issues on the forefront of analysis of mitigation options and integrated assessment. The vibrancy of this work is illustrative of the excitement in the field of integrated assessment. The papers are pushing for advances in many directions in IAMs. These papers offer significant new insights in themselves and hint at the promise of this rapidly advancing area of research.

15

Bottom-up modeling of energy and greenhouse gas emissions: approaches, results, and challenges to inclusion of end-use technologies

Jayant A. Sathaye

15.1 Introduction

Two general approaches have been used for the integrated assessment of energy demand and supply – the so-called "bottom-up" and "top-down" approaches. The bottom-up approach focuses on individual technologies for delivering energy services, such as household durable goods and industrial process technologies. For such technologies, the approach attempts to estimate the costs and benefits associated with investments in alternative fuels and technologies, and increased energy efficiency, often in the context of reductions in greenhouse gas (GHG) emission or other environmental impacts. The top-down method assumes a general equilibrium or macroeconomic perspective, wherein costs are defined in terms of changes in economic output, income, or GDP, typically from the imposition of energy or emissions taxes.

A fundamental difference between the two approaches is in the perspective each typically takes on consumer and firm behavior, and the performance of markets for energy efficiency. The bottom-up approach typically assumes that various market factors ("barriers") prevent consumers from taking actions that would be in their private self-interest, that is, would result in the provision of energy services at lower cost. These market barriers include lack of information about energy efficiency opportunities, lack of access to capital to finance energy efficiency investment, and misplaced incentives which separate responsibilities for making capital investments and paying operating costs. In contrast, the top-down approach typically assumes that consumers and firms correctly perceive, and act in, their private self-interest (are utility and profit maximizers), and that unregulated markets serve to deliver optimal investments in energy efficiency as a function of prevailing prices. In this view, any market inefficiencies pertaining to energy efficiency result solely from the presence of environmental externalities that are not reflected in market prices.

To the extent that these perspectives prevail, and as we note later both types of models are making an effort to remove these disadvantages, an assessment carried out using the bottom-up approach will therefore typically show significantly lower costs for meeting a given objective – e.g., a limit on carbon emissions – than will one using a top-down approach. To some extent, the differences may lie in a failure of bottom-up studies to account accurately for all costs associated with implementing specific actions. Top-down methods, on the other hand, can fail to account realistically for consumer and producer behavior by relying too heavily on aggregate data (Krause et al., 1993). In addition, some top-down methods sacrifice sectoral and technology detail in return for being able to solve for general equilibrium resource allocations. Finally, top-down methods often ignore the fact that economies depart significantly from the stylized equilibria represented by the methods (Boero et al., 1991). Each approach, however, captures costs or details of technologies, consumer behavior, or impacts that the other does not. Consequently, a comprehensive assessment should combine elements of each approach to ensure that all relevant costs and impacts are accounted for.

Following are some key questions that an analyst would seek to answer from a bottom-up assessment:

1 What is the economic cost of providing energy services in a baseline or policy scenario? What is the incremental cost

Human-induced Climate Change: An Interdisciplinary Assessment, ed. Michael Schlesinger, Haroon Kheshgi, Joel Smith, Francisco de la Chesnaye, John M. Reilly, Tom Wilson and Charles Kolstad. Published by Cambridge University Press. © Cambridge University Press 2007.

between scenarios? What are the capital and foreign exchange implications of pursuing alternative scenarios?
2. What is the economic cost of pursuing particular mitigation (policy and technology) options, such as high-efficiency lighting or a renewable technology, and what are its local and global environmental implications?
3. What are the costs of reducing emissions to a predetermined target level? (Target may be annual or cumulative.)
4. What is the shape of the marginal cost curve for reducing carbon emissions? How do alternative technologies and/or practices rank in terms of their carbon abatement potential?

In this chapter, we report on the typical structure and types of bottom-up assessments, and some of the key findings from these studies. Given the broad scope of bottom-up assessments, we focus primarily on the key challenges posed in the inclusion of demand-side technologies in bottom-up studies, and we discuss recent findings that shed light on their resolution. These challenges include: (1) aggregation over time, regions, sectors, and consumers, (2) representing decision rules used by consumers, (3) not accounting for transaction costs, (4) incorporating technical change, and (5) inclusion of secondary or productivity benefits.

15.2 Bottom-up assessment structure and models

The energy sector comprises the major energy demand sectors (industry, residential and commercial, transport, and agriculture), and the energy supply sector, which consists of resource extraction, conversion, and delivery of energy products. GHG emissions occur at various points in the sector, from resource extraction to end-use, and accordingly, options for mitigation exist at various points.

The bottom-up approach involves the development of scenarios based on energy end-uses and evaluation of specific technologies that can satisfy demands for energy services. One can compare technologies based on their relative cost to achieve a unit of GHG reduction and other features of interest. This approach gives equal weight to both energy supply and energy demand options. A variety of screening criteria, including indicators of cost-effectiveness as well as non-economic concerns, can be used to identify and assess promising options, which can then be combined to create one or more mitigation scenarios. These are evaluated against the backdrop of a baseline scenario, which simulates the events assumed to take place in the absence of mitigation efforts. Mitigation scenarios can be designed to meet specific emission reduction targets or to simulate the effect of specific policy interventions. The results of a bottom-up assessment can then be linked to a top-down analysis of the impacts of energy sector scenarios on the macroeconomy.

As discussed earlier, it is common to divide energy models into two types, so-called bottom-up and top-down, depending upon their representation of technology, markets, and decisionmaking. In practice, there is a continuum of models, each combining technological and economic elements in different ways. At one extreme are pure bottom-up energy models, which focus upon fuels and energy conversion or end-use technologies and treat the rest of the economy in an aggregated fashion. At the other extreme are pure top-down models, which treat energy markets and technologies in an aggregated manner and focus instead upon economy-wide supply–demand relations and optimizing behavior.

Between these two cases are a number of models that combine elements of both extremes with varying degrees of emphasis and detail. Such models integrate data on energy demand and supply. The models can use this information for determining an optimal or equilibrium mix of energy supply and demand options. The various models use cost information to different degrees, and provide for different levels of integration between the energy sector and the overall economy.

15.3 Accounting models: salient results

Energy accounting models reflect an engineering or input–output conception of the relations among energy, technology, and the services they combine to produce. This view is based on the concept of energy services that are demanded by end-users. With some accounting models, the evaluation and comparison of policies is performed by the analyst external to the model itself. Accounting models are essentially spreadsheet programs in which energy flows and related information such as carbon emissions are tracked through such identities. The interpretation of the results, and the ranking of different policies quantified in this manner, is external to the model and relies primarily on the judgment of the analyst. More recently accounting (or simulation) models have introduced behavioral parameters to try to reproduce real-world technology selection. Models such as ISTUM, CIMS, and Energy 2020 have market-splitting features based on the probabilistic behavior of investors. The estimation of behavioral parameters is based on discrete choice theory. In contrast to optimization models, accounting models cannot easily generate a least-cost mitigation solution. They can be used, however, to represent cost-minimizing behavior as determined by the analyst. They tend to require fewer data and less expertise, and are simpler and easier to use than optimization models.

The Kaya identity forms the basis for the approach used in such models. Using the identity, carbon emissions at an aggregate economy-wide level may be expressed as:

$$CO_2 = P \times GDP/P \times E/GDP \times CO_2/E \qquad (15.1)$$

where P = population, GDP = gross domestic product, and E = primary energy use.

Likewise, the identity may be used at the sectoral and end-use level in a similar manner. Knowing the historical values of the above parameters, the identity may be used to decompose the contribution of each factor to carbon emissions (Schumacher and Sathaye, 1999). Using a similar approach at the sectoral level, other studies have projected carbon emissions for both developing and developed countries by sector

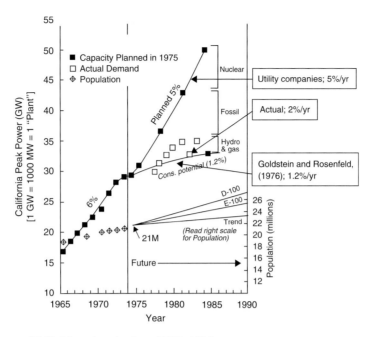

Figure 15.1 California peak power (1965–74) and projections (1975–1984).
Source: Rosenfeld (1999). *Annual Review of Energy and the Environment* **24**, 33–82.

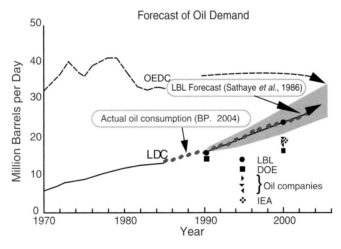

Figure 15.2 Less-developed country (LDC) oil demand projections: bottom-up projections may be more accurate! LBL: Lawrence Berkeley Lab.

and for the overall economy (Sathaye and Ketoff, 1991; Schipper *et al.*, 1992).

Bottom-up models typically project lower costs than top-down models for achieving the same level of emissions reductions. In part, this arises because bottom-up models often have lower reference case projections since they incorporate a higher potential for energy efficiency improvements in the model. This artifact has characterized the approach from the very beginning. Figure 15.1 shows a graph of the California demand for peak power from 1965 to 1974, and the projected growth in demand to 1984. Using a straight line approach, utility companies at the time projected a growth rate of 5% per year. Goldstein and Rosenfeld (1976) projected a growth in demand of only 1.2% per year, and the growth rate turned out to be much closer to their estimate at about 2% per year. Indeed, peak power demand in California increased at a continued low rate of about 2.2% a year between 1985 and 2000 (Brown and Koomey, 2002).

Figure 15.2 shows a counterexample of using the bottom-up accounting approach. It shows the crude oil demand in developing countries as estimated by Sathaye *et al.* (1986) based on sectoral studies of energy demand in the developing world (Sathaye *et al.*, 1987). Unlike the Goldstein and Rosenfeld (1976) report, the crude oil demand in this case was projected to be *higher* than that reported by several oil companies, the International Energy Agency, and the US Energy

Information Administration. The actual crude oil demand turns out to have been about the same as that projected in 1986 by LBL for the years 1990 and 2000.

A key feature of the accounting models that permits relatively accurate short- to medium-term projections is the reliance on the counting of physical assets and technologies, a good understanding, quantitative and otherwise, of the energy performance of current and future technologies, and penetrating insights gathered from the studies of individual energy markets in the study regions.

15.4 Other bottom-up models: costs and carbon emissions projections

In *engineering optimization* models, the model itself provides a numerical assessment and comparison of different policies. These models use mathematical programming, linear and otherwise, in which the most basic criterion is total cost of providing economy-wide energy services under different scenarios. In engineering optimization models, macroeconomic factors enter in two ways. First, they are used to construct forecasts of useful energy demands. Second, they can be introduced as constraints. For example, the overall cost-minimization can be constrained by limits on foreign exchange or capital resources. In both cases, the models do not provide for the representation of feedbacks from the energy sector to the overall economy.

Partial equilibrium models incorporate the dynamics of market processes related to energy via an explicit representation of market equilibrium, that is, the balancing of supply and demand. These are used to model a country's total energy system, and do not explicitly include an economy model integrated with the energy system model. Thus, macroeconomic factors enter the model exogenously, as in the previous model types discussed. Prices and quantities are then adjusted iteratively until equilibrium is achieved. This iterative approach makes it much easier to include non-competitive-market factors in the system. Recent improvements to the optimization approach also permit the representation of a partial equilibrium within the model structure (MARKAL/TIMES family of models).

In *hybrid* models, the basic policy measure is the maximization of the present value of the utility to a representative consumer through the model planning horizon. Constraints are of two types: macroeconomic relations among capital, labor, and forms of energy, and energy system constraints. In this type of model, energy demands are endogenous to the model rather than imposed exogenously by the analyst. Specifically, in accounting and engineering optimization models, these data – on GDP, population growth, capital resources, etc. – enter essentially in the underlying constructions of the baseline and policy scenarios. In the hybrid model, however, such data enter in the macroeconomic relations (technically, the aggregate production function) as elasticities and other parameters. Within this model framework, changes in energy demand and supply can feedback to affect macroeconomic factors. It should be noted that, despite their inclusion of engineering optimization subcomponents, these models typically do not contain as much detail on specific end-use technologies as many purely engineering models.

Figure 15.3 shows a compilation of country results using bottom-up studies of emissions reductions that may be achieved using a carbon tax or comparable instrument. The models are largely partial equilibrium models (e.g., AIM: JAPAN 1 and 2; NEMS: EIA; MARKAL: LOUKA; PRIMES: CAP-99; and MARKAL-MACRO, a simple general equilibrium model. Negative (positive) values on the horizontal axis indicate future emissions higher (lower) than the 1990 base year. For example, at a marginal cost of $100 per tonne C, the models show emissions that range from about 20% higher compared with those in 1990 for the US using the EIA model to as much as 55% lower for the EU using the GIEPIE model. The differences arise through different energy endowments, economic growth rates, energy intensity levels, and other country-specific conditions.

A recent hybrid model, AMIGA, incorporates detailed technology characteristics in a general equilibrium framework and permits evaluation of technology-specific efficiency improvements (Hanson and Laitner, 2004). It notes that a technology-led investment strategy can secure substantial domestic reductions of carbon emissions at a net positive impact on the US economy. Another hybrid model, CIMS, shows higher cost estimates when run in the hybrid mode compared to running it as a purely bottom-up model (Jaccard et al., 2003).

In addition to these studies, bottom-up models also include national and regional sectoral and cross-sectoral studies that assume large potential for energy efficiency gains, in other words large no-regrets benefits. Krause et al. (1999) show that 2030 European emissions could be reduced by up to 50% below 1990 levels at negative cost. Likewise, the five-lab studies (Brown et al., 1997; Interlaboratory Working Group, 2000) indicate that the Kyoto reduction target could be achieved at negative overall cost ranging between −US $7 billion and −US $34 billion. Using a modification of the NEMS model, Koomey et al. (1998) estimate that 60% of the Kyoto target could be reached while achieving an increase in US GDP. Using the MARKAL model for Canada, Loulou et al. (2000) and Kanudia and Loulou (1998) show that with efficiency measures a positive cost of US $20 billion turns into negative cost of −$26 billion for Canada to reach its Kyoto target. Sector-specific studies similarly stress the availability of a large negative cost potential in the forestry sector in tropical countries (Sathaye et al., 2001b).

15.5 Key challenges in the bottom-up modeling approach

The aforementioned bottom-up studies show a wide variety of cost estimates for achieving similar reductions in carbon emissions. While a complete explanation supported by empirical data is lacking, one can point to several reasons for these differences. These include: (1) representing decision rules used by

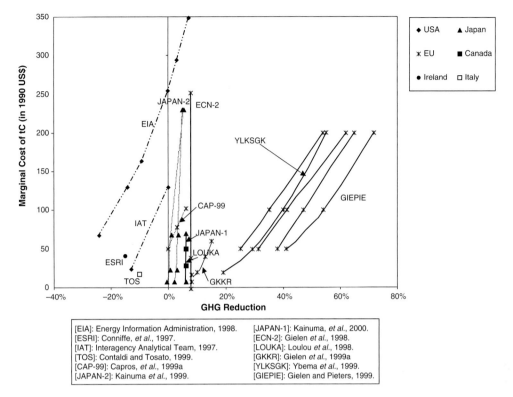

Figure 15.3 Bottom-up results: marginal costs of reducing emissions relative to 1990, single region, no trading.
Note: Negative GHG reduction implies an increase of emissions from 1990 level. This increase implies a positive marginal cost when it corresponds to a lower increase than would have occurred in the absence of climate policy.
Source: Hourcade and Shukla (2001).

consumers, (2) not accounting for transaction costs, (3) incorporating technical change, (4) inclusion of secondary or productivity benefits, and (5) aggregation over time, regions, sectors, and consumers. Prior to reporting some of the empirical analyses that highlight the importance of these factors, it is useful to provide a conceptual framework for the differences in cost estimates.

15.5.1 Conceptual framework: factors, potentials, and transaction costs

Earlier reports have enumerated lists of several factors (barriers) affecting the penetration of energy-efficient devices by customer class or tariff category, region and/or sector (Reddy, 1991; Golove and Eto, 1996; Eto et al., 1997; Sathaye et al; 2001). These include lack of information, lack of access to capital, misplaced incentives, flaws in market structure, performance uncertainties, decisions influenced by custom and habits, inseparability of features, heterogeneity of consumers, hidden costs, transaction costs, bounded rationality, product unavailability, externalities, and imperfect competition. The extent of their inclusion affects both costs and the mitigation potential of a technology or a mix of technologies. Sathaye et al. (2001), following on the work of Jaffe and Stavins (1994), classify factors into two categories. The first category refers to factors that economists may typically classify as "market failures," the second to factors that are manifestations of consumer preferences, custom, cultural traits, habits, and lifestyles.[1]

Associated with each category is the concept of potentials for GHG mitigation (Figure 15.4). Each concept of the potential represents a hypothetical projection that might be made today regarding the extent of GHG mitigation into the future. The leftmost line, labeled *market potential*, indicates the amount of GHG mitigation that might be expected to occur under forecast market conditions, with no changes in policy or implementation of measures whose primary purpose is the mitigation of GHGs. At the other extreme, the *technical potential* describes the maximum amount of GHG mitigation achievable through technology diffusion. This is a hypothetical projection of the extent of GHG mitigation that could be achieved over time if all technically feasible technologies were used in all relevant applications, without regard to their cost or user acceptability.

By definition, we can say that whatever physical, cultural, institutional, social, or human factors are preventing us from reaching the technical potential are factors affecting the

---

[1] Whether government policy can, may or will be used to remove these barriers is a question for political economists, and we do not attempt to address it here.

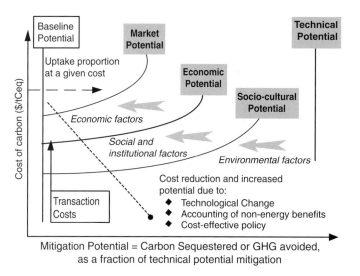

Figure 15.4 Penetration of mitigation technologies: a conceptual framework.

mitigation of GHG via technology diffusion. Since, however, our ultimate goal is to understand policy options for mitigation, it is useful to group these factors in a way that aids in understanding the kinds of policies that would be necessary to overcome them. As we create these different categories of factors, we correspondingly create intermediate conceptions of the potential for GHG mitigation. Starting at the left, we can imagine addressing factors (often referred to as market failures) that relate to markets, public policies, and other institutions that inhibit the diffusion of technologies that are (or are projected to be) cost-effective for users without reference to any GHG benefits they may generate. Amelioration of this class of *market and institutional imperfections* would increase GHG mitigation towards the level that is labeled as the *economic potential*. The economic potential represents the level of GHG mitigation that could be achieved if all technologies that are cost-effective from the consumer's point of view were implemented. Because economic potential is evaluated from the consumer's point of view, we would evaluate cost-effectiveness using market prices and the private rate of time discounting, and also take into account consumers' preferences regarding the acceptability of the technologies' performance characteristics.[2]

Many of the market failures listed above can be broadly grouped together as cognitive factors affecting product diffusion. By this, we mean that there are limitations to consumers' ability to gather and process information. Before any consumer can decide to adopt a technology, he or she must as a minimum be aware of its existence. Once aware, a consumer needs to make some effort to gather the information needed to make an informed decision about whether a given technology provides more benefits than it costs. To do this, an individual needs the analytic capacity to quantify fairly accurately the benefits and costs. Even an aware, informed, capable consumer must ultimately make the effort to assess benefits and costs before making the decision to adopt.

A consumer who has made this decision needs to find a vendor for the product in question. Relatively new technologies are likely to be less widely available than their more standard counterparts. Thus, limitations on cognitive resources are described by factors such as performance uncertainty, information costs, and bounded rationality.

Of course, elimination of all of these market and institutional factors would not produce technology diffusion at the level of the *technical potential*. The remaining factors, which define the gap between economic potential and technical potential, are usefully placed in two groups separated by a socio-cultural potential. The first group may be further subdivided into two sets of factors. The first set defines the difference between private and social economic potential. The socio-economic potential represents the level of GHG mitigation that would be achieved by implementing all technologies that are cost-effective on the basis of using a social, rather than a private, rate of discount, including externalities. The second set of factors is derived from people's *preferences and other social and cultural factors* affecting the diffusion of technology. That is, even if market and institutional factors were removed, some GHG-mitigating technologies might not be widely used simply because people did not like them or because existing social and cultural forces operated against their acceptance. If, in addition to overcoming market and institutional factors, this second set of factors could be overcome, we would achieve what we label as the *socio-cultural*

---

[2] The identification of "economic potential" with implementation of technologies that are cost-effective from the consumer's point of view adopts, in effect, the economist's view that economic potential corresponds to the elimination of market failures. Other analysts have used the phrase "economic potential" to incorporate a broader conception, similar to what we have dubbed "socio-economic potential" (Jaffe and Stavins, 1994).

potential, a level attained without regard to concerns about some performance characteristics or to social and cultural obstacles to their use.

Finally, even if all market, institutional, social, and cultural factors were removed, some technologies might not be widely used because they are too expensive after including the cost of policy intervention. That is, the definition of socio-cultural potential includes the requirement that technologies be cost-effective. Elimination of this requirement would therefore take us to the level of *technical potential*, the maximum technologically feasible extent of GHG mitigation through technology diffusion. Moving from right to left, the figure shows that factors increase costs and reduce the savings potential of a mitigation technology.

Figure 15.4 presents a snapshot in time of the factors and potentials for the penetration of technologies. Over time technological progress, discoveries of new resources and/or technologies, and cost-effective government policies and programs could eliminate some of the factors and hence move the potential lines to the right and down, thereby increasing the savings from a mitigation option. The snapshot also refers to an immature or nascent market that has higher transaction costs. As the market matures, the decline in transaction costs caused by learning by doing and standardization will push the cost curve lower, which will increase penetration of technologies at a given cost of competing energy supply options.

15.5.2 Empirical evidence of the influence of factors

Figure 15.4 above conceptually illustrates the influence of many factors that act as barriers to and opportunities for technology penetration. Empirical research to validate the existence of these factors and to gage their influence on costs and potentials, while limited, provides useful insights into the way these affect not only the penetration levels but also the ranking of options. This has both policy and programmatic implications.

Representing decision rules used by consumers
A key characteristic of bottom-up models is the use of engineering cost estimates of new technologies. However, these may leave out the costs of bringing technologies to the market, and reliance of a model on such engineering estimates could severely underestimate their real market costs. Likewise, technological potential may be overestimated if the model does not properly account for access to capital to purchase such technologies, or their limited availability and applicability in current market settings. This facet of technological potential was illustrated as early as 1994 (Brown, 1993). In his thesis, Brown illustrates that at the 1990 electricity price of 7.8 c/kWh the US potential estimated for electricity efficiency improvements in 2010 declines by over 50% from 410 TWh to 185 TWh, relative to estimates that relied on engineering costs only. Another way to interpret this result is that to achieve a reduction of 185 TWh, costs increase from about 3.8 c/kWh to 7.8 c/kWh. These changes are caused by re-estimation of the space conditioning potential which had assumed some unrealistically high operating conditions for space conditioning and appliance use, consideration of a takeback or rebound effect, and lack of savings persistence.

Accounting for transaction costs
Transaction costs arise in the trade of technologies, and may include costs of programs to increase their market penetration. As noted above, engineering cost estimates may leave out the routine costs of bringing technologies to the market and promoting them. In addition, transaction costs also arise in emissions trading to the extent the intent of using these technologies is to reduce GHG emissions. Both allowance trading and project-based emissions trading programs incur costs. The UK program is one example of an allowance trading program. The UK National Audit Office reports that brokers have charged a fee of 2% per transaction (National Audit Office, 2004). A 2% fee would amount to a charge of $0.33 – $1.65 per t Ceq. for brokerage services alone. Transaction costs of the latter type of trading program have been reported to range from as low as a few cents per t C to as high as $200 per t C, but with large proportion of projects report costs between tens of cents and about $20 per t C (Michaelowa *et al.*, 2003). The costs of such emissions trading programs need to be explicitly factored into the total costs of carbon reduction using a bottom-up model.

Accounting for technological change
While not all models may account for technological change, those that include this factor do it primarily in two different ways. One way is to include an exogenous parameter that may be based on historical evidence about technological improvements and accounts for change (often decline) in the cost of the technology over time. Nakicenovic *et al.* (1998), for instance, have published curves showing the decline in costs of electricity supply technologies over time. Not unexpectedly, photovoltaic technology demonstrates a faster decline in cost than a more conventional coal power plant. Another approach is to make the technological costs endogenous to the model such that costs decline over time as the penetration of technology increases, i.e., through so-called learning by doing, and if the model chooses to not increase the penetration of a particular technology, costs remain higher. Costs of both demand- and supply-side technologies may be modeled in this manner. Laitner and Sanstad (2004) show for instance that a mature technology, such as the magnetic ballast, shows a low learning rate of 0.04, indicating a slow rate decline. By contrast, an electronic ballast, which is an advanced technology, shows a learning rate of 0.18. The learning rate indicates the rate of cost reduction with an increase in cumulative output. Yet another approach is to incorporate "endogenous technology learning" by which

technology learning is linked to investment cost, whereby additional investment results in additional learning which results in additional investment and so on (de Feber, 2003; Gielen et al., 2004).

Inclusion of non-energy benefits
Most bottom-up models are either accounting, optimization, or partial equilibrium models of the energy sector. The models may be restricted to one or more of the end-use sectors, to a particular geographic region, or to an energy form. Since the focus is on an element of the energy sector, these models adequately account for energy flows and the changes in capital costs, and some, such as MARKAL, MESSAGE, etc., account for changes in operation and maintenance costs. Because their representation of demand-side technologies is limited, however, even these models may miss the significant effects of non-energy benefits.

Models may not routinely account for changes in other costs that might arise owing to a change in resource requirement that accompanies a change in energy end-use. Clothes washers are a technology whose use of energy has been studied, and standards have been proposed to regulate the penetration of energy-inefficient washers. On the basis of the reduced use of electricity in the washer and that used to heat the water, the standards would be too stringent, but when water savings are accounted for the standards become cost-effective (Sathaye and Murtishaw, 2004). Accounting for such "hidden benefits" requires that bottom-up models look beyond the energy markets and examine the cost considerations in light of their impact on other resource markets. A similar example is reported by Worrell et al. (2003) for the US iron and steel industry. They report a cost-effective annual primary energy savings of 1.9 GJ/tonne of output for this sector due to the implementation of an array of 47 measures. Inclusion of labor and material cost savings during the operation of an efficient iron and steel plant, however, increases the potential to 3.8 GJ/tonne of output at the same cost. More importantly, the ranking of technologies to implement changes dramatically; an oxy-fuel burner ranked no. 41 when only energy cost savings are included becomes the no. 1 technology to implement. Inclusion of all resource benefits thus is crucial to understanding the full cost impacts of a technology. This may be particularly relevant to end-use energy efficiency technologies whose main goal often is not providing or saving energy but providing some other form of service or producing an industrial product.

Aggregation over time, regions, sectors, and consumers
Perhaps the most fundamental formulation issue is the level of aggregation at which costs and benefits are calculated. Aggregation bias arises when interdependencies between mitigation options are not accounted for, opportunities for carbon trading across regions are ignored, changes in characteristics of mitigation options over time are not accounted for, and/or diversity among agents is overlooked.

In decomposition analysis, the level at which decomposition is conducted can have a large influence on the proportion of changes in energy savings (and carbon emissions) that are attributed to changes in structure vs. intensity. Fischer-Vanden et al. (2004) show that sectoral shift contributes to an increase in China's energy use, which is offset by a decline in energy intensity at the one-digit level, while at the three-digit level, both factors contribute to the decline in energy use, although the contribution of energy intensity is still 50% higher.

An obvious source of error that has been noted by many authors is not accounting for the interactions between two energy efficiency options when constructing a supply curve (Stoft, 1995). Adding efficient lighting as a mitigation option reduces electricity demand for lighting but it can also reduce the air conditioning load of a building. Most models take such interactions into consideration, but it is important to note that such interactions reduce the potential for energy savings.

Both bottom-up and top-down models illustrate that costs of reducing greenhouse gas emissions decline when trading of emissions rights is permitted over larger geographic regions (where flexibility). The IPCC (2001) report notes that the mean value of cost of carbon emissions reduction, which was estimated to be between \$178–331 per t C with domestic trading only, declines by 50% or more to \$68 per t C with global trading. The decline is caused by lower mitigation cost in both the developing countries and in the economies in transition.

Prima facie, this seems appropriate since the latter countries and economies have lower incomes per capita and labor costs. A recent study, however, challenges this conventional wisdom by examining the actual cost of emissions reduction from a natural gas combined cycle plant when compared with a coal-fired one (Sathaye and Phadke, 2004). The study compares the cost of reducing carbon emissions from this mitigation option in the United States and India. It concludes that while components that contribute to the total cost of a power plant such as capital, operations and maintenance, plant heat rate, and lifetime are very similar for coal power plants in India, they are uniformly higher for combined cycle units in India. As a consequence, the cost of carbon reduction is much higher in India than in the United States. A key observation from this study is that labor costs make up a small proportion of the construction costs and hence the labor component of the cost of technologies in developing countries, which typically have lower wage rates for comparable skill categories, can be more than offset by higher capital and fuel costs. In a project-based trading regime that requires that projects meet emissions additionality conditions (the condition that emissions reductions must be in addition to what would have happened in the absence of a program), new technologies will need to be deployed and their costs may be higher in developing than in developed countries.

Furthermore, aggregation across income classes even within a region overlooks the equity implications of climate change costs. Most models tend not to report on the sectoral impacts of mitigation options, although from a policy perspective alleviating these inequities may prove critical in gaining acceptance for climate policies. Krause *et al.* (2003) show for instance that the impact of mitigation will be borne disproportionately by the coal sector compared with the oil and gas sectors of the US economy.

15.6 Summary

The distinguishing feature of a bottom-up approach is that it focuses on individual technologies and attempts to estimate the costs and benefits associated with investments in increased penetration of the technologies. Since it focuses on physical assets, it serves as a useful tool for ensuring that technology penetration is not over- or under-estimated. Historically, such approaches have served as an important means to provide reliable, and often controversial, projections of energy and oil demand. In often relying on engineering cost and potential estimates, the approach can be overly optimistic about penetration and costs of mitigation technologies. In this paper, we have suggested a conceptual framework that can improve consistency in the estimation of costs and potentials, and have reported on empirical results that validate these concepts.

References

Boero, G., Clarke, R. and Winters, L. (1991). *The Macro-economic Consequences of Controlling Greenhouse Gases: A Survey*. UK Department of the Environment.

Brown, R. (1993). *Estimates of the Achievable Potential for Electricity Efficiency Improvements in US Residences*. Unpublished ERG Masters thesis, University of California Berkeley.

Brown, R. and Koomey, J. (2002). Electricity use in California: past trends and present usage patterns. *Energy Policy* **31**(9), 849–864.

de Feber, M. A. P. C., Schaeffer, G. J., Seebregts, A. J. and Smekens, K. E. L. (2003). *Enhancements of Endogenous Technology Learning in the Western European MARKAL Model: Contributions to the EU SAPIENT project*. ECN-C-03–032, April 2003.

Eto, J., Prahl, R. and Schlegel, J. (1997). *A Scoping Study on Energy-efficiency Market Transformation by California Utility DSM Programs*. Lawrence Berkeley National Laboratory.

Fischer-Vanden, K., Jefferson, G., Liu, H. and Tao, Q. (2004). What is driving China's decline in energy intensity? *Resource and Energy Economics* **26** (1), 77–97.

Gielen, D., Unander, F., Mattsson, N. and Sellers, R. (2004). Technology learning in the ETP model. Paper presented at the 6th IAEE European Conference, Zurich, Switzerland.

Goldstein, D. and Rosenfeld, A. (1976). *Projecting an Energy-efficient California*. LBL-3274/EEB 76–1. Lawrence Berkeley National Laboratory.

Golove, W. H. and Eto, J. H. (1996). *Market Barriers to Energy Efficiency: A Critical Reappraisal of the Rationale for Public Policies to Promote Energy Efficiency*. Lawrence Berkeley National Laboratory.

Hanson, D. and Laitner, J. A. S. (2004). An integrated analysis of policies that increase investments in advanced energy-efficiency/low-carbon technologies. *Energy Economics* **26**(4), 739–755.

Hourcade, J. and Shukla, P. (2001). Global, regional, and national costs and ancillary benefits of mitigation. In *Climate Change 2001: Mitigation. Contribution of Working Group III to the Third Assessment Report of the Intergovernmental Panel on Climate Change*, ed. B. Metz, O. Davidson, R. Swart and J. Pan. Cambridge: Cambridge University Press.

Interlaboratory Working Group (2000). *Scenarios for a Clean Energy Future*. Oak Ridge, TN: Oak Ridge National Laboratory; and Berkeley, CA: Lawrence Berkeley National Laboratory), ORNL/CON-476 and LBNL-44029, November.

IPCC (2001). *Climate Change 2001: Synthesis Report. Third Assessment Report of the Intergovernmental Panel on Climate Change*, ed. R. T. Watson. Cambridge: Cambridge University Press.

Jaccard, M., Nyboer J., Bataille, C. and Sadownik, B. (2003). Modeling the cost of climate policy: distinguishing between alternative cost definitions and long-run cost dynamics. *The Energy Journal* **24**(1), 49–73.

Jaffe, A. and Stavins, R. (1994). The energy efficiency gap: what does it mean? *Energy Policy* **22**(10), 804–810.

Kanudia, A. and Loulou, R. (1998). Robust response to climate change via stochastic MARKAL: the case of Quebec. *European Journal of Operations Research* **106**, 15–30.

Koomey, J., Richey, C., Laitner, S., Markel, R. and Marnay, C. (1998). *Technology and GHG Emission: An Integrated Analysis Using the LBNL-NEMS Model*. Berkeley, CA: LBNL Report 42054.

Krause, F., Haites, E., Howarth, R. and Koomey, J. (1993). *Cutting Carbon Emissions: Burden or Benefit? The Economics of Energy-tax and Non-price Policies*. El Cerrito, CA: International Project for Sustainable Energy.

Krause, F., Koomey, J. and Olivier, D. (1999). *Cutting Carbon Emissions while Saving Money: Low Risk Strategies for the European Union*. Executive Summary in Energy Policy in the Greenhouse. Vol. II, Part 2IPSEP, International Project for Sustainable Energy Paths, El Cerrito CA. November 33 p.

Krause, F., De Canio, S., Hoerner, J. A. and Baer, P. (2003). Cutting carbon emissions at a profit (Part II): Impacts on US competitiveness and jobs. *Contemporary Economic Policy*, **21** (No. 1), 90–105.

Laitner, J. A. S. and Sanstad, A. H. (2004). Learning-by-doing on both the demand and the supply sides: implications for electric utility investments in a heuristic model. *International Journal of Energy Technology and Policy* **2**(1–2), 142–152.

Loulou, L., Kanudia, A., Labriet, M., Margolick, M. and Vaillancourt, K. (2000). *Integration of GHG Abatement Options for Canada with the MARKAL Model*. Report prepared for the National Process on Climate Change and the Analysis and Modeling Group, Canadian Government.

Michaelowa, A., Stronzik, M., Eckermann, F. and Hunt, A. (2003). Transaction costs of the Kyoto mechanisms. *Climate Policy* **3** (3), 261–226.

Nakicenovic, N., Grubler, A. and McDonald, A. (eds.) (1998). *Global Energy Perspectives.* Cambridge: Cambridge University Press.

National Audit Office (2004). *The UK Emissions Trading Scheme: A New Way to Combat Climate Change.* Ordered by the House of Commons.

Reddy, A. K. N. (1991). Barriers to improvements in energy efficiency. *Energy Policy* **19**, 953–961.

Rosenfeld, A. (1999). The art of energy efficiency: protecting the environment with better technology. *Annual Review of Energy and the Environment* **24**, 33–82.

Sathaye, J. and Ketoff, A. (1991). CO_2 emissions from major developing countries: better understanding the role of energy in the long term. *The Energy Journal* **12**(1), 161–171.

Sathaye, J. and Murtishaw, S. (2004). *Market Failures, Consumer Preferences, and Transaction Costs in Energy Efficiency Purchase Decisions.* CEC-500–2005–020/LBNL-57318. Lawrence Berkeley National Laboratory for the California Energy Commission, PIER Energy-Related Environmental Research.

Sathaye, J. and Phadke, A. (2004). Cost and carbon emissions of coal and combined cycle power plants in India: implications for costs of climate mitigation projects in a nascent market. *Energy Policy* **34**, 1619–1629.

Sathaye, J., Schipper L. and Levine, M. (1986). *Growth in LDC Oil Demand: Implications and Opportunities for the US.* Presentations to William F. Martin, Deputy Secretary, US Department of Energy, Washington DC.

Sathaye, J., Ghirardi, A. and Schipper, L. (1987). Energy demand in developing countries: a sectoral analysis of recent trends. *Annual Review of Energy* **12**, 253–281.

Sathaye, J., Bouille D., Biswas, D. *et al.* (2001a). Barriers, opportunities, and market potential of technologies and practices. In *Climate Change 2001: Mitigation. Contribution of Working Group III to the Third Assessment Report of the Intergovernmental Panel on Climate Change,* ed. B. Metz, O. Davidson, R. Swart and J. Pan. Cambridge: Cambridge University Press.

Sathaye, J., Maundi, W., Andrasko, K. *et al.* (2001b). Carbon mitigation potential and costs of forestry options in Brazil, China, India, Indonesia, Mexico, the Philippines, and Tanzania. *Mitigation and Adaptation Strategies for Global Change* **6** (3–4), 185–211.

Schipper, L., Meyers, S., Howarth, R. and Steiner, R. (1992). *Energy Efficiency and Human Activity: Past Trends, Future Prospects.* Cambridge: Cambridge University Press.

Schumacher, K. and Sathaye, J. (1999). Carbon emissions trends for developing countries and countries with economies in transition. *Quarterly Journal of Economic Research* **4**, Special Issue: Energy Structures Past 2000.

Stoft, S. (1995). The economics of conserved-energy supply curves. *The Energy Journal* **16** (4), 109–137.

Worrell, E., Laitner, J. A., Ruth, M. and Finman, H. (2003). Productivity benefits of industrial energy efficiency measures. *Energy* **28** (11), 1081–1098.

16

Technology in an integrated assessment model: the potential regional deployment of carbon capture and storage in the context of global CO₂ stabilization

J. A. Edmonds, J. J. Dooley, S. K. Kim, S. J. Friedman and M. A. Wise

16.1 Introduction

Stabilizing the concentration of greenhouse gases implies stabilizing the concentration of carbon dioxide (CO_2), the most important greenhouse gas. As Wigley et al. (1996) showed, this goal implies a peak in emissions of carbon to the atmosphere followed by a decline in global carbon emissions that goes on indefinitely thereafter. Carbon dioxide from fossil fuel burning is the dominant anthropogenic emission to the atmosphere.[1] A central question in limiting emissions is therefore how to provide the energy services that are at present delivered predominantly using fossil fuels, without concurrently releasing CO_2 into the atmosphere. Prior analysis, including that of Edmonds et al. (2004), has shown that carbon dioxide capture and storage (CCS) has the potential to reduce emissions and limit emissions mitigation costs. Furthermore, earlier surveys of aggregate global CO_2 storage capacity such as Edmonds et al. (2001) indicate that potential gross global reservoir capacity is larger – potentially many times larger – than the scale of gross global CO_2 capture from energy and industrial processes over the course of the twenty-first century. However, this simplification does not reflect the reality that CO_2 would be captured at point sources and deposited into specific reservoirs.

Work has begun to consider local source–reservoir relationships within a specific country (see for example Dooley et al. [2004]). The analysis we present here is the first to examine the implications of regionally heterogeneous CO_2 storage on a global scale.

To accomplish this task, we

1. developed estimates of maximum theoretical potential capacity of geologic storage reservoirs for five reservoir types disaggregated into four resource grades in 14 regions including coastal zones constituting the world;
2. incorporated these region-specific estimates of potential CO_2 storage capacity into a substantially enhanced version of our integrated assessment model, MiniCAM, which allowed us to treat CO_2 storage as a regionally specific, graded resource for which storage costs depend on resource grade; and
3. examined CO_2 storage in the context of other technology options against a background of alternative CO_2 concentrations and the availability of suitable geologic storage capacity.

16.2 A regionally disaggregated CO₂ storage potential

The first order of business is to estimate the regional availability of suitable CO_2 storage reservoirs. While estimates exist of total global capacity, for example Herzog et al. (1997) and Freund and Ormerod (1997), no studies report the geographic distribution of that capacity across the globe. In this

---

[1] Land-use change is the other major source of anthropogenic carbon emissions. Fossil fuel carbon emissions are estimated to have been 6.5 PgC/yr in 1999 (CDIAC, 2004), while land-use change emissions are estimated to have been 1.7 PgC/yr with substantial uncertainty, 0.6 PgC/yr to 2.5 PgC/yr (CDIAC, 2004).

Human-induced Climate Change: An Interdisciplinary Assessment, ed. Michael Schlesinger, Haroon Kheshgi, Joel Smith, Francisco de la Chesnaye, John M. Reilly, Tom Wilson and Charles Kolstad. Published by Cambridge University Press. © Cambridge University Press 2007.

Table 16.1 First-order estimates of geologic CO_2 storage potential, all resource grades, by region (petagrams of carbon (PgC)).

Region	Unminable coal deposits	Depleted oil plays	Gas basins	Deep saline formation on-shore	Deep saline formation off-shore	Total
USA	16	3	10	745	248	1022
Canada	1	0	1	273	68	344
Western Europe	1	2	11	20	39	73
Japan	0	0	0	0	0	0.3
Australia and New Zealand	8	0	3	56	130	196
Former Soviet Union	5	6	70	101	378	560
China	4	1	2	90	9	106
Middle East	0	9	52	61	4	126
Africa	2	4	17	32	63	118
Latin America	1	4	13	51	15	85
Southeast Asia	7	1	8	33	49	97
Eastern Europe	1	0	2	29	3	35
Korea	0	0	0	0	0	0.3
India	2	0	2	51	51	105
Total global capacity	48	31	190	1540	1057	2867

study we consider five candidate-types of geologic storage reservoir:

1. On-shore deep saline formations;
2. Off-shore deep saline formations;
3. Depleted oil fields;
4. Depleted gas fields; and
5. Unminable coal deposits.

Our analysis disaggregates the world into the 14 geopolitical regions listed in Table 16.1. For each of these regions we estimate maximum potential capacity (MPC) for each of the five classes of geologic CO_2 reservoirs listed in Table 16.1. Each of these five reservoir classes is in turn partitioned into four resource grades. Estimates of the capacity of candidate reservoirs by region and resource grade were derived from the technical literature supplemented by our own technical judgment. Sedimentary basins were counted broadly as targets, while cratonal areas were excluded.[2] This assessment is documented in Dooley and Friedman (2004).

In general, the surveyed literature contains information about CO_2 storage potential in certain classes of geologic reservoirs and in certain regions of the world. For example, there are numerous estimates of the CO_2 storage capacity for depleted oil and gas fields and coal basins, but relatively little information on the storage potential of on-shore deep saline reservoirs and even less on that of off-shore deep saline reservoirs. As a general rule, the CO_2 storage potential of geologic formations that do not hold potentially valuable hydrocarbons and that lie in developing nations received the least attention in the literature. Therefore, the precision and accuracy underlying the estimates in Table 16.1 vary considerably. The data in Table 16.1 are thus considered a first-order preliminary assessment of the storage potential in these regions. It is our intent to update this data set as new information becomes available and to bring this new knowledge into the evolving O^bjECTS.MiniCAM model.[3] As such, these numbers serve as a departure point for storage capacity estimates, and not as a definitive compendium of reservoir storage volumes.

Aggregating the regional estimates yields a global total. Our estimates of the global total are within the range of estimates reported in earlier studies of potential aggregate storage capacity such as Freund and Ormerod (1997) and Herzog et al. (1997),

Table 16.2 Comparison with earlier estimates of global CO_2 storage potential (PgC).

Carbon storage reservoir	Herzog et al. (1997)	Freund and Ormerod (1997)	This study*
Oil plays	41	191	31
Gas reservoirs	136	300	190
Deep saline reservoirs	87	2727	2597
Coal basins	20	>20	48
Total	284	3218	2866

*Dooley and Friedman (2004).

---

[2] A craton is the relatively stable nucleus of a continent comprised of crystalline, igneous rocks. As such, cratons are not good targets for CO_2 storage.

[3] The complete set of literature consulted and the assumptions used to transform the literature's estimates of CO_2 storage capacity into the form needed by O^bjECTS.MiniCAM (e.g., in particular the subdivision of aggregate capacity to the four resource grades) is described fully in Dooley and Friedman (2004).

but closer to the high end of the range than the low end (Table 16.2).

16.3 Analysis cases

In this analysis, CO_2 capture and storage is considered in the context of alternative global emissions limitation regimes that stabilize the concentration of atmospheric CO_2 at different levels. Our approach is to develop a counterfactual case, which assumes that the world develops throughout the period from the present to 2095 without consideration of climate change. This counterfactual case, which is often referred to as the "business as usual" or "reference" case, is then contrasted with three illustrative scenarios in which the concentration of CO_2 is limited to 450, 550, and 650 ppm, three reference concentrations common in the literature.[4] These concentrations are used for illustrative purposes only, as science cannot at present determine the appropriate level at which to stabilize CO_2 concentrations.

Atmospheric CO_2 stabilization is accomplished by constraining net emissions to rates prescribed by Wigley et al. (1996). Concentrations are stabilized by charging a price for each tonne of carbon released to the atmosphere in the model. The same carbon emissions price is charged in all 14 regions. Each region therefore faces the same marginal cost of a ton of emissions and the resulting geographic distribution of mitigation activities is consistent with an idealized global emissions limitation regime. Global costs of stabilizing atmospheric CO_2 concentrations are therefore minimized subject to the technology available in the scenario. Of course, policies that improve the set of available technologies – for instance a policy that promoted R&D and enhanced technology transfer to developing countries – would result in lower costs.

To explore the potential deployment of carbon capture and storage technology, we examine five cases in which the maximum assumed storage is constrained to a fraction of the MPC as the concentration of atmospheric CO_2 is stabilized.

- **Case 0: Unlimited storage**. Assumes that the cost of CO_2 storage is a constant across space and time, which is equivalent to assuming that CO_2 storage reservoirs are not meaningfully limited, as has been the assumption in previous studies.[5]
- **Case 1: 100% MPC**. Stabilization with 100% of the maximum potential reservoir capacity identified in Table 16.1 being available.
- **Case 2: 50% MPC**. Stabilization with 50% of the maximum potential reservoir capacity identified in Table 16.1 being available.
- **Case 3: 10% MPC**. Stabilization with 10% of the maximum potential reservoir capacity identified in Table 16.1 being available.
- **Case 4: CCS is unavailable**. Stabilization with 0% of the maximum potential reservoir capacity identified in Table 16.1 being available.

These five cases were designed to reflect the uncertainty in the total resource that will be available for CO_2 storage. In particular, these cases are meant to account for the uncertainty associated with the data contained in Table 16.1. As there has yet to be a systematic global survey of potential geologic CO_2 storage reservoirs, there is bound to be significant uncertainties in these kinds of estimates. We therefore implement this limitation proportionally across all reservoir types and grades in all regions. At a secondary level, these cases are also designed to help account for other uncertainties such as political, economic, and technological factors (e.g., a region may decide to adopt strict rules as to which formations can be used for CO_2 storage, which would limit the effective available storage capacity). Political or technical constraints might imply non-constant proportionality limitations across reservoir types and grades. As yet, we have no basis upon which to formulate specific cases and thus have made no attempts to explore such cases in this analysis.

16.4 Modeling tools

CCS must compete with other technologies in the global energy system. Its deployment will depend on a variety of factors including the scale and pattern over time of the economy, the policy environment, and the costs and availability of other energy/carbon management technologies. Assessing the potential role of CCS in this context requires a long-term, global energy–technology–economy model with an integrated carbon cycle model. In this analysis, we employ the equation structure of the MiniCAM model, described in Edmonds et al. (2004). That equation structure has been embedded in an object-oriented model implementation environment that we refer to as O^{bj}ECTS.MiniCAM.

We employ O^{bj}ECTS.MiniCAM version 2004.04, which disaggregates the world into the 14 regions listed in Table 16.1. The model is long-term, examining energy–economy interactions over the period 1990 to 2095 in 15-year time steps. It tracks energy production, transformation, and use.[6] In this modeling framework, energy production, transformation, and end-use potentially employs more than 100 alternative energy

---

[4] See for example, Wigley et al. (1996) and IPCC (2001).
[5] It is important to note that we are not implying that others actually believe that CO_2 storage reservoirs are effectively unlimited. However, CO_2 storage reservoirs are often modeled as if this were the case.

[6] The model tracks nine primary energy forms (oil, natural gas, coal, biomass, nuclear fission, nuclear fusion, land-based solar, wind, and space-based solar). Fossil fuel resources are disaggregated by grade, while biomass energy is disaggregated by type (traditional fuels, waste recycling, and commercial energy crops). Biomass energy is derived from plants growing on land that in turn is modeled in an economic setting which allocates land resources to competing ends.

Table 16.3 Minimum cost of carbon storage by grade and resource grade (1990 $/TC).

Region	Coal basins	Depleted oil plays	Gas basins	Deep saline formation on-shore	Deep saline formation off-shore
Grade 1	<$1	<$1	$18	$18	$92
Grade 2	$7	$7	$44	$44	$220
Grade 3	$37	$37	$110	$110	$550
Grade 4	$183	$183	$367	$367	$1,100

technology options. Energy demands are established within the model and are determined by such factors as the size and age distribution of regional populations, the rate of labor productivity increase, the resources of fossil fuels, uranium, and other energy resources,[7] availability of energy technology options, energy market prices, and policy.

O^{bj}ECTS.MiniCAM version 2004.04 includes regional representations for the capacity and cost of CO_2 storage in the five reservoir types listed in Table 16.1. Each of these types is in turn disaggregated into four discrete grades of the storage resource. Each grade is described by a minimum cost of bringing the first ton of CO_2 from the site at which it was captured to the storage reservoir and injecting it into that reservoir grade, and a maximum extent of potential storage. This minimum cost by reservoir grade and reservoir class is given in Table 16.3. The marginal cost of storage is assumed to rise linearly between reservoir grades. Thus, the grade of the resource is defined by its cost of production and its extent is determined by the geology and geography of reservoir sites.

In each region where CO_2 is captured, it must be stored in a regional repository. We assume no inter-regional shipments of CO_2 for storage in extra-regional reservoirs.[8] We further assume that within a region competition drives the marginal cost of carbon storage to be equal across the five reservoir types: coal basins, depleted oil plays, gas basins, deep saline formations on-shore, and deep saline formations off-shore. Thus, regional resource availability and capacity differences mean that the marginal cost of storage can vary from region to region, even in a world in which emissions mitigation is assumed to be perfectly efficient. In this paper, we do not consider the potential CO_2 storage capacity of deep, saline-filled basalt flows (McGrail et al., 2003) or black shales (Nuttall et al., 2004). We also do not consider storage of CO_2 in the oceans. If included, these other reservoir types would surely increase our estimated CO_2 storage capacities.

In this analysis CCS is disaggregated into two components: CO_2 capture and CO_2 storage. We thus model the cost of CCS as the sum of the costs of capture and storage. CCS is deployed against a background of technology competition. For any activity in the model (e.g. electric power generation), a portfolio of technology options is deployed. The mix of technologies used in any region depends on the costs of options available in that region. CCS is a technology that is assumed available for electric power generation and energy transformation (e.g. hydrogen production from carbonaceous feedstocks) in this analysis. The regional cost of generating electric power (or transforming energy) using CCS in a specific region in a specific period is the sum of the cost of generating power (or transforming energy) using a technology that captures CO_2 plus the cost of transporting and storing the CO_2 in a suitable reservoir. Each energy technology option in the model has an efficiency of transformation and levelized cost of non-energy resources. Electric power generation or energy transformation infrastructure assets that employ CO_2 capture are assumed to have lower efficiency of transformation and higher levelized non-energy costs than similar alternative technology options without CO_2 capture. Characteristics of fossil fuel power generation technologies with and without CCS are given in Table 16.4.

The cost of CO_2 storage in turn depends on the reservoirs available in the region, the amount of the reservoirs' capacity previously filled, and the grade of the storage reservoirs being filled in that period. Reservoirs are assumed to be filled so that the marginal cost of adding the last ton of CO_2 to the last grade of each reservoir is equal across all reservoir types. That is, within any given region the cheapest reservoirs and reservoir grades are filled first. As time proceeds, reservoirs fill. Poorer grades of the reservoirs (e.g. reservoirs that lie further from the CO_2 source) come into play, and costs rise. The degree to which this occurs depends critically on the regional availability of reservoirs in each grade and the suite of reservoirs available. If some reservoirs are unavailable, then the cost of CCS will become supply-constrained. When all reservoirs in a region are filled, no additional CO_2 capture and storage is allowed. Once CO_2 is stored in a reservoir it is assumed to reside there indefinitely. That is, we do not consider the

---

[7] Oil, gas, and coal are treated as exhaustible, graded resources, subject to increasing cost as resources are depleted and decreasing costs as technologies improve. Uranium resources are treated similarly. Biofuels are treated differently. They are produced on land and must therefore compete with other land uses. Like other crops, they must compete on the basis of expected profitability. Hydroelectric power is treated exogenously as its deployment is constrained by non-energy considerations. Wind and solar power also compete based on economic performance, but must be paired with auxiliary power or power-storage options as their penetration exceeds 15% of total generation.

[8] While the exporting of CO_2 generated in one country to be sequestered in another country is already happening with the Weyburn enhanced oil recovery project where CO_2 from the United States is being transported into Canada, we treat projects like Weyburn as exceptions over the course of this century. We also show in this paper that the geologic CO_2 reservoir capacity in these two regions is sufficiently large that regional capacity is not a meaningful constraint for either the United States or Canada.

Table 16.4 Technology assumptions for carbon dioxide capture.

Technology	Parameter	2020	2050	2095
Advanced fossil electric without CO_2 capture				
Coal integrated gasification combined cycle	Efficiency	34%	34%	55%
	Non-energy costs[a]	$3.90	$3.90	$3.90
Natural gas combined cycle	Efficiency	38%	43%	60%
	Non-energy costs[a]	$2.00	$2.00	$2.00
Oil combined cycle	Efficiency	38%	41%	47%
	Non-energy costs[a]	$2.00	$2.00	$2.00
Fossil CO_2 capture technology incremental resource requirements				
Coal integrated gasification combined cycle	Efficiency penalty	25%	14%	11%
	Incremental non-energy costs[a]	$0.88	$0.59	$0.49
	Capture rate	90%	90%	90%
Natural gas combined cycle	Efficiency penalty	13%	9%	7%
	Incremental non-energy costs[a]	$0.89	$0.67	$0.53
	Capture rate	90%	90%	90%

[a] Capital plus operations and maintenance costs (1975 constant USD/GJ).

possible impermanence of storage (i.e. leakage) in this analysis.[9] Our focus here is solely on the question of the effect of a regionally distributed availability of storage reservoirs on the regional and global cost and deployment of CCS technologies.

As shown in Table 16.4, we have considered technological progress in CO_2 capture in terms of lower incremental cost and a smaller efficiency penalty. We have not explicitly considered technological progress in our representation of storage. Technological progress in storage would lead to more storage capacity at lower cost. The scenarios we have constructed with 0, 10, 50, and 100% of MPC can be thus seen as sensitivities to assumptions about technology, as moving from 0 to 100% of MPC leads to more storage at lower costs. One reason for only 10 or 50% of MPC being available could be that the remaining capacity presents technological challenges that are not overcome. More realistic scenarios might simulate a gradual increase on capacity supply over time (technological progress) or a decrease in capacity supply (leakage or environmental concerns develop that limit capacity in the future compared with what we expect today).

16.5 The reference scenario

For comparison purposes we begin with a reference scenario that places no limits on the release of CO_2 into the atmosphere other than those already in effect,[10] and in which CCS is not used as a carbon management technology.[11] This scenario is constructed for comparison purposes only – to be compared with other scenarios in which explicit limits on CO_2 emissions and explicit assumptions regarding CO_2 storage are introduced.

In the reference scenario global population grows to approximately 9 billion in the year 2050 and eventually stabilizes at a level of 9.5 billion people at the end of the twenty-first century. Populations in Europe, Japan, Korea, and China peak in the first half of the century and decline thereafter. Total primary energy grows from approximately 420 exajoules (EJ) per year in 2000 to approximately 1275 EJ per year in 2095 (Figure 16.1a). That energy is provided by fossil fuel supplemented by biomass, nuclear, solar, wind, and other renewables. That in turn leads to continued growth in fossil fuel CO_2 emissions which grow to approximately 22 000 TgC/yr by 2095 (Figure 16.1b). The geographic distribution of emissions shifts over time so that an increasing fraction of emissions is associated with non-Annex I nations.[12] Several of the regions in which populations peak and decline have emissions that peak and decline in the absence of policies to limit fossil fuel carbon emissions, including Europe, Japan, and Korea. Other regions exhibit growing emissions over most or all of the twenty-first century.

---

[9] We have explored this issue elsewhere. See, for example, Dooley and Wise (2003).

[10] While a variety of climate-motivated policies are already in effect in the real world, explicit limits, such as those embodied in the Kyoto Protocol, are assumed to expire and not be renewed in the reference case. Thus, this scenario should not be viewed as either a forecast or prediction. The function of this scenario is purely to create a well defined, if artificial, benchmark against which to compare other scenarios.

[11] In the reference case CCS is limited to niche applications such as tertiary recovery of oil and coal bed methane recovery.

[12] Annex I refers to a list of nations in the Framework Convention on Climate Change (United Nations, 1992): Australia, Austria, Belarus, Belgium, Bulgaria, Canada, Croatia, Czech Republic, Denmark, Estonia, European Community, Finland, France, Germany, Greece, Hungary, Iceland, Ireland, Italy, Japan, Latvia, Lithuania, Luxembourg, The Netherlands, New Zealand, Norway, Poland, Portugal, Romania, Russian Federation, Spain, Sweden, Switzerland, Turkey, Ukraine, United Kingdom of Great Britain and Northern Ireland, and the United States of America.

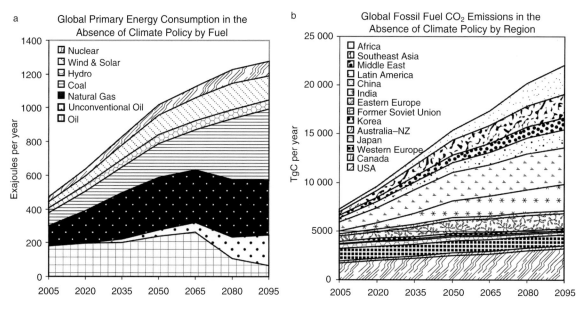

Figure 16.1 Primary energy and fossil fuel emissions in the non-climate policy reference scenario.

16.6 Carbon dioxide concentrations and the global value of carbon

In the absence of external limitations, CO_2 concentrations rise to more than 700 ppm in the reference case. Limiting concentrations to levels ranging from 450 to 650 ppm therefore require limits on emissions. In this analysis, those limits are achieved by placing a value on carbon emissions. The value of carbon is set so as to trace out an emissions trajectory consistent with Wigley et al. (1996). The marginal value of carbon faced by the global economy across this century is shown in Figure 16.2. Marginal carbon values are shown for Cases 0, 1, 2, 3, and 4 for the three concentration limitations, 450, 550, and 650 ppm.

For all three of the CO_2 concentrations examined, we note that the marginal value of carbon is generally higher when regionally explicit geologic resources are considered than when storage is assumed to be unlimited. This is not universally the case. The improved parameterization of the CO_2 storage resource presented in this paper allows for the fuller representation of the range of CO_2 storage costs likely to be encountered in practice including the representation of so-called value-added CO_2 storage reservoirs (i.e., CO_2-driven enhanced oil recovery and CO_2-driven enhanced coal bed methane recovery). The inclusion of these value-added CO_2 storage reservoirs reveals its importance by depressing the marginal value of carbon at low carbon prices. That is, not only does the fixed marginal cost of storage parameterization (Case 0) underestimate long-term CO_2 storage costs but it also misses critical low-cost early deployment opportunities for CCS.

Note also that for any CO_2 concentration, marginal carbon values or carbon prices are universally lower when CCS is available than when it is not. By the year 2095 the marginal value of carbon or carbon emission price is ordered, lowest to highest, Case 0, Case 1, Case 2, Case 3, and Case 4. That is, unlimited storage is associated with the lowest marginal cost of carbon in all cases regardless of the CO_2 concentration limit. All regionally specific storage resource cases are more expensive than the unlimited Case 0 and those in turn ordered by the amount of the maximum potential resource assumed available. In general, for any of the cases in which we have explicitly modeled CO_2 storage reservoirs as a region-specific, finite and graded resource, i.e. Cases 1, 2, or 3, the further the scenario proceeds into the future, the greater the marginal value of carbon or carbon price compared with Case 0 (CO_2 storage is an infinitely available resource available at a fixed price). Yet even in cases with relatively constrained amounts of CO_2 storage opportunities (e.g. Cases 2 and 3), the marginal cost of carbon remains significantly below Case 4, in which CCS systems are not able to deploy as part of society's response to climate change.

The marginal value, relative to Case 0, shown in Figure 16.2 differs between Cases 1, 2, and 3 for any given year. For Case 1 the increase in the marginal value of carbon relative to Case 0 ranges from 4% to 10% when 100% of the estimated reservoir capacity is available. While costs are somewhat higher in the Case 2 than in Case 1, a more significant increase in marginal value of carbon is observed in Case 3 when the assumed resource availability declines to 10%. With only 10% of the maximum potential estimated reservoir capacity available, the marginal value of carbon rises between 28% and 36%, relative to Case 0, in Case 3, depending on the concentration limitation.

Figure 16.2b shows the corresponding cumulative storage of CO_2 measured in TgC in geologic reservoirs. Values range from more than 80 PgC when the maximum assumed storage

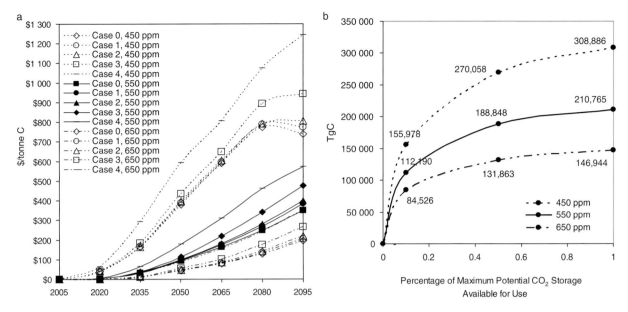

Figure 16.2 a, Time profile of marginal value of carbon for three stabilization concentrations: 450 ppm, 550 ppm, and 650 ppm, for Case 0 (unlimited storage), Case 1 (100% maximum potential capacity (MPC) available), Case 2 (50% MPC available), and Case 3 (10% MPC available), and Case 4 (CCS is unavailable). b, Cumulative storage 1990 to 2095.

is only 10% of the MPC and the CO_2 concentration is 650 ppm to 300 PgC when the maximum assumed storage is 100% of the MPC and the CO_2 concentration is 450 ppm. Clearly the relationship between CCS and the fraction of the MPC available is non-linear. When the maximum assumed storage is 10% of the MPC, the deployment of CCS is more than half of the value when the maximum assumed storage is 100% of the MPC for any CO_2 concentration. This finding emphasizes the value of CCS as an element in society's portfolio of energy–technology options even if only a fraction of its full potential is realized. If CO_2 storage reservoirs are available and if policies, rules and regulations are in place that allow those reservoirs to be used cost effectively, CCS systems will deploy more extensively in stringent CO_2 stabilization scenarios.

16.7 The regional marginal cost of storage

The heterogeneous distribution of the global geologic CO_2 storage resource, coupled with regionally heterogeneous energy systems and varying hypothetical limits on atmospheric CO_2 concentrations, results in a wide range in the marginal cost of storage and annual rates of CO_2 injection into reservoirs around the world. For the three cases 1, 2, and 3, in which we explicitly represent CO_2 storage as a regionally specific, graded, and finite natural resource as outlined in Table 16.1, three general patterns present themselves.

1. Regional CO_2 storage is effectively unlimited and is never a meaningful constraint;
2. Regional CO_2 storage is finite and becomes a meaningful constraint around the middle of the twenty-first century; and
3. Regional CO_2 storage is a meaningful constraint before the middle of the twenty-first century.

These three circumstances reflect the interplay of national energy systems, potential regional reservoir capacity and the CO_2 concentration limit. Figures 16.3 and 16.4 show these patterns for nine cases examined – Cases 1, 2, and 3 in combination with CO_2 concentrations of 450, 550, and 650 ppm. Figure 16.3 shows the regional patterns of storage costs while Figure 16.4 shows patterns of annual regional storage rates in the same nine cases.

Figure 16.3 shows that for any region costs rise monotonically with time. This reflects the assumption that low-cost reservoirs will be filled before higher-cost reservoirs. Costs are lowest in Case 1 with a CO_2 concentration of 650 ppm. Conversely they are highest for Case 3 with a CO_2 concentration of 450 ppm. For any region, the marginal cost of storage rises as available storage diminishes. Thus, for any CO_2 concentration the regional marginal cost of storage in any year is lower in Case 1 than in Case 2 and is lower in Case 2 than in Case 3.

Similarly, Figure 16.3 shows that for any case, the marginal cost of storage is higher the lower the CO_2 concentration. That is, for any case, the marginal cost of storage experienced in any region in any year for the 550 ppm CO_2 concentration is higher than for the 650 ppm CO_2 concentration and lower than for the 450 ppm CO_2 concentration. This follows from the fact that allowable emissions in any year are lower for lower CO_2 concentrations, and that there is a consequent greater deployment of CCS technology for lower CO_2 concentration. This in turn accelerates reservoir fill rates and raises regional costs as the region moves through its endowment of

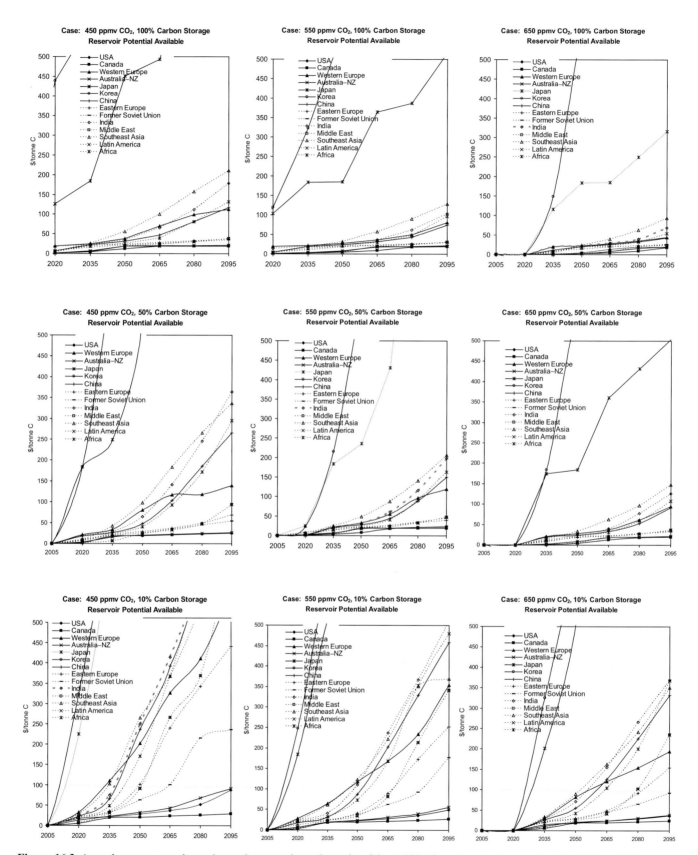

Figure 16.3 Annual storage costs by region and year at three alternative CO_2 stabilization concentration levels: 450, 550, and 650 ppm.

Technology in an integrated assessment model

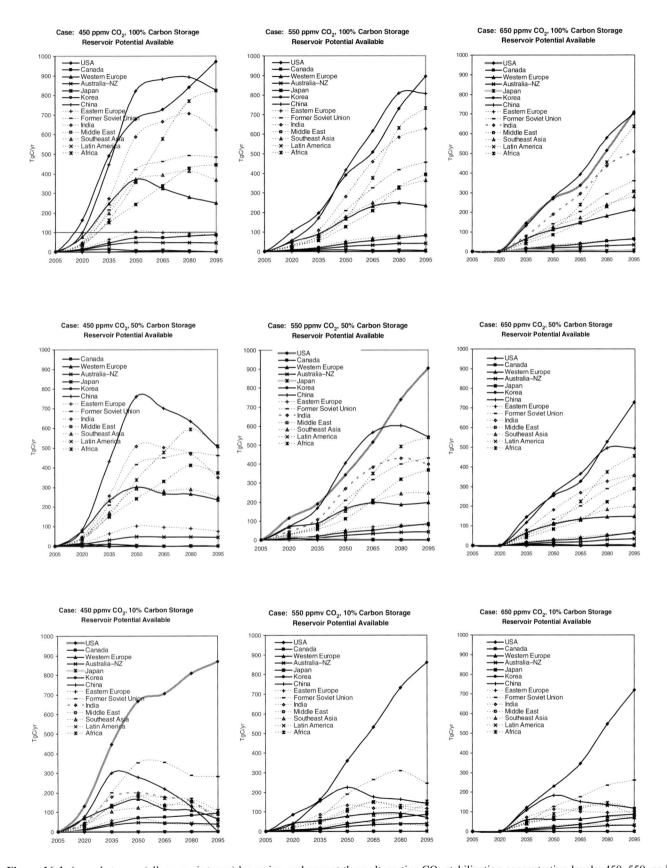

Figure 16.4 Annual storage (all reservoir types) by region and year at three alternative CO_2 stabilization concentration levels: 450, 550, and 650 ppm.

lower prices, 'value-added' CO_2 storage grades, and on to the non-value-added CO_2 storage reservoirs that are likely to be the dominant reservoirs over the course of this century and in certain regions perhaps well beyond the twenty-first century.[13]

Some regions have high rates of carbon emission and relatively limited potential storage reservoir capacity and always exhibit a willingness to pay for storage opportunities that exceed $200/tonne C, regardless of the concentration at which atmospheric CO_2 is stabilized and the degree to which maximum potential geologic storage is assumed to be available.[14] Only three regions, Australia–New Zealand, Canada, and the United States, have sufficiently abundant storage capacity relative to their emissions that the marginal cost of storage never exceeds $100/tonne C regardless of the CO_2 stabilization concentration and the degree to which maximum potential geologic storage is assumed to be available. Costs in other regions range from less than $100/tonne C to $500/tonne C or more depending on the degree to which maximum potential geologic storage is assumed to be available and the CO_2 stabilization target.

The interaction between reservoir availability and emissions mitigation can also be seen playing out in Figure 16.4, which shows regional patterns of physical storage over time for the nine combinations of Cases 1, 2, and 3 with concentrations 450 ppm, 550 ppm, and 650 ppm. Three general patterns of annual storage are reflected in Figure 16.4. For some regions, storage capacity is effectively unconstrained, prices remain low, and use grows throughout the century regardless of CO_2 stabilization concentration. Other regions follow this pattern in some cases but not others. As Figure 16.4 shows, there are many instances in which the rates of CO_2 storage grows for some time, but then peaks and begins to decline in the second half of the century. This is the direct consequence of the depletable resource nature of CO_2 storage reservoirs. As the cheapest grades of the resource are exploited, costs rise and reduce the fill rates. A third pattern of reservoir usage is one of severely limited capacity relative to regional emissions resulting in rapidly escalating prices of storage and usage that quickly peaks and declines to low levels. These three categories largely reflect the abundance of potential storage relative to unconstrained emissions.

16.8 The regional pattern of cumulative CO_2 storage over the twenty-first century

Figure 16.5 shows cumulative CO_2 storage by region in TgC for Cases 1, 2, and 3 at 450, 550, and 650 ppm. Figure 16.6 organizes the information into panels showing Cases 1, 2, and 3 in each panel and successive panels showing CO_2 concentrations for 450, 550, and 650 ppm. In each panel there is a clear interaction between maximum assumed storage capacity and CCS deployment. It is interesting to note that in Case 1 (in which each region is able to make full use of the CO_2 storage potentially outlined in Table 16.1), regardless of CO_2 concentration considered, China is the most intense user of CCS technologies, when measured as the total amount of CO_2 stored over the course of the century. The United States and India are the next two most intense users of CCS technologies. However, when CO_2 storage capacity is assumed to be only 50% of the amounts listed in Table 16.1 (Case 2) this order changes, with the United States, China and India in that order being the most intensive users of CCS regardless of CO_2 concentration. In Case 3 where only 10% of the storage capacity listed in Table 16.1 is available, the order shifts again. The United States is the largest user of CCS technologies owing to its large potential storage capacity and its assumed large demand for CO_2 storage driven by the size of its economy. The Former Soviet Union and China are the next two most intensive users of CCS with India dropping to fourth regardless of CO_2 concentration. In all cases Japan and Korea are the least intensive users of CCS, highlighting the relative paucity of potential CO_2 storage opportunities within their borders and under nearby coastal waters.

Figure 16.6 organizes the same information by case and shows CCS relative to the fraction of maximum assumed capacity available in that scenario. Figure 16.6c shows Case 3 where the maximum assumed storage is limited to 10% of the values listed in Table 16.1. In this case, *global average utilization* of the potentially available resource never reaches 50%, even in the more stringent Case 3–450 ppm scenario. However, in the Case 3–450 ppm scenario 9 of the 14 regions have filled more than 75% of their maximum assumed storage and 7 of those 9 regions have filled more than 90% of their maximum assumed storage. This pattern of usage as a percentage of the maximum assumed storage attests to the very heterogeneous distribution of the CO_2 storage resource across

---

[13] See for example Dooley *et al.* (2004) which shows for the case of North American how so-called value-added reservoirs are consumed in early periods of a mitigation regime with the CO_2 that was once going towards these value-added reservoirs shifting in later periods to the region's massive deep saline formations and depleted gas fields, both of which are non-value-added reservoirs.

[14] It is important to stress once again that a significant advance in the modeling of CCS technologies is being presented in this paper through our ability to model CO_2 storage reservoirs as a graded resource. The lowest-cost grades, which are very low-cost, are intended to be representative of situations in which large stationary anthropogenic CO_2 sources are located close to so-called value-added reservoirs in which it is hoped that the proceeds from the CO_2-driven advanced hydrocarbon recovery (oil or natural gas) can be used to offset the cost of CO_2 transport, injection into a suitable geologic reservoir, and measurement, monitoring, and verification. The highest-cost grades, in contrast, are meant to be indicative of situations in which CO_2 must be transported over great distances to remote and perhaps hard to reach storage reservoirs (e.g. a deep saline formation located beneath the seafloor some distance from the shore). This is not to say that we necessarily believe that all CO_2 storage opportunities will be fully exploited (including those in the higher-cost categories) but rather that CCS technologies appear to be so valuable that at the end of this century when carbon prices are in excess of a couple of hundred dollars per tonne of carbon it appears as if CO_2 capture coupled with these higher-cost grades of storage opportunities could be cheaper than other available abatement options.

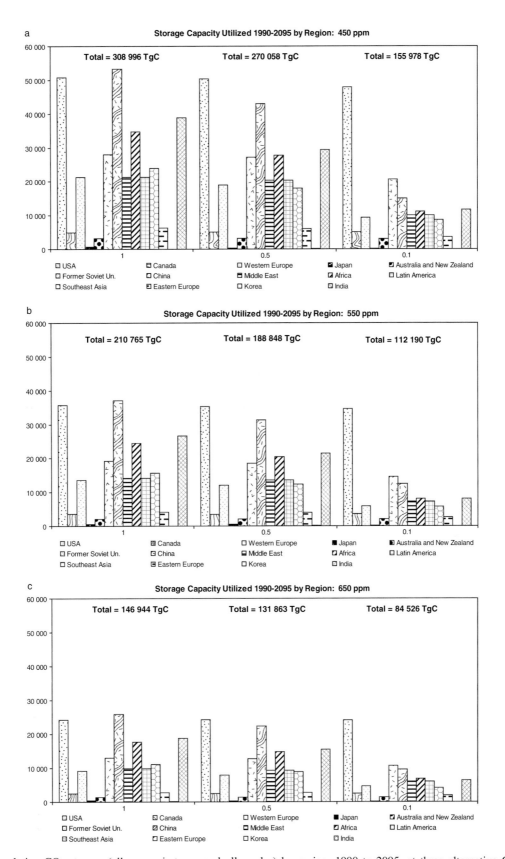

Figure 16.5 Cumulative CO_2 storage (all reservoir types and all grades) by region, 1990 to 2095, at three alternative CO_2 stabilization concentration levels: 450, 550, and 650 ppm and three alternative levels of availability relative to the maximum potential capacity of Table 16.1.

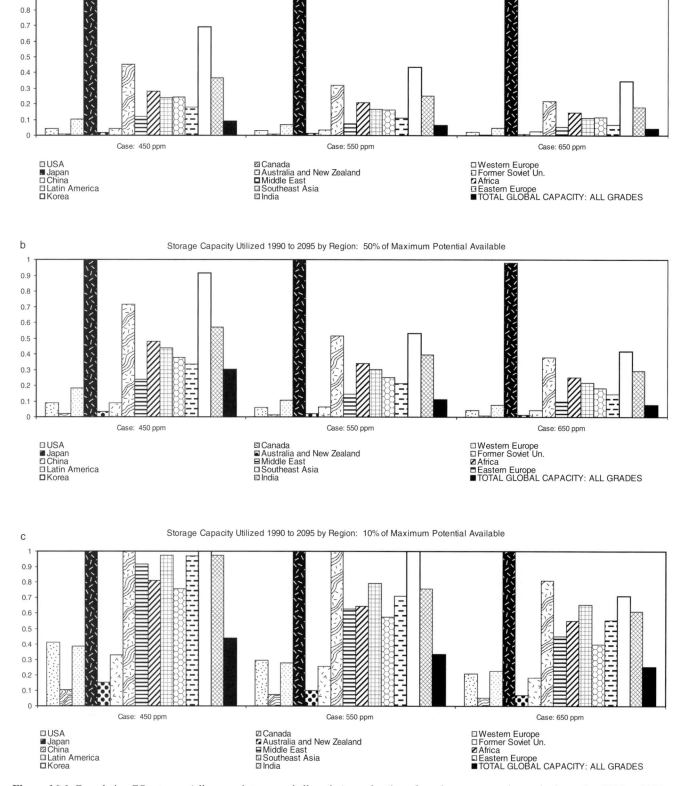

Figure 16.6 Cumulative CO_2 storage (all reservoir types and all grades) as a fraction of maximum assumed capacity by region 1990 to 2095 at three alternative CO_2 stabilization concentration levels: 450, 550, and 650 ppm.

the globe and the equally heterogeneous demand for CO_2 storage driven by differing growth rates, population, and natural resource endowments across the globe over the course of this century. Even though Japan's available CO_2 storage resource appears to be quite limited, CCS is such a valuable tool for addressing climate change that even in the most generous case (Case 1 in which 100% of the maximum potential of Table 16.1 is available) and with CO_2 concentrations stabilized at 650 ppm, 88% of Japan's available capacity is exploited by 2095.

The situation in Figure 16.6a is very different. This panel represents Case 1 in which 100% of the capacity outlined in Table 16.1 is available. Here average global utilization of CO_2 storage reservoirs over the period 1990 to 2095 never exceeds 10% of the maximum assumed storage, even for the most stringent concentration level. Only two regions, Japan and Korea, use more than 50% of potential storage capacity even for the 450 ppm CO_2 concentration.

Case 2, in which 50% of the storage capacity listed in Table 16.1 is available, is shown in Figure 16.6b. Case 2 lies between Cases 1 and 3, but is more similar to Case 1 than Case 3. With the exceptions of Japan and Korea, 30% or more of the maximum assumed storage (in this case 50% of the MPC) remains available in 2095. For most regions more than half of their maximum assumed storage remains available in 2095.

16.9 Technology choice and regional storage

Creating regional markets for CO_2 storage can potentially lead to different regional technology choice when compared with results from models that assumed a fixed cost of universally available unlimited CO_2 storage. To explore this difference, we compare Case 0 with Cases 1, 2, and 3. Case 0 places no limits on storage, but assesses a fixed flat storage fee, $15/tC. This assumption characterized earlier analysis, as for example in Edmonds et al. (2002). Not surprisingly Case 0 results in greater global total cumulative deployment of CCS technologies over the period 1990 to 2095 than in Cases 1, 2, or 3, as the cost for CO_2 transport and storage is inappropriately never allowed to escalate beyond $15/tC no matter how intensively CCS systems are deployed. Case 0 scenarios show approximately double the cumulative storage of Case 3 scenarios over the period 1990 to 2095, and approximately 15% greater cumulative storage than Case 1 scenarios.

The precise pattern varies from region to region and from case to case. Figure 16.7 compares Cases 0 and 2 at 550 ppm CO_2 concentration in four panels. Panel a shows global carbon values for the period 2020 to 2095. As noted earlier, the value of carbon is lower in Case 0 than in Case 2. Panel b shows the corresponding cumulative global CO_2 capture and storage. Cumulative global CCS is higher for Case 0 than for Case 2. Panel c shows the regional pattern of storage costs for Japan and the United States for these two cases, and Panel d shows cumulative storage in Japan and the United States. In Japan, where geologic storage capacity appears to be quite limited, the marginal cost of storage is higher in 2020 in Case 2 than in Case 0 (fixed price case) and escalates in succeeding periods. This leads to a significantly lower deployment of CCS in Japan in Case 2 than in Case 0 as the more realistic parameterization of CO_2 storage reservoirs in Case 2 results in other carbon management technologies appearing to be more competitive than they would be in the unrealistic assumptions of Case 0. The United States exhibits a very different pattern. Here, cumulative storage is actually greater in Case 2 than in Case 0. In the periods before 2050, storage costs are lower in the United States in Case 2 than in Case 0. The combination of a globally common marginal value of carbon and a regionally heterogeneous availability of storage sites in Case 2 acts to shift the deployment of CCS between regions as compared with Case 0. Even after 2050 United States cumulative emissions remain greater in Case 2 than in Case 0, because the global carbon value is higher in Case 2 than in Case 0 by more than the regional marginal cost of storage.

As would be expected, the degree of availability of suitable CO_2 storage reservoirs exerts an influence on the relative long-term contribution of CCS in overall emissions mitigation. We define emissions mitigation as the difference between reference case emissions and emissions in the case where controls are applied. The contribution of CCS to total mitigation is therefore defined as the ratio of CCS deployment to total mitigation. In the near term, the ratio is strongly influenced by the CO_2 concentration level. Before the middle of the century, the contribution of CCS is higher in all of the 450 ppm cases than in others. Of course, the greater the availability of potential storage, the greater is its contribution. In Case 1, for example, more than 45% of emissions mitigation is accounted for by CCS in 2035. And, even in Case 3, more than one-third of emissions mitigation is accounted for by CCS with the 450 ppm CO_2 concentration. Over time the influence of the availability of potential regional CO_2 storage options increases. By the end of the century, the contribution of CCS to cumulative mitigation is largely shaped by storage availability. CCS accounts for approximately 20% of cumulative mitigation in all instances of Case 3, approximately 30% of cumulative mitigation in all instances of Case 2, and approximately 35% of cumulative mitigation in all instances of Case 1.

These differences in regional CCS patterns between Cases 0 and 1, 2, or 3 reflect differences in the underlying global energy system. Japan, for example, deploys significant fossil-fuel-using capacity coupled with CCS in power generation as a response to the hypothetical emissions mitigation regimes examined here when CO_2 storage is assumed to be infinitely available at a fixed fee (Case 0). This is shown in Figure 16.8. However, in Case 2, little CCS-enabled fossil-fired generation is deployed as an emissions mitigation option, as the potential CO_2 storage capacity in and near Japan appears to be too limited and too expensive. The shift from Case 0 to Case 2 is accompanied by a shift from CCS to nuclear power,

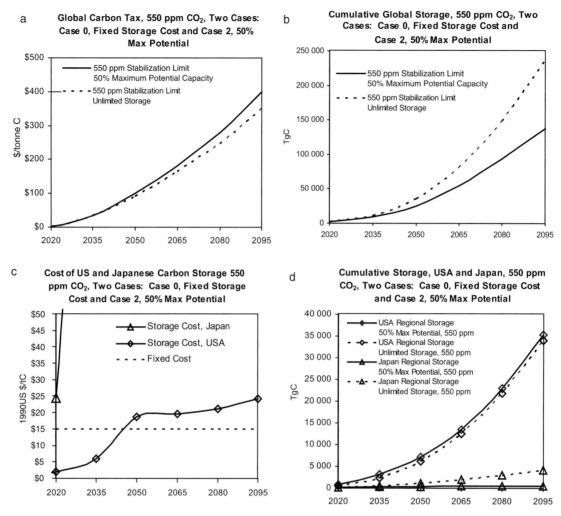

Figure 16.7 Global and regional patterns of costs and deployment of CCS.

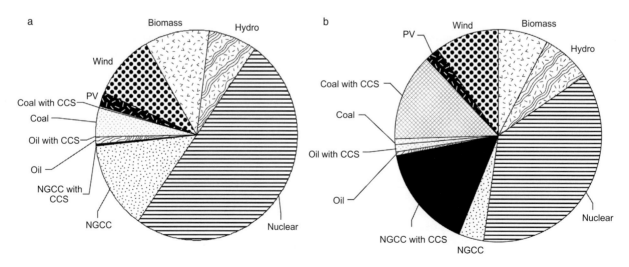

Figure 16.8 Japanese power generation in 2095 by source of power: 550 ppm CO_2 concentration. a, Case 2, regional CCS 50% maximum potential; b, Case 0, fixed cost storage with unlimited storage.

Table 16.5 Present discounted cost of stabilizing the concentration of CO_2 (discounted at 5% per year 2095 to 2005, trillions of 1990 US dollars).

CO_2 concentration	Case 0 Unlimited storage	Case 1 100% MPC	Case 2 50% MPC	Case 3 10% MPC	Case 4 No storage
450 ppm	$10.2	$10.3	$10.5	$11.6	$15.4
550 ppm	$1.8	$1.9	$2.0	$2.4	$3.2
650 ppm	$0.6	$0.6	$0.7	$0.8	$1.2

commercial biomass fuels, and natural gas combined cycle (NGCC) turbines.[15]

16.10 The economic value of CCS

The introduction of regionally specific CO_2 storage reservoir endowment has a marginal effect on cumulative deployment relative to Case 0, where it is assumed to be universally available in unlimited supply at a fixed cost. To explore the effect of regionally disaggregated CO_2 storage reservoir endowment we have computed the present discounted cost of stabilizing the concentration of CO_2 for all cases. These are reported in Table 16.5. Note that these values should be interpreted as minimum values as a common global value of carbon is assumed which escalates at an efficient rate. Actors are assumed to anticipate carbon values fully and act appropriately. Costs would be higher if carbon values were set at non-optimum rates or if some regions of the world did not participate from the outset. Table 16.5 is useful for comparative purposes.[16] That is, it is useful to explore relative effects of technology on cost. The table shows that there is relatively little difference (single digit percentage changes) in the cost of stabilization between Cases 0 and 1, regardless of the CO_2 concentration. As the available CO_2 storage potential decreases, the cost of meeting the CO_2 concentration rises. The change is non-linear with the fraction of assumed available CO_2 storage capacity. Reducing available storage from 100% to 50% raises costs less than reducing available CO_2 storage capacity from 50% to 10%, and so forth.

The difference in cost between Case 4 and Cases 0, 1, 2, and 3 can be interpreted as the value of having CCS available, given the suite of other available technologies. The value of CCS plotted against the fraction of MPC is given in Figure 16.9. The relationship is highly non-linear. The value changes relatively little between Cases 1 and 2. There is a more significant change between Cases 2 and 3. However, most of the change occurs between Cases 3 and 4. The implication is that even 10% of our estimated MPC provides a significant reduction in the cost of meeting a CO_2 concentration limit.

This analysis reaffirms the value of development and deployment of CCS technology. Furthermore, it finds that the value of CCS is robust against a variety of available scales of deployment. Of course, this analysis does not address the many important steps that must be taken for that potential to be realized. But it does suggest that the regional distribution of CCS does not significantly alter the basic finding of earlier analyses, that CCS has significant potential to help control the cost of stabilizing the concentration of atmospheric CO_2. However, the presence or absence of CO_2 storage opportunities in any given region can have a significant impact on the future evolution of the region's energy infrastructure in a carbon constrained world.

16.11 Final remarks

This analysis examines the global potential of CCS in more detail than previously. Specifically, this study treats CCS as a large, but finite, graded resource that is available regionally. This study both confirms important previous results and provides insights that were not heretofore available. First, the value of CCS in addressing climate change is reaffirmed. Regardless of the CO_2 concentration, the reduction in cost associated with CCS availability was denominated in hundreds of billions to trillions of present discounted dollars. Even if only 10% of MPC is available, the economic value of the technology remains strong. Furthermore, deployment of the technology is large across all of the cases and concentrations examined. Even if the CO_2 concentration is 650 ppm and only 10% of MPC is available, 312 $PgCO_2$ (85 PgC) were captured and stored over the period to 2095. In cases with lower CO_2 concentrations or a higher percentage of MPC available, cumulative storage can rise above 1100 $PgCO_2$ (300 PgC).

Our estimates of the geographic distribution of candidate CO_2 storage reservoirs indicate that they are not distributed evenly over the globe. This CO_2 storage resource appears to be available in great abundance relative to future cumulative carbon emissions over the course of this century in some regions, e.g. the United States, Canada, the Former Soviet Union, and Australia–New Zealand, while severely limited in others, e.g. Japan and Korea. Most regions of the world find

---

[15] Potential expansion of wind and photovoltaic electric power is tempered in this analysis by the requirement for electricity storage capacity for deployments beyond 15% of total generation. With greater experience in integrating large wind systems into the grid, these values may expand relative to other non-carbon sources.

[16] As noted earlier, assumptions such as the commonly applied price for carbon emissions administered uniformly in all regions in all periods bias the cost estimates downward.

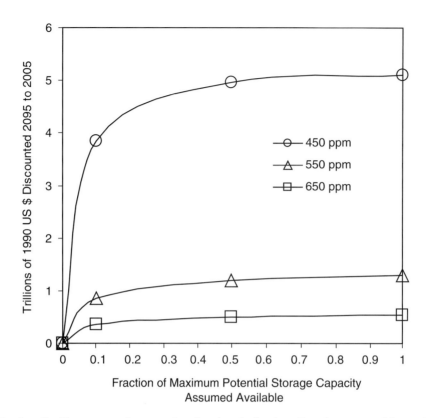

Figure 16.9 The value of carbon dioxide capture and storage plotted against the fraction of maximum potential capacity. The present discounted economic cost for four cases were computed for each stabilization level (650 ppm, 550 ppm, and 450 ppm): Cases 1, 2, 3, and 4. For each stabilization level and case, costs were computed for each year over the period 2005 to 2095. Each year's costs were discounted from the year in which they occurred to their 2005 equivalent using a 5%/yr discount rate. Discounted costs for all years were then summed to give a present discounted cost. Costs are reported in year 1990 inflation-adjusted US dollars. The value of CCS for a given case and stabilization level is computed by subtracting the present discounted cost for the given case and stabilization level from the present discounted cost of Case 4 at that stabilization level.

themselves endowed with CO_2 storage reservoir resources that lie somewhere between these two extremes. Our estimates are based on the available literature, and for many important regions of the world current knowledge about the extent, quality, and location of candidate CO_2 storage reservoirs is limited. For regions such as China and India, which fall into this middle category, and for which the amount of CO_2 that might be stored in their reservoirs over the course of this century could be large, improved knowledge of these reservoirs would be a high priority for assessing the value of the CCS option in these regions. The heterogeneous distribution of potential CO_2 storage reservoirs was shown to result in relatively greater deployment of CCS technology in those regions with relatively more abundant endowments of potential storage reservoirs.

We recognize the preliminary nature of this work. A great deal of additional research remains to be undertaken both to improve the present analysis and to address a broader set of questions associated with CCS including, for example, the implications of cross-border CO_2 trade. That is, there is no inherent reason that CO_2 could not be captured in one region and shipped to another, either in pipelines or ships, for storage. Similarly, future work is needed to investigate the implications of an expanded suite of reservoirs, reservoir retention, and the potential role of technological change for CCS in general and storage technology in particular.

Acknowledgements

The authors would like to express their appreciation to John Reilly for his many helpful editorial comments. If he makes many more contributions we will have to make him a co-author. We are also indebted to Brian Flannery for comments and suggestions on an earlier draft of this paper. In addition, the authors benefited from comments, consultations, and counsel with Gerry Stokes, Hugh Pitcher, Sonny Kim, Steve Smith, and John Weyant as well as feedback received when these results were first presented to the Tenth Annual Meeting on the Integrated Assessment of Climate Change in Snowmass, Colorado, in August, 2004.

References

CDIAC (Carbon Dioxide Information Analysis Center) (2004). http://cdiac.esd.ornl.gov/

Dooley, J. J. and Friedman, S. J. (2004). *A Regionally Disaggregated Global Accounting of CO_2 Storage Capacity: Data and*

Assumptions. Battelle Pacific Northwest Division Technical Report Number PNWD-3431.

Dooley, J. J. and Wise, M. A. (2003). Potential leakage from geologic sequestration formations: allowable levels, economic considerations, and the implications for sequestration R&D. In *Greenhouse Gas Control Technologies: Proceedings of the Sixth International Conference on Greenhouse Gas Control Technologies*, Vol. 1, ed. J. Gale and Y. Kaya. Elsevier Science, pp. 2373–2378.

Dooley, J. J., Dahowski, R. T., Davidson, C. L. *et al.* (2004) A CO_2 storage supply curve for North America and its implications for the deployment of carbon dioxide capture and storage systems. In *Proceedings of seventh International Conference on Greenhouse Gas Control Technologies. Volume 1: Peer-Reviewed Papers and Plenary Presentations*, ed. E. S. Rubin, D. W. Keith and C. F. Gilboy. Cheltenham, UK: IEA Greenhouse Gas Programme, pp. 593–604. http://uregina.ca/ghgt7/PDF/papers/peer/282.pdf

Edmonds, J., Freund, P. and Dooley, J. J. (2002). The role of carbon management technologies in addressing atmospheric stabilization of greenhouse gases. In Williams DJ, Durie RA, McMullan P, Paulson CAJ and Smith YA editors. *Greenhouse Gas Control Technologies. Proceedings of the Fifth International Conference on Greenhouse Gas Control Technologies*, ed. R. A. Durie, D. J. Williams, A. Y. Smith *et al.* Collingswood, VIC: CSIRO, 46–51.

Edmonds, J., Clarke, J., Dooley, J., Kim, S. H. and Smith, S. J. (2004). Stabilization of CO_2 in a B2 world: insights on the roles of carbon capture and storage, hydrogen, and transportation technologies. *Energy Economics (Special Issue)* **26** (4), 517–537.

Freund, P. and Ormerod, W. G. (1997). Progress toward storage of carbon dioxide. *Energy Conversion and Management* **38**, Supplement, pp. S199–S204.

Herzog, H., Drake, E. and Adams, E. (1997). *CO_2 Capture, Reuse, and Storage Technologies for Mitigation Global Climate Change*. Cambridge MA: MIT Energy Laboratory.

IPCC (2001). *Climate Change 2001: The Scientific Basis. The Contribution of Working Group I to the Third Assessment Report of the Intergovernmental Panel on Climate Change*, ed. J. T. Houghton, Y. Ding, D. J. Griggs *et al.* Cambridge: Cambridge University Press.

McGrail, B. P., Ho, A. M., Reidel, S. P. and Schaef, H. T. (2003). Use and features of basalt formations for geologic sequestration. In *Proceedings of the Sixth International Conference on Greenhouse Gas Control Technologies*, ed. J. Gale and Y. Kaya. Elsevier Science Ltd., Oxford, pp. 1637–1640.

Nuttall, B. C., Eble, C., Bustin, R. M. and Drahovzal, J. A. (2004). Analysis of Devonian black shales in Kentucky for potential carbon dioxide sequestration and enhance natural gas production. In *Proceedings of the Third Annual Conference on Carbon Sequestration*. Washington DC: Monitor Exchange Publications and Forums.

United Nations (1992). *Framework Convention on Climate Change*. New York: United Nations.

Wigley, T. M. L., Richels, R. and Edmonds, J. A. (1996). Economic and environmental choices in the stabilization of atmospheric CO_2 concentrations. *Nature* **379**, 240–243.

17

Hydrogen for light-duty vehicles: opportunities and barriers in the United States

James L. Sweeney

17.1 Underlying energy policy issues

Three fundamental issues are now explicit in energy policy – at least in US energy policy – and have been explicit or implicit in energy policy at least since the very early 1970s:

- Reducing environmental impacts of energy production, distribution, use;
- Providing security against supply system disruption;
- Supplying and using plentiful energy at a reasonable cost.

This paper relates these issues to the quest for a new energy carrier – molecular hydrogen – which might take a place comparable to that of electricity. The paper focuses attention on molecular hydrogen for use in light duty vehicles.

The United States has learned or is learning to deal with most of the worst environmental impacts of energy use. But there is one problem we have not learned to control – carbon dioxide (CO_2) releases from combustion of fossil fuels.

Internationally, the Kyoto Protocol is a response but it has been rejected in the United States. The protocol tells what commitments are expected by various countries but does nothing to make such changes economically viable. To meet the goals requires not simply institutional and economic changes, it needs technological advances.

Figure 17.1 provides 2003 data on US sources of CO_2 emissions (EPA, 2005). The largest two sources are coal in electricity generation and petroleum in transportation, particularly in light duty vehicles. Technologies that reduce CO_2 from electricity generation or from transportation vehicles could, in principle, be very important for addressing this serious environmental problem.

In the United States, public policy on energy security goes back at least to the taskforce appointed by President Eisenhower to examine energy security implications of increasing oil imports. The 1973 Arab OPEC reduction in oil production led to a worldwide depression and indirectly to worldwide inflation.

There has been much progress since 1973. But recent events suggest that we need renewed attention on oil supply vulnerability. World oil demand has been growing for the past decade at about 1.5% per year, rapidly enough to put significant pressure on the world oil market, while oil reserves have not been keeping up. World oil markets may be tight for decades to come. The tighter the market, the greater the price jump that would stem from an oil supply disruption and the more damaging would be the impacts on the world economy.

Thus, the challenge is to create technologies that allow the United States – and all nations – the availability of plentiful energy at reasonable cost, while sharply reducing atmospheric CO_2 releases and reducing the vulnerability of the oil supply system. Molecular hydrogen used in light duty vehicles has the potential to meet that challenge if fundamental barriers can be overcome.

17.2 Hydrogen: an emerging energy carrier?

There are similarities between molecular hydrogen and electricity. Those similarities suggest the potential of hydrogen as a parallel to electricity for the energy system.

First, molecular hydrogen, like electricity, is not a primary energy source, but must be produced from primary sources. Thus hydrogen, like electricity, is an "energy carrier."

Human-induced Climate Change: An Interdisciplinary Assessment, ed. Michael Schlesinger, Haroon Kheshgi, Joel Smith, Francisco de la Chesnaye, John M. Reilly, Tom Wilson and Charles Kolstad. Published by Cambridge University Press. © Cambridge University Press 2007.

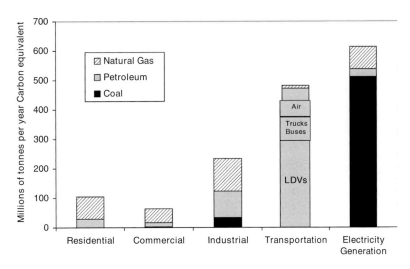

Figure 17.1 United States CO_2 emissions: 2003.

Second, there is a potential that, like electricity, hydrogen could ultimately be usable to provide virtually all energy services. With hydrogen fuel cells that could economically generate electricity using hydrogen at the point of electricity use, hydrogen and electricity would compete as energy carriers. And with economical hydrogen storage, hydrogen could satisfy all uses to which electricity is put and it could be stored and used for mobile purposes, particularly transportation.

Third, use of hydrogen to generate electricity or direct combustion of hydrogen is carbon-dioxide free *at the point of use*. And such electricity generation does not release criteria pollutants; it simply releases water and heat after it combines with oxygen.

Fourth, hydrogen, like electricity, can be produced using any primary energy resources: coal, natural gas, oil, hydropower, nuclear, solar energy, wind, biomass, and geothermal energy. Coal or biomass can be gasified to produce hydrogen. Natural gas can produce hydrogen through steam shift reforming. High-temperature nuclear reactors might provide the capability to dissociate water into hydrogen and oxygen. And any electricity can create hydrogen through electrolysis.

Given the wide variety of potential feedstocks, hydrogen, like electricity, allows the *potential* for any nation to harness any available primary energy resources. Since many primary energy sources are broadly distributed around the world, use of hydrogen as an energy carrier has the potential of sharply reducing the security risk of highly geographically concentrated supplies of hydrocarbons.

Fifth, although CO_2-free at the point of use, hydrogen, like electricity, may or may not be CO_2-free at the point of production. Hydro-, solar, nuclear, and wind power are inherently CO_2-free, but fossil fuels include carbon. However, there is the potential to separate CO_2 from the gas stream and store it permanently, in spent oil and gas reservoirs, in coal beds, or in salt water aquifers. And biomass-based hydrogen offers the possibility of fixing atmospheric CO_2 and permanently storing it.

In principle, with technological advances, ultimately hydrogen could be a second broadly used energy carrier competing with electricity. Both carriers could allow use of a broad variety of primary energy sources and both could allow abundant energy with no carbon dioxide release at the point of production or the point of energy use. In this vision, the different physical properties of electricity and hydrogen would help determine which would be used for various energy needs. Electricity could be used in all-electric vehicles, but only if battery technology advances greatly. Hydrogen, which is storable on vehicles and allows for quick refueling, could be the more attractive alternative. For heating and lighting, grid-delivered electricity is likely to be more economical than hydrogen generating electricity on site. But backup generators based on fuel cells could convert electricity to hydrogen and hydrogen back to electricity if backup power were needed.

In what follows I discuss opportunities for and barriers to using hydrogen for one use: fueling light duty vehicles. Hydrogen could reduce emissions and improve energy security in this large component of the petroleum-reliant transportation sector. I next summarize two chapters from the NRC (2004) study, *The Hydrogen Economy: Opportunities, Costs, Barriers, and R&D Needs*.[1] In what follows I will refer to this as the "NRC hydrogen study" and to its authors as "the Committee."

17.3 Hydrogen for light duty vehicles: the opportunity

In the NRC hydrogen study, the Committee examined environmental, security, natural resource, and economic implications of the United States moving toward widespread use of hydrogen to fuel light duty vehicles. The Committee developed spreadsheet models of unit impacts of producing hydrogen: impacts per kilogram (kg) of hydrogen. Models were developed for

---

[1] NRC (2004). James Sweeney, as a member of that committee, took the lead in drafting chapters from which the following material was derived. Interested readers are directed to that report.

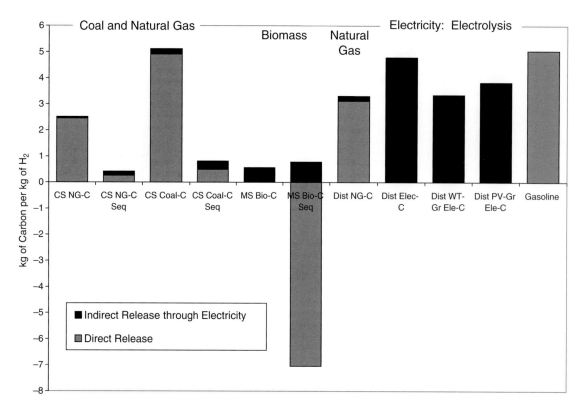

Figure 17.2 Carbon dioxide released per kg of hydrogen: current technologies. CS: Central station hydrogen production; MS: Medium sized hydrogen production plant; Dist: Distributed hydrogen production; NG: Natural gas feedstock; Coal: Coal feedstock; Bio: Biomass feedstock; Nu: Nuclear reactor; Elec: Electrolysis, electricity from grid; WT-Gr-Ele: Electrolysis, electricity from wind turbine with grid backup; WT-Ele: Electrolysis, electricity from wind turbine without grid backup; WT-PV-Ele: Electrolysis, electricity from photovoltaics with grid backup; -C: Current technology; -F: Future technology.

combinations of technologies and feedstocks from which molecular hydrogen could be produced.

Based on unit-impact estimates, the Committee developed scenarios describing a possible path of hydrogen adoption into the vehicle fleet. The successful-hydrogen scenario is based on the assumption that fuel cell vehicles (FCVs) could be greatly reduced in cost and increased in durability and that the difficulties of on-board storage of hydrogen could be overcome. Unit-impact estimates, combined with these market penetration scenarios, provide quantitative estimates of economic, environmental, and security implications of large-scale adoption of hydrogen for light duty vehicles.

17.3.1 Unit carbon dioxide releases of hydrogen production technologies

Figure 17.2 shows the NRC-developed estimates of CO_2 released per kg of hydrogen for the various technologies and feedstocks.[2] In addition, a gasoline efficiency adjusted (GEA) estimate is included for comparison: the CO_2 released by an equivalent hybrid gasoline-fueled vehicle, driven the same number of miles as the FCV could be driven with a kilogram of hydrogen. Figure 17.2 includes both the direct CO_2 releases (quantities released at the point of hydrogen production) and the indirect CO_2 releases through electricity (quantities released at the point of generation of electricity required to produce hydrogen).[3]

The first four bars on the left show CO_2 releases if hydrogen is produced using fossil fuels – coal and natural gas – in large central station plants. Higher bars show estimated impacts under the assumption that the CO_2 is vented into the atmosphere. The two lower bars labeled Seq show estimated impacts under the assumption that the CO_2 is captured, compressed, transported, and permanently stored (CCS). Without CCS, production of hydrogen from coal would release as much CO_2 into the atmosphere as would an equivalent hybrid vehicle. With CCS, CO_2 emissions could be reduced by 85%. Without CCS, production of hydrogen from natural gas would reduce the CO_2 emissions by 50%; with CCS, use of natural gas would reduce CO_2 emissions by over 90%.

---

[2] The basic models were developed by Dale Simbeck and Elaine Chang of SFA Pacific. The models were subsequently further developed and model parameters were developed, often subjectively, by members of the Committee.

[3] Figures 17.2 through 17.10 and Figures 17.13 through 17.16 originally appeared in the NRC hydrogen study.

The next two bars (with and without sequestering) are based upon biomass grown for hydrogen production, gasified, and converted to hydrogen. Biomass would eliminate direct releases of CO_2 because CO_2 released at the point of hydrogen production would be of the same magnitude as the CO_2 fixed from the atmosphere as the biomass was growing. But production of hydrogen requires electricity; generation of that electricity itself releases CO_2 into the atmosphere. With CCS, over 80% of the CO_2 drawn from the atmosphere would be permanently removed from the atmosphere.

The seventh bar estimates the impacts of hydrogen produced at the fueling station using a steam shift reformation of natural gas. Unit CO_2 emissions would be roughly 40% lower than for the equivalent gasoline-fueled hybrid vehicle.

The final three hydrogen bars represent generation of hydrogen through electrolysis, using three alternative electricity sources. The first includes distributed electrolysis at the filling station with electricity drawn from the grid, using electricity generation consistent with marginal impacts of additional electricity.[4] Although there would be no direct release of CO_2 at the point of electrolysis, generation of the electricity would release almost as much CO_2 as would a hybrid vehicle.

The next two bars represent electrolysis at the filling station using electricity generated from two renewable sources of electricity – photovoltaics and wind turbines. Photovoltaics and wind turbines produce electricity only intermittently. With such units, their operators could use the electrolyzers only when the wind was blowing or the Sun was shining and could leave them idle otherwise. Or operators could use electrolyzers constantly, backing up the wind-derived or photovoltaics-derived electricity with grid electricity. The capital cost of electrolyzers is now very high and it would be more profitable for operators to follow the latter strategy. We have assumed this profit-maximizing behavior. Thus use of these renewables would not greatly reduce CO_2 released.

Committee members made their best judgements about possible advancement of hydrogen technologies *if* industry and government pursued a serious and sustained technology development program. These technologies might become available in several decades. Following the NRC terminology, I refer to these as "future technologies" or "possible future technologies." These possible future technologies would have, for the most part, similar CO_2 releases; most progress would be in cost reduction.

Figure 17.3 shows CO_2 estimates for possible future technologies. Most bars are similar, with two differences. First, this includes a technology not currently possible: very-high-temperature direct dissociation of water from a new generation of nuclear power. This technology would release no CO_2.

The second difference is in hydrogen produced through electrolysis, with electricity generated using wind turbines. With advances in fuel cells there could be equivalent advances in electrolysis, sharply reducing capital costs. Large enough reductions imply that the strategy of electrolyzer use would change to one of installing large electrolyzers, generating all electricity with wind turbines, and leaving capacity idle when the wind was not blowing. Under this mode, generating hydrogen by electrolysis with electricity generated from wind turbines would release no CO_2 into the atmosphere.

17.3.2 Unit costs of hydrogen production technologies

The NRC hydrogen study also developed estimates of unit costs of various hydrogen production technologies. The spreadsheets build up unit cost estimates through estimation of various cost elements: capital cost, operating and maintenance cost, feedstock costs, electricity costs. An imputed carbon cost, equal to $50 per tonne of carbon released into the atmosphere, was included in the costs.[5] Costs included normal taxes on earnings, but no road use or "gasoline" taxes. Details of the cost calculations appear in the NRC hydrogen study.[6] Figure 17.4 provides cost estimates for all technologies considered in the NRC hydrogen study, using current technologies. These are the same technologies represented in Figure 17.2. The ordering of the bars in the chart is identical to the ordering in Figure 17.2.

Figure 17.4 shows that using current technologies, with large central hydrogen production facilities, hydrogen can be produced at costs similar to the retail cost of gasoline. However, these cost estimates assume supply from large central station hydrogen-generating facilities, each large enough to produce fuel for 2 million vehicles, located relatively close (within 100 kilometers) to fueling locations for these vehicles.[7] These units would not be built until sufficient vehicles were operating. In the interim, fueling technologies will probably be based on natural gas generating hydrogen at fueling stations. With current technologies these costs are about 80% greater than the unit costs of gasoline.

Figure 17.4 also shows that the cost of generating hydrogen using renewable primary energy sources or electrolysis greatly

---

[4] Specifically, the analysis assumes grid-delivered electricity will, at the margin, release 0.32 kg of CO_2 per kWhr of electricity. This can be compared with a current estimated average of 0.75 kg of CO_2 per kWhr of electricity.

[5] A cost of $50 per tonne of carbon was chosen to roughly account for the carbon externality. Appropriately quantifying this externality is the subject of continual debate. Results are not highly sensitive to the $50 per tonne assumption.

[6] For these cost estimates it is assumed that the wellhead price of crude oil is $30 per barrel, lower than the current crude oil price. Natural gas is being sold at market locations near wellheads at a price of $4.50 per million BTU.

[7] Some observers have suggested that the Committee underestimated the distribution cost of hydrogen from central stations, arguing that more pipeline capacity would be needed, either because the plants could not be built so close to the point of hydrogen use or because the pipeline system would need to be more complicated than we have assumed.

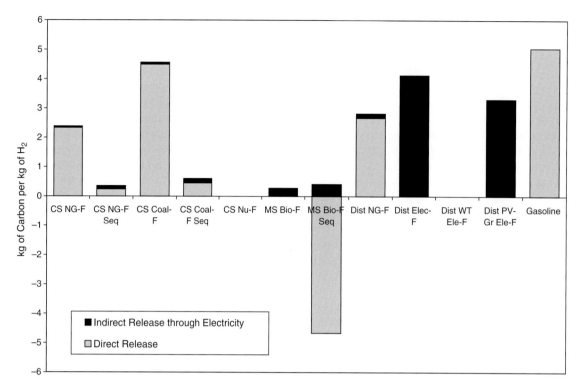

Figure 17.3 Carbon dioxide released per kg of hydrogen: possible future technologies.

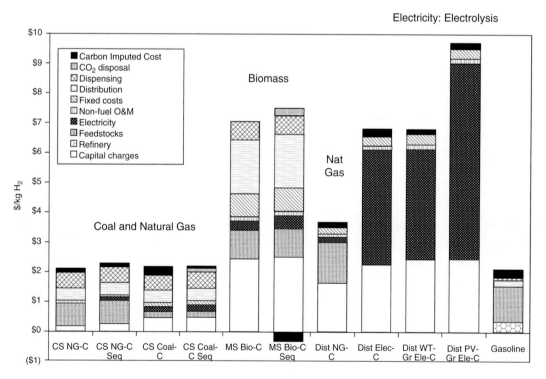

Figure 17.4 Hydrogen cost estimates: all current technologies.

exceeds the cost of generating hydrogen using fossil fuels. Therefore, with current technological options, is unlikely that a competitive hydrogen industry could be based upon either biomass or electrolysis.

In addition, Figure 17.4 includes details of the various estimated cost components. Note that for biomass and electrolysis-based generation, capital costs alone are estimated to exceed the cost of gasoline, even before any other costs are

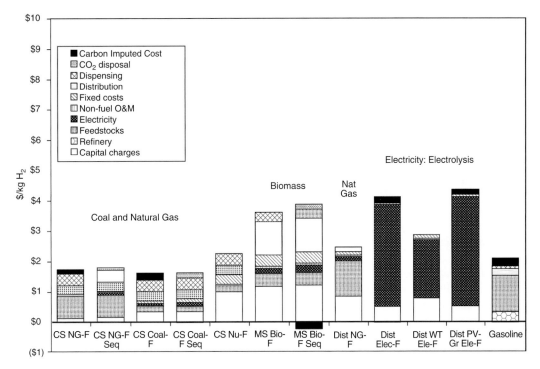

Figure 17.5 Hydrogen cost estimates: potential future technologies.

considered. The cost of electricity purchased from the grid or generated using wind turbines alone amounts to between $3 and $4 per kilogram of hydrogen.[8] The electricity costs of photovoltaics under current technologies exceed $6 per kilogram of hydrogen.[9]

Figure 17.4 is based on technologies that, while currently feasible, would typically require significant engineering before implementation. With a concerted serious technology development program, the Committee believes that the cost of these technologies could be greatly reduced. Figure 17.5 provides highly subjective estimates of possible costs after a successful technology development program.

Costs of coal and natural gas facilities could be expected to decrease somewhat but not by much, primarily because these processes are already well understood. Costs for electrolysis-based systems could decrease greatly if the capital cost of electrolyzers and the conversion efficiency were to greatly improve. Assuming that electrolysis efficiency improves, less electricity would be needed to produce hydrogen. We assumed that delivered cost of electricity from wind turbines could be decreased to four cents per kilowatt hour. We assumed great progress in photovoltaics and the cost of electricity generated from photovoltaics.

We assumed that productivity of land for growing biomass would increase sharply and that the biomass gasifiers would improve in efficiency and capital costs. With these assumptions, biomass would cost around $4 per kilogram. A new generation of very efficient nuclear technologies, directly dissociating water, might have competitive costs. Distributed generation of hydrogen using steam reforming of natural gas could be reduced to almost competitive levels.

Thus in the judgement of Committee members, successful R&D efforts could allow molecular hydrogen to be supplied using several technologies at costs roughly competitive with gasoline costs. Increases in crude oil prices above our assumed $30 per barrel would make these technologies more competitive; increases in natural gas costs would make natural gas-based technologies less competitive.

The unit impacts estimates and unit cost estimates described above are next used to estimate possible impacts on CO_2 emissions and on fueling system costs,[10] again following the NRC study.

---

[8] The analysis for current conditions assumes that grid-based electricity is available all of the time at a delivered price of $0.07/kWh, photovoltaic-derived electricity is available 20% of the time at an average cost of $0.32/kWh, and wind-turbine-generated electricity is available 30% of the time at an average cost of $0.06/kWh.

[9] Note also that the distribution cost of hydrogen from biomass facilities significantly exceeds distribution cost from large central stations, primarily because biomass plants could not be built in as large a scale as could central station natural gas plants. Since the production facility would be significantly smaller, it is likely that pipeline distribution would not be viable and distribution using more costly truck transportation would be necessary.

[10] In addition, the NRC report provides other unit impact estimates of hydrogen use, of coal use, of land use associated with the various technologies and feedstocks.

17.3.3 Three scenarios of vehicle technology adoption

Fuel-cell vehicles are still extremely costly and the fuel cells have relatively short lives. Storage of hydrogen on board vehicles, for adequate range, is challenging. Research vehicles being driven now are estimated to cost between $1 million and $4 million each.

Major advances in fuel cells and in on-board storage of hydrogen are necessary before the vision for hydrogen can be implemented. And success cannot be assured. Without reduction in FCV cost enough to compete with conventional vehicles, diesel vehicles, and hybrid vehicles, there will be little or no diffusion of this technology and little or no energy security and environmental benefits.

Light duty vehicles in the three scenarios
Given this uncertainty about FCVs, the Committee developed three scenarios, which I describe as:

- "H_2 and hybrid vehicles." Hydrogen FCVs and hybrid vehicles are both successful.
- "Hybrids, no H_2." FCVs are not successful, hybrid vehicles ultimately become dominant.
- "Without hydrogen or hybrids." Neither FCVs or hybrids capture significant market share.

The three scenarios are illustrated in Figure 17.6. Figure 17.6a shows the assumed fractions of new vehicle sales yearly, showing the new conventional, hybrid, and FCVs for the various scenarios. Figure 17.6b shows estimated fractions of the total fleet of vehicles yearly.

In the "Without hydrogen or hybrids" scenario, neither hybrid vehicles nor hydrogen vehicles capture any market share. The scenario has not been drawn in Figure 17.6, but can be envisioned as a horizontal line at 100% of conventional vehicles for all years and 0% fraction for hybrids and FCVs for all years.

In the "Hybrids, no H_2" scenario, hybrid vehicles replace conventional vehicles over time, ultimately capturing 100% market share in the year 2035. Hydrogen vehicle technology never advances sufficiently for hydrogen FCVs to become competitive in the marketplace.

In the "H_2 and hybrid vehicles" scenario, FCV technology successfully advances. Hydrogen FCVs are first sold in significant numbers in the year 2015, and subsequently, the market share of new hydrogen vehicles continues to grow, ultimately reaching 100% market share in 2038. Hybrid vehicle sales initially grow less rapidly than in the second scenario, but ultimately peak in the year 2025, declining afterwards, as FCV sales grow.

This third scenario represents a view of the maximum speed that hydrogen vehicles could expand if technological barriers are overcome and under very strong legislative mandates. It is assumed in this last scenario that FCVs take market share from hybrids, but not from conventional vehicles, based on the idea that FCVs would initially appeal to the same market segments as do hybrid vehicles.

In each scenario, fuel efficiency would differ significantly between the three vehicle classes. The NRC assumed that hybrid vehicles would be 45% more fuel-efficient (45% higher miles per gallon) than conventional vehicles and that hydrogen FCVs would be 66% more efficient than hybrid vehicles.[11] Fuel economy of each vehicle type is assumed to increase by 1% per year,[12] shown in Figure 17.7.

In the scenarios we have taken the simplest assumption on vehicle miles traveled (VMT): continued growth at 2.3% per year.

Fuel use by light duty vehicles in the three scenarios
These assumptions – fractions of various vehicles, fuel efficiency of various vehicles, vehicle miles traveled, and type of fuel – together provide estimates of how much gasoline and hydrogen would be used under the three scenarios. Gasoline estimates are plotted in Figure 17.8. The scenarios show the great potential for hydrogen in a transition from oil. But they also underline the long time required to accomplish this transition, even if hydrogen FCVs are completely successful. The analysis suggests that even with hydrogen success, during the next 20–25 years the United States would use more gasoline than it currently uses. Only later would gasoline consumption be lower than current levels. But by 2050 in the "H_2 and hybrid vehicles" scenario the United States could entirely eliminate use of gasoline in light duty vehicles.

The analysis also shows the important role in the near term of hybrid vehicles. Twenty-five years from now the reduction in gasoline as a result of a growth in hybrids could be two-thirds as large as the reduction resulting from the combination of hybrids and hydrogen vehicles.

In the long term, however, hybrid vehicles could only delay growth in gasoline consumption, reversing it only temporarily; hydrogen vehicles could entirely eliminate use of gasoline in light duty vehicles.

Carbon dioxide emissions by light duty vehicles in the three scenarios
If hydrogen FCVs are successful, reductions in CO_2 will depend on the rate of their adoption and on the particular technologies and feedstocks used to generate the hydrogen. The NRC Committee estimated CO_2 emissions in the three scenarios; for the "H_2 and hybrid vehicles" scenario, the Committee provided estimates of CO_2 emissions for each

---

[11] These assumptions are based on averages from Wang's (2002) review of other studies.
[12] One percent a year is highly uncertain. Recent history has seen shifts in the light duty fleet to larger and heavier vehicles, with a corresponding decline in average fleetwide fuel economy. But this shift in fleet composition appears to be nearly complete, implying that a moderate improvement in fuel economy is plausible.

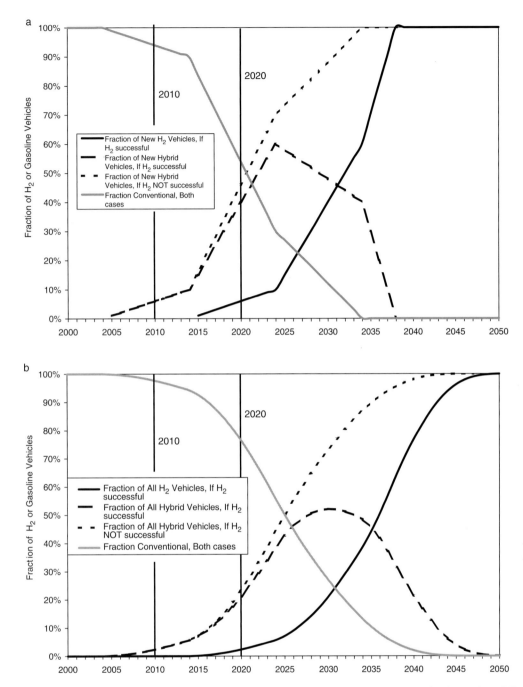

Figure 17.6 Scenarios for hydrogen and hybrid vehicle adoption. a, Fractions of new vehicle sales; b, fractions of total fleet.

technology/feedstock combination, assuming that hydrogen would be produced exclusively using one technology/feedstock combination.[13]

The CO_2 emissions in the three scenarios for the various feedstock combinations are shown in Figures 17.9 and 17.10. These figures use estimates based on the *future* technologies, assuming that concerted R&D were successfully pursued and that the technical barriers in vehicles were overcome. Figure 17.9 shows estimates based upon coal, natural gas, and nuclear power; Figure 17.10 provides estimates based on renewable feedstocks and/or electrolysis. Similar graphs are shown in the NRC hydrogen study for current technologies, but not repeated here. The differences are relatively small.

Figures 17.9 and 17.10 show that the growth of hybrid vehicles alone could reduce CO_2 emissions from light duty vehicles by around 200 million tonnes of carbon per year

---

[13] In fact, if hydrogen technologies are successful, a variety of different technologies would be used. Such combinations can be readily estimated as combinations of the pure estimates from the NRC hydrogen study.

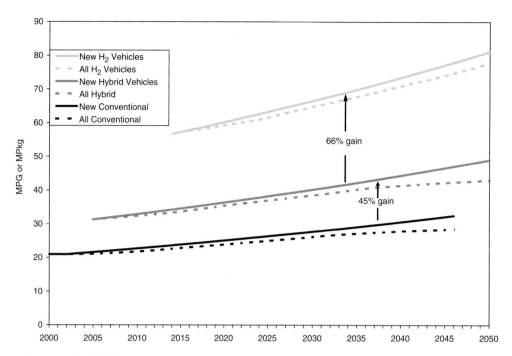

Figure 17.7 Assumed average fuel efficiency: conventional, hybrid, hydrogen light duty vehicles.

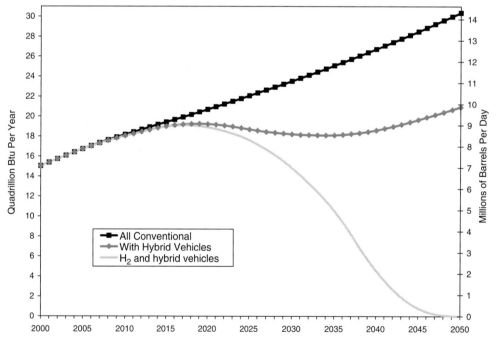

Figure 17.8 Gasoline use estimated for three scenarios.

from 2040 on, from the constantly growing level of emissions which could be expected in the absence of hybrid vehicles.

In addition to emissions reductions from hybrids, Figures 17.9 and 17.10 show that hydrogen vehicles could save an additional 500 million tonnes of carbon per year by 2050 if hydrogen was generated from either coal or natural gas with CCS, nuclear energy, biomass, or wind turbines. Without CCS, hydrogen from natural gas would save less, about 200 million tonnes per year. Without CCS, hydrogen from coal would have very little, if any, additional savings. Hydrogen generation through electrolysis could lead to some carbon savings, but if a large fraction of electricity were generated using CO_2-emitting technologies, electrolysis would have relatively little incremental impacts on CO_2 emissions.

Hydrogen for light-duty vehicles 207

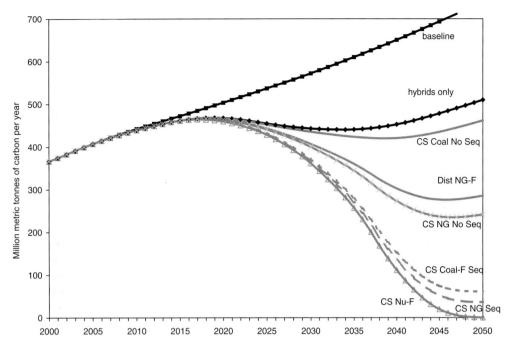

Figure 17.9 Carbon dioxide releases for various scenarios and feedstocks: future technologies.

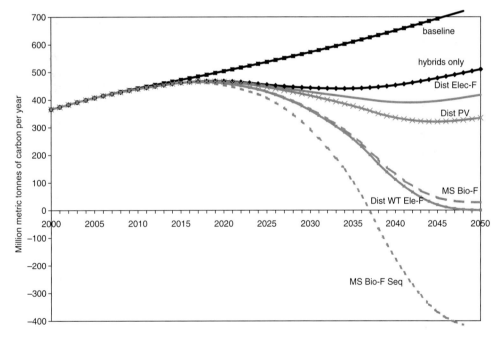

Figure 17.10 Carbon dioxide releases, various scenarios and feedstocks: future technologies.

17.4 Hydrogen for light duty vehicles: the barriers

The vision for hydrogen offers much promise. But whether this vision becomes a reality within the envisioned time-frame is highly uncertain because of the formidable barriers to implementation of hydrogen-based transportation technologies. I turn here to a more discussion of the barriers.

17.4.1 Demand-side technology barriers in vehicles

First and foremost is a technological barrier – the technology of hydrogen FCVs. Fuel cells are still far too expensive and have too-short lives for economically competitive implementation in light duty vehicles. Proton exchange membrane (PEM) fuel cells need platinum or other noble metals for catalysis. The cells deteriorate long before the 5000 hours needed for a normal automobile life. Adequate storage of hydrogen on board vehicles is a technological problem and, for very high pressure tanks, could be a safety problem. Internal combustion of hydrogen is technically feasible, but combustion would not have the efficiency gains anticipated for FCVs. Lower efficiency implies shorter range between refueling for a given hydrogen storage or needs for more storage. Lower efficiency also implies higher fuel cost per mile of driving.

Unless the technological problems of vehicles are solved, the potential benefits of hydrogen in light duty vehicles will not be obtained. Therefore, advances in hydrogen fuel cells and on-board hydrogen storage technologies are essential for hydrogen success.

For FCVs to capture a significant market share, FCV costs must decrease greatly to levels similar to those of conventional or hybrid vehicles. This is the case even if gasoline costs were to increase significantly. Although fueling costs could be lower for FCVs, high initial costs could more than compensate for any fuel savings.

This can be illustrated by examining the total fuel used (gallons of gasoline or kg of hydrogen) over an assumed 140 000-mile lifetime of a hybrid vehicle or a FCV, using 2015 average fuel economies from Figure 17.7. Over the life of a hybrid vehicle, the vehicle would consume 4000 gallons of gasoline, while the equivalent hydrogen fuel cell vehicle would consume about 2400 kg of hydrogen. If hydrogen and gasoline prices, including taxes, were $2.50 per gallon of gasoline or $2.50 per kg of hydrogen, then the fuel savings of a hydrogen vehicle over its entire lifetime would be $3900. But discounting future cash flows, the present value of the savings would be around $2300 per vehicle, depending on the annual driving pattern and discount rate.[14] Therefore, if a hydrogen FCV were more expensive than a hybrid by more than $2300 per vehicle, fuel cost savings would not make up for this initial cost differential. If the price of gasoline were to increase by $2.00 per gallon, the cost of fuel over the entire lifetime of a hybrid vehicle would increase by $8000. The total fuel cost differential would be $11,900, or $7,500 once discounted. Therefore, even with this gasoline price increase, if a hydrogen FCV were more expensive than a hybrid by more than $7,500 per vehicle, fuel cost savings would not make up for this capital cost differential.

A hydrogen vehicle might have physical advantages or disadvantages over a hybrid vehicle of equivalent size and performance. The ability of a hydrogen vehicle to generate electricity even when not running could be an advantage. But hybrid vehicles would have electricity stored in batteries and thus could provide much of the equivalent functionality. Repair costs of a hydrogen vehicle could differ from costs for a hybrid.

The often alleged benefits of plugging one's home into a vehicle to provide home electricity is a fallacy. Fuel cells on vehicles are likely to have a limited life. If the fuel cell life in a vehicle were 5000 hours and the vehicle were always used as a source of electricity to the home when not being driven, then the entire lifetime of the vehicle would be about seven months. The high capital cost makes the idea of plugging one's home into the vehicle for regular use a very unattractive option.

Thus it is not clear that there would be great consumer benefits to a FCV not available in a hybrid vehicle. But if a FCV were significantly more costly than a hybrid vehicle, that capital cost differential would be a great barrier to the widespread introduction of hydrogen vehicles.

Only if these vehicular technology problems are solved, resulting in only small cost differentials, will the additional barriers become relevant. While significant, I believe that each of these other barriers can be solved if the technological and economic problems of vehicles are solved.

17.4.2 Supply-side technology barriers

As discussed above, there currently exist several technologies to produce hydrogen from coal or from natural gas.

There are other supply-side barriers. If hydrogen were produced a long distance from fueling stations, transportation to the refueling station would be needed. Although long-distance transportation is possible using pipelines, tube trucks, or cryogenic trucks, it is costly. Therefore, under current hydrogen transportation technologies, either the hydrogen must be produced close to the point of fueling (or at the fueling station) or the costs of hydrogen in vehicles must include a large cost of transporting that hydrogen to the filling station. Particularly before there is a large concentration of hydrogen vehicles or for sparsely populated regions of the country, the high cost of transportation can be an important technology barrier to hydrogen. Research and development directed toward sharply reducing the cost of hydrogen transportation might lower this barrier.

The reduced-carbon-dioxide gain of hydrogen depends on either CO_2 being captured and permanently stored from fossil fuel hydrogen generation or hydrogen being produced without net release of CO_2. There are technology issues in both of these situations and these issues could be important barriers to the potential environmental gain.

In principle, CO_2 could be permanently stored in unminable coal seams, in spent oil and gas reservoirs, or in saltwater aquifers. Deep ocean storage may be possible, but environmental damages may be great, even if invisible. There is some experience with CO_2 storage. The Sleipner gas field in the North Sea produces high-CO_2 natural gas; CO_2 is separated from the natural gas and is injected into a deep saline aquifer. Carbon dioxide from a coal-to-methane plant in North Dakota is piped to Saskatchewan and injected into an oil field to enhance oil recovery and for long-term storage.

---

[14] This estimate assumes that a car is driven 14 000 miles per year for 10 years and that there is an annual 10% discount rate. The exact estimate in this case is $2320. Note that the model results are sensitive to this highly uncertain parameter.

But whether these are viable long-term CCS options is not clear. There is relatively little experience in saltwater aquifer storage; whether carbon would remain permanently or would leak must be determined. Spent oil and gas reservoirs seem like ideal storage locations, but only if the seals on the geological formations remain intact. Injection for enhanced oil recovery may provide only a relatively limited potential for storage. These are all areas of active research. Until there is confidence in the ability to store the CO_2 safely and dependably for hundreds of years, then we cannot count on CCS; until then these technical issues will be barriers to the environmental potential of hydrogen.

Hydrogen from non-fossil-fuel based sources has its own technological, and thus economic, barriers.

Hydrogen from biomass is likely to remain too costly, in the absence of technologies not yet at the development stage, such as genetically modified algae or other microbes. In Figures 17.4 and 17.5, hydrogen production from biomass is among the most expensive options. Land constraints may also make hydrogen based on gasified biomass economically not viable, since large-scale biomass production would eventually have to compete with food crops for the finite agricultural land supply.

Another option for carbon-free production of hydrogen is electrolysis, with the electricity generated from photovoltaics or from wind. For wind, the primary barrier is the cost of the electrolyzer. If the wind cost were as low as 4 cents per kilowatt hour, this hydrogen source would still be uneconomical because of the currently high cost of electrolyzers.[15] Only with technology breakthroughs that greatly reduce the capital cost of electrolyzers can this become economic. For photovoltaics, the barrier is both the current cost of electrolyzers and the cost of generating electricity from photovoltaics. Even with a technology breakthrough that greatly reduces the cost of electrolzyers, the cost of the photovoltaic-generated electricity would keep this source uneconomic.

Some renewables might ultimately become carbon-dioxide-free sources of hydrogen. Hydrogen can be produced from a variety of feedstocks, allowing it to take advantage of whichever new technology breakthrough occurs. But, unless and until there are such technology breakthroughs, the technology difficulties of advanced biomass, electrolyzers, and photovoltaics provide economic barriers to generating hydrogen from carbon-dioxide-free sources.

17.4.3 Fueling cost barriers hydrogen to production

Figure 17.4 and 17.5 show that for large central hydrogen production stations based on coal or natural gas, hydrogen fueling cost per mile in FCVs and gasoline fueling cost per mile in hybrid vehicles are very similar. Thus if the same road-use taxes were imposed on hydrogen as on gasoline, the cost at the filling station would be similar and there would be no fueling cost barrier to hydrogen once the system developed enough that such large central plants were setting the competitive hydrogen price.

However, in a transition period of at least several decades, there would not be enough vehicles to justify large central station plants. Smaller, distributed generation at the filling station is more likely. But distributed generation leads to a higher cost of hydrogen production. Thus, without subsidies, hydrogen would be significantly more costly than gasoline. Consumers would not tend to buy hydrogen FCVs. With few vehicles, hydrogen producers would continue to avoid building central station plants. The price differential would persist, and the potential benefits of moving to a widespread hydrogen fueling system would be lost.

To get past this barrier, governments could provide subsidies for hydrogen fuel generated at distributed stations. Figure 17.4 shows that hydrogen based upon distributed natural gas reforming would cost roughly $1.40 per kilogram more than the gasoline equivalent cost. Since road-use taxes ("gasoline taxes") are relatively small, it would be necessary to eliminate all road-use taxes on hydrogen vehicles and offer further direct subsidies. With advance in technologies for distributed steam reforming of natural gas, the cost differential could be narrowed considerably, down to about $0.20 per kilogram and direct subsidies could be avoided.

To estimate the needed magnitude of the direct subsidy or road-use tax reduction, I will assume that the financial burden would be covered through an increase in taxes on gasoline. Figure 17.11 plots the gasoline tax increase that would be just sufficient to subsidize hydrogen in order to make the hydrogen price equivalent to the gasoline price at the filling station.[16] Figure 17.11, based on current technologies, includes only hydrogen produced using fossil fuel feedstocks.

With initially few hydrogen vehicles on the road, total subsidy cost would be relatively small and thus the gasoline tax would initially be very small. Figure 17.11 shows that with current technologies, the requisite gasoline tax would remain below $0.05 per gallon through 2025. However, this low gasoline tax rate would not be sustainable once there were large numbers of hydrogen vehicles. By 2030 the requisite gasoline tax would have increased to $0.17 per gallon and by 2035 to $0.45 per gallon. The smaller gasoline tax during early periods of the transition might be politically feasible, but the tax is unlikely to be politically sustainable once the requisite gasoline tax increase became large.

Figure 17.11 also shows that with existing technology, much smaller gasoline taxes would be required if production

---

[15] These estimates are in Figure 17.4 and Figure 17.5.

[16] The gasoline tax is calculated as the difference between the hydrogen unit cost and the equivalent gasoline cost divided by 1 plus the ratio of gasoline quantity sold to hydrogen quantity sold. This is derived by solving two equations. One sets the unit cost of gasoline plus gasoline tax equal to the unit cost of hydrogen minus the subsidy. The other sets the product of hydrogen quantity and hydrogen subsidy equal to the product of gasoline quantity and gasoline tax.

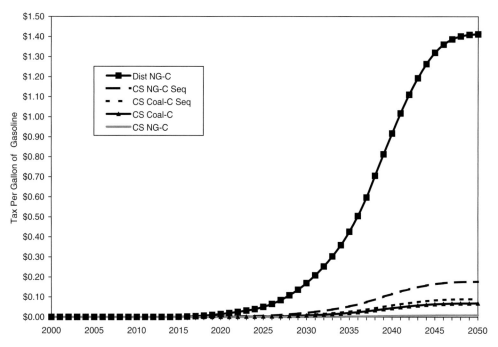

Figure 17.11 Gasoline tax increase to subsidize hydrogen: current technologies.

were based upon large central station facilities. For example, using either coal or natural gas as a feedstock and permanently storing the CO_2 would allow a $0.10 per gallon gasoline tax increase to be more than adequate through the year 2035, based upon central station costs.

However, if successful R&D reduces hydrogen production costs, costs of hydrogen from fossil-fuel-based central stations would be lower than the gasoline cost and no subsidy would be needed once there were sufficient vehicles. With distributed steam reforming of hydrogen, a $0.07 per gallon gasoline tax increase would be sufficient for the first two decades of the transition – through 2035. But even with such technological improvements, the requisite gasoline tax increase would increase to $0.22 per gallon in subsequent years in the absence of large central station facilities.

It should be noted, however, that there would be very little CO_2 reduction while the system was dependent upon distributed generation of hydrogen from natural gas, since CCS from such distributed facilities is very unlikely.

To reduce CO_2 emissions during the transition period, absent large central station units, would require renewable technologies such as solar wind or biomass. Figure 17.12 shows a similar tax analysis, but for renewable technologies. This figure includes cost estimates for technologies based on successful R&D programs that greatly reduce the cost of the renewable technologies; with current technologies, requisite taxes would be much larger than those shown in Figure 17.12. With the potential future wind technologies the requisite gasoline tax increase could remain below $0.10 per gallon to 2030. But in 2030, the requisite gasoline tax for a system based entirely on other renewables would be between $0.18 and $0.20 per gallon and would rapidly increase in subsequent years. These high requisite gasoline taxes are likely to be politically untenable.

17.4.4 Fueling cost barriers: hydrogen retailing/other infrastructure

This issue discussed above is similar to the much postulated "chicken and egg" problem for hydrogen retailing. The "chicken and egg" problem is often posed as: suppliers of hydrogen would be unwilling to invest in filling stations until there were sufficient hydrogen vehicles on the road; automobiles buyers would be unwilling to purchase FCVs until there were sufficient filling stations on the road; therefore nothing happens.

In the cost analyses summarized above (Figures 17.4 and 17.5), costs of the retail distribution infrastructure and of the long-distance transportation infrastructure have been included. But they have been included based upon normal capacity use of that equipment. However, if capital costs must be incurred but capacity use is very low, actual unit costs would be higher than those estimated above. The "chicken and egg" problem reduces to the assertion that in the transition the retail infrastructure and perhaps the hydrogen transportation infrastructure would be used at well below capacity and that therefore suppliers would not provide infrastructure unless they could sell the product at a higher than equilibrium price. Automobile buyers in turn would expect inconvenience in purchasing fuels and would be willing to buy hydrogen fuel vehicles only if they would pay less for fuel than the equilibrium price.

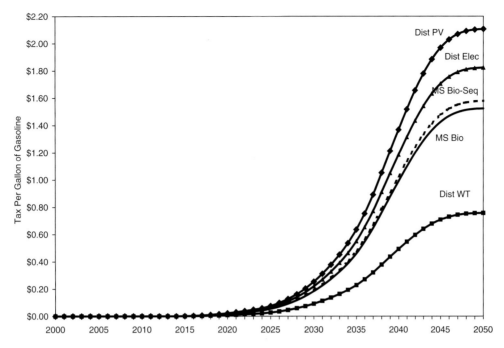

Figure 17.12 Gasoline tax increase to subsidize hydrogen: future technologies.

One solution for this problem would be analogous to that of the high cost of distributed natural gas generation. Fueling stations could be subsidized so more would be built; if more stations were built, consumers would not be inconvenienced and would purchase more hydrogen vehicles. And hydrogen vehicles could be subsidized so more would be purchased. Like the issue of high-cost distributed generation versus low-cost central station generation, such subsidies would be needed in a transition period only. That is fundamentally the "Hydrogen Highway" plan developed for Governor Schwarzenegger in California.

However, the infrastructure "chicken and egg" problem seems less difficult than the issue of high production costs from distributed generation. The total subsidy required would be smaller because retail distribution costs are only a fraction of the total cost of hydrogen. Only the fixed costs of the retail distribution system – not the variable costs – would require subsidy.[17] In addition, the subsidy need not necessarily be paid for by governments. Suppliers themselves may find it in their economic interests initially to subsidize their own retail activities in order to establish a network of distribution stations and thereby to build brand identification and brand loyalty. Translating brand identification into brand loyalty can provide a competitive advantage as the hydrogen market develops more fully. And companies that build early networks of hydrogen distribution stations will gather technical, managerial, and marketing information from those early experiences. For these reasons corporations – particularly large corporations – may find it economically attractive to internally subsidize an early network of fueling stations.

Thus, in my judgement, the "chicken and egg" problem, while requiring small subsidies either from corporations or from government entities for a complete solution, will not impose an important barrier to the development of hydrogen as a widespread automotive fuel.

In addition, a dual fueling system – gasoline and hydrogen – will be needed for decades after the widespread commercialization of molecular hydrogen for vehicles. Therefore some retail distribution stations may be operated at less than full capacity, but only in areas of low-vehicle population density. But this situation does not seem fundamentally very different from the current situation in which gasoline of many different brands is sold, but each at stations that sell only one brand of gasoline. Each station may operate at somewhat lower capacity than it would if there were fewer stations, each dispensing any brand of gasoline. But this has not been an essential problem limiting sales of gasoline and I do not believe it will provide a fundamental barrier to hydrogen. Similarly, assuring that there is enough local competition among fueling stations that retailers cannot exercise excessive market power will itself increase cost. But again, this is likely to be at a small cost issue, not an important barrier to the development of a hydrogen system.

17.4.5 Resource limitations

Other possible supply-side barriers stem from possible limitations on availability of resources used as inputs to hydrogen

---

[17] Of course, the large upfront subsidy for the infrastructure capital outlay may be difficult for political reasons, and this also holds for central station generation.

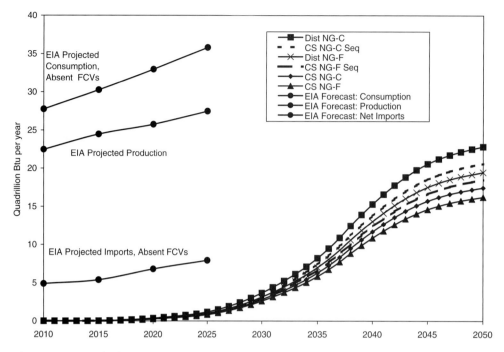

Figure 17.13 Natural gas to generate hydrogen.

production. While these barriers are unlikely to be relevant across all potential sources of hydrogen, they can be relevant for particular sources.

Natural gas supply and demand
Natural gas markets in the United States remain tight, with natural gas prices recently increasing greatly. The NRC hydrogen study examined the potential usage of natural gas as a feedstock and compared that usage with the current and projected supply, demand, and imports of natural gas.

Figure 17.13, based on an assumption that natural gas is used to provide feedstock for all hydrogen production, shows use of natural gas for this purpose and compares these amounts with US Energy Information Administration (EIA) projections about US natural gas production, consumption, and imports. In the first 10 years of the transition, the United States could rely upon natural gas to generate hydrogen, increasing imports of natural gas by only a small fraction of current imports. However, if the United States were to rely upon natural gas beyond the first 20 years, natural gas imports would more than double. The natural gas infrastructure impacts would be great.

Such large imports may transform the problem of oil security into a problem of security of supply of natural gas. Such a shift, rather than reduction, in the source of energy supply disruption risks implies that one of the important benefits of moving to hydrogen would be lost.

The increased demand for natural gas in the United States could be expected to increase natural gas prices worldwide, motivating other nations to use less natural gas and more coal or oil. I have not examined worldwide impact on natural gas supply and demand balances. But induced fuel substitution may ultimately blunt the expected reduction in global CO_2 emissions.

Resources for geological storage
A second possible resource limitation is availability of sites for geological storage of CO_2 if hydrogen is produced from fossil fuels. The NRC hydrogen study addressed this problem, estimating cumulative quantities of CO_2 that would require storage if fossil fuels were used for producing hydrogen.

Figure 17.14 estimates cumulative storage of CO_2 through 2050, based upon future technologies for central station coal, central station natural gas, or mid-size biomass facilities. Storage needs would be massive, requiring a tremendously large new infrastructure. For coal-based natural gas production, the United States would need to capture and permanently store 20 million metric tons of CO_2 by 2050. And another 20 million metric tons would need to be stored roughly every 15 years.

There have been only very rough estimates of the geological resources. One estimate, in a review by Anderson and Newell (2004), is that one could permanently store between 5 and 10 billion tonnes in unminable US coal seams and another 25 to 30 or more billion tonnes in depleted US oil and gas reservoirs. The potential for long-term storage in saltwater aquifers is many times larger, but largely unmapped. If the technological barriers of CCS can be solved, there appears to be sufficient storage resources to satisfy needs to the end of this century, even without saltwater aquifers. Therefore this potential resource limitation does not appear to be a barrier.

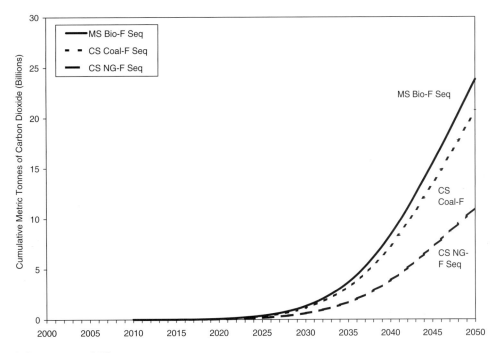

Figure 17.14 Cumulative storage of CO_2.

Land for biomass

In addition to the problems of cost, use of biomass (gasified to produce hydrogen) requires land to grow the biomass. The NRC hydrogen study estimated land use if all hydrogen were produced using biomass. Figure 17.15 provides these estimates, based on both current and possible future hydrogen production technologies. This graph should be compared with the current use of land for agricultural purposes in the United States: 700 000 square miles of crop land and 900 000 square miles of pasture land.

Figure 17.15 suggests that, were the United States to rely upon biomass as the primary source for hydrogen, it would need an unacceptably large fraction of agricultural land to grow the energy crops.[18]

These estimates do not imply that no hydrogen could be produced viably from biomass. Forest and agricultural wastes might be used as feedstocks for hydrogen production, without significant impact on agricultural lands. And this analysis does not rule out the possibility that alternatives involving direct production of hydrogen from genetically modified microorganisms would require far less land.

However, in the absence of entirely new biomass technologies, analysis of land requirements suggests that the lack of available land for energy crop production would be an important barrier to widespread use of biomass for hydrogen production.

Coal industry expansion

Similar to the analysis of natural gas, the NRC hydrogen study examined impacts on US coal production. Figure 17.16 provides those estimates. A reliance upon coal as a feedstock for hydrogen would require a large expansion of the coal industry: by 30% greater than it would otherwise over the next 30 years and by 80% through 2050. However, coal resources in the United States are massive and the industry seems quite capable of expanding, given this long time period.[19] Therefore, this does not appear to be a barrier to the use of coal as the basis for hydrogen production.

17.4.6 Other barriers to consumer adoption

As with any new technology competing with existing technologies, consumer perceptions of the desirability of the technology can significantly influence its rate of adoption and its ultimate success. Such perceptions may help or hinder hydrogen as a fuel.

Of these perceptions, one that might become a barrier is safety. Hydrogen burns with an invisible flame, cannot be seen or smelled, will combust over a wide range of concentrations in air, and can be easily ignited. Thus potential risks exist. The perception of risk from very high pressure hydrogen canisters

---

[18] This analysis assumes that growing of biomass becomes more productive with genetic engineering and other advances; possible future technologies assume that 50% more biomass crop could be grown per acre of land. Note, however, that for large-scale biomass production, yields may decrease if more marginal land is put into production because of the exhaustion of better land. This effect would contribute to preventing biomass from ever reaching the levels found in Figure 17.15.

[19] If this became the case, other environmental consequences of coal industry expansion (e.g., watershed pollution) would have to be addressed.

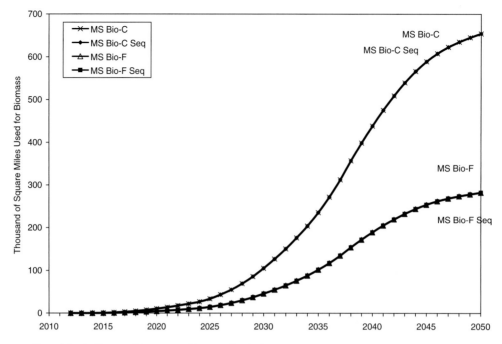

Figure 17.15 Square miles of land (thousands) to grow biomass for hydrogen production.

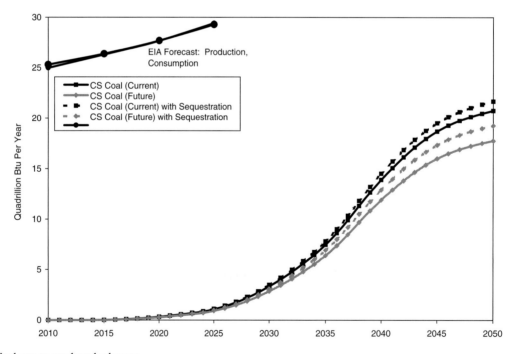

Figure 17.16 Coal use to produce hydrogen.

in vehicles could be enough to discourage consumer purchases. To help alleviate both real and perceived risk, hydrogen sensors and standards for hydrogen handling and storage will be important, especially because humans cannot smell leaking hydrogen.

But there are also risks using gasoline. We have assured gasoline safety through standards, safeguards embedded in the technology, and education programs. Similar efforts would be needed for hydrogen. Hydrogen sensors exist and can be used to detect leaks. There is already long experience for industrial hydrogen use. Standard setting is already under way, including joint government/industry groups. Therefore although safety is an important issue, it is unlikely to be an important barrier to hydrogen.

17.4.7 Competitive technologies

Finally is the issue of competition from existing technologies. Gasoline-fueled vehicles will continue to improve in many dimensions desired by consumers. Hybrid vehicles, now rapidly improving, allow greater fuel economy. Hybrid SUVs are now available and mid-size vehicles will soon be available. Improvements in diesel-fueled light duty vehicles could also provide significant competition. But such vehicles would still be fueled by petroleum. If petroleum costs were to increase dramatically over time, as would be possible if world oil supply starts to decline, then even large gains in efficiency of petroleum-fueled vehicles would make such vehicles economically unattractive, relative to vehicles fueled by non-petroleum sources.

Although advances in battery technology have been disappointing, we might expect a breakthrough at some point. If low-weight, low-cost, high-capacity batteries could be developed, then dedicated electric vehicles could become attractive. And it is likely that plug-in hybrids, allowing the vehicle to be driven for shorter ranges powered only by the electricity, but including a gasoline fuel engine for long trips, could be attractive alternatives even before dedicated electric vehicles were economical. Such battery advances could allow electricity to become the main energy source for light duty vehicles. In such a case there may be no market niche that could be filled by hydrogen-fueled-vehicles.

Thus improvements of these competing technologies could ultimately be the greatest barrier to the widespread implementation of hydrogen vehicles, and could ultimately preclude any significant market acceptance.

17.5 In summary

Hydrogen holds great promise as an energy carrier that (1) could be produced by a range of domestic primary energy sources, (2) releases no CO_2 or any criteria pollutants at the point of use, (3) could be produced with little or no CO_2 release at the point of production, and (4) could facilitate a transition away from petroleum products. The opportunities are great.

But barriers are formidable. Most fundamental is a technological barrier: unless fuel cell costs dramatically drop and their lifetimes greatly increase, hydrogen will not become a viable competitor for use in light duty vehicles. Second, without improvement in on-board hydrogen storage, the range of hydrogen vehicles may be unacceptably short and the passenger safety risk may be significant. And advances in competitive technologies could make those technologies dominant, leaving no market opportunity for molecular hydrogen as a vehicle fuel. Although additional barriers abound, they pale in comparison with the technological barriers in vehicles.

To address the barriers, R&D is necessary. Some will be sponsored by government agencies but much of it will be undertaken by industry. Demonstration programs, standard-setting, and education programs will all be important parts of overcoming barriers. But because the most fundamental problem is basic technology, the most fundamental solution must involve technological development.

In any event, if the transition to hydrogen light duty vehicles is successful, it will not be quick. As the most optimistic scenario suggests, the transition could be possible by the middle of this century, if technological and other barriers are overcome. Otherwise the transition will be slower, if it happens at all. Thus the vision of hydrogen – no matter how promising – must not be allowed to get in the way of shorter-run fuel economy improvements in light duty vehicles.

Acknowledgements

The material in Section 17.3 draws heavily on two chapters from the National Research Committee study, (NRC, 2004). In addition, I would like to thank Kenny Gillingham for assistance in the development of this paper.

References

Anderson, S. and Newell, R. (2004). Prospects for carbon capture and storage technologies. *Annual Review of Environment and Resources* **29**, 109–142.

Energy Information Agency (EIA) (2003). *Annual Energy Outlook 2003 with Projections to 2025*. Washington, DC: US Department of Energy.

Environmental Protection Agency (EPA) (2005). *Inventory of US Greenhouse Gas Emissions and Sinks: 1990–2003*. Washington, DC: EPA, April 2005.

National Research Council (NRC) (2004). *The Hydrogen Economy: Opportunities, Costs, Barriers, and R&D Needs*. Committee on Alternatives and Strategies for Future Hydrogen Production and Use. Washington, DC: National Academy of Engineering.

Wang, M. (2002). Fuel choices for fuel cell vehicles: well-to-wheels energy and emissions impacts. *Journal of Power Sources* **112**, 307–321.

18

The role of expectations in modeling costs of climate change policies

Paul M. Bernstein, Robert L. Earle and W. David Montgomery

18.1 Introduction

Policies to address climate change in the European Union and the United States, as well as international negotiations under the UN Framework Convention on Climate Change, have come to share a surprising feature. All define carbon limits over only a very short time-frame compared with either the timescales characteristic of climate processes or the time horizon over which investments in mitigation measures must be evaluated. The Kyoto Protocol set national greenhouse gas emission targets only through 2012, and did not allow negotiations on targets for the Second Commitment period from 2013 to 2017 to begin until the protocol enters into force. In the European Union, a pilot emission trading system has been set up for the years 2005–2007, but targets and allocations for the First Commitment period limits were not specified. In the United States, proposed legislation to limit US greenhouse gas emissions has been specific about limits to 2020 but purposely ambiguous about limits further in the future.

The short time horizons arise at least in part from the fundamental impossibility of binding future governments (through legislation or treaty) to policies that must be implemented and enforced over the indefinite future. Short policy horizons pose very difficult problems for assessing costs because silence on long-term limits leaves considerable ambiguity over long-term expectations about policies not yet enacted. This ambiguity becomes more than a footnote because these long-term expectations can be crucially important in determining what near-term market responses – and therefore near-term mitigation costs – will be. This is not a new theme in economics. The "Lucas critique" that led to the reformulation of modern macroeconomics arose from the same realization that expectations of the long-term consequences of current policy actions have a profound effect on the immediate market response to policy changes (Lucas, 1976; Lucas and Sargent, 1981).

In this chapter, we perform two analyses to understand the role of expectations in modeling the costs of climate change policies. First, we use a dynamic computable general equilibrium (CGE) model of the US economy to illustrate the importance of the specification of long-term policy expectations on near-term actions and costs. We find that policies that are identical through 2020 have welfare impacts prior to 2020 that are substantially lower when the policy is assumed to end in 2020 than when the policy is expected to tighten or continue unchanged.

We follow up this first analysis, which is based on comparing alternative scenarios in which all agents are certain about policies beyond the policy horizon, with our second analysis, which considers the true situation in which agents are uncertain about future policies, although that uncertainty may be conditioned on current policy actions. This uncertainty leads to questions of how enactment of one particular current policy affects expectations about future policies, and whether behavior in the presence of future policy uncertainty is accurately characterized by means of scenario analysis. To examine these questions, the second analysis makes use of a stochastic optimization model of the utility sector to compare decisions under alternative scenarios of policy certainty with a sequential decision problem in which policy uncertainties are progressively resolved over time. Incorporating uncertainty in decisions leads to higher costs of achieving carbon limits than a comparable scenario analysis, suggesting that the

Human-induced Climate Change: An Interdisciplinary Assessment, ed. Michael Schlesinger, Haroon Kheshgi, Joel Smith, Francisco de la Chesnaye, John M. Reilly, Tom Wilson and Charles Kolstad. Published by Cambridge University Press. © Cambridge University Press 2007.

approach typically used in integrated assessments is likely to underestimate mitigation cost.

The rest of the chapter is organized into three sections as follows. In the first section, "Modeling with perfect foresight," we discuss our general equilibrium model and use it to analyze representative climate change policies. This section concludes with a description of how the results from modeling depend on long-term horizon expectations. In the second section, "Modeling uncertainty: three approaches," we turn to the nature of policy uncertainty. Here, we examine the relationship between various modeling approaches and the characteristic effects on the measurement of policy costs. Then we illustrate some of these effects with our stochastic partial equilibrium model. The third section, "Conclusions," offers some implications for integrated assessment (IA) modeling practice.

18.2 Modeling with perfect foresight

18.2.1 Basic structure of the multi-region national model

For the analysis in this section, we use our Multi-Region National model (MRN).[1] We have used it in numerous studies to analyze the economic costs of the climate change policies. MRN is a computable general equilibrium (CGE) model of region-specific impacts and regional interactions in the US economy. The model solves for income, production levels, relative prices, trade, and consumption by accounting for behavioral as well as technological responses to changes in policy. The equilibrium is fully dynamic, meaning that investment decisions determine the future capital stock, which in turn determines future income and consumption, and the agent in the model has perfect foresight of future equilibrium prices and endowments. Investment today requires forgoing consumption of current output in order to devote more of current GDP to investment. Consumer decisions maximize utility, which implies that an optimal tradeoff is made between consumption today and consumption in the future.

18.2.2 Data

Data that characterize the interrelationships of commodities, labor, and capital within the economy are of primary importance in quantifying the impacts from alternative carbon abatement policies. We use a Social Accounting Matrix (SAM) developed for each state by the Minnesota IMPLAN Group, Inc. (MIG). The IMPLAN database represents the activities in 530 sectors for all 50 states and the District. Adjustments to the original energy data were necessary to bring them in line with the Energy Information Agency's (EIA) state level energy data,

Table 18.1 MRN model's energy and non-energy sectors.

Energy sectors	Non-energy sectors
Coal extraction	Agriculture
Gas distribution	Energy intensive sectors
Oil and gas extraction	Manufacturing
Oil refining/distribution	Motor vehicles
Electricity generation	Services

which are more accurate than the corresponding IMPLAN data. In addition, the SAM completes the circular flow with an account of factor incomes, household savings, trade, and institutional transfers.

18.2.3 Benchmarking

The SAM is a snapshot of the economy along a dynamic growth path. Calibration of the dynamic equilibrium is completed by incorporating growth forecasts for industries, population, and carbon emissions. Although many dynamic growth models assume a balanced growth path, this is not necessary with MRN. Since the model's purpose is analysis of policy changes, the baseline is calibrated to match an exogenous forecast, in this case the forecast for GDP growth and carbon emissions in the EIA's *Annual Energy Outlook* 2001. The benchmarking procedure involves working backwards from the benchmark price and quantity path to solve for free parameters of the model that will make that benchmark path the equilibrium solution of the model. There is no need for the baseline to involve balanced growth, although we do generally let the model converge to balanced growth after 2030.

18.2.4 Sectoral disaggregation

Aggregation of regions and sectors in MRN is completely flexible. Since this chapter considers uniform nationwide programs, we aggregate all the regions of our modified IMPLAN SAM into one US region for this analysis.

All the important energy sectors contained in the detailed SAM are represented in MRN. We then aggregate the remaining non-energy sectors into five categories (see Table 18.1). We break out motor vehicles separately so that we can correctly account for individuals' responses to higher fuel costs caused by carbon abatement policies. Therefore, the model is run with the ten sectors listed in Table 18.1.

To incorporate impacts through international trade, we have built a hierarchical structure of models in which relevant international information is passed down from a more complete geographic model to MRN, which is the more detailed regional model.[2]

---

[1] We use our international CGE model (Multi-Sector Multi-Region Trade model) to provide international prices to MRN, but all reported economic impacts come from MRN.

[2] For this analysis it is not the regional or sectoral detail that necessitates this decomposition, but rather the special MRN features involved in modeling taxes and personal travel.

Our international trade model is the Multi-Sector Multi-Region Trade (MS-MRT) model. For this analysis, we used a version of MS-MRT that incorporates nine regional trade blocs and nine of the ten commodities in MRN.[3] MS-MRT predicts changes in the prices of US imports and exports. Given this information, an open-economy model of the United States alone can be run independently of the other regions. Since we incorporate terms of trade impacts from MS-MRT, MRN results become consistent with the international trade implications of the assumed policy scenario.

18.2.5 Time horizon

To correctly account for the long-term impacts, specifically consumption and investment decisions, of GHG abatement policies, we extend MRN's model horizon out to 2070. With this long horizon, the minimum time steps while maintaining computational feasibility are 5 years.

18.2.6 Policy instruments

Currently, MRN only tracks carbon emissions and therefore does not capture the other GHGs. To incorporate carbon emissions in the model, a constructed emissions permit is tracked for each of the three fossil fuel inputs (coal, natural gas, and refined petroleum products). In the baseline equilibrium these permits are not scarce (their price is zero), and the quantity of permits demanded equals the projected baseline emissions. The carbon permit is required at the fuel's point of purchase according to the carbon content of the purchase. Limiting the number of permits available imposes an emissions constraint, and the permit price reflects the marginal cost of abatement. This method of incorporating emissions, via permits, is convenient in terms of providing a number of policy instruments that involve emissions trading or specific wedges between abatement costs across geographic regions or sectors.

18.2.7 Representation of production and consumption decisions

MRN is a standard intertemporal Arrow–Debreu model, in which macro-level outcomes are driven by the self-interested decisions of consumers and producers. Consumers are represented by a single agent (the household sector) that maximizes utility subject to endowments of primary factors and available production technologies that transform factors and intermediates into commodities. Household utility is defined by a constant elasticity of substitution (CES) infinite sum of discounted transitory utility. Utility in a given time period is the CES composite of consumption and leisure. The budget constraint equates the present value of consumption to the present value of income earned in the labor market and the value of the initial capital stock minus the value of post-terminal capital.[4] The representative agent optimally distributes wealth over the horizon by choosing how much output in a given period to consume and how much to forgo for investment.

Two primary factors are supplied by the household sector for production: labor, which grows exogenously, and capital. The capital stock depreciates geometrically but can be augmented in each period through an investment activity. We model adjustment costs in the capital stock through a partial putty-clay production structure and equilibrium unemployment. In addition to labor and capital, the model is extended to include primary resource factors specific to the extraction of crude oil and natural gas, and extraction of coal.

Production sectors are assumed to be competitive, exhibiting constant returns to scale (except the natural resource extracting sectors). A nested CES structure is employed for production in the non-resource extraction sectors that use new capital. The CES process combines material (intermediate) inputs of non-energy commodities with capital, labor, and energy to produce final goods for consumption and intermediate goods for other sectors. The production nesting is intended to accommodate fuel substitution that might result from carbon abatement.

18.2.8 Representation of international trade

The model described above was further extended to an open economy with interstate and international trade. We use an intertemporal balance-of-payments constraint that dictates no change in net indebtedness over the horizon. Capital markets are otherwise unrestricted. Trade is specified such that all goods (except for crude oil) are differentiated by their origin; this is the popular Armington formulation (Armington, 1969). We assume that crude oil is a homogeneous world good (i.e., the Armington elasticity is infinite). An Armington aggregate good, which is either consumed or used as an intermediate in production, is the CES composite of imports of the good and goods produced within the United States. Similarly, a constant elasticity of transformation is defined between output destined for home consumption and output for international markets.

18.2.9 MRN's personal automobile use component

Analogous to the capital stock that supports production, a motor vehicle earns rents for its owner as it is combined with gasoline and other support commodities to produce vehicle miles traveled. We treat automobile purchases as an investment activity that augments a stock of automobiles. This is a unique view of the social accounting matrix (SAM) and

---

[3] For a full description of MS-MRT, see Bernstein et al. (1999).

[4] Consistent with a formulation of equilibrium unemployment, the wage is net of the premium paid to workers matched to a job. This is the correct rate at which to measure the labor–leisure trade off (equals the marginal benefit of leisure).

Role of expectations in modeling costs 219

requires a unique set of adjustments, for although the SAM fully tracks the purchases of automobiles and gasoline, it does not capture the joint product that they produce – personal transportation via automobile usage. Similar to the calibration of the social accounts to a dynamic equilibrium, assumptions about rates of return and depreciation, the level of investment, and a given benchmark equilibrium trajectory imply a value of capital stock. This formulation allows us to capture consumers' trade-offs between alternative methods of reducing fuel use, by buying more fuel-efficient vehicles or driving fewer miles to reduce emissions.

18.2.10 Tax instruments

The model takes into account the wedges between prices received by factor owners and marginal products of those factors, and the marginal costs of production and market prices, caused by the inclusion of taxes. The taxes represented in the model include: FICA (or labor taxes), corporate income tax, property taxes, indirect business taxes (or output and sales taxes), and personal income taxes.

18.2.11 Welfare measurement

Since MRN is based on intertemporal utility maximization, the best measure of the costs of a climate change policy is the change in representative household's intertemporal utility function, expressed as the loss in income that would produce the same change in welfare.[5] This measure gives a single number that takes into account all changes in impacts over the model horizon, discounts future impacts based on the representative agents rate of discount, and includes an adjustment for welfare losses that occur after the end of the model horizon (Bernstein *et al.*, 2003).

Policy studies more commonly use annual measures of impacts, such as the change in GDP or the change in the value of macroeconomic consumption. These annual measures are the source of much of the difficulty that arises because of uncertainty about future policies. They also have some deficiencies as true measures of economic welfare, in particular because they fail to include a consistent measure of how changes in investment affect the value of future consumption opportunities.

18.3 Defining policy scenarios for the long term

18.3.1 Background

We use as an example of a short-term policy intervention the amended Climate Stewardship Act of 2003 (S.139) introduced in 2003 by US senators John McCain and Joseph Lieberman (alternatively referred to as the McCain–Lieberman bill or M/L). This policy caps aggregate national GHG emissions at their 2000 levels by 2010 and maintains the cap until 2020. The policy is silent on emission targets after 2020.

18.3.2 Three alternative extensions of the McCain–Lieberman Phase I cap

In modeling McCain–Lieberman or any other particular piece of short-term climate legislation, it is reasonable to ask what comes next. In the case of McCain–Lieberman, we considered three potential extensions beyond 2020.[6] They are:

1. The Phase I cap persists forever (M/L Forever);
2. The cap is eliminated after 2020 (M/L to 2020); and
3. The cap is tightened after 2020 (M/L + Tighten).

Under all three scenarios, all other assumptions are identical. Specifically, the backstop technology is assumed to sequester carbon at $300/tonne for all time.

18.4 MRN results of three alternative extensions of McCain–Lieberman

We first discuss the welfare impacts of the McCain–Lieberman proposal under each of these long-term extensions to understand how different long-term expectations affect near-term costs. Then we return to a discussion of how these scenarios might be derived from assumptions about policy expectations, and how different treatments of long-term expectations fit with the policy analyst's goal of providing the most informative characterization of a particular near-term policy.

Figure 18.1 illustrates the time profile of carbon permit prices that result under these three scenarios. For M/L to 2020, the permit price drops to zero once the cap is removed (after 2020). In the other cases, the permit price rises to the cost of removal using the backstop technology or $300/tonne. The price rises more rapidly under M/L + Tighten because the caps are more binding.

Table 18.2 provides results for our measures of economic welfare. Since MRN includes an income–leisure trade-off, the relevant measure of welfare is what is called "full consumption" and measured as the loss in income equivalent to the change in consumption plus leisure time.

Logically, the lower the long-term carbon cap, the greater will be the percentage loss in welfare over a very long time horizon. Losses in welfare rise from 0.1% if the policy is terminated after 2020, to 1.1% if the carbon limits decline

---

[5] See Varian (1992) for the definition of equivalent variation and the conditions under which this can be calculated as the change in the value of the 'money metric utility function.'

[6] The long time horizon of MRN, although relatively short for integrated assessment models that run to a time at which concentrations and temperature can be stabilized, is much longer than the time horizon typically considered in legislation.

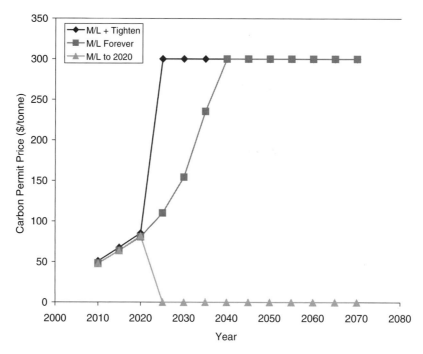

Figure 18.1 Carbon prices in three scenarios.

continuously after 2020. Figure 18.1 shows that carbon prices are identical through 2020, so that the remarkable finding is that the corresponding measure of an annual loss in welfare in the years before 2020 differs substantially across the long-term policy scenarios.

There are two reasons why an expectation of future carbon limits affects current behavior in a model and, therefore, produce changes in measures of economic welfare before those future limits take effect:

- Current investment decisions are made in light of expectations about all future market conditions and investment opportunities, leading to investment in anticipation of future requirements;
- Individuals adjust their consumption and savings choices over time, taking into account expected changes in future economic conditions attributable to a policy.

When the policy is not expected to continue, the loss in full consumption (consumption plus leisure) in 2005 is negligible, whereas if the policy is expected to become more stringent the loss in full consumption is substantial and nearly the same in every time period through 2020. This is an example of the "consumption smoothing" typical of intertemporal utility maximization. The household reallocates its savings and work effort to avoid having years with high consumption and years with low consumption – a direct consequence of the declining marginal utility of income that the standard utility function used in MRN implies.

The reallocation of consumption is so dramatic because under the cost assumptions that produce the permit prices

Table 18.2 Welfare impacts under different long-term policy expectations.

Inf. Horizon Welfare %	M/L to 2020	M/L Forever	M/L + Tighten
	−0.10	−0.71	−1.1

	Consumption (% change)			
Scenario	2005	2010	2015	2020
M/L to 2020	−0.18	−0.79	−0.98	−1.0
M/L Forever	−0.38	−1.0	−1.3	−1.6
M/L + Tighten	−0.35	−1.0	−1.4	−1.8

	Consumption + Leisure (% change)			
Scenario	2005	2010	2015	2020
M/L to 2020	−0.02	0.01	−0.09	−0.09
M/L Forever	−0.47	0.42	−0.50	−0.55
M/L + Tighten	−0.72	−0.71	−0.80	−0.83

assumed in MRN, welfare losses increase over time at a rate sufficiently high that, unlike normal present value calculations, discounting does not make losses in the distant future vanish. We can see this effect of long-term costs on near-term costs by adjusting the difference in long-term costs between the M/L Forever and M/L + Tighten policies. We accomplish this analysis by varying the cost of the backstop technology

Table 18.3 Welfare impacts with different backstop technologies.

Backstop cost ($/tonne of C)	Discounted PV of welfare through 2020		Difference	
	M/L Forever	M/L + Tighten	(%)	(billions of $s)
100	−127	−136	7%	9
300	−257	−322	20%	65
600	−296	−478	38%	183

and examining the effect of the different backstop prices on welfare (see Table 18.3). With a low enough backstop price, long-term costs for the two policies are nearly equivalent as the price of carbon permits reaches the backstop price at the same time under both policies. Therefore, the influence of policy expectations for the period after 2020 on welfare before 2020 goes away. As the backstop price rises, long-term costs rise and the difference in costs between the two policies grows.

18.5 Implications for long-term expectations in policy analysis

These results reveal several issues associated with long-term expectations. The first is that expectations beyond normal policy horizons matter, both for net present value measures of cost and for costs during the initial period when policies are defined. What is assumed about the continuation of a policy can produce a wider range of estimates for near-term costs than changing the near-term policy while holding longer-term expectations fixed. The second issue is that policy expectations are different from market expectations. The assumption of "perfect foresight" in deterministic models or of "rational expectations" in models that incorporate uncertainty is really an equilibrium condition. It is the requirement that all decisions be based on the equilibrium values of prices and quantities (or the equilibrium probability distributions of prices and quantities) over all time periods.

This concept of an equilibrium condition on expectations needs to be broadened to apply to policy expectations. Government is not represented as one of the agents in a standard general equilibrium model, and in modeling climate change policies, the actions of governments are normally specified by defining some scenario in which taxes or emission limits are imposed. But since governments cannot commit themselves irrevocably to future actions, this scenario must also be specified as some set of expectations about future policies. The obvious choice is to examine the consequences of economic agents assuming that an announced policy will be carried out. This literal approach is not always possible. In some cases, the policy will not be defined for the entire model horizon, and in other cases there may be reasons to doubt that economic agents will believe the announcement. Therefore, under a deterministic framework, multiple scenarios must be analyzed to capture the range of expectations that agents will have about a policy that has not been fully defined for all time.

In analyzing the McCain–Lieberman bill, we considered the possible range of policy extensions: less stringent caps (or termination), continuation of the policy as defined, and more stringent caps. The three different scenarios examined in our analysis of S.139 can be associated with different beliefs of economic actors about policies under investigation.

18.5.1 Scenario 1: Phase I lasts forever – no change in policy

This assumption may be inconsistent with everyone's expectations as a constant cap at 2000 levels has no support as an optimal long-term policy under any assumption about the costs and benefits of reducing greenhouse gas emissions. To those who are not convinced action is required, the cap is too tight, and yet for those who believe GHG concentrations must be stabilized to avoid impacts to climate, the cap is not tight enough.

18.5.2 Scenario 2: Phase I cap is loosened

In the case of termination of the cap, this scenario represents the discovery that "it was all a mistake," and could represent the expectations of economic actors not convinced that climate change is occurring. This scenario is relevant, but difficult to interpret.

Decisionmakers in the private sector may not be sufficiently confident that the mistaken policy will be reversed in a timely manner, and therefore will make long-term investments in order to avoid some of its perceived future costs. At a minimum, the possibility that a mistaken policy will lead to changed long-term behavior should be included in the analysis so that it is possible to identify the full potential consequences of a policy mistake.

18.5.3 Scenario 3: cap is tightened after 2020

The third scenario assumes the cap is tightened after 2020. This represents the "Watershed" view of policy on the part of private sector decisionmakers, in which consumers and firms are looking for a signal of where the policy process is going, and once they see any action, assume that climate policy will move toward an ultimate goal of stabilizing concentrations and global temperatures.

18.6 Policy expectations and policy analysis

Policy studies compare a "policy" scenario to a "no policy" scenario to measure costs and effects. When we examine policy scenarios in which long-term economic expectations differ from expectations in the no policy case, we are implicitly

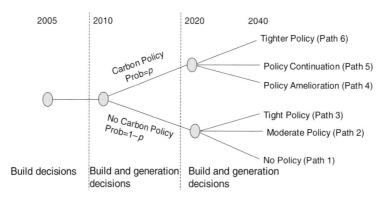

Figure 18.2 Representation of uncertainty in the carbon tax policy.

assuming that the passage of legislation *changes* expectations about future policy. Indeed, that is about all legislation can do when a later legislature can change anything done by an earlier legislature unless, under the US Constitution, that action imposes *ex post* penalties or takes property without due process. In our world of scenarios, the only way that we can address this function of legislation is by constructing different scenarios for the extension of policies. It is not clear that there are verifiable empirical procedures for ascertaining how policies change expectations.

Alternatively, the policy analyst could postulate an optimal long-term policy and compare all proposals to that policy – which dooms every proposal except the optimal one. We have seen no explicit examples of this approach, possibly because few analysts are sufficiently confident that they know the optimal long-term policy – or that the relevant decisionmakers in the private sector could be convinced that they are right. But this approach would have a great deal of appeal as a theory of how expectations are formed if it were possible to assume that in the long run an optimal policy would be followed. Unfortunately, every model recognizes that the optimal policy is very uncertain. Therefore one problem of a theory of policy expectations is with the certainty framework.

18.7 Modeling uncertainty: three approaches

The natural alternative is to model behavioral responses to policies as the outcome of a process of decisionmaking under uncertainty, in which actors are not certain of what future policies will be. This approach still takes government policies as essentially exogenous variables, fixed for all time, but recognizes that they are uncertain. We therefore develop a relatively simple positive model of decisionmaking under uncertainty in power generation, one of the most important sources of greenhouse gas emissions and a clear target for most policies. We contrast the implications of this model for near-term behavior and costs to the results of deterministic scenario analysis that is based on the assumption that actors are certain that the future will unfold as the scenario implies. This comparison suggests that there are material differences in near-term decisions when future policies are recognized as uncertain, and that these differences and their implications for cost cannot be captured in scenario analysis. We also reach some conclusions about whether near-term policies can in fact motivate the kind of long-term investments required for cost-effective mitigation if future policies must remain uncertain.

In this section, we consider legislation similar to the amended S.139 in structure. That is, there is some probability that a carbon tax policy will be enacted in 2010, and then in 2020, there is another decision point for regulators and a probability associated with possible policy extensions from 2020 onward. One way to represent this policy uncertainty is through a tree structure as illustrated in Figure 18.2.

In the figure, new information about the government's policy is revealed in 2010 and 2020. In 2010, with probability p, a carbon policy is introduced, and with probability $1-p$, no policy is introduced. Then in 2020, there is a revision to the government's approach with six possible outcomes each with their own probability.[7] Given this representation of possible outcomes, a rational firm will make decisions in 2005 that take into account the different possibilities of policy change in 2010 and 2020. That is, stochastic optimization (SO) only makes decisions for the first point in the tree, with subsequent decisions being random variables dependent on information revelation. Under such a circumstance, the decisions taken in 2005 will differ from those taken if the exact path of policy decisions going forward were certain.

This approach for policy analysis contrasts with the common approach of scenario analysis (SA) used by policymakers. Under SA, a deterministic framework is assumed, but a variety of possible policies or scenarios is examined. These scenarios are typically reducible to the paths as labeled in the figure, where path 6 might be considered the "worst case scenario" and path 1, the "best case scenario."[8] Scenario analysis then would treat every path on the tree as a separate

---

[7] The labels on the end states are meant to be illustrative. As discussed below a number of cases were run that explore how changes in conditional expectations affect firm behavior.

[8] The terms "worst case" and "best case" are not meant to be commentaries on policy approaches but reflections on policy costs.

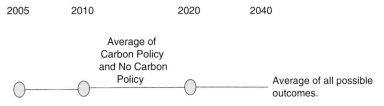

Figure 18.3 Expected value approach.

problem with all decisions made at the beginning. As a result, scenario analysis systematically underestimates policy costs. This is not only true if one takes the values given by each path and averages them accordingly to the probabilities of each path, but also if one considers each path in turn. In particular, the so-called worst case of scenario analysis underestimates what the worst case costs would be since it does not take uncertainty into account.

Another typical approach to planning is the expected value problem (EVP) approach. This approach takes the expected values of the parameters (in this case the level of the carbon tax) to construct a single path as illustrated in Figure 18.3 and then optimizes.

Costs developed through an expected value approach overstate policy costs, however. Thus the relationship between the approaches is given as:

$$\mathrm{SA} \leq \mathrm{SO} \leq \mathrm{EVP}. \qquad (18.1)$$

18.8 Description of the stochastic model

Under any of the analyses the tradeoff faced by the firm is between (1) relatively high capital cost choices with lower variable costs and (2) lower capital cost outlays but higher variable costs. In the stochastic optimization problem, in contrast to scenario analysis or the expected value problem, a firm can both delay its decisions for future periods whilst waiting to see what the policy outcomes will be, and reverse previously made decisions. In the language of option theory, there may be value to the firm in waiting to make decisions if it does not have to pay too high a price to do so. Moreover, the option may have value if certain states of the world are realized. For instance, because of the possibility of a high carbon price outcome, a natural gas plant that might in expectation be uneconomic, could be the right plant to build.

In order to explore these concepts we used our partial equilibrium model of the electric utility sector. This model is formulated as a mathematical program in the form of minimizing costs while meeting electricity demand and operating limitations. The results as presented here use a linear representation of costs and constraints in the form of a linear programming model.[9]

The model consists of variable and fixed costs for generating units as well as emissions costs, which are simply equal to the policy's imposed carbon tax times the level of emissions. A variety of generating technologies are represented including nuclear, coal (several different technologies), and natural gas (both combustion turbine and combined cycle). Each technology type has its own heat rate and therefore fuel costs based on the underlying heat rate. Variable costs, then, consist of fuel costs and variable operating and maintenance costs which also vary across technology types. Fixed costs are based on capital expenditures needed to build a unit as well as fixed operating and maintenance expenses. Costs were generally taken from the EIA's 2004 *Annual Energy Outlook*. Emissions costs in the results presented in this chapter are in the form of a $/tonne tax on carbon emissions from generation. We vary this tax from $0/tonne in the lowest case up to $250/tonne.

Figure 18.4 illustrates the probability structure for the results presented here. We solve a three-stage stochastic program with six possible end states. There are multiple years in some of the stages, but we only represent the "stochastic decision points" in the figure. Both costs and demand changed from one stage to another with a small growth in demand, increasing fuel costs, and decreasing technology costs (especially for new technologies such as integrated gasification combined cycle, IGCC). These costs (excluding emissions costs), however, did not vary from one scenario (state) to another within a stage. So, for instance, variable and fixed costs for generation and demand were the same for both nodes in Figure 18.4 for the year 2020. Scenarios, however, did differ by carbon taxes, which evolved over time, but also varied from scenario to scenario. Our base-case carbon taxes are shown in Figure 18.4.

We examine two sets of cases to explore the theme of policy uncertainty. The first is a set of cases in which expectations about the final policies are not influenced by current decisions, but only the degree of uncertainty about future policies varies. These are the balanced cases. On the other hand, it might be reasonable to expect that under some conditions the adoption of a policy might give some information about the likelihood of the final policy end state. This assumption describes our second set of cases.

---

[9] We also developed non-linear versions of the model to take into account the fact that many decisions are not continuous, but rather are discrete. For example, in deciding whether to build a generating plant or not, the practical reality is one is limited to some set of discrete plant sizes, rather than building just exactly the MW size of plant that would be optimal. The results for this mixed integer program were essentially the same as the linear results presented here.

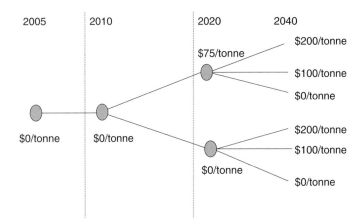

Figure 18.4 Base-case carbon tax scenarios

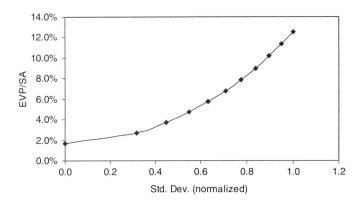

Figure 18.5 Standard deviation of carbon tax vs. error in analysis – balanced cases.

18.9 Comparison of results under balanced cases

In order to explore the concepts discussed above, we solve this model by examining cases which are "balanced" in the sense that the conditional expected value of the third-stage carbon tax level is the same in both cases whether an initial policy is enacted or not. The cases are then varied by changing the probabilities of the end states, but keeping the expected value of the tax at the last stage the same after both second-stage outcomes.[10] Figure 18.5 illustrates how as the standard deviation increases, the spread between the total costs under scenario analysis and the expected value problem increases (reported as the percentage increase in total costs under EVP over the total costs under SA). A similar relationship obtains for the relationship between SA and SO.[11] In other words, as uncertainty, as measured by the standard deviation in the carbon tax at the end state, increases, SA and EVP perform worse and worse. Intuitively, one expects the deviation among the approaches to increase as the variance in possible policies increases. That is, if there were only one policy that was known with certainty, then the three approaches would yield the same result.

Importantly, differences in generating capacity choices are significant across the cases. For SA, conventional coal tends to be chosen for the paths with less stringent policies (paths 1 to 4), and combined cycle for the more stringent regimes (paths 5 and 6) (Fig. 18.2). EVP chooses the highest-cost solution more often: the combined cycle. SA, on the other hand, hedges by choosing advanced coal in the first stage. Then again, sensitivity cases run with further delay in policy implementation suggest that to keep options open, a measure with high operating costs and low capital costs may be preferred to a measure with low operating costs and high capital costs even if the latter measure has lower cost per tonne emissions avoided (in NPV terms over the life of the capital).

---

[10] Because this is a partial equilibrium model the assumption is that the carbon taxes are exogenous in each scenario reflecting the resulting value of carbon permits from a nationwide emission reductions policy. The question explored is the reaction to changes in expectation in policy. Another question, for further research, is the degree to which the same reductions in emissions are achieved under varying expectations. This would imply an ability of a government to commit to a policy, at least in the sense of reducing the variability of possible future policies.

[11] Note that the percentage difference of total costs is plotted, and the difference in marginal costs would be greater.

18.10 The importance of changes in legislation on expectations

The next results that we present address what happens when the initial policies in 2010 give a signal as to later policies. There are two possibilities. When initial policies signal continuation, then costs are lower, and the spread amongst all three approaches decreases compared with cases where initial policies signal reversal. The results illustrate, again, the phenomenon that increased uncertainty about end states increases costs and decreases the accuracy of SA and EVP measures.

Our partial equilibrium model then shows that there is an immediate problem with regulators using SA to estimate policy costs. Costs will in fact be higher, both in the short run and in a long-run present value sense, when future policies are correctly modeled as uncertain. Total net present value (NPV) costs of meeting a fixed cap will be greater because capital investments will be delayed relative to the schedule that would be optimal under certainty, in order to buy time to learn more about future policies. Thus we conclude that there is a bias toward underestimating costs in IA models that assume certainty about future policies.

We see very different changes in near-term behavior when legislation is "informative," in the sense that the probabilities assigned to future carbon prices differ in the branches with and without legislation passed in period 2. This leads naturally to the practical question, "Changed from what?" We also find that when passage of legislation leads to a downward revision of beliefs about future carbon prices the effects on behavior are very different from the case in which it leads to upward revision. Some legislation may be so weak that its passage would lead to the reaction, "If that's all they can pass, I can stop worrying so much about tight carbon caps," or, in the specific context of the United States, passage of relatively mild Federal legislation could through pre-emption modify expectations that some states could adopt much more stringent limits. The first step toward using this insight may have to be development of an empirical approach to measuring current market expectations of future carbon prices.

18.10.1 How are market expectations changed?

What, then, is the process by which adoption of policies changes expectations (Newbery, 1990)? Putting an explicit time horizon into a policy with some set of criteria for decisions to be made at that point may help to narrow future possibilities. A sunset provision signals that the need for a policy will be re-evaluated, but is normally used with programs that serve an inherently temporary purpose.[12] That is not the case for limits on carbon emissions, save in the case when it is assumed that a zero-carbon technology comes to dominate all fossil fuel technologies at some point in the future.

There are few relevant limits in the US constitutional system on what a future Congress can do to change actions of a previous Congress, as long as it does not establish criminal penalties for actions taken before passage of the law or take property without due process. Thus in a sense all policies, no matter what they say, become increasingly uncertain looking further out in the future.

The central issues in debates about climate legislation have been whether climate change is a problem that needs to be addressed now through emission limits, and what the cost of such limits will be. That is how the issue of long-term policy expectations, and how near-term costs are affected by policy expectations, arose. Some approaches to climate policy give more certainty to future costs than others. The original version of S.139, had it passed, could have been more informative than the amended version, because the explicit provision to ratchet the carbon limit down in 2016 was more specific than the indefinite implementation of Phase I, which left almost any follow-on policy possible. Negotiations over a "safety valve" which would put an upper limit on the price of a carbon permit under S.139, and therefore on the costs that could be imposed by S.139, are consistent with the expectation that a policy with explicitly limited future costs will be more easily passed and more likely to remain unchanged (or if changed, more likely changed in only one direction). A policy that used revenues from carbon taxes or permit auctions to create a trust fund with an organized constituency would also tend to have greater long-run certainty, although standard public finance principles teach that such earmarking is also likely to lead to highly uneconomic uses of the funds.

18.11 Conclusions

Issues about expectations and their implications for lessened incentives and more costly compliance arise from policies that provide incentives for current investments by changing expected prices and costs over the life of the investments. Such policies are by their nature reversible, although some may be viewed as less likely to be reversed than others. Policies with low and slowly rising prices are least likely to be reversed, policies with high prices or dependent on the actions of sovereign countries under an international agreement that has no penalties for non-participation or non-compliance are most likely. Unfortunately, the former policies also do little to bring about needed investments, whereas the latter would. The key policy instruments most often analyzed in studies of climate change policies – international agreements on targets and timetables that support emission trading and reliance on announced future policies to motivate R&D – are not credible because of fundamental dynamic inconsistencies.[13] Therefore targets and timetables are not likely to be accepted

---

[12] Such as trade restrictions designed to give time for domestic industry to restructure or adapt.

[13] See, for example, Barrett (2003), and Montgomery and Smith, Chapter 27 in this volume.

either as a certainty or even as the most probable among possible paths.

There are alternative policy approaches not subject to these dynamic inconsistencies brought about by measures that are easily reversible. A different class of policies, involving current actions by government to bring about irreversible changes in the economy, would avoid the problem. Fiscal and regulatory measures are reversible by sovereign governments, so that incentives based on expectations of future carbon prices supported by such measures – especially ones tied to international agreements such as the Kyoto Protocol – will be diluted. But once an investment in R&D or physical capital is made, it will not be reversible. Thus policies that stimulate such investments through up-front incentives can indeed have long-lasting effects, even if there is uncertainty about whether they will continue.

There are a number of implications of the foregoing analysis for both modelers and policymakers. First, for modelers, regardless of the framework for the analysis it is important that the model horizon extends far beyond the policy time-frame. Actions taken as a result of the policy could have important impacts that occur beyond the policy time-frame that need to be taken into account in policy analysis. Also, expectations of the policy beyond the time-frame envisioned in a given policy framework will affect the actions taken by economic agents. This leads to a second implication for modelers. Assumptions about future policies can lead to widely differing results, so that explicit treatment of these assumptions is important. A third implication, along with a way to analyze the impacts of assumptions about future policies, is that stochastic optimization can both improve the answers given by scenario analysis, and yield insights as to how future expectations affect actions here-and-now. A final implication for the modeling community is the need to be clear about which problems we are analyzing. In the optimal policy world, scenario analysis will underestimate costs because there is real uncertainty about the direction of optimal policy. While some uncertainty may be resolvable with increased knowledge of the science and economics of the problem, some of the uncertainty may be irreducible.

In addition to these implications for modelers, there are some implications for policymakers as well. First, it is important that policies clearly convey the regulators' expectations of the direction of future policies so that agents' expectations will be in line with those of regulators. Second, to the extent that policy uncertainty is driven by expectations of the private sector that a policy will change erratically, this raises the costs of compliance. It seems warranted, then, that rather than simply continuing to focus on the question of the optimal response to a given policy, more effort is need to judge how policy uncertainty from any source can drive costs. Policy uncertainty may indeed come from many sources, some more or less resolvable through more science and better economics, and some may be inherent stemming from both domestic and global political uncertainties.

References

Armington, P. S. (1969). *A Theory of Demand for Products Distinguished by Place of Production*. IMF Staff Papers 16, pp. 159–176.

Barrett, S. (2003). *Environment and Stategraft: The Strategy of Enviromental Treaty-Making*. Oxford: Oxford University Press.

Bernstein, P. M., Montgomery, W. D. and Rutherford, T. F. (1999). Effects of restrictions on international permit trading: the MS-MRT model. *The Energy Journal*, Kyoto Special Issue, 221–256.

Bernstein, P., Montgomery, W. D. and Smith, A. (2003). *Analysis of Welfare Measures*. Submitted to US Department of Energy.

Lucas, R. E. (1976). Econometric policy evaluation: a critique. *Carnegie-Rochester Conference Series on Public Policy* **1**, 19–46.

Lucas, R. E. and Sargent, T. (1981). After Keynesian macroeconomics. In *Rational Expectations and Econometric Practice*. Minneapolis: University of Minnesota Press, pp. 295–319.

Newbery, D. M. (1990). Cartels, storage, and the suppression of futures markets. *European Economic Review* **34** (No. 5, July) 1041–1060.

Varian, H. (1992). *Microeconomic Theory*, 3rd edn. New York: Norton, pp. 160–171.

19

A sensitivity analysis of forest carbon sequestration

Brent Sohngen and Robert Mendelsohn

19.1 Introduction

This paper examines the sensitivity of estimates of both the baseline of carbon in forests and the efficacy of carbon sequestration programs to the supply of land for forests, timber demand, and technology change in forestry. We explore how changes in these parameters affect estimates of the amount of carbon that forests will store in the absence of carbon sequestration programs. We also explore how changes in these parameters will alter how much an efficient carbon sequestration program will store. Although we have previously estimated expected outcomes for the baseline (Sohngen and Sedjo, 2000) and for efficient carbon sequestration programs (Sohngen and Mendelsohn, 2003), the model used for this analysis has been updated from these earlier versions, and this sensitivity analysis is new. Given that the parameters we are exploring in this analysis are highly uncertain, especially at the global level, it is important to see how these adjustments affect the conclusions of earlier studies.

In this study, we follow the definition of an efficient carbon sequestration program established in an earlier paper (Sohngen and Mendelsohn, 2003). We assume that carbon sequestration programs must be embedded in the overall effort to control greenhouse gases. The marginal cost of sequestering carbon in forests should be equated to the marginal cost of sequestering carbon in other resources such as agriculture and the marginal cost of preventing carbon emissions in the energy sector at each moment in time. It is interesting to contrast the results found in this paper with the chapters on agriculture and forestry sequestration (McCarl et al., Chapter 20, this volume) and energy mitigation precisely for this reason. Because the marginal cost of carbon sequestration must be equated with the marginal cost of abatement in the energy sector, the sequestration program must be tied to the optimal path of marginal prices of carbon predicted by an optimal control model of greenhouse gases (e.g. Nordhaus and Boyer, 2000). For a forest carbon sequestration program to be efficient, it must have the correct dynamic intertemporal incentives and it must capture the substitutions possible in the global forest sector across space and across management intensity (see Sohngen and Mendelsohn, 2003).

While the paper ties estimates of carbon sequestration to carbon prices from an integrated assessment model, the estimates do not directly account for potential climate change impacts in forests. Sohngen et al. (2001) show how climate change itself could increase productivity in the world's forests, and influence forest management and market prices. Such changes could have important influences on the efficiency of carbon sequestration, but we have not incorporated these effects in this study. Section 19.2 provides a detailed literature review of economic models of timber and agricultural markets that have been applied to estimate carbon sequestration costs. Section 19.3 describes the model and analysis, Section 19.4 displays the results, and Section 19.5 reviews the conclusions and policy implications.

19.2 Literature review

Many authors have projected the amount of carbon that forests will hold in the absence of any carbon sequestration programs. Between 2.0 and 2.2 Pg/yr ($1 Pg = 1 \times 10^{15}$ g) of carbon (C) have been lost from forests because of changes in land use in the past few decades (Houghton, 2003). The Intergovernmental Panel on Climate Change (IPCC, 2000) predicts that forests will continue to be converted to agriculture in tropical countries, causing large carbon emissions over the next few

Human-induced Climate Change: An Interdisciplinary Assessment, ed. Michael Schlesinger, Haroon Kheshgi, Joel Smith, Francisco de la Chesnaye, John M. Reilly, Tom Wilson and Charles Kolstad. Published by Cambridge University Press. © Cambridge University Press 2007.

decades (1–2 Pg/yr). At the same time, timber stocks have increased throughout temperate regions (IPCC, 2001), there have been large increases in the area of fast-growing timber plantations in many regions of the world (FAO, 2003; Haynes, 2003), and the technology of growing trees and harvesting forests has improved (Sedjo, 1999; 2004a, b). Sohngen and Mendelsohn (2003) estimate that in the baseline scenario, forests emit 292 Tg C per year on average over the next century. The IPCC Special Report on Emission Scenarios (2000) suggests similar total land-use emission levels over the entire century, but they allocate a large proportion of the emissions to earlier periods.

In addition to baseline studies, numerous studies have been conducted to assess the potential of programs to sequester carbon in forests. One of the most comprehensive recent reviews of existing studies is that of Richards and Stokes (2004), who suggest that it may be possible to store up to 2 PgC per year for $10–$150 per tonne C. This is an exceptionally wide range of cost estimates. One reason for this is that many studies analyze only specific regions or countries. Costs may vary considerably across regions. Further, while regional studies often benefit from having detailed spatial data, they potentially ignore two important issues necessary for calculating net carbon benefits: leakage and substitution. Leakage is a problem because efforts to sequester carbon in one place may cause forest owners to release carbon in another place. Ignoring leakage can cause studies to underestimate the total costs of carbon sequestration (Murray et al., 2004; Sohngen and Brown, 2004). Global substitution is important because forestry can shift harvesting from one region to another and from one level of management intensity to another. In order to design an effective global sequestration program, one needs to understand which forests are most suitable for sequestration and which forests should be used for production. The recent widespread increase of plantations in subtropical forests is an important example of the global supply response required to make forests renewable.

Another issue contributing to the wide range of results shown in Richards and Stokes (2004) is that some studies have failed to make the distinction between setting aside carbon for a year and setting aside carbon permanently. Lengthening a rotation increases carbon storage for a limited period of time. Setting aside old growth in a conservation area permanently stores carbon if that area would have been harvested otherwise but does nothing at all if that area would never have been harvested. The value of holding 1 tonne C for a year is the rental rate of carbon. The value of permanently setting carbon aside is the cost of carbon. It is the cost of carbon that is equivalent to the marginal cost of abatement in the energy sector. The rental rate of carbon can be determined from the cost of carbon using appropriate discounting techniques. Finally, many studies reviewed by Richards and Stokes (2004) rely on static models. Static models cannot handle changing conditions such as changes in demand and technical change, and they specifically cannot address how much carbon sequestration might occur if incentives for carbon sequestration are changing over time.

As an alternative, a number of authors have applied dynamic forestry models, or forestry and agricultural models, to estimate carbon sequestration costs. For example, Adams et al. (1999) examine potential sequestration in the United States using a dynamic model of the forestry and agricultural sectors, and McCarl and Schneider (2001) update this analysis by including several additional options like agricultural soil carbon sequestration and biomass energy production. By accounting for the equilibrium in land markets between many different uses, these two approaches implicitly capture leakage, and not surprisingly, suggest higher costs than many of the local studies examined in Richards and Stokes (2004). In other regions, such as Europe, work has been conducted to assess future baseline carbon under alternative scenarios using economic models (Nabuurs et al., 2003). Regional studies such as these often benefit from using fairly specific and well-known local data sources. They potentially miss interactions that occur across country or regional boundaries. In the context of carbon sequestration analysis, where large-scale changes in land-use may occur throughout the world, capturing interactions across countries can be important.

Several models have been applied at the global level to assess carbon sequestration. For example, the IMAGE model includes both agriculture and forestry, and van Vuuren et al. (2006) have used IMAGE to assess carbon sequestration policy. The IMAGE model, however, focuses on the spatial distribution of forests and agriculture rather than stock adjustments in forestry, and it does not fully exploit economic processes in allocating land to different sectors. An alternative approach is the intertemporal computable general equilibrium approach used in the D-FARM model (Ianchovichina et al., 2001). This study clearly shows that technology change in agriculture can have potentially important implications for future land-use change, but the authors do not monitor the stock of forests, and consequently cannot assess carbon sequestration.

Sohngen and Mendelsohn (2003) use a dynamic optimization model of timber markets to assess global carbon sequestration potential across a range of prices. Because the model is dynamic, it can capture important adjustments in the forest stocks (and consequently carbon stocks) resulting from changes in demand for agricultural land, timber demand, and technology. However, the demand for agricultural land is captured through exogenously specified land rental functions, making it difficult to account directly for changes in demand for agricultural products or technological change in the agriculture sector. A more recent addition to the modeling literature is the GCOMAP model described in Sathaye et al. (2006). This model incorporates highly detailed data on land supply in specific regions of the world, but as with the model described in Sohngen and Mendelsohn (2003), it does not model the agricultural–forestry interface endogenously.

Table 19.1 Alternative baseline scenarios.

Scenario	Assumptions
Baseline	Demand growth = 0.5% per year, declining 5% per year
	Land supply elasticity = 0.25
	Subtropical plantation Yield = 0.25%/yr
	Harvesting costs = constant
	Plantation establishment costs = constant
High demand growth	Demand growth = 2.0% per year, declining 5% per year
High land supply elasticity	Land supply elasticity = 0.38
Low land supply elasticity	Land supply elasticity = 0.13
Low yield growth	Subtropical plantation yield = 0.05%/yr
Lowering harvesting costs	Harvesting costs = −0.5%/yr
Lowering plantation establishment costs	Plantation establishment costs = −0.5%/yr

19.3 Analysis and results

This paper uses the global timber market model described in Sohngen and Mendelsohn (2003) to assess global carbon sequestration potential. For this analysis, the model has been extensively updated to include additional regions and timber types, alternative land rental function specifications, and new yield functions and biomass expansion factors for many forests. These changes, as well as a brief mathematical exposition on the model, are described in the appendix to this chapter.

For the present study, we rely on sensitivity analysis to address the implications of demand for agricultural land, demand for timber, and technology change in the forestry sector on potential carbon sequestration in forests. The full set of sensitivity analyses is shown in Table 19.1. The first scenario examines the potential that demand for industrial wood increases more rapidly over time than our baseline assumption. Although global wood harvests have been relatively constant for the past 20–30 years (FAOSTATS, 2003), it is possible that rising population and income in Asia, India, and other developing countries could substantially expand the demand for wood products. The baseline assumes that industrial wood demand grows at 0.5% per year, with growth falling by 5% per year. The sensitivity analysis assumes that industrial wood demand grows at 2% per year initially, with the growth rate falling by 5% per year.

The second and third scenarios change the assumed price elasticity of the supply of land from agriculture. The baseline assumes that this elasticity is 0.25 initially, suggesting that a 1% increase in forest values would increase forestlands by 0.25%. These results are consistent with recent estimates of land supply elasticity in the United States (Hardie and Parks, 1997; Plantinga et al., 1999), but they have not been estimated for other regions of the world. It is entirely plausible that land supply could be more or less "elastic" in other regions. A higher (lower) price elasticity would suggest that more (less) agricultural land would shift into forestry as forest prices increase. With current data sets, it is difficult to identify the proper elasticity estimate to use for each individual region, so we have assumed that all regions have the same elasticity initially. For the sensitivity analysis, we assess both higher and lower land supply elasticity, and apply these adjustments to each region of the world uniformly (Table 19.1). Alternatively, one could shift the land rental functions without changing their price elasticity. For example, rising populations or income could shift the rental functions up. We have not conducted such analyses in this study, although this would be a promising issue to explore in future research.

The fourth scenario explores the influence of technological change on the yield of timber from high value plantations. In recent years, a number of technological advances have occurred in many fast-growing plantation species (see Sedjo, 2004a, b). The baseline scenario assumes that these advances continue and that they are adopted relatively quickly. As an alternative, we assume that technical change has a smaller beneficial effect on plantation yields, so that yields increase at 0.05% per year, rather than the 0.25% per year increase assumed in the baseline.

The fifth scenario assesses change in harvesting technology. As forest industries in Europe and the United States have adapted to changing environmental constraints, they have tended to adopt new and improved methods for extracting timber from forests. Widespread adoption of these new methods has potentially improved productivity in the forestry sector (Sedjo, 1999). To model adjustments in harvesting technology, we assume that harvesting costs in all regions decline by 0.5% per year throughout the modeling horizon.

Finally, we examine the role of fast-growing timber plantations. One important trend in forestry in recent years has been the expansion of subtropical plantations in regions such as Chile, Argentina, Brazil, South Africa, Indonesia, Australia, and New Zealand. Many of these expansions involve conversion of agricultural land or existing natural forests to intensive plantation management. As an alternative scenario, we assume that the costs of establishing these plantations decline over time by 0.5% per year. This accounts for the possibility that the new technologies above that improve plantation yields will make these sites even easier to establish in the future. These changes in costs are applied only to the area of high-value subtropical plantations, which initially account for only a small proportion of total world forests

(73 million hectares out of 3594 million hectares in forests in total).

In addition to considering how these alternative baseline assumptions influence future carbon flux, we also analyze how these assumptions affect the efficacy of efficient carbon sequestration programs. The carbon price scenarios are drawn from Sohngen and Mendelsohn (2003), where a forestry model was linked with the DICE model described in Nordhaus and Boyer (2000). For this analysis, we treat the price path of carbon as exogenous, and two sets of carbon price paths are considered, a low and a high damage case. The low damage case assumes that greenhouse gases will have only a modest effect on global warming and future impacts. The high damage case takes into account possible extreme scenarios such as the melting of Antarctic glaciers or the loss of the thermohaline circulation (Nordhaus and Boyer, 2000). Carbon prices in the low damage case are initially $7.14 per tonne C in 2005, and rise to $61.36 per tonne C by 2105. For the high damage case, carbon prices are $21.82 in 2005, and rise to $187.63 in 2105. For each alternative baseline scenario, we compare and contrast potential carbon storage.

19.4 Results

19.4.1 Baseline

The baseline for this analysis assumes that demand grows at a declining rate over time. Growth starts at 0.5% per year but declines 5% per year. The elasticity of demand is assumed to be unitary, 1.0, in the initial period. The elasticity of land supply is assumed to be 0.25 for every timber-producing region. Technical change is assumed to increase yields in subtropical plantations at 0.25% per year. The yield functions for all other species are assumed to remain constant over time. The baseline also assumes that harvesting costs and plantation establishment costs are constant over time. There is no carbon sequestration program in the baseline.

The model predicts that the area of forestland and global carbon storage will decline over the next century in the baseline (see Table 19.2). During the period 2005–15, the net annual reduction in forest area is approximately 10 million hectares per year, and global carbon emissions from land-use change are approximately 1.9 Pg per year, which are consistent with estimates from other authors (Dixon et al., 1994; Houghton, 2003). However, the model predicts that this deforestation rate will slow over time so that the rate is closer to 267 000 ha/yr in the second half of the century and the amount of carbon being lost will drop to 0.5 Pg/yr. Subtropical plantations are projected to expand during the period of analysis, rising from 73 million hectares in 2005 to nearly 136 million hectares by 2105. This relatively robust increase in plantations occurs largely in South America, China, and Oceania. The area of inaccessible temperate and boreal forests declines from 1121 million hectares in 2005 to approximately 938 million hectares. Most of these losses occur in Canada and Russia.

The changes in subtropical plantation forests and inaccessible temperate and boreal forests have important implications for carbon sequestration. Within North America, for example, the baseline estimates in this study suggest that the carbon pool will be relatively stable, and even declining slightly over the coming century. The area of forests within the United States increases over time, but these increases do not offset carbon losses from harvests in forests that currently are natural

Table 19.2 Baseline scenario: carbon storage in forests and forest areas.

	Carbon (PgC stored by year)			Total forest area (million hectares)			Inaccessible temperate and boreal forest area (million hectares)			Subtropical plantation area (million hectares)		
	2015	2055	2105	2015	2055	2105	2015	2055	2105	2015	2055	2105
United States	49.6	49.8	50.5	205.5	206.1	206.0	53.0	37.2	34.7	13.5	14.2	14.6
Canada	131.2	131.6	132.0	420.4	420.4	420.0	286.2	260.4	258.6	0.0	0.0	0.0
S. America	206.3	194.4	190.6	833.7	732.0	715.4	0.0	0.0	0.0	15.8	24.4	27.4
C. America	11.2	9.2	10.1	49.5	37.9	37.1	0.0	0.0	0.0	3.0	3.9	3.9
Europe	28.5	29.3	30.3	189.0	192.5	195.1	2.6	0.6	0.2	4.5	7.7	8.3
Russia	252.8	249.8	250.1	839.7	839.7	839.8	692.9	648.8	644.6	0.0	0.0	0.0
China	27.8	28.4	29.2	159.9	166.7	168.7	18.2	2.6	1.4	13.6	20.4	22.9
India	9.9	9.4	9.3	48.7	45.3	45.6	0.0	0.0	0.0	4.7	5.7	5.8
Oceania	25.4	25.9	26.1	195.9	180.1	164.9	0.0	0.0	0.0	7.4	11.5	12.9
SE Asia	51.3	39.5	36.1	184.3	127.8	116.8	0.0	0.0	0.0	12.3	14.9	15.9
Central Asia	6.3	6.2	6.2	38.4	39.8	40.0	0.0	0.0	0.0	2.5	3.7	3.8
Japan	4.1	4.3	4.9	23.5	24.1	24.6	0.0	0.0	0.0	10.4	10.6	10.7
Africa	75.8	62.6	58.6	305.0	202.7	190.1	0.0	0.0	0.0	5.3	8.1	9.3
Total	880.2	840.6	833.9	3493.7	3215.2	3164.0	1053.0	949.6	939.5	92.9	125.1	135.5

Table 19.3 Sensitivity analysis: projected changes in total carbon and total forest area for alternative scenarios relative to the baseline in 2105.

	Carbon (PgC stored above baseline by 2105)						Forest area (million hectares above baseline in 2105)					
	High demand	High elasticity	Low elasticity	Low harvest cost	Low plt. cost	Low yield chg.	High demand	High elasticity	Low elasticity	Low harvest cost	Low plt. cost	Low yield chg.
United States	1.2	0.3	−0.3	−0.3	−0.1	−0.4	11.5	3.1	−2.7	1.4	−0.8	0.0
Canada	−0.3	0.2	−0.2	−0.7	0.0	−0.3	6.1	1.6	−1.5	0.7	−0.4	0.3
S. America	−2.7	3.9	−3.7	−2.4	0.3	−0.7	4.2	34.0	−33.9	1.1	3.8	−2.4
C. America	−0.1	0.1	−0.2	−0.1	0.0	0.0	−0.2	1.8	−1.7	−0.2	0.0	0.0
Europe	0.8	0.1	−0.1	0.1	−0.1	−0.4	17.2	1.0	−2.8	2.0	−1.6	−1.0
Russia	−2.9	0.0	−0.1	−2.6	0.1	−0.3	2.5	0.7	−1.0	0.4	−0.3	0.6
China	0.6	−0.4	0.4	−0.2	−0.8	0.5	9.6	1.2	−0.8	0.9	4.7	−2.1
India	0.5	0.1	−0.1	0.1	0.1	0.0	5.5	2.2	−2.1	0.3	2.0	0.3
Oceania	0.4	0.1	−0.2	−0.3	0.2	−0.3	1.5	5.2	−7.3	−4.8	2.9	−1.6
SE Asia	−1.9	1.0	−1.4	−1.5	0.1	−0.3	3.1	5.3	−9.1	−1.6	2.2	−0.7
C. Asia	0.0	0.0	0.0	−0.1	0.0	0.0	1.4	−0.2	0.3	0.3	1.2	0.3
Japan	0.6	0.0	0.0	0.0	0.0	−0.4	3.1	0.3	−0.2	0.2	0.0	−0.6
Africa	−2.8	0.7	−1.1	−2.3	0.1	−0.4	−0.5	7.1	−9.3	−0.6	1.2	−1.0
Total	−6.6	6.3	−7.0	−10.3	0.1	−2.9	64.9	63.1	−72.1	0.1	14.8	−7.9

Chg. = change
Plt. = plantation

and conversions from natural forests to plantations, particularly in the south. Mature natural forests are assumed to have more carbon per hectare, mostly because they are much older on average. We assume that species conversions that enhance merchantable timber production do not also enhance carbon sequestration. Within Canada, continued harvests in inaccessible regions of the boreal forests offset many of the gains in carbon that occur with afforestation. Similar results occur for Russia over the coming century. Note that for both Canada and Russia, these results are contingent on assuming that inaccessible forests are mature and that they are not accumulating additional carbon.

19.4.2 Sensitivity analysis with no carbon sequestration program

Many of the alternative scenarios in the sensitivity analysis have little effect on total forestland and carbon storage in the world's forests. Higher demand for forest products increases the price of forest products (timber prices are approximately 48% higher in 2105). These higher prices do increase the overall area of land in forests by approximately 65 million hectares in 2105 (Table 19.3), but this represents a gain of only about 2% globally. The largest gains occur in the European Union, the United States, China, and Canada.

Global carbon storage is projected to decline by 6.6 Pg under the high timber demand assumption by 2105, despite the projected increase in forestland area. As with total forestland area, this represents a small proportional deviation in total global carbon storage in 2105, approximately 1%. The reduction in carbon results from more intensive harvests of inaccessible forests and more intensive harvesting of all forests. Regions that have large areas of inaccessible forests, the boreal and tropical regions, experience a decline in carbon sequestration. These additional harvests do not lead to land-use change, but they do reduce carbon intensity in inaccessible forests. Canada, South and Central America, Russia, southeast Asia, and Africa would be likely to hold less carbon with higher demand, while the United States, Europe, China, and Japan are projected to hold more carbon.

Even changing the elasticity of land supply has little relative effect. The more (less) elastic land supply, the larger (smaller) is the forest area and the more (less) carbon is sequestered. Globally, the high elasticity assumption leads to approximately 63 million additional hectares in forestry, with the largest gains occurring in tropical regions. Currently, there are approximately 1.5 billion hectares of rain-fed agricultural land globally (Fischer *et al.*, 2002). If the entire 63 million hectares comes from rain-fed agricultural land, this implies a 4% reduction in agricultural land (and a 1.7% increase in forestland). The carbon consequences of these changes amount to less than a 1% change in total carbon stored globally. Each 1% increase or decrease in rain-fed agricultural land leads to approximately a 0.25% change in carbon (i.e. 1 additional hectare of land in rain-fed agriculture reduces carbon storage by 80 to 141 Mg C).

The final three scenarios, low harvesting costs, low plantation establishment costs, and low technological change in plantation yields, have even smaller impacts on total carbon storage and forestland. Low harvesting costs have the largest effects on carbon storage, lowering total carbon in forests in 2105 by 10 PgC, or around 1.3%. These losses result mainly from larger incursions into inaccessible forest areas in boreal and tropical regions. The lower costs have little effect on land area in forests. Although lower plantation establishment costs do increase the area of subtropical plantations in regions like South and Central America, Oceania, southeast Asia, and Africa, lower timber prices lead to offsetting reductions in forestland area in other regions. The net effects on forestland area amount to approximately 23 million additional hectares of land in plantations. The carbon consequences of these changes are quite small, with a less than 1 Pg gain in carbon. High-valued timber plantations, while quickly producing timber for markets, do not sequester substantial carbon. Finally, the low technology change scenario has a small effect on both land area and carbon sequestration.

19.4.3 Sensitivity analysis of carbon sequestration programs

The baseline scenario suggests that by 2105, a global carbon sequestration program could cause the area of forests to increase by 232 and 796 million hectares in the low and high damage scenarios respectively (Table 19.4). This will in turn result in an additional 52.9 and 125.8 PgC being stored in forests. Compared with the results in Sohngen and Mendelsohn (2003), the new model suggests approximately 50% more carbon will be stored in the low damage scenario and approximately 15% more carbon in the high damage scenario by the end of the century. Several modeling changes explain this increase in carbon sequestration projections. First, the new model captures the inaccessible boreal forests in Canada and Russia more carefully. The model now recognizes that the inaccessible forests in these regions are initially old-growth forests when first harvested but they will be slow to re-grow to that status. This increases the potential to use boreal forests to sequester additional carbon. Second, carbon intensity in Asia (including China and southeast Asia) and Africa increases in this model relative to the earlier model. These forests will consequently be able to store more carbon at the same cost. A final change is that the carbon sequestration functions used in this study for most regions have been updated. While carbon storage levels for mature forests are consistent with the earlier study, the conversion parameters used in this study suggest more carbon in early periods of forest growth than in the earlier study.

Under the alternative scenarios, the carbon sequestration program stores between 47.1 and 64.1 PgC globally for the low damage case, and 102.0 to 149.5 PgC for the high damage case (Tables 19.5 and 19.6). The results are most sensitive to the alternative land supply elasticity assumptions. Higher land supply elasticity leads to approximately 19–21% more carbon than estimated under the baseline scenario, and lower land supply elasticity leads to 11–19% less carbon storage globally.

Table 19.4 Projected effect of low and high carbon price, baseline carbon sequestration programs.

	Carbon (additional PgC stored)		Forest area (million additional hectares)		Inacc. temp. and boreal forest area (million additional hectares)		Subt. plant. (million additional hectares)	
	Low price	High price	Low price	High price	Low price	High price	Low price	High price
United States	2.0	6.8	12.5	67.5	3.2	8.7	0.7	4.5
Canada	1.6	4.9	14.2	52.9	7.9	17.4	0.0	0.0
S. America	8.8	22.3	54.9	175.9	0.0	0.0	1.6	5.7
C. America	1.2	2.5	4.5	15.6	0.0	0.0	0.0	0.0
Europe	2.3	9.7	8.7	61.6	0.3	0.9	0.0	0.0
Russia	3.9	7.9	10.0	44.8	34.4	67.8	0.0	0.0
China	3.4	9.7	21.2	67.5	6.1	13.9	5.1	6.3
India	1.0	3.3	8.1	21.9	0.0	0.0	0.5	2.9
Oceania	0.4	1.3	8.4	41.1	0.0	0.0	0.4	3.3
SE Asia	16.7	36.0	33.6	85.1	0.0	0.0	1.2	4.8
C. Asia	0.7	1.4	3.0	12.3	0.0	0.0	0.8	1.7
Japan	0.1	1.1	2.8	11.3	0.0	0.0	0.2	0.7
Africa	10.8	18.9	49.8	138.6	0.0	0.0	1.7	2.7
Total	52.9	125.8	231.6	795.9	51.9	108.8	12.1	32.5

Inacc. = inaccessible
Temp. = temperate
Subt. plant. = subtropical plantation

Table 19.5 Sensitivity analysis for low carbon price sequestration scenario: comparison of alternative scenarios relative to baseline.

	Baseline	High demand		High elasticity		Low elasticity		Low harvest cost		Low plt. est. cost		Low tech. chg.	
	PgC	PgC	%	PgC	%	PgC	%	PgC	%	PgC	%	PgC	%
United States	2.0	1.8	−12	2.3	17	1.4	−32	2.1	4	1.8	−9	2.1	2
Canada	1.6	1.7	5	2.4	52	1.6	4	2.0	27	2.0	29	1.7	9
S. America	8.8	9.2	4	11.7	32	6.8	−23	10.8	23	9.2	4	9.2	5
C. America	1.2	0.8	−32	1.4	17	0.9	−20	1.2	5	1.2	−1	1.2	−1
Europe	2.3	2.6	10	3.0	28	2.6	11	2.9	24	2.5	8	2.2	−5
Russia	3.9	5.2	32	4.4	13	4.2	7	6.5	66	4.2	7	4.5	16
China	3.4	1.9	−44	5.4	56	3.4	−3	4.5	29	5.2	50	3.4	0
India	1.0	0.4	−61	1.6	63	0.7	−27	0.7	−32	1.1	9	1.0	−2
Oceania	0.4	0.4	11	0.5	41	0.4	0	0.5	45	0.4	19	0.4	10
SE Asia	16.7	15.4	−7	18.4	10	15.2	−9	18.4	10	17.1	2	17.2	3
C. Asia	0.7	0.3	−52	0.7	0	0.6	−18	0.6	−19	0.6	−12	0.6	−14
Japan	0.1	0.0	−72	0.2	109	0.0	−50	0.1	45	0.1	6	0.2	108
Africa	10.8	12.1	12	12.1	12	9.4	−13	13.1	22	11.0	2	11.2	4
Total	52.9	51.8	−2	64.1	21	47.1	−11	63.4	20	56.3	7	55.0	4

Plt. est. = plantation establishment
Tech. chg. = technological change

Table 19.6 Sensitivity analysis of the high carbon price sequestration scenario: comparison of alternative scenarios relative to baseline in 2105.

	Baseline	High demand		High elasticity		Low elasticity		Low harvest cost		Low plt. est. cost		Low tech. chg.	
	PgC	PgC	%	PgC	%	PgC	%	PgC	%	PgC	%	PgC	%
United States	6.8	6.7	−3	8.8	29	4.8	−29	7.2	5	6.9	1	6.7	−2
Canada	4.9	5.3	8	6.6	34	3.6	−26	6.1	25	5.0	3	4.8	−3
S. America	22.3	24.3	9	28.8	29	16.1	−28	24.7	11	22.0	−1	22.6	1
C. America	2.5	2.6	1	3.1	22	1.9	−25	2.6	5	2.5	0	2.5	1
Europe	9.7	8.0	−17	10.5	8	8.2	−16	8.9	−9	9.8	1	8.6	−12
Russia	7.9	10.3	31	8.6	9	7.1	−10	10.4	32	7.9	0	8.1	3
China	9.7	7.3	−25	14.5	50	6.7	−30	8.5	−12	12.0	24	7.3	−24
India	3.3	2.4	−27	4.4	31	2.9	−14	3.2	−2	3.3	−2	3.3	−1
Oceania	1.3	1.5	16	1.7	26	1.1	−18	1.7	29	1.4	5	1.3	−1
SE Asia	36.0	26.5	−26	38.4	7	32.5	−10	37.4	4	36.0	0	36.1	0
C. Asia	1.4	1.4	−3	1.6	15	1.2	−14	1.5	7	1.4	−4	1.4	−1
Japan	1.1	0.6	−43	1.4	32	0.7	−38	1.3	23	1.0	−4	1.1	−2
Africa	18.9	21.1	12	21.1	12	15.2	−20	21.2	12	18.7	−1	19.2	2
Total	125.8	118.0	−6	149.5	19	102.0	−19	134.7	7	127.9	2	122.8	−2

Higher (lower) land supply elasticity reduces (increases) the slope of the land supply function, and reduces (increases) the costs of adding land to increase carbon sequestration.

The other sensitivity analyses have a relatively small effect on carbon sequestration globally. A larger increase in demand for forest products over time reduces potential sequestration 2% or 6% respectively in the high and low price scenarios. Lower harvesting costs and lower plantation establishment costs both increase potential carbon sequestration. Both these changes tend to increase the value of land in forests, and therefore increase the value of establishing forests for carbon and for timber. Slower yield growth reduces potential sequestration in the low carbon price scenario, but increases it slightly in the high carbon price scenario. The effect is relatively small, only around 1%. Globally, sequestration programs are most sensitive to factors that influence the supply of land to forests.

Although the global results may not be sensitive to many of the alternative scenarios, some regions are very sensitive. For instance, under the high demand scenario, the United States sequesters less carbon under the low price scenario, whereas Canada is projected to sequester more carbon in forests.

Higher demand has little effect on the carbon sequestration program in the United States because many of the low-cost opportunities for adding carbon are captured in the baseline case. Further, the higher timber prices cause land to shift into intensive plantations that only reduce carbon. In Canada, the sequestration program can keep some of the inaccessible forests from being used to satisfy the high demand for timber. This turns out to be a relatively inexpensive sequestration option and so the Canadian program sequesters more carbon. Canada is consequently more sensitive to the tested assumptions than the United States.

19.5 Discussion and conclusion

This paper presents a sensitivity analysis of carbon sequestration in global forests using a global forestry model (Sohngen and Mendelsohn, 2003). The model is updated with new inventory data and new information on potential carbon sequestration in global forests. Several analyses are presented with the new model. First, the model is used to see how much carbon would be stored globally without any carbon sequestration programs. We calculate an expected baseline estimate and then several alternative estimates using different assumptions about timber demand, land supply, and technological change in the forestry sector. Second, we examine two efficient global sequestration programs. The two programs are based on low and high paths for carbon prices that have been calculated from optimal mitigation models. A sensitivity analysis of these carbon sequestration programs is examined using the same alternative model assumptions discussed above.

The forestry model predicts that 430 million ha of forestland would be converted to other land-uses (primarily agriculture) resulting in the global loss of 64 PgC of carbon over the next century. The rate that forestland is lost and the rate that carbon is lost falls gradually over the century. Higher demand for forest products increases the cumulative century-long loss by another 7 PgC as more inaccessible forests are harvested. A lower elasticity for land supply results in shifting another 72 million ha out of forestland over the coming century and leads to a loss of approximately 7 PgC. Lower harvest costs would result in the loss of another 10 PgC, a relatively large effect, because they influence harvests of inaccessible forests, which tend to have high carbon content. In contrast, other technical changes in the forestry sector (i.e., those that would reduce plantation regeneration costs or increase plantation yields) have a very small net effect on carbon sequestration.

The paper also explores the implementation of two global forest sequestration programs. The two programs rely on dynamic price paths that have been generated by optimal mitigation models (Nordhaus and Boyer, 2000). The low price path (from \$7/t C to \$61/t C) reflects a low estimate of climate damage and the high price path (from \$22 to \$188/t C) reflects a high estimate of the damages from climate change. By tying sequestration to these price paths, the analysis equates the marginal benefits and the marginal costs of sequestration over time. The resulting sequestration program is a dynamic and efficient policy. With the low set of prices, the forestry model predicts the carbon sequestration program could sequester 53 PgC, and with the high set of prices, it will sequester 126 PgC. These new estimates are higher than projections in Sohngen and Mendelsohn (2003) with the same carbon prices because of new inventory data in regions like Russia, China, and the United States, as well as updated information on carbon intensity in regions like southeast Asia and Africa.

The carbon sequestration program is sensitive to alternative assumptions about land supply elasticity. Increasing the elasticity of land supply by 50% leads to approximately 20% more carbon sequestered globally, whereas reducing the elasticity by 50% leads to approximately 19% less carbon globally. The other scenarios have a relatively small effect on global carbon storage, with the exception of harvesting costs. With lower harvesting costs, a carbon sequestration program can save inaccessible forests at a relatively low cost thus increasing the amount of carbon stored at a given price per tonne.

The research indicates that deforestation could add about 64 PgC of carbon back into the atmosphere by the end of the next century. Efficient carbon sequestration programs that pay forest owners to store carbon can pull a substantial amount of that carbon back into the forest. If the damages from climate change turn out to be low, a carbon sequestration program could sequester between 47 and 64 PgC and if the damages are high, the program could sequester between 102 and 149 PgC by 2100.

These large estimates suggest that carbon sequestration in forests is a low-cost way to slow greenhouse gas concentrations from rising over time. An efficient program to control greenhouse gases should include forest carbon sequestration. However, creating an efficient sequestration program is admittedly a big challenge. The research in this paper explores global sequestration programs. Programs that targeted only limited regions would have problems with leakage. That is, increases in carbon in one region may be offset by reductions in carbon in other regions outside the program. To enhance efficiency, programs consequently must equate the marginal cost of sequestration across all regions.

The analysis also suggests that it is critical to design dynamic policies that increase the incentives to sequester over time in concert with the price of carbon (the benefit of sequestration). Programs that are too aggressive too soon will be unnecessarily expensive. For example, only one-third of the carbon that should be stored this century should be set-aside by 2050 (Sohngen and Mendelsohn, 2003). In this exercise, we relied on paying forest owners worldwide a price for each unit of carbon they stored. That price increased over time along the "known" price path of carbon. This is one way to design an efficient sequestration program, but there may well be other alternatives that have equivalent efficient properties.

The paper reviews the impact of carbon sequestration programs on carbon, forestland, and timber markets. However, it does not examine what sequestration programs would do to habitat, water resources, or forest recreation. Although one may be able to associate changes in natural forestland with changes in habitat, water, and recreation, the precise effects would depend on the spatial pattern of these changes and the species mix. Although timber plantations are likely to fall short of natural forests as non-timber habitat, the role of plantations on non-timber services also needs to be explored. Because of the importance of all of these non-timber impacts, future studies should try to expand the analysis to include them.

Acknowledgements

This research was conducted while B. Sohngen was on leave at RTI International, as part of the Stanford Energy Modeling Forum EMF-21 study. Funds for the work were provided by US Environmental Protection Agency Climate Analysis Branch. The authors appreciate comments from participants in the Energy Modeling Forum Snowmass Meeting.

References

Adams, D. M., Alig, R. J., McCarl, B. A., Callaway, J. M. and Winnett, S. M. (1999). Minimum cost strategies for sequestering carbon in forests. *Land Economics* **75**, 360–374.

Dixon, R. K., Brown, S., Houghton, R. A. *et al.* (1994). Carbon pools and flux of global forest ecosystems. *Science* **263**, 185–190.

Fang, J., Chen, A., Peng, C., Zhao, S. and Ci., L. (2001). Changes in forest biomass carbon storage in China between 1949 and 1998. *Science* **292**, 2320–2322.

FAO (2003). *State of the World's Forests 2003*. Rome: United Nations Food and Agricultural Organization.

FAOSTATS (2003). *FAOSTATS Database of Global Forest Production*. Rome: United Nations Food and Agricultural Organization.

Fischer, G., van Velthuizen, H., Shah, M. and Nachtergaele, F. (2002). *Global Agro-ecological Assessment for Agriculture in the 21st Century: Methodology and Results*. Research Report RR-02-02. Vienna: International Institute for Applied Systems Analysis (IIASA).

Hardie, I. W. and Parks, P. J. (1997). Land use with heterogeneous land quality: An application of an area base model. *American Journal of Agricultural Economics* **79**, 299–310.

Haynes, R. (2003). *An Analysis of the Timber Situation in the United States: 1952–2050*. General Technical Report, PNW-GTR-560. Portland, Oregon: US Department of Agriculture, Forest Service, Pacific Northwest Research Station.

Houghton, R. A. (2003). Revised estimates of the annual net flux of carbon to the atmosphere from changes in land use and land management 1850–2000. *Tellus* **55B**, 378–90.

Ianchovichina, E., Darwin, R. and Shoemaker, R. (2001). Resource use and technological progress in agriculture: a dynamic general equilibrium analysis. *Ecological Economics* **38** (2), 275–291.

IPCC (2000). *Emissions Scenarios: A Special Report of Working Group III of the Intergovernmental Panel on Climate Change*, ed. N. Nakicenovic and R. Swart. Cambridge: Cambridge University Press.

IPCC (2001). *Climate Change 2001: Mitigation. Contribution of Working Group III to the Third Assessment Report of the Intergovernment Panel on Climate Change*, ed. B. Metz, O. Davidson, R. Swart and J. Pan. Cambridge: Cambridge University Press.

McCarl, B. A. and Schneider, U. A. (2001). Greenhouse gas mitigation in United States agriculture and forestry. *Science* **294**, 2481–2482.

Murray, B. C., McCarl, B. A. and Lee, H. (2004). Estimating leakage from forest carbon sequestration programs. *Land Economics* **80** (1), 109–124.

Nabuurs, G. J., Päivinen, R. and Schelhaas, M. J. (2003). *Development of European Forests until 2050*. Research Report 15. Helsinki: European Forestry Institute.

Nordhaus, W. and Boyer, J. (2000). *Warming the World: Economic Models of Global Warming*. Cambridge, MA: MIT Press.

Plantinga, A. J., Mauldin, T. and Miller, D. J. (1999). An econometric analysis of the costs of sequestering carbon in forests. *American Journal of Agricultural Economics* **81**, 812–824.

Richards, K. R. and Stokes, C. (2004). A review of forest carbon sequestration cost studies: a dozen years of research. *Climatic Change* **63**, 1–48.

Sathaye, J., Makundi, W., Dale, L. and Chan, P. (2006). GHG mitigation potential, costs and benefits in global forests: a dynamic partial equilibrium approach. *Energy Journal* **27**, 127–163.

Sedjo, R. (1999). Land use change and innovation in US forestry. In *Productivity in Natural Resource Industries*, ed. R. D. Simpson. Washington, DC: Resources for the Future, pp. 141–174.

Sedjo, R. (2004a). The potential economic contribution of biotechnology and forest plantations in global wood supply and forest conservation. In *The Bioengineered Forest: Challenges for Science and Society*, ed. S. H. Strauss and H. D. Bradshaw. Washington, DC: Resources for the Future.

Sedjo, R. (2004b). *Transgenic Trees: Implementation and Outcomes of the Plant Protection Act*. Discussion Paper DP-04-10. Washington, DC: Resources for the Future.

Smith, J. E., Heath, L. S. and Jenkins, J. C. (2003). *Forest Volume-to-biomass Models and Estimates of Mass for Live and Standing Dead Trees of United States Forests*. General Technical Report, NE-298. Radnor, PA: US Department of Agriculture, Forest Service, Northeastern Research Station.

Sohngen, B. and Brown, S. (2004). Measuring leakage from carbon projects in open economies: a stop timber harvesting project as a case study. *Canadian Journal of Forest Research* **34**, 829–839.

Sohngen, B. and Mendelsohn, R. (2003). An optimal control model of forest carbon sequestration. *American Journal of Agricultural Economics* **85**(2), 448–457.

Sohngen, B. and Sedjo, R. (2000). Potential carbon flux from timber harvests and management in the context of a global timber market. *Climatic Change* **44**, 151–172.

Sohngen, B., Mendelsohn, R. and Sedjo, R. (1999). Forest management, conservation, and global timber markets. *American Journal of Agricultural Economics* **81**(1), 1–13.

Sohngen, B., Mendelsohn, R. and Sedjo, R. (2001). A global model of climate change impacts on timber markets. *Journal of Agricultural and Resource Economics* **26** (2), 326–343.

van Vuuren, D. P., Eickhout, B., Lucas, P. L. and den Elzen, M. G. J. (2006). Long-term multi-gas scenarios to stabilise radiative forcing: exploring costs and benefits within an integrated assessment framework. *Energy Journal* **27**, 201–234.

Winjum, J. K., Brown, S. and Schlamadinger, B. (1998). Forest harvests and wood products: sources and sinks of atmospheric carbon dioxide. *Forest Science* **44**, 272–284.

Appendix A19.1 Global forestry model

The forestry model used in this analysis is built upon the model described in Sohngen et al. (1999), and used by Sohngen and Mendelsohn (2003) to analyze global sequestration potential. The model used in this analysis differs from the earlier version in several important ways. First, the model has been expanded to contain 146 distinct timber types in 13 regions. Second, several important regions have been split in this version. Canada and the United States have been split apart; Central and South America have been split; Asia Pacific has been split into Southeast Asia, Central Asia, and Japan; the Former Soviet Union (FSU) has been split into Russia only; areas of the FSU in Europe are now in the European area, and areas of the FSU in Central Asia have been included in the new Central Asia region. Third, new data on age class inventories and yield functions for several major regions have been incorporated, including the United States, Russia, China, Australia, and New Zealand. Fourth, yield functions for subtropical plantations have been updated, and technological change in these plantations has been directly modeled. Technological change for plantations is predicted to increase timber yields by 0.25% annually. Fifth, carbon conversion parameters in the United States (Smith et al., 2003) and China (Fang et al., 2001) have been updated. Carbon parameters in other regions have been adjusted so that the carbon content of forests corresponds to biomass estimates from the UN Food and Agricultural Organization (FAO, 2003).

For the purposes of describing the model, each of the 146 timber types modeled can be allocated into one of three general types of forest stocks. Stocks S^i are moderately-valued forests, managed in optimal rotations, and located primarily in temperate regions. Stocks S^j are high-value timber plantations that are managed intensively. Subtropical plantations are grown in the southern United States (loblolly pine plantations), South America, southern Africa, the Iberian Peninsula, Indonesia, and Oceania (Australia and New Zealand). Stocks S^k are relatively low-value forests, managed lightly if at all, and located primarily in inaccessible regions of the boreal and tropical forests. The inaccessible forests are harvested only when timber prices exceed marginal access costs.[1]

The forestry model maximizes the present value of net welfare in the forestry sector. Formally, this is:

$$\text{Max} \sum_0^\infty \rho^t \left\{ \begin{array}{l} \int_0^{Q^*(t)} \{D(Q_t, Z_t) - C_{H^i}(\cdot) - C_{H^j}(\cdot) - C_{H^k}(\cdot)\} \\ \times \, dQ(t) - \sum_{i,k} C_G^{i,k}(G_t^{i,k}, m_t^{i,k}) \\ - \sum_j C_N^j(N_t^j, m_t^j) - \sum_{i,j,k} R^{i,j,k}(L_t^{i,j,k}) + CC_t \end{array} \right\} \quad (19.1)$$

where $Q_t = \sum_{i,j,k} \left(\sum_a H_{a,t}^{i,j,k} V_{a,t}^{i,j,k}(m_{t0}) \right)$.

In Equation (19.1), $D(Q_t, Z_t)$ is a global demand function for industrial wood products given the quantity of wood, Q_t, and income, Z_t. The quantity of wood depends upon $H^{i,j,k}$, the area of land harvested in the timber types in i, j, or k, and $V_a^{i,j,k}(m_{t0})$, the yield function of each plot. The yield per hectare depends upon the species, the age of the tree (a), and the management intensity at the time of planting (m_{t0}). $C_H(\cdot)$ is the cost function for harvesting and transporting logs to mills from each timber type. Marginal harvest costs for temperate and subtropical plantation forests (i and j) are constant, while marginal harvest costs for inaccessible forests rise as additional land is accessed. $C_G^{i,k}(\cdot)$ is the cost function for planting land in temperate and previously inaccessible forests, and $C_N^j(\cdot)$ is the cost function for planting forests in subtropical plantation regions. $G_t^{i,k}$ is the area of land planted in types i and k, N_t^j is the area of land planted in plantation forests, and L_t^j is the total area of land in forest type j. The planting cost functions are given as:

$$C_G^{i,k}(\cdot) = p_m^{i,k} m_t^{i,k} G_t^{i,k} \quad (19.2)$$

$$C_N^j(\cdot) = p_m^j m_t^j N_t^j + f(N_t^j, L_t^j) \quad (19.3)$$

where $m_t^{i,j,k}$ is the management intensity of those plantings purchased at price p_m^i, p_m^j, or p_m^k. $f(N_t^j, X_t^j)$ is a function representing establishment costs for new plantations. The cost function for establishing new plantations rises as the total area of plantations expands.

The yield function has the following properties typical of ecological species: $V_a > 0$ and $V_{aa} < 0$. We assume that management intensity is determined at planting. The following two conditions hold for trees planted at time t_0 and harvested "a" years later $(a + t_0) = t_{ai}$:

$$\frac{dV^i(t_{a_i} - t_0)}{dm^i(t_0)} \geq 0 \quad \text{and} \quad \frac{d^2 V^i(t_{a_i} - t_0)}{dm^i(t_0)^2} \leq 0. \quad (19.4)$$

---

[1] In this study, forests in inaccessible regions are harvested when marginal access costs are less than the value of the standing stock plus the present value of maintaining and managing that land as an accessible forest in the future.

The total area of land in each forest type is given as $L_t^{i,j,k}$ ($= \sum_a X_{a,t}^{i,j,k}$). $R^{i,j,k}(\cdot)$ is a rental function for the opportunity costs of maintaining lands in forests. Two forms of the rental function are used:

$$R(L_t^{i,j,k}) = \alpha L_t^{i,j,k} + \beta (L_t^{i,j,k})^2 \quad \text{for temperate and boreal regions}$$

$$R(L_t^{i,j,k}) = \alpha (L_t^{i,j,k})^2 + \beta (L_t^{i,j,k})^3 \quad \text{for tropical regions.} \quad (19.5)$$

The marginal cost of additional forestland in tropical forests is assumed to be non-linear to account for relatively high opportunity costs associated with shifting large areas of land out of agriculture and into forests. The rental functions represent a weakness of the current model because it is likely that the land supply function will vary a great deal across ecosystems and regions, and we have only considered two potential functional forms. The parameters of the rental function are chosen so that the elasticity of land supply is 0.25 initially, the reported relationship between forests and agriculture in the United States (Hardie and Parks, 1997; Plantinga et al., 1999). This elasticity implies that the area of forests could increase by 0.25% if forests can pay an additional 1% rental payment per year. The same elasticity estimate is applied globally.

The stock of land in each forest type adjusts over time according to:

$$X_{a,t}^i = X_{a-1,t-1}^i - H_{a-1,t-1}^i + G_{a=1,t-1}^i \quad i = 1 - I \quad (19.6a)$$

$$X_{a,t}^j = X_{a-1,t-1}^j - H_{a-1,t-1}^j + N_{a=1,t-1}^j \quad j = 1 - J \quad (19.6b)$$

$$X_{a,t}^k = X_{a-1,t-1}^k - H_{a-1,t-1}^k + G_{a=1,t-1}^k \quad k = 1 - K \quad (19.6c)$$

Stocks of inaccessible forests in S^k are treated differently depending on whether they are in tropical or temperate/boreal regions. All inaccessible forests are assumed to regenerate naturally unless they are converted to agriculture. In tropical regions, forests often are converted to agriculture when harvested, so that $G_{a=1}^k$ is often 0 for tropical forests in initial periods when the opportunity costs of holding land in forests are high. As land is converted to agriculture in tropical regions, rental values for remaining forestland declines, and land eventually begins regenerating in forests in those regions. This regeneration is dependent on comparing the value of land in forests versus the rental value of holding those forests. In this study, we do not track the type of agriculture to which forests are converted, i.e. crops or grazing. Inaccessible forests in temperate/boreal regions that are harvested are converted to accessible timber types so that $G_{a=1}^k$ is set to 0. The stock of inaccessible forests in S^k is therefore declining over time if these stocks are being harvested. Each inaccessible boreal timber type has a corresponding accessible timber type in S^i, and forests that are harvested in inaccessible forested areas in temperate/boreal regions are converted to these accessible types. Thus, for the corresponding timber type, we set $G_{a=1}^i \geq H_{a-1}^k$. Note that the area regenerated, $G_{a=1}^i$, can be greater than the area of the inaccessible timber type harvested because over time, harvests and regeneration occurs in forests of the accessible type.

The term CC_t represents carbon sequestration rental payments. Rental payments are made on the total stock of carbon in forests, thus, the form for CC_t is given as:

$$CC_t = CR_t \sum_{i,j,k} \gamma_{i,j,k} \sum_a \left\{ V_{a,t}^{i,j,k}(m^{i,j,k}(t_0)) \right\} X_{a,t}^{i,j,k}$$
$$+ P_t^C \sum_{i,j,k} \theta_{i,j,k} \sum_a \left\{ V_{a,t}^{i,j,k}(m^{i,j,k}(t_0)) \right\} H_{a,t}^{i,j,k} - E_t^b, \quad (19.7)$$

where CR_t is the annual rental value on a tonne of carbon, P_t^C is the price of a tonne of carbon, $\gamma_{i,j,k}$ is a conversion factor to convert forest biomass into carbon, $\theta_{i,j,k}$ is a conversion factor to convert harvested biomass into carbon stored in products, and E_t^b is baseline carbon sequestration. For this model, we assume that product storage in long-lived wood products is 30% of total carbon harvested (Winjum et al., 1998).

The model is programmed into GAMS and solved in 10-year time increments. Terminal conditions are imposed on the system after 150 years. These conditions were imposed far enough into the future not to affect the study results over the period of interest. For the baseline case, $P_t^C = 0$, there is no sequestration program, and the term CC_t has no effect on the model. Baseline carbon sequestration is then estimated, and used for E_t^b in the carbon scenarios. The sequestration program scenarios are based on an independently estimated path for P_t^C.

20

Insights from EMF-associated agricultural and forestry greenhouse gas mitigation studies

Bruce A. McCarl, Brian C. Murray, Man-Keun Kim, Heng-Chi Lee, Ronald D. Sands and Uwe A. Schneider

20.1 Introduction

Integrated assessment modeling (IAM) as employed by the Energy Modeling Forum (EMF) generally involves a multi-sector appraisal of greenhouse gas emission (GHGE) mitigation alternatives and climate change effects, typically at the global level. Such a multi-sector evaluation encompasses potential climate change effects, and mitigative actions within the agricultural and forestry (AF) sectors. In comparison with many of the other sectors covered by IAM, the AF sectors may require somewhat different treatment owing to their critical dependence upon spatially and temporally varying resource and climatic conditions. In particular, in large countries like the United States, forest production conditions vary dramatically across the landscape. For example, some areas in the southern United States present conditions favorable to production of fast-growing, heat-tolerant pine species, while more northern regions often favor slower-growing hardwood and softwood species. Moreover, some lands are currently not suitable for forest production (e.g., the arid western plains). Similarly, in agriculture, the United States has areas where citrus and cotton can be grown and other areas where barley and wheat are more suitable. This diversity across the landscape causes differential GHGE mitigation potential in the face of climatic changes and/or responses to policy or price incentives.

It is difficult for a reasonably sized global IAM to reflect the full range of sub-national geographic AF production possibilities alluded to above. AF response in the face of climate change alterations in temperature precipitation regimes plus mitigation incentives will be likely to involve region-specific shifts in land use and agricultural/forest production. This chapter addresses AF sectoral responses in climate change mitigation analysis. Specifically, we draw upon US-based studies of AF GHGE mitigation possibilities that incorporate sub-national detail drawing largely on a body of studies done by the authors in association with EMF activities. We discuss characteristics of AF sectoral responses that could be incorporated in future IAM efforts in climate change policy.

20.2 Types of insights to be discussed

The insights discussed here arise from various studies of the economically optimal portfolio of GHGE mitigation strategies as it varies across different incentive or program target levels. The studies examine GHGE mitigation from either a static or a dynamic perspective, looking at the implications for agricultural and forest sector goods production, the incidence of non-GHG co-effects, regional heterogeneity, and to a limited extent discounting. These findings are drawn primarily from the dissertations by Schneider (2000), Lee (2002) and related work under projects sponsored by US EPA, USDA, and DOE. One caveat: the studies were done over time and were not simultaneously redone for this paper. Thus, some results are based on variations in the price per metric tonne of carbon while others are based on price per metric tonne of carbon

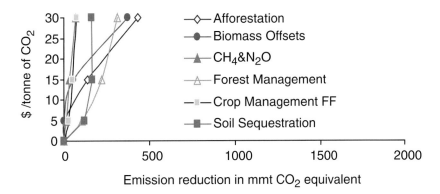

Figure 20.1 NPV portfolio of mitigation strategies (mmt, million metric tonnes).

dioxide. The results also arise across various generations of the models involved.

20.2.1 Mitigation strategy portfolio

The first set of insights arise from results on the quantity of GHGE "offsets" produced at alternative prices (Figure 20.1). There the AF offset quantities represent the extent to which sequestration can increase, emissions can be reduced, or substitute products embodying lower offsets can be produced. The price/offset quantity schedule reflects the amount that would be produced at different price signals where the price represents the amount that would need to be paid to cause that level of AF offset production.

In our modeling frame work AF offsets can be produced by GHGE mitigation alternatives that can be broadly characterized into six categories:

- Afforestation – offsets that arise due to establishment of new forest on agricultural crop and pasture lands.
- Biofuels – offsets that arise from substitution of AF produced commodities for fossil fuel energy replacing ethanol or coal in power generation. The substitution is dominated by replacement of coal for electricity generation with switch grass and short-rotation poplar (Schneider and McCarl, 2003).
- CH_4 and N_2O – methane (CH_4) and nitrous oxide (N_2O) emission reductions through altered agricultural management. This includes manipulations involving livestock manure, livestock enteric fermentation, rice production, and nitrogen fertilization (McCarl and Schneider [2000] discuss the full set of alternatives).
- Forest Management – GHGE offsets generated by altering management of existing forests principally through adoption of longer rotations or use of more intensive practices that promote forest (and carbon) growth.
- Crop Management FF – the GHGE offset generated by altering agricultural management practices to conserve the amount of fossil fuel used.
- Soil Sequestration – the GHGE offset generated by altering agricultural tillage practices and by changing land use from cropping to pasture or grasslands.

Such data were first developed in the thesis by Schneider (2000) and later by McCarl and Schneider (2001). The concept was subsequently generalized for a multi-period case in Lee's (2002) thesis and associated presentations (Adams et al., 2005; Lee et al., 2005) to a portrayal of the net present value (NPV) of the GHGE offset as it arises over time. Following the conceptual approach discussed in Richards (1997), the time-dependent contributions generated by a multi-period model of sectoral response are discounted back to the present using a 4% discount rate.

A representative graphic of these results appears in Figure 20.2 and summarizes dynamic results from the FASOMGHG (Adams et al., 1996, 2005; Lee, 2002) model. The offset quantities in Figure 20.2 report offsets on a carbon dioxide equivalent basis where the 100-year global warming potential (GWP) is used to aggregate results across alternative GHGs.

We now turn our attention to insights.

Multi-strategy nature

Figure 20.1 shows that the AF sectors can produce offsets from a multitude of sources. The four biggest classes of contributions arise from agricultural soil sequestration, management of existing forest, biofuels, and afforestation. The non-CO_2 strategies and the offsets arising from alterations in agricultural fossil fuel usage patterns make smaller but still sizable contributions. This shows agriculture and forestry can play a role considerably beyond the often discussed sequestration role and leads to:

Insight 1

The agricultural and forestry potential greenhouse gas emission mitigation portfolio is diverse and composed of a number of alternatives.

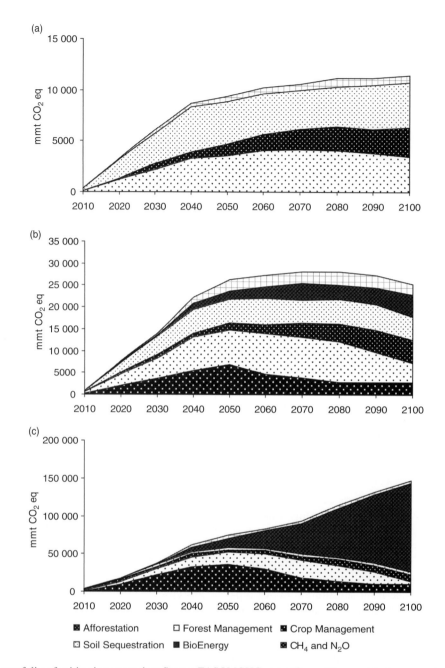

Figure 20.2 Dynamic portfolio of mitigation strategies; Source FASOMGHG. (mmtCe: million metric tonnes carbon equivalent.)

Portfolio price dependency

Figure 20.1 indicates that the importance of the various portfolio elements depends upon the magnitude of the offset price. In particular, at low levels of offset prices, the cost-effective strategies are those that are complementary to existing production – agricultural soil sequestration and existing forest management. However, when the offset prices become higher, then substitution possibilities such as biofuel production and afforestation take over. The reason for the shift involves basic economics. At low prices, the GHGE mitigation strategies employed continue to produce traditional agricultural and forestry commodities while also producing GHGE offsets. Such strategies yield relatively low per acre mitigation rates but are complementary to existing production so can occur at low prices. At higher prices, the biofuel and afforestation strategies take over. These strategies produce larger per acre amounts of mitigation but divert production from traditional commodities.

For example, West and Post (2002) indicate that agricultural soil sequestration generates roughly a quarter of a tonne of carbon offset per acre while afforestation and biofuel generate offsets in excess of one tonne per acre but require giving up traditional production. This leads to:

> **Insight 2**
>
> The mix of "best" GHGE mitigation strategies depends on the magnitude of the greenhouse gas emission offset (carbon) price.

20.2.2 Portfolio variation over time

While the above results show what happens in a static or net present value sense, it is also important to examine how mitigation effects change over time. In particular, biological phenomena underlying some of the agricultural and forestry GHGE mitigation alternatives (largely the sequestration ones) have characteristics that influence their potential dynamic contribution. Strategies such as forest management, afforestation, and soil sequestration add carbon to the ecosystem but ecosystems exhibit what has been often called "saturation." Carbon accumulates in the ecosystem under a particular management régime until the rate of carbon addition equals the rate of carbon decomposition. At that time, a new carbon equilibrium is achieved, and sequestration ceases. West and Post (2002) argue this happens in 15 to 20 years when changing agricultural tillage while Birdsey (1996) presents data that show carbon ceases accumulating in an undisturbed southern pine plantation after about 80 years. Under either circumstance, the duration of the GHG increments is finite, furthermore if the practice is reversed the carbon is released. This dynamic issue has been widely discussed under the topic of "permanence" in the international GHGE mitigation dialogue (see discussion in Marland et al. [2001], Kim [2004], or Smith et al., [2007]). On the other hand, biofuel and emission control strategies do not exhibit saturation.

Such characteristics are manifest in the Figure 20.2 dynamic results drawn from Lee (2002) and subsequent papers (Lee et al., 2005, McCarl et al., 2005). There we portray the cumulative contribution by each and all strategies as they arise over time for three different offset prices. Generally, the cumulative offsets from the sequestration strategies rise quickly but level off after 30 to 40 years and may even diminish over the longer run (the short time to effective saturation for forest occurs because harvest disturbances begin). However, the non-CO_2 emissions and biofuel strategies continue to accumulate over time. Further, at high prices the biofuel strategy dominates, although it takes some time to penetrate the market due to capital stock considerations. This leads to the collective observation that in the near term and at low prices the sequestration strategies are employed whereas in the longer term and at higher prices the emission control and biofuel strategies dominate. This leads to:

> **Insight 3**
>
> The cost-effectiveness and desirability of strategies vary with time largely owing to the limited life involved with sequestration strategies and the market penetration time for biofuels.

20.2.3 Policy design dependency

The above results largely assume that the market price for all offsets is the same. However, as widely discussed in the international process there may be discounts that arise owing to permanence, additionality, leakage and uncertainty or other factors (see Kim [2004] and Smith et al. [2007] for a literature review). In actuality this has led to a degree of market discounting when considering for example non-permanent alternatives. McCarl et al. (2001) looked at the effect of permanence-related sequestration discounts on portfolio share using a static AF sector model. In that study a 50% discount was applied to the price paid for agricultural soil carbon sequestration relative to that paid for non-sequestration offsets. Simultaneously a 25% emission reduction was applied to carbon sequestration arising from forest management and afforestation. The results (Figure 20.3) show a portfolio composition shift with less sequestration and more biofuels. Kim has shown that discounts can be substantial in the case of AF offsets:

> **Insight 4**
>
> AF offsets may not all be priced on an equal footing and that would shift the portfolio of "best" mitigation strategies.

20.2.4 Portfolio variation over space

Now we turn our attention to regional issues. Figure 20.4 portrays results arising from Lee's work on the major regional GHGE mitigation activity choices across the set of US regions (for regional definitions see the FASOM documentation; Adams et al. [1996]; McCarl et al. [2005]). These data show the not unexpected result that portfolio composition varies across the landscape reflecting climate and other variations. In particular, if one looks at the data, one finds that agricultural activities dominate in major agricultural

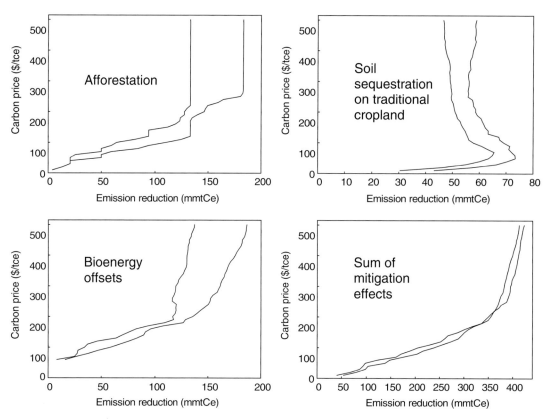

Figure 20.3 Effects of permanence discounting on optimal portfolio composition. Dashed lines represent no discounting; solid lines represent discounting with 50% on soil carbon credits, 25% on afforestation credits, and 0% on biofuel offsets.

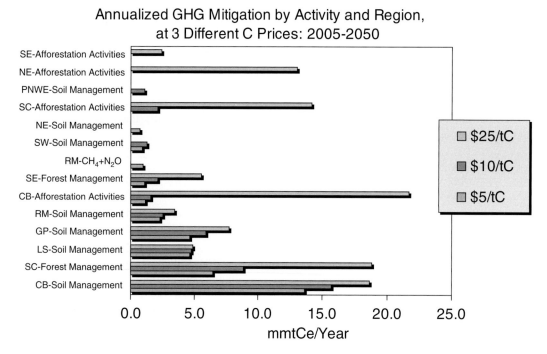

Figure 20.4 Regional NPV portfolio of mitigation strategies.

regions (e.g., CornBelt) and that forestry activities dominate in important forestry regions (e.g., Southeast) and leads to:

> **Insight 5**
>
> The optimal mitigation portfolio varies across space with different regions preferring different strategies depending on resource endowments and opportunity costs.

This indicates the importance of depicting sub-national areas in obtaining a reasonable set of GHGE mitigation responses for incorporation in an IAM.

20.2.5 Individual item potential and portfolio role

It is common in the AF GHGE mitigation related literature to find studies that address only single strategies (e.g. the book by Lal *et al.* [1998] on cropland-based agricultural soil carbon sequestration or virtually all of the IPCC [2001] estimates). Such treatments frequently deal with the strategy in isolation and report a total potential GHGE offset quantity for that strategy. However, the approach of evaluating options in isolation can produce biased results of mitigation potential. Figure 20.5 illustrates strategy potential in the context of agricultural soil carbon offsets drawn from the work of Schneider (2000) and McCarl and Schneider (2001). Two types of biases can be discussed.

First, when the mitigation potential estimate does not consider resource availability or economic costs of production it will overstate, perhaps substantially, the absolute mitigation potential for that strategy. In the vertical line to the far right is a US-wide estimate of soil agricultural carbon offset potential based on data in the Lal *et al.* (1998) book – we call this "technical potential." The monotonically increasing line to the left of the vertical line arises from an economic model that examines strategies at various GHGE offset prices when agricultural soil sequestration is the only available strategy – we call this the "economic potential." These data show that as higher offset prices are paid, the technical potential is approached but never attained. Such results indicate that the physical estimate of potential can substantially overstate the economic estimate:

> **Insight 6**
>
> Omitting consideration of the opportunity cost of their use can overstate the importance of individual strategies.

The second form of bias can come from ignoring the interaction between mitigation actions that compete for the same resources. This is especially important for the AF sectors, where land is fixed and must be allocated across competing uses. This notion of "competitive potential" is illustrated in Figure 20.5 as the inner line depicting the mitigation initially rising then falling. The falling part shows the influence of resource competition. Namely, soil carbon sequestration, biofuels, and afforestation all share common resources (in this case principally the land base) and at higher prices the chosen offset strategies move land to afforestation and biofuels diminishing agricultural soil carbon sequestration. Hence the "backward" bend in the curve represents the competitive potential. This leads to the insight:

> **Insight 7**
>
> Omitting consideration of the complete set of strategies and resource competition between strategies can also lead one to overstate the importance of individual strategies and total mitigation potential.

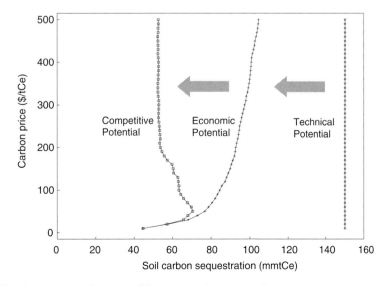

Figure 20.5 Estimates of soil carbon sequestration potential under varying assumptions.

20.2.6 Mitigation portfolio: dynamics and economy-wide role

The implications of Insights 6 and 7 are that it is not appropriate to examine the potential of AF GHGE mitigation alternatives in isolation. By logical extension, then, AF alternatives should not be examined in isolation but rather in a full economy context in comparison with energy sector and other alternatives via IAMs. However, some have argued that, because of their impermanence and potential for reversal, AF sequestration activities should not be allowed or should be viewed with extreme skepticism (Greenpeace, 2003). However, a preliminary dynamic investigation by Sands and McCarl (2005) shows a potentially important role (Figure 20.6). Those results show the AF contribution is quite important in the near term constituting initially more than 50% of potential mitigation across all sectors. Later the AF share diminishes as investments in energy sector mitigation and technological developments in carbon capture–storage emerge. This indicates the desirability of future dynamic studies on the potential relevance of AF GHGE mitigation alternatives, perhaps based in part on data coming from extensions of the work discussed herein, and leads to:

> **Insight 8**
>
> While agricultural and forestry activities may not have unlimited duration, they may be very important in a world that requires time and technological investment to develop low-cost greenhouse gas emission offsets.

20.3 Mitigation activities: effects on traditional production

Another important factor in AF approaches to GHGE mitigation is the interrelationship between the GHGE mitigation alternatives and the levels of traditional sectoral production. Figures 20.7 and 20.8 portray the relationship over time between total production indices and GHGE offset prices over time.

On the agricultural side, these data show competition between traditional crop production and GHGE offset production. In particular, the data in Figures 20.7 and 20.8 show that as the offset price gets larger, agricultural crop production generally decreases. This occurs because of resource substitution. Namely, as offset prices get larger more and more land is diverted from traditional agricultural crops to biofuels and afforestation. While not portrayed here, an index of total livestock production also shows declines, although to a smaller extent. This leads to:

> **Insight 9**
>
> Employment of agricultural mitigation activities generally involves reductions in production of traditional agricultural products.

On the forestry side, the commodity production effects of GHG mitigation are somewhat more complex as shown in

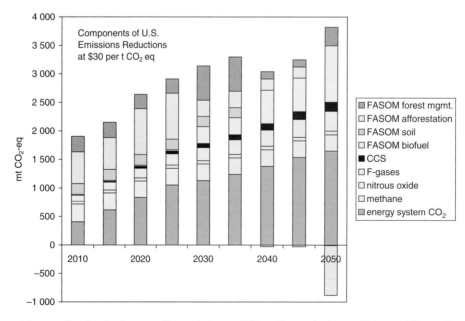

Figure 20.6 Economy-wide annualized reductions or offsets relative to US baseline emissions at $30 per t CO_2-eq. Bars are in order listed in the key, except for 2050 where result for afforestation is negative.
Notes: cumulative offsets from FASOMGHG are converted to an annual basis for comparison to other greenhouse gas mitigation options. CCS represents CO_2 capture and storage. Carbon prices begin in year 2010 and are held constant thereafter.

Insights from agricultural and forestry studies 245

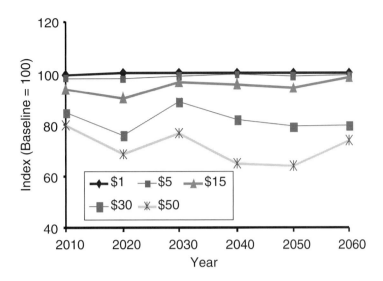

Figure 20.7 Agricultural crop production over time as a function of GHGE offset prices (prices are in $/tonne CO_2 equivalent).

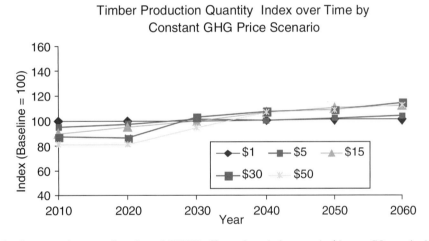

Figure 20.8 Timber production over time as a function of GHGE offset prices (prices are in $/tonne CO_2 equivalent).

Figure 20.8. There one sees short-term substitution but longer-term complementarity. In the short term, when GHGE offset prices rise, one finds that timber rotation lengths get longer (i.e., harvesting is delayed) thereby reducing forest product supply initially. However, afforestation is also occurring and will generate harvestable timber at some point. Consequently, in the longer run, harvest deferral and afforestation combine so that both forest carbon and timber volume are accumulating and then harvesting begins to occur. The long-run increase in harvesting activity takes into account diminishing sequestration rates and the fact that some carbon will be retained post-harvest in lumber and other products. This leads to:

> **Insight 10**
>
> Employment of forestry mitigation activities generally involves short-run substitution with traditional production, but a longer-run complementary relationship arises.

20.4 Environmental co-effects

A number of agricultural mitigation activities not only generate emission reductions or sequestration gains, but also

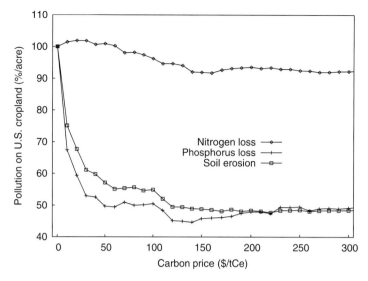

Figure 20.9 Co-effects of increasing carbon prices on non-GHG pollution.

exhibit environmental and economic byproduct effects as shown in Figure 20.9. Such effects are generally called co-benefits or, considering that cases may arise where things have costs, co-effects. Such co-effects arise in several arenas. For example, Schneider (2000) and Plantinga and Wu (2003) show substantial aggregate reductions in erosion when GHGE mitigation strategies were used. Schneider (2000) shows reductions in phosphorus and nitrogen runoff. In turn, Pattanayak et al. (2005) show this leads to improvements in regional water quality. Others indicate such actions affect species diversity and hunting opportunities (Matthews et al., 2002; Plantinga and Wu, 2003).

Economic effects have also been shown in terms of increases in producer income and decreases in governmental income support (Callaway and McCarl, 1996; McCarl and Schneider, 2001) as shown in Fig. 20.10. This occurs since the availability of profitable GHGE mitigation alternatives expands producer opportunities to sell goods and, in turn, generate income. On the other hand, findings indicate a worsening of the foreign trade balance (net exports), foreign welfare and domestic consumers' welfare. This leads to:

> **Insight 11**
>
> Implementation of forestry and agricultural mitigation activities leads to co-effects.

One should be careful not to overemphasize the role of co-benefits based on studies confined to the AF sector, as mitigation options in other sectors may also have co-benefits worthy of consideration. For example, AF sector GHGE offsets may allow the energy sector to increase GHGE when GHGE are fixed at an economy-wide cap. If this allows, for instance, more coal-fired generation, there may be health and other co-effects due to particulate emissions of NO_x and SO_2 (see discussion in Burtraw et al. [2003] and Elbakidze and McCarl [2004]) that should be considered to determine the net impacts of the AF offsets.

20.5 Concluding comments

The role of agricultural and forest sector GHGE mitigation activities remains a controversial topic in climate policy. On the positive side, AF activities seem to provide relatively inexpensive options for GHGE mitigation compared with other sectors. Moreover, the mitigation technologies are endowed by natural forces and are available now to meet the challenge. In contrast, mitigation actions in the energy and industrial sectors may require substantial time for development and commercial adoption. Therefore, AF activities have been viewed by some as a potential bridge to the future when low-emitting energy sources pervade the landscape. The bridge analogy is particularly apt, because the opportunities to mitigate from AF activities may be temporary. Planted forests reduce the rate at which they accumulate carbon as they age, and may be harvested in the future, thereby re-introducing CO_2 to the atmosphere. This raises some concerns that AF sequestration is merely a quick fix and ultimately solves nothing. These concerns may be warranted if the new, cheaper mitigating technologies are not developed in the interim (while the bridge is open) and AF activities merely serve to delay the inevitable. We would argue that a two-prong approach is required: AF activities in the short run and low-emitting technologies in the long run.

With this as background, we draw from our collective research experience and those of others addressing AF sector mitigation activities to offer some insights that we hope can inform both policy and future IAM efforts. Ours is by no

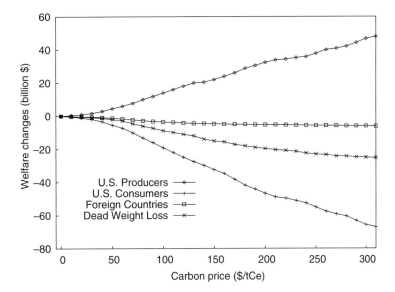

Figure 20.10 Sectoral welfare changes as carbon prices increase.

means the final word on this, and new findings are modifying the conventional wisdom all the time. We believe it is critical that the modeling community build on the features of and information generated by sector-and region-specific models – in the AF sector and other key sectors – as well as general equilibrium models and global scale models to better inform the policy community of the complex and potentially important feedback effects that different policy initiatives can produce.

Acknowledgements

Note that while McCarl is senior author, all others contributed equally. This work benefited from support largely from US EPA, but with contributions from USDA and DOE.

References

Adams, D. M., Alig, R. J., Callaway, J. M., McCarl, B. A. and Winnett, S. M. (1996). *The Forest and Agricultural Sector Optimization Model (FASOM): Model Description*. USDA Forest Service Report PNW-RP-495.

Adams, D. M., Alig, R. J., McCarl, B. A. *et al.* (2005). *FASOMGHG Conceptual Structure, and Specification: Documentation*. http://agecon2.tamu.edu/people/faculty/mccarl%2Dbruce/FASOM.html.

Birdsey, R. A. (1996). Carbon storage for major forest types and regions in the conterminous United States. In *Forests and Global Change*, Vol. 2: *Forest Management Opportunities for Mitigating Carbon Emissions*, ed. R. N. Sampson and D. Hair. Washington DC: American Forests.

Burtraw, D., Krupnick, A. Palmer, K. *et al.* (2003). Ancillary benefits of reduced air pollution in the US from moderate greenhouse gas mitigation policies in the electricity sector. *Journal of Environmental Economics and Management* **45**, 650–673.

Callaway, J. M. and McCarl, B. A. (1996). The economic consequences of substituting carbon payments for crop subsidies in US agriculture. *Environmental and Resource Economics* **7**, 15–43.

Elbakidze, L. and McCarl, B. A. (2004). Should we consider the co-benefits of agricultural GHG offsets. *Choices*, 3rd Quarter, www.choicesmagazine.org/2004-3/climate/2004-3-10.htm.

Greenpeace (2003). Sinks in the CDM: After the climate, biodiversity goes down the drain. http://elonmerkki.net/GP_CDMSinks_analysi.pdf.

IPCC (2001). *Climate Change 2001: Mitigation. Contribution of Working Group III to the Third Assessment Report of the Intergovernmental Panel on Climate Change*, ed. B. Metz, O. Davidson, R. Swart and J. Pan. Cambridge: Cambridge University Press.

Kim, M. (2004). *Economic Investigation of Discount Factors for Agricultural Greenhouse Gas Emission Offsets*. Unpublished Ph.D. thesis, Department of Agricultural Economics, Texas A&M University.

Lal, R., Kimble, J. M., Follett, R. F. and Cole, C. V. (1998). *The Potential of U.S. Cropland to Sequester Carbon and Mitigate the Greenhouse Effect*. Chelsea, MI: Sleeping Bear Press Inc..

Lee, H.-C. (2002). *The Dynamic Role for Carbon Sequestration by the U.S. Agricultural and Forest Sectors in Greenhouse Gas Emission Mitigation*. Unpublished Ph.D. thesis, Department of Agricultural Economics, Texas A&M University.

Lee, H.-C., McCarl, B. A., Gillig, D. and Murray, B. C. (2005). U.S. agriculture and forestry based greenhouse gas emission mitigation: an economic exploration of time dependent effects. In *Rural Lands, Agriculture and Climate beyond 2015: Usage and*

Management Responses, ed. F. Brouwer and B. A. McCarl. Dordrecht: Kluwer Press.

Marland, G., Fruit, K. and Sedjo, R. (2001). Accounting for sequestered carbon: the question of permanence. *Environmental Science and Policy* **4**, 259–268.

Matthews, S., O'Connor, R. and Plantinga, A. J. (2002). Quantifying the impacts on biodiversity of policies for carbon sequestration in forests. *Ecological Economics* **40**(1), 71–87.

McCarl, B. A. and Schneider, U. A. (2000). Agriculture's role in a greenhouse gas emission mitigation world: an economic perspective. *Review of Agricultural Economics* **22**(1), 134–159.

McCarl, B. A. and Schneider, U. A. (2001). The cost of GHG mitigation in U.S. agriculture and forestry. *Science* **294**, 2481–2482.

McCarl, B. A. and Spreen, T. H. (1996). Applied mathematical programming using algebraic systems. Available at http://agecon.tamu.edu/faculty/mccarl.

McCarl, B. A., Murray, B. C. and Schneider, U. A. (2001). Jointly estimating carbon sequestration supply from forests and agriculture. Paper presented at Western Economics Association Meetings, San Francisco, July 5–8, 2001.

McCarl, B. A., Gillig, D. Lee, H.-C. *et al.* (2005). Potential for biofuel-based greenhouse gas emission mitigation: rationale and potential. In *Agriculture as a Producer and Consumer of Energy*, ed. J. L. Outlaw, K. Collins and J. Duffield. Cambridge, MA: CABI Publishing, pp. 300–316.

Pattanayak, S. K., McCarl, B. A., Sommer, A. J. *et al.* (2005). Water quality co-effects of greenhouse gas mitigation in US agriculture. *Climatic Change* **71**, 341–372.

Plantinga, A. J. and Wu, J. (2003). Co-benefits from carbon sequestration in forests: evaluating reductions in agricultural externalities from and afforestation policy in Wisconsin. *Land Economics* **79**(1), 74–85.

Richards, K. R. (1997). The time value of carbon in bottom-up studies. *Critical Reviews in Environmental Science and Technology* **27**, 279–292.

Sands, R. D. and McCarl, B. A. (2005). Competitiveness of terrestrial greenhouse gas offsets: are they a bridge to the future? Paper presented at the USDA Greenhouse Gas Symposium, Baltimore, Maryland, March 22–24, 2005.

Schneider, U. (2000). *Agricultural Sector Analysis on Greenhouse Gas Emission Mitigation in the United States*. Unpublished Ph.D. thesis, Department of Agricultural Economics, Texas A&M University.

Schneider, U. A. and McCarl, B. A. (2003). Economic potential of biomass based fuels for greenhouse gas emission mitigation. *Environmental and Resource Economics* **24**(4), 291–312.

Smith, G. R., McCarl, B. A. Li, C. S. *et al.* (2007). *Quantifying Greenhouse Gas Emission Offsets Generated by Changing Management*. Durham, NC: Duke University Press.

West, T. O. and Post, W. M. (2002). Soil organic carbon sequestration rates by tillage and crop rotation: a global data analysis. *Soil Science Society of America Journal* **66**, 1930–1946.

Williams, J. R., Jones, C. A. and Dyke, P. T. (1984). A modeling approach to determining the relationship between erosion and soil productivity. *Transactions of the ASAE* **27**, 129–144.

Appendix 20.1 Basic structure of some agricultural and forest-related models

A number of studies have been done within the agricultural and forestry sectors regarding the competitiveness of alternative agriculture and forestry-based GHGE mitigation strategies under different market conditions and time. The fundamental sources for these findings are from Schneider (2000), Lee (2002), and Kim (2004) and follow-up works. In this section we will illustrate the basic structure of the models used in associated works.

A20.1.1 Agricultural sector and mitigation of greenhouse gas (ASMGHG) model

This section documents the essential structure of the US agricultural sector and mitigation of greenhouse gas (ASMGHG) model. Here, we focus on the general model structure, which is not affected by data updates or model expansion toward greater detail. Data and a generalized algebraic modeling system (GAMS) version of a regionally aggregated ASMGHG version is available on the Internet. The aggregated model can be used to examine and verify the model structure and data and to qualitatively replicate the results presented in this article. In representing ASMGHG's mathematical structure, we will use summation notation because it corresponds very closely to the ASMGHG computer code.

ASMGHG is designed to emulate US agricultural decision making along with the impacts of agricultural decisions on agricultural markets, the environment, and international trade. To accomplish this objective, ASMGHG portrays the following key components: natural and human resource endowments, agricultural factor (input) markets, primary and processed commodity (output) markets, available agricultural technologies, and agricultural policies. Because of data requirements and computing feasibilities, sector models cannot provide the same level of detail as do farm-level or regional models. Therefore, ASMGHG depicts only representative crop and livestock enterprises in 63 aggregated US production regions rather than individual farms' characteristics. International markets and trade relationships are portrayed in 28 international regions.

Agricultural technologies in the United States are represented through Leontief production functions specifying fixed quantities of multiple inputs and multiple outputs. Producers can choose among several alternative production technologies. Specifically, alternative crop production functions arise from combinations of three tillage alternatives (conventional tillage, conservation tillage, and zero tillage), two irrigation alternatives (irrigation, dryland), four alternative conservation measures (none, contour plowing, strip cropping, terracing), and three nitrogen fertilization alternatives (current levels, a 15% reduction, and a 30% reduction) specific to each US region, land, and crop type.[1] Alternative livestock production functions reflect different production intensities, various manure treatment schemes, alternative diets, and pasture management for 11 animal production categories and 63 US regions. Processing functions identify first or higher level processing opportunities carried out by producers.

ASMGHG is set up as a mathematical programming model and contains more than 20 000 individual variables and more than 5000 individual equations. All agricultural production activities are specified as endogenous variables such as crop management variables, land-use transformation, livestock raising, processing, and production factor (input) supply variables. Additional variables reflect the dissemination of agricultural products with US domestic demand, US interregional and international trade, foreign region excess supply, foreign region excess demand, emissions, and emission reduction or sequestration variables.

ASMGHG consists of an objective function, which maximizes total agricultural welfare including both US and foreign agricultural markets, and a set of constraining equations, which define a convex feasibility region for all variables. Feasible variable levels for all depicted agricultural activities range from zero to an upper bound, which is determined by resource limits, supply and demand balances, trade balances, and crop rotation constraints.[2] Solving ASMGHG involves the task of finding the "optimal" level for all endogenous variables subject to compliance with all constraining equations. By means of ASMGHG's objective function, optimal levels of all endogenous variables are those levels which maximize agricultural sector-based welfare, which is computed as the sum of total consumers' surplus, producers' surplus, and governmental net payments to the agricultural sector minus the total cost of production, transportation, and processing. Basic economic theory demonstrates that maximization of the sum of consumers' plus producers' surplus yields the competitive market equilibrium as reviewed by McCarl and Spreen (1996). Thus, the optimal variable levels can be interpreted as equilibrium levels for agricultural activities under given economic, political, and technological conditions.

Supply and demand balance equations for agricultural commodities form an important constraint set in ASMGHG, which link agricultural activities to output markets. Specifically, the total amount of commodities disseminated in a US region through domestic consumption, processing, and exports cannot exceed the total amount of commodities supplied through crop production, livestock raising, or imports.

The most fundamental physical constraints on agricultural production arise from the use of scarce and immobile resources. Particularly, the use of agricultural land, family labor, irrigation water, and grazing units is limited by given regional endowments of these private or public resources.

In ASMGHG, trade activities by international region of destination or origin are balanced through trade equations which force a foreign region's excess demand for an agricultural commodity not to exceed the sum of all import activities into that particular region from other international regions and from the United States. Similarly, the sum of all commodity exports from a certain international region into other international regions and the United States not to exceed the region's excess supply activity.

A set of constraints addresses aggregation-related aspects of farmers' decision processes. These constraints force producers' cropping activities to fall within a convex combination of historically observed choices. Based on decomposition and economic duality theory, it is assumed that observed historical crop mixes represent rational choices subject to weekly farm resource constraints, crop rotation considerations, perceived risk, and a variety of natural conditions. ASMGHG contain the observed crop mix levels for the past 30 years by historical year and region, whose level will be determined during the optimization process. The mix of livestock production is constraint in a similar way as crop production. Particularly, the amount of regionally produced livestock commodities is constraint to fall in a convex combination of historically observed livestock product mixes.

Agricultural land owners do not only have a choice between different crops and different crop management strategies, they can also abandon traditional crop production altogether in favor of establishing pasture or forest. Equivalently, some existing pasture or forest owners may decide to convert suitable land fractions into cropland. In ASMGHG, land-use conversions are portrayed by a set of endogenous variables. Certain land conversion can be restricted to a maximum transfer whose magnitude was determined by geographical information system (GIS) data on land suitability.

The assessment of environmental impacts from agricultural production as well as political opportunities to mitigate negative impacts is a major application area for ASMGHG. To facilitate this task, ASMGHG includes environmental impact accounting. For each land management, livestock, or processing activity, environmental impact coefficients contain the absolute or relative magnitude of those impacts per unit of activity. Negative values of greenhouse gas account

---

[1] We use representative crop production budgets for 63 US regions, 19 crops (cotton, corn, soybeans, 4 wheat types, sorghum, rice, barley, oats, silage, hay, sugar cane, sugar beets, potatoes, tomatoes, oranges, grapefruits), six land classes (low erodible cropland, medium erodible cropland, highly erodible cropland, other cropland, pasture, and forest).

[2] Crop rotation constraints force the maximum attainable level of an agricultural activity such as wheat production to be equal to or below a certain fraction of physically available cropland.

coefficients, for example, indicate emission reductions. A detailed description of environmental impact categories and their data sources is available in Schneider (2000).

A20.1.2 Forest and agricultural sector optimization and mitigation of greenhouse gas (FASOMGHG) model

FASOMGHG solves a multi-period, multi-market optimization problem by maximizing the present value of aggregated consumers' and producers' surpluses in the agricultural and forest sectors subject to resource constraints. The solutions reveal the prices and quantities of agricultural and forest markets in each period under the assumption that producers and consumers have perfect knowledge of market responses at the beginning of the modeling period. The basic structure of FASOMGHG follows the formulation in McCarl and Spreen (1996) in which the life of the activities, such as forest, is determined endogenously and production activities adjust over time. The model includes 48 primary agricultural, 54 secondary agricultural commodities, and 8 forest products produced in 11 geographical regions. The agricultural sector activities are based on the agricultural sector mitigation of greenhouse gas (ASMGHG) model (Schneider, 2000).

Responding to the projected population growth and technology improvements over time, demands and supplies in timber and agricultural product markets vary dynamically. Likewise, the diminishing projected land availability to the agricultural and forest sectors reflects the land loss to infrastructure and urbanization. However, resource endowments such as water and labor remain the same over the modeling period. In addition, agricultural product demand and foreign supply/demand change over time. Time-dependent yield improvements also keep agricultural productivity growing over time.

The carbon sector in FASOMGHG accounts for terrestrial carbon in (1) forest ecosystems on existing forest stands, (2) regenerated and afforested ecosystems, (3) carbon losses in forest non-commercial carbon pools after harvest, (4) carbon loss in timber products over time, and (5) carbon in agricultural soil. For the carbon fate in merchantable timber products, harvested stumpage is transformed into four merchantable products, saw logs, pulpwood, fuel wood, and milling residue. All fuel wood is burned, within a decade of harvesting. Carbon in burned wood is divided into two categories, a substitute of fossil fuel or emissions to the atmosphere by fixed fractions. The carbon content kept in wood products decays over time as specified by a set of coefficients. In addition, the terrestrial carbon accounting in the agricultural sector is distinguished by region, crop type, tillage and other chosen management, as well as the time length of chosen tillage management that has been practiced. Therefore, in FASOMGHG, two underlying assumptions regarding soil sequestration in the agricultural sector are made: (1) tillage practices and all other soil management practices will affect the capability of soil sequestration, and (2) saturation condition is specified.

Tree planting is the major forest GHG mitigation strategy in FASOMGHG. In addition to afforestation, reforestation, reducing deforestation size, and changing forest management on existing forest stands, such as intensifying management and extending rotation age, can also enhance carbon sequestration in the forest sector.

Since agriculture is a major source of US CH_4 and N_2O emissions, strategies associated with these two GHGs are included in the FASOMGHG mitigation domain. Methane in agriculture arises from enteric fermentation, livestock manure, and rice cultivation. Fertilizer application and manure management are the main sources for N_2O. In addition, land management in agriculture is the major CO_2 source in the agricultural sector although its magnitude is relatively small compared to the emissions from other sectors. However, changing cropland management or converting cropland to pastureland can cause agricultural soil to become a potential sink to absorb atmospheric carbon. In addition, direct and indirect emissions from fossil fuel burning in cropping processes or the process of manufacturing agricultural production inputs are considered as sources of the agricultural CO_2 emissions. Similarly, the CO_2 emissions for energy used in irrigation, drying, and tillage are also counted as CO_2 emission sources.

Ethanol and biofuel are treated as fossil fuel substitutes. In addition to CO_2 emission offsets, the use of biofuel will also reduce the CH_4 emissions, but increase N_2O emissions compared with conventional fossil-powered plants.

All agricultural GHG emission data are converted from Schneider (2000). He developed the Agricultural Sector and Greenhouse Gas Model (ASMGHG), incorporating a number of mitigation options in the agricultural sector. Most emission data in ASMGHG are defined in 63 US regions, and FASOMGHG aggregate data in ASMGHG into 11 FASOMGHG regions.

All emission coefficients are assumed to be the same across all modeling years for a given production activity, except for agricultural terrestrial carbon sources and sinks. For example, emissions generated from fuel usage for planting one acre of corn in the Corn Belt will be fixed over time under each tillage practice, land type, fertilizer application, and irrigation condition. As for agricultural soil, carbon emissions or savings are accounted for only in the first three decades right after the tillage operation is switched. Therefore, the net carbon emission for a specific cropland will be zero if its current tillage practice has lasted for more than 30 years.

For biofuel production and other agricultural not sequestration-related activities, credits for GHG emission reductions are generated whenever such activities occur. Terrestrial carbon sinks are capable of accumulating carbon, but limited by ecosystem capability. Moreover, the carbon stored by the sinks still exists on the ground or biomass in a potentially volatile form. Therefore, GHG emission reductions by sinks can result in potential future GHG emission increases. FASOMGHG assumes when a piece of cropland switches its tillage practice, there could be carbon gain/loss, depending on the tillage management switching to a more or less disturbing one, but it stops after the first 30 years. This 30-year limit assumption is based on the previous tillage studies (Lal *et al.*, 1998; West and Post, 2002), opinions of soil scientists, and ASMGHG data, whose agricultural soil carbon sequestration rates are an average of 30-year EPIC simulation results (Williams *et al.*, 1984). Furthermore, the sequestering tillage practice has to remain in use even after its soil carbon content reaches equilibrium, otherwise the carbon would be released.

FASOMGHG assumes that soil carbon saturation occurs 30 years after a specific change of tillage operation is made. Once tillage management is changed, it cannot change again for 30 years. Therefore, 30 years after continuing in some specific tillage management, a farmer has to decide whether to maintain such tillage practice, switch to another tillage management, or even retire the land.

There are three tillage systems, conventional, conservation, and no-till, included in FASOMGHG. The conservation practice in FASOMGHG refers to practices that leave the residue on the field at levels between conventional and no-till operation. No-till operations offer the highest cropland use capacity to accumulate carbon, followed by conservation tillage, then conventional tillage.

FASOMGHG further differentiates carbon gain/loss on cropland according to its previous tillage practice. A different previous tillage practice indicates a different equilibrium level of carbon content to start off, but carbon content will reach another stable level, depending on tillage practice switching, 30 years later. Cropland previously in conventional tillage practice will exhibit carbon gains if its tillage management changes to either conservation or no-till operation. The carbon gain will be greater in the case where its tillage management switches to no-till rather than to conservation tillage. Carbon gains or losses in FASOMGHG are spread out linearly over 30 years. The initial tillage distribution of cropland follows the 1998 tillage information available through The Purdue University Conservation Technology Information Center.

The land used for tree planting has the highest carbon content, followed by grazing use, then cropland use. When land transfers between cropland, pastureland, and forestland, there are two steps with regard to soil carbon change accounting in FASOMGHG. If a pastureland or forestland is changed to crop planting, its carbon content will first drop to a base level instantly. Then its carbon content will continue to decrease for the future 30 years if it uses conventional tillage management. Its carbon content will increase steadily in the future 30 years if conservation or no-till operations are used. As cropland converts to pastureland or forestland, FASOMGHG assumes the carbon content in cropland will start from the base level and be followed by a carbon gain, whose magnitude depends on switching to grassland or forestland.

When cropland converts to grassland or is converted from grassland, carbon gains and losses occur for 30 years. Agricultural soil carbon can enter emission or sequestration accounting since the cropland can be either a CO_2 emission source or sink.

The carbon accounting sums over emissions/sequestration from tillage change in previous periods. We assume that the new cropland carbon equilibrium after changing tillage practice will be reached in 30 years. Furthermore, the soil carbon content of the cropland is assumed to be in equilibrium at the beginning of the modeling period. Therefore, unless there is a tillage change on existing cropland, there is neither soil carbon flux for emission nor sequestration accounting. A detailed description of FASOMGHG is available in Lee (2002).

21

Global agricultural land-use data for integrated assessment modeling

Navin Ramankutty, Tom Hertel, Huey-Lin Lee and Steven K. Rose

21.1 Introduction

Human land-use activities, while extracting natural resources such as food, fresh water, and fiber, have transformed the face of the planet. Such large-scale changes in global land use and land cover can have significant consequences for food production, freshwater supply, forest resources, biodiversity, regional and global climates, the cycling of carbon, nitrogen, phosphorus, etc. In particular, land management and changes in land use can affect fluxes of greenhouse gases including carbon dioxide, nitrous oxide, and methane.

Recently, Hannah *et al.* (1994) estimated that roughly 50% of the planet's land surface (75% of the habitable area) has been either moderately or severely disturbed; only the core of the tropical rainforests and boreal forests, deserts, and ice-covered surfaces are still relatively untouched by humans. Moreover, Vitousek *et al.* (1997) estimated that around 40% of the global net primary productivity is being co-opted by humans, while Postel *et al.* (1996) estimated that over 50% of the available renewable freshwater supply is being co-opted.

The major mode of human land transformation has been through agriculture. Since the invention of agriculture, ~10 000 years ago, humans have modified or transformed the land surface; today, roughly a third of the planet's land surface is being used for growing crops or grazing animals (National Geographic Maps, 2002; Foley *et al.*, 2003). The pace of land clearing for cultivation has been particularly rapid in the past 300 years, in which Richards (1990) estimated that we have lost 20% of forests and woodlands, and 1% of grasslands and pastures (although most grasslands were converted to pastures), and croplands expanded by 466%.

It has been estimated that roughly a third of the total emissions of carbon into the atmosphere since 1850 has resulted from land-use change (and the remainder from fossil fuel emissions) (Houghton, 2003). For example, in the 1990s, 6.4 GtC/yr was emitted to the atmosphere from industrial activities and 2.2 GtC/yr was emitted from tropical deforestation. In addition, agricultural activities are responsible for approximately 50% of global atmospheric inputs of methane (CH_4), and agricultural soils are responsible for 75% of global nitrous oxide emissions (Scheehle and Kruger, 2006; USEPA, 2006) as well as soil carbon fluxes.

While the clearing of tropical forests today results in a source of carbon to the atmosphere, natural ecosystems are also sinks of carbon. It is known, from attempts to close the global carbon budget, that roughly 3 GtC/yr is taken up by terrestrial ecosystems, and all evidence points to a significant uptake in the high latitude regions of the northern hemisphere (Tans *et al.*, 1990; Prentice *et al.*, 2001). Many hypotheses for the uptake have been proposed including CO_2 fertilization, nitrogen fertilization, climate-driven uptake, and carbon uptake by vegetation recovering from past disturbances (Prentice *et al.*, 2001). It is becoming increasingly clear that terrestrial ecosystems can be a source or sink of carbon depending on how they are managed. The notion of terrestrial *carbon sequestration* emerged from this understanding, wherein terrestrial ecosystems are actively manipulated, through afforestation, soil management, etc., to withdraw carbon dioxide from the atmosphere (IPCC, 2000). This notion has become so widespread that in the Kyoto Protocol, terrestrial carbon sequestration was included as a possible means for nations to reduce their net emissions of carbon.

Human-induced Climate Change: An Interdisciplinary Assessment, ed. Michael Schlesinger, Haroon Kheshgi, Joel Smith, Francisco de la Chesnaye, John M. Reilly, Tom Wilson and Charles Kolstad. Published by Cambridge University Press. © Cambridge University Press 2007.

Appropriate management of forests and agricultural lands can reduce emissions of CO_2, CH_4, and N_2O, and can further sequester CO_2. Biomass substitutes for fuels can reduce emissions that would otherwise result from burning fossil fuels. Numerous estimates have been made by the integrated assessment community of the cost of abating greenhouse gas emissions through land-use change (Richards and Stokes, 2004). Early estimates suggest that mitigating emissions from land-use management could be less expensive than mitigation from traditional industrial activities (Lee, 2005). Therefore, numerous partial and general equilibrium models are being developed to study climate change policy and the role of land-use change in abating GHG emissions. Furthermore, integrated assessment models are being developed to assess the impacts of climate change on land-use practices, and the subsequent feedbacks to land-use mitigation. However, lack of a consistent global land resource database linked to underlying economic activity, and to emissions and sequestration drivers, has hindered these analyses.

This research provides such a database within the Global Trade Analysis Project (GTAP) at Purdue University.[1] In addition to allocating cropland and production to specific commodities, the database represents biophysical heterogeneity to better capture varying land-use potential. In this paper we describe the methodology used to derive land-use information for the GTAP database and provide insights into various features of the data that improve global emissions modeling.

21.2 The GTAP database

The GTAP database includes detailed bilateral trade, protection, and energy data characterizing economic linkages among regions, as well as individual country input–output tables that account for inter-sectoral linkages within regions. GTAP version 6 disaggregates 87 regions of the world economy (a mix of individual countries and groups of countries) along with 57 sectors in each of these regions.[2] GTAP presents a baseline database of the world at a particular point in time – version 6 presents data for the year 2001, but the database is constantly updated to reflect the most recent data available.

GTAP began as a database and modeling framework to assess the global implications of trade policies. However, over the past decade, through a series of grants from the US Department of Energy and the US Environmental Protection Agency, GTAP has become increasingly central to analyses of the global economic consequences of attempts to mitigate greenhouse gas emissions. The first step in this direction involved integrating the International Energy Agency's database on fossil fuel consumption into GTAP. When coupled with CO_2 emission coefficients, this permitted researchers to estimate more accurately changes in economic activity and fossil-fuel-based emissions in the wake of policies aimed at curbing CO_2 emissions. More recent work has extended the GTAP database to include data on non-CO_2 greenhouse gas emissions and carbon sequestration. Most importantly, these emissions and sequestration are linked to the underlying economic drivers of emissions and sequestration, which are faithfully represented in the core GTAP database. However, the treatment of land use in the standard GTAP model has been rudimentary at best, with no differentiation across land types, and no link to the global land-use databases. In response to the increasing emphasis on land-use change as a factor in greenhouse gas abatement, GTAP has been actively working to develop a new land-use database which has the capability to support the associated land-use modeling activities, particularly in agriculture and forestry (www.gtap.agecon.purdue.edu/databases/projects/Land_Use_GHG/default.asp). This has involved close collaboration with researchers at SAGE, the Center for Sustainability and the Global Environment, based at the University of Wisconsin, as well as at the Ohio State University. This paper documents the methodology used to derive land-use information for the GTAP database and draws out the subsequent implications for climate change mitigation analysis.

21.3 Global land-use data from SAGE

The Center for Sustainability and the Global Environment (SAGE) at the University of Wisconsin has been developing global databases of contemporary and historical agricultural land use and land cover. SAGE has chosen to focus on agriculture because it is clearly the predominant land-use activity on the planet today, and provides vital services – food and other environmental services – for human society.

SAGE has developed a "data fusion" technique to integrate remotely sensed data on the world's land cover with administrative-unit-level inventory data on land use (Ramankutty and Foley, 1998, 1999). The advent of remote sensing data has been revolutionary in providing consistent, global, estimates of the patterns of global land cover. However, remote sensing data are limited in their ability to resolve the details of agricultural land cover from space. Therein lies the strength of the ground-based inventory data, which provide detailed estimates of agricultural land-use practices. However, inventory data are limited in not being spatially explicit, and are plagued by problems of inconsistency across administrative units. The "data fusion" technique developed by SAGE exploits the strengths of both the remotely sensed data and the inventory data.

Using SAGE's methodology, Ramankutty and Foley (1998) [RF98 hereafter] developed a global data set of the world's cropland distribution for the early 1990s (Figure 21.1). This was accomplished by integrating the Global Land Cover Characteristics (GLCC; Loveland et al., 2000) database at

---
[1] See www.gtap.agecon.purdue.edu/databases/projects/Land_Use_GHG/default.asp.
[2] See GTAP website, www.gtap.agecon.purdue.edu/databases/v6/default.asp for a list.

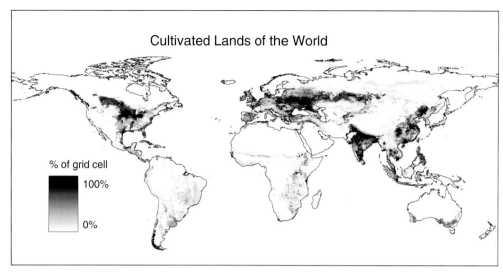

Figure 21.1 The global distribution of croplands *c.* 1992 from Ramankutty and Foley (1998).

1 km resolution (derived from the Advanced Very High Resolution Radiometer instrument, AVHRR), with comprehensive global inventory data (at national and sub-national levels) of cropland area. The resulting data set, at a spatial resolution of 5 min (~10 km) in latitude by longitude, describes the percentage of each 5 min grid cell that is occupied by croplands. Leff *et al.* (2004) further disaggregated the RF98 data set to derive the spatial distribution of 19 crop types of the world (18 major crops and one "other crop" type; maps of individual crops not shown – see Leff *et al.* [2004] for detailed maps).

The SAGE data sets described above are being used for a wide array of purposes, including global carbon cycle modeling (McGuire *et al.*, 2001), analysis of regional food security (Ramankutty *et al.*, 2002), global climate modeling (Bonan, 1999, 2001; Brovkin *et al.*, 1999; Myhre and Myhre, 2003), and estimation of global soil erosion (Yang *et al.*, 2003). They also formed part of the BIOME300 effort, initiated by two core projects, LUCC (Land Use and Land Cover Change) and PAGES (Past Global Changes), of the International Geosphere–Biosphere Programme (IGBP). In other words, they are a widely recognized and widely used data set of global agricultural land use.

The SAGE agricultural land-use data form the core of the GTAP land-use database. In addition to the SAGE data, to derive information on crop yields and irrigation, some ancillary data were obtained from the Food and Agriculture Organization (FAO). In the next section, we describe the procedure used to adapt the SAGE data and the ancillary data to derive land-use information for GTAP.

21.4 Incorporating SAGE agricultural land-use data into GTAP

In the standard GTAP model, land is assumed to be transformable within a nation (or regional grouping of nations) between uses of crop growing, livestock breeding, or timber plantation, or between different crops, regardless of climatic or soil constraints. However, most crops can only grow on lands with particular temperature, moisture, soil type, land form, etc. Lands that are suitable for growing wheat may not be good for rice cultivation, even under transformation at a reasonable cost. In order to account for this difference in crop suitability within a nation, we introduced the concept of the agro-ecological zone (AEZ). AEZs are homogenous units within each country characterized by similar growing conditions (see Section 21.4.1 below for more details). The recognition of various AEZs within each nation results in a more realistic approach in modeling land-use change in GTAP, where land is mobile between crops *within, but not across, AEZs*. The introduction of AEZs in GTAP helps to better inform the issue of land mobility and analysis of competition among competing land uses.

The key land-use variable in the GTAP input–output database that requires specification is the land rent, which is the payment that would be made by a normal tenant for the use of the land. In the database, national-level land rents are specified for eight crop sectors.[3] We partitioned the national-level rents into 18 AEZs (or land endowments) within each nation, in proportion to the AEZ-specific crop production shares – yield times harvested area – from the SAGE land-use data. The spatially explicit nature of the SAGE data allows us to derive the sub-national AEZ distribution of crops. These form the basis for the disaggregation of total crop production in each GTAP region into 18 (potential) production outcomes, which, when other costs are deducted, yield land rents by AEZ. In this paper we focus our attention on how the crop production shares by AEZ are derived.

---

[3] The 18 SAGE crops are translated into the 8 GTAP sectors. That translation is not discussed further here.

Table 21.1 GTAP land-use matrix.

	Land-use types in region r								
AEZ[a]	$Crop_1$	...	$Crop_N$	$Livestock_1$	...	$Livestock_H$	$Forest_1$	...	$Forest_V$
AEZ_1									
...									
...									
AEZ_M									
Total									

[a] Agro-ecological zone
M indicates total number of AEZs, N indicates number of GTAP crop sectors, H indicates total number of GTAP livestock sectors, and V indicates numbers of GTAP forest sectors.

To supply the necessary data for this specification of GTAP, the spatially explicit land-use data sets from SAGE needed to be aggregated along the lines suggested in Table 21.1. Table 21.1 shows the format of the GTAP land-use data. We identify land located in each agro-ecological zone, and this land is allocated down the rows in Table 21.1. The land uses (sectors or activities) are in turn allocated across the columns in Table 21.1. In the GTAP database, each country has its endowment of agricultural land, used in the crops, livestock, and forestry sectors. This total land endowment is represented by the sum over all of the cells in Table 21.1.

The following developments were required in order to complete this database: (1) development of global agro-ecological zones (GAEZs) for deriving sub-national information on land endowments; (2) mapping SAGE data into the GAEZs within nations of the world; and (3) deriving yield (and production) data for the crop sectors. These developments are described in detail below.

21.4.1 Development of global agro-ecological zones

In constructing the GTAP land-use database, we adopt the FAO/IIASA convention of AEZs. The FAO was one of the first organizations to develop the concept of AEZs. Recently, FAO, in collaboration with the International Institute for Applied Systems Analysis (IIASA), developed a global database of AEZs that is considered state of the art (Fischer et al., 2000, 2002). For our purposes, we adopted the "length of growing period" (LGP) data from the IIASA/FAO GAEZ database. Fischer et al. (2000) derived the length of growing period by combining climate, soil, and topography data with a water balance model and knowledge of crop requirements. LGP describes the number of days during a growing season with adequate temperature and moisture to grow crops.

We first derived six global LGPs by aggregating the IIASA/FAO GAEZ data into six categories of ~60 LGP days each: (1) LGP1: 0–59 days; (2) LGP2: 60–119 days; (3) LGP3: 120–179 days; (4) LGP4: 180–239 days; (5) LGP5: 240–299 days; (6) LGP6: > 300 days. These six LGPs roughly divide the world along humidity gradients, and are generally consistent with previous studies in global agro-ecological zoning (Alexandratos, 1995). We also subdivided the world into three climatic zones – tropical, temperate, and boreal – using criteria based on absolute minimum temperature and "Growing degree days", as described in Ramankutty and Foley (1999). By overlaying the six categories of LGPs with the three climatic zones, we developed a global map of 18 AEZs (see Figure 21.2 in the colour plates section and Table 21.2).

In the future, we plan to simulate shifts in AEZs as a function of changing climate. Furthermore, one could potentially define a suite of land uses within each AEZ that although infeasible under current conditions could become feasible under future conditions.

21.4.2 Deriving crop production data

The SAGE land-use data provides information on global crop distributions (Leff et al., 2004) [LEFF04 hereafter]. The global distribution of 18 major crops were derived by compiling crop harvested area statistics from national and sub-national sources, estimating the proportions of harvested area of each crop to total harvested area, and then redistributing it using the RF98 croplands map. These global maps of crop area were superimposed on the GAEZ map to derive crop harvested area for 160 countries of the world, for 18 AEZs within each nation. Several data adjustments and assumptions were made during this procedure. These are described in detail in Appendix A21.1.

SAGE does not have complementary crop yield data accompanying the LEFF04 harvested area information. To derive yield information, we obtained a database from the FAO consisting of harvested area, yield, and production, for 94 developing countries, for several FAO agro-ecological zones. This data was developed based on primary data obtained in the 1970s, and has been periodically updated since then based on observed aggregates (Jelle Bruinsma, personal communication, 2003). So, while the FAO data are not strictly reliable for direct use, they are the only available data on yields by AEZ. We have therefore chosen to use the FAO data as a provisional measure, until improved estimates become available.

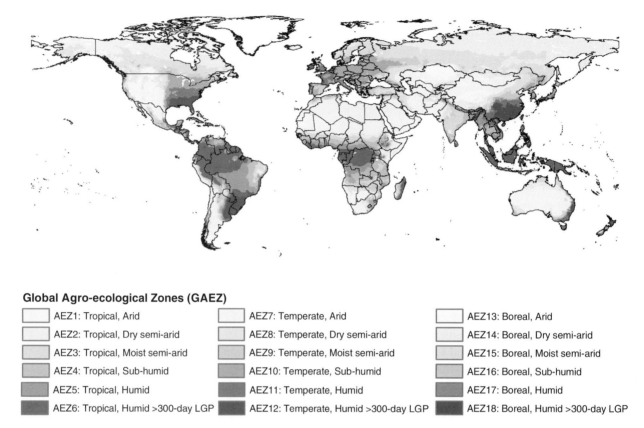

Figure 21.2 The global distribution of global agro-ecological zones (AEZ) from this study, derived by overlaying a global data set of length of growing periods (LGP) over a global map of climatic zones. The LGPs in green shading are in tropical climatic zones, LGPs in yellow-to-red shading lie in temperate zones, while LGPs in blue-to-purple lie in boreal zones. LGPs increase as we move from lighter to dark shades.

Table 21.2 Definition of global agro-ecological zones used in GTAP.

LGP in days	Moisture regime	Climate zone	GTAP class
LGP1: 0–59	Arid	Tropical	AEZ1
		Temperate	AEZ7
		Boreal	AEZ13
LGP2: 60–119	Dry semi-arid	Tropical	AEZ2
		Temperate	AEZ8
		Boreal	AEZ14
LGP3: 120–179	Moist semi-arid	Tropical	AEZ3
		Temperate	AEZ9
		Boreal	AEZ15
LGP4: 180–239	Sub-humid	Tropical	AEZ4
		Temperate	AEZ10
		Boreal	AEZ16
LGP5: 240–299	Humid	Tropical	AEZ5
		Temperate	AEZ11
		Boreal	AEZ17
LGP6: >300 days	Humid > 300-day growing season	Tropical	AEZ6
		Temperate	AEZ12
		Boreal	AEZ18

Adopting the FAO yield data for GTAP involved the following procedures: (1) mapping from FAO's agro-ecological zones to the GTAP AEZs; (2) estimating data for the countries for which FAO did not provide yield data; and (3) merging the crop harvested area and yield data and ensuring consistency between the two. The details of these procedures can be found in Appendix A21.1.

Finally, we derive crop production by AEZ by calculating the product of harvested area and yield data by AEZ. The crop production shares were used by GTAP to partition total production, as well as national land rents into 18 AEZs within each nation.

21.5 Results

21.5.1 Overview of land use and production by AEZ

Our final product is data on crop harvested area, yield, and production, for 160 countries, for 18 AEZs within each country, benchmarked to the year 2001. The uniqueness of this product is the partitioning of data by AEZs within each nation;

the national-level statistics on area, yield, and production are available directly from FAOSTAT. Therefore, here we do not discuss the patterns of agricultural land-use data across the various countries of the world, but rather discuss the patterns across the AEZs.

We have calculated the global total harvested area and production for our 18 major crops for each AEZ (Figure 21.3). The different distribution of warm season and cool season crops is clearly visible in this global summary. Among the cereals (top panels of Figure 21.3), rice, sorghum, and millet are clearly tropical crops, while barley, rye, and wheat are predominantly temperate crops. Maize is equally prevalent in both tropical and temperate zones. Cassava, sugarcane, pulses, oilpalm, and groundnuts are predominantly grown in tropical climates, while potatoes, sugarbeets, rapeseed, and sunflowerseed are grown in temperate climates. Cotton is grown in both tropical and temperate climates. Very little cultivation occurs in the boreal zones – some barley, wheat, and potatoes are grown in the boreal climates of Russia, Kazakhstan, China, and Canada.

In terms of the growing period lengths, the global summary shows that rice, cassava, sugarcane, oilpalm, and soybeans dominate in humid climates, while millet, sorghum, and cotton dominate in semi-arid climates, and the rest of the crops are grown in semi-arid to humid climates. Wheat is the crop with the greatest global extent and is grown across a variety of growing period regimes, but mostly in temperate climates. Soybean is a globally important fast-expanding new crop; the data show that it is mostly grown in temperate climates, but also in the tropical AEZ5, which is consistent with the large extent of soybeans grown in Brazil. Oilpalm is an important crop in southeast Asia (predominantly Malaysia), and this is shown strikingly by its predominant production in AEZ6.

Our analysis also provides insights on the agro-climatic regimes where production of each crop is biophysically feasible (Figure 21.3). For example, wheat is ubiquitous and can be grown in any climatic zone and all but three AEZs – the long growing period boreal AEZs in which acreage of any kind is scarce. Conversely, millet, rice, sorghum, cassava, sugarcane, groundnuts, and oilpalm are mainly tropical crops that cannot be grown in boreal climates. Barley, rye sugarbeets, and rapeseed predominate in temperate climates.

The agro-ecological zone with the largest extent of crops is AEZ10, followed by AEZ9, AEZ3 and AEZ8.[4] AEZ10 and AEZ9 are both temperate AEZs, in sub-humid and moist semi-arid LGP regimes respectively. Indeed, of the 18 major crops, seven of them – barley, maize, rye, wheat, potatoes, sugarbeet, and soybeans – find both their largest extent and production in AEZ10. None of the 18 crops finds its largest extent in AEZ9, even though it comes a close second in many cases. Four crops – sorghum, pulses, groundnuts, and cotton – find their largest extent in AEZ3, the tropical moist semi-arid zone. For each crop, we also queried for the AEZ and country in which the greatest production occurs (Table 21.3). China's AEZ12 and India's AEZ3 are some of the most productive places on the planet, followed by AEZs10/12 in the United States, and AEZ10 in Germany. The database is also consistent with the other well-known crop-growing regions – sunflowers in Argentina, sugarcane in Brazil, cassava in Indonesia, oilpalm in Malaysia, and rye in Poland.

We looked at the distribution of yields across LGPs (Figure 21.4). In general, yields increase with growing period length as expected. However, yields peak by LGP4 (sub-humid) for most crops, with the exception of rice, sorghum, cassava, sugarbeets, groundnuts, oilpalm, and sunflowerseed that have their highest yields in LGP6 (humid; >300 days growing season). Maize, in particular, has much higher yields in LGP4 relative to the rest, while oilpalm has much higher yields in LGP6 relative to the rest.

We finally examined the top five nations of the world ranked in terms of total harvested area, the harvested area of the top five crops within each nation and the AEZs they are grown in (Figure 21.5 in colour plates section. India and China have the largest extent of harvested area in the world, mostly devoted to the cultivation of rice. In India, rice is grown mainly in tropical AEZ2 and AEZ3 (dry and moist semi-arid), while in China, rice is grown in temperate AEZ12 (humid LGP). In the United States, soybeans now exceed maize in terms of harvested area. They are grown in very similar climates, and therefore compete with each other. Russia grows wheat and barley in some temperate AEZs, and in boreal AEZ15 (moist semi-arid). Brazil's growing dominance in soybeans is indicated by the large extent grown in AEZ5 – Brazil is the second largest producer of soybeans in the world, next to the United States. India grows a lot of millet and sorghum.

While the actual location of crop production reflects biophysical feasibility as well as different production structures (i.e. yields), resource availability, market conditions, available technologies, policy, and cultural circumstances, being able to identify the technically viable crops will allow modelers to define the set of cropping alternatives and model changes in the other factors, e.g. trade liberalization or technological change. Over long time horizons, e.g. 50–100 years, AEZs may geographically migrate, expand, or contract with changing land productivity due to evolving temperatures, precipitation, and atmospheric fertilization and deposition. Being able to define biophysically viable land-use alternatives will be crucial for modeling the evolution of the land resource base.

21.5.2 Implications for the cost of mitigation

Thus far we have focused on the use of the AEZ database for descriptive purposes. However, as noted in the introduction, the main motivation for this project has been to provide a better

---

[4] We examine area rather than production here because summing production over different crops is skewed by the fact that some crops have much larger yield than others (e.g., sugarcane has very large yields and will skew the total production, even if it has small extent).

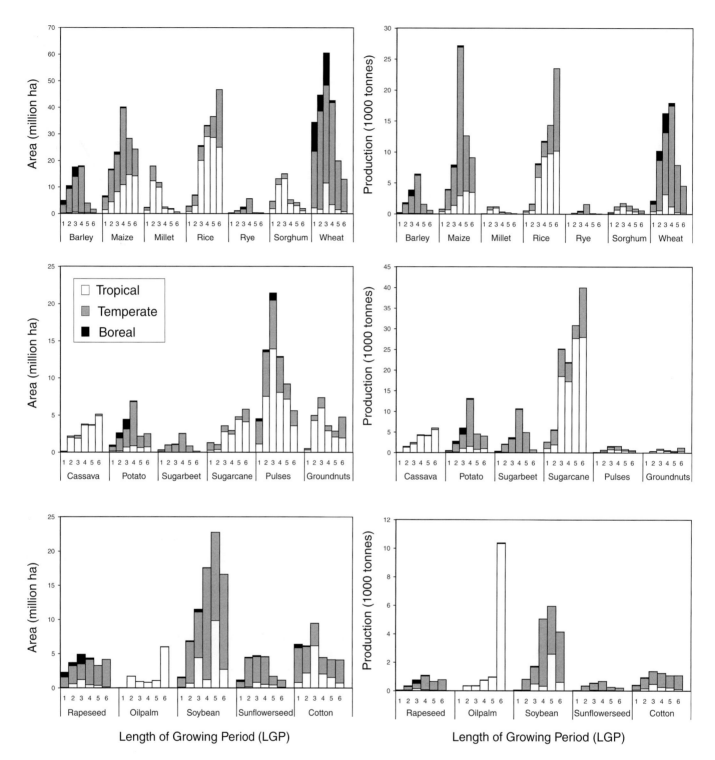

Figure 21.3 Distribution of global total harvested area (left panels) and global total production (right panels) across AEZs for the 18 major crops.

underpinning for the analytical modeling of the cost of climate change mitigation. By providing a more accurate representation of the potential for competition among alternative land uses, the crop production by AEZ database is a critical element of any serious analysis of the role of land-use change in climate change mitigation.

To illustrate this point, we draw on the work of Lee (2005) who calculates the cost of reducing greenhouse gas emissions in the USA and China, taking into account both CO_2 and non-CO_2 gases. She initially computes the cost of achieving various national abatement targets in the absence of the AEZ specification, essentially assuming that there is no mobility of

Table 21.3 The country and AEZ of greatest production for each major crop.

Sunflowerseed	Argentina	AEZ12
Sugarcane	Brazil	AEZ5
Rice	China	AEZ12
Potato	China	AEZ12
Groundnuts	China	AEZ12
Rapeseed	China	AEZ12
Barley	Germany	AEZ10
Sugarbeet	Germany	AEZ10
Millet	India	AEZ3
Sorghum	India	AEZ3
Wheat	India	AEZ3
Pulses	India	AEZ3
Cassava	Indonesia	AEZ6
Oilpalm	Malaysia	AEZ6
Rye	Poland	AEZ10
Maize	USA	AEZ10
Soy	USA	AEZ10
Cotton	USA	AEZ12

land across alternative uses within agriculture and between agriculture and forestry. She then compares this cost to that obtained when the potential to move land across uses, within AEZs, is taken into account. For a marginal (5%) reduction in emissions from the baseline, Lee (2005) finds that the failure to account for land mobility within AEZs results in an overstatement of the marginal cost of mitigation by 41% in the case of the United States and 32% in the case of China. Clearly this addition to the analytical framework is very important.

21.6 Conclusions

In this study, we have combined global land-use data sets to derive land-use information at the national level, by 18 different AEZs (Table 21.4) for use in the GTAP database. In particular, we have used the following global land-use/land-cover data sets developed by SAGE: spatially explicit maps of 19 crops (18 major plus "other crops"), agro-ecological zones, and national boundaries. We also used the following data developed by FAO: harvested area, yield, and production for 94 developing nations, for 34 crops, by 6 AEZs within each nation, and FAOSTAT national statistical data for all crops and countries for 2001. These data sets were synthesized, adjusted for consistency, and calibrated to the year 2001.

The results from our synthesis seem reasonable, and broadly consistent with general knowledge about the patterns of global agricultural production systems. The database shows that the traditional wheat belts of the world are prevalent in the United States, Russia, Argentina, India, and China. However, soybeans are becoming a dominant crop, with the fourth largest extent in the world behind wheat, rice, and maize; soybeans are currently the largest crop in the United States, Brazil, and Argentina.

The geo-referenced cropping database supports spatial location of GHG emissions sources and mitigation opportunities. For instance, higher N_2O from nitrogen fertilizer use will coincide with wheat, maize, and rice production, which are responsible for approximately 50% of global fertilizer use (FAO, 2000). Similarly, nitrogen-fixing crops such as soybeans are another identified driver of nitrogen oxide emissions (IPCC, 1997), while paddy rice is a significant source of global methane emissions (Scheehle and Kruger, 2006; USEPA, 2006).

In summary, the SAGE land-use data sets, integrated with ancillary information from FAO, have been customized for the GTAP database. The database provides crop harvested area and production data (Table 21.4), for 160 nations of the world, and for 18 agro-ecological zones within each nation. The data will be valuable for global integrated assessment models examining the role played by agricultural land use in GHG mitigation strategies, and stabilization modeling needing a base year characterization of land to assess land responses to climate changes. Preliminary results indicate that for a 5% reduction in emissions from the baseline, failure to account for land mobility within AEZs results in an overestimation of the marginal cost of mitigation by 41% in the United States and 32% in China.

Future directions in this work involve meeting numerous challenges, including the consideration of multiple cropping, differentiating between planted and harvested area, accounting for fallow lands, and the consideration of land heterogeneity within AEZs.

Acknowledgements

We are grateful for the support of this research through funding from the Office of Biological and Environmental Research, US Department of Energy (Award No. DE-FG02–0IER63215). The authors would like to thank participants at two workshops that focused on land use and integrated assessment modeling – one held in September 2002 at MIT and the other in May 2004 at the US EPA – for their ideas and useful feedback.

Appendix A21.1 Key assumptions and procedures

A21.1.1 Crop harvested area

The SAGE land-use data provide information on crop areas (Leff et al., 2004) [LEFF04 hereafter]. The global distributions of major crops were derived by compiling crop harvested area statistics from national and sub-national sources, estimating the proportions of harvested area of each crop to total harvested area, and then redistributing it using the RF98 croplands map.

The RF98 croplands and LEFF04 crop area maps represents a notional "physical cultivated area" on the ground, i.e. multiple cropped areas were counted only once. However, crop *harvested* area, wherein multiple cropped areas are counted as many times as they are multi-cropped, and total production is more representative than physical cultivated area of the actual

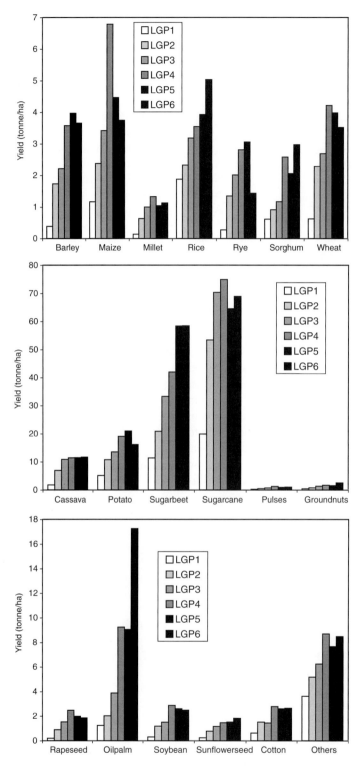

Figure 21.4 Distribution of global average yields across length of growing period (LGP) regimes for the 18 major crops. Here, the climatic zone distinctions (tropical, temperate, boreal) are not drawn out because those distinctions are not made when translating yields from the FAO AEZs to the GTAP AEZs (see Table 21.7); therefore, the same yield pattern across LGPs will hold for all three climatic zones.

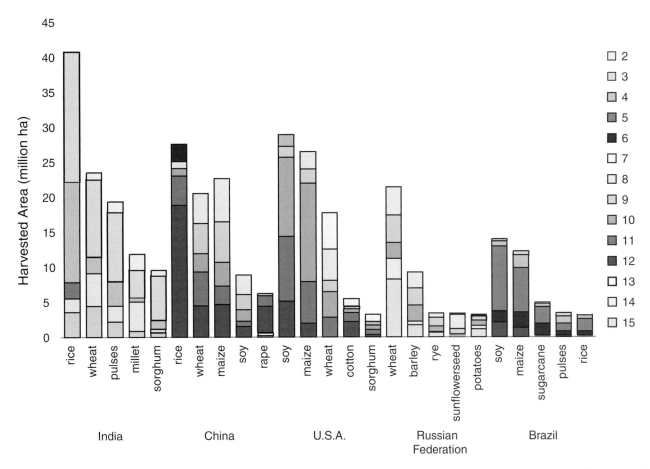

Figure 21.5 The top five nations of the world, in terms of total harvested area, the top five crops and their harvested areas within each nation, and the AEZs they are grown in. Tropical AEZs are shown in green, temperate in yellow-to-red, and boreal in blue-to-purple.

Table 21.4 Summary of the SAGE agricultural land-use data set provided to GTAP.

Items	Variables & units	Specifications	Reference period
19 LEFF04 crops	Harvested area (ha); Yield (kg/ha)	160 countries, 18 AEZs within each country	2001

land-use activity on the ground, and reflective of the true revenue from the land. Land inputs such as irrigation and fertilization, as well as land-based emissions such as CH_4 emissions from paddy rice cultivation, are also more directly related to harvested area.[5] Therefore, to convert the LEFF04

data from physical area to harvested area, we recalibrate the LEFF04 data against the national crop harvested area statistics from FAOSTAT (FAO, 2004).

Let $A'_{\text{LEFF}}(i, mc)$ be the LEFF04 crop area (in units of ha) for pixel i and major crop mc. Note that the original LEFF04 data sets are gridded, at 0.5 degree resolution in latitude by longitude. We recalibrate the LEFF04 data to the FAOSTAT harvested area data $A_{\text{FAO}}(l, mc, t_{\text{ref}})$ (ha) as follows:

$$A_{\text{LEFF}}(i, mc) = A'_{\text{LEFF}}(i, mc) \times \frac{A_{\text{FAO}}(l, mc, t_{\text{ref}})}{\sum_{i \in l} A_{\text{LEFF}}(i, mc)}, \quad (21.1)$$

where l represents countries in FAOSTAT, $i \in l$, and t_{ref} is the reference time period 2001 (for consistency with GTAP version 6.0).

The recalibrated LEFF04 data are then overlain with: (1) global AEZ map; and (2) political boundaries, and aggregated to derive harvested areas of 19 crops for all nations of the world, for 18 AEZs within each nation.

Let this aggregated data be represented by $A_{\text{LEFF}}(l, mc, z)$, where l is the country, mc is one of 19 LEFF04 major crops, and z is one of 18 AEZs.

---

[5] Planted area would probably be a better indicator of land-based emissions from agriculture, but global data on planted area are not easily available except for a few countries of the world. The standard FAO database only provides harvested area and production – variables that the market sees.

A21.1.2 Estimating crop yields from FAO data

The FAO provided GTAP with estimates of harvested area, yield, and production, for 94 developing countries, for several FAO agro-ecological zones (including an FAO AEZ labeled "irrigated"). As described in Section 21.4.2, these data were developed based on primary data obtained in the 1970s, and periodically updated based on observed aggregates (Jelle Bruinsma, personal communication, 2003). Although not strictly reliable for direct use, they are the only available data on yields by AEZ within countries. Here we describe how we adopted the FAO data for our purpose.

Deriving yields from FAO data for 94 developing countries
The FAO data were provided for six different agro-ecological zones (Table 21.5; see Alexandratos [1995]), defined slightly differently from our AEZs, and for 34 different crops. We therefore had to match the FAO AEZs and crops with GTAP's 18 AEZs and 19 LEFF04 crops.

FAO reports yields separately for four rain-fed AEZs (AT1, AT2, AT3, AT4+AT5), one AEZ with fluvisol/gleysol soils (AT6+AT7; naturally flooded soils), and one irrigated AEZ (denoted "Irrigated land"). In other words, FAO has separated out irrigation and the occurrence of naturally flooded soils into separate AEZs. In this study, we choose to treat AEZs as a *climate only* constraint (including the influence of soil moisture), and therefore irrigation and/or fluvisols/gleysols are considered to potentially occur within each AEZ. As a result, we needed to repartition the irrigated and AT6+AT7 yields into the rain-fed zones to estimate the total yields for each AEZ. This procedure is described below.

We first mapped from the 34 FAO crops to the 19 LEFF04 major crops, based on the mapping given in Table 21.6 (harvested area weighted averages were calculated when multiple FAO crops mapped into one LEFF04 crop).

Let $Y_{FAO,RF}(n, mc, fz)$ be the FAO reported yield (in units of kg/ha) for the four rain-fed AEZs fz, for nation n, and crop mc, where mc = one of 19 LEFF04 major crops, fz = FAO AEZs AT1, AT2, AT3, AT4+AT5 (Table 21.5), and RF refers to rain-fed.

Let $Y_{FAO}(n, mc)$ be the national yield from FAO for each crop (harvested area weighted average of all six zones).

We calculated *national rain-fed* yields for each crop,

$$Y_{FAO,RF}(n,mc) = \frac{\sum_{fz=AT1}^{AT4+AT5} Y_{FAO,RF}(n,mc,fz) \times A_{FAO,RF}(n,mc,fz)}{\sum_{fz=AT1}^{AT4+AT5} A_{FAO,RF}(n,mc,fz)},$$

(21.2)

where $A_{FAO,RF}(n, mc, fz)$ = harvested area data from FAO, corresponding to the yield data.

Table 21.5 Definition of FAO AEZs.

FAO AEZ	Moisture regime (LGP in days)	Description
AT1	75–119	Dry semi-arid
AT2	120–179	Moist semi-arid
AT3	180–269	Sub-humid
AT4	270+	Humid
AT5	120+	Marginally suitable land in moist semi-arid, sub-humid, humid classes
AT6	Naturally flooded	Fluvisols/gleysols
AT7	Naturally flooded	Marginally suitable fluvisols/gleysols
Irrigated land	Irrigated	Irrigated

Table 21.6 Mapping from FAO crops to LEFF04 crops.

No.	FAO crops	No.	LEFF04 crops
1	WHEA	19	Wheat
2	RICE	12	Rice
3	MAIZ	5	Maize
4	BARL	1	Barley
5	MILL	6	Millet
6	SORG	14	Sorghum
7	OTHC	13	Rye
8	POTA	9	Potato
9	SPOT	8	Others
10	CASS	2	Cassava
11	OTHR	8	Others
12	PLAN	8	Others
13	BEET	16	Sugarbeet
14	CANE	17	Sugarcane
15	PULS	10	Pulses
16	VEGE	8	Others
17	BANA	8	Others
18	CITR	8	Others
19	FRUI	8	Others
20	OILC	8	Others
21	RAPE	11	Rape
22	PALM	7	Oilpalm
23	SOYB	15	Soy
24	GROU	4	Groundnuts
25	SUNF	18	Sunflower
26	SESA	8	Others
27	COCN	8	Others
28	COCO	8	Others
29	COFF	8	Others
30	TEAS	8	Others
31	TOBA	8	Others
32	COTT	3	Cotton
33	FIBR	8	Others
34	RUBB	8	Others

Then, we estimated total yield (rain-fed plus irrigated plus fluvisol/gleysol) for each of the FAO AEZs, AT1, AT2, AT3, and AT4+AT5, by applying the ratio of national total yield to national rain-fed yield to each AEZ,

$$Y''_{FAO}(n,mc,fz) = Y_{FAO,RF}(n,mc,fz) \times \frac{Y_{FAO}(n,mc)}{Y_{FAO,RF}(n,mc)}, \text{ if } Y_{FAO,RF}(n,mc) > 0. \quad (21.3)$$

As an average across all countries, the national total yield is 70% greater than rain-fed yields for rice. This is reasonable because paddy rice is heavily irrigated, and irrigated yields are higher than rain-fed yields. The lowest ratio of national total yield to rain-fed yield occurs for cassava and oilpalm, which are not irrigated at all.

This yield is then adjusted to match FAOSTAT national statistics,

$$Y_{FAO}(n,mc,fz) = Y''_{FAO}(n,mc,fz) \times \frac{\overline{Y}_{FAO}(n,mc)}{Y_{FAO}(n,mc)}, \text{ if } Y_{FAO,RF}(n,mc) > 0, \quad (21.4)$$

where $\overline{Y}_{FAO}(n,mc)$ = FAOSTAT national statistic on crop yield.

If total rain-fed yield is zero (i.e. FAO reports that for a particular crop and country, the crop is entirely irrigated or found in the gleysol/fluvisol AEZ), then we simply repartition the national-level FAOSTAT yields using an estimated global average of the proportion of yield in each AEZ to total yield. This is described in greater detail below (note that the estimation of average yields is executed prior to the calculation below for zero total rain-fed yields).

$$Y_{FAO}(n,mc,fz) = \overline{Y}_{FAO}(n,mc) \times \frac{1}{N}\sum_{n=1}^{N} \frac{Y_{FAO}(n,mc,fz)}{Y_{FAO}(n,mc)},$$
$$\text{if } Y_{FAO,RF}(n,mc) = 0, \quad (21.5)$$

where N = total number of countries with $Y_{FAO}(n,mc,fz) > 0$ and $Y_{FAO}(n,mc) > 0$. The summation in the above equation is only performed when both numerator and denominator are non-zero.

Estimating crop yields by AEZ for countries without FAO data

As FAO data were available for only 94 countries, we estimated information on yield variation across AEZ for the remaining countries (66) using averages calculated over the 94 countries and applying them to the national statistics for

Table 21.7 Mapping from FAO AEZs to GTAP AEZs.

FAO AEZs	GTAP AEZs
Estimated (see text)	AEZ1, AEZ7, AEZ13
AT1	AEZ2, AEZ8, AEZ14
AT2	AEZ3, AEZ9, AEZ15
AT3	AEZ4, AEZ10, AEZ16
AT4 + AT5	AEZ5, AEZ11, AEZ17
AT4 + AT5	AEZ6, AEZ12, AEZ18
AT6 + AT7	No separate AEZ (see text)
Irrigated land	No separate AEZ (see text)

the remaining countries (figure not shown).[6] Note that we did not average the yields themselves, but rather the proportion of yield in each AEZ to national yields.

For each country m, without FAO data by AEZ,

$$Y_{FAO}(m,mc,fz) = \overline{Y}_{FAO}(m,mc) \times \frac{1}{N}\sum_{n=1}^{N} \frac{Y_{FAO}(n,mc,fz)}{Y_{FAO}(n,mc)}, \quad (21.6)$$

where $\overline{Y}_{FAO}(m,mc)$ = FAOSTAT national statistic on crop yield, and N = total number of countries with $Y_{FAO}(n,mc,fz) > 0$ and $Y_{FAO}(n,mc) > 0$. The summation in the above equation is only performed when both numerator and denominator are non-zero. The yields are set to zero if the LEFF04 crop harvested areas are zero for that country.

Merging the data and adjusting for consistency with SAGE crop harvested area

The yields from the 94 countries are merged with the estimated yields for the remaining countries,

$$Y_{FAO}(l,mc,fz) = Y_{FAO}(n,mc,fz) \cup Y_{FAO}(m,mc,fz). \quad (21.7)$$

FAO does not report yields for GTAP AEZ1, AEZ7, and AEZ13 (see Table 21.7). Furthermore, often the FAO yield data and the recalibrated LEFF04 harvested area data are inconsistent, with FAO reporting non-zero yields even though recalibrated LEFF04 reports zero harvested areas, and conversely, FAO reporting zero yields while LEFF04 reports non-zero harvested area. In all of these cases, we adjusted the FAO yield data to match the recalibrated LEFF04 harvested area data.

We first mapped the FAO yield data from FAO's AEZs to GTAP AEZs based on Table 21.7. To fill in gaps in FAO yield data (i.e. zero reported yields when recalibrated LEFF04 harvested area is non-zero), we estimate yields using a regression across all countries and all crops of yields in each

---

[6] Averaging across all 94 countries may introduce biases. For example, the 94 countries are developing countries, and not representative of developed country yield variations across AEZ. In future versions, proxy data for averaging may be selected on the basis of similarity in climates, as well as socio-economic conditions.

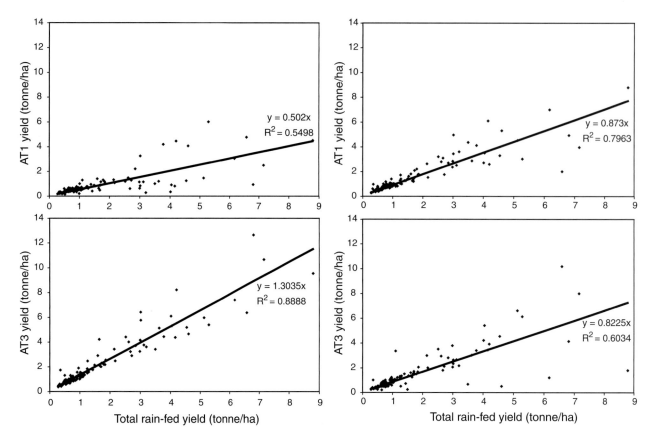

Figure 21.6 A regression across all countries and all crops of yields in each rain-fed AEZ to total rain-fed yields. This regression is used to fill gaps (see text) in FAO yield data (i.e. when yield is not reported for a particular crop while LEFF04 harvested area is non-zero).

rain-fed AEZ to total rain-fed yields (Figure 21.6).[7] For GTAP AEZ1, AEZ7, and AEZ13 (0–60 day LGP, with no data reported by FAO), we assumed that yield is one-tenth of the total rain-fed yield for the corresponding crop and country (the value of 0.1 is arbitrary, but meant to represent a small yield in these arid AEZs; because not much is grown in these AEZs (see Figure 21.3), this assumption should not have significant influence on the final results). In other words,

$$Y_{\text{FAO}}(l,mc,z) \begin{cases} \leftarrow Y_{\text{FAO}}(l,mc,fz), \text{ based on Table 21.7} \\ = 0, \text{ if } A_{\text{LEFF}}(l,mc,z) = 0 \end{cases}$$
(21.8)

If $(Y_{\text{FAO}}(l,mc,z) = 0 \quad A_{\text{LEFF}}(l,mc,z) \neq 0)$, then

$$Y_{\text{FAO}}(l,mc,z) = \alpha_z Y_{\text{FAO,RF}}(l,mc) \frac{Y_{\text{FAO}}(l,mc)}{Y_{\text{FAO,RF}}(l,mc)},$$
(21.9)

where $\alpha_z = \alpha_{fz} = 0.10$ for AEZ1, and AEZ13;

= 0.50 for AT1;
= 0.87 for AT2;
= 1.30 for AT3; and
= 0.82 for AT4 + AT5 (based on Figure 21.6, and Table 21.7).

Recalibrating the yield data to year 2001
Finally, because the recalibrated harvested area data by AEZ are derived from LEFF04, and the yield data are from FAO, the re-calculated national yields will change. Also, the yields need to be calibrated to the reference period of 2001. We do this as follows:

$$Y(l,mc,z) = Y_{\text{FAO}}(l,mc,z) \times \frac{P_{\text{FAO}}(l,mc,t_{\text{ref}})}{\sum_z Y_{\text{FAO}}(l,mc,z) \times A_{\text{LEFF}}(l,mc,z)},$$
(21.10)

where $P_{\text{FAO}}(l, mc, t_{\text{ref}})$ = the FAOSTAT national production for $t_{\text{ref}} = 2001$.

References

Alexandratos, N. (1995). *World Agriculture Towards 2010*. Rome: Food and Agriculture Organization of the United Nations.

---

[7] This averaging is done across all crops to maintain a sufficiently large sample size for the regression. For now, this is used here simply for consistency, and therefore will not bias the final results very much. Future versions should consider establishing this relationship for individual crop types.

Bonan, G. B. (1999). Frost followed the plow: impacts of deforestation on the climate of the United States. *Ecological Applications* **9**, 1305–1315.

Bonan, G. B. (2001). Observational evidence for reduction of daily maximum temperature by croplands in the Midwest United States. *Journal of Climate* **14**, 2430–2442.

Brovkin, V., Ganopolski, A., Claussen, M., Kubatzki, C. and Petoukhov, V. (1999). Modelling climate response to historical land cover change. *Global Ecology and Biogeography* **8**, 509–517.

FAO (2000). *Fertilizer Requirements in 2015 and 2030*. Rome: Food and Agriculture Organization of the United Nations.

FAO (2004). *FAOSTAT Data*. Food and Agriculture Organization of the United Nations. http://apps.fao.org.

Fischer, G., van Velthuizen, H., Nachtergaele, F. and Medow, S. (2000). *Global Agro-Ecological Zones (Global – AEZ)*. Food and Agricultural Organization/International Institute for Applied Systems Analysis (FAO/IIASA), CD-ROM and website www.fao.org/ag/AGL/agll/gaez/index.html.

Fischer, G., van Velthuizen, H., Medow, S. and Nachtergaele, F. (2002). *Global Agro-Ecological Assessment for Agriculture in the 21st Century*. Food and Agricultural Organization/International Institute for Applied Systems Analysis (FAO/IIASA), CD-ROM and website www.iiasa.ac.at/Research/LUC/SAEZ/index.html.

Foley, J. A., Costa, M. H., Delire, C., Ramankutty, N. and Snyder, P. (2003). Green surprise? How terrestrial ecosystems could affect Earth's climate. *Frontiers in Ecology and the Environment* **1**, 38–44.

Hannah, L., Lohse, D., Hutchinson, C., Carr, J. L. and Lankerani, A. (1994). A preliminary inventory of human disturbance of world ecosystems. *Ambio* **23**, 246–250.

Houghton, R. A. (2003). Revised estimates of the annual net flux of carbon to the atmosphere from changes in land use and land management 1850–2000. *Tellus Series B – Chemical and Physical Meteorology* **55**, 378–390.

IPCC (1997). *Revised 1996 IPCC Guidelines for Greenhouse Gas Emissions Inventories*. Cambridge: Cambridge University Press.

Lee, H.-L. (2005). *The GTAP Land Use Data Base and the GTAPE-AEZ Model: Incorporating Agro-Ecologically Zoned Land Use Data and Land-based Greenhouse Gases Emissions into the GTAP Framework*. www.gtap.agecon.purdue.edu/resources/res_display.asp?RecordID=1839.

Leff, B., Ramankutty, N. and Foley, J. (2004). Geographic distribution of major crops across the world. *Global Biogeochemical Cycles*, **18** GB1009, doi:10.1029/2003 GB002108.

Loveland, T. R., Reed, B. C., Brown, J. F. et al. (2000). Development of a global land cover characteristics database and IGBP DISCover from 1 km AVHRR data. *International Journal of Remote Sensing* **21**, 1303–1330.

McGuire, A. D., Sitch, S., Clein, J. S. et al. (2001). Carbon balance of the terrestrial biosphere in the twentieth century: analyses of CO_2, climate and land use effects with four process-based ecosystem models. *Global Biogeochemical Cycles* **15**, 183–206.

Myhre, G. and Myhre, A. (2003). Uncertainties in radiative forcing due to surface albedo changes caused by land-use changes. *Journal of Climate* **16**, 1511–1524.

National Geographic Maps (2002). *A World Transformed, Supplement to National Geographic September 2002*. Washington DC: National Geographic Society.

Postel, S. L., Daily, G. C. and Ehrlich, P. R. (1996). Human appropriation of renewable fresh water. *Science* **271**, 785–788.

Prentice, I. C., Farquhar, G., Fashm, M. et al. (2001). The carbon cycle and atmospheric carbon dioxide. In *Climate Change 2001: The Scientific Basis. Contribution of Working Group I to the Third Assessment Report of the Intergovernmental Panel on Climate Change*, ed. J. T. Houghton, Y. Ding, D. J. Griggs et al. Cambridge: Cambridge University Press, pp. 183–237.

Ramankutty, N. and Foley, J. A. (1998). Characterizing patterns of global land use: an analysis of global croplands data. *Global Biogeochemical Cycles* **12**, 667–685.

Ramankutty, N. and Foley, J. A. (1999). Estimating historical changes in global land cover: croplands from 1700 to 1992. *Global Biogeochemical Cycles* **13**, 997–1027.

Ramankutty, N., Foley, J. A. and Olejniczak, N. J. (2002). People on the land: changes in population and global croplands during the 20th century. *Ambio* **31** (3), 251–257.

Richards, J. F. (1990). Land transformation. In *The Earth as Transformed by Human Action*, ed. B. L. Turner, W. C. Clark, R. W. Kates et al. New York: Cambridge University Press, pp. 163–178.

Richards, K. and Stokes, C. (2004). A review of forest carbon sequestration cost studies: a dozen years of research. *Climatic Change* **63**, 1–48.

Scheehle, E. A. and Kruger, D. (2006). Global anthropogenic methane and nitrous oxide emissions. *The Energy Modeling Journal* Special Issue No. 3: *Multi-Greenhouse Gas Mitigation and Climate Policy*.

Tans, P. P., Fung, I. Y. and Takahashi, T. (1990). Observational constraints on the global atmospheric CO_2 budget. *Science* **247**, 1431–1438.

USEPA (2006). *Global Emissions of Non-CO_2 Greenhouse Gases: 1990–2020*. Washington, DC: Office of Air and Radiation, US Environmental Protection Agency (USEPA); www.epa.gov/nonco2/econ-inv/international.html.

Vitousek, P. M., Mooney, H. A., Lubchenco, J. and Melillo J. M. (1997). Human domination of Earth's ecosystems. *Science* **277**, 494–499.

Yang, D. W., Kanae, S., Oki, T., Koike, T. and Musiake, K. (2003). Global potential soil erosion with reference to land use and climate changes. *Hydrological Processes* **17**, 2913–2928.

22

Past, present, and future of non-CO$_2$ gas mitigation analysis

Francisco C. de la Chesnaye, Casey Delhotal, Benjamin DeAngelo, Deborah Ottinger-Schaefer and Dave Godwin

22.1 Introduction

"Other greenhouse gases" (OGHGs) and "non-CO$_2$ greenhouse gases" (NCGGs): these are terms that are now much more familiar to the climate modeling community than they were a decade ago. Much of the increased analytical relevance of these gases, which include methane, nitrous oxide, and a group of fluorinated compounds, is due to work conducted under the Stanford Energy Modeling Forum (EMF) and facilitated by meetings at Snowmass, Colorado, going back to 1998.

The two principal insights from over five years of analysis on NCGGs are (1) the range of economic sectors from which these emissions originate is far larger and more diverse than for carbon dioxide (CO$_2$); and (2) the mitigation costs for these sectors and their associated gases can be lower than for energy-related CO$_2$. Taken together, these two factors result in a larger portfolio of potential mitigation options, and thus more potential for reduced costs, for a given climate policy objective. This is especially important where carbon dioxide is not the dominant gas, on a percentage basis, for a particular economic sector and even for a particular region.

This paper provides an analytical history of non-CO$_2$ work and also lays out promising new areas of further research. There are five sections following this introduction. Section 22.2 provides a summary of non-CO$_2$ gases and important economic sectors. Section 22.3 covers early efforts to estimate non-CO$_2$ emissions and mitigation potential. Section 22.4 covers recent work focusing on mitigation. Section 22.5 covers new work on improvements at estimating technological change and greater sectoral coverage. And finally, Section 22.6 points to future research needed to improve NCGG emission and mitigation analysis further.

22.2 Summary of non-CO$_2$ gases and sources

Comparing and aggregating GHGs can be an analytical challenge. This is because NCGGs are emitted from a variety of sources, have different atmospheric lifetimes, and exert different climate forcings (an outside, imposed perturbation in the energy balance of the Earth's climate system; Hansen et al., [1997]) compared with one another and with CO$_2$. However, it is possible to quantify the radiative forcings exerted by the various anthropogenic GHGs at their past and current atmospheric concentrations. These gases are estimated to have produced an enhanced greenhouse effect of 2.9 W/m^2 for the period 1850–2000, where the combined NCGGs account for about 40% of that effect (Hansen and Sato, 2001); see Figure 22.1.[1]

---

[1] Note that the "F-gases" in Figure 22.1 include chlorofluorocarbons, hydrochlorofluorocarbons, and other ozone-depleting substances (ODSs) in addition to hydrofluorocarbons (HFCs), perfluorocarbons (PFCs), and sulfur hexafluoride (SF$_6$). The ODSs contributed the vast majority of the climate forcing shown for the F-gases in Figure 22.1. In contrast, the "F-gases" in Figure 22.2 include only HFCs, PFCs, and SF$_6$. Because they are regulated under the Montreal Protocol, the ODSs are excluded from the UNFCCC and from most GHG mitigation analyses.

Human-induced Climate Change: An Interdisciplinary Assessment, ed. Michael Schlesinger, Haroon Kheshgi, Joel Smith, Francisco de la Chesnaye, John M. Reilly, Tom Wilson and Charles Kolstad. Published by Cambridge University Press. © Cambridge University Press 2007.

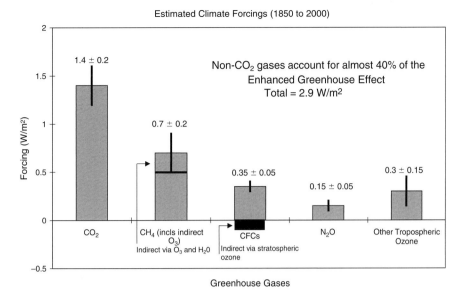

Figure 22.1 Estimated climate forcings 1850 to 2000 (Source: Hansen *et al.*, 2000).

In the area of climate mitigation and GHG emission inventories, the common practice is to compare and aggregate emissions by using global warming potentials (GWPs). Here emissions of all NCGGs are converted to a carbon dioxide equivalent (CO_2e) basis using GWPs (21 for methane and 310 for nitrous oxide) as published by the Intergovernmental Panel on Climate Change (IPCC) and recognized by the United Nations Framework Convention on Climate Change. GWPs used here and elsewhere are calculated over a 100-year period, and vary because of both the gases' ability to trap heat and their atmospheric lifetime compared with an equivalent mass of CO_2.[2] These equivalents are then converted to carbon equivalents (Ce) by multiplying by the weight ratio of C to CO_2 (12/44). Although the GWPs have been updated by the IPCC in the Third Assessment Report, estimates of emissions in this report continue to use the GWPs from the Second Assessment Report, in order to be consistent with international reporting standards under the UNFCCC (IPCC, 1997; 2001). By using the GWP metrics, a recent EMF study estimates total anthropogenic GHG emissions, including CO_2 from energy and land use and NCGGs, at about 11 300 million metric tons of carbon equivalent (MtCe) (Weyant and de la Chesnaye, 2006); see Figure 22.2. Carbon-equivalent emission aggregations of all GHGs can also help illustrate the variety and range of economic sectors, globally, that emit GHGs; see Table 22.1.

22.2.1 Methane

Among the NCGGs, methane has the largest total radiative forcing but a relatively short atmospheric lifetime (roughly a decade before natural sinks consume it [IPCC, 2001]). This relatively short lifetime makes methane an excellent candidate for mitigating the near-term increases in total climate forcing. Methane emissions from anthropogenic and natural sources are shown in Table 22.2.

In a recent study, methane emissions were estimated at 1553 MtCe in 1990 increasing to 1618 MtCe in 2000 and projected to 2153 MtCe in 2020 (Scheehle and Kruger, 2006). The slow growth in methane emission in the past 15 years is due to various factors; the most important include emission reduction programs, effects of non-climate regulations on waste management, and economic and sectoral restructuring in Eastern Europe and the Commonwealth of Independent States (CIS). For more details on methane emissions by sector and region, see Scheehle and Kruger (2006).

22.2.2 Nitrous oxide

Nitrous oxide has a much longer lifetime than methane (about 100 years; IPCC [2001]). As is the case for methane, anthropogenic emissions of nitrous oxide were estimated to have increased over the period 1990 to 2000, from 865 MtCe to 947 MtCe, and they are projected to increase to 1259 MtCe by 2020. There is a wide range in the estimates of natural emissions of nitrous oxide given by IPCC (IPCC, 2001). Both global natural and human-generated emissions by sector are provided in Table 22.3.

---

[2] One factor that complicates comparisons among the gases is that GWPs calculated over 20- or 500-year time horizons are generally different from GWPs calculated over a 100-year horizon. In general, the shorter-lived a gas, the higher is its 20-year GWP (and the lower its 500-year GWP) relative to its 100-year GWP. For example, methane, with a lifetime of about a decade, has a 20-year GWP that is almost three times as large as its 100-year GWP. On the other hand, CF_4, with a lifetime of 50 000 years, has a 20-year GWP that is only two-thirds as large as its 100-year GWP.

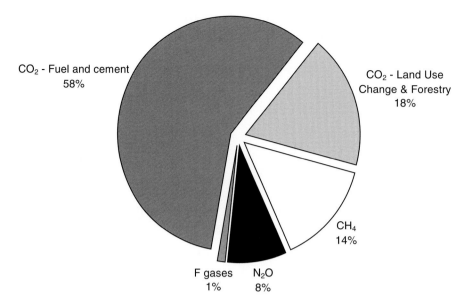

Figure 22.2 Global net GHG emissions for 2000 (Weyant and de la Chesnaye, 2006).

Table 22.1 Global GHG emissions for 2000 (MtCe).

Sectors	Sub-sectors	CO_2	Methane	N_2O	F-gases
Energy	Coal	2218	123		
	Nat. gas	1309	244		
6843	Petroleum	2857	17		
60%	Stationary/mobile sources		16	61	
	LUCF (net)	2081			
Land-use change, forestry,	Soils			711	
agriculture	Biomass		134	51	
3691	Enteric fermentation		476		
32%	Manure management		61	56	
	Rice		177		
	Cement	226			
	Adipic and nitric acid production			43	
Industry	HFCs				26
408	PFCs				29
4%	SF_6				15
	Substitution of ODS				52
Waste	Landfills		213		
388	Wastewater		154	22	
3%	Other		3	3	
Total GHG		11 378	8691	1618	947	122
			76%	14%	8%	1%

Source: EPA, EMF21, CDIAC

22.2.3 Fluorinated gases

The fluorinated gases, including hydrofluorocarbons (HFCs), perfluorocarbons (PFCs), and sulfur hexafluoride (SF_6), are emitted from a number of processes and products, the most significant of which are detailed in Table 22.4. Over the past 100 years, atmospheric concentrations of most of these gases have risen from zero to the parts-per-trillion level. Although the gases make up a very small fraction of greenhouse gas emissions by mass, they are responsible for a more significant amount of climate forcing because of their high GWPs, which range from approximately 120 to 22 200 (the 100-year GWPs of HFC-152a and SF_6, respectively) (IPCC, 2001). In addition, in the absence of voluntary or regulatory action to reduce emission rates, emissions of HFCs, PFCs, and SF_6 would be expected to rise quickly because of (1) the rapid growth rate of some emitting

Table 22.2 Global methane emissions for 2000 (MtCe).

Anthropogenic methane			Natural methane	
Energy			Wetlands	659–1357
Coal mining	123	8%		
Natural gas & oil systems	261	16%	Termites	114
Fossil fuel combustion	16	1%		
Biofuel combustion	59	4%	Ocean	57–85
Agriculture			hydrates	29–57
Livestock manure management	61	4%		
Livestock enteric fermentation	476	29%		
Rice cultivation	177	11%		
Biomass burning:	76	5%		
Agricultural residue burning				
Prescribed burning of savanna				
Waste				
Solid waste management	213	13%		
Wastewater treatment	154	10%		
Other	3	0.2%		
Total	1618	100%		

Note: totals may not sum because of rounding.
Source: Scheehle and Kruger (2006); IPCC (2001).

Table 22.3 Global nitrous oxide emissions for 2000 (MtCe).

Anthropogenic nitrous oxide			Natural nitrous oxide	
Energy				
Fossil fuel combustion	61	6%	Ocean	132–757
Biofuel combustion	10	1%		
Industry			Atmosphere	40–159
Adipic and nitric acid production	43	5%	(NH_3 oxidation)	
Agriculture				
Agricultural soils	711	75%	All soils	438–1315
Livestock manure management	56	6%		
Other agricultural sources:	41	4%		
Agricultural residue burning				
Prescribed burning of savanna				
Waste				
Human sewage	22	2%		
Other non-agricultural	3	0.3%		
Solvent use				
Waste incineration				
Total	947	100%		

Note: totals may not sum because of rounding. Percentages are % of global anthropogenic N_2O emissions.
Source: Scheehle and Kruger (2006); IPCC (2001).

industries (e.g. semiconductor manufacture, and magnesium production and processing), and (2) the replacement of ozone-depleting substances (ODSs) with HFCs (and other chemicals) in a wide range of applications. Finally, the PFCs and SF_6 have very long atmospheric lifetimes, ranging from 3200 years for SF_6 to 50 000 years for CF_4. Thus, emissions of PFCs and SF_6 today will affect the climate for millennia.

22.3 Early work on non-CO_2 GHGs

Seminal work on identifying mitigation options for NCGGs, primarily methane, was done mainly in the United States (see Hogan, 1993a,b), the Netherlands (see de Jager and Blok, 1993), and the UK (see IEAGHG, 1999). This work has also been expanded to include nitrous oxide and some F-gases

Table 22.4 Sources and emissions of HFCs, PFCs, and SF_6.

Source category	Emitted chemical(s)	2000 emissions (MtCe)	Reason emitted
1. ODS substitutes			
Air conditioning and refrigeration	HFC-32, HFC-125, HFC-134a, HFC-143a, HFC-152a, HFC-236fa	31.5	Refrigerant emitted from system leaks and during system service and disposal.
Foams	HFC-134a, HFC-152a, HFC-245fa, HFC-365mfc	1.7	Blowing agent released during foam manufacture, use, and disposal.
Solvents	HFC-4310mee, various PFCs and PFPEs (perfluoropolyethers)	11.6	Solvent leaks from cleaning equipment and evaporates from open-air uses.
Metered-dose inhalers (MDIs)	HFC-134a, HFC-227ea	2.7	Propellant released to deliver medicine to lungs.
Aerosols (non-MDI)	HFC-134a, HFC-152a	3.4	Propellant released to deliver product.
Fire suppression	CF_4, HFC-23, HFC-227ea, HFC-236fa	0.9	Suppression agent emitted from system leaks and to extinguish fires.
2. Industrial			
HCFC-22 production	HFC-23	26.0	Byproduct resulting from incidental overfluorination of HCFC-22.
Aluminum production	CF_4, C_2F_6	17.7	Byproducts formed during transient process disturbances.
Semiconductor manufacture	CF_4, C_2F_6, C_3F_8, C_4F_8, SF_6, HFC-23, NF_3	10.3	Fluorine sources for etching circuits, cleaning chambers. Incompletely reacted.
Magnesium production and casting	SF_6	2.4	Cover gas to prevent oxidation of magnesium. Incompletely reacted; leaks from blending/distribution system.
Electrical equipment manufacture	SF_6	2.9	Insulating gas for equipment.
Electrical equipment use	SF_6	9.8	Insulating gas emitted from equipment leaks and during equipment service and disposal.

(see van Ham et al., 1994). For the most part, these bottom-up, engineering-economic analyses were focused on the cost and performance characteristics of mitigation options. However, information on precisely how the options would be applied to national and regional emissions baselines was not supplied. This made it difficult to integrate these early mitigation estimates into regional and international climate economic models. The data that were critically needed included differentiation of potential reductions and mitigation costs by sector, by region, and over time. The first step in developing these more detailed cost data was met by two independent studies. The first, published by the US Environmental Protection Agency, detailed costs and revenue of individual technologies for each of the sectors in the US from 1990 to 2020 (USEPA 1999; 2001a). The second study was conducted by the European Commission and evaluated the average cost and revenue of particular technologies for an "average firm" in the 15 countries of the European Union (EU-15) from 1990 to 2010 (EC, 2001).

From all of these engineering-economic analyses, climate-economic modelers were able to conduct their own first analysis of the effects of mitigating more than CO_2 from energy. This early work had to use the existing regional mitigation analyses, e.g. US, EU, or Dutch, as a proxy for mitigation costs in other regions, which could lead to an under- or overestimation of mitigation potentials depending on how similar a particular region's make-up of NCGG sources was to that of these regions. Notwithstanding this limitation, all the studies demonstrated that expanding mitigation options to include NCGGs (a multi-gas abatement strategy) reduces overall costs (see Brown et al., 1999; Hayhoe et al., 1999; Reilly et al., 1999; Tol, 1999; Burniaux, 2000; Manne and Richels, 2001).

22.4 Recent work on non-CO_2 GHGs

Over the past two years, a major international modeling effort has tackled the challenge of improving mitigation estimates of NCGGs and of incorporating them into state-of-the-art climate economic and integrated assessment models. This effort was organized by the Stanford Energy Modeling Forum (EMF-21) and grew to include over 25 international modeling teams. This section provides a summary of the NCGG mitigation analysis for the EMF-21 study by gas and sector. For more details on the NCGGs analyses and results of all the international modeling teams in EMF-21 see Weyant and de la Chesnaye (2006).

22.4.1 Methane from the energy and waste sectors

In the EMF-21 study, Delhotal *et al.* (2006) focused on the major sources of methane from energy and waste sectors, solid waste management, natural gas production, processing and transmission/distribution (T&D), and coal mining. The work furthered previous marginal abatement cost studies conducted in the United States (USEPA, 1999, 2000a) and the EU (EC, 2001) by developing new marginal abatement curves (MACs) for 21 countries and regions, with worldwide coverage. Generally, the mitigation options fall into several broad categories including capture and use as natural gas (landfills, coal mining), capture and conversion to electricity (landfills, coal mining), flaring (landfills, natural gas), and reduction of leakage (natural gas). The two data sets, representing over 100 mitigation options, were combined by calculating the average cost and revenue of particular methane abatement technology options. The total list of technologies used in this analysis reflects both US and EU technologies currently available on the market.

Costs include capital or one-time costs, operation and maintenance costs, and labor costs. Each of these costs is average, or in other words, they represent the average cost of the mitigation option to an average-sized firm. The labor costs are adjusted by the average manufacturing wage in each region. Capital costs are not adjusted in this analysis because of the difficulty of calculating domestic production costs of a technology compared to the importation of a US or EU constructed technology. Operations and maintenance (O&M) costs are also adjusted according to labor rates in the country or region.

Benefits or revenues from installing methane mitigation options are calculated based on the efficiency of the technology to capture and convert methane to a useful energy source. The efficiency is a percentage of total methane that a specific technology can capture and use. The percentage is applied to the country or regions total baseline to estimate an absolute reduction. The amount of electricity generated or the amount of natural gas equivalent produced is converted into a dollar amount using local electricity and natural gas prices.

Applying a standard net present value formula that equates country-specific costs and benefits of each technology, the break-even carbon price can be calculated over the lifetime of the technology for various discount and tax rates. The result is a stepwise marginal abatement cost curve where each point on the curve represents the full implementation of a given technology.

The entire set of sector-specific mitigation technologies or a subset of the technologies is applied to each country or region Some options are not applied to particular regions, on the basis of expert analysis of conditions in the region. The selective omission of options represents a "static" view of the region's socio-economic conditions. Limited use of region-specific data and the static view of the potential socio-economic conditions in the region make the cost estimates in this analysis conservative. The analysis does not account for technical change over time, which would reduce the cost of abatement, increase the efficiency of the abatement options, and encourage domestically produced mitigation options for sale on the market. The analysis also does not incorporate firm-level data, relying on average firm sizes and average costs.

The results for the solid waste, natural gas and coal sectors are presented as a series of tables by country or region, where results vary with discount rates, tax rates and energy prices. Discount rates range from the "social" discount rate of 4% to the "industry perspective" discount rate of 20%. Tax rates are either 0% or 40%, with 40% reflecting the corporate income tax rate. The discount and tax rate combinations available are 4%/0%; 5%/0%; 10%/40%; 15%/40%; and 20%/40%. Energy prices range from negative 50% to 200% of the 1999 energy price for that country or region. Table 22.5 provides an illustrative example of the data provided in the EMF-21 study.

Results are in partial equilibrium and do not reflect perfectly functioning markets. Inclusion of market equilibrium effects in the modeling would change the amount of potential methane because of changes in prices. Results also do not include transactions costs, information barriers, policy barriers or other hidden costs. As a result, the analysis shows "no-regret" costs. No-regret options are defined as options where the revenue or cost savings from an option is greater than the cost of the option over the lifetime of the technology used. The no-regret options in this analysis range from options which are practical, but knowledge within the sector is limited, to options where hidden costs actually make the option impractical. The presence and abundance of no-regret options illustrate where policy change may have the most impact on reducing methane.

22.4.2 Methane and nitrous oxide from other non-agricultural sectors

Other sources of methane and nitrous oxide were considered in the EMF-21 study. These included methane emissions from the oil sector and nitrous oxide emissions from adipic and nitric acid production. Mitigation in the industrial and municipal wastewater sector is not covered in the EMF-21 study.

Oil extraction is a minor source of methane emissions. Emissions, and emission reduction potential, are concentrated in the Middle East, Mexico, Canada, and the United States which have the strongest oil production industries. Methane released during oil extraction can be flared, turned into liquefied natural gas, or captured and converted to electricity for local use. However, the costs of mitigation are relatively high compared with other sectors, starting at $40/tCe. The EMF-21 analysis estimated that the mitigation potential was small with only 8 MtCe available for reduction worldwide.

Adipic and nitric acid production is a small source of nitrous oxide emissions. Few countries have significant emissions of nitrous oxide from adipic and nitric acid production, with the

Table 22.5 Marginal abatement for methane-emitting sectors in 2020 at $200/tCe (MtCe).

	Solid waste		Natural gas		Coal		Oil		Total	
	abslt	**prct**	**abslt**	**prct**	**abslt**	**prct**	**abslt**	**prct**	**abslt**	**prct**
Africa	13.3	53%	4.7	44%	2.2	86%	0.2	37%	20.4	52%
ANZ	3.3	37%	1.8	38%	6.5	83%	0.0	14%	11.6	54%
Brazil	2.8	53%	2.3	37%	0.0	0%	0.0	27%	5.2	43%
Canada	2.4	37%	2.1	30%	0.0	0%	1.2	27%	5.6	31%
China	28.2	53%	1.2	44%	63.9	84%	0.4	38%	93.7	71%
CIS	2.8	53%	11.6	36%	1.6	58%	0.0	33%	16.0	40%
Eastern Europe	5.9	53%	4.1	34%	5.4	73%	0.0	13%	15.5	50%
EU-15	7.5	37%	2.6	29%	2.1	41%	0.0	12%	12.2	35%
India	2.2	53%	10.1	43%	8.4	84%	0.0	18%	20.8	55%
Japan	0.3	47%	0.2	35%	0.3	98%	0.0	0%	0.8	52%
Latin America	9.8	53%	20.8	44%	0.0	0%	0.7	32%	31.3	44%
Mexico	2.9	53%	3.0	44%	0.8	86%	1.3	35%	8.0	47%
Middle East	9.0	53%	24.3	37%	0.0	86%	1.7	39%	35.1	40%
Non-EU Europe	1.1	86%	0.1	37%	0.0	50%	0.0	38%	1.2	79%
Russia	5.4	53%	28.3	35%	6.0	84%	0.1	31%	39.9	41%
South and SE Asia	9.3	53%	9.9	44%	9.8	84%	0.1	24%	29.1	56%
South Korea	0.5	37%	1.3	43%	0.0	0%	0.0	2%	1.9	38%
Turkey	1.4	37%	0.8	43%	0.3	24%	0.0	11%	2.5	37%
Ukraine	4.4	53%	2.7	38%	4.3	85%	0.0	16%	11.4	56%
United States	25.3	53%	12.2	31%	17.2	86%	1.0	18%	55.8	49%
Total	138.1	51%	144.2	38%	128.7	80%	6.8	29%	417.8	50%

Notes: data are calculated for 20% discount rate and 40% tax rate. abslt = absolute reductions. prct = percent reductions from baseline. ANZ = Australia + New Zealand.
Non-EU Europe includes Andorra, Cyprus, Iceland, Leichtenstein, Malta, Monaco, Norway, Switzerland.
EU-15 includes Austria, Belgium, Denmark, Finland, France, Germany, Greece, Ireland, Italy, Luxembourg, Netherlands, Portugal, Spain, Sweden, United Kingdom.
Eastern Europe includes Albania, Bosnia and Herzvegovina, Bulgaria, Croatia, Czech Republic, Estonia, Hungary, Latvia, Lithuania, Macedonia, Poland, Romania, Slovak Republic, Slovenia, Former Yugoslavia.
Source: Delhotal *et al.* (2006).

largest emissions coming from the EU, the United States, and China. The EMF-21 study considers several mitigation options for these sectors, compiled from a US report on nitrous oxide abatement (USEPA, 2001b) and an IEAGHG report (IEAGHG, 2000). Thermal destruction, which destroys the off-gases from the boiler, was the only mitigation option considered for adipic acid production. For nitric acid production, either high-temperature or low-temperature catalytic reduction methods, or a non-selective catalytic reduction method, can be used to mitigate nitrous oxide. These methods reduce nitrous oxide emissions by 89 to 98%. The analysis estimates that 48 MtCe are available for reduction worldwide, with approximately 30 MtCe coming from the nitric acid production sector. All reductions are implemented at $5/tCe.

22.4.3 Methane and nitrous oxide from agriculture

Global agriculture accounts for a significant share of total GHG emissions, and these emissions are projected to increase for the foreseeable future owing to population growth, increases in per capita caloric intake, and changing diet preferences (e.g. greater proportion of meat consumption). Of the world's total anthropogenic N_2O and methane emissions, agriculture is responsible for almost 90% and over 50%, respectively.

Agriculture presents unique challenges when assessing GHG mitigation at the aggregated national and international scales. First, there is a high degree of spatial and temporal heterogeneity in farming practices and resultant emissions (often from one farm to the next), which are rarely directly monitored. Second, regionally specific cost data (for example including any changes in labor requirements) to implement identifiable mitigation options are scarce. And third, estimating the expected level of adoption of mitigation options in response to financial incentives (e.g. carbon price) is difficult, as there are barriers, such as cultural preferences and lack of information that are often unknown and difficult to quantify. The agricultural mitigation analyses in EMF-21 (DeAngelo *et al.*, 2006), present results from an ongoing project to assess the mitigation potential of the largest agricultural GHG sources – N_2O emissions from cropland soils, and methane emissions from management and rice cultivation – with data and methods that are generally consistent across emission sources and regions. Changes in GHG emissions other than the primary GHG emitted from each category (e.g. N_2O emissions from rice cultivation) are not currently included in the analysis.

In addition, estimates of options that increase soil carbon, reduce fossil fuel carbon dioxide (CO_2) emissions from agricultural operations, and use bioenergy crops as fossil fuel offsets are currently not included (see IPCC [2000], chapter 4).

Application of nitrogen (N)-based fertilizers to croplands is a key determinant of N_2O emissions – the largest source of agricultural GHGs – as excess nitrogen not used by the plants is subject to gaseous emissions, as well as leaching and runoff. Therefore, a key N_2O mitigation strategy is to identify where and when N-based fertilizers can be applied more efficiently (i.e. not in excess of crop needs) without sacrificing yields. Total global mitigation for soil N_2O emissions is estimated to be almost 16 MtCe at negative or zero cost, over 42 MtCe at $200/tCe (Figure 22.3), and 63 MtCe at much higher costs. The maximum global N_2O mitigation represents a reduction in 2010 baseline emissions of roughly 8%. Negative costs result from options where there are net cost savings due to lower applications of fertilizers while maintaining yields, whereas high-cost options are those where revenues decline as yields decline in response to sub optimal fertilizer applications. A greater portion of mitigation is estimated at the higher cost range owing to the assumption that most mitigation options result in crop yield declines; however, it seems feasible that equivalent levels of mitigation could be estimated at much lower costs, since a different set of mitigation options, or ones tailored more to local conditions, could reduce N_2O with less adverse effects on yields.

Enteric methane emissions (Figure 22.4) occur through microbial fermentation in the digestive system of livestock. Mitigation options include improvements in food conversion efficiency, increased animal productivity through the use of natural or synthetic compounds, feed supplements to combat nutrient deficiencies, and changes in herd management. Total global mitigation for enteric CH_4 is estimated to be 29 MtCe at negative or zero cost, almost 31 MtCe at $200/tC, and 34 MtCe at even higher costs. The maximum mitigation here represents a reduction in the 2010 global baseline of 6%. This is a conservative estimate because many of the mitigation options are not considered feasible, or are assumed to have limited penetration – at least by the 2010 time-frame – in regions where livestock

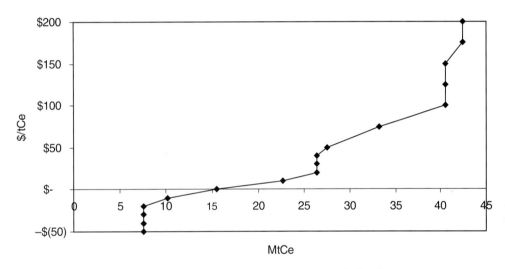

Figure 22.3 Global marginal abatement cost curve for N_2O emissions from cropland soils in 2010.

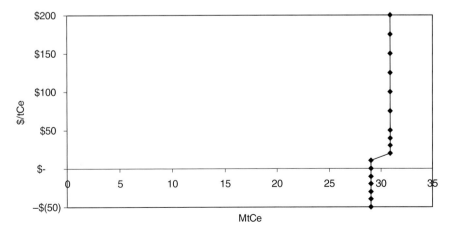

Figure 22.4 Global marginal abatement cost curve for CH_4 emissions from enteric fermentation in 2010.

herds are large and there is significant room for improved efficiency (e.g. Africa, Brazil, China, and India). For nearly all enteric CH_4 options, costs are estimated to be negative because of increased efficiency for an assumed constant level of regional milk or beef production (i.e. methane reductions are assumed to occur through aggregate reductions in livestock populations made possible through efficiency gains on per animal).

Rice emissions of methane occur through the anaerobic decomposition of organic matter in flooded rice fields. All the mitigation options here involve a change in water management regime (to reduce anaerobic conditions), fertilizers or amendments (to inhibit methanogenesis), or rice variety. Total global mitigation for rice methane is estimated to be 19 MtCe at negative or zero cost, almost 56 MtCe at $200/tCe (Figure 22.5), and 61 MtCe at higher costs. The maximum mitigation estimate is a reduction in the global 2010 baseline of 33% (non-Asian regions were not included in the DeAngelo et al. analysis, 2006). These estimates can be considered high-end values because, unlike the enteric methane category, there were no quantitative data available to quantify adoption feasibility, and therefore the estimates represent full-scale adoption.

Manure methane emissions occur through the anaerobic decomposition of the manure. Anaerobic conditions are created when livestock or poultry manure is treated and stored as a liquid in lagoons, ponds, tanks or pits. Anaerobic digesters that cover and capture the methane emitted from collected manure, and then flare it (converting it to the less-effective global warming gas, CO_2) or use it as an on-farm energy source, represent a major mitigation option. Total global mitigation for manure methane is estimated to be roughly 0.4 MtCe at negative or zero cost, almost 10 MtCe at $200/tCe (Figure 22.6), and over 11 MtCe at even higher prices. This maximum estimated mitigation represents a reduction of 17% of the global 2010 baseline.

Work is continuing in the agricultural sector to improve (1) representation of the *net* GHG effects of the mitigation

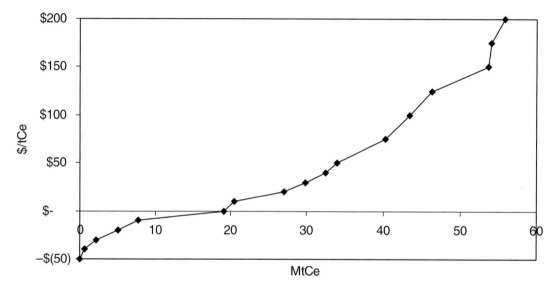

Figure 22.5 Global marginal abatement cost curve for CH_4 emissions from rice cultivation in 2010.

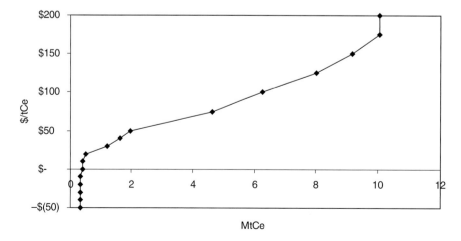

Figure 22.6 Global marginal abatement cost curve for CH_4 emissions from manure management in 2010.

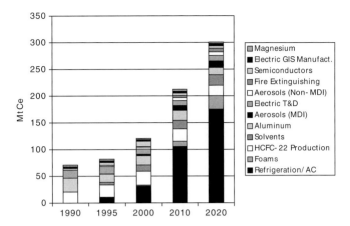

Figure 22.7 Fluorinated gas emissions by sector, 1990–2020. T&D = transmission and distribution; AC = air condition; GIS = manufacturing of gas-insulated switchgear.

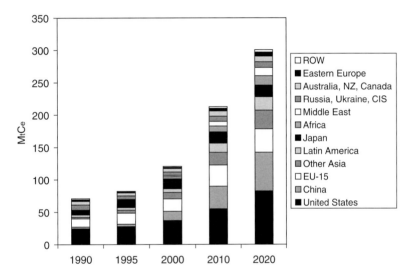

Figure 22.8 Fluorinated gas emissions by region, 1990–2020.

options; (2) regional characterization of baseline practices; (3) regional adoption feasibility of the identified mitigation options; and (4) potential market effects due to adoption of the mitigation practices.

22.4.4 Fluorinated gases

Because few countries have published estimates of their present and future HFC, PFC, and SF_6 emissions on an industry-specific basis, the emission baselines for the EMF-21 analysis, Schaefer et al. (2006), were based on previous analyses performed by EPA and Ecofys (USEPA, 2001c; Harnisch et al., 2001, 2002). Schaefer et al. built on both of these previous analyses to develop country-by-country and industry-by-industry projections of emissions using projections of activity data, emission factors, or other data related to emissions (Schaefer et al., 2006). To reflect the global emission reduction commitments that have been made by some industries that emit fluorinated gases (which have already resulted in decreased emission rates), Schaefer et al. assumed that emission rates will fall in several industries. At the same time, the authors adjusted the MACs to account for the adoption of some reduction options in some regions in the baseline. As a consequence of these assumptions, emissions from the industrial sectors are projected to decline by 20% between 2000 and 2020. However, this is not the case for the sectors where HFCs and PFCs are substituting for ozone-depleting substances (ODSs), whose emissions are projected to increase by nearly a factor of five over the same period.

Figure 22.7 displays sector-by-sector emissions of fluorinated gases. Growing from essentially zero emissions in 1990, air conditioning and refrigeration accounts for 58% of total emissions by 2020, seven times as much as the second largest source, foams. HCFC-22 production is the third largest source in 2020 and the second largest source in 2010.

Figure 22.8 shows emissions by region. In general, developing countries make up an increasing share of total emissions over time, reflecting two anticipated trends: (1) robust

Table 22.6 Fluorinated gas emission reduction technologies yielding largest low-cost (< $50/tCe) reductions (2010).

Technology	Gas(es)	Cost ($/tCe)	Reduction (million tCe)	Share of baseline
Thermal oxidation (HCFC-22 production)	HFC-23	0.80	22	10%
Refrigerant recovery	Short-lived HFCs	5.40	14	7%
Distributed refrigeration (commercial ref.)	Short-lived HFCs	(21)	5.5	3%
Major retrofit (aluminum)	PFCs	(14)	4.8	2%
A/C&R leak repair	Short-lived HFCs	4.40	3.6	2%

Numbers in parentheses are negative costs.

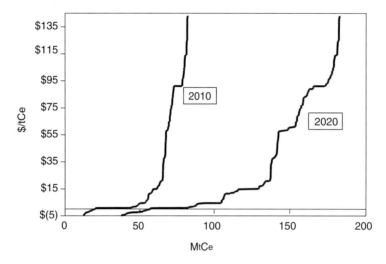

Figure 22.9 Global MACs for fluorinated gases, 2010 and 2020.

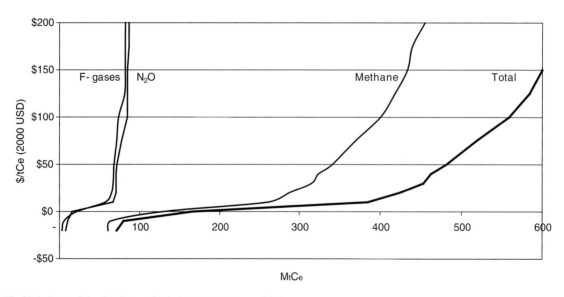

Figure 22.10 Global non-CO_2 GHG marginal abatement curves, 2010.

economic growth in developing countries, and (2) greater implementation of emission reduction measures in developed countries.

Potential reductions are strongly related to projected emissions. In 2010, refrigeration and air conditioning yields the largest reductions (30 MtCe) of any sector. HCFC-22 production follows with 22 MtCe of reductions. Table 22.6 shows the largest low-cost (< $50/tCe) reductions from a range of technologies in 2010.

Figure 22.9 displays the 2010 and 2020 global MAC curves for all sources of fluorinated gases, using a 4% discount rate and 0% tax rate. The 2020 MAC is larger than the 2010 MAC for two reasons. First, emissions are larger in 2020 than in 2010 owing to the continuing growth of ODS substitutes. Second, the percentage of emissions abated is higher in 2020 than in 2010 because many of the abatement options for air conditioning and refrigeration take time to penetrate the stock of equipment.

22.4.5 Summary of NCGG mitigation results

Figure 22.10 provides global reduction potentials by gas or gas group, and shows that the overall NCGG mitigation potential for 2010 is about 500 MtCe at $100/tCe with methane the largest contributor.

22.5 New directions in NCGG mitigation analysis

22.5.1 Improved cost estimates for the energy and waste sectors

New estimates of international MACs were developed with the objective of improving upon the limitations of the EMF-21 analysis (Gallaher and Delhotal, 2004; Delhotal and Gallaher, 2005). The newer analysis (1) incorporates a statistical distribution of firm sizes in each country or region allowing for firm-level cost analysis, (2) estimates the transformation from importing US or EU produced technologies toward domestic production of technologies, (3) allows for country-specific changes in labor productivity over time, and (4) accounts for the reduction of costs and increases in benefits over time by technology. The results presented are a set of MACs by 10-year interval, by country, and by sector. Only the top methane-emitting countries were included in the analysis because of the amount of regional data required for the firm-level analysis, representing between 48 and 75% of total future emissions by sector.

MACs for each country and sector are estimated for 10-year intervals from 2010 to 2030. Figures 22.11 and 22.12 show the MACs for 2010, 2020, and 2030 for the US coal mining and the Mexico landfill sectors, respectively. The magnitude of the shifts reflects both changes in costs and benefits of abatement technologies and technical changes, such as increased reduction efficiency and trends in production. For example, after 2020, the shift in the US MACs slows because underground mine production is projected to decrease slightly. However, the main driver for technology shift is the predicted rate of technology improvement, which continues to drive down costs.

In contrast, the MACs for landfills in Mexico shift out and downward steadily, reflecting the decreasing technology costs and the projected growth in landfills. It should be noted that these future reductions could be overstated depending on landfills becoming subject to non-climate regulations. For example, if Mexico were to enact a rule similar to the US or EU landfill rules that control volatile organic compounds (VOCs) through flaring, baseline emissions would be lower and thus there would be lower potential for estimated reductions in the MAC. The shift outward of the curve is also illustrative of the assumptions in the model of technology transfer. The model assumes that over time more and more of the capital and all of the labor will be domestic instead of imported from the United States. It should also be noted that the results reflect a financial analysis and do not capture all the factors influencing the adoption of mitigation technologies. If unobservable transition costs or institutional or informational barriers to adoption were included, we would not expect the "no-regrets" options to increase so dramatically over time.

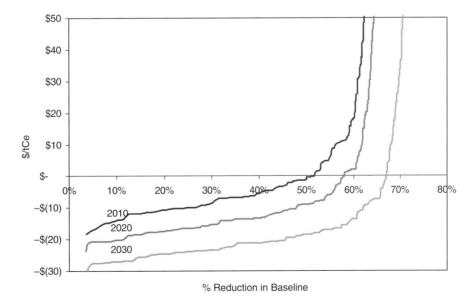

Figure 22.11 Shift in the United States' MAC for the coal mining sector over 30 years.

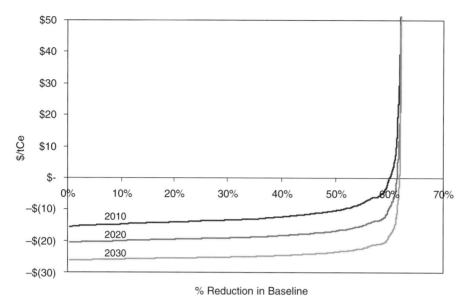

Figure 22.12 Shift in Mexico's MAC for landfill sector over 30 years.

Countries other than the United States have greater decreases in costs because this analysis assumes that they are currently importing most inputs, but will increase the use of significantly lower-cost domestic capital, labor, and materials over time. This results in greater downward shifts in these countries' curves over time relative to the United States. The changes in costs are also a function of each country's relative prices. For example, the percentage change in annual costs is not as great in Russia and Poland as in China, because China has lower wages than these countries. Therefore, China experiences a greater decrease in annual costs when switching to domestic labor rather than importing inputs.

22.5.2 Comparison with EMF-21 MACs

The effect of the above approach can be illustrated by comparing the curves in the later paper (referred to as "enhanced" curves) with the MACs produced for the EMF-21 multi-gas study as described above. Table 22.7 shows the percentage methane abatement in 2020 at different carbon prices for both the EMF-21 and enhanced projections (Delhotal and Gallaher, 2005).

Generally, the enhanced approach leads to more abatement at lower break-even prices owing the use of entity-level data and reduced costs (or increased efficiency) of technologies over time. Using the US coal mining sector as an example, the EMF-21 estimates project no abatement at a negative $10/tCe whereas the enhanced approach suggests 48% abatement. However, at higher break-even prices, the EMF-21 estimates project a larger abatement percentage, 86% at $50/tCe, compared with 66% at $50/tCe for the enhanced approach. This is because the enhanced approach incorporates different assumptions about technical applicability or potential market penetration. In this example, the enhanced MACs use mine-level data to determine if ventilation air methane (VAM) reduction technology is technically applicable for each mine in 2020, instead of assuming all mines are eligible.

Similar trends are seen in the natural gas and solid waste sectors. "No regrets" abatement, or abatement where revenues from energy recovery are greater than the costs of mitigation, is approximately 15% for natural gas and approximately 60% for landfills using the enhanced approach in 2020. At higher break-even prices in the natural gas and solid waste sectors, differences in assumptions regarding technical potential lead to differences in the maximum abatement potential. As with the coal sector, the enhanced approach yields less abatement potential in the natural gas sector at higher break-even prices compared with the EMF-21 estimates. Again, the difference is due to more conservative assumptions being used for technical applicability of natural gas mitigation options in the enhanced approach. In contrast, the enhanced approach for the solid waste sector yields a higher maximum potential, reflecting increases in reduction efficiency by 2020.

Two important issues need further consideration when considering long-term methane mitigation potential. The first is estimating the adoption of mitigation options over time. The above analysis does not incorporate estimates of actual implementation of mitigation options by sector by country over time; it only provides an estimate of what is available for adoption. Inclusion of adoption estimates would reduce the emission projection baseline, thus reducing the implied emission factor over time. Historically, no-regrets options implemented in the baseline significantly reduce the baseline estimates. Because baselines do not currently include future adoption of any options including no-regret options, the baselines may be overestimated.

Table 22.7 Percentage reduction from the baseline of methane emissions at a given carbon price: comparison with EMF-21 MAC curves.

Sector	Break-even prices ($/tCe)							
	−$20	−$10	$0	$10	$20	$30	$40	$50
Coal								
United States								
EMF-21	0.0%	0.0%	49.2%	49.2%	66.5%	86.0%	86.0%	86.0%
Enhanced	11.7%	47.8%	58.0%	61.9%	62.6%	63.5%	63.9%	64.4%
China								
EMF-21	0.0%	0.0%	0.0%	0.8%	49.7%	84.5%	84.5%	84.5%
Enhanced	0.0%	35.8%	72.3%	81.8%	85.5%	86.7%	88.9%	89.8%
Natural gas								
United States								
EMF-21	1.9%	5.6%	14.5%	14.5%	18.9%	18.9%	19.2%	19.2%
Enhanced	8.4%	9.3%	9.4%	10.6%	11.9%	12.1%	12.3%	12.4%
Russia								
EMF-21	0.0%	0.0%	3.8%	9.2%	25.0%	26.6%	26.6%	26.9%
Enhanced	0.0%	9.6%	14.4%	14.6%	14.7%	14.7%	14.7%	14.7%
Landfill								
United States								
EMF-21	0.0%	0.0%	10.0%	10.0%	31.4%	31.4%	42.1%	42.1%
Enhanced	28.2%	54.1%	61.7%	65.1%	67.1%	67.5%	68.3%	68.6%
Mexico								
EMF-21	0.0%	0.0%	10.0%	20.7%	31.4%	42.1%	42.1%	42.1%
Enhanced	13.6%	54.1%	55.9%	56.4%	56.5%	56.6%	56.6%	56.7%

Source: Delhotal and Gallaher (2005).

The second issue is the short-term nature of the study. The technical change estimates presented in the above study only run to 2030, limiting the usefulness of the study to climate-economic and integrated assessment modelers. Future work will focus on how to translate technical change, including the transformation of the basic infrastructure in these sectors, into long-term estimates of technical change. Long-term estimates can take on two forms. The first form may be a sector-specific index allowing for the exogenous representation of technical change in economic models. The second would be the endogenous representation of these sectors and technical change in long-term climate-economic and integrated assessment models.

22.5.3   Improvements to fluorinated gas analyses

The EMF-21 study illuminated some important uncertainties and sensitivities in projected emissions and reductions of fluorinated gases. The most important of these include (1) emissions from ODS substitutes in developing countries, (2) the growth rates of both emitting activities and emission factors, particularly more than 10 years into the future, (3) the costs and reductions associated with technologies that have not yet been widely implemented (such as recovery or destruction of the HFC blowing agent during the disposal of rigid foams),

and (4) regional technical and cost data, particularly for developing countries. Future analyses can address many of these uncertainties and sensitivities, both by further researching poorly understood areas and by generating multiple scenarios to understand the range of possible futures. Among the most important questions to address are those involving the emissions and abatement costs of developing countries, because these are expected to account for a large fraction of global emissions and abatement opportunities by 2020.

22.6   Conclusion

Inclusion of NCGGs in climate-economic modeling is critical to understanding the full range of policy options for reducing climate change. Until recently, little work had been done on including these gases in the overall analysis of climate policies. Advances in analyses on specific gases and sectors have enabled modeling teams to include marginal cost estimates of these gases exogenously. Results have shown that including NCGGs in climate analysis can reduce costs. However, including these gases has proved to be complicated because of the diverse nature of the gases and the processes emitting these gases. Further work is needed to endogenize these gases and their mitigation technologies fully, particularly in long-run models.

Disclaimer

The views of the authors do not necessarily represent the views of the US Government or the Environmental Protection Agency.

References

Brown, S., Kennedy, D., Polidano, C. et al. (1999). *Assessing the Economic Impacts of the Kyoto Protocol: Implications of Accounting for the Three Major Greenhouse Gases*. ABARE Research Report 99.6, Canberra, Australia.

Burniaux, J. M. (2000). *A Multi-Gas Assessment of the Kyoto Protocol*. Working Paper # 270. Paris: OECD Economics Department, www.oecd.org/pdf/M00002000/M00002105.pdf.

CDIAC (2004). Carbon Dioxide Information Analysis Center, http://cdiac.esd.ornl.gov/home.html.

de Jager, D. and Blok, K. (1993). *Cost-Effectiveness of Emission Reducing Measures for Methane in the Netherlands*. Utrecht: Ecofys Research and Consultancy.

DeAngelo, B., de la Chesnaye, F., Beach, R., Sommer, A. and Murray, B. (2006). Methane and nitrous oxide mitigation in agriculture. *Energy Journal: Special Issue on Multigas Mitigation and Climate Policy*.

Delhotal, K. C. and Gallaher, M. (2005). Estimating technical change and potential diffusion of methane abatement technologies for the coal-mining, natural gas, and landfill sectors. In *IPCC Expert Meeting on Industrial Technology Development, Transfer and Diffusion* (peer reviewed conference proceedings). http://arch.rivm.nl/env/int/ipcc/pages_media/itdt.html.

Delhotal, K. C., de la Chesnaye, F. C., Gardiner, A., Bates, J. and Sankovski, A. (2006). Mitigation of methane and nitrous oxide emissions from waste, energy and industry. *Energy Journal* (in press).

EC (European Commission). (2001). *Economic Evaluation of Sectoral Emission Reduction Objectives for Climate Change*. Brussels. http://europa.eu.int/comm/environment/enveco/climate_change/sectoral_objectives.htm

Gallaher, M. and Delhotal, K. C. (2004). Modeling the impact of technical change on emissions abatement investments in developing countries. *The Journal of Technical Change* **30**, 211–225.

Hansen, J., Sato, M. and Ruedy, R. (1997). *Journal of Geophysical Research* **102**, 6831–6864.

Hansen, J., Sato, M., Ruedy, R., Lacis, A. and Oinas, V. (2000). Global warming in the twenty-first century: an alternative scenario. *Proceedings of the National Academy of Sciences* **97**, 9875–9880.

Harnisch, J., Stobbe, O., de Jager, D. (2001). *Abatement of Emissions of Other Greenhouse Gases: Engineered Chemicals*. Report Number PH3/35, undertaken by Ecofys (Utrecht) for the International Energy Agency Greenhouse Gas R&D Programme (IEA GHG). Cheltenham.

Harnisch, J., Stobbe, O., Gale, J. and de Jager, D. (2002). Halogenated compounds and climate change: future emission levels and reduction costs. *Environmental Science Policy Research* **9** (6), 369–375.

Hayhoe, K., Jain, A., Pitcher, H. et al. (1999). Costs of multigreenhouse gas reduction targets in the USA. *Science* **286**, 905–906.

Hogan, K. B. (ed.) (1993a). *Options for Reducing Methane Emissions Internationally.* Vol I: *Technological Options for Reducing Methane Emissions*. EPA 430-R-93–006. Washington, DC: US Environmental Protection Agency.

Hogan, K. B. (ed.). (1993b). *Options for Reducing Methane Emissions Internationally.* Vol II: *International Opportunities for Reducing Methane Emissions*. EPA 430-R-93–006 B. Washington, DC: US Environmental Protection Agency.

IEAGHG (1999). *Technologies for the Abatement of Methane Emissions*, Vol 2. IEA Greenhouse Gas R&D Programme, UK.

IEAGHG (2000). *Abatement of Emissions of Other Greenhouse Gases: Nitrous Oxide*. IEA Greenhouse Gas R&D Programme, UK.

IPCC (1997). *Revised 1996 IPCC Guidelines for National Greenhouse Gas Inventories*, ed. J. T. Houghton, L. G. Meira Filho, B. Lim et al. IPCC/OECD/IEA. Bracknell: UK Meteorological Office.

IPCC (2000). *Land Use, Land-Use Change, and Forestry 2000: Special Report of the Intergovernmental Panel on Climate Change*, ed. R. T. Watson, I. R. Nobel, B. Bolin et al. Cambridge: Cambridge University Press.

IPCC (2001). *Climate Change 2001: The Scientific Basis. Contribution of Working Group I to the Third Assessment Report of the Intergovernmental Panel on Climate Change*, ed. J. T. Houghton, Y. Ding, D. J. Griggs et al. Cambridge: Cambridge University Press.

Manne, A. and Richels, R. (2001). An alternative approach to establishing trade-offs among greenhouse gases. *Nature* **410**, 675–677.

Reilly, J., Prinn, R., Harnisch, J. et al. (1999). Multi-gas assessment of the Kyoto Protocol. *Nature* **401**, 549–555.

Schaefer, D., Godwin, D. and Harnish, J. (2006). Estimating future emissions and potential reductions of HFCs, PFCs, and SF_6. *Energy Journal: Special Issue on Multigas Mitigation and Climate Policy*.

Scheehle, E. and Kruger, D. (2006). Methane and nitrous oxide baselines and projections. *Energy Journal: Special Issue on Multigas Mitigation and Climate Policy*.

Tol, R. S. J. (1999). Kyoto, efficiency, and cost-effectiveness: applications of FUND. *Energy Journal: Special Issue on the Costs of the Kyoto Protocol: A Multi-Model Evaluation*, 130–156.

USEPA (1999). *US Methane Emissions 1990–2020: Inventories, Projections, and Opportunities for Reductions*. Washington DC: US Environmental Protection Agency, www.epa.gov/methane.

USEPA (2001a). *Addendum Update to U.S. Methane Emissions 1990–2020: Inventories, Projections, and Opportunities for Reductions*. Washington DC: US Environmental Protection Agency, www.epa.gov/methane.

USEPA (2001b). *U.S. Adipic Acid and Nitric Acid Nitrous Oxide Emissions 1990–2020: Inventories, Projections and Opportunities for Reductions*. Washington DC: US Environmental Protection Agency, www.epa.gov/methane.

USEPA (2001c). *Non-CO₂ Greenhouse Gas Emissions from Developed Countries: 1990–2010*. EPA-430-R-01–007. Washington DC: Office of Air and Radiation, US Environmental Protection Agency.

van Ham, J., Janssen, L. H. J. M., and Swart, R. J. (eds.) (1994). *Non-CO₂ Greenhouse Gases: Why and How to Control?* Proceedings of an International Symposium, Maastricht, The Netherlands, 13–15 December 1993. Dordrecht: Kluwer Academic Publishers.

Weyant, J. and de la Chesnaye, F. (eds.) (2006). Multigas mitigation and climate change. *Energy Journal: Special Issue on Multigas Mitigation and Climate Policy.*

23

How (and why) do climate policy costs differ among countries?

Sergey Paltsev, John M. Reilly, Henry D. Jacoby and Kok Hou Tay

23.1 Introduction

There have been many studies of the cost to Annex B countries of meeting Kyoto Protocol commitments. Unfortunately for these analyses, the Protocol has proved to be a moving target in terms of its interpretation and likely implementation. In addition, the economic performance and future expectations for some parties also are changing, and with these changes come revisions in reference emissions, which have a strong influence on the projected cost of meeting Protocol requirements. Looking back across these studies, the progression of work can be divided into three broad phases. The first studies were conducted soon after the Protocol was signed in 1997, and they focused on carbon emissions from fossil fuels. Often they assumed an idealized system of harmonized carbon taxes, or cap-and-trade among all the Annex B parties, contrasting such systems with implementation without international permit trade but with an idealized trading system operating within each country (see, for example, Weyant and Hill [1999]). These studies showed a high cost of the Protocol with autarkic compliance, but huge benefits of international trading because it made Russian "hot air" (potentially tradable emission quotas in excess of their anticipated emissions) accessible to other Annex B parties.

A second phase of studies followed the final negotiations in Marrakech in 2001 (Manne and Richels, 2001; Babiker et al., 2002; Bohringer, 2002). By that time, the United States had withdrawn from the Protocol, and the potential contribution of Article 3.4 carbon sinks had been defined for each party. Progress in economic modeling also made it possible to consider the economic cost and contribution of non-CO_2 greenhouse gases (GHGs). These changes – US withdrawal, added sinks, and the consideration of non-CO_2 GHGs – led to downward revisions in the cost of meeting the Protocol, particularly if idealized trading among the remaining Annex B parties was assumed (Babiker et al., 2002). In fact, many analyses concluded that credits from Russian hot air, and to lesser extent from Eastern European Associates of the EU, could be sufficient to meet Kyoto targets without additional effort.

We are now in a third phase of this work, where studies seek to estimate the cost under a more realistic representation of how the target reductions might actually be achieved. These studies seek to understand the cost of a mixed set of policies that almost certainly will differ among Kyoto parties and across economic sectors, and that may thus be more costly than the idealized systems studied earlier. Under some conditions costs may be less, if the policy is structured to avoid exacerbating pre-existing distortions. Some elements of such "imperfect" implementation have been considered in past studies. Examples include the potential for the exercise of monopoly or monopsony power in the permit market (e.g., Ellerman and Sue Wing, 2000; Bernard et al., 2003), the impact of pre-existing tax distortions on labor and capital (e.g., Shackleton et al., 1993; Goulder, 1995; Fullerton and Metcalf, 2001) or on energy (Babiker et al., 2000a), and policies focused on particular technologies or sectors (e.g., Babiker et al., 2000b, 2003a). In this analysis, we extend and interpret this work.

To study these latter aspects of cost realistically requires attention to the exact set of policies and measures likely to be implemented, and representation of ways they will interact with existing tax policies. Because circumstances can differ among countries and across sectors, a result that holds in one country may not hold in another. For example, much analysis

Human-induced Climate Change: An Interdisciplinary Assessment, ed. Michael Schlesinger, Haroon Kheshgi, Joel Smith, Francisco de la Chesnaye, John M. Reilly, Tom Wilson and Charles Kolstad. Published by Cambridge University Press. © Cambridge University Press 2007.

of the double-dividend (from the use of carbon revenue to reduce distorting capital and labor taxes) has been conducted in the United States, and these studies have found evidence of at least a weak double-dividend. But recent work has shown a much smaller double-dividend effect in other Annex B countries (Babiker et al., 2003b). This result occurs because reducing labor taxes within a particular fiscal structure may further distort relative prices by widening the divergence between labor rates and energy prices (tax inclusive). Where this happens, the distorting effect may offset other potential gains of revenue recycling.

Furthermore, the presence of varying distortions and fragmented policy implementation across countries raises questions about the degree to which a party can take advantage of the various international flexibility mechanisms, and if they do whether such trading is in fact welfare enhancing. Without comparable trading systems that can be linked, and without the willingness of parties to permit unrestricted cross-border trading, international flexibility may exist only on paper. Recent studies emphasize that extreme care is needed in the design of policies to make them economically efficient, and they highlight the difficulty of realistically achieving the low costs found under idealized cap-and-trade systems. Thus, the Protocol could end up being relatively costly for some parties but, without the United States involved, it will not achieve the environmental benefit imagined when it was signed in Kyoto. While international emissions trading has been shown to be beneficial for all parties *when trading includes the hot air from Russia and Eastern European Associate nations*, this result has not held up in more limited trading scenarios. In these cases trading may not be beneficial to all parties, even if they enter into the system voluntarily – a result that can be traced to the interaction of the carbon policy with existing energy taxes (Babiker et al., 2004).

As the set of issues addressed by these economic modeling studies has become richer and more complex, the definition of economic cost itself and how it is estimated in economic models has presented a puzzle, particularly to those outside the economic modeling community. Depending on the study, cost results may be reported in various ways, e.g. in terms of the carbon or carbon-equivalent price of GHGs, as the integrated area under an abatement curve, or applying broader measures of economy-wide welfare such as the reduction in consumption or GDP. Adding to the confusion thus created is the fact that the costs of a policy as estimated by different concepts, even using a single model, often appear inconsistent. For example, early studies examining Kyoto costs without international emissions trading showed Japan to have the highest carbon price, followed by the United States and other OECD regions and the EU. But among this group the cost in percentage loss of welfare was lower in Japan than in the EU. The apparent inconsistency of these results, where the ranking of cost by one measure is reversed when another is used, emphasizes the importance of distinguishing among these cost concepts.

In this paper we explore these concepts and the differences among them. We consider sector-specific policies, using the example of Japan to explore the implication of alternative domestic implementation strategies. The results reported here are in the domain of the third phase of studies outlined above, where more complex and more realistic policies (not necessarily designed for economic efficiency) are the focus of analysis. To be sure, even after Kyoto's entry into force much remains in doubt, with the domestic policy details of most parties still not specified, so possible costs to Japan or any other party remain speculative. Nevertheless, this work can pave the way toward a likely fourth phase of studies where it will be necessary to investigate the costs and effectiveness of policies actually implemented – whether intended to achieve the Kyoto targets or other goals.

We begin the analysis with an overview of the EPPA model used to perform the simulations. In Section 23.3 we lay out a set of policies designed to study possible implementation of the Kyoto Protocol in Japan and elsewhere, and compare these costs across countries and regions using two of the more common measures of cost. This Kyoto example yields some seemingly paradoxical results about the variation in cost among countries, the difference in relative burdens depending on the measures used and the effects of emissions trading on welfare. Therefore in Section 23.4 we construct a diagnostic case where precisely the same constraint is imposed across countries and use this example to resolve some of these puzzles about the factors that contribute to welfare cost in any country. Finally, Section 23.5 reflects back on the Kyoto assessment and draws some lessons for future analyses of issues of this type.

23.2 The EPPA model

The EPPA model is a recursive-dynamic multi-regional general equilibrium model of the world economy, which is built on the GTAP data set (Hertel, 1997) and additional data for greenhouse gas (CO_2, CH_4, N_2O, HFCs, PFCs and SF_6) and urban gas emissions (Mayer et al., 2000). The version of EPPA used here (EPPA 4) has been updated in a number of ways from the model described in Babiker et al. (2001). The updates are presented in Paltsev et al. (2005). The various versions of the model have been used in a wide variety of policy applications (e.g., Jacoby et al., 1997; Jacoby and Sue Wing, 1999; Reilly et al., 1999; Paltsev et al., 2003). Compared with the previous version, EPPA 4 includes (1) greater regional and sectoral disaggregation, (2) the addition of new advanced technology options, (3) updating of the base data to the GTAP 5 data set (Dimaranan and McDougall, 2002) including newly updated input–output tables for Japan, the United States, and the EU countries, and rebasing of the data to 1997, and (4) a general revision of projected economic growth and inventories of non-CO_2 greenhouse gases and urban pollutants.

EPPA 4 aggregates the GTAP data set into 16 regions and 10 sectors shown in Table 23.1. The base year for the EPPA 4 model is 1997. From 2000 onward, it is solved recursively at 5-year intervals. To focus better on climate policy the model is

Table 23.1 Regions and sectors in the EPPA model.

Country/Region	Sectors
Annex B	**Non-energy**
United States (USA)	Agriculture(AGRI)
Canada (CAN)	Services (SERV)
Japan (JPN)	Energy intensive products(EINT)
European Union + [a] (EUR)	Other industries products (OTHR)
Australia/New Zealand (ANZ)	Industrial transportation (TRAN)
Former Soviet Union (FSU)	Household transportation (HTRN)
Eastern Europe[b] (EET)	**Energy**
Non-Annex B	Coal (COAL)
India (IND)	Crude oil (OIL)
China (CHN)	Refined oil (ROIL)
Indonesia (IDZ)	Natural gas (GAS)
Higher income East Asia[c] (ASI)	Electric: fossil (ELEC)
Mexico (MEX)	Electric: hydro (HYDR)
Central and South America (LAM)	Electric: nuclear (NUCL)
Middle East (MES)	**Advanced Energy Technologies**
Africa (AFR)	Electric: biomass (BELE)
Rest of world[d] (ROW)	Electric: natural gas combined cycle (NGCC)
	Electric: NGCC with sequestration (NGCAP)
	Electric: integrated coal gasification with combined cycle and sequestration (IGCAP)
	Electric: solar and wind (SOLW)
	Liquid fuel from biomass (BOIL)
	Oil from shale (SYNO)
	Synthetic gas from coal (SYNG)

[a] The European union (EU-15) plus countries of the European Free Trade Area (Norway, Switzerland, Iceland).
[b] Hungary, Poland, Bulgaria, Czech Republic, Romania, Slovakia, Slovenia.
[c] South Korea, Malaysia, Philippines, Singapore, Taiwan, Thailand.
[d] All countries not included elsewhere: Turkey, and mostly Asian countries.

disaggregated beyond that provided in the GTAP data set for energy supply technologies and for transportation, and a number of supply technologies are included that were not in use in 1997 but could take market share in the future under some energy price or climate policy conditions. All production sectors and final consumption are modeled using nested constant elasticity of substitution (CES) production functions (or Cobb–Douglas and Leontief forms, which are special cases of the CES). The model is written in the GAMS software system and solved using the MPSGE modeling language.

The regional disaggregation of EPPA 4 includes a breakout of Canada from Australia/New Zealand, and a breakout of Mexico to focus better on North America. Regional groupings of developing countries were altered to create groups that were geographically contiguous. New sectoral disaggregation includes a breakout of services (SERV) and transportation (TRAN) sectors. These were previously aggregated with other industries (OTHR). This further disaggregation allows a more careful study of the potential growth of these sectors over time and the implications for an economy's energy intensity. In addition, the sub-model of final consumption was restructured to include a household transportation sector. This activity provides transportation services for the household, either by purchasing them from TRAN or by producing them with purchases of vehicles from OTHR, fuel from ROIL, and insurance, repairs, financing, parking, and other inputs from SERV. While the necessary data disaggregation for TRAN is included in GTAP 5 (Dimaranan and McDougall, 2002), the creation of a household transportation sector required augmentation of the GTAP data as described in Paltsev et al. (2004a).

23.3 Policy scenarios

23.3.1 Cases studied

In order to study the potential cost of the Kyoto Protocol a number of assumptions are required. Exactly how countries will attempt to meet their commitments has not been fully determined, and it is not yet known which of the Kyoto Protocol's international flexibility mechanisms will in fact be

How (and why) do climate policy costs differ?

Table 23.2 Scenarios.

NoTrad	No emissions trading among parties. Reference average annual growth in GDP for Japan of 1.7%, the US of 3.2%, and the EU of 2.7% over the period 1997 to 2010. We assume an economy-wide cap-and-trade without CDM credits. The US is constrained to meet the Bush intensity target (18% emissions intensity improvement from 2000 to 2010).
NoTradR	NoTrad but with rapid economic recovery in Japan, with GDP growing at a rate of 2.6%/yr for 1997 to 2010.
ExTran	NoTrad but with Japan's transportation (TRAN) sector exempted from the cap.
N-30	NoTrad but with Japan's nuclear capacity reduced by 30% to reflect temporary shutdowns.
ExtEU	Extended EU bubble that includes EU expansion countries.[a] Japan and other regions continue to meet Kyoto targets with domestic, economy-wide cap-and-trade systems but do not trade with each other or ExtEU.
EUJ	Extended EU bubble includes Japan.
FullTrd	Full trade among Annex B parties to the Protocol, without US participation.

[a] Modeled here as including the EET EPPA regional group, dominated by countries that will become part of the expanded EU.

available. For example, European policy regarding international trading of emissions permits is not yet fully specified, and even then the EU trading system does not cover all sectors but omits transportation and households and focuses only on large point sources of CO_2. Like Europe, other countries may adopt policies that differentiate among sectors, limiting domestic trading system to point sources. Government-to-government transfers of permits outside a market trading system may be considered, but the outcome of those negotiations cannot be known at this time.

With our focus on domestic implementation in Japan, we also consider other economic and energy changes that affect the estimation of economic costs of a greenhouse gas target. Japan's economy has not recovered as strongly as was projected a few years ago, substantially reducing its likely future emissions levels. On the other hand, early plans to achieve the emissions commitment put heavy reliance on increasing the contribution of nuclear power (e.g., Babiker *et al.*, 2000c). Even when originally proposed, these plans seemed difficult to achieve because they would have required licensing and building many new reactors within a decade, when the planning and construction of some recent capacity additions stretched to 20 years. To have any chance of bringing this capacity on line by the Kyoto commitment period, Japan would have to begin a massive reactor construction program immediately and seek ways to speed up licensing and construction. Even more troublesome, a large share of the existing nuclear capacity was recently shut down temporarily. Thus not only is additional nuclear capacity unlikely to provide a large contribution to meeting the Kyoto commitment, but any recurring plant closures will lead to even more emissions from electricity production.

Recognizing these uncertainties about the precise economic conditions during the first Kyoto commitment period, and uncertainties about access to flexibility mechanisms, we construct a set of scenarios in order to explore the range of possible outcomes. They are summarized in Table 23.2. These scenarios include a case without emissions trading across regions (NoTrad), and one with international trading among all Kyoto parties with full access to Russian hot air (FullTrd). Scenarios with idealized emissions trading systems operating in each Annex B party serve as a basis for comparison with other implementation options. In all the cases studied in this part of our analysis we assume that the United States does not return to its original Kyoto target, but only meets the GHG intensity target set out by the Bush Administration.[1] The implied change in emissions is shown in Figure 23.1 (White House, 2002). While the United States Administration anticipates meeting its target with voluntary programs, we find that meeting it will require a positive carbon-equivalent price. We achieve the intensity target with an economy-wide cap-and-trade system covering all greenhouse gases. We also assume that Protocol members make full use of the agreed sinks allocated under the final agreement at Marrakech (see Babiker *et al.*, 2002), and that both Protocol members and the United States include in their cap-and-trade systems all of the non-CO_2 greenhouse gases identified in the Protocol.

We then consider a number of implementation variants. Focusing first on Japan, we consider a high economic growth or "Recovery" scenario (NoTradR) that is consistent with expectations for growth when Kyoto was signed but now looks unlikely considering recent experience. Given 1997–2003 economic performance, to achieve a 2010 GDP level like that projected only a few years ago would require a large and immediate economic turn-around and continued rapid growth over the remaining 6 years. Next we consider the implications of omitting the transportation sector from the Japanese emissions cap (ExTran), and of the effect of a 30% reduction of nuclear capacity (N–30), as might be realized if some level of shutdown were to recur during the Kyoto commitment period.

The next set of scenarios considers trading schemes focused on the EU. To approximate planned EU expansion, we create an extended EU bubble including the EET (ExtEU) region in EPPA. This trading system immediately allows the EU to access hot air in the EET (Eastern European economies in transition). We then consider the implications for Japan if it can trade within this system (ExtEUJ).

---

[1] A different set of assumptions is made in Section 23.4 where we decompose the factors contributing to national cost.

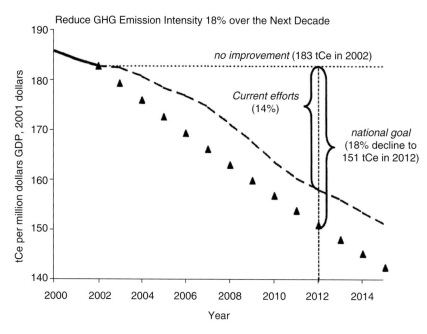

Figure 23.1 Bush administration intensity target for the United States (White House, 2002). tCe, metric tonnes carbon equivalent.

23.3.2 Simulation results

We focus our presentation of results on four regions, the United States, EU, Japan, and Canada, and on the effects in 2010. The top half of Table 23.3 shows the carbon-equivalent prices in each region for each scenario in $1997 per tonne C. Notably, Japan's carbon-equivalent price is much higher than the United States or the EU. Canada's carbon price is the highest of the group. Canadian emissions grew strongly from the mid-1990s up to the present, and our reference forecast for Canada continues this growth. The United States intensity target leads to a far smaller reduction than would its Kyoto target, but it still requires a $12 carbon price in the United States in 2010 given our EPPA reference growth in emissions. The US price varies only slightly across the scenarios because we assume that the United States does not engage in emissions trade with other regions. Similarly, Canada's carbon price varies little because we do not consider a case where Canada participates in emissions trading. (The policy variants do affect the United States and Canada through trade in other goods, yielding only a small effect on the carbon price.) The EU price in the NoTrad case is roughly half that of Japan while Canada's price is one-third higher. Note, however, that in the ExtEU case, which is intended to simulate the result of EU enlargement to include several Eastern European Associates, the EU's carbon-equivalent price falls to $21 per tonne. Thus, in carbon-price terms the EU Kyoto-target, with EU enlargement, requires only slightly more effort than the Bush intensity target in the United States. However, Japan's effort, as measured by the carbon-equivalent price, is five to ten times that in the EU or the United States, and Canada's is still higher.

The scenario variants for Japan show that a partial cap-and-trade system that (1) excludes transportation (ExTran), or (2) must be carried out with less than full nuclear capacity in operation (N-30), or (3) is implemented under conditions of more rapid economic growth (NoTradR) could cause Japan's carbon-equivalent price to increase by 50–100%. In some of the cases the carbon price exceeds that of Canada. Of course, uncertainties and implementation considerations are also present for the EU, the United States, and Canada. By focusing on Japan we can identify very specific issues facing Japan. At the same time, these issues are illustrative of the uncertainties any of these regions will face.

Access to international emissions trade, even if limited to the extended EU bubble, brings Japan's carbon-equivalent price down to $29. We estimate that full emissions trading, including full access to Russian hot air, would reduce the price in the Annex B parties to essentially zero. The very low price under full trading among the parties, absent the United States, is a finding of previous work (e.g., Babiker et al., 2002).

We turn now to the results in terms of change in economic welfare in 2010, measured in terms of consumption and stated in percentage terms, also shown in the bottom half of Table 23.3. This measure shows a very different picture of economic burden. The NoTrad case results in a somewhat greater burden on the EU than on Japan, with Canada's cost by far the highest. The percentage consumption loss for Japan rises under the conditions described by the ExTran, N-30, and NoTradR cases, but still does not reach that of Canada. Of course the EU costs, even as a percentage, fall under ExtEU, and are thus much below Japan's costs in all scenarios based on NoTrad.

Table 23.3 Carbon-equivalent prices and consumption loss, 2010.

		NoTrad	**NoTradR**	**ExTran**	**N-30**	**ExtEU**	**EUJ**	**FullTrd**
Carbon-equivalent prices ($/tonne C)	Japan	100	217	141	186	99	29	0
	EU	52	52	52	52	21	29	0
	USA	12	12	12	12	12	12	9
	Canada	135	134	135	135	134	134	0
Consumption loss (%)	Japan	0.45	1.04	0.57	0.98	0.45	0.13	0
	EU	0.54	0.52	0.54	0.54	0.20	0.31	0
	USA	0.01	0.01	0.01	0.01	0.01	0.01	0
	Canada	1.32	1.33	1.32	1.32	1.29	1.29	0.01

Beyond the very different carbon price and percentage loss effects, there are two additional paradoxical sets of results in Table 23.3. The first is that the United States has virtually no consumption loss in any of these cases – a result attributable to changes in goods markets (explored below). The second is that the costs of climate policy *rise* for the EU when all of Europe trades with Japan (ExtEUJ), compared with the EU costs when there is trading only within Europe (ExtEU). This result contradicts the common expectation that trading benefits both buyers and sellers of permits. Here, because the ExtEU bubble has a lower carbon price than Japan, expanding the bubble to include Japan will mean that ExtEU will be a net seller and Japan a net buyer of permits. Babiker *et al.* (2004) found a similar result: in a study of the implications for the EU parties of trading among themselves, the selling countries lost from entering a trading regime. In order to understand these paradoxical results we turn now to a discussion of what lies behind these different cost concepts.

23.4 Cost concepts and why countries differ

23.4.1 An equal-reduction comparison

To investigate the various factors that can lead to differences in cost among countries we first create a new set of scenarios that allows a focus on the energy efficiency question and other factors affecting the energy markets, isolating them from differences in the emissions growth and from the effects of non-CO_2 GHGs and sinks. Cost differences among studies have often been traced to different assumptions about emissions growth (e.g., Weyant and Hill, 1999). While many of the Kyoto Parties adopted very similar reductions from the 1990 level, it was clear even as the Protocol was signed that some regions were on a growth path that would lead to their emissions being as much as 30% above this target in 2010 (e.g. the United States) whereas others were likely to be only 10 or 15% (or less) above the target (e.g. the EU). Not surprisingly analysts who assumed rapid growth for some countries usually found higher costs than those who assumed lower growth.

We would like to look beyond these obvious reasons for cost differences and thus we conduct a new set of simulations where each region (including the United States) is assumed to reduce its emissions by an equal 25% from the reference level in 2010 (the projected level in the absence of any action). We also focus only on carbon, excluding the non-CO_2 greenhouse gases and any consideration of sinks. The non-CO_2 GHGs often offer inexpensive abatement options (Babiker *et al.*, 2002), and emissions of these gases relative to CO_2 vary among countries. Their influence is not large enough to explain the cost phenomena summarized in Table 23.3, however, and their omission simplifies the analysis. As can be seen in a comparison with the emissions price of the NoTrad case in Table 23.3, the 25% reduction is much larger than that implied by the Bush intensity target. The cut is somewhat more stringent than Kyoto for Japan and is roughly equivalent for the EU and Canada.

One of the important ways these countries differ is in their emissions intensity (emissions per dollar of GDP). Japan, which relies less on coal and is very energy efficient, has a much lower intensity than the United States or Canada. Table 23.4 shows how energy efficiency can be responsible for high costs (in terms of carbon price) or low costs (in terms of consumption loss). In terms of carbon price (Column 1), from highest to lowest the order is Japan followed by the United States, then the EU, and finally Canada. Note that for a comparable percentage reduction in emissions from reference, Canada has a lower price than the others, whereas in the Kyoto results of Table 23.3 (excluding the cases showing Japan with higher growth or various restrictions) Canada's carbon price was much higher. This comparison shows that the carbon price in Canada under Kyoto is high because of the high reference growth in emissions, rather than from a lack of technological options.

In terms of percentage consumption loss (Column 2), on the other hand, the United States cost is by far the lowest, less than 25% of that in the EU. Japan's percentage consumption loss falls about midway between these two extremes. Canada's loss is very similar to Japan's. Using market exchange rates (1997 US$), the economies of the United States and EU are of comparable size, with that of the United States somewhat larger, and thus the absolute consumption loss for the EU is roughly four times larger than for the United States (Column 3). Japan's economy in 2010 is approximately 60% of the EU or the United States. Even with the smaller economy, the absolute consumption loss is larger than in the United States.[2] Canada's economy is smaller still: less than one-fifth that of Japan.

In fact, energy efficiency – or more specifically the difference in carbon emissions intensity (Column 4) resulting from a combination of lower energy intensity and less reliance on coal in Japan than in the United States or the EU – can explain this difference in consumption loss. Japan's carbon intensity is less than half that of the United States, and about 70% of Europe. On one hand, Japan's already emissions-efficient economy is the main reason its carbon price is much higher. For example, whereas the United States, the EU, and Canada can fuel-switch from coal to natural gas, reducing CO_2 emissions for the same energy output, that option is limited in Japan. On the other hand, the absolute reduction required in million metric tonnes (mmt) of carbon is 70% larger in the United States than in Europe even though the economies are comparable in size. In contrast, Japan's reduction is only 20% of that required in the United States and only 32% of that required in Europe. Thus, while the carbon price is higher in Japan, any measure of total cost (cost per tonne times the number of tonnes) will be proportionally less because of the smaller number of tonnes reduced. Canada is the most GHG-intensive of the four regions shown. This fact is reflected in a required reduction in mmt that is more than one-half that of Japan even though the economy is less than one-fifth as big. Thus the carbon price level, often casually used as an indicator of relative cost among countries, is an exceptionally poor indicator.

There is a still deeper paradox, shown in Column (6). The *average* consumption loss per tonne is the total consumption loss (Column 3) divided by the number of tonnes reduced (Column 5). By this measure the cost turns out to be very similar in Japan and the EU, but this value is about six times the cost in the United States. Canada's social cost is roughly twice that in the United States but not nearly as high as that in the EU and Japan.

23.4.2 The influences on national cost

To explore these differences we need to consider in more detail how these different measures of cost are defined. First, the carbon price is a marginal cost, and an average cost will differ from the marginal. We can see this divergence with reference to a marginal abatement curve (MAC), shown in Figure 23.2, for Japan and the EU. (These curves are derived using the EPPA model, again focusing only on carbon.) We derive the MAC by running the model with successively tighter emissions constraints and plotting the emissions reduction on the horizontal axis and the carbon price on the vertical axis.[3] A dashed horizontal line is drawn at the price in Table 23.3 associated with a 25% cut from reference for each country. We then draw a vertical line from the point where it intersects that MAC to the horizontal axis to confirm that this approach can accurately estimate the abatement quantity for each. Thus, such curves are a useful way to summarize the relationship between the carbon-equivalent price and emissions reduction (Ellerman and Decaux, 1998).

In a partial equilibrium analysis (focusing only on the energy sector and ignoring other effects) it is possible under some conditions to consider the area under the MAC curve, up to the required reduction, to be the total cost of the policy. The area is the sum of the marginal cost of each tonne, the first tonnes costing very little (approaching zero) and the cost of the last tonnes approaching the carbon price. We can take this total cost as estimated and divide by the number of tonnes to get an average cost per tonne. Because the very highest cost reduction is the last tonne of removal required to meet the target, the average cost derived in this way will necessarily be lower than the carbon price. If the MAC is bowed as in Figure 23.2, as it typically is when estimated from models of this type, then the average cost will be somewhat less than one-half of the marginal cost. Without an estimate of the full MAC, a rough approximation of the total cost of a policy can be constructed by assuming the MAC is not bowed in this way but is simply a straight line. The area is then the formula for a triangle:

$$\text{Total cost} = \frac{1}{2}PQ, \quad (23.1)$$

where P is the price of carbon, and Q is the quantity of carbon abated. Dividing by Q on both sides of Eq. (23.1) shows that, in this case, the average cost is exactly $1/2P$. For the United States, Table 23.3 shows that indeed the average consumption cost per tonne is somewhat less than one-half of the carbon price. The consumption cost is thus roughly consistent with an integrated area under the abatement curve.

---

[2] It is important to note that these cross-country comparisons of the absolute cost (or size of the economy) depend heavily on the exchange rate. Given that the GTAP base year data are for 1997, and we must balance trade flows with capital accounts, we use 1997 market exchange rates (MERs). The US dollar has fallen against the Yen and Euro since 1997, so these absolute comparisons would change, widening further the relative reduction cost in absolute dollars between the United States and the EU. For comparing the "real" cost purchasing power parity (PPP) conversions are often used and these are at least more stable than MERs. In emissions trading scenarios we assume international trades of permits are at MERs, and to the extent that the price of permits do not represent the "real" cost to a domestic purchaser or seller, this presents a further distortion between equalized marginal cost at MER and the real social cost of the trade.

[3] One complicating factor that immediately emerges is how to construct a MAC when multiple countries are involved in the policy. One can simultaneously tighten the policy gradually on all parties; one can assume that all other parties are meeting a policy constraint and alter only the policy in the country for which one is constructing the MAC; or one can assume no policy in any party except for the party for which one is constructing the MAC. Because a policy can have spillover economic effects, the existence of a policy or not, and its severity, outside the country of interest will affect the emissions of that country and these different approaches to estimating the MAC will give somewhat different results. Here we have constructed the MACs assuming other parties are meeting their Kyoto commitments and that the United States meets the Bush intensity target.

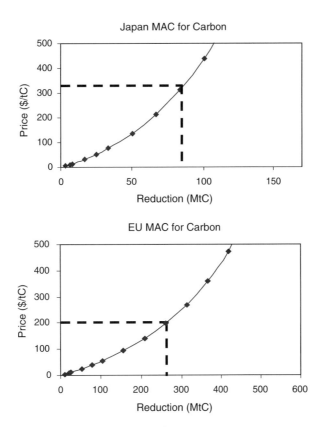

Figure 23.2 Marginal abatement curves for carbon, Japan and the EU.

Table 23.4 Measures of cost in 2010: 25% reduction from reference of carbon only.

	C-price, 1997$/tonne	Consumption loss,[a] %	Consumption loss, billions of 1997$	Carbon intensity, tonne/million 1997$[b]	Carbon reduction MtCe	Average consumption cost, 1997$ tonne
Japan	323	1.5	48	68	89.6	538
EU	205	2.2	160	96	275.1	586
USA	231	0.5	43	155	461.2	93
Canada	127	1.1	7	199	49.0	137

[a] Macroeconomic consumption loss as a percentage of total consumption.
[b] GDP intensity: carbon emissions/GDP.

But this result does not hold true either for Japan or for the EU. In fact the average consumption cost per tonne is *larger* than the carbon price, and thus clearly cannot be an average of the marginal cost of each tonne as represented in the MAC. Similarly, Canada's average consumption cost per tonne is also above the carbon price. Integrating under the MACs for Japan and Europe, up to a 25% reduction, and dividing by the tonnes reduced, we find the average cost derived is $150 for Japan and $100 for Europe. As expected, the average costs calculated from the MAC are slightly less than one-half the carbon price. Thus, there are other considerations, not captured in the MAC, that increase the consumption cost of climate policy in these regions.

Before continuing our explanation of this difference it now can be seen why, under these conditions, a region that sells permits can be made worse off, as happens to the EU in the ExtEUJ case. Using the carbon prices from Table 23.4 a private firm at the margin in the EU sees the direct additional cost of reducing emissions to be $205, whereas a Japanese firm at the margin sees the cost as $323. The EU firm is willing to undertake reductions that cost $205 or more (let us assume $205) and sell them to the Japanese firm, and the Japanese firm would at the margin be willing to pay as much as $323 for these credits to avoid making the reductions themselves. We would expect an equilibrium market price to fall somewhere in between $205 and $323; let us suppose it is $250. The EU firm

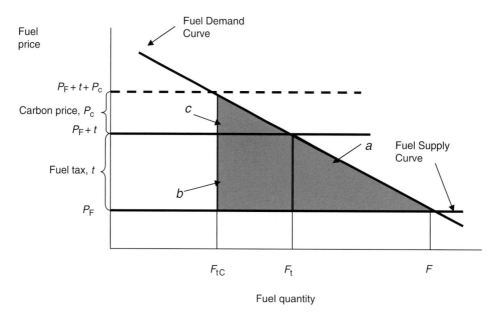

Figure 23.3 Effects of existing fuel taxes on the cost of carbon policy. The economic cost of the fuel tax (t) added to the fuel price (P_F) is given by area a. Adding a carbon constraint, with a price of carbon (P_C), raises the total fuel price, inclusive of the tax and carbon price, to $P_F + t + P_C$. Without the fuel tax, the economic cost of the carbon policy would be just the area labeled c. But, the pre-existing tax means that the cost represented by the area b, in excess of the actual cost of the fuel, is also an added economic cost of the policy. A marginal abatement curve (MAC) for carbon will include only area c.

thus profits $45 per permit by selling at $250 a reduction that cost it only $205. The Japanese firm benefits $73 per permit purchased by paying only $250 for reductions that would have cost it $323. But the *average social cost* of that tonne in the EU is $586, so the EU has a net loss from the trade, on average, of $586 − $250 = $336. On the other hand, Japan gains not only the avoided direct cost but the avoided social cost of $538 − $250 = $288. In this example, the result of trading, summing across the two regions, is to reduce total consumption because the gain of $288 in Japan is less than the loss of $336 in the EU.

The trading result for the ExtEUJ case depends on the divergence of the direct cost, as measured by the marginal abatement cost, from the social cost as measured by the consumption loss. Two factors are mainly responsible for this phenomenon. One is the interaction of the GHG policy with existing taxes (or subsidies) and the other is change in the terms of trade.[4] A major difference for the United States compared with the EU and Japan is that US fuel taxes are quite low, and thus we would immediately suspect that the fuel tax distortion effect is a major contributor to the high average consumption loss in Europe and Japan.[5]

Figure 23.3 graphically depicts how existing energy taxes can increase the cost of a carbon policy. Here we represent the demand for a carbon-containing fuel and its supply. The economic cost of a tax is the lost value of consumption of the good, less the cost of producing it. It is the area under the demand curve from the market equilibrium without the tax up to the market equilibrium point with the tax. The cost of the fuel tax is thus the shaded area, labeled a. Adding a carbon price results in additional loss of the area labeled c, the direct economic loss associated with the carbon constraint, plus the area labeled b, the increase in the loss due to the original fuel tax.[6]

If we consider that there is an area, here labeled c, under the demand for each fuel for each sector (and final consumption), the sum of all of these areas for an economy will approximately equal the area under the MAC for carbon.[7] Here we

---

[4] A country's terms of trade is the ratio of the weighted average of its export prices divided by the weighted average of the prices of its imports. An increase in the ratio (more imports from the same physical quantity of exports) will increase welfare, and vice versa.

[5] There can also be a further interaction with capital and labor taxes, but this does not occur in EPPA because of the particular formulation of the model. Labor supply is fixed in each period, and thus taxes or changes in the wage rate do not affect the quantity supplied. Thus, there is no deadweight loss with

labor taxes directly represented in EPPA, and so with changes in the wage rate there is no additional deadweight loss from this source. Similarly, the formulation of EPPA as a recursive-dynamic model means that capital in a period is a fixed supply, determined by the previous period's capital, less depreciation, plus investment, with investment equal to the previous period's level of savings.

[6] To some degree the fuel tax may represent other externalities (e.g., congestion) that are being managed using this instrument. We do not compensate for this potential effect of the tax, as discussed below.

[7] To imagine this, consider a single fuel and a single market for that fuel. We could measure the fuel, instead of in gallons or exajoules, in terms of the tonnes of carbon contained in the fuel. We could also compute the fuel price, instead of in dollars per gallon or exajoule, as the cost of the fuel per tonne of carbon in it. The demand curve so transformed and plotted as price rose from the initial market price without the carbon constraint would be exactly the

have drawn the demand curves as strictly linear, but as usually represented it will be convex with respect to the origin.[8] This non-linear relationship of demand to price gives rise to the curvature of the MAC. The important implication, however, is that if there are pre-existing energy taxes, the area under the MAC will underestimate the cost of a climate policy by a significant amount because the excluded cost is a rectangle whereas direct cost is approximately a triangle.

Paltsev *et al.* (2004b) develop a decomposition analysis and apply it to these EPPA scenarios. They show that in the cases evaluated in this section, for Japan the distortion cost is about 2.5 times the direct cost measured by the area under the MAC; for the EU it is over 4.5 times the direct cost; and for Canada it is about the same size, whereas there is no distortion loss for the United States. However, there is an offsetting oil market benefit due to changing terms of trade of between approximately 3.5 and 6% for Japan, the EU, and the United States, whereas Canada suffers in terms of trade loss in the oil market because it is a net exporter. How can these extra distortion costs be so large for the EU and Japan? The answer is that fuel taxes are extremely high. A numerical example can illustrate the importance of these distortions. Given the carbon content of gasoline, a $100/tonne carbon charge equates to $0.34/gallon of gasoline. Thus a $100 carbon price is the equivalent of a $0.34 per gallon gasoline tax. Gasoline taxes in the EU ranged from $2.80 to $3.80 per gallon as of 2004 (OECD/IEA, 2004). That is the equivalent of an $800 to $1200/tonne carbon tax. In the gasoline market, then, a $100 carbon tax would cause a direct cost loss of approximately $1/2 * \$100 = \$50$/tonne at the margin, but the distortion cost would be the full rectangle of $800 to $1200 per tonne, 16 to 24 times the direct cost. Other fuels in the EU are not taxed at as high a rate, so the average social cost in Table 23.3 is an average across the economy, weighted by how much of the reduction occurs in each sector with a different fuel tax rate.

The existence of these high fuel taxes in transportation raises several issues. A first issue is whether these taxes are correcting some other external effects of fuel use such as issues of dependence on foreign fuel sources, urban congestion, or injury to others in automobile accidents. As reviewed in Babiker *et al.* (2004), analysts who have looked carefully at these taxes do not believe a good case can be made that they are optimally correcting for such externalities. If they were, the levels and changes in them would take some account of the marginal benefit in terms of externality reduction, which is not the case. Furthermore, for the most important externalities associated with vehicle use, a fuel tax would be an extremely blunt instrument. For example, all fuel users in a country pay the tax but congestion occurs only in urban areas during particular times of day. In the United States the gasoline tax revenue is designated to the Federal Highway Trust Fund that builds and maintains highways, and is best treated as a justified user fee. The high fuel taxes in Europe and Japan have a complex history but are now treated mainly as a source of government revenue. The high taxes through their influence on driving probably have some indirect effect of lowering externalities. We have not assessed the magnitude of these indirect benefits. However, because fuel taxes only at best bluntly correct the various externalities associated with automobile use, there will remain a substantial distortionary effect.

A second issue is how policies may be adjusted to deal with the influence of these existing distortions. We show that international emissions trading can decrease welfare compared with a non-trading situation, at least for the permit selling region and possibly for the entire trading block in a Pareto sense. That is, gains in those regions that are net purchasers may be insufficient to compensate losses in selling regions.[9] The proposed EU trading system currently exempts transportation and thus would avoid exacerbating the effects of high fuel taxes. However, avoiding the distortionary effect by exempting those sectors with fuel taxes means that the burden of reduction for the country must fall on the non-exempt sectors The EU trading system as so far specified is a test program designed to operate prior to the first Kyoto commitment period, and the cap levels are within 1 or 2% of the expected reference level of emissions. Discussions of "adequacy" of these caps in moving toward the goals of the first commitment period of the Kyoto Protocol appear to focus on whether the capped sectors are bearing an appropriate proportion of the country-wide reduction. But if, as national targets are tightened, the exemptions are to remain in place in order to avoid exacerbating distortions, emissions will need to be disproportionally reduced in the capped sectors. The problem is alleviated only if the cap on non-transport sectors can be met in some other way that does not exacerbate fuel tax distortions, perhaps through Joint Implementation or CDM (Clean Development Mechanism) credits.

Finally is the issue of whether existing models consider all the pre-existing distortions in the economy that might be affected by the climate policy or the introduction of emissions trading. The short answer is no. In recent work, some success toward including these externalities in the EPPA model has been made (Yang *et al.*, 2005). Uncontrolled externalities or externalities controlled very bluntly by a fuel tax may be worsened by removing the fuel tax completely. These externalities must be controlled in another way or the taxes partially removed (if the tax is an efficient means of dealing

---

MAC plotted as a "demand for carbon", rather then the inverse – the willingness to do without carbon (i.e., abate).

[8] A fuel demand curve is not represented directly in a CGE model like EPPA, but is indirectly represented by the representation of the production technology, usually as a constant elasticity of substitution (CES) production function. To obtain the MAC from just the individual fuel demand functions, we required an uncompensated demand function that incorporates the changes in demand for downstream goods whose prices change.

[9] How to aggregate losses and gains in different regions depends, however, on having an accurate intercountry comparison if market exchange rates fail to accurately reflect real income comparisons.

with some of the externalities) to be sure that a country is indeed better off. Moreover, to the extent taxes are removed, new sources of revenue may need to be identified to make up for lost revenue, and these sources may themselves have a distortionary effect. A careful assessment of each country's domestic situation and how the climate policy will interact with it is needed. No doubt these details of country structure complicate climate policy negotiations. One country may see its policies as interventions to correct another problem, whereas another country may see them as an attempt to skew emissions trading or competitiveness.

In the end, while an efficiency case can be made for exemptions or other deviations from a country-wide cap-and-trade system, experience in trade negotiations would suggest that in practice the efficiency case for the policy deviation is often weak or non-existent. Instead, the exemption policy may represent a further distortion. Thus, caution is warranted in believing too easily either that equalizing the marginal cost of carbon across regions is necessarily a move toward improved efficiency, or that every case for deviation or exemption is legitimately based on efficiency grounds.

23.5 Conclusions

As implementation of the Kyoto Protocol or policies such as the Bush Administration's intensity target proceeds, economic analysis of mitigation policies must begin to deal with more realistic implementation, representing national economies with the varying levels of existing taxes and economic distortions. As the set of issues addressed by these economic modeling studies has become richer and more complex, the definition of economic cost itself, and how it is estimated in economic models, has presented a puzzle, particularly to those outside the economic modeling community. Results can seem paradoxical: focusing on carbon-equivalent price, Japan or Canada looks to be the high-cost region, and Europe the low-cost. But in terms of welfare loss Japan appears low-cost, Europe higher-cost, and Canada higher-cost still. These cost estimates are not independent of the specific policies by which targets will be achieved, and these policies are far from settled.

The diagnostic case developed in Section 23.4 illuminates the implications of the use of different measures of cost, and explains why the relative carbon price among Kyoto parties may not be an accurate measure of efficiency or of which country is bearing the largest cost burden. Relative emissions intensity, different levels of distortions in the form of fuel and other taxes, and different effects through trade are shown to be important factors. We show that existing distortions, particularly fuel taxes, greatly increase the cost of an economy-wide cap-and-trade applied in the EU or Japan. The United States has much lower fuel taxes so these distortions are negligible. Canada's costs are higher than the direct mitigation cost because it loses energy market export revenue owing to policy-induced reductions in world producer prices of energy. The United States gains from this fuel market effect because it is a net importer of fuel,

particularly oil. These examples illustrate why integrating under a MAC captures only the direct cost, and thus can be highly misleading in the presence of large tax distortions or other important general equilibrium effects.

We consider comparable reductions across regions (25% from reference in 2010) to isolate the effect of difference reference growth in emissions on these cost calculations. We also consider the cases of the EU, Japan, and Canada meeting their Kyoto targets and the United States meeting the Bush intensity target. For a country with a low emissions intensity, as in Japan, the absolute reduction in tonnes is small relative to the total size of the economy, and this fact means that welfare loss is a smaller share of total welfare. Low emissions intensity (high energy efficiency) also means the economy has fewer options to reduce emissions further, resulting in a higher carbon price. Energy efficiency thus pushes in both directions, lowering the number of tonnes that need to be reduced but raising the direct cost per tonne. On net this lowers costs in Japan relative to the EU.

A country's economic cost of complying with Kyoto or pursuing any climate mitigation policy will depend on a number of characteristics, including its economic growth, how it implements its climate policies, and interactions with other taxes and uncontrolled externalities. Early studies of the Kyoto Protocol focused on the huge benefits of international emissions trading, but those benefits depended on access to Russian hot air and to a lesser extent on hot air and low-cost reductions from Eastern European Associates of the EU. Russian economic growth has been more rapid than some imagined, however, and Russia is concerned about the Protocol becoming a constraint on its economic growth. That worry and the difficulty of creating and enforcing a cap-and-trade system in Russia may prevent many of these permits from being available. With entry into the Union, the Eastern European Associates are under the EU bubble, and this extended EU region has thus become a source of low-cost permits for Japan should a trading system be established among these parties. Economic distortions come into play in this situation, however. While the EU's private cost of emissions reduction is lowered when the bubble is extended to include Japan, its average social cost is increased. That result raises the question of whether the EU will in fact create such a trading system with Japan.

The coverage of the European Trading System is different from the example represented here, as it includes only large point sources and, importantly, excludes the household sector and highly taxed transport sector. Thus the implications for the EU and Japan of opening trading in emissions might be different from the calculations shown here. Only when the cap for the EU's covered sectors and the mechanisms used to restrain other emissions have been specified in full, and similar details are determined for Japan, will it be possible to gain an accurate picture of Kyoto compliance and the effects of emissions trade. But one lesson is clear: caution is warranted in assessing the welfare effects of policies based on estimated greenhouse gas prices given the presence of pre-existing taxes and other distortions in the countries involved.

References

Babiker, M. H., Reilly, J. M., and Jacoby, H. D. (2000a). The Kyoto Protocol and developing countries. *Energy Policy* **28**, 525–536.

Babiker, M. H., Bautista, M., Jacoby, H. D. and Reilly, J. M. (2000b). *Effects of Differentiating Climate Policy by Sector: A United States Example*. Cambridge MA: MIT Joint Program on the Science and Policy of Global Change, Report 61.

Babiker, M. H., Reilly, J. M. and Ellerman, A. D. (2000c). Japanese nuclear power and the Kyoto agreement. *Journal of the Japanese and International Economies* **14**, 169–188.

Babiker, M. H., Reilly, J. M., Mayer, M. *et al.* (2001). *The MIT Emissions Prediction and Policy Analysis (EPPA) Model: Revisions, Sensitivities, and Comparisons of Results*. Cambridge MA: MIT Joint Program on the Science and Policy of Global Change, Report 71.

Babiker, M. H., Jacoby, H. D., Reilly, J. M. and Reiner, D. M. (2002). The evolution of a climate regime: Kyoto to Marrakech. *Environmental Science and Policy* **5**(3), 195–206.

Babiker, M. H., Criqui, P., Ellerman, A. D., Reilly, J. M. and Viguier, L. L. (2003a). Assessing the impact of carbon tax differentiation in the European Union. *Environmental Modeling and Assessment* **8**(3), 187–197.

Babiker, M. H., Metcalf, G. E. and Reilly, J. M. (2003b). Tax distortions and global climate policy. *Journal of Environmental Economics and Management* **46**, 269–287.

Babiker, M. H., Reilly, J. M. and Viguier, L. (2004). Is international emissions trading always beneficial? *The Energy Journal* **25**(2), 33–56.

Bernard, A., Paltsev, S., Reilly, J. M., Vielle, M. and Viguier, L. (2003). *Russia's Role in the Kyoto Protocol*. Cambridge MA: MIT Joint Program on the Science and Policy of Global Change, Report 98.

Bohringer, C. (2002). Climate politics from Kyoto to Bonn: from little to nothing? *The Energy Journal* **23**(2), 51–71.

Dimaranan, B. and McDougall, R. (2002). *Global Trade, Assistance, and Production: The GTAP 5 Data Base*. West Lafayette, IN: Center for Global Trade Analysis, Purdue University.

Ellerman, A. D. and Decaux, A. (1998). *Analysis of Post-Kyoto CO_2 Emissions Trading Using Marginal Abatement Curves*. Cambridge MA: MIT Joint Program on the Science and Policy of Global Change, Report 40.

Ellerman, A. D. and Sue Wing, I. (2000). Supplementarity: an invitation to monopsony? *The Energy Journal* **21**(4), 29–59.

Fullerton, D. and Metcalf, G. E. (2001). Environmental controls, scarcity rents, and pre-existing distortions. *Journal of Public Economics* **80**, 249–267.

Goulder, L. H. (1995). Environmental taxation and the "Double Dividend:" a reader's guide. *International Tax and Public Finance* **2/2**, 157–183.

Hertel, T. (1997). *Global Trade Analysis: Modeling and Applications*. Cambridge: Cambridge University Press.

Jacoby, H. D., Eckhaus, R. S., Ellerman, A. D. *et al.* (1997). CO_2 emissions limits: economic adjustments and the distribution of burdens. *The Energy Journal* **18**(3), 31–58.

Jacoby, H. D. and Sue Wing, I. (1999). Adjustment time, capital malleability, and policy cost. *The Energy Journal Special Issue: The Costs of the Kyoto Protocol: A Multi-Model Evaluation*, 73–92.

Jacoby, H. J. and Ellerman, A. D. (2004). The safety valve and climate policy. *Energy Policy* **32**(4), 481–491.

Manne A. and Richels, R. (2001). *US Rejection of the Kyoto Protocol: The Impact on Compliance Cost and CO_2 Emissions*. AEI-Brookings Joint Center for Regulatory Studies, Working Paper 01–12.

Mayer, M., Hyman, R., Harnisch J. and Reilly, J. M. (2000). *Emissions Inventories and Time Trends for Greenhouse Gases and Other Pollutants*. Cambridge, MA: MIT Joint Program on the Science and Policy of Global Change, Technical Note 1.

OECD/IEA (2004). *Energy Prices and Taxes, Quarterly Statistics*, 2nd Quarter. Paris: Organization for Economic Cooperation and Development/International Energy Agency.

Paltsev, S., Reilly, J. M., Jacoby, H. D., Ellerman, A. D. and Tay, K. H. (2003). *Emissions Trading to Reduce Greenhouse Gas Emissions in the United States: The McCain-Lieberman Proposal*. Cambridge MA: MIT Joint Program on the Science and Policy of Global Change, Report 97.

Paltsev, S., Viguier, L., Reilly, J. M., Tay, K. H. and Babiker, M. H. (2004a). *Disaggregating Household Transport in the MIT-EPPA Model*. Cambridge MA: MIT Joint Program on the Science and Policy of Global Change, Technical Note 5.

Paltsev, S., Reilly, J. M., Jacoby, H. D. and Tay, K. H. (2004b). *The Cost of Kyoto Protocol Targets: The Case of Japan*. Cambridge MA: MIT Joint Program on the Science and Policy of Global Change, Report 112.

Paltsev, S., Reilly, J., Jacoby, H. *et al.* (2005). *The MIT Emissions Prediction and Policy Analysis (EPPA) Model: Version 4*. Cambridge MA: MIT Joint Program on the Science and Policy of Global Changes, Report 125.

Reilly, J. M., Prinn, R. G., Harnisch, J. *et al.* (1999). Multi-gas assessment of the Kyoto Protocol. *Nature* **401**, 549–555.

Shackleton, R. M., Shelby, M., Cristofaro, A. *et al.* (1993). *The Efficiency Value of Carbon Tax Revenues*, Stanford CA: Energy Modeling Forum EMF Working Paper 12.8.

Weyant, J. P. and Hill, J. N. (1999). *The Costs of the Kyoto Protocol: A Multi-Model Evaluation. The Energy Journal Special Issue*: Introduction and overview, vii–xliv.

White House (2002). *Global Climate Change Policy Book*. Washington DC: White House, February.

Yang, T., Reilly, J. M. and Paltsev, S. (2005). Air pollution health effects: toward an integrated assessment. In *The Coupling of Climate and Economic Dynamics*. ed. A. Haurie and L. Viguier. Dordrecht: Springer.

24

Lessons for mitigation from the foundations of monetary policy in the United States

Gary W. Yohe

24.1 Introduction

Many analysts, (including Pizer [Chapter 25], Keller *et al.* [Chapter 28], Webster [Chapter 29] and Toth [Chapter 30] in this volume, as well as others like Nordhaus and Popp [1997], Tol [1998], Lempert and Schlesinger [2000], Keller *et al.* [2004] and Yohe *et al.* [2004]) have begun to frame the debate on climate change mitigation policy in terms of reducing the risk of intolerable impacts. In their own ways, all of these researchers have begun the search for robust strategies that are designed to take advantage of new understanding of the climate systems as it evolves – an approach that is easily motivated by concerns about the possibility of abrupt climate change summarized by, among others, Alley *et al.* (2002). These concerns take on increased importance when read in the light of recent surveys which suggest that the magnitude of climate impacts (see, for example, Smith and Hitz [2003]) and/or the likelihood of abrupt change (IPCC, 2001; Schneider, 2003; Schlesinger *et al.*, 2005) could increase dramatically if global mean temperatures rose more than 2 or 3 °C above pre-industrial levels. Neither of these suggestions can be advanced with high confidence, of course, but that is the point. Uncertainty about the future in a risk-management context becomes *the fundamental reason* to contemplate action in the near term even if such action cannot guarantee a positive benefit–cost outcome either in all states of nature or in expected value.

Notwithstanding the efforts of these and other scholars to reflect these sources of concern in their explorations of near-term policy intervention, the call for a risk-management approach has fallen on remarkably deaf ears. Indeed, uncertainty is frequently used by many in the United States policy community and others in the consulting business as *the fundamental reason* not to act in the near term. For evidence of this tack, consider the policy stance of the Bush Administration that was introduced in 2002. In announcing his take on the climate issue, the President emphasized that "the policy challenge is to act in a serious and sensible way, *given our knowledge. While scientific uncertainties remain*, we can begin now to address the factors that contribute to climate change" (www.whitehouse.gov/ new/releases/2002/02/climate change.html; my emphasis). Indeed, the *New York Times* reported on June 8th, 2005 that Philip Cooney, the White House Council on Environmental Quality, repeatedly inserted references to "significant and fundamental" uncertainties into official documents that describe the state of climate science even though he has no formal scientific training (Revkin, 2005). In large measure responding to concerns about uncertainty among his closest advisers, the President's policy called for more study, for voluntary restraint (by those motivated to reduce their own emissions even though others are not), and for the development of alternatives to current energy-use technologies (because reducing energy dependence is otherwise a good idea even though promising advancements face enormous difficulty penetrating the marketplace).

If productive dialogue is to resume, advocates of risk-based near-term climate policy will have to express its value in terms that policymakers will understand and accept. The case must be made, in other words, *that a risk-management approach to near-term climate policy would be nothing more than the application of already accepted policy-analysis tools and principles to a new arena*. This paper tries to contribute to this process by turning for support to recent descriptions of how monetary policy is conducted in the United States. Opening remarks offered by Alan Greenspan, Chairman of the Federal Reserve Board, at a symposium that was sponsored by the Federal Reserve Bank of Kansas City in August 2003 are a great place to start (Greenspan, 2003). In his attempt to motivate three days of intense conversation among policy experts, Chairman Greenspan observed:

> For example, policy A might be judged as best advancing the policymakers' objectives, conditional on a particular model of the economy, but might also be seen as having relatively severe adverse consequences if the true structure of the economy turns out to be other than the one assumed. On the other hand, policy B might be somewhat less effective under the assumed baseline model ... but might be relatively benign in the event that the structure of the economy turns out to differ from the baseline. *These considerations have inclined the Federal Reserve policymakers toward policies that* **limit the risk** *of deflation even though the baseline forecasts from most conventional models would* **not project such an event.** (Greenspan (2003), p. 4; my emphasis).

The Chairman expanded on this illustration in his presentation to the American Economic Association (AEA) at their 2004 annual meeting in San Diego:

> ... the conduct of monetary policy in the United States has come to involve, at its core, crucial elements of risk management. This conceptual framework emphasizes understanding as much as possible the many sources of risk and uncertainty that policymakers face, quantifying those risks *when possible*, and assessing the costs associated with each of the risks........ This framework also entails, in light of those risks, a strategy for policy directed at maximizing the probabilities of achieving over time our goals ... Greenspan (2004), p. 37; my emphasis).

Clearly, these views are consistent with an approach that would expend some resources over the near term to avoid a significant risk (despite a low probability) in the future. Indeed, the Chairman used some familiar language when he summarized his position:

> As this episode illustrates (the deflation hedge recorded above), policy practitioners under a risk-management paradigm may, at times, be led to undertake actions intended to provide *insurance against especially adverse outcomes*. (Greenspan (2004), p. 37; my emphasis).

So how did the practitioners of monetary policy come to this position? By some trial and error described by Greenspan in his AEA presentation, to be sure; but the participants at the earlier Federal Reserve Bank symposium offer a more intriguing source. Almost to a person, they all argued that the risk-management approach to monetary policy evolved most fundamentally from a seminal paper authored by William Brainard (1967); see, for example, Greenspan (2003), Reinhart (2003), and Walsh (2003).

This paper is crafted to build on their attribution by working climate into Brainard's modeling structure in the hope that it might thereby provide the proponents of a risk-based approach to climate policy some access to practitioners of macroeconomic policy who are familiar with its structure and its evolution since 1967. It does so even though the agencies charged with crafting climate policy in the United States (the Department of State, the Department of Energy, the Environmental Protection Agency, the Council for Environmental Quality, etc.) are not part of the structure that crafts macroeconomic policy (the Federal Reserve Board, the Treasury, the Council of Economic Advisors, etc.). The hope, therefore, is really that the analogy to monetary policy will spawn productive dialogue between the various offices where different policies are designed and implemented, even as it provides the environmental community with an example of a context within which risk-management techniques have informed macroscale policies.

The paper begins with a brief review of the Brainard (1967) structure with and without a climate policy lever and proceeds to explore the circumstances under which its underlying structure might lead one to appropriately ignore its potential. Such circumstances can and will be identified in Sections 24.2 and 24.3, but careful inclusion of a climate policy lever makes it clear that they are rare even in the simple Brainard-esque policy portfolio. In addition, manipulation of the model confirms that the mean effectiveness of any policy intervention, the variance of that effectiveness and its correlation with stochastic influences on outcome are all critical characteristics of any policy. Section 24.4 uses this insight as motivation when the text turns to describing some results drawn from the Nordhaus and Boyer (2001) DICE model that has been expanded to accommodate profound uncertainty about the climate's temperature sensitivity to increases in greenhouse gas concentrations. Concluding remarks use these results, cast in terms of comparisons of several near-term policy alternatives, to make the case that creative and responsive climate policy can be advocated on the basis of the same criteria that led the Federal Reserve System of the United States to adopt a risk-management approach to monetary policy.

24.2 The Brainard model

The basic model developed by Brainard (1967) considers a utility function on some output variable y (read GDP, for example) of the form:

$$V(y) = -(y - y^*)^2, \quad (24.1a)$$

where y^* represents the targeted optimal value. The function $V(y)$ fundamentally reflects welfare losses that would accrue if actual outcomes deviate from the optimum. The correlation between y and some policy variable P (read a monetary policy indicator such as the discount rate, for example) is taken to be linear, so

$$y = aP + \varepsilon.$$

In specifying this relationship, a is a parameter that determines the ability of policy P to alter output and ε is an unobservable random variable with mean μ_ε and variance σ_ε^2. The expected value of utility is therefore

$$\begin{aligned}E\{V(y)\} &= -E\{y^2 - 2yy^* + (y^*)^2\}\\ &= -\{E(y^2) - 2y^*E(y) + (y^*)^2\}\\ &= -\{\mu_y^2 - \sigma_y^2 - 2y^*\mu_y + (y^*)^2\}\\ &= -\{\sigma_y^2 + (\mu_y - y^*)^2\}.\end{aligned} \quad (24.1b)$$

In this formulation, of course, σ_y^2 and μ_y represent the variance and mean of y, respectively, given a policy intervention through P and the range of possible realizations of a and ε.

If the decisionmaker knew the value of parameter $a = a_o$ with certainty, then $\sigma_y^2 = \sigma_\varepsilon^2$. Moreover, prescribing a policy P_c such that $\mu_y = y^*$ would maximize expected utility. In other words,

$$P_c = \{y^* - \mu_\varepsilon\}/a_o \quad (24.2a)$$

In an uncertain world where a is known only up to its mean μ_a and variance σ_a^2, however,

$$\mu_y = \mu_a P + \mu_\varepsilon \text{ and}$$

$$\sigma_y^2 = P^2 \sigma_a^2 + \sigma_\varepsilon^2$$

under the assumption that a and ε are independently distributed. Brainard focused his attention primarily on estimation uncertainty, but subsequent applications of his model (see, for example, Walsh [2003]) have also recognized many of the other sources that plague our understanding of the climate system – model, structural, and contextual uncertainties, to name just three.

The first-order condition that characterizes the policy P_u that would maximize expected utility in this case can be expressed as

$$\begin{aligned}\partial E\{V(y)\}/\partial P &= -\{2P_u\sigma_a^2 + 2(\mu_y - y^*)\mu_a\}\\ &= -\{2P_u\sigma_a^2 + 2(\mu_a P + \mu - y^*)\mu_a\} = 0\end{aligned}$$

Collecting terms,

$$\begin{aligned}P_u &= \{(y^* - \mu_\varepsilon)\mu_a\}/\{\sigma_a^2 + \mu_a^2\}\\ &= P_c/\{(\sigma_a^2/\mu_a^2) + 1\}\end{aligned} \quad (24.2b)$$

under the assumption that the distribution of a is anchored with $\mu_a = a_o$. Notice that $P_u = P_c$ if uncertainty disappears as σ_a^2 converges to zero. If σ_a^2 grows to infinity, however, policy intervention becomes pointless and P_c collapses to zero. In the intermediate cases, Brainard's conclusion of caution – the "principle of attenuation" to use the phrase coined by Reinhart (2003) – applies. More specifically, policy intervention should be restrained under uncertainty about its effectiveness, at least in comparison with what it would have been if its impact were understood completely.

Reinhart (2003) and others have noted that considerable effort has been devoted to exploring the robustness of the Brainard insight in a more dynamic context where the loss function associated with deviations from y^* is not necessarily symmetric. They note, for example, that the existence of thresholds for $y < y^*$ below which losses become more severe at an increasing rate can lead to an intertemporal hedging strategy that pushes policy further in the positive direction in good times even at the risk of overshooting the targeted y^* with some regularity. Using such a strategy moves μ_ε higher over time so that the likelihood of crossing the troublesome threshold falls in subsequent periods. In the realm of monetary policy, for example, concerns about deflation have defined the critical threshold; in the realm of climate, the possibility of something sudden and non-linear such as the collapse of the Atlantic thermohaline circulation comes to mind as a critical threshold to be avoided by mitigation.

Practitioners of monetary policy have also worried about avoiding states of nature where the effectiveness of P can be eroded, and so they have found a second reason to support the sort of dynamic hedging just described. In these contexts, for example, central bankers have expressed concern that the ability of reductions in the interest rate to stimulate the real economy can be severely weakened if rates have already fallen too far. In the climate arena, decisionmakers may worry that it may become impossible to achieve certain mitigation targets over the long run if near-term interventions are too weak. This point is illustrated in Yohe et al. (2004) when certain temperature targets become infeasible if nothing is done over the next 30 years to reduce greenhouse gas emissions.

Both of these concerns lie at the heart of the Greenspan comparison of two policies, of course. However, neither confronts directly the question at hand: *under what circumstances (if any) can the effects of climate change on the real economy be handled by standard economic interventions without resorting to direct mitigation of the drivers of that change?*

24.3 Extending the model to include a climate module

To address this question, we add a climate module to the Brainard model so that we can search for conditions under which it would make sense for policymakers who have their hands on the macro-policy levers (the P in the basic model) to ignore climate policy when they formulate their plans. To that end, we retain the symmetric utility function recorded in Eq. (24.1a), but we add a new policy variable C (read mitigation for the moment) to the output relationship so that

$$y = aP + cC + \varepsilon.$$

The error term ε now includes some reflection of climate risk to the output variable. Since the expected value of utility is preserved, perfect certainty about a and now c still guarantees that $\sigma_y^2 = \sigma_\varepsilon^2$ so that prescribing a policy P_c such that $\mu_y = y^*$ would still maximize expected utility depicted by Eq. (24.1b).

In other words,

$$P_c = \{y^* - \mu_\varepsilon\}/a_o$$

would persist and the optimal intervention could be achieved without exercising the climate policy variable. In this certainty case, clearly, climate policy could be set equal to zero without causing any harm.

In an uncertain world where a and c are known only up to means (μ_a and μ_c) and variances (σ_a^2 and σ_c^2), however, we now have

$$\mu_y = \mu_a P + \mu_c C + \mu_\varepsilon \text{ and } \sigma_y^2 = P^2\sigma_a^2 + C^2\sigma_c^2 + \sigma_\varepsilon^2$$

under the assumption that a, c and ε are all independently distributed. We already know that this sort of uncertainty can modify the optimal policy intervention, but does it also influence the conclusion that the climate policy lever could be ignored?

24.3.1 The climate lever in an isolated policy environment

To explore this question, note that Eq. (24.2b) would still apply for setting policy P if the policymaker chose to ignore the climate policy variable; i.e.,

$$P_{uo} = P_u = P_c/\{(\sigma_a^2/\mu_a^2) + 1\}.$$

As a result, the first-order condition characterizing the policy C_{uo} that would maximize expected utility can be expressed as

$$\partial E\{V(y)\}/\partial C = -\{2C_{uo}\sigma_c^2 + 2(\mu_\varepsilon - y^* + \mu_a P_u + \mu_c C_{uo})\mu_c\} = 0$$

Collecting terms,

$$C_{uo} = \{[\sigma_a^2\mu_c^2]/[D_a D_c]\}\{(y^* - \mu_\varepsilon)/\mu_c\}, \quad (24.3)$$

where

$$D_a \equiv \{\sigma_a^2 + \mu_a^2\} \text{ and } D_c \equiv \{\sigma_c^2 + \mu_c^2\}.$$

Notice that $C_{uo} = 0$ if uncertainty about the effectiveness of P disappeared as σ_a^2 converged to zero. The climate policy lever could therefore still be ignored even in the context of uncertainty drawn from our understanding of the climate system, in this case. This would not mean, however, that climate change should be ignored. The specification of P_{uo} would recognize the effect of climate through its effect on μ_ε.

If σ_a^2 grew to infinity, however, then l'Hospital's rule shows that policy intervention through C would dominate. Indeed, in this opposing extreme case,

$$C_{uo} = \{(y^* - \mu_\varepsilon)/\mu_c\}/\{(\sigma_c^2/\mu_c^2) + 1\} \quad (24.4)$$

so that policy intervention through C would mimic the original intervention through P while P_{uo} collapsed to zero. In the more likely intermediate cases in which the variances of both policies are non-zero but finite, the optimal setting for climate policy would be positive as long as $\mu_c > 0$.

It follows, from consideration of the intermediate cases, that bounded uncertainty about the effectiveness of both policies can play a critical role in determining the relative strengths of climate and macroeconomic policies in the policy mix. *Put another way, uncertainty about the effectiveness of either or both policies becomes the reason to diversify the intervention portfolio by undertaking some climate policy even if the approach taken in formulating other policies remains unchanged. Moreover, the smaller the uncertainty about the link between climate policy C and output, the larger should be the reliance on climate mitigation.*

24.3.2 The climate lever in an integrated policy environment

These observations fall short of answering the question of how best to integrate macroeconomic and climate policy in an optimal intervention portfolio. Maximizing expected utility if both policies were considered together in a portfolio approach would produce two first-order conditions:

$$\partial E\{V(y)\}/\partial P = -\{2P_u T \sigma_a^2 + 2(\mu_\varepsilon - y^* + \mu_a P_{uT} + \mu_c C_{uT})\mu_a\} = 0 \text{ and}$$

$$\partial E\{V(y)\}/\partial C = -\{2C_u T \sigma_c^2 + 2(\mu_\varepsilon - y^* + \mu_a P_{uT} + \mu_c C_{uT})\mu_c\} = 0.$$

In recording these conditions, P_{uT} and C_{uT} represent the jointly determined optimal choices for P and C, respectively. Solving simultaneously and collecting terms,

$$P_{uT} = \{[\sigma_c^2\mu_a^2]/[D_a D_c - \mu_a^2\mu_c^2]\}\{(y^* - \mu_\varepsilon)/\mu_a\}, \text{ and} \quad (24.5a)$$

$$C_{uT} = \{[\sigma_a^2\mu_c^2]/[D_a D_c - \mu_a^2\mu_c^2]\}\{(y^* - \mu_\varepsilon)/\mu_c\}. \quad (24.5b)$$

Table 24.1 shows the sensitivities of these policies to extremes in the characterizations of the distributions of the parameters a and c. Notice that the policy specifications recorded in Eq. (24.5a and b) collapse to the certainty cases for C and P if the variance of c or a (but not both) collapses to zero, respectively. The policies also converge to the characterizations in Eq. (24.2b) or (24.4) if the variances of a or c grow without bound, respectively (again by virtue of l'Hospital's rule). In between these extremes, Eq. (24.5a and b) show how ordinary economic and climate policies can be integrated to maximize expected utility. In this regard, it is perhaps more instructive to contemplate their ratio:

$$\{C_{uT}/P_{uT}\} = \{[\sigma_a^2\mu_c]/[\sigma_c^2\mu_a]\}. \quad (24.6)$$

Equation (24.6) makes it clear that climate policy should be exercised relatively more vigorously if the variance of its effectiveness parameter falls or if its mean effectiveness

Table 24.1 Integrating policies in the extremes.

Limiting case	C_{uT}	P_{uT}
$\sigma_c^2 \to \infty$ with $0 < \sigma_a^2 < \infty$ (i.e., $D_c \to \infty$)	$C_{uT} \to 0$	$P_{uT} \to \{(y^* - \mu)/\mu_a\}/\{(\sigma_a^2/\mu_a^2) + 1\}$
$\sigma_c^2 \to 0$ with $0 < \sigma_a^2 < \infty$ (i.e., $D_c \to \mu_c^2$)	$C_{uT} \to \{(y^* - \mu_\varepsilon)/\mu_c\}$	$P_{uT} \to 0$
$\sigma_a^2 \to \infty$ with $0 < \sigma_c^2 < \infty$ (i.e., $D_a \to \infty$)	$C_{uT} \to \{(y^* - \mu_\varepsilon)/\mu_c\}/\{(\sigma_c^2/\mu_c^2) + 1\}$	$P_{uT} \to 0$
$\sigma_a^2 \to \infty$ with $0 < \sigma_c^2 < \infty$ (i.e., $D_a \to \mu_a^2$)	$C_{uT} \to 0$	$P_{uT} \to \{(y^* - \mu_\varepsilon)/\mu_a\}$
$\sigma_c^2 = 0$ and $\sigma_a^2 = 0$	Undefined	Undefined

increases. In addition, comparing Eq. (24.2b) and (24.5a) shows that

$$\{P_u/P_{uT}\} = \{D_c/\sigma_c^2\} + \{\mu_a^2 \mu_c^2 / \sigma_c^2 D_a\} > 1;$$

i.e., an integrated approach diminishes the role of ordinary economic policy in a world that adds climate to the sources of uncertainty to which it must cope as long as σ_c^2 is bounded.

It is, of course, possible to envision responsive climate policy that corrects itself as our understanding of the climate system evolves – ramping up (or damping) the control if it became clear that damages were more (less) severe than expected and/or critical thresholds were closer (more distant) than anticipated. In terms of the Brainard model, this sort of properly designed responsive policy would create a negative covariance between the effectiveness parameter c and the random variable ε. Since the variance of output is given by

$$\sigma_y^2 = P^2 \sigma_a^2 + C^2 \sigma_c^2 + \text{cov}(c; \varepsilon) + \sigma_\varepsilon^2,$$

in this case, repeating the optimization exercise reveals that

$$\begin{aligned} P'_{uT} &= \{[\sigma_c^2 \mu_a^2]/[D_a D_c - \mu_a^2 \mu_c^2]\}\{(y^* - \mu_\varepsilon)/\mu_a\} \\ &\quad + \{\text{cov}(c; \varepsilon)/D_a\} \\ &= P_{uT} + \{\text{cov}(c; \varepsilon)/D_a\} < P_{uT} \end{aligned} \quad (24.7a)$$

and

$$\begin{aligned} C'_{uT} &= \{[\sigma_a^2 \mu_c^2]/[D_a D_c - \mu_a^2 \mu_c^2]\}\{(y^* - \mu_\varepsilon)/\mu_c\} \\ &\quad - \{\text{cov}(c; \varepsilon)/D_c\} \\ &= C_{uT} - \{\text{cov}(c; \varepsilon)/D_c\} > C_{uT} \end{aligned} \quad (24.7b)$$

As should be expected, the ability of responsive climate policy to deal more effectively with worsening climate futures would increase its emphasis in an optimizing policy mix at the expense of ordinary economic policy intervention.

24.3.3 Discussion

Equation (24.6) shows explicitly that an integrated policy portfolio would ignore climate policies at its increasing peril, especially if the design of the next generation of climate policy alternatives could produce smaller levels of implementation uncertainty. Targeting something closer to where impacts are felt in the causal chain (like shooting for a temperature limit rather than trying to achieve emissions pathways whose associated impacts are known with less certainty) could, for example, be preferred in the optimization framework if the technical details of monitoring and reacting could be overcome. As in any economic choice, however, there are tradeoffs to consider. Moving to the impact end of the system should reduce uncertainty on the damages side of the implementation calculus (if monitoring, attribution, and response could all be accomplished in a timely fashion, of course), but it could also increase uncertainty on the cost side.

In any case, Eq. (24.7a and b) show that the potential advantage of climate policy could turn on the degree to which its design could accommodate a negative correlation. They support consideration of a comprehensive climate policy that could incorporate mechanisms at some level by which mitigation could be predictably adjusted as new scientific understanding of the climate system, climate impacts, and/or the likelihood of an abrupt or non-linear change became available (much in the same way that the rate of growth of the money supply can be predictably adjusted in response to changes in the overall health of the macroeconomy).

In addition, the same caveats discovered by the practitioners of monetary policy certainly apply to the climate side of the policy mix. Considering combined policies in a dynamic context, that includes critical thresholds beyond which abrupt, essentially unknown but potentially damaging impacts could occur, would still support more vigorous intervention; and climate policy should be particularly favored for this intervention if it becomes more effective in avoiding those thresholds when crossing their boundaries becomes more likely. Indeed, the Greenspan warning can be especially telling in these cases.

24.4 The hedging alternative under profound uncertainty about climate sensitivity

The Brainard structure is highly abstract, to be sure, and so conclusions drawn from its manipulation beg the question of its applicability to the climate policy question as currently formulated. This section confronts this question directly by exercising a version of the Nordhaus and Boyer (2001) DICE integrated assessment model that has been modified to accommodate wide uncertainty in climate sensitivity *and* the

Lessons for mitigation from US monetary policy

Table 24.2 Calibrating the climate module.

Climate sensitivity	Likelihood	Alpha-1 calibration
1.5 degrees	0.30	0.065742
2 degrees	0.20	0.027132
3 degrees	0.15	0.014614
4 degrees	0.10	0.011550
5 degrees	0.07	0.010278
6 degrees	0.05	0.009589
7 degrees	0.03	0.009157
8 degrees	0.03	0.008863
9 degrees	0.07	0.008651

Source: Yohe et al. (2004).

problem of setting near-term policy with the possibility of making "midcourse" adjustments sometime in the future.[1] It begins with a description of uncertainty in our current understanding of climate sensitivity. It continues to describe the modifications that were implemented in the standard DICE formulation, and it concludes by reviewing the relative efficacy, expressed in terms of expected net present value of gross world (economic) product (GWP), of several near-term policy alternatives.

24.4.1 A policy hedging exercise built around uncertainty about climate sensitivity

Andronova and Schlesinger (2001) produced a cumulative distribution of climate sensitivity based on the historical record. Table 24.2 records the specific values of a discrete version of this CDF that was used in Yohe et al. (2004) to explore the relative efficacy of various near-term mitigation strategies. There, the value of hedging in the near term was evaluated under the assumption that the long-term objective would constrain increases in global mean temperature to an unknown target. Calibrating the climate module of DICE to accommodate this range involved specifying both a climate sensitivity and an associated parameter that reflects the inverse thermal capacity of the atmospheric layer and the upper oceans in its reduced-form representation of the climate system. Larger climate sensitivities were correlated with smaller values for this capacity so that the model could match observed temperature data when run in the historical past. The capacity parameter was defined from optimization of the global temperature departures, calculated by DICE, and calibrated against the observed temperature departures from Jones and Moberg (2003) for the prescribed range of the climate sensitivities from 1.5 through 9 °C.

It is widely understood that adopting a risk-management approach means that near-term climate policy decisions should, as a matter of course, recognize the possibility that adjustments will be possible as new information about the climate system emerges. The results that follow are the product of experiments that recognize this understanding. Indeed, they were produced by adopting the hedging environment that was created under the auspices of the Energy Modeling Forum in Snowmass to support initial investigations of the policy implications of extreme events; Manne (1995) and Yohe (1996) are examples of this earlier work. They were, more specifically, drawn from a policy environment in which decisionmakers evaluate the economic merits of implementing near-term global mitigation policies that would be in force for 30 years beginning in 2005 under the assumption that all uncertainty will be resolved in 2035. These global decisionmakers would, therefore, make their choices with the understanding that they would be able to "adjust" their interventions in 2035 when they would be informed fully about both the climate sensitivity and the best policy target. In making both their initial policy choice and their subsequent adjustment, their goal was taken to be maximizing the expected discounted value of GWP across the range of options that would be available at that time.

The hedging exercise required several structural and calibration modifications of the DICE model in addition to changes in the climate module that were described above. Since responding to high sensitivities could be expected to put enormous pressure on the consumption of fossil fuel, for example, the rate of "decarbonization" in the economy (reduction in the ratio of carbon emissions to global economic output) was limited to 1.5% per year. Adjustments to mitigation policy were, in addition, most easily accommodated by setting initial carbon tax rates in 2005 and again in 2035. The initial and adjusted benchmarks appreciated annually at an endogenously determined return to private capital so that "investment" in mitigation was put on a par with investment in economic capital. Finally, the social discount factor for GWP included a zero pure rate of time preference in deference to a view that the welfare of future generations should not be diminished by the impatience of earlier generations for current consumption.

24.4.2 Some results

Suppose, to take a first example of how the critical mean, variance, and covariance variables from the Brainard foundations might be examined, that global decisionmakers tried to divine "optimal" intervention given the wide uncertainty about climate sensitivity portrayed in Table 24.2. Table 24.3 displays the means and standard deviations of the net value, expressed in terms of discounted value through 2200 and computed across the discrete range of climate sensitivities recorded in Table 24.2, for optimal policies that would be chosen if each of the climate sensitivities recorded in

---

[1] Climate sensitivity is defined as the increase in equilibrium global mean temperature associated with a doubling of greenhouse gas concentrations above pre-industrial levels, expressed in terms of CO_2 equivalents.

Table 24.3 Exploring the economic value of deterministic interventions in the modified DICE environment. Mean returns in billions of 1995$ with the standard deviations in parentheses.

Policy context	1.5°	2.0°	3.0°	4.0°	5.0°	6.0°	7.0°	8.0°	9.0°
Economic value	68.34 (36.33)	86.36 (53.97)	95.38 (92.13)	84.21 (115.3)	72.38 (129.5)	62.59 (139.2)	54.05 (146.0)	44.65 (153.2)	44.65 (153.2)

Table 24.4 Exploring the economic value of the mean climate policy contingent on climate sensitivity in the modified DICE environment. Return in billions of 1995$ and maximum temperature change in °C.

Climate sensitivity	1.5°	2.0°	3.0°	4.0°	5.0°	6.0°	7.0°	8.0°	9.0°	Mean	Standard deviation
Economic value	0	52	125	164	186	200	209	215	219	96.62	81.94
Max ΔT	2.71	3.46	4.69	5.62	6.32	6.85	7.25	7.57	7.83	4.55	1.76

Table 24.5 Exploring the economic value of the responsive climate policy contingent on climate sensitivity in the modified DICE environment. Return in billions of 1995$ and maximum temperature change in °C.

Climate sensitivity	1.5°	2.0°	3.0°	4.0°	5.0°	6.0°	7.0°	8.0°	9.0°	Mean	Standard deviation
Economic value	25	58	126	177	212	236	256	271	281	117.82	90.11
Max ΔT	2.80	3.53	4.67	5.53	6.18	6.67	7.03	7.32	7.57	4.53	1.64

Table 24.2 were used to specify the uncontrolled baseline. The mean returns of these interventions peak for the policy associated with a 3°C climate sensitivity, but the standard deviations grow monotonically with the assumed sensitivity. Selecting the mean of these interventions produces a net expected discounted value of $96.62 billion with a standard deviation of $81.94 billion. The first row of Table 24.4 shows the distribution of the underlying net values for this policy across the range of climate sensitivities, and the second row displays the associated maximum temperature increases that correspond to each policy.

Now suppose that decisionmakers recognized that a policy adjustment would be possible in 2035, but they could not tell in 2005 which one would be preferred. The first row of Table 24.5 displays the corresponding net values under the assumption that climate policy could be adjusted in the year 2035 to reflect the results of 30 years of research into the climate system that would produce a complete understanding of the climate sensitivity. In other words, the policy intervention would respond to new information in 2035 to follow a path that would then be optimal. The expected value of this responsive policy, computed now with our current understanding as depicted in Table 24.2, climbs to $117.82 billion (nearly a 22% increase), but the standard deviation also climbs to $90.11 billion (nearly a 21% increase in variance). Looking at Eq. (24.6) might suggest almost no change in the policy mix, as a result, but comparing the second rows of Tables 24.4 and 24.5 would support, instead, an increased emphasis on a climate-based intervention because the negative covariance of such a policy and possible climate-based outcomes has grown in magnitude (the relative value of climate policy has grown significantly in the upper tail of the climate sensitivity distribution). Notice, though, that these adjustments have little effect on the mean temperature increase; indeed, only the standard deviation seems to be affected.

Given the wide range of temperature change sustained by either "optimal" climate intervention, we now turn to exploring how best to design a Greenspan-inspired hedge against a critical threshold. If, to construct another example, a 3°C warming were thought to define the boundary of intolerable climate impacts, then the simplified DICE framework under the median assumption of a 3°C climate sensitivity would require a climate policy that restricted greenhouse gas concentrations to roughly 550 parts per million (in carbon dioxide equivalents). Adhering to a policy targeted at this concentration limit would, however, fall well short of guaranteeing that the 3°C threshold would not be breached. As shown in the first row of Table 24.6, in fact, focusing climate policy on a concentration target of 550 ppm would produce only a distribution of temperature change across the full range of climate sensitivities with nearly 40% of the probability anchored above 3°C. The associated discounted economic

Table 24.6 Exploring the economic value of a concentration-targeted climate policy contingent on climate sensitivity in the modified DICE environment.
Return in trillions of 1995$ and maximum temperature change in °C.

Climate sensitivity	1.5°	2.0°	3.0°	4.0°	5.0°	6.0°	7.0°	8.0°	9.0°	Mean	Standard deviation
Economic value	−3.04	−2.49	−1.58	−0.99	−0.61	−0.35	−0.17	−0.03	0.08	−1.81	1.12
Max ΔT	1.83	2.31	3.00	3.45	3.79	4.06	4.29	4.48	4.65	2.86	0.96

Table 24.7 Exploring the economic value of the responsive temperature-targeted climate policy contingent on climate sensitivity in the modified DICE environment.
Return in trillions of 1995$ and maximum temperature change in °C.

Climate sensitivity	1.5°	2.0°	3.0°	4.0°	5.0°	6.0°	7.0°	8.0°	9.0°	Mean	Standard deviation
Economic value	−0.01	−0.81	−1.58	−1.03	−0.56	−0.25	−0.02	0.13	0.24	−0.54	0.60
Max ΔT	2.87	3.00	3.00	3.00	3.00	3.00	3.00	3.00	3.00	2.96	0.06

Table 24.8 Exploring the economic value of the responsive optimization after a temperature-targeted hedge contingent on climate sensitivity in the modified DICE environment.
Return in billions of 1995$.

Climate sensitivity	1.5°	2.0°	3.0°	4.0°	5.0°	6.0°	7.0°	8.0°	9.0°	Mean	Standard deviation
Economic value	0	35	114	165	205	249	270	296	311	106.15	106.19

values of this policy intervention (given the DICE calibration of damages) are recorded in the second row, and they are not very attractive. Indeed, the concentration target policy would produce a positive value only if the climate sensitivity turned out to be 9 °C and the expected value shows a cost of $1.807 trillion (with a standard deviation of more than $1.1 trillion).

Table 24.7 shows the comparable statistics for a responsive strategy of the sort described above; it focuses on temperature and not concentrations, so it operates closer to the impacts side of the climate system. In this case, the policy is adjusted in 2035 to an assumed complete understanding of the climate sensitivity so that the temperature increase is held below the 3 °C threshold (barely, in the case of a 9 °C climate sensitivity). In this case, the reduced damages associated with designing a policy tied more closely to impacts dominates the cost side and reduces the expected economic cost of the hedge to a more manageable $535 billion with a standard deviation of nearly $600 billion. Moreover, we know from the first section that beginning this sort of hedging strategy early not only reduces the cost of adjustment in 2035, but also preserves the possibility of meeting more restrictive temperature targets should they become warranted and the climate sensitivity turn out to be high.

Finally, Table 24.8 illustrates what would happen if it were determined in 2035 that the 3 °C temperature target was not required so that adjustment to an optimal deterministic policy would be best. Notice that hedging would, in this eventuality, produce non-negative economic value regardless of which climate sensitivity were discovered. Indeed, the mean economic value (discounted to 2005) exceeds $100 billion. Even though the variance around this estimate is high (caused in large measure because the value of the early hedging would be very large if a high climate sensitivity emerged), this is surely an attractive option.

24.5 Concluding remarks

The numerical results reported in Section 24.4 are surely model dependent, and they ignore many other sources of uncertainty that would have a bearing on setting near-term policy. They are not, however, the point of this paper. The point of this paper is that decisionmakers at a national level are already comfortable with approaching their decisions from a risk-management perspective. As a result, they should welcome climate policy to their arsenal of tools when they come to recognize climate change and its potential for abrupt and intolerable impacts as another source of stress and uncertainty with which they must cope. In this context, the numbers are important because they are evidence that currently available methods can provide the information that they need. Moreover, they are also important because they provide evidence from the climate arena to support the insight drawn from a manipulation of the Brainard framework (where uncertainty about the effect of policy is recognized) that doing nothing in the near term is as much of a policy decision as doing something.

Acknowledgements

I gratefully acknowledge the contributions and comments offered by participants at the 2004 Snowmass Workshop and two anonymous referees; their insights have improved the paper enormously. I would also like to highlight the extraordinary care in editing and the insightful contributions to content provided by Francisco de la Chesnaye; Paco's efforts were "above and beyond". Finally, the role played by William Brainard, my dissertation adviser at Yale, in supporting this and other work over the years cannot be overestimated – although I expect that Bill will be surprised to see his work on monetary policy cited in the climate literature. Remaining errors, of course, stay at home with me.

References

Alley, R. B., Marotzke, J., Nordhaus, W. *et al.* (2002). *Abrupt Climate Change: Irreversible Surprises.* Washington DC: National Research Council.

Andronova, N. G. and Schlesinger, M. E. (2001). Objective estimation of the probability density function for climate sensitivity. *Journal of Geophysical Research* **106** (D190), 22 605–22 611.

Brainard, W. (1967). Uncertainty and the effectiveness of monetary policy. *American Economic Review* **57**, 411–424.

Greenspan, A. (2003). Opening remarks, *Monetary Policy and Uncertainty: Adapting to a Changing Economy.* Federal Reserve Bank of Kansas City, pp. 1–7.

Greenspan, A. (2004). Risk and uncertainty in monetary policy. *American Economic Review* **94**, 33–40.

IPCC (2001). *Climate Change 2001: Impacts, Adaptation and Vulnerability. Contribution of Working Group II to the Third Assessment Report of the Intergovernmental Panel on Climate Change*, ed. J. J. McCarthy, O. F. Canziani, N. A. Leary, D. J. Dokken and K. S. White. Cambridge: Cambridge University Press.

Jones, P. D. and Moberg, A. (2003). Hemispheric and large-scale surface air temperature variations: an extensive revision and an update to 2001. *Journal of Climate* **16**, 206–223.

Keller, K., Bolker, B. M. and Bradford, D. F. (2004). Uncertain climate thresholds and optimal economic growth. *Global Environmental Change* **48**, 723–741.

Lempert, R. and Schlesinger, M. E. (2000). Robust strategies for abating climate change – an editorial essay. *Climatic Change* **45**, 387–401.

Manne, A. S. (1995). *A Summary of Poll Results: EMF 14 Subgroup on Analysis for Decisions under Uncertainty.* Stanford University.

Nordhaus, W. D. and Boyer, J. (2001). *Warming the World: Economic Models of Global Warming.* Cambridge, MA: MIT Press.

Nordhaus, W. D. and Popp, D. (1997). What is the value of scientific knowledge? An application to global warming using the PRICE model. *Energy Journal* **18**, 1–45.

Reinhart, V. (2003). Making monetary policy in an uncertain world. In *Monetary Policy and Uncertainty: Adapting to a Changing Economy.* Federal Reserve Bank of Kansas City.

Revkin, A. (2005). Official played down emissions' links to global warming. *New York Times*, June 8th.

Schlesinger, M. E., Yin, J. Yohe, G. *et al.* (2005). Assessing the risk of a collapse of the Atlantic thermohaline circulation. In *Avoiding Dangerous Climate Change.* Cambridge: Cambridge University Press.

Schneider, S. (2003). *Abrupt Non-linear Climate Change, Irreversibility and Surprise.* ENV/EPOC/GSP(2003)13. Paris: Organization for Economic Cooperation and Development.

Smith, J. and Hitz, S. (2003). *Estimating the Global Impact of Climate Change.* ENV/EPOC/GSP(2003)12. Paris: Organization for Economic Cooperation and Development.

Tol, R. S. J. (1998). Short-term decisions under long-term uncertainty. *Energy Economics* **20**, 557–569.

Walsh, C. E. (2003). Implications of a changing economic structure for the strategy of monetary policy. In *Monetary Policy and Uncertainty: Adapting to a Changing Economy.* Federal Reserve Bank of Kansas City.

Yohe, G. (1996). Exercises in hedging against extreme consequences of global change and the expected value of information. *Global Environmental Change* **6**, 87–101.

Yohe, G., Andronova, N. and Schlesinger, M. E. (2004). To hedge or not to hedge against an uncertain climate future. *Science* **306**, 416–417.

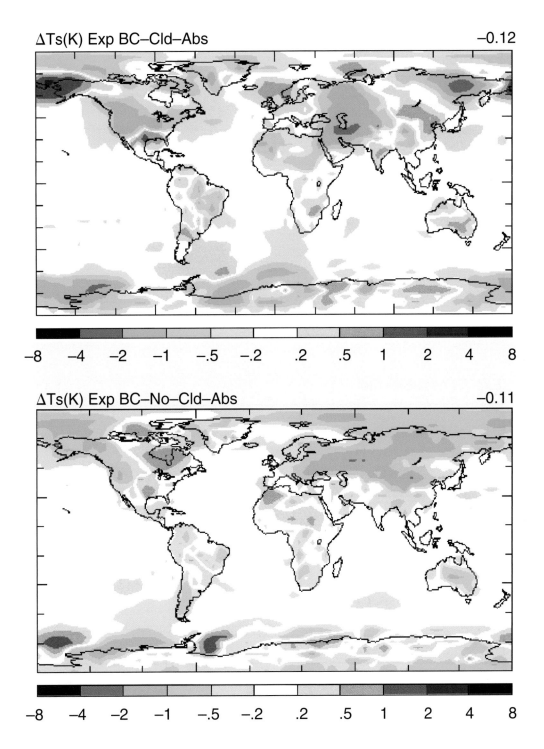

Figure 3.1 Model simulated annual surface temperature change (K) for year 2000 − Year 1850 for simulations that account for BC absorption in-cloud (top panel) and that do not account for BC (bottom panel).

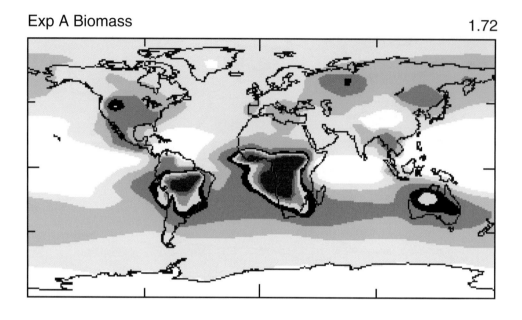

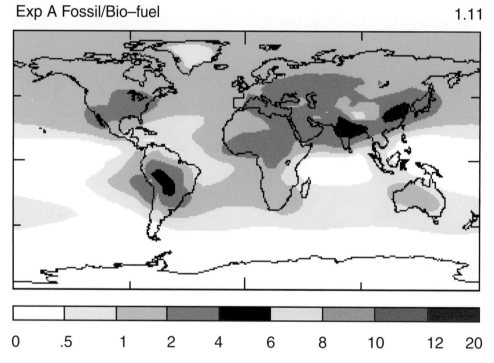

Figure 3.2 Annual values of carbonaceous aerosol column burden distribution (mg/m^2) from biomass (top panel and fossil- and biofuel sources (bottom panel). Global mean values are on the right-hand side of the figure.

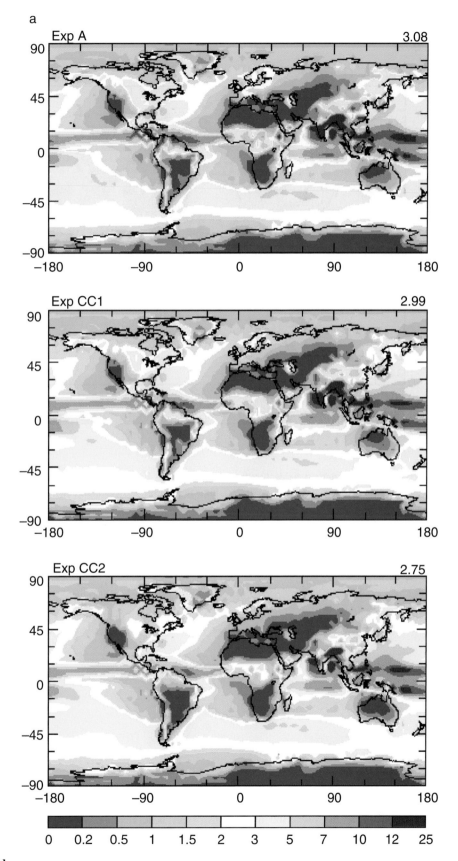

Figure 3.3 Continued

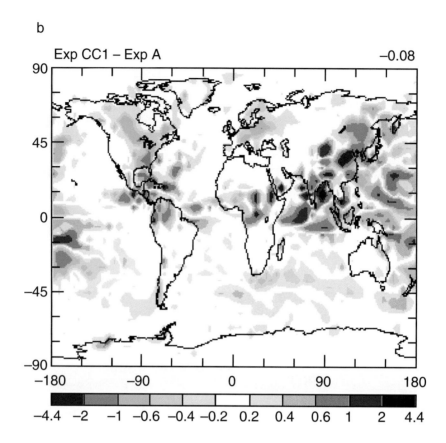

Figure 3.3 June–July–August precipitation (mm/day) fields for the year 2000 from Exp A, Exp CC1 and Exp CC2 (a), and change in precipitation between Exp CC1 and Exp A (b). Global mean values are indicated on the right-hand side.

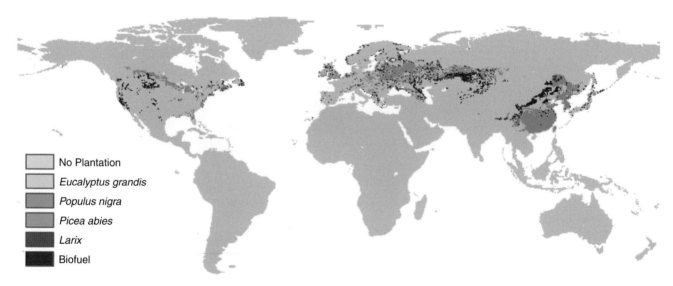

Figure 6.2 Carbon-plantation tree types for the year 2100 in the IM-C experiment. Because of the extra surplus-NPP constraint on C-plantations and bioclimatic limits, the total area of these is smaller than that of biomass plantations. The additional area of biomass plantations in the IM-bio experiment is indicated in red. Land-cover changes for regions other than the northern hemisphere regions selected for the sensitivity experiments in this paper are not shown here.

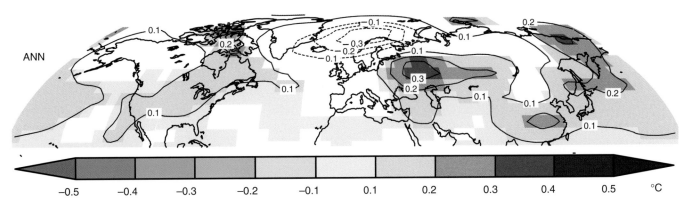

Figure 6.8 Difference in annual-mean surface-air temperature in 2071–2100 (°C) of carbon-plantation with respect to biomass-plantation ensemble mean, including albedo effects. Contours are plotted for all model grid cells. Colored are the grid cells for which the difference between the two ensemble means is significant above the 95% level (2-tailed t-test).

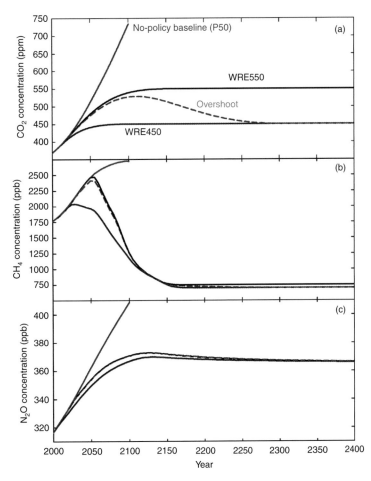

Figure 7.1 (a) Revised WRE and a new overshoot concentration stabilization profile for CO_2 compared with the baseline (P50) no-climate-policy scenario. (b) Methane concentrations based on cost-effective emissions reductions (Manne and Richels, 2001) corresponding to the WRE450, WRE550, and overshoot profiles for CO_2. The baseline (P50) no-climate-policy scenario result is shown for comparison. (c) Nitrous oxide concentrations based on cost-effective emissions reductions (Manne and Richels, 2001) corresponding to the WRE450, WRE550, and overshoot profiles for CO_2. The baseline (P50) no-climate-policy scenario result is shown for comparison.

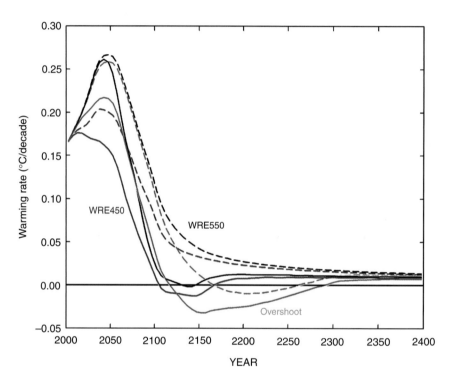

Figure 7.4 Rates of change of global-mean temperature (°C/decade) for the temperature projections shown in Figure 7.3a.

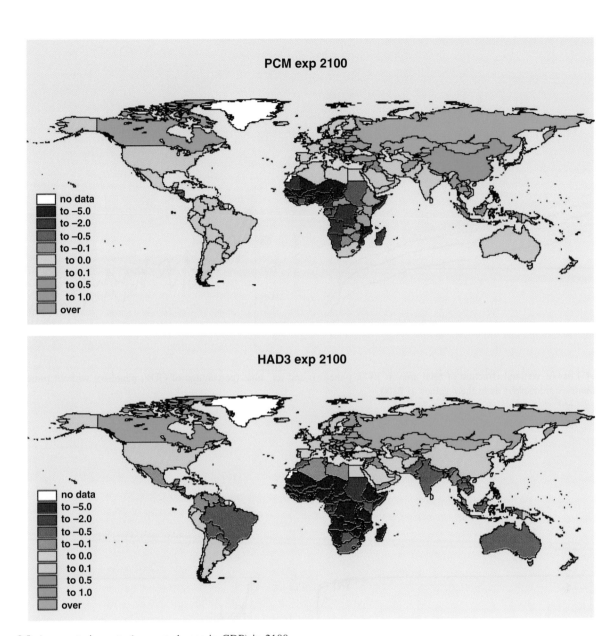

Figure 9.8 Aggregate impacts (percent change in GDP) in 2100.

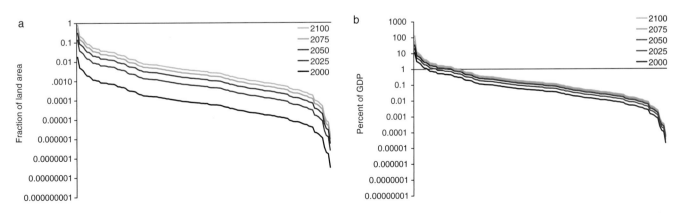

Figure 10.3 Loss of dryland (fraction of total area in 2000; panel(a)) and its value (percentage of GDP; panel(b)) without protection. Countries are ranked as to their values in 2100.

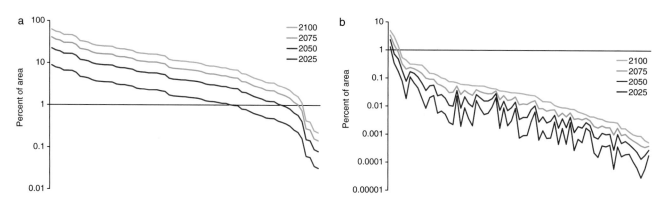

Figure 10.4 Loss of wetland (fraction of total area in 2000; panel(a)) and its value (percentage of GDP; panel(b)) without protection (left panels). Countries are ranked as to their values in 2100.

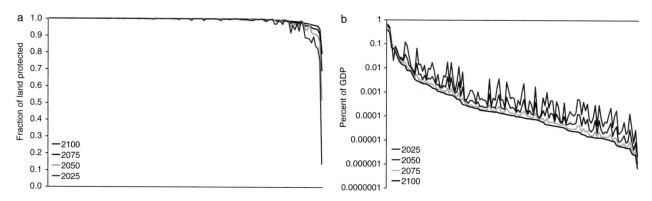

Figure 10.5 Protection level (fraction of coast protected; left panel) and the costs of protection (percent of GDP; right panel). Countries are ranked as to their protection level.

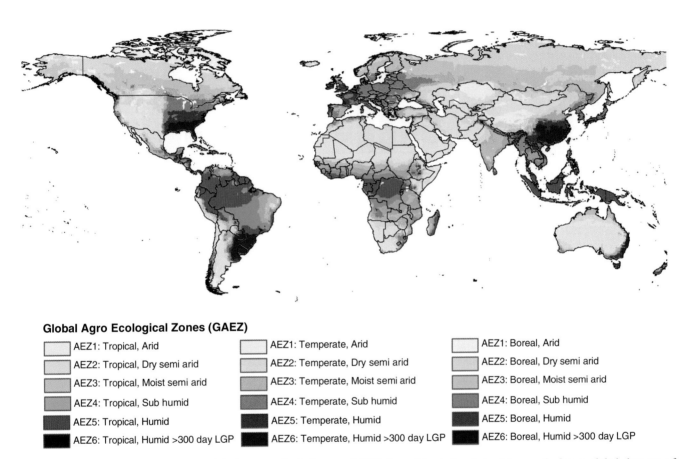

Figure 21.2 The global distribution of global agro-ecological zones (AEZ) from this study, derived by overlaying a global data set of length of growing periods (LGP) over a global map of climatic zones. LGPs in green shading are in tropical climatic zones, LGPs in yellow-to-red shading lie in temperate zones, while LGPs in blue-to-purple lie in boreal zones. LGPs increase as we move from lighter to dark shades.

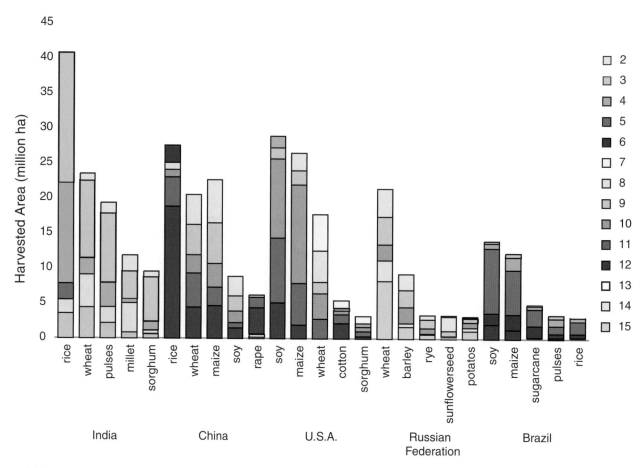

Figure 21.5 The top five nations of the world, in terms of total harvested area, the top five crops and their harvested areas within each nation, and the AEZs they are grown in. Tropical AEZs are shown in green, temperate in yellow-to-red, and boreal in blue-to-purple.

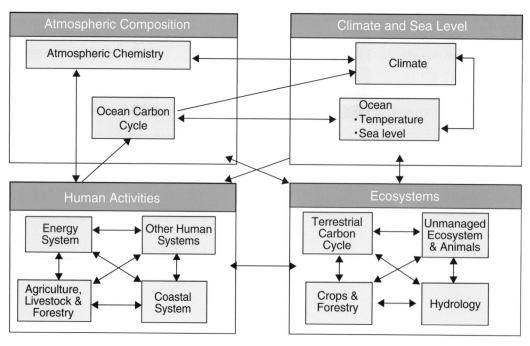

Figure 31.1 A schematic diagram of the major components of a climate-oriented integrated assessment model.

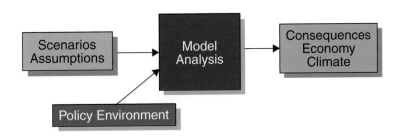

Figure 31.2 A schematic diagram of the elements of an IA model-based policy process.

Part IV

Policy design and decisionmaking under uncertainty

Charles Kolstad and Tom Wilson, coordinating editors

Introduction

The ultimate goal of nearly all climate research, and certainly the motivation behind government investments in climate research, is to develop effective and efficient approaches for addressing climate change. Climate change is not just an environmental issue – it is an energy issue and an economic development issue. Global in extent, it is characterized by its century scale as well as substantial uncertainties in science (highlighted in earlier chapters of this book) and economics.

This section examines a number of important issues in climate policy. While not comprehensive, the chapters tackle many of the critical factors that are likely to drive climate policies.

In Chapter 25, Billy Pizer addresses the fundamental question of how climate policy should change given real uncertainty about climate impacts and in fact, most dimensions of the climate problem. Does the presence of uncertainty imply different policy prescriptions than if uncertainty were absent or modest?

In Chapter 26, Mikiko Kainuma and colleagues present a suite of disaggregated models, collectively known as the Asia–Pacific Integrated Model, used to formulate climate policy in the Asian region, particularly Japan. The chapter demonstrates the usefulness of the model in looking at specific policies to mitigate greenhouse gas emissions.

In Chapter 27, W. David Montgomery and Anne Smith take a closer look at policies to induce innovation in technologies for abating carbon or otherwise reducing emission of greenhouse gases. The authors argue that imposing a cost on emitting carbon is, by itself, ineffective at catalyzing fundamental technological advances. The authors suggest other incentive schemes for inducing the innovation needed to make mitigation cheaper and easier.

Klaus Keller and colleagues focus on a different issue in Chapter 28: threshold effects of climate change. Threshold effects are major abrupt and non-marginal changes in the climate from a marginal increase in forcing. Examples are the disintegration of the West Antarctic ice sheet and the shut down of the gulf stream. The authors look at the interesting question of what kinds of policies can be used to avoid such abrupt changes.

In Chapter 29, Mort Webster invokes the well-known metaphor of boiling a frog by placing it in cold water and slowly raising the temperature. Webster suggests that integrated assessment models that suggest very modest current mitigation may end up with mankind never taking steps to control climate change. He suggests that this myopia may end up with major non-marginal changes in the climate and nothing having been done to avert the problem. By using a more sophisticated model of sequential decisionmaking under uncertainty, he shows that this is in fact non-optimal.

In Chapter 30, Ferenc Toth examines directly the problem of preventing dangerous climate change in a context of uncertainty. He examines the meaning of Article 2 of the Framework Convention on Climate Change (FCCC) which stipulates the goal of preventing such dangerous change. He then looks at various methods and models for structuring policy to deal with this goal.

Hugh Pitcher and colleagues take a step back in Chapter 31, looking at the past 10 years of Snowmass workshops on Climate Change Impacts and Integrated Assessment. They review what has been learned and what remains to be learned about integrated assessment of climate.

In Chapter 32, Richard Richels, the late Alan Manne and Tom Wigley look at the problem of limiting temperature increases, rather than just limiting atmospheric concentrations of greenhouse gases. This chapter is important, not only because of the substance of the chapter but also because one of its authors, Alan Manne, passed away between the time of the 2004 workshop where the paper was originally presented and the publication of the book. Alan was one of the great analysts of the climate change problem, providing innovative analyses and original insights that have benefited many. Alan is greatly missed. Similar to the goals in Chapter 30, the authors in this chapter argue that focusing on temperature increases rather than concentration increases allows one to better implement the avoidance of dangerous climate change embodied in the FCCC.

Taken together, the papers in this section provide a wide variety of insights into important problems facing the policy community as it seeks solutions to the problem of climate change.

25

Climate policy design under uncertainty

William Pizer

25.1 Introduction

The uncertainty surrounding the costs and benefits associated with global climate change mitigation creates enormous obstacles for scientists, stakeholders, and especially policy-makers seeking a practical policy solution. Scientists find it difficult to accurately quantify and communicate uncertainty; business stakeholders find it difficult to plan for the future; and policymakers are challenged to balance competing interests that frequently talk past each other.

Most emissions trading programs to date have focused on absolute caps that either remain fixed or decline over time. Examples include the US SO_2 trading program and NO_x Budget Program, the EU Emissions Trading Scheme (EU ETS), Southern California's NO_x RECLAIM program, and a host of other regional pollutant trading schemes in the United States.[1] Even the Kyoto Protocol, by most accounts, is viewed as a first step in capping emissions that must then lead to even lower levels in subsequent periods.

Yet the uncertainty surrounding climate change suggests that such an approach to regulating greenhouse gas emissions is problematic. On the one hand, we are unsure about what atmospheric concentrations need to be in the long run to prevent dangerous interference with the climate system.[2] And regardless of the stabilization target, considerations of the global economic system and its dependence on fossil fuels suggests that optimal global emissions trajectories will continue to grow for some time (Wigley *et al.*, 1996; Manne and Richels, 1999). On the other hand, economic analysis suggests that for most plausible assumptions, price-based policies – rather than quantitative caps – are more efficient for addressing the problem of climate change in the face of uncertainty (Pizer, 2002; Newell and Pizer, 2003).

From an economic point of view, then, emissions caps – and in particular non-increasing emissions caps – may be less desirable in the near term. Other arguments arise in the non-economic rhetoric against caps, positing that we do not know enough to proceed with limits on carbon dioxide emissions, that such limits would otherwise impede economic growth, and that the most prudent action is technology policy.

With concern over simple emissions caps and the consequences of uncertainty as a backdrop, this paper explores how two mechanisms – an intensity-based emissions target and a price-based safety valve – can attenuate some of the economic downsides of emission caps. Importantly, the discussion does not start and end with a neo-classical economics perspective. The most simple and elegant solution may not, in

---

[1] Fuel taxes do exist (e.g., gasoline) but are typically motivated by revenue concerns. There is also a jurisdictional problem that tax-based environmental regulation would probably fall under the purview of the Finance/Ways and Means Committees, rather than Environment and Public Works/Environment and Natural Resource Committees. Perhaps most importantly, viewing environmental regulation as a tax is clearly seen as a negative – and is used that way by stakeholders seeking to diminish support for such regulations (National Mining Association, 2005).

[2] This is the language used in the UN Framework Convention on Climate Change to describe the eventual goal of global cooperation on climate change mitigation.

Human-induced Climate Change: An Interdisciplinary Assessment, ed. Michael Schlesinger, Haroon Kheshgi, Joel Smith, Francisco de la Chesnaye, John M. Reilly, Tom Wilson and Charles Kolstad. Published by Cambridge University Press. © Cambridge University Press 2007.

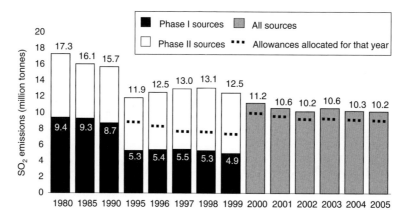

Figure 25.1 Allocations and use of SO_2 allowances.
Note: The increase in allocation in 2000 arose from inclusion of Phase II source allocations in that year and thereafter.
Source: US EPA (2006).

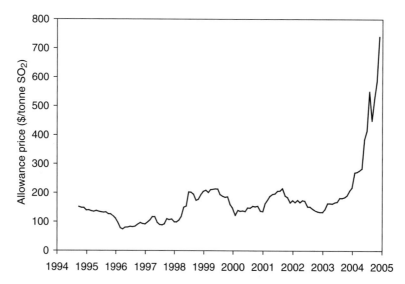

Figure 25.2 Allowance price of SO_2.
Source: Cantor Fitzgerald Environmental Brokerage.

fact, satisfy key stakeholder concerns. Without trying to explain these concerns fully, the paper considers how policy design can disarm the non-economic rhetoric.

25.2 Background on market-based programs: experience with cost uncertainty

Market-based programs have a long and varied history in the United States. While the SO_2 trading program is perhaps the best-known example, trading programs have been used to regulate NO_x emissions, halibut and sablefish fisheries, lead in gasoline, and even pollution from heavy-duty engines.[3] Some, including the regulation in lead in gasoline and pollution from heavy-duty engines, are examples of tradable performance standards. That is, pollution allowances are given out in relation to production; more production means more allowances and more pollution – an idea we will come back to a little later.

Historically, major policy actions typically follow after a convergence of opinions about costs and benefits. In the case of the 1990 Clean Air Act Amendments creating the SO_2/acid rain trading program, there was a remarkable alignment of science on both the cost and benefits of controls (Kete, 1993). Both suggested that a 50% reduction in emissions was scientifically justified and economically reasonable. On top of that, there was a remarkable determination within the White House to implement these reductions using a flexible cap-and-trade program.

An important feature in the SO_2 cap-and-trade program was its banking provision: allowances that remained unused at the end of any compliance period could be saved for use

---
[3] See Chapter 6 of Council of Economic Advisers (2002).

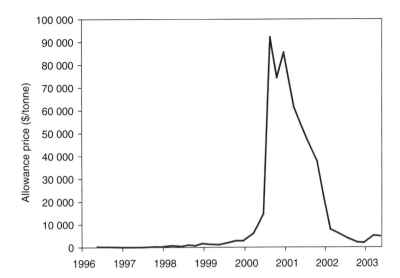

Figure 25.3 RECLAIM allowance price.
Source: Cantor Fitzgerald Environmental Brokerage.

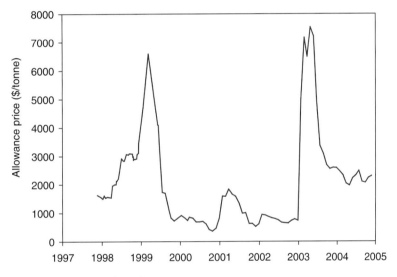

Figure 25.4 OTC Current vintage allowance price for NO_x.
Source: Cantor Fitzgerald Environmental Brokerage.

in any future compliance period without any limitations. Figure 25.1 shows the pattern of allowance use and banking through 2005. Most analyses find that this feature accounted for a large part of cost-savings (Carlson *et al.*, 2000; Ellerman *et al.*, 2000). In addition to smoothing out the path of emissions between the two phases of the program, allowance banking also created a large reserve that could accommodate annual fluctuations in emissions levels. Starting with more than 3 million allowances after the first year and rising to more than 10 million in 1999, there were 8.6 million allowances in the bank at the end of 2003 (US EPA, 2004a).

Figure 25.2 shows SO_2 allowance prices from the beginning of the program until 2005. Despite large (10%) fluctuations in annual emissions, allowance prices remained in a fairly narrow range of $100–$200 per tonne until 2005, when expectations of further reduction requirements under the Clean Air Interstate Rule (CAIR) began to incite additional abatement and banking.[4] The same cannot be said for other trading programs in which banking is not allowed or has not developed.

For example, Figure 25.3 shows the path of allowance prices in the RECLAIM market in the South Coast Air Quality Management District surrounding Los Angeles. The RECLAIM program included one-year compliance periods with no banking. Because the compliance periods were

---

[4] Under the proposed rule, current allowances could be banked and used to comply with tighter emissions limits in the future.

overlapping for different sources – some ended in December and others in June – it would have been possible to create a bank through trading. However, no such bank developed, and when the California energy crisis struck in the fall of 2000 and NO_x-emitting, gas-fired electricity generation increased to fill the gap caused by abnormally low hydroelectric generation, prices skyrocketed from about $3000 per tonne to more than $70 000.

Figure 25.4 shows the price of allowances in the NO_x Ozone Transport Commission (OTC) market and subsequent State Implementation Plan (SIP) call market (after 2003; jointly referred to as the NO_x Budget Program). The NO_x market allows use of banked allowances up to 10% of annual allocations without penalty; after 10%, banked allowances must be retired 2:1 for every tonne emitted (referred to as "flow control"). This occurred, for example, after a large number of allowances were banked in 1999. Also, when the OTC program was subsumed into the SIP call program in 2003, banked allowances could only carry over in 2003 with considerable restrictions and limits. This banking constraint, coupled with concerns about high natural gas prices, explains why prices jumped in 2003. The jump in 1999, the first compliance year, occurred because of uncertainty about allocation and participation as the program started – again without a bank.[5]

The lesson from all of these market-based programs regarding cost uncertainty is fairly clear: a large and unrestricted bank substantially buffers prices, as it has in the SO_2 program. Prices in that program remained in the $100–$200 range for nearly 10 years, until expectations of a tighter program in the future drove prices up to $700, close to their expected level under a future program. In contrast, a restricted or absent bank creates circumstances where prices can spike up by a factor of 5, 10, or even 20 times in the face of a shortage.

25.3 Uncertainty about climate change mitigation benefits

Among the many features that distinguish climate change mitigation from conventional pollution control, two loom large. One is the timescale: whereas SO_2 and NO_x controls are focused on immediate concentration levels related to current emissions, greenhouse gas controls are focused on long-term concentration levels related to accumulated emissions over many decades. The other is uncertainty: we have relatively strong epidemiological evidence on fine particulates that has allowed us to estimate benefits of the order of $500–$2300 for NO_x and $3700–$11 000 for SO_2.[6] At the same time, a recent summary of various studies of climate change mitigation found estimated benefits of between $3 and $20 per tonne of CO_2 – a slightly higher high-to-low estimate ratio of six compared with the 3–5 range associated with SO_2 and NO_x.[7] Yet broader studies find wider ranges for CO_2; Tol (2005) finds estimates as high as $450.

Distinct from these cost–benefit analyses, decisionmakers and scientists can alternatively describe a "safe" level of greenhouse gases in the atmosphere and seek an emissions trajectory that preserves it. This is the approach suggested by the United Nations Framework Convention on Climate Change (UNFCCC), whose goal is to stabilize concentrations at levels that avoid "dangerous anthropogenic interference with the climate system," and in the United States Clean Air Act, which sets air quality standards for conventional pollutants that "are requisite to protect the public health."[8] Unfortunately, this adds further hurdles to climate policy design: not only must a safe level be chosen, but also the corresponding emissions must be allocated across nations and time.[9]

This uncertainty about mitigation benefits has been the major reason why the Bush administration has avoided mandatory emissions controls for greenhouse gases. On February 14, 2002, President Bush noted that compared with conventional pollution of SO_2 and NO_x, "the science [of climate change] is more complex, the answers are less certain, and the technology is less developed" (White House, 2002b). Earlier, in a letter on March 13, 2001, to Senator Hagel, President Bush stated, "I do not believe ... that the government should impose on power plants mandatory emissions reductions for carbon dioxide, which is not a 'pollutant' under the Clean Air Act ... This is especially true given the incomplete state of scientific knowledge of the causes of, and solutions to, global climate change and the lack of commercially available technologies for removing and storing carbon dioxide" (White House, 2001).

Uncertainty has had exactly the opposite effect in Europe (Majone, 2002) where, in addition to ratifying the Kyoto Protocol, the European Union initiated its Emissions Trading Scheme in January 2005, creating the largest emissions trading market in the world (Kruger and Pizer, 2004). While the reduction requirements, allowance price, and overall stringency remain unclear, there is no arguing that the EU has taken the question of immediate emissions reductions more seriously than the United States. The July 2005 price of allowances in the European market was 30 per tonne of CO_2 (Point Carbon, 2005).

---

[5] Here, a key question was whether Maryland, an expected net seller, would participate in the program. With the risk they would not participate, prices spiked, reflecting the expected loss of market supply; once it was clear they would be part of the program, prices returned to normal.

[6] See page 46 of OMB (2000).

[7] See Table 13 of Gillingham et al. (2004).

[8] Article 2, UNFCCC, and Section 109(b)(1), Clean Air Act.

[9] While cost–benefit analysis may involve debate over the correct measurement of benefits (and costs), it remains a well-defined paradigm. The determination of a "safe" level requires an initial debate over what constitutes "safety" before then connecting concentration (and eventually emission levels) to that definition. See also discussion in Kopp (2004).

25.4 Price-based approaches

Environmental policies, despite appearances to the contrary, typically revolve around economic costs. Even where the law requires standards set to protect the public health, without regard to cost, some notion of reasonableness enters the analysis. Often, this may be an attempt to identify the "knee" of the cost curve – the point where further environmental controls quickly become more expensive (Sagoff, 2003); or there is some basic notion of what society will bear in terms of dislocation and burden.

As President Bush noted above, there are no proven technologies for controlling greenhouse gas emissions from fossil fuel combustion; the only alternative is to burn less fuel.[10] This means switching from coal to gas, or gas to nuclear and/or renewables, or using less energy overall. It also means that emissions reductions are necessarily modest – there are no scrubbers or catalysts to reduce emissions with otherwise modest changes in behavior. Indeed, as countries have been forced to enact actual emissions limits in Europe, their National Allocation Plans have focused on levels at, or slightly above, current levels (Forrester, 2004). The proposal by the National Commission on Energy Policy similarly suggests a gradual deflection from an otherwise business-as-usual emissions path (NCEP, 2004).

The problem with seeking such modest emissions reductions is that they do not naturally allow creation of the kind of allowance bank that has arguably been so important in many existing programs. In the SO_2 program, the bank was equal to half the annual emissions limit after the first year. In the NO_x OTC program, the bank was about 20% of the annual limit after the first year.[11] Modeling of modest CO_2 policies suggest a bank of perhaps 2–3% after the first year.[12] While an eventual bank of 10% or more is possible – a range that could be adequate given annual fluctuations in baseline CO_2 emissions from weather, economic business cycles, and other events – it is unlikely to happen immediately.

This raises the question of whether and how one might avoid the risk of price spikes seen in the RECLAIM and NO_x Budget Programs. Some degree of temporarily high prices may be acceptable and even desirable as a way to preserve a given expected allowance price and to induce desired actions. However, price spikes up to 20 times the typical level are hard to justify. The highest estimate of NO_x benefits was $2300 per tonne – yet prices went to $7000 in the Budget Program and $90 000 in the RECLAIM program.

A transparent way to address these risks is to provide additional allowances at a fixed price. This so-called "safety valve" limits price spikes in much the same way that governments might intervene in currency and bond markets to limit fluctuations. The difference here is that the government is intervening in a market that was, itself, created by the government.[13] Such mechanisms have already been used in markets for individual fishing quotas and renewable portfolio standards. They were also proposed in the Bush administration's Clear Skies Initiative to reduce NO_x, SO_2 and mercury (Hg) emissions and included in the recommendations of the National Energy Commission to control greenhouse gas emissions. In some cases, the price is set at a level that is expected to be achieved, in which case the safety valve is effectively setting the market price; in other cases, the price is set at a level somewhat above the expected market price and functions more as an emergency mechanism.

As noted in Pizer (2002) and in Newell and Pizer (2003), there are clear economic arguments for favoring a price-based approach to regulating greenhouse gas emissions. Emissions accumulate gradually over time, making particular emissions levels in a given year less important than long-term downward pressure. Empirical estimates of damage functions are also hard-pressed to justify strict emissions limits. Yet more convincing, in many ways, is the common sense argument that preserving the cap at all costs is simply not worth it. Two questions logically follow: how likely is it that the cost will rise to unbearable levels, and should there be an automatic mechanism to deal with it? While allowance banking often provides a cushion that significantly reduces the likelihood of such events, the fact that a bank may be slow to develop raises concerns about whether that particular cushion will arise and how quickly. Price-based mechanisms provide an obvious and transparent alternative.

25.5 Intensity targets: disarming long-term concerns

A safety-valve mechanism can either be set at a level that will become the market price or at a level that reduces its chances of being activated. For example, in the Clear Skies Initiative, the safety-valve prices for NO_x, SO_2, and Hg were set at about three times the estimated market price of allowances. In contrast, the National Commission on Energy Policy recommended a safety-valve price for CO_2 that would quickly become the market price under the reference case modeling.

This raises an interesting question: what is the best way to set the safety-valve level and/or the associated quantitative emissions target? With well-defined costs and benefits, it is straightforward to determine the welfare-maximizing

---

[10] Capture and sequestration is theoretically possible at central power stations but at estimated costs of $50–70 per tonne of CO_2 (Anderson and Newell, 2004). Despite small projects and trials, large-scale projects have yet to demonstrate the practicality of such options.
[11] Some of this reflects early reduction credits given out for activities prior to the first year of the program; during the first few years of the program, abatement continued to be 5 to 10% below the required emissions level, and the bank continued to grow (US EPA, 2004b).
[12] See Table B20 in EIA (2004) and Figure 5 in NCEP (2004).

[13] See Newell et al. (2005) and Pizer (2002).

solution. This was the general question examined by Weitzman (1978) and Roberts and Spence (1976), while Pizer (2002) focused specifically on climate change. Pizer concluded that the global welfare-maximizing choice for climate change set the safety valve at the expected marginal benefit and the emissions target substantially below the otherwise "optimal" target without a safety valve – that is, a fairly aggressive emissions target that will trigger the safety valve most of the time.

Of course, economic optimization is not the only criterion for policy. One reason for preferring an emissions trading program with a safety valve over a carbon tax is that the government can give out most of the emissions rights rather than making emitters pay for them.[14] This is itself not without economic consequence, as the revenues could be used to reduce the burden of the existing tax system (Parry, 1995). However, the interest among emitters in receiving those rights is quite strong (Braine, 2003) and efforts to use them in other ways creates additional opposition to the proposed policy (Samuelsohn, 2003). This is true despite studies suggesting that only a fraction of the allowances are needed to make key emitting sectors equally well off (Goulder and Bovenberg, 2002).

This further suggests that setting a reasonable emission target may be just as important as setting a reasonable safety valve. In addition to this concern over allowance allocation, there are two other reasons the target is important. First, even if a safety valve is proposed, stakeholders may be concerned that the safety-valve mechanism will be removed – either before the policy is finalized or phased out at some point in the future (Peabody Energy, 2005). Indeed, as pointed out all along, the presence of a sufficiently large bank could serve the same insulating purpose, making the safety valve less important in the future. Second, as a negotiating tool, the emissions target has important symbolic and practical consequences, given uncertainty about future targets. Namely, it becomes a natural focal point for discussions of both future targets and, in an international setting, targets in other nations. Were there no uncertainty about mitigation costs and benefits and the likely level/path of future emissions reductions, this would be much less important.

As it is, there is tremendous uncertainty surrounding future reductions. Part of this uncertainty is over future technological developments and economic growth, and the concern is whether an emissions limit, however modest, will in time become unduly onerous. Such concern underlies the Byrd–Hagel resolution, reflecting the US Senate's reluctance to consider any climate treaty that would adversely affect the US economy (US Senate, 1997). Similarly, a common theme among critics of any action, however modest, is that it is simply a vehicle to force much more aggressive reductions in the future – the proverbial camel's nose under the tent (Payne, 2001). If past policies for SO_2 and NO_x are any guide, that tendency to reduce, rather than slow the growth of, emissions would certainly seem plausible. Yet most economic analysis suggests optimal trajectories with rising emissions for the next decade or two (Wigley et al., 1996; Manne and Richels, 1999). Note that the issue here is not whether historic emissions trajectories for SO_2 and NO_x were or were not optimal, or whether future emissions trajectories for greenhouse gases will be optimal – the issue is that existing policy tools inevitably carry with them the history and associations related to their past use.

This is where an intensity target may be able to help. An intensity target, analogous to the tradable performance standards used in some domestic pollution programs, would target emissions levels in relation to aggregate economic activity. As the economy grows, an intensity target would naturally allow more emissions. In fact, because economies typically grow faster than their use of energy and emissions of greenhouse gases, an intensity target would probably come in the form of a rate of decline, rather than an absolute level. For example, the Bush administration called for an 18% decline in emissions intensity over 10 years (White House, 2002a). The National Commission on Energy Policy recommended a 2.4% annual decline (NCEP, 2004b, p. 14).

Unlike a cap, an intensity target does not automatically become more onerous as time passes: rather, a conscious choice must be made to lower the target. Thus, while an intensity target can achieve the same results as an ordinary emissions cap, it does not have the same natural tendencies to do so and, in fact, tends to be less aggressive. In an uncertain world where targets are being set and then revised, this changes expectations about what future targets might look like and could assuage critics' fears about a modest policy evolving into an aggressive one. It is worth noting that these uncertainties and concerns about the consequences of an ordinary emissions cap, as well as the potential advantages of an intensity target, are only amplified for developing countries.

It is also worth noting that an intensity approach is in no way an alternative to the safety valve in terms of ameliorating short-term price spikes. There is a tendency, after recognizing that some amount of cost uncertainty hinges on fluctuations in annual economic growth, to imagine emissions intensity targets as a natural way to hedge against such fluctuations. Unfortunately, while emissions growth is tied to economic growth, so is improvement in technology – so much so that intensity targets could lead to higher price spikes. Worse, those price spikes occur in the face of low growth rather than high growth (Ellerman and Sue Wing, 2003; Pizer, 2004). For these reasons, it makes sense to think about intensity targets as a way to frame long-term goals but not as a way to accommodate short-term economic fluctuations.

Thinking about intensity targets simply as a way to frame long-term goals implies that they can, at a practical level,

---

[14] Note that this would not be true for an extremely aggressive cap – say one tending toward zero – where there are no emissions rights to distribute. For CO_2, however, even an aggressive cap implies only a 10% or so reduction from business-as-usual forecasts in the near term (see Weyant [1999] for a summary of the relationship between emissions prices and abatement levels).

be implemented as an ordinary emissions trading program. That is, the intensity target is applied to forecasts of economic activity over a fixed horizon (say 10 years). The resulting emissions levels then become the target in an ordinary trading program. This is exactly the formulation under a recent proposal by Senator Jeff Bingaman (Senate Energy Committee, 2005).

Finally, whether one views intensity targets as a good thing hinges on both positive and normative arguments – none of which are as clear from an economics point of view as the price/quantity distinction. On the positive side, the argument is that equivalent intensity and emissions targets are likely to lead to different targets in the future. The Bush administration, for example, has a target of reducing intensity by 18% over 2002–2012 to 150 tonnes per million dollars of GDP. That is equivalent to an absolute target of 2173 million tonnes (White House, 2002a). Ignoring for a moment the voluntary nature of the target, the point is that framing the 2012 target as 150 tonnes per million dollars of GDP is likely to lead to a different choice of targets after 2012 compared with a target described as 2173 million tonnes – even though they are equivalent in 2012.

To a dispassionate analyst, this may sound crazy: why would the way the initial target is described – given that either achieves precisely the same emissions and cost outcome – lead to a different target choice in the future? The answer is that this choice of description may, in fact, *not* lead to a different target choice. But if stakeholders *believe* it might make a difference, in much the way that people may believe certain types of advertising make a difference, it can affect their support for the policy. In particular, stakeholders may be concerned that to show progress, the numbers "need to go down." Because GDP is growing, emissions can go up while intensity goes down. The same is not true for an ordinary emissions target.

If one agrees with this positive argument that there is a difference, or at least people believe there is a difference, the question becomes practical – which formulation brings together the necessary people? For those convinced that emissions need to start going down immediately, perhaps because of preferences for a particularly low concentration goal, it would be absolute targets. For those concerned more about protecting the economy, the ability to fine-tune the target without jumping to emission decreases makes intensity targets appealing.

25.6 Conclusions

The uncertainty surrounding the costs and benefits of climate change mitigation make it particularly challenging for policymakers: we are unsure about the appropriate response, we are unsure how the appropriate response will change over time, and we are unsure how much any particular response – viewed in terms of an emissions level – will cost. While early work in the area of uncertainty and climate change focused on the role of sequential decisionmaking and learning (Nordhaus, 1994; Kelly and Kolstad, 1999), this paper has emphasized more recent thinking about the policy design at a particular decision point.

One key observation has been that the economics and politics of greenhouse gas mitigation argue for policies based on allowance prices rather than allowance quantities. Harking back to Weitzman (1974), marginal benefits are relatively flat compared with costs. But more to the point, stakeholders are unlikely to stomach substantial price risk regardless of academic debates about relative slopes. While banking has proved to be a useful mechanism to hedge price risks in many emissions trading markets, a bank may be less effective in the climate change context, at least initially, because the small abatement levels make accumulation of a bank more difficult. This is true even with the relatively generous allocation schemes considered in recent proposals. One transparent solution would be a price-based safety valve – that is, a price at which the government would sell addition emissions permits in order to keep allowance prices and costs from rising above acceptable levels.

A second issue is that while emissions caps have historically been used to reduce emissions levels rather than slowing emissions growth, in the case of climate change, optimal emissions paths typically involve a period of slowed growth before absolute levels decline. Popular criticism, as well as criticism from developing countries, focuses on the related fear that caps will become a limit to growth and pose an undue economic burden on domestic economies. While a safety valve technically addresses the issue by allowing emissions to increase above any specified cap if costs are too high, there is still political uncertainty about whether the safety valve will always remain in place and, in turn, whether current or future emissions caps will become more relevant. Targets based on improvements in aggregate intensity – emissions per volume of economic activity – might be more suitable for trajectories that initially slow, rather than halt, emissions growth and could attenuate this fear by more naturally accommodating emissions growth alongside economic growth.

When thinking about the uncertainty surrounding climate change, many people focus on the uncertainties surrounding the consequences of greenhouse gas emissions – temperature and precipitation, sea-level rise, agricultural impacts, vector-borne diseases, etc. However, it is arguably not those uncertainties (at least entirely) that have limited action in many countries. Uncertainty about costs, as well as whether targets over the next decade or two should slow growth or reduce absolute emissions levels, implies that the use of traditional (non-growth) emissions cap-and-trade programs is not an innocuous design choice. Economic theory is not entirely silent on the normative consequences of these choices. Perhaps more importantly, this paper posits that a practical obstacle to market-based domestic climate policies could be concern about whether a reasonable, meaningful step is possible in response to the current threat without risking excessive near-term costs or landing on a slippery slope to more costly and burdensome efforts in the future.

This paper has suggested modifications to the traditional cap-and-trade approach that reduce the uncertainty surrounding immediate costs as well as the possibility (or stakeholder belief in the possibility) that modest near-term policies could promote unduly onerous efforts in the future. To economists, most normative work on the tradeoff between cost and benefit uncertainty as well as optimal emissions trajectories supports such modifications. To those more concerned about achieving environmental improvements, these modifications may seem counterproductive: they introduce the possibility of above-target emissions via the safety valve as well as a weaker position for proposing further reductions based on emissions intensity. However, from a practical political standpoint, these or similar modifications may be necessary to reach an agreement in the near term. The question then becomes, is getting started quickly a more effective long-term strategy than waiting for a consensus for stronger action?

References

Anderson, S. and Newell, R. (2004). Prospects for carbon capture and storage technologies. *Annual Review of Environment and Resources* **29**, 102–142.

Braine, B. (2003). Sold! *Electric Perspectives*, January/February, 20–30.

Carlson, C., Burtraw, D., Cropper, M. and Palmer, K. L. (2000). SO_2 control by electric utilities: what are the gains from trade. *Journal of Political Economy* **108** (6), 1292–1326.

Council of Economic Advisers (2002). *Economic Report of the President*. Washington: Government Printing Office.

Ellerman, A. D. and Sue Wing, I. (2003). Absolute versus intensity-based emission caps. *Climate Policy* **3** (2), S7–S20.

Ellerman, A. D., Joskow, P. L., Schmalensee, R., Montero, J. P. and Bailey, E. M. (2000). *Markets for Clean Air: The US Acid Rain Program*. Cambridge: Cambridge University Press.

Energy Information Administration (EIA) (2004). *Analysis of Senate Amendment 2028, the Climate Stewardship Act of 2003*. Washington DC: EIA.

Forrester, D. (2004). *Supplemental Report on EU National Allocation Plans*. Washington: National Commission on Energy Policy, Research Compendium.

Gillingham, K., Newell, R. and Palmer, K. (2004). *Retrospective Examination of Demand-Side Energy Efficiency Policies*. Washington: RFF.

Goulder, L. and Bovenberg, L. (2002). *Addressing Industry-Distributional Concerns in US Climate-Change Policy*. Stanford: Stanford University.

Kelly, D. L. and Kolstad, C. D. (1999). Bayesian learning, growth, and pollution. *Journal of Economic Dynamics and Control* **23** (4), 491–518.

Kete, N. (1993). *The Politics of Markets: the Acid Rain Control Policy in the 1990 Clean Air Act Amendments*. Unpublished Ph.D. thesis, The Johns Hopkins University, Baltimore.

Kopp, R. (2004). *Near-term Greenhouse Gas Emissions Targets*. Discussion Paper 04–41. Washington: Resources for the Future.

Kruger, J. and Pizer, W. (2004). Greenhouse gas trading in Europe: the new grand policy experiment. *Environment* **46** (8), 8–23.

Majone, G. (2002). What price safety? The precautionary principle and its policy implications. *Journal of Common Market Studies* **40** (1), 89–109.

Manne, A. and Richels, R. (1999). The Kyoto Protocol: a cost-effective strategy for meeting environmental objectives? *Energy Journal Special Issue*, 1–23.

National Commission on Energy Policy (NCEP) (2004). *Economic Analysis of Commission Proposals*. Washington: NCEP.

National Mining Association (2005). *Bingaman Climate Proposal Rations and Taxes US Coal Production and Consumption*. Washington: National Mining Association.

Newell, R. and Pizer, W. (2003). Regulating stock externalities under uncertainty. *Journal of Environmental Economics and Management* **45**, 416–432.

Newell, R., Pizer, W. and Zhang, J. (2005). Managing permit markets to stabilize prices. *Energy and Resource Economics*, **31**, 133–157.

Nordhaus, W. D. (1994). *Managing the Global Commons*. Cambridge: MIT Press.

Office of Management and Budget (OMB) (2000). *Report to Congress on the Costs and Benefits of Regulation*. Washington: US OMB.

Parry, I. W. H. (1995). Pollution taxes and revenue recycling. *Journal of Environmental Economics and Management* **29** (3), S64–77.

Payne, H. (2001). (Mmph) What camel? *Detroit News*, October 29.

Peabody Energy (2005). *NCEP Climate Plan Cuts Coal Production*. Washington: Peabody Energy.

Pizer, W. A. (2002). Combining price and quantity controls to mitigate global climate change. *Journal of Public Economics* **85** (3), 409–434.

Pizer, W. (2004). *The Case for Intensity Targets*. Washington: Resources for the Future.

Point Carbon (2005). Midday Market Update. *Carbon Market News*, July 8.

Roberts, M. J. and Spence, M. (1976). Effluent charges and licenses under uncertainty. *Journal of Public Economics* **5** (3–4), 193–208.

Sagoff, M. (2003). Cows are better than condos, or how economists help solve environmental problems. *Environmental Values* **12** (4), 449–470.

Samuelsohn, D. (2003). Trading plans highlight differences between Clear Skies, competing bills. *Greenwire*, May 7.

Senate Energy Committee, Minority Staff (2005). *Climate and Economy Insurance Act*, draft legislation. Washington: Senate Energy Committee.

Tol, R. S. J. (2005). The marginal costs of carbon dioxide emissions. *Energy Policy* **33** (16), 2064–2075.

US Environmental Protection Agency (EPA) (2004a). *Acid Rain Program: 2003 Progress Report*. Washington: US EPA.

US EPA (2004b). *Progressive Flow Control in the OTC NOx Budget Program: Issues to Consider at the Close of the 1999 to 2002 Period*. Washington: US EPA.

US EPA (2006). *Acid Rain Program: 2005 Progress Report*. Washington: US EPA.

US Senate (1997). *A Resolution Expressing the Sense of the Senate Regarding the Conditions for the United States Becoming a Signatory to any International Agreement on Greenhouse Gas Emissions under the United Nations Framework Convention on Climate Change*. Senate reports 105–54, S 98.

Weitzman, M. L. (1974). Prices vs. quantities. *Review of Economic Studies* **41** (4), 477–491.

Weitzman, M. L. (1978). Optimal rewards for economic regulation. *American Economic Review* **68** (4), 683–691.

Weyant, J. P. and Hill, J. (1999). The costs of the Kyoto Protocol: a multi-model evaluation. Introduction and Overview. *Energy Journal Special Issue*.

White House (2001). *Text of a Letter from the President to Senators Hagel, Helms, Craig, and Roberts*. Available from www.whitehouse.gov/news/releases/2001/03/20010314.html. [cited July 7 2005].

White House (2002a). *Global Climate Change Policy Book*. White House 2002, Available from www.whitehouse.gov/news/releases/2002/02/climatechange.html. [cited August 23 2004].

White House (2002b). *President Announces Clear Skies & Global Climate Change Initiatives*. Available from www.whitehouse.gov/news/releases/2002/02/20020214-5.html. [cited July 7 2005].

Wigley, T. M. L., Richels, R. and Edmonds, J. A. (1996). Economic and environmental choices in the stabilization of atmospheric CO_2 concentrations. *Nature* **379**, 240–243.

26

Climate policy assessment using the Asia–Pacific Integrated Model

Mikiko Kainuma, Yuzuru Matsuoka, Toshihiko Masui, Kiyoshi Takahashi, Junichi Fijino and Yasuaki Hijioka

26.1 Introduction

It is predicted that global climate change will have significant impacts on the society and economy of the Asia–Pacific region, and that the adoption of measures to tackle global climate change will impose a large economic burden on the region. Also, if the Asia–Pacific region fails to adopt such countermeasures, it has been estimated that its emissions of greenhouse gases (GHG) will increase to over 50% of total global emissions by 2100.[1] To respond to such a serious and long-term threat, it is critical to establish communication and evaluation tools for policymakers and scientists in the region. The Integrated Assessment Model provides a convenient framework for combining knowledge from a wide range of disciplines, and is one of the most effective tools to increase the interaction among groups.

The Asia–Pacific Integrated Model (AIM) is one of the most frequently used models in the Asia–Pacific region (Kainuma *et al.*, 2003; Shukla *et al.*, 2004). The distinctive features of AIM are: (1) it involves Asian country teams from Japan, China, India, Korea, Thailand, and so on; (2) it has detailed description of technologies; and (3) it uses information from a detailed geographic information system to evaluate and present the distribution of impacts at local and global levels. Besides preparing country models for evaluation at the state and national level, we have also developed global models to analyze international economic relationships and climate impacts in order to evaluate policy options from a global viewpoint.

Some analyses using AIM are presented in this chapter. Section 26.2 shows how different AIM models are integrated to assist climate policy assessment. Section 26.3 estimates climate change impacts on crop productivity under different scenarios and shows the effectiveness of GHG mitigation policies. Section 26.4 estimates economic impacts of reducing GHG emissions, and Section 26.5 assesses Japanese policy to comply with the Kyoto Protocol.

26.2 Integrated modeling framework

The AIM family of models consists of more than 20 models which have been used for assessing climate policy options, classified into emission models, climate models, and impact models. These models have been used independently or in combination depending on specific policy requirements.

The AIM models used in this chapter are AIM/Impact, AIM/Impact [Policy], AIM/CGE [Global], AIM/Material, and AIM/Enduse. These models are used interactively to analyze policy options for emissions pathways to stabilization of climate change and impacts in an integrated framework. Further, the global-level analysis is integrated with country-level models to evaluate economic impacts and technology options specific to each country (see Figure 26.1).

AIM/Impact is a process model for assessing detailed regional-level climate impacts on vegetation, health, and other socio-physical indicators. It is also used to provide adaptation details. Its outputs are used by a simplified global-level assessment model, AIM/Impact [Policy], which estimates

---

[1] In this paper, we use the term GHG to denote all greenhouse gases.

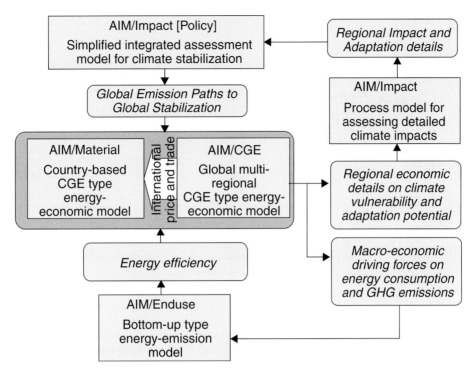

Figure 26.1 Interaction among AIM modules.

global emission paths to climate stabilization. The estimated global emission paths are then allocated regionally, and the regional allocations are used by AIM/CGE [Global]. It estimates regional economic impacts and regional macro-economic driving forces, including inter-regional trade, on energy consumption and GHG emissions. Global and regional estimates are integrated with AIM/Material and AIM/Enduse which are country-level models. The former is a top-down model used to estimate economic impacts and analyze economic instruments, whereas the latter is a bottom-up type energy-emission model for assessing GHG reduction potential from the perspective of technology development. The calculated energy efficiency improvement and additional investment are aggregated and put into AIM/Material for analysis of their economic effects. Thus, on the one hand, this assessment framework integrates analyses at the levels of climate, economy, vegetation, health, and other socio-physical effects. On the other hand, it integrates global-level macro-analysis with detailed country-level analysis.

26.3 Climate change impacts on crop productivity

26.3.1 Methods and scenarios

How much will we suffer from the adverse impact of climate change if there is no policy intervention on GHG emission, and how much can we abate the adverse impact through GHG mitigation policies? In order to understand the policies, potential crop productivity was estimated under BaU (SRES-B2), GHG-500ppm and GHG-600ppm scenarios.[2] Impacts on the potential productivity of rice and wheat in the current major producing countries were estimated using climate change projections of the six Atmosphere–Ocean General Circulation Models (AO-GCMs) distributed publicly at the IPCC Data Distribution Centre (IPCC-DDC).[3] The six AO-GCMs are CCSR/NIES (Nozawa et al., 2001), CGCM2 (Flato and Boer, 2001), CSIRO-mk2 (Hirst et al., 1997), ECHAM4 (Roeckner et al., 1996), GFDL-R30 (Manabe and Stouffer, 1996), and HADCM3 (Gordon et al., 2000). GHG emission paths to achieve concentration stabilization targets of 500 ppmv and 600 ppmv were calculated by the emission module of AIM/Impact [Policy] described in Section 26.3.3.

Country-averaged change in potential productivity of each cereal crop was estimated by combining (1) changes in country-averaged annual mean temperature and precipitation supplied by AO-GCM simulation results, and (2) sensitivity coefficients of potential crop productivities which have been calculated from sensitivity analysis using the AIM/Impact model (a process model with spatial resolution of 0.5°; Takahashi et al., 1998). A shortcoming of this procedure is that the heterogeneity of climate change on a finer spatial scale than country-average is ignored. In other words, a similar spatial pattern of climate change is assumed in each country. Another shortcoming is that the seasonal difference of climate

---

[2] GHG-500ppm and GHG-600ppm correspond to cap on GHG concentration of 500 ppmv and 600 ppmv, respectively. BaU represents "business as usual".
[3] IPCC-DDC (http://ipcc-ddc.cru.uea.ac.uk/ipcc_ddc.html).

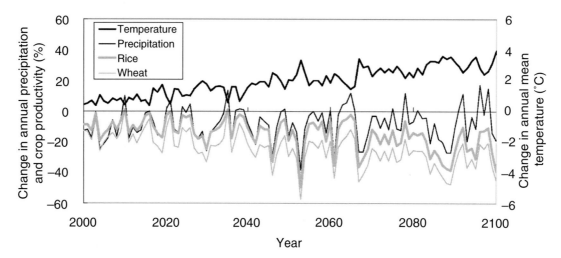

Figure 26.2 Changes of annual mean temperature (°C), precipitation (%) and potential productivity of crops (rice and wheat, %) in India from the present mean conditions (SRES-B2).

change cannot be represented, since only annual mean values are considered. Despite these drawbacks we adopted this procedure since it requires a small computer resource to accomplish a large number of simulations in a permissibly short time. Such a feature is indispensable for time-series analysis of many GHG emission scenarios.

26.3.2 Key gaps in impact assessment studies

Inter-annual fluctuation of projected future climate and its impact on potential crop productivity
Figure 26.2 shows the changes in annual mean temperature (°C), precipitation (%), and potential productivity of two crops (rice and wheat, %) in India from the present average conditions (30-year mean from 1961 to 1990) under the climate scenario based on the CSIRO-SRES-B2 experiment. In Figure 26.2, inter-annual fluctuation is not omitted in order to show the possible existence of years with extremely severe climate. Even if the decadal mean of the decrease in potential productivity is estimated to be about 20%, it is possible that the decrease would often reach 40%.

Range of uncertainty in future climate projection by AO-GCMs
Figure 26.3 shows the temperature change and percentage precipitation change in India from the present average conditions (defined as above) under the SRES-B2 scenario (11-year moving average). The black and grey lines show temperature change and precipitation change respectively. The thick lines show the mean value of simulation results projected by six different GCMs, while the thin lines show the range of results projected by the GCMs. The uncertainty range for the long-term future caused by model differences is larger than that for the short-term future. The mean estimate of temperature change from the present condition to the end of the twenty-first century is about 3 °C, while the highest and the lowest estimates are 4 °C and 2 °C respectively. In the case of precipitation change, although the mean estimate shows a monotonous increasing trend in the twenty-first century, the lowest estimate shows a decrease from the present mean condition.

Figure 26.4 shows the potential productivity changes for rice and wheat (%) in India from the present mean conditions under the SRES-B2 scenario (11-year moving average). The black and grey lines show the productivity changes for rice and wheat, respectively. The thick lines show the mean value of impact assessment results under the six climate scenarios based on the different GCMs, while the thin lines show the range of results. The mean estimate of rice productivity change from the present condition to the end of the twenty-first century is −18%, while the largest and the smallest estimates are −35% and −10%, respectively. In the case of wheat, the mean estimate of productivity change from the present condition to the end of the twenty-first century is about −25%, with uncertainty range from −21 to −30%.

Below, we discuss the effectiveness of GHG stabilization policy to mitigate adverse impacts of climate change, using the graphs of the 11-year moving average of the mean value of assessment results under six climate scenarios based on the different GCMs.

26.3.3 Climate change impacts mitigated by GHG stabilization policy

Table 26.1 shows the top 20 countries for rice and wheat production in the world (5-year average from 1996 to 2000; FAO, 2004) and Figures 26.5a and b show the change in potential productivity of rice and wheat in these countries for each crop under the SRES-B2 scenario (11-year moving average; mean value of impact assessment results under six climate scenarios based on the different GCMs).

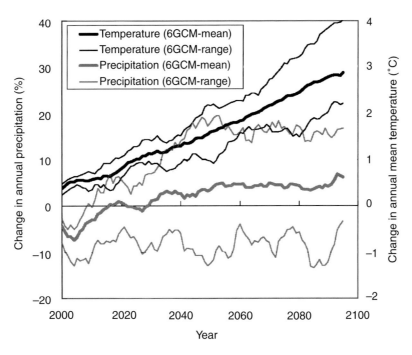

Figure 26.3 Temperature change (°C) and precipitation change (%) in India (SRES-B2).

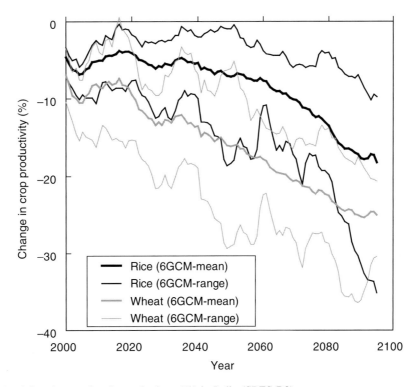

Figure 26.4 Potential productivity changes for rice and wheat (%) in India (SRES-B2).

From Figure 26.5a, India, Brazil, and Thailand are identified to be current big producers of rice that will suffer significant decreases in potential productivity under the expected climate change. Here we investigate the effectiveness of GHG stabilization policy to mitigate the adverse impact on rice potential productivity in the three countries. In similar manner, from Figure 26.5b, we investigate the effectiveness of GHG stabilization policy to mitigate the adverse impact on wheat potential productivity in India, Australia, and Germany.

Figure 26.6 shows the change of potential rice productivity in the three major rice-producing countries identified as climate-sensitive, from the present through to the end of the twenty-first century under the SRES-B2, GHG-600ppm and GHG-500ppm scenarios (11-year moving average).

In India, the potential productivity of rice will be rather stable in the first half of the twenty-first century, but it will begin to decrease significantly after 2050 to the end of the century under the SRES-B2 baseline scenario. The change from the present to the end of the century will be about −18%. By introducing the GHG emission mitigation policy targeting GHG-600ppm stabilization, the decrease in the latter half of the century is reduced and the change from present to the end of the century will be about −8%. By introducing the more radical GHG mitigation policy targeting GHG-500ppm stabilization, the productivity decrease in the latter half of the twenty-first century can be almost avoided, although the productivity decrease from the baseline period (1961–90) to the year 2000 is not regained.

In Brazil and Thailand, potential rice productivity will monotonously decrease from early in the twenty-first century to the end of the century under the SRES-B2 scenario. The mitigation policies will relieve the decrease in the latter half of the century, although they are not adequate to recover the productivity decrease occurring in the first half of the century.

Figure 26.7 shows the change of potential wheat-productivity in the three major climate-sensitive wheat-producing countries from present to the end of the twenty-first century under SRES-B2, GHG-600ppm, and GHG-500ppm scenarios (11-year moving average). In India, wheat potential productivity will decrease severely throughout the century under SRES-B2 scenario. The GHG mitigation policies will relieve the rate of decrease after 2050, but the change of productivity from the baseline period (1961–90) to the end of the twenty-first century will be about −18% even under the radical mitigation policy targeting GHG-500ppm stabilization. In Australia, climate change affects wheat productivity positively in the first decade of the century, but it will drop significantly after that and will decrease by 25% from the baseline period to the end of the the twenty-first century under the SRES-B2 scenario. In Germany, potential wheat productivity will decrease at a constant rate through the century under the SRES-B2 scenario, although its rate of decrease is slower than that in India and Australia.

Table 26.1 Top 20 countries for rice and wheat annual production in the world (unit: 1000 tonne/year).

Rice		Wheat	
Country	Production	Country	Production
China	198 119	China	111 420
India	127 456	India	69 090
Indonesia	50 496	USA	64 434
Bangladesh	31 684	France	36 802
Viet Nam	29 398	Russian Fed.	34 327
Thailand	23 876	Canada	26 325
Myanmar	18 572	Australia	22 380
Japan	11 999	Germany	20 035
Philippines	11 056	Turkey	19 441
Brazil	9 502	Pakistan	18 237
USA	8 490	UK	15 628
Korea Rep.	7 088	Argentina	15 084
Pakistan	6 981	Ukraine	14 134
Egypt	5 333	Iran	9 755
Nepal	3 820	Poland	8 772
Cambodia	3 679	Kazakhstan	8 339
Nigeria	3 248	Italy	7 658
Sri Lanka	2 542	Egypt	6 118
Madagascar	2 511	Spain	5 706
Iran	2 425	Romania	4 916

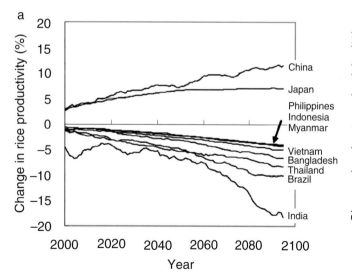
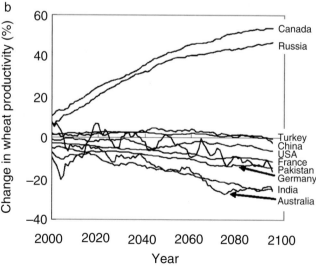

Figure 26.5 Change of potential productivity of (a) rice and (b) wheat in the top 10 countries under the SRES-B2 GHGs emission scenario.

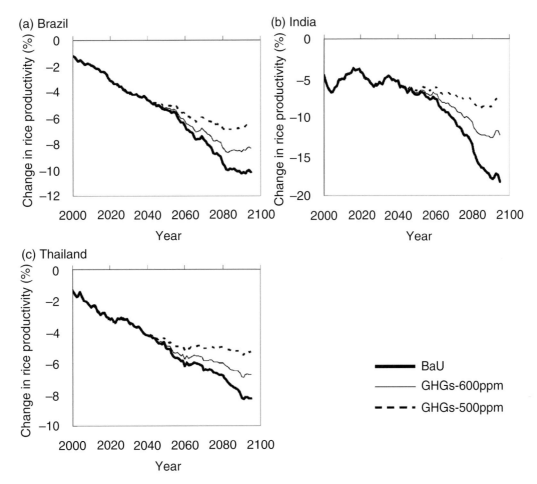

Figure 26.6 Change in rice potential productivity in the climate-sensitive three major rice producer countries under the SRES-B2, GHG-600ppm and GHG-500ppm scenarios.

The mitigation policy targeting GHG-500ppm stabilization can avoid the decrease in the latter half of the twenty-first century.

From the above analysis, it is evident that unmitigated GHG emission (SRES-B2 scenario) will significantly reduce potential crop productivity in some current major producer countries. The GHG mitigation policy targeting GHG-500ppm stabilization is expected to relieve the rate of productivity decrease in the latter half of the twenty-first century, but it is not effective enough to regain the decrease during the first half of the century. This implies that we need to achieve the GHG-500ppm stabilization target to avoid the trend of sharp decrease in crop productivity throughout the twenty-first century in the climate-sensitive countries.

26.4 Multi-gas analysis of stabilization scenarios

26.4.1 Global emission pathways

The dynamic optimization model included in AIM/Impact [Policy] provides global GHG emissions paths under different socio-economic scenarios with constraints on GHG emissions, temperature increases, rates of temperature increases, and rises in sea levels.

The model provides a framework to assist policymakers in meeting the UNFCCC's ultimate objective: "stabilization of GHG concentrations in the atmosphere at a level that would prevent dangerous anthropogenic interference with the climate system." To discuss the time-frame and volume of GHG reductions, we consider four cases as described below:

- SRES-B2: Business as usual
- CO_2-450ppm: 450 ppmv cap on CO_2 concentration
- GHG-500ppm: 500 ppmv cap on total GHG concentrations
- GHG-550ppm: 550 ppmv cap on total GHG concentrations

With respect to the cases other than SRES-B2, constraint optimization calculations were carried out in which CO_2 or total GHG concentrations do not exceed the constraint levels from 1990 to 2200. Total GHG concentrations were calculated based on their global warming potential as reported by the IPCC. Figure 26.8 compares carbon emissions for the four cases. Figure 26.9 shows the results of global mean temperature

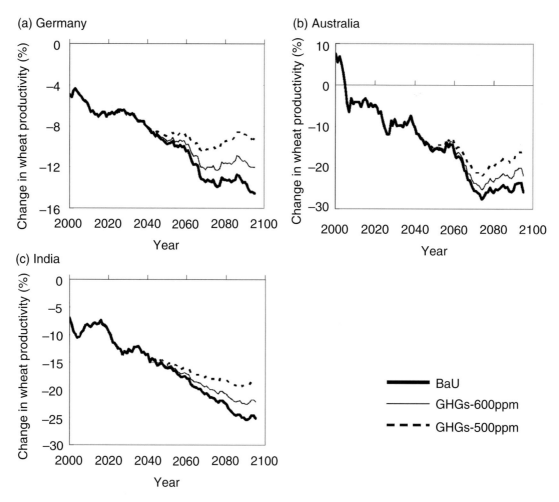

Figure 26.7 Change in wheat potential productivity in the climate-sensitive three major wheat producer countries under the SRES-B2, GHG-600ppm and GHG-500ppm scenarios.

changes from 1990 for the four cases. Under the SRES-B2 case, CO_2 emissions continue to rise till 2050, and in 2150 the CO_2 concentration is almost three times its 1990 level. The temperature increases by nearly 2 °C in 2060 and by 4.5 °C in 2150.

The German Advisory Council on Global Change (WBGU, 2003) suggested from a review of the available literature that "above 2 °C the risks increase very substantially, involving potentially large extinctions or even ecosystem collapses, major increases in hunger and water shortage risks as well as socioeconomic damage, particularly in the developing countries."

The temperature increases for the cases involving restrictions on CO_2 concentrations surpass 2 °C in 2150. This is because there are no constraints on the emissions of GHG other than CO_2. In the CO_2-450ppm case, the temperature increases by 2.7 °C in 2150.

In the cases involving constraints on GHG concentrations, GHG-500ppm indicates a temperature increase of 2.2 °C relative to 1990 in 2150. The GHG reductions required to achieve a 500 ppmv cap on total GHG concentrations are 1.9–3.0 gigatonnes CO_2 equivalent, GtCe (CO_2 reduction: 1.6 GtC) in 2020 and 6.5 GtCe (CO_2 reduction: 4.5 GtC) in 2030, compared with the BaU case.

26.4.2 Burden sharing and its economic impact using AIM/CGE [Global]

This section focuses on allocation of regional GHG emission caps using global emission trajectories considering GHG-500ppm constraint simulated by AIM/Impact [Policy] and its regional economic impacts. Although several burden sharing schemes, such as C&C (contraction and convergence) (Meyer, 2000), the Brazilian Proposal (including historical emissions), Multi-stage, and Intensity target, have already been proposed, quantitative simulation studies on this subject are scant (Den Elzen and Lucas, 2003). We have developed a burden sharing model that allocates regional emission caps using global emission constraints and socio-economic conditions. This analysis is combined with that of the AIM/CGE [Global] model which evaluates economic impacts using the computable general equilibrium (CGE) method (Kainuma *et al.*, 2003; Fujino *et al.*, 2004).

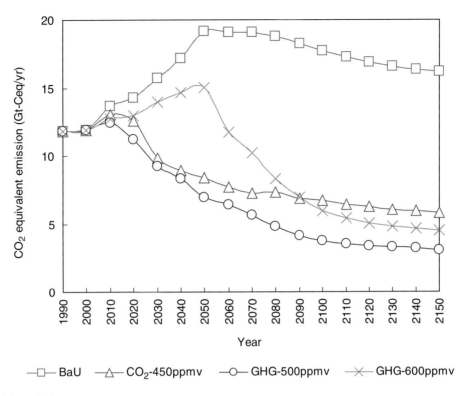

Figure 26.8 Global CO_2 emissions.

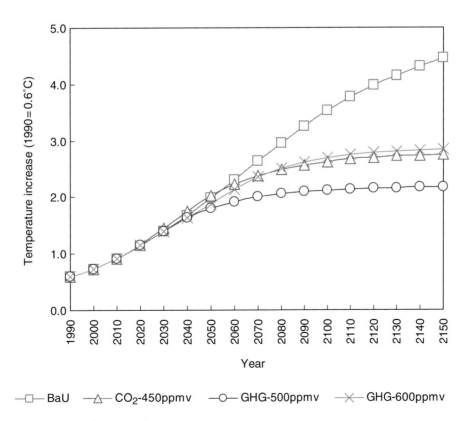

Figure 26.9 Global mean temperature changes.

Regional allocation cap using burden sharing model

In this analysis we adopted the C&C scheme, a simple and clear scheme for regional allocation. It needs current regional emission data, projection of regional population, and assumption of convergence year. It assumes that per capita emission paths in different regions will converge after the convergence year. For this analysis we prepare B2-based country-level population projection data and set the convergence year as 2050. Figure 26.10 shows that per capita emission will converge to 0.53 tCe/yr in 2050 because the GHG-500ppm constraint itself requires considerable reduction globally. Canada and the United States are required to reduce their per capita emission by more than 5 tCe/yr from 2000 to 2050, whereas a few regions such as Indonesia, India, other South Asian regions, and Africa are allowed to increase their per capita emissions.

Regional economic impacts using AIM/CGE [Global]

The multi-region, multi-sector, multi-gas model AIM/CGE [Global] has been developed to analyze long-term stabilization scenarios. This model is a recursive dynamic CGE model based on GTAP-EG structure. It is programmed with GAMS/MPSGE. GTAP version 5 database (base year 1997) and IEA energy statistics are used for the economic database and energy database respectively. AIM/CGE [Global] is a long-term model with time horizon 1997 to 2100, and includes 21 world regions and 14 economic sectors. With this model, we calculated three cases. They are BaU; regional GHG emission constraint without trading; and regional GHG emission constraint with trading. Regional GHG emission caps are calculated with the C&C scheme.

Figure 26.11 shows regional allocation caps and estimated economic impacts in 2020 and 2030. Under emission constraint, global GHG emissions are reduced by 27% and 38% from their level in the BaU case, in 2020 and 2030 respectively. Many developed regions are required to reduce their GHG emissions by more than 50% in 2030, as compared with the BaU case. Although global GDP loss is estimated as 0.7% and 1.5% for 2020 and 2030 respectively as compared with the BaU, the impact varies across different regions (Figure 26.12). Several developing regions gain GDP because of the advantage of producing excess energy-intensive goods with a moderate GHG allocation cap. On the other hand, Russia and Korea will face some loss of GDP. Although those countries have already developed their economies close to developed regions, they are mostly reliant on energy-intensive industries such as iron and steel, chemical, energy import industries, and so on. Regional carbon prices in 2020 and 2030 are relatively large because the model does not consider back stop technologies, and the severe GHG reductions force changes in regional economic and energy systems. In the emission trading case, global GDP loss will be lessened to 0.2% and 0.5%, and carbon price will be 79 and 155$/tCe, in 2020 and 2030 respectively (Figure 26.13).

These results indicate that (1) C&C cap constraint on the GHG-500ppm case does not allow most regions to increase

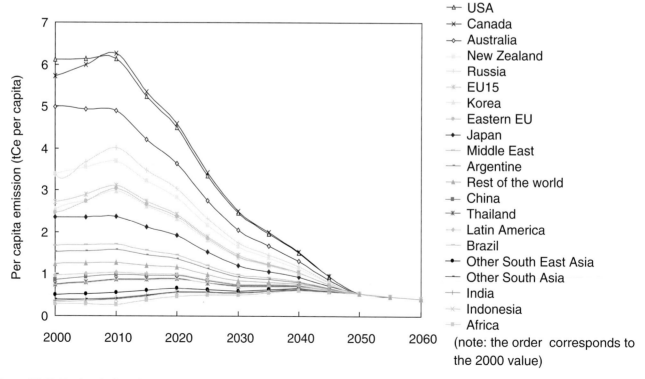

Figure 26.10 Regional allocation cap using C&C scheme under the GHG-500ppm case.

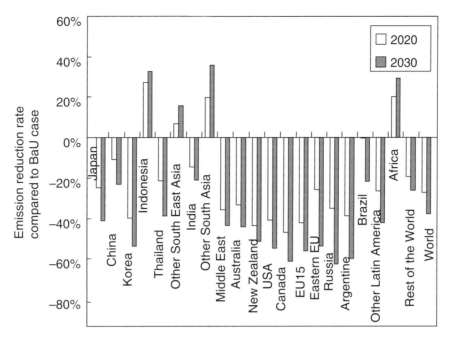

Figure 26.11 Emission reduction rate under C&C cap constraint with GHG-500ppm, as compared with BaU (SRES-B2).

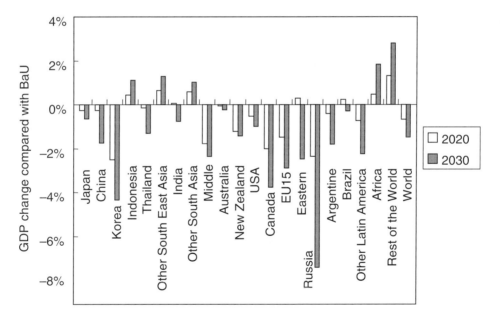

Figure 26.12 GDP change without trading under C&C cap constraint with GHG-500ppm, as compared with the BaU (SRES-B2).

per capita emission in 2020 and 2030; (2) GDP loss without emission trading will be 1.5% in 2030, and will be lessened to about 1% in the case of global emission trading; and (3) the impact on socio-economic structure will vary across regions.

26.5 Reduction target for CO_2 in Japan and its cost

In order to realize the long-term target, what can be done in the short term? According to the Kyoto Protocol, GHG emissions in Japan in the first commitment period (from 2008 to 2010) should be reduced by 6% of those in 1990. Moreover, in order to reduce the GHG emissions, the Government of Japan adopted the New Climate Change Policy Programme in March 2002 (Government of Japan, 2002). Under this program, the quantitative target of CO_2 emissions reduction in Japan related to energy consumption in the first commitment period is set to 2% of that in 1990. The carbon sink by reforestation, emissions trading, and non-CO_2 emissions reduction will cover the rest. On the other hand, the

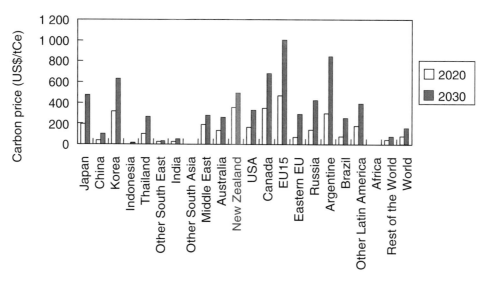

Figure 26.13 Carbon price without trading under C&C cap constraint with GHG-500ppm, as compared with the BaU (SRES-B2).

CO$_2$ emissions in 2000 increased by 10% more than those in 1990. As a result, this reduction of CO$_2$ emissions is equivalent to more than 10% of emissions in 2000. This target can be considered as the minimum target for long-term climate stabilization. In this section, we adopted this CO$_2$ target, which should be achievable with domestic measures alone.

AIM/Enduse and AIM/Material models were applied to estimate the feasibility of the reduction target, its marginal cost and the economic impact, and the appropriate policy mix to realize the target. AIM/Enduse model is a bottom-up model and AIM/Material model is a top-down model. The linkage of these different type models can lead to the more comprehensive estimate. First of all, by using the AIM/Enduse model and the future energy service demand, potential CO$_2$ emissions reduction from technology development is evaluated under several counter measures such as carbon taxation and so on. At the same time, sector-wise aggregated energy efficiency improvement and the required additional investment are also calculated, and these results are fed into the AIM/Material model which estimates the economic impact of introduction of the carbon tax in Japan.

26.5.1 Technologies to achieve Kyoto Protocol

Figure 26.14 shows the results of the carbon tax rate and CO$_2$ emissions in Japan as estimated by the AIM/Enduse model. If technology improvements do not take place and technology shares remain unchanged after 2002, CO$_2$ emissions will increase along the line A in Figure 26.14. When the total system cost (including initial investment and running cost) of technology selection is minimized, the CO$_2$ emissions will follow the line B. In this case, the CO$_2$ emissions will be stabilized at the level of year 2000. In order to reduce the CO$_2$ emissions to achieve the target, a carbon tax policy is introduced after 2005.

Since the energy price increases by introducing the carbon tax, the differences of the total cost between ordinary devices and expensive energy-saving devices will reduce. When the tax rate exceeds a certain level, the energy-saving technology devices are selected, and consequently, the CO$_2$ emissions will decrease. In the case of 29 US$ tax at 2000 price per tonne carbon (US$/tC) (line C) and 290 US$/tC (line D), the CO$_2$ emissions will decrease, but not achieve the target. In order to achieve the target by carbon tax alone, the tax rate has to be 430 US$/tC (line E). This, however, implies a severe tax burden.

In order to realize the CO$_2$ emissions reduction and low burden of tax payment simultaneously, we consider a combination of carbon tax and subsidy to lower the marginal cost of CO$_2$ reduction countermeasures. Figure 26.15 shows the idea of this policy mix. Instead of the high carbon tax, a low carbon tax is imposed, and the tax revenue is used to subsidize the countermeasures to reduce emissions. From the AIM/Enduse model, the countermeasures to reduce emissions as shown in Table 26.2 are selected. In this case, 33 US$/tC of carbon tax is required.

26.5.2 Economic impact of Kyoto Protocol

The economic impacts are calculated for these carbon tax scenarios and for this combined carbon tax and subsidy scenario by using the AIM/Material model. The reference scenario has been designed based on the service demand projections provided by various expert agencies in Japan. In the reference scenario, the average GDP growth rate from 2000 to 2012 is about 1.4%per year. As shown in Figure 26.16, when only carbon tax is imposed to reduce CO$_2$ emissions, the average GDP loss in the first commitment period will be 0.16% compared with the GDP in the reference scenario. When a combined tax and subsidy policy is introduced, the average

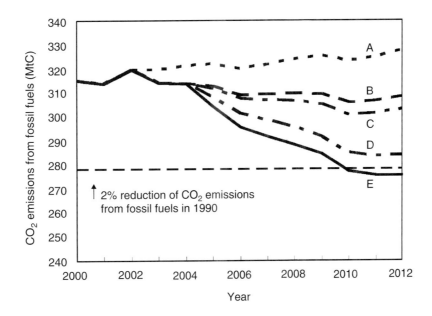

A: Case of unimproved technology and fixed technology share
B: Case of cost minimized technology selection without carbon tax
C: Case of cost minimized technology selection with 29 US$/tC of carbon tax
D: Case of cost minimized technology selection with 290 US$/tC of carbon tax
E: Case of cost minimized technology selection with 430 US$/tC of carbon tax or 33 US$/tC of carbon tax with subsidy

Figure 26.14 CO_2 emissions trajectories in Japan.

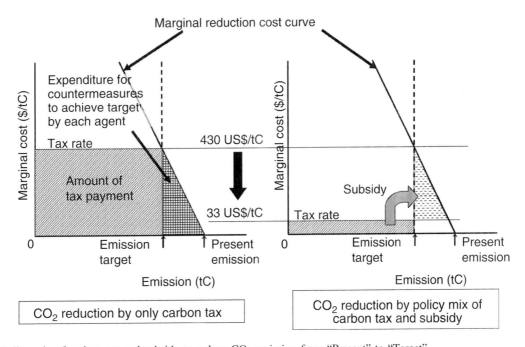

Figure 26.15 Policy mix of carbon tax and subsidy to reduce CO_2 emission from "Present" to "Target".

GDP loss in the same period will be 0.061% of the GDP in the reference scenario. The GDP loss in the first commitment period is not severe. On the other hand, the economic impact on sectors will be quite different as follows:

1. Production in the thermal power generation sector, especially coal power plant, will drastically decrease.
2. Production in the sectors that supply fossil fuels will also decrease.

Table 26.2 Countermeasures, additional investment and carbon tax rate required to reduce CO_2 emissions by 2% of emissions in 1990 in the first commitment period in Japan.

Sector	Introduced measures and devices	Additional investment (million US$/year)
Industrial sector	Boiler conversion control; high performance motor; high performance industrial furnace; waste plastic injection blast furnace; LDF with closed LDG recovery; high efficiency continuous annealing; diffuser bleaching device; high efficiency clinker cooler; Biomass power generation	978
Residential sector	High efficiency air conditioner; high efficiency gas stove; solar water heater; high efficiency gas cooking device; high efficiency television; high efficiency VTR; latent heat recovery type water heater; high efficiency illuminator; high efficiency refrigerator; standby electricity saving; insulation	3417
Commercial sector	High efficiency electric refrigerator; high efficiency air conditioner; high efficiency gas absorption heat pump; high efficiency gas boiler; latent heat recovery type boiler, solar water heater; high efficiency gas cooking device; high frequency inverter lighting with timer; high efficiency vending machine; amorphous transformer; standby electricity saving; heat pump; insulation	1878
Transportation sector	High efficiency gasoline private car; high efficiency diesel car; hybrid commercial car; high efficiency diesel bus; high efficiency small-sized truck; high efficiency standard-sized track	1029
Forest management	Plantation; weeding; tree thinning; multilayered thinning; improvement of natural forest	1889
Total		9191
Tax rate to appropriate required subsidiary payments (US$/tC)		33

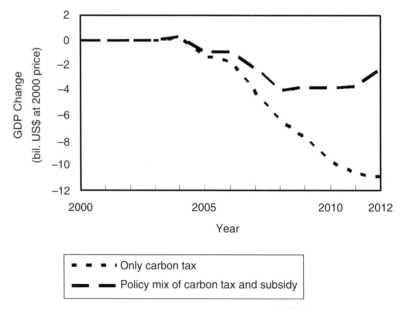

Figure 26.16 GDP change compared with the reference scenario.

3 On the other hand, manufacturing of electric machinery, equipment and supplies and transportation equipment will increase because of the increase in demand for energy-saving devices.

26.6   Concluding remarks

The followings are the major findings from global, regional, and country-level assessment using AIM.

1 Unmitigated GHG emission (SRES-B2 scenario) will cause a significant reduction in potential crop productivity in some of the current major producer countries. The mitigation policy targeting 500 ppmv GHG stabilization is expected to relieve the rate of productivity decrease in the latter half of the twenty-first century.
2 The GHG reductions required to achieve a 500 ppmv cap on total GHG concentrations are 1.9 GtCe (CO_2 reduction: 1.7 GtC) in 2020 and 5.2 GtCe (CO_2 reduction: 4.6 GtC) in 2030, compared with the BaU case (business as usual; the case without taking any measures).
3 To follow the GHG 500 ppmv constraint and to allocate global emissions under the C&C constraint (convergence in 2050), many developed regions are required to reduce their GHG emissions by more than 50% in 2030, as compared with the BaU case. This will impose severe economic losses. The economic losses would be reduced by mixed policy options such as multi-gas reduction policies and global trading.
4 To realize the CO_2 emissions reduction and low burden of tax payment simultaneously, a combination of low carbon tax and subsidy to lower the marginal cost of CO_2 reduction countermeasures is an effective policy mix.

Active debates have been taking place in the process of climate policy planning. While a consultative process involving interactions among different stakeholders such as policymakers, local communities, scientists, and analysts, is critical for arriving at implementable and desirable decisions, quantitative modeling tools can provide useful inputs to such a process. An integrated assessment framework like the one illustrated in this chapter plays a crucial role by analyzing complex interactive linkages between different specialized themes like economy, energy, vegetation, and health on the one hand, and macro-economic and technological systems on the other. Simulation results from such models enormously enrich the climate policy debates.

Acknowledgements

The AIM model development was inspired by the late Professor Tsuneyuki Morita. We acknowledge his guidance of our climate policy studies. This research was supported by the Global Environment Research Fund by the Ministry of Environment Japan.

References

Den Elzen, M. G. J. and Lucas, P. (2003). *FAIR 2.0: A Decision-support Model to Assess the Environmental and Economic Consequences of Future Climate Regimes*. RIVM Report 550015001.

FAO (2004). FAOSTAT. Available at www.fao.org/

Flato, G. M. and Boer, G. J. (2001). Warming asymmetry in climate change simulations. *Geophysical Research Letters* **28**, 195–198.

Fujino, J., Masui, T., Nair, R., Kainuma, M. and Matsuoka, Y. (2004). Development of computer model to analyze GHG reduction including non-CO_2 gas. *Proceedings of the 20th Conference on Energy, Economy, and Environment*. Tokyo, Japan, pp. 87–90.

Gordon, C., Cooper, C., Senior, C. A. *et al.* (2000). The simulation of SST, sea ice extents and ocean heat transports in a version of the Hadley Centre coupled model without flux adjustments. *Climate Dynamics* **16**, 147–168.

Government of Japan (2002). *The New Climate Change Policy Programme*. Available at www.env.go.jp/en/topic/cc/ 020319.pdf.

Hirst, A. C., Gordon, H. B. and O'Farrell, S. P. (1997). Response of a coupled ocean–atmosphere model including oceanic eddy-induced advection to anthropogenic CO_2 increase. *Geophysical Research Letters* **23**(23), 3361–3364.

Kainuma, M., Matsuoka, Y. and Morita, T. (eds.). (2003). *Climate Policy Assessment: Asia-Pacific Integrated Modeling*. Tokyo: Springer. www-iam.nies.go.jp/aim/index.htm.

Manabe, S. and Stouffer, R. J. (1996). Low frequency variability of surface air temperature in a 1,000-year integration of a coupled atmosphere–ocean–land surface model. *Journal of Climate* **9**, 376–393.

Meyer, A. (2000). *Contraction and Convergence, The Global Solution to Climate Change*. Bristol: Green Books for the Schumacher Society.

Nozawa, T., Emori, S., Numaguti, A. *et al.* (2001). Projections of future climate change in the 21st century simulated by the CCSR/NIES CGCM under the IPCC SRES Scenarios. In *Present and Future of Modeling Global Environmental Change*, ed. T. Matsuno and H. Kida. Tokyo: Terrapub, pp. 15–28

Roeckner, E., Arpe, K., Bengtsson, L. *et al.* (1996). *The Atmospheric General Circulation Model ECHAM-4: Model Description and Simulation of Present-day Climate*. Report No. 218. Hamburg: Max-Planck Institute for Meteorology.

Shukla, P. M., Rana, A., Garg, A., Kapshe, M. and Nair, R. (2004). *Climate Policy Assessment for India: Applications of Asia-Pacific Integrated Model* (AIM). Delhi: University Press.

Takahashi, K., Matsuoka, Y. and Harasawa, H. (1998). Impacts of climate change on water resources, crop production and natural ecosystem in the Asia and Pacific region. *Journal of Global Environment Engineering* **4**, 91–103.

WBGU, (2003). *Special Report 2003: Climate Protection Strategies for the 21st Century. Kyoto and Beyond*. Berlin: WBGU.

27

Price, quantity, and technology strategies for climate change policy

W. David Montgomery and Anne E. Smith

27.1 Introduction

Cost-effective market approaches to environmental management all entail the use of some form of price signal, and they can generally be grouped as quantity-based or tax-based regulations. The quantity-based approach, in the form of "cap-and-trade" programs, has gained an especial popularity in the past two decades. The political acceptance of cap-and-trade over the tax approach has its roots in a favorable set of political attributes (Hahn and Stavins, 1991) and a successful experience with the Title IV SO_2 cap. Its support among economists comes from its theoretical capability to achieve any stated cap at minimum cost. Most importantly, it has been accepted by those with a greater concern for environmental outcomes than for program costs, because the idea of a cap provides certainty of particular emissions levels in return for uncertainty about the costs of achieving those emissions levels.

Cap-and-trade was such a popular policy prescription during the 1990s that almost the entire discussion about climate change policy during that time was cast in terms of how to design a greenhouse gas (GHG) trading program, and how to build international agreements and institutions that could allow emission trading to function even where caps could not be directly imposed. (Joint implementation and the extensive set of rules and procedures surrounding the Clean Development Mechanism are examples of the latter.) Alternative approaches were nearly impossible to discuss in any climate change policy forum, despite papers that were written from the earliest days of GHG policy discussions warning that GHG cap-and-trade programs would pose much greater challenges to implement than for other typical types of air emissions (Smith 1991; Smith *et al.*, 1992; Montgomery, 1993, 1996).

Early papers on the difficulty of applying cap-and-trade to GHGs focused primarily on issues of multiple types of GHGs, extremely numerous and small sources, the timescales and uncertainties involved in climate change, and scale of financial commitment. Many different forms of hybrid trading designs were proposed and complex institutional structures (such as the Clean Development Mechanism) were developed to try to address some of these concerns. A cottage industry of potential brokers promoting policy based on emission trading appeared. Integrated assessment tools were enhanced to address the implications of these alternative and more complex formulations of the basic cap-and-trade design (Bernstein *et al.*, 1999; Smith *et al.*, 1999).

By coupling economic models with models of the climate change process, integrated assessment studies have made it possible to analyze how to design caps on GHG emissions that would lead to stabilization of atmospheric concentrations. These studies have led to a general recognition among analysts that the timescales over which policies must operate in order to deal effectively with climate change are measured in centuries. Additionally, they generally find that time paths for emissions that achieve concentration targets at minimum cost demand relatively small emission reductions in the next few decades, and much larger emission reductions in the more distant future (Wigley *et al.*, 1996). These emission reductions in the distant future start after technologies that can provide energy without greenhouse emissions, but do not exist today, are expected (and assumed) to become available; without assuming such future technologies, no goal of stabilization of atmospheric concentrations can be achieved without ending and even reversing economic growth (Hoffert et al., 2002).[1]

Modelers call this now unavailable, and possibly unknown, technology the "backstop technology" because ultimately it meets all incremental demand for energy services in order that stabilization be achieved.[2] Modelers have become uncomfortably aware of the fact that the estimated cost of any GHG stabilization policy tends to be more or less driven by the cost at which the backstop technology has been assumed to be available. That cost affects the timing of when it becomes the GHG reduction method of choice, and that in turn determines the amount of control effort that the model expends to meet the GHG target imposed in the first few years of a program.

The enormous importance of the backstop technology assumption is not just a modeling inconvenience. In fact, it is a signal coming from the analysis that the cap-and-trade approach is fundamentally a misfit for this environmental problem. It is time for researchers to turn to the question of why this is the case, so that more effective prescriptions can start to be developed.

27.2 Why climate change is different from previous externality problems

There are three fundamental differences between GHGs and other forms of emissions that alter the potential effectiveness of cap-and-trade when applied to climate change policy (Montgomery, 1996; Jacoby et al., 2004).

1. The long timescales over which climate change policy must be defined;
2. The necessity of developing new technologies to make stabilization of concentrations economically feasible; and
3. The requirement of a complete changeover of the capital stock to embody those technologies.[3]

These features limit the need for and value of immediate emission reductions in GHGs, which is the primary and well-proven advantage of the cap-and-trade approach. At the same time, these features demand a policy prescription that is effective in stimulating future technology development without imposing inefficiently high costs in the near-term stages of the policy. Previous applications have not shown that the cap-and-trade approach can perform in this manner, nor has this been explored adequately in modeling or in theory.

Cap-and-trade has performed well in controlling emissions such as SO_2 and NO_x, because, for those emissions, the policy problem is to find the least-cost set of *near-term* control actions from a much larger set of currently possible control actions. In those applications, an array of retrofit technologies and process changes are already known and readily identified by each emitting company. However, the cost of the options often varies substantially when applied at different sources (usually at different companies), such that overall policy costs can be greatly reduced if the controls are selectively applied where the cost per unit of emissions reduction is greatest. Identifying this least-cost set of controls is almost impossible for a regulator with imperfect knowledge, but the price signals created by a market for a limited number of emissions allowances allow the companies to self-select into the least-cost set of controllers without any informational demands on the regulator.[4]

Thus, the main problem that cap-and-trade resolves so handily is how to find and induce use of the best controls from a large array of options. In practice it has almost entirely focused on existing technology, and making reductions that are economically feasible to achieve right now. In doing so, cap-and-trade applications have also demonstrated an ability to motivate innovation that further reduces the costs of meeting caps.[5] However, this incentive for innovation has only been demonstrated in situations where innovation could lead immediately to cost savings (because the policies in question imposed the full force of the intended emissions reductions within five years or less from program initiation), and where the ultimate cap would be achievable at a socially acceptable cost using existing technologies, even if the innovations were not to have materialized.

It is much less clear what incentives for innovation a cap-and-trade policy would engender when the cap would not be lowered to levels anywhere near what is necessary to have material impact on the externality until *after* the innovations

---

[1] Hoffert et al. note that within the next 50 years, the world will require 15–30 TW of carbon-free energy to meet stabilization targets of 550 to 350 ppm, which is more than double the approximately 12 TW of energy consumed today (85% of which is fossil-fueled).

[2] In some models, such as CRA's MRN model and Manne and Richels' MERGE model, the backstop is identified only as a technology that provides an unlimited supply of carbon-free energy at constant marginal cost, leaving to the reader's imagination what energy sources and processes might be involved. In other cases, such as the AIM model, it is assumed that a number of technologies supporting provision of energy services from carbon-free sources become available at specific future times and at specific costs. We include all these approaches in our notion of a backstop, since they all share the essential feature of assuming that carbon-free technologies with defined costs usable on a large scale become available at some future date, and that singly or in combination these technologies eventually satisfy all growth in demand for energy services.

[3] There are additional difficulties in applying a cap-and-trade system at the international level, owing to the impossibility of defining adequate property rights in future emissions, as discussed in Barrett (2003). In this paper we concentrate on the reasons why cap-and-trade policies cannot work as climate policy that apply even within an individual country.

[4] The cap-and-trade approach has many other merits that are of less direct relevance to the point we wish to make here. These include the way the program provides an automatic mechanism by which the higher-cost emitters pay their share of the controls incurred by other companies via their purchases of allowances from controlling companies.

[5] The impact of this incentive to innovate under a cap-and-trade policy is sometimes called the *dynamic* cost reduction component, while the cost reductions that occur without any additional innovation are called the *static* element of cost reductions attributable to cap-and-trade. (See Burtraw, 1996; Ellerman and Montero, 1998.)

that the policy is supposed to motivate will have occurred. The latter situation poses a "dynamic inconsistency" in incentives. We define dynamic inconsistency more formally later, but it is generally reflected in the following dilemma in the case of GHGs. Environmentally meaningful GHG reductions cannot be mandated before research and development (R&D) has been successful, while any GHG cap whose imposition could be economically justified before that time would not impose high enough costs to motivate the technological improvements necessary to bring the cost of meaningful reductions down to acceptable levels. Thus the incentive for R&D has to take the form of a credible threat to impose a high future cost of control, which will provide economic returns to the innovator of a superior and lower-cost technology. The problem is that a policy that imposes sufficiently high costs to provide a return on the investment in R&D – as opposed to covering the variable costs of using the technology – is not credible. If future caps are set on the assumption that a low-cost backstop technology will be available, it is not credible that such a tight cap will be maintained if the innovation does not appear.[6] Thus, there will be no private consequences from failure to invest in (or be successful at) R&D. At the same time, the private sector does risk negative financial consequences from investing in R&D. This is because the fixed cost of performing the R&D is large, yet under a future stringent cap, a competitive emissions market will result in an emissions price that would reflect only the *forward* costs of installing and operating the new technology for reducing emissions.

These incentive problems have not been present in other cap-and-trade applications, but they are the defining characteristic for GHG policy. The need for technological advancement undermines the usefulness of cap-and-trade for achieving near-term reductions, and dynamic inconsistency of incentives undermines the ability of a near-term cap-and-trade program to produce the necessary incentives for innovation. After a closer look at the underlying types of technological change required, and the literature on induced technological change, we will demonstrate that this problem of incentives applies to emissions taxes and to "safety valve" proposals,[7] as well as to pure cap-and-trade programs – effectively all of the standard tools for addressing environmental externalities that have been prescribed in the past.[8]

The dynamic inconsistency of policies that announce future carbon prices is not the only reason cap-and-trade or carbon tax policies cannot motivate private sector investment in R&D. Studies of the inherent instability of international agreements without mechanisms for enforcing participation or compliance imply that cap-and-trade systems based on such agreements cannot create a credible incentive for long-term R&D (Schelling, 2002; Barrett, 2003). It is well established in the literature on innovation that unless the results of R&D are fully appropriable, the incentive effect of even a certain future carbon price will be diluted or eliminated.

The dynamic inconsistency analyzed in this paper is related to the instability of international agreements, but exists even when sovereign national governments announce an intention to put a future price on carbon, and makes it impossible for announced future carbon prices to provide a credible incentive even if the results of R&D are fully appropriable. To focus on this dynamic inconsistency, we address in this study a case in which the results of R&D are fully appropriable, so that the developer can expect to receive all of the quasi-rents created by an innovation. We note that if this is not the case – for example, if innovation takes the form of a series of incremental improvements, for which it is impossible to claim any share of the quasi-rents earned by resulting new energy technologies when they are deployed – then the inadequacy and irrelevance of future carbon prices as incentive for R&D is established without reference to dynamic inconsistency. In intermediate cases, in which some but not all of the quasi-rents generated by R&D can be captured, then inappropriability diminishes the incentive effects of announced future prices, and dynamic inconsistency diminishes them further.

27.3 What kind of new technology is required?

What is the nature of the future technology that might serve as the solution? Even this is unknown at present. Hoffert *et al.* (2002: p. 981) argue that "the most effective way to reduce CO_2 emissions with economic growth and equity is to develop revolutionary changes in the technology of energy production, distribution, storage and conversion." They go on to identify an entire portfolio of technologies, suggesting that the solution will lie in achieving advances in more than one of the following categories of research:

- wind, solar, and biomass
- nuclear fission
- nuclear fusion
- hydrogen fuel cells
- energy efficiency
- carbon sequestration

---

[6] Government will try to equate marginal benefits to marginal cost of emissions reduction by setting the future cap at a level that would be too costly today but viable after R&D produces a backstop technology. If the backstop is not available, the cap will have to be much less stringent if it is to be set at a point where the marginal cost of emissions reduction equals marginal benefit.

[7] Recently proposals that the government should offer to sell an unlimited number of permits at a 'safety valve' price to remove the cost uncertainty characteristic of emission caps have been made by Resources for the Future and the National Commission on Energy Policy. The idea originated with Roberts and Spence (1976).

[8] Command-and-control forms of regulation fare even worse, since no regulatory requirement to adopt currently available technology over a very

short time horizon can provide an effective long-term incentive to develop a revolutionary new technology.

Although the list may be familiar, the authors point out that there is not enough supply of low-cost reductions currently available to meet the GHG challenge. Developing that supply will require basic science and fundamental breakthroughs in a number of disciplines. The magnitude of possible reductions in the next decade or two is dwarfed by the magnitude of reductions that successful innovation would supply through these routes.[9]

Thus, except for harvesting low-cost opportunities in fuel switching, consumer behavior, or investments in technologies that enhance options for future reductions, larger near-term reductions in emissions are not only unnecessary in the context of a long-term cumulative emissions problem, but are not even desirable. They are highly expensive, while not contributing significantly to the long-run goal of stabilization. The economic resources spent on these low-return activities could limit national willingness to make the large R&D investments that are required for this problem.

27.4 Climate policy must focus on creating and adopting new technology

When new technology and new capacity investments are the issue, the only policy strategies that matter immediately are those that will increase incentives to invest in R&D, and direct the R&D toward technologies that will create a much larger supply of carbon-free energy alternatives at acceptable costs. Therefore, the only attribute of a cap-and-trade program that could serve to create an incentive for R&D will be the future course of the cap and its implications for future permit prices. This raises three problems that we turn to now.

1. Unless set at a level that requires only small emission reductions using existing technology, near-term caps on emissions impose unnecessarily high costs, since the same benefits to future GHG concentrations can be achieved with emission reductions made after low-cost technology becomes available in the future.
2. Incentives for new technology will be inadequate when they are provided only by unenforceable expressions of an *intention* to cause permit prices to rise over time.
3. A near-term carbon cap will place an unnecessary tax on existing technology (which cannot and need not respond to the price signal) in its effort to move future technology to a different place.

These problems stand in contrast to the goal of forcing near-term reductions in emissions on existing technology. The benefit of the latter actions is a reduced near-term contribution to future accumulated concentrations of GHG in the atmosphere. The cumulative nature of the GHG risk implies that emission reductions in any year can be substituted for emission reductions in any other year without changing subsequent greenhouse gas concentrations.[10] This implies that the right way to evaluate a reduction in emissions is to compare its cost with the cost of the same reduction taken in the future when lower-cost, new technologies might be available. Thus, the benefit of early emission reductions is the discounted present value of the avoided cost of future emission reductions. If R&D can make future controls possible at even a moderate cost, this discounted present value will be small.[11]

It is without question worthwhile to adopt current measures to achieve emission reductions that pass this cost-effectiveness test.[12] These measures might include modest carbon caps or taxes in the near term, or might avoid the cost of creating new regulatory or tax-collection mechanisms by taking other approaches that would target cost-effective near-term reductions. In particular, there are likely to be opportunities to reduce emissions growth in developing countries through measures that provide overall economic benefits (Bernstein *et al.*, in press). However, the relevant criterion for cost-effectiveness is the present value of the cost of avoided future emission reductions that would come from the future technologies, once they become available.

27.5 Induced technological change in economic models

A cap-and-trade system can certainly promote use of the low-cost opportunities in fuel switching or consumer behavior, but these are likely to be a very small part of the long-term solution for climate change. The question then is whether a cap-and-trade system can motivate the significant development of basic science and technology needed without

---

[9] For example, if all of the existing US natural gas-fired combined cycle generating capacity were suddenly to be fully utilized, we estimate based on our models of the US power sector that current annual US CO_2 emissions would be reduced by about 80 million metric tonnes C – about a 4% reduction in total US GHG emissions – and it would come at a cost of about $80/metric tonne C, even if gas prices would not be inflated by the sudden surge in natural gas demand.

[10] The rate of natural removal of greenhouse gases in the atmosphere makes the actual calculation of the tradeoff between current and future emission reductions a bit more complicated. Edmonds has suggested that earlier actions to reduce emissions contribute less to stabilization of concentrations than later actions, because a significant share of current emissions will have been removed from the atmosphere by natural processes before the date at which stabilization will be achieved.

[11] Various integrated assessments have placed that present value at around $10/tonne carbon. The exception is for concentration goals sufficiently low – e.g. below 450 ppm – that they are impossible to achieve unless emission reductions start immediately. The question then is whether achieving such tight concentration targets is cost-effective, a topic on which there is now clearly no consensus but considerable doubt. See, for example, Nordhaus and Boyer (2000).

[12] No matter what concentration level might be chosen as a goal, the least-cost sequence of emission reductions to achieve that concentration will have the property that the marginal cost of mitigation must rise over time at the rate of discount. What the acceptable marginal cost will be on any given date depends on the concentration goal chosen, as well as on the cost characteristics of all the mitigation options available throughout the time horizon.

imposing unnecessarily high near-term costs. Relying on a cap-and-trade system to bring about innovation implies that the caps to be set in the future, and not the caps that are set for the present, will determine the nature and pace of R&D to create the backstop technology.

This is recognized explicitly in the literature that has arisen on "induced technological change" (ITC). In models of ITC, the cap-and-trade system leads to a change in the rate and direction of technological change, resulting in a progressive reduction in the cost of technologies that can be used to reduce GHG emissions.[13] Models of ITC have been used to reach a number of conclusions about how optimal climate change policy should be designed. However, the conclusions vary depending on what the model assumes is the underlying basis for the ITC. One key distinction is between technological change that is induced by the policy causing an increase in investment in research and development ("R&D-based ITC"), and technological change that occurs because the policy induces increased use of existing but higher-cost production methods, which in turn results in falling process costs due to "learning by doing" (LBD).

Goulder (2004) has surveyed this literature and concludes that the optimal time path of emission caps depends on which type of ITC occurs. With R&D-based ITC, Goulder agrees with the analysis presented thus far that emission reductions should largely be delayed until the technology is available. Goulder also subscribes to the belief that an "announcement effect" will induce this R&D even if no near-term targets are imposed.[14] He contrasts R&D-based ITC to the case of LBD, in which the cost of new technologies is driven down as a function of cumulative output using those technologies. LBD leads to the policy recommendation for early emission caps to drive adoption and learning.

Goulder (2004) concludes that a two-pronged approach is required: "To promote ITC and reduce GHG emissions most cost-effectively, two types of policies are required: policies to reduce emissions and incentives for technological innovation." The need for a two-pronged approach would hold true if LBD were important, and it would also hold true under R&D-based ITC, *if the announcement effect exists*. However, as we will next discuss, the LBD hypothesis is unlikely to be valid for the situation of GHG reductions. Additionally, the announcement effect is unlikely to exist under the R&D-based ITC case because of the dynamic inconsistencies that we demonstrate in the next section. There may be little harm in the prescription to announce a future cap simultaneously with implementing an R&D strategy. However, there *is* potential harm in the prescription of near-term caps in the hope that LBD is the possible mode of ITC. Therefore, it is important to look closely at the available evidence on the case for LBD.

27.6 Learning by doing?

"Learning by doing" is simply not the right model for the types of innovation that are needed in the case of GHGs. Near-term reductions imply primarily greater natural gas utilization, but greater use of natural gas is notably absent from the portfolio of technologies that Hoffert *et al.* identified as potential solutions in their extensive review of technology paths to global climate stability. Even if it were not a dead-end path for GHG reductions, there is also little potential learning to be done in the efficient use of natural gas. In short, there is no reason to believe that any relevant technological evolution (learning) might be achieved regarding how to solve the GHG problem by the types of actions that would result from forcing near-term reductions.

Nordhaus (2004) has discussed the lack of empirical support for the assumption that LBD will be a substantial part of ITC. His discussion of LBD in airframe production (which was the example that brought the phenomenon to the attention of economists), semiconductor production, and other examples lead him to several conclusions. To quote (emphasis in the original):

- There is clear *structural* evidence of learning.
- The *mechanism* by which learning occurs is *complex*, including worker experience, forgetting, investment, R&D, and perhaps but not clearly cumulative output.
- *Spillovers* differ greatly depending upon the technology and whether they apply to workers, firms, specific technologies, and countries. Given current structure, it would be *folly* to rely upon LBD to rationalize a costly or critical component of climate change policy.

Nordhaus concludes that the historical evidence is that LBD is largely firm-specific, that it occurs largely within a single generation of technology, and that it is unclear that there is any association between LBD and cumulative output with a technology. In the case of climate change, the needed technologies simply do not exist today, so that LBD with existing technology will not contribute to development or

---

[13] Most models embody technological change even in their baselines, in forms such as autonomous energy efficiency improvement (AEEI) rates. ITC might be modeled as an increase in the baseline rate of the AEEI that occurs as a result of imposition of a policy. This is more generic than the types of technological change that are envisioned as needed for GHG reduction, where the new low-emitting technologies that will become available will change the mix of input factors, rather than simply reduce the quantity of energy used to produce a unit of output. For example, if the combination of IGCC with sequestration were to become a cost-effective control option, its adoption would increase the use of coal while reducing emissions of carbon, whereas in most models of technological change, including AEEI and ITC, increased coal burning always produces increase carbon emissions.

[14] The "announcement effect" would occur if the anticipation that a challenging emissions cap will be imposed at a specific future date would motivate the desired R&D before that date, in order for private entities to reduce the erstwhile costs of meeting that cap. We will make the case in the next section that the problem of dynamic inconsistency eliminates the announcement effect in the case of GHG policy.

reduction in the cost of the needed technologies. It is interesting that LBD is not even mentioned in an extensive recent survey of the ITC literature in environmental economics (Jaffe et al., 2003).

27.7 The basic economics of R&D

Private sector R&D is motivated by profits to be earned from successful innovation. These profits (also called "quasi-rents" because they are earned for temporary exclusive ownership of intellectual property) can be realized by selling or licensing technology or by using technology breakthroughs to create new products or lower costs. Earning the profits requires protection of intellectual property so that imitation does not drive market price down to cost of production with the new technology.

To the extent that climate R&D must focus on fundamental science and requires many incremental steps of technology development before a mature new energy source is achieved, it may be impossible to appropriate the benefits of that R&D. This is a problem in all fundamental R&D. However, climate R&D is clearly different in another way: the very existence of economic benefits from low- or zero-carbon technology that make it possible to profit from its adoption depends on the decisions of governments to put a price on carbon emissions. Market forces do not establish that price in the absence of government intervention in the form of a tradable permit system or carbon taxes. Even if benefits are otherwise appropriable through protection of intellectual property, the ability to take profits depends on the price of carbon, and that price is set by government policy. Therefore, how the government reacts to successful (or unsuccessful) R&D is key to the prospect of making profits from the fruits of R&D for GHG-reducing technologies.

27.8 Emission trading and R&D-based ITC

If it is not possible to rely on LBD to produce the requisite future technologies for reducing GHG emissions, and it is in fact necessary to stimulate R&D in order to induce technological change, then an important question is how a cap-and-trade system can create these incentives. The literature on ITC has recognized that it is the expectation of future policies that motivates R&D, and that emission caps put in place before innovations resulting from R&D can be deployed have no effect as incentives. Indeed, the literature emphasizes the "announcement effect" of future carbon limits. What the literature has failed to recognize is the impossibility of creating a credible announcement of a future carbon limit or carbon price at a level sufficient to motivate R&D investment by the private sector. This difficulty is so fundamental that it undercuts the very basis for the theoretical demonstration that emission trading can achieve environmental targets at minimum cost (Montgomery, 1972).

The clear definition and enforcement of caps is central to the efficient operation of the emission trading system (Bernstein et al., Chapter 18, this volume). The idea of emission trading is to circumvent the "public goods" problem of climate change by setting up a system of emission rights that can be traded, by first placing a cap on total emissions, and then dividing up that total allowable amount (emission cap) and enforcing limits on greenhouse gases from every individual source. Once these limits and the required monitoring and enforcement are set up, allowing those well-defined emission rights to be traded makes it possible for market forces to determine where emissions should be reduced, so that the aggregate limit is achieved at the lowest possible cost.[15] Emission trading will achieve this result, but only if all the relevant markets exist and operate reasonably effectively. The two relevant principles of all markets are that:

- Competitive markets lead to efficient resource use only if there are "complete" markets for everything that matters; and
- These markets must extend over space and time and allow for trades that are conditional on how uncertain events turn out (Debreu, 1959; Arrow, 1964; Arrow and Hahn, 1971).

These conditions also apply to the creation of new markets through emission trading: emissions markets can be expected to lead to efficient use of resources only if they are also complete, cover all the relevant sources and time periods, and allow participants to lay off risks through use of appropriate securities or other instruments (Montgomery, 1972; Mäler, 1974).

In theory, emission trading would put a price on carbon emissions, thereby creating value for investments and technologies that can bring about reductions in carbon emissions at unit costs less than the price of permits. Expectations about the future price of carbon permits would be the motivating force to change long-term investment behavior, and would stimulate R&D. This concept of regulation-induced R&D implies a policy prescription where the optimal amount of control is less now, and more later.[16]

This is a very attractive picture. Not only does it promise a route to solving the climate policy problem, but it also allows decisions to be made by the scientists, inventors, entrepreneurs, and investors in the private sector who, one surmises, know what they are doing. And it keeps government agencies,

---

[15] This leaves out the question of how the limit is chosen. The operation of competitive markets can achieve the limit efficiently under the conditions stated, but the only way to tell if a limit is justified is to compare its overall costs and benefits. No one has ever devised a form of emission trading that could use market forces to reveal where the overall limit should be set and, indeed, the limits defined by the Kyoto Protocol and similar agreements have been primarily political in nature.

[16] Alternatively, it implies that the optimal profile of externality taxes would be lower relative to that of a situation where the regulation does not induce R&D. See Jaffe et al., (2003), p. 484.

who have demonstrably failed to conduct successful energy R&D programs, out of the picture.

27.9 The fundamental impossibility

The problem is that it won't work. All of the really important actions by the private sector have to be motivated by price expectations far in the future. Creating that motivation requires that emission trading establish not only current but future prices, and also establish a confident expectation that the future prices will remain high enough to justify the desired current R&D and investment expenditures. This requires that clear, enforceable property rights in emissions be defined far into the future – so that emission permits for 2030, for example, can be traded today in confidence that they will be valid and enforceable on that future date.

The time period over which emission rights are defined is the most critical question in evaluating whether emission trading can be a viable approach to climate change. What motivates R&D is the expectation that future costs will be high – not the observation that current costs are high. It is unnecessary to bear high costs of reducing emissions today to induce R&D, and it is not even clear that doing so would have any effect on R&D at all. Near-term reductions may even undermine long-term incentives (e.g., by creating such high costs that there is a general expectation that the regulatory system will be dismantled before any R&D could come to fruition).

Moreover, there is another theme that has emerged both from the integrated assessment literature and from the efforts of the IPCC to summarize climate economics and climate science. It is that both the costs of controlling emissions and the relationship between emissions and ultimate global temperatures are highly uncertain, and that the impacts of global temperatures on human and natural systems and therefore the damages from climate change are even harder to predict. There remains great uncertainty about what costs are worth bearing to mitigate climate change, and therefore what the ultimate concentration goal should be.

This scientific uncertainty translates into policy uncertainty. Emission caps can be justified based on one view of costs and one view of the ultimate concentration goal. But both current views of costs and current views on an appropriate concentration goal will almost certainly need to be revised as more experience with costs of control and better climate science appear in the future. As more is learned about costs and climate science, emission caps should be revised in light of the changed balance between costs and benefits. Thus no government with the objective of maximizing social welfare should commit at this point irrevocably to a particular set of emission caps for the future, because today's choice will almost certainly turn out to be suboptimal.

Further, even if R&D is successful, and the initial control level chosen by the government is correct, the price of carbon permits will be lower than their total cost unless there is a monopolistic supplier of the future emissions reduction technology. This is because there is a fixed cost to the R&D investment, but the resulting methods will be subject to competition in their deployment in the emissions market. The competition to achieve least-cost emissions reductions will drive the price of carbon permits down to just the forward costs of control – the cost of building and operating the new technology. This market price generally will not provide any quasi-rents that will allow those who performed the R&D to recoup their investment in the emissions market.[17]

27.10 Carbon taxes also are ineffective for R&D-based ITC

The natural reaction of environmental economists to the sorts of uncertainties that we have outlined above for the cap-and-trade approach has been to suggest that emissions taxes must be the appropriate alternative. There is a fundamental economic principle at work here: when costs are very uncertain and there are no threshold effects in damages, it is better to use price signals, rather than binding caps, to determine the quantity of new emissions (Weitzman, 1974). Since there is a fundamental cost uncertainty in new technology for reducing greenhouse gas emissions, this principle indicates by itself that caps are not the way to go (and the cumulative or "stock" nature of the GHG externality reinforces this conclusion; Newell and Pizer, 2003).

This realization has been embodied in proposals that a harmonized carbon tax, starting at a relatively low value, should be adopted (Cooper, 1998; Nordhaus, 2002), or that if a cap-and-trade system is put in place there should be a safety valve, in the form of a commitment by the government to make additional permits available at a fixed price to limit costs (NCEP, 2004). But taxes or safety valves present problems of their own, and exploiting the feature that taxes limit costs makes taxes insufficient to stimulate R&D. This is strengthened by examining the credibility of announcements of a tax or a safety valve price: any tax low enough to be credible will be insufficient to stimulate R&D, and any tax sufficient to stimulate R&D will not be credible.

If the tax is set so low that it does not produce significant emission reductions (e.g., below a level sufficient to cause switching from coal to gas for power generation), a tax today serves little purpose but to raise revenues. Allowing for a gradually rising tax at some reasonable rate of increase means that the tax must either be too high in the near-term, or it will not be high enough later – at the time when the R&D results

---

[17] If the technology is successful, then unless innovators are able to tie up the technology in patents and successfully restrict licensing, competitive pressures will drive permit prices down to the level of forward costs allowing no return on R&D. The purpose of an ideal patent system is to prevent this problem, but since the innovations needed for reducing the cost of GHG emission control require basic science and a complex of supporting technological developments, such exclusion is not likely to be possible.

can be expected or needed.[18] This leaves one viable alternative, a very low tax now that will rise rapidly at some point. Included in this idea would be no tax today, but the announcement that a rather large tax will be imposed at some specified date in the relatively distant future such as 10 or 20 years out.

Unfortunately, as with the announcement of a future set of challenging emission caps, a high future safety valve price or carbon tax also cannot create a credible incentive for private sector R&D. The reasons are slightly different. In the case of cap-and-trade, the fundamental flaw was the knowledge that the competitive market price of carbon permits would be insufficient to produce returns on the R&D investment. In the case of a tax or safety valve, the problem lies in an expectation of the future actions of the government. Governments will always have an incentive to fall back to a lower carbon price after innovations occur – a price that will motivate companies to adopt the new control technology, but not the much higher one that was necessary to motivate the large R&D investment (Kremer, 2000).[19] This leads to a fundamental dynamic inconsistency that makes any effort to announce future carbon prices sufficient to stimulate R&D not credible.

27.11 Dynamic inconsistency in using emissions taxes to stimulate ITC

In order to stimulate investment in R&D, the future safety valve price or carbon tax must reach a level that is sufficient to provide a return on the capital invested in R&D. This price will necessarily exceed the price required to induce deployment of the technology once it is developed, because to induce deployment the carbon price need only exceed the amount by which the average variable cost of the new technology exceeds the average variable cost of the next best alternative. But to motivate R&D the price must exceed the difference in variable cost by enough to create quasi-rents large enough to provide the required return on investment.

Even if the government announces a future emission tax that creates such an incentive, the government will prefer reneging on the level of that tax once the technology is developed. As in the case of patents, there is a tradeoff between efficiency in resource allocation and providing an incentive for R&D. A carbon price above the level necessary to induce adoption of the new technology will cause avoidable deadweight losses as all energy supply and use decisions are distorted. The government can avoid these deadweight losses by limiting the carbon price to the level just necessary to get the technology adopted, and not high enough to provide a return on R&D. Since private investors can understand this is the optimal strategy for government – and indeed would probably be skeptical of the political ability of any government to proceed with what will look like "corporate welfare" – they will not be motivated to invest in R&D by any announcement of future climate policy.

The fundamental problem for emissions taxes, then, is that there is no equilibrium solution to the R&D game in climate change. The government controls the key policy variables, and not even a perfect system of intellectual property rights can give innovators an expectation that they will earn positive quasi-rents if their innovations are successful.[20] First we characterize the R&D game, then we show that it has no equilibrium solution when government controls carbon prices.[21]

To simplify the R&D game we consider two actors, the government and a private sector that carries out innovation and investment, and two time periods, now and future. We assume the government takes into account how the private sector will respond to its actions, but the private sector plays competitively, taking the government's best response as given. In the first time period, the government announces a future price for carbon and the private sector decides whether to innovate. In the future period, there will be a (known) function that relates marginal damages from climate change to the reduction of emissions from baseline levels, and a marginal

---

[18] The nature of the climate problem, which arises from concentrations of greenhouse gases in the atmosphere based on cumulative GHG emissions, suggests that the carbon tax should rise at the rate of interest. Proposals to allow "banking" of credits will cause prices to rise at no more than the rate of interest.

[19] Kremer has identified this same problem with government intervention in pricing in the case of pharmaceutical R&D. Malaria, tuberculosis, and the strains of HIV common in Africa kill approximately 5 million people each year, yet research on vaccines for these diseases remains minimal. Kremer makes the case that this is largely because potential vaccine developers fear that they would not be able to sell enough vaccine at a sufficient price to recoup their research expenditures. This is because the marginal cost of manufacturing the vaccines is very small relative to the cost per dose that would be necessary for companies to recoup their R&D investment. Governments in countries that need these vaccines, usually poor countries, may refuse to pay more than the marginal manufacturing costs. Anticipation of this response of governments leads to underinvestment in vaccine research.

[20] The analogous problem with emission caps is that the competitive emissions price in the future, once the R&D has succeeded, will never reflect the sunk costs associated with the R&D unless those who have funded the R&D would be able to become monopolistic suppliers of the new emissions control technology. In cap-and-trade, the problem does not stem from the lack of equilibrium in a game between government and private actors, but rather it is the result of announcing that competitive market pricing of emissions will determine the rewards for private sector R&D.

[21] This policy problem is very similar to the difficulty in the analysis of macroeconomic policy that was demonstrated by Lucas (1976), who showed that only unexpected changes in monetary policy can have any effect on real output. In line with the Lucas critique, Kydland and Prescott demonstrated that there is a dynamic inconsistency in monetary policy, because governments always have an incentive to create a surprise increase in the money supply to stimulate output. The only case in which private agents' expectations of inflation will be realized is one in which the government announces such a high rate of inflation that further surprise increases would not be beneficial to the economy. Thus the equilibrium outcome when government cannot pre-commit to an inflation policy is one that creates greater inflation than optimal. Our demonstration of the dynamic inconsistency when governments cannot pre-commit to a future carbon tax follows the proof given by Kydland and Prescott. See Lucas (1976), and Kydland and Prescott (1977, pp. 473–91).

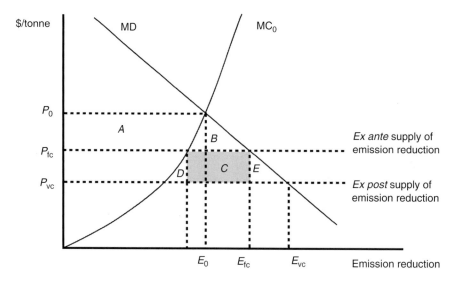

Figure 27.1 Optimal carbon taxes before and after innovation.

cost curve for achieving emission reductions from existing capital and technologies. If the R&D is successful, a carbon-free technology will be available. The carbon price that will cover variable costs of using the carbon-free technology to replace fossil fuel used in the baseline is P_{vc}. The carbon price that would cover the variable costs of the new carbon-free technology plus provide quasi-rents adequate to motivate *ex ante* the investment required to produce the innovation is P_{fc}.

Figure 27.1 describes the government's problem of choosing a carbon price in the future. The curve labeled MC_0 is the marginal cost of reducing emissions without the future technology, and MD reflects the marginal damages curve.[22] With no innovation, the optimal emission reduction is E_0 and the price of carbon is P_0.

The *ex post* supply curve of emission reductions (after the carbon-free technology has been innovated) coincides with MC_0 up to the price P_{vc}, at which point it becomes flat indicating unlimited emission reductions available at that price.[23] The optimal future emission reduction assuming successful R&D is given by the intersection of the marginal damage curve with this *ex post* supply curve – the socially optimal carbon price would be P_{vc}, and the optimal emissions reduction would be E_{vc}.

But now consider the same problem at a time prior to the development of new technology. The minimum price that will act as incentive for the innovation supporting the backstop technology is P_{fc}, higher than P_{vc}. As seen in Figure 27.1, P_{fc} would produce more emission reduction from the existing capital stock, higher costs attributable to these emission reductions, and less adoption of the future technology in the

---

[22] To keep Figure 27.1 simple enough to clarify the key distinction between the *ex ante* and *ex post* supply of emission reduction, we use the emission reduction observed in a single year as indicative of the change in emissions for the entire time horizon. We need to simplify the relationship between emissions and damages because damages are a function of concentrations, which in turn depend on cumulative emissions rather than emissions in a single year. Therefore the emission reductions measured on the horizontal axis should be thought of as being taken from a representative year in a program that reduces emissions over the entire time horizon. Thus the emission rate E_0 is taken from a program that leads to concentrations for which marginal damages are P_0; likewise E_{vc} is taken from a program leading to a lower stabilized concentration and hence a lower marginal damage, P_{vc}. By construction this means that Figure 27.1 refers to a representative year in the future period in which large emission reductions become available – i.e. after zero-carbon technologies could in principle be deployed. The time path of emissions during the transitional period in which new technologies are being deployed and the date at which a net zero contribution to concentrations is achieved will determine the level at which concentrations are stabilized. Different cumulative emissions during this transition period will lead to different ultimate concentrations and to a downward sloping marginal damage curve. This implies that the lower is the cost of reducing emissions during this transitional period, the lower will be the concentration justified by an overall cost–benefit analysis.

[23] We also note that the argument does not depend on the carbon-free technology being a backstop in the sense that it can supply *unlimited* amounts of energy at constant cost. Our conclusions only depend on the assumption that one of the critical new technologies necessary to enable stabilization will be on the margin relative to the *ex ante* price P_{fc}. There could well be a series of smaller cost reductions or improvements in existing technologies resulting from R&D, each with limited applicability or resource bases. These multiple innovations would create differences at many points along the base supply curve, MC_0. For these new but inframarginal technologies, there would be no dynamic inconsistency problem, but achievement of such modest technology improvements along the way would not by definition provide the breakthroughs needed to enable carbon stabilization. It is hard for us to imagine breakthroughs of the sort that would enable actual stabilization not including at least one such critical, large-scale technology that would be on the margin. (Inframarginal new technologies would, however, provide useful additional cost-reductions from a price signal, and thus make the benefits of a price signal larger than can be calculated now.)

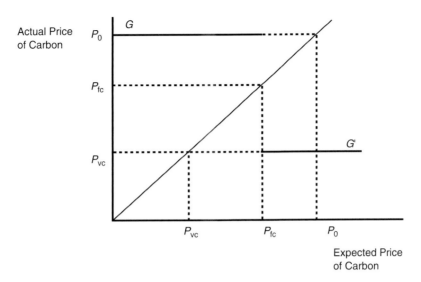

Figure 27.2 Equilibrium in the R&D game.

optimum. If the optimal emission reduction is determined based on the cost of R&D plus deployment of the backstop technology, it will be less than the optimal emission reduction based solely on the variable cost of deploying the technology (the amount $E_{vc} - E_0$), further reducing the benefits of the future technology.

Economic benefits are the difference between benefits and costs. In the case of no innovation at all, this is the large area A between MD and MC_0 from the origin to E_0. A policy of pre-committing to a carbon price just sufficient to induce innovation that produces the backstop technology provides additional benefits equal to the triangular area B. By construction, the shaded area C is the cost of R&D that must be covered by the difference between the price of carbon and the cost of deploying the backstop technology.[24]

However, this policy (P_{fc}, E_{fc}) is not dynamically consistent. The policy that maximizes social welfare *after the R&D needed to create the advanced technology has been successful* is to set the price of carbon equal to P_{vc}.[25] To reduce the market price of the backstop technology to P_{vc} and obtain these benefits, the government must renege on its original announcement that the carbon price will be P_{fc}.

But it is clear from the outset that such is the future incentive of the government because the government cannot irrevocably commit itself *ex ante* to a future price higher than P_{vc}. The private sector can fully expect that it will only be able to charge a future price of P_{vc} for that technology. Thus the market expectation that the government will maintain carbon prices high enough to provide a return on innovation is not credible. This leads to the conclusion that the R&D game does not have an equilibrium in which the private sector is motivated to produce R&D by the government's announcement of a future carbon price.[26]

Figure 27.2 illustrates the equilibrium problem. The horizontal axis represents the carbon price expected by the private sector, and the vertical axis represents the carbon price chosen by the government. On the 45° line, private expectations are fulfilled in that the government chooses the price expected by the private sector. Three prices are drawn in the figure: P_{fc}, P_{vc}, and the price at which the marginal cost and marginal benefit of reducing GHG emissions are equalized when no backstop technology is available, P_0. None is an equilibrium price. If the private sector expects any price greater than or equal to P_{fc} (including P_0), then innovation will take place and the government's best reply will be to set price at P_{vc}. Thus the expectation of a price of P_0, or any price greater than or equal to P_{fc}, will not be fulfilled and cannot be an equilibrium. If the private sector expects any price less than P_{fc}, including P_{vc}, there will be no innovation, so that the government will set the price at P_0. Thus, the expectation of prices less than P_{fc} are also not in equilibrium.

---

[24] The area C equals the amount by which the price of permits exceeds the variable cost of deploying the backstop technology, times the amount of emission reduction achieved by the backstop technology at E_{fc}.

[25] Reducing the price of carbon to P_{vc} would produce an additional social gain equal to the triangular areas D and E, which are the savings on unnecessary emission reductions from the existing capital stock and the benefits of additional reduction of global warming achievable by reducing the market price of the backstop technology to P_{vc}. It would also, however, transfer the costs of the R&D, C, onto those who made the investment.

[26] An analogous situation exists for a pure cap-and-trade system. We have noted already that if the innovator cannot maintain a monopoly then the permit price in the market will fall to P_{vc}. If, however, the innovator can preserve a monopoly, then the government will have strong incentives to maintain a cap at the level E_{vc} once the technology to achieve such reductions at P_{vc} exists. This suggests that the government will pursue whatever other policy tools it has available to prevent the monopolist from restricting use of its technology, including antitrust proceedings that would prevent coordination among patent holders to raise the price of technology to the levels they would need to recoup their investment costs.

This outcome of dynamic inconsistency is also known as a lack of sub-game perfect equilibrium. The government's strategy that would be optimal if it could be made a binding commitment prior to innovation becomes no longer optimal once innovation occurs. The private actors will realize at the beginning that announcing and then reneging is the optimal dynamic strategy for the government; they will therefore be willing to undertake R&D in response to the announcement only if the government can bind itself to impose the higher carbon tax. But, as we have discussed, it is impossible for the government to make such a commitment if the policy is to involve a tax that will create a future burden on the economy at large. Therefore, announcing future carbon limits or prices cannot in principle provide an adequate incentive for private investment.

27.12 Implications for technology policy

Our proof of the dynamic inconsistency of announced carbon prices when R&D is required does not depend on the assumption that there is a single large innovation or "silver bullet." All it requires is the assumption that a large innovation is on the margin *ex ante*, so that the full cost of the new technology sets the relevant carbon price.

There are two separate issues about the incentives for R&D into climate technology. One is the general problem of appropriability. At an early stage in research into enabling technologies, standard problems of appropriability are likely to appear, exacerbated by the possibly long time and indirect connection between scientific achievement and the actual production of low-carbon energy. It is necessary to assume a sequence of licenses and participation in final profits if emission trading is to stimulate this kind of R&D. If such participation is not possible, non-appropriability will eliminate the incentive for R&D independently of the identified dynamic inconsistency. The second issue is that even with such a sequence to provide potential rewards, the rewards also depend on the credible announcement of a carbon tax high enough to make the resulting intellectual property worth something.

At a later stage in R&D, rewards of innovation are likely to become more readily appropriable. It is here that the dynamic inconsistency of announced carbon prices eliminates what could otherwise be a sufficient incentive for induced technological change. But it does not matter if it is one or many technologies that are capable of being deployed.

For the case of multiple technologies being deployed together or over time, the proof of dynamic inconsistency need only note that the government will observe the forward-looking marginal cost of producing low- or zero-carbon energy – ignoring sunk costs of R&D – and be able to compare this with the announced price of carbon permits.[27] The optimal policy, looking forward from any point after innovations have begun to appear, is then to set a carbon price equal to the marginal cost of achieving desired reductions in carbon emissions. This marginal cost will not contain the sunk cost of R&D required to develop the marginal technology, and therefore will eliminate any quasi-rents to reward innovation at the margin.

Likewise, the effect of this dynamic inconsistency on the incentive to invest in R&D also remains if there are a number of separate potential innovations that could become economic at different carbon prices, as well as in the simple case illustrated in Figure 27.1 above. An investor in R&D will not know ahead of time whether an innovation will be on the margin or not. The prospect that an innovation may be on the margin is sufficient to make the announced carbon price a less than credible incentive. The larger and more important an innovation is expected to be, the greater the likelihood it will be on the margin and the greater will be the diminution of incentives due to the dynamic inconsistency of policy.

Moreover, reversing an announced carbon price may be even easier if there are many innovations and innovators that could profit from a high carbon price, because it will make the action appear more general and less as a taking from a single innovator.

Thus, no matter whether one's view of the backstop technology is one of a "silver bullet" or of a stream of incremental improvements, the inability of governments to pre-commit to a future carbon price will lead to the expectation that there will not be sufficient profits available to make all the requisite R&D profitable. The solution is not to write a constitutional amendment setting a carbon tax in place for all time. As has been pointed out in the macroeconomic literature (Friedman and Kuttner, 1996), there is a legitimate role for policy flexibility. This is especially true when the costs and benefits of a policy are both uncertain and likely to be updated over time, and this certainly applies to the situation for carbon policy.

This then suggests another approach, which is that the government provides compensation for the innovation separately from the price of carbon, that is, outside the emission trading or carbon tax system. Direct, up-front incentives would also allow innovations to be deployed optimally, without a period of time in which deployment is restricted to provide quasi-rents to innovators. By avoiding this inevitable consequence of a patent regime, provision of up-front R&D incentives would be superior even to the outcome if a government could pre-commit to keep carbon prices high enough to induce innovation. We take this up in the final section of this chapter.

27.13 The fundamental problem is the need for effective R&D

Emission taxes and emission caps are the two basic tools that economists traditionally prescribe for managing environmental externalities. There is an extensive literature indicating

---

[27] Assuming, as we pointed out earlier, that some new technology is on the margin.

conditions under which the price or the quantity approach is more appropriate. However, we find that neither the price nor the quantity approaches appear to work effectively in the case of the GHG externality. Why is this? The fundamental problem for cost-effective GHG controls keeps coming back to the externality associated with R&D, and the unique degree to which solving the GHG externality depends on effective technological change. Until the R&D externality is resolved, there is almost no case to be made for starting to control the environmental externality.

The main arguments that have been advanced for attempting to pair a direct emission control strategy with an R&D strategy, and *in advance of any R&D success*, lie in the belief that either there will be meaningful learning-by-doing from near-term reductions, or that there will be some added incentive for R&D-based ITC in the form of an announcement effect. The LBD hypothesis is simply inconsistent with the types of control action that are necessary for large GHG reductions. The announcement effect is incapacitated by dynamic inconsistencies. Dynamic inconsistencies apply to both cap-and-trade and tax approaches, although for somewhat different reasons. Thus, implementing policies for addressing the environmental externality before the fruits of R&D are available would be of minimal, if any, value. If made too stringent (i.e., sufficient to motivate more action than just harvesting the very lowest-cost actions viable today), direct emissions policies might even be counterproductive. Taxes or carbon prices could become relatively high before they would motivate any significant action, and are more punitive than motivational for companies that have few near-term options to change their emissions.

The lesson learned about technology from the past decade of integrated assessment is that there are no ways that we can achieve the needed GHG cuts with what we know how to do today.

The analysis tells us that what we must focus on now is how to devise effective R&D policies that will overcome the dynamic inconsistencies of incentives, and the other externalities associated with R&D. This is central, and we have been diverting ourselves for too long in search of workable cap-and-trade, tax, or hybrid systems that are inherently unsuited to this need. The only function of a price on current greenhouse gas emissions is to motivate emission reductions through changes in use of the existing capital stock, or new capital investments using existing technologies, that can be achieved at a cost no greater than the discounted present value of the cost at which future emission reductions can be achieved if R&D is successful. This could justify a relatively low carbon tax at present, but picking any specific number for a carbon tax requires an assessment of what the cost of emission reductions will be in the future when zero-carbon technologies are available on a near-universal scale. Therefore, even calculating what the carbon tax should be in, for example, the year 2010 requires more thought about the design and goals of a long-term R&D program.

Thus the central question becomes one of how R&D can be motivated effectively and in a cost-effective fashion (Wright, 1983). Even deciding on where to set a current carbon tax logically requires that this question be addressed first. We do not have answers, but conclude with some general discussion of the options that need to be explored more aggressively for GHG policy.

27.14 Patent rights

Simply awarding extended patent rights cannot be a sufficient answer because, as we have shown, governments always have an incentive to reduce the announced carbon price in a way that reduces either license fees or quasi-rents from use of the patented technology. This action would remove the value of the patents without constituting a taking, because it in no way affects the validity of a patent. Patents convey a right to property, but not a right to a particular price. Even though patents allow exclusive use, the law is clear that there is no guarantee they will have value. Thus patents need to be combined with some other way of addressing the dynamic inconsistency of policy announcements as an incentive for R&D.

27.15 Grants and tax incentives

Grants and tax incentives face an intractable tradeoff. Either they are so narrowly circumscribed as to what qualifies that government ends up making all the choices about the direction of research, or they are so broad that opportunistic behavior is encouraged – so funding goes to research that would have been done anyway, leading to either excessive budgetary cost for the amount of progress made or for a tendency for research to drift into areas that are not the most productive for developing climate change technologies. It is also noteworthy that these traditional "push" incentives are inefficient because although they motivate R&D, they do not target R&D that is most likely to be successful. "The extra R&D carried out because of tax breaks or government grants might lead to more researchers being hired, more scientific papers being published and more discoveries with lucrative applications that can be spun off from research into vaccines, but it does not necessarily produce any new vaccines" (Crooks, 2001). Design of R&D grants or tax incentives need to work around these problems, perhaps in the manner designed for the direction of the Global Climate and Energy Project (GCEP) at Stanford University.

27.16 Prizes

Michael Kremer of Harvard University has proposed the use of prize competition in a very similar R&D policy problem, that of creating vaccines for AIDS, tuberculosis, and malaria for developing countries (Kremer, 2000). Kremer proposes a "pull" program that rewards the companies once vaccines have been discovered. He suggests cash prizes for successful development or giving companies that develop vaccines extended patents for some of their other drugs. The best method, he finds, is for governments to guarantee to buy a promised quantity of the vaccines at a set price per person immunized. The companies would then know that if their research was successful there would be a guaranteed payoff.

Adapting this idea to technologies needed to reduce GHG emissions would take some thought. A malaria vaccine has readily definable properties and would be the direct result of one company's R&D. A climate change technology is likely to be the product of a number of steps taken by different companies, ultimately embodied in a complex of processes known as a "technology." Thus to have prizes it would be necessary either to break up the awards into prizes for smaller steps, as has been done by NASA in its Centennial Challenge awards program, or to define a generic technology (e.g. something that will generate electricity at zero-carbon emissions) and let the "winner take all." But the structure of the award proposed by Kremer, which is a specific cash payment for each unit sold (which could be a kWh of generating capacity) would then allow the inventor to compete at variable cost.

A question of credibility also arises for prizes, since again this entails a future commitment by the government. Governments could possibly commit themselves, in this case, through contracts or creating of trust funds to pay the prizes. Given the appropriation process in the United States, and the problems it has caused for even NASA's tiny program, these problems of having the funds available to fulfill commitments and effective instruments for creating an enforceable obligation to pay the prize should not be ignored.[28]

27.17 R&D consortia

It appears that carbon sequestration technology may be a special case in which formation of a vertical R&D consortium involving the coal and power industries would be sufficient to achieve the rewards of innovation in a manner not subject to the dynamic inconsistency in the setting of carbon prices. For a coal producer, developing a technology that uses coal without GHG emissions has value if there is a credible expectation that policies will be adopted that will make it impossible to use coal without the technology. It is an established principle that it is possible to capture the rewards of an innovation without patent protection by purchasing the assets whose value is increased (Hirshleifer, 1971). A firm that owns coal can get value from an innovation that allows continued profitable sales in the event that a climate policy is adopted.

There is still a free rider problem for the coal industry, since no company has an individual incentive to invest in order to make the technology available. This makes carbon sequestration technology no different from any other technology, because it will be impossible to profit from the technology unless carbon prices exceed the variable cost of using the technology by more than the required return on R&D investment. If the carbon price just equals the cost of using this technology, then the innovator cannot get his return, because any coal user will prefer burning coal and buying permits to using the technology to sequester. All the government has to do is set carbon price at just above the variable cost of using technology, and the innovator will be better off accepting that minimal profit than taking nothing. But if the innovator does not own all the coal resources, under the free rider problem the innovator will not have an incentive to perform the R&D.

However, it may not even be in the interest of the entire coal industry to undertake the R&D, if the profits available from continuing to sell coal are not enough to provide an adequate return to the investment in R&D. Then it would be necessary to expand the consortium, for example by including some portion of the utility industry that likewise has assets whose value would be diminished by a carbon price that makes coal uneconomic as an energy source.

This commonality of interests between buyers and sellers is the classic case for industry R&D consortia, and in particular vertical consortia (Congressional Budget Office, 1990). If all pool their resources, then they will all be in a position to use the technology without an issue of paying for the innovation. Both the R&D and the diffusion incentives will be right.

The open question is whether the prospect of carbon policy, and the potential benefits of a new technology in reducing compliance costs and retaining a market for coal, are sufficient to motivate the requisite investment in R&D. This in turn depends on whether anticipated policies that would raise the cost of using coal would also cause sufficient harm to some segments of the coal and utility industry to make investment in creating this coal-based, low-carbon emission technology attractive.

If that is the case, then there is at least some prospect for a cooperative solution involving the coal and utility industries in R&D consortia. Individual companies would join based on their assessment of the cost and likelihood of success of the planned R&D program. But this solution puts those decisions in the hands of the private sector parties who will develop and use the technology, solving the general problems of government direction and researcher opportunism. Whether these

---

[28] Definition, judging, and credibility have been a problem in the entire history of prizes. The first record of a major technology prize competition was the offer by the British government in 1714 of a prize £20 000 ($12 million today) for the discovery of an accurate method of determining longitude. Parliament appropriated the funds, thus dealing with one aspect of credibility. The review committee, which included Sir Isaac Newton, disallowed the claim of the ultimate winner, who in fact created the key piece of technology still used to determine longitude, because it did not fit with the preconceptions of the leading scientists of the day about how the problem could be solved (their reluctance has also been attributed to the vagueness of the criteria stated by Parliament for the award and to lobbying by established members of the scientific community who wanted the prize for themselves). The inventor was a carpenter and clockmaker, who did not agree with the scientists that it was impossible to make a sufficiently accurate chronometer that could operate under conditions at sea. See J. Dash (2000) *The Longitude Prize*. Every aspect of this history is informative.

coal and utility companies have or can obtain the technical competence to best solve the R&D problems of economical coal gasification and sequestration is another question.

27.18 The challenge for climate change policy analysis

The three apparently viable possibilities of prizes, consortia, and grant administration along the lines of Stanford University's GCEP are surely not the only ones. Many more ideas are needed. This is where the fresh ideas and economic research need to be focused in the next decade.

References

Arrow, K. J. (1964). The role of securities in the optimal allocation of risk-bearing. *Review of Economic Studies* **31**, 91–96.

Arrow, K. J. and Hahn, F. H. (1971). *General Competitive Analysis*. San Francisco: Holden-Day.

Barrett, S. (2003). *Environment & Statecraft: The Strategy of Environmental Treaty-Making*. New York: Oxford University Press.

Bernstein, P. M., Montgomery, W. D. and Rutherford, T. (1999). Effects of restrictions on international permit trading: The MS-MRT model. *The Energy Journal, Kyoto Special Issue*, June, 221–256.

Bernstein, P. M., Montgomery, W. D. and Tuladhar, S. (in press). Potential for reducing carbon emissions from non-Annex B countries through changes in technology. *Energy Economics*.

Burtraw, D. (1996). The SO_2 emissions trading program: cost savings without allowance trades. *Contemporary Economic Policy* **14**, 79–94.

Congressional Budget Office (1990). *R&D Consortia*, Background Paper. Washington DC.

Cooper, R. N. (1998). Toward a real global warming treaty. *Foreign Affairs*, March/April.

Crooks, E. (2001). Prizes for saving lives: drug companies will not develop new vaccines unless they are rewarded for it. *Financial Times*, April 25.

Dash, J. (2000). *The Longitude Prize: The Race Between the Moon and the Watch-Machine*. Farrar, New York: Strauss and Giroux.

Debreu, G. (1959). *The Theory of Value*. New Haven, CT: Yale University Press.

Ellerman, A. D. and Montero, J.-P. (1998). The declining trend in sulfur dioxide emissions: implications for allowance prices. *Journal of Environmental Economics and Management* **36**, No. 1, July, 26–45.

Friedman, B. M. and Kuttner, K. (1996). A price target for U.S. monetary policy? Lessons from the experience with money growth targets. *Brookings Papers on Economic Activity*, No. 1, 77–125.

Goulder, L. H. (2004). *Induced Technological Change and Climate Policy*. Washington DC: Pew Center on Global Climate Change.

Hahn, R. W. and Stavins, R. N. (1991). Incentive-based environmental regulation: a new era from an old idea? *Ecology Law Quarterly* **18**, 1–42.

Hirshleifer, J. (1971). The private and social value of information and the reward to inventive activity. *The American Economic Review* **61**, No. 4, 561–574.

Hoffert, M. I., Caldeira, K., Benford, G. *et al.*, (2002) Advanced technology paths to global climate stability: energy for a greenhouse planet. *Science* **298**, 981–987.

Jacoby, H. D., Reilly, J. M., McFarland, J. R. and Paltsev, S. (2004). *Technology and Technical Change in the MIT EPPA Model*. Report No. 111. Cambridge, MA: MIT Center for the Science and Policy of Climate Change.

Jaffe, A. B., Newell, R. G. and Stavins, R. N. (2003). Technological change and the environment. In *Handbook of Environmental Economics*, Vol. 1, ed. K.-G. Mäler and J. R. Vincent, Amsterdam: Elsevier Science.

Kremer, M. (2000). *Creating Markets for New Vaccines* Parts I and II. NBER Working Paper 7716 and 7717.

Kydland, F. E. and Prescott, E. C. (1977). Rules rather than discretion: the inconsistency of optimal plans. *Journal of Political Economy* **85**, 473–491.

Lucas, R. E. (1976). Econometric policy evaluation: a critique. *Carnegie-Rochester Conference Series on Public Policy* **1**, 19–46.

Mäler, K.-G. (1974). *Environmental Economics: A Theoretical Inquiry*, Washington DC: Resources for the Future.

Montgomery, W. D. (1972). Markets in licenses and efficient pollution control programs. *Journal of Economic Theory* **5**, No. 6 (December), 395–418.

Montgomery, W. D. (1993). Interdependencies between energy and environmental policies. In *Making National Energy Policy*, ed. H. Landsberg. Washington, DC: Resources for the Future.

Montgomery, W. D. (1996). Developing a frame work for short- and long-run decisions on climate change policies. In *An Economic Perspective on Climate Change Policies*, ed. C. Walker, M. Bloomfield, and M. Thorning. Washington, DC: American Council for Capital Formation.

National Commission on Energy Policy (NCEP) (2004). *Ending the Energy Stalemate: A Bipartisan Strategy to Meet America's Energy Challenges*. Washington DC: NCEP.

Newell, R. G. and Pizer, W. A. (2003). Regulating stock externalities under uncertainty. *Journal of Environmental Economics and Management* **45**, 416–432.

Nordhaus, W. D. (2002). *After Kyoto: Alternative Mechanisms to Control Global Warming*. Paper prepared for the meetings of the American Economic Association and the Association of Environmental and Resource Economists, January 2002.

Nordhaus, W. D. (2004). Economic modeling of climate change: where have we gone? Where should we go? *Anniversary Edition of Climate Change Impacts and Integrated Assessment (CCI/IA)*. Snowmass, Colorado.

Nordhaus, W. D. and Boyer, J. (2000). *Warming the World: Economic Models of Global Warming*, Cambridge, MA: MIT Press.

Roberts, M. and Spence, M. (1976). Effluent charges and licenses under uncertainty. *Journal of Public Economics* **5**, No. 3–4, 193–208.

Schelling, T. C. (2002). What makes greenhouse sense? *Foreign Affairs* **81**, No. 3 (May/June) 2.

Smith, A. E. (1991). *Issues in Implementing Tradeable Allowances for Greenhouse Gas Emissions*. Paper 91–169.4. 84th Annual Meeting of the Air & Waste Management Association, Vancouver, British Columbia, June 16–21.

Smith, A. E., Gjerde, A. R., Dehain, L. I. and Zhang, R. R. (1992). *CO₂ Trading Issues: Choosing the Market Level for Trading*. Final report by Decision Focus Incorporated to US Environmental Protection Agency, Contract No. 68-CO-0021.

Smith, A. E., Montgomery, W. D., Balistreri, E. J. and Bernstein, P. M. (1999). *Analysis of the Reduction of Carbon Emissions Through Tradable Permits or Technology Standards in a CGE Framework*. AERE/Harvard Workshop on Market-Based Instruments for Environmental Protection, Cambridge, MA, July 18–20.

Weitzman, M. L. (1974). Prices vs. quantities. *Review of Economic Studies* **41**, 477–491.

Wigley, T. M. L, Richels, R. and Edmunds, J. A. (1996). Economic and environmental choices in the stabilization of atmospheric CO_2 concentrations. *Nature* **379**, 240–243.

Wright, B. D. (1983). The economics of invention incentives: patents, prizes, and research contracts. *American Economic Review* **73**, No. 4, 691–707.

28

What is the economic value of information about climate thresholds?

Klaus Keller, Seung-Rae Kim, Johanna Baehr, David F. Bradford and Michael Oppenheimer

28.1 Introduction

The field of integrated assessment of climate change is undergoing a paradigm shift towards the analysis of potentially abrupt and irreversible climate changes (Alley et al., 2003; Keller et al., 2007). Early integrated studies broke important new ground in exploring the relationship between the costs and benefits of reducing carbon dioxide (CO_2) emissions (e.g., Nordhaus, 1991; Manne and Richels, 1991; or Tol, 1997). These studies project the climate response to anthropogenic CO_2 emissions to be relatively smooth and typically conclude that the projected benefits of reducing CO_2 emissions would justify only small reductions in CO_2 emissions in a cost–benefit framework. The validity of the often-assumed smooth climate response is, however, questionable, given how the climate system has responded to forcing in the geological past. Before the Anthropocene, the geological time period where humans have started to influence the global biogeochemical cycles considerably (Crutzen, 2002), the predominant responses of the climate system were forced by small changes in solar insolation occurring on timescales of thousands of years (Berger and Loutre, 1991). Yet this slow and smooth forcing apparently triggered abrupt climate changes – a threshold response where the climate system moved between different basins of attraction (Berger, 1990; Clement et al., 2001). Anthropogenic forcing may trigger climate threshold responses in the future (Alley et al., 2003; Keller et al., 2007). Examples of such threshold responses include (i) a collapse of the North Atlantic meridional overturning circulation (Stommel, 1961; Rahmstorf, 2000), (ii) a disintegration of the West Antarctic Ice Sheet (Mercer, 1978; Oppenheimer, 1998), (iii) abrupt vegetation changes (Claussen et al., 1999; Scheffer et al., 2001), or (iv) changes in properties of the El Niño-Southern Oscillation, ENSO (Fedorov and Philander, 2000; Timmermann, 2001). Here we focus on the first two examples: a possible collapse of the North Atlantic meridional overturning circulation and a possible disintegration of the West Antarctic Ice Sheet.

Our analysis addresses three main questions. (i) What underlying mechanisms define a climate threshold? (ii) What are the key scientific uncertainties in predicting whether these thresholds may be crossed in the future? (iii) What might be reasonable order-of-magnitude estimates of the expected economic value of reducing these uncertainties? Our analysis suggests that climate strategies designed to reduce the risk of crossing climate thresholds have to be designed in the face of large uncertainties. Some of the key uncertainties (e.g., the sensitivity of the meridional overturning circulation to anthropogenic greenhouse gas emissions) can be reduced by observation systems. We conclude by identifying future research needs.

28.2 What defines a climate threshold?

Slow and smooth forcing is capable of inducing abrupt and persistent changes in the climate system – in other words, the system exhibits a threshold response. What, then, is a climate threshold and how can crossing a climate threshold cause

Human-induced Climate Change: An Interdisciplinary Assessment, ed. Michael Schlesinger, Haroon Kheshgi, Joel Smith, Francisco de la Chesnaye, John M. Reilly, Tom Wilson and Charles Kolstad. Published by Cambridge University Press. © Cambridge University Press 2007.

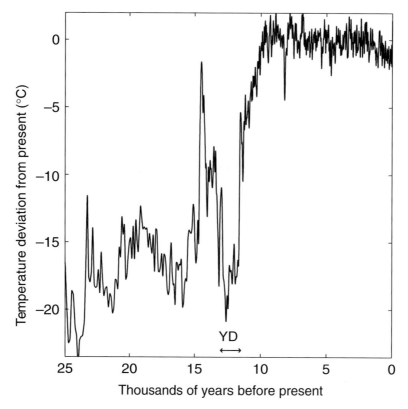

Figure 28.1 Geological evidence for the potential of abrupt climate change as recorded in the Greenland ice core. Data from Meese *et al.* (1994a), Stuiver *et al.* (1995) using a 20-year running mean. The conversion between the oxygen isotope and the temperature signal is based on Cuffey *et al.* (1995).

abrupt climate change? Broadly speaking, a threshold is the level of perturbation below which the response of the climate system is proportional to the perturbation, and above which the response to a small perturbation might lead to an abrupt and persistent change.

An analysis of the temperature deviations in central Greenland over the past 25 000 years illustrates the possible implications of climate thresholds (Figure 28.1) (Meese *et al.*, 1994b; Stuiver *et al.*, 1995). A key conclusion from this record is that climate change during the Holocene (the current interglacial, approximately the past 10 000 years) was relatively smooth compared with the pre-Holocene record. One salient example of abrupt climate change is the climate period known as the Younger Dryas (YD). During the YD, temperatures changed by close to 10 °C within a few decades (Alley, 2000). Crossing climate thresholds has probably caused abrupt climate change in the past. In the following sections, we briefly discuss basic mechanisms that could drive potential threshold responses of the North Atlantic meridional overturning circulation and the West Antarctic Ice Sheet.

28.2.1 The North Atlantic meridional overturning circulation

One key hypothesis to explain the Younger Dryas is associated with sudden changes in the meridional overturning circulation (MOC) in the North Atlantic (Alley *et al.*, 2002). The meridional overturning circulation is technically defined as the zonally integrated north–south flow in an ocean basin. In the North Atlantic, surface currents (e.g., the Gulf Stream) transport warm and salty waters to the high latitudes in the North Atlantic, where cooling increases the sea-surface density. Eventually, cooling-induced deep convection and sinking takes place; the surface waters sink and flow southward as cold deep currents (Dickson and Brown, 1994). The MOC is part of the global deepwater circulation system, which is sometimes called the "conveyor belt" (Broecker, 1991). In the Atlantic, the MOC accounts for most of the oceanic heat transport (Hall and Bryden, 1982). One effect of this northward heat transport through the MOC is seen in the relatively mild climate of Western Europe. Paleoclimatic records suggest that the MOC has undergone rapid changes in the past. Evidence derived from oceanic sediment and model simulations suggest that the above-mentioned YD was a global scale event (Denton and Hendy, 1994; Yu and Wright, 2001) and associated with a collapse of the North Atlantic MOC (Manabe and Stouffer, 1995; Teller *et al.*, 2002; McManus *et al.*, 2004). Furthermore, numerical simulations suggest that anthropogenic CO_2 emissions may cause a weakening or even a collapse of the MOC (Cubasch and Meehl, 2001; Gregory *et al.*, 2005; Schmittner *et al.*, 2005). Such an effect would entail a reduction in the North Atlantic heat transport, which in

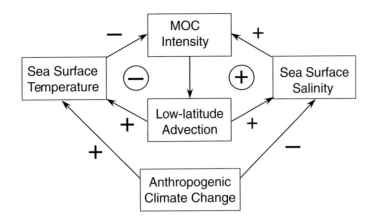

Figure 28.2 Conceptual diagram of key positive and negative feedbacks affecting the intensity of the North Atlantic meridional overturning circulation (MOC). Adapted from Stocker *et al.* (2001).

turn would result in considerable and global-scale climate changes (Vellinga and Wood, 2002).

In the Atlantic, the MOC comprises both a buoyancy-forced contribution and wind-driven meridional transports. The density-driven part of the MOC is maintained by density contrasts in the ocean due to differences in the heat and salt content of the water (and hence often referred to as thermohaline circulation). Changes in the surface fluxes of heat and freshwater influence the sea-surface density contrasts. The warm and salty surface waters transported by the MOC to the convection regions have opposite effects on the density of the surface waters. In principle, the import of warm water acts to reduce the density, whereas the import of saline water acts to increase the density. In addition, at high latitudes the ocean generally loses heat and gains freshwater from continental river runoff and precipitation, which again have opposite effects on density. Predicting MOC changes thus requires a detailed understanding of the various driving processes and the combined effect of the associated feedback mechanisms.

Two specific feedbacks are thought to be key drivers of the MOC response to anthropogenic forcing (Figure 28.2) (Stocker *et al.*, 2001). One feedback acts through the interactions between sea surface temperature, MOC intensity, and the low-latitude advection. This is a negative feedback, acting to stabilize the MOC. Consider a perturbation to the system temporarily weakening the MOC. The weakened MOC causes a slowdown in the advection towards the deepwater formation site. This slowed advection increases the time the surface waters are exposed to the colder atmosphere and increases the heat loss. The resulting decrease in sea surface temperature increases the density of the surface waters and acts to strengthen the MOC. Hence, this feedback acts to stabilize the MOC.

Why, then, is the MOC potentially unstable? The key to this question lies in the complex positive feedbacks, imposed, for example, through the interactions between sea surface salinity and MOC strength. The North Atlantic is an area of net freshwater input (Baumgartner and Reichel, 1975). Hence, the more the transit of surface waters across the North Atlantic slows down, the more sea surface salinity decreases. Decreasing sea surface salinities act to decrease the density of the surface waters. As a result, a perturbation temporarily weakening the MOC decreases the advection. The decreased advection causes a decreased sea surface salinity. The decreased sea surface salinity then causes a weakening MOC, reinforcing the initial perturbation. This positive feedback amplifies an initial perturbation and makes the systems potentially unstable. Anthropogenic greenhouse gas emissions can cause a MOC collapse by decreasing the density gradient between surface and deep waters. This occurs through two main mechanisms: the increase in sea surface temperatures and the decrease in sea surface salinities at the current deep-water formation sites (Schmittner and Stocker, 1999) due to changes in the hydrological cycle resulting from anthropogenic CO_2 emissions. These processes decrease the water densities at the surface faster than the densities of deeper water, hence they decrease the density gradient.

28.2.2 The West Antarctic Ice Sheet

The West Antarctic Ice Sheet (WAIS) may disintegrate in response to anthropogenic greenhouse gas emissions (Mercer, 1978; Oppenheimer, 1998). The consequences of WAIS disintegration could include a global sea level rise of about 5 m and disruption of global oceanic circulation patterns (Oppenheimer, 1998). The WAIS may disintegrate abruptly (i.e., on multi-century timescales) as a result of positive feedbacks deep in the ice sheet that trigger a threshold response to climate change, as in the case of the MOC. The positive feedbacks might be activated in part as a dynamical response to the rapid disintegration of floating ice shelves at the ice sheet periphery (Oppenheimer and Alley, 2005). It is also plausible that WAIS would disintegrate only gradually and exhibit no threshold behavior, and perhaps only for a very large warming (Huybrechts and de Wolde, 1999; Thomas *et al.*, 2004).

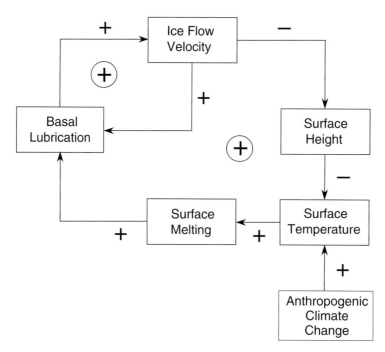

Figure 28.3 Conceptual diagram of key processes affecting the stability of the West Antarctic Ice Sheet. See Holland et al. (2003) and Oppenheimer (1998) for details.

Three positive feedback mechanisms that have received considerable attention are (i) the slip-rate–temperature feedback (Macayeal, 1992, 1993), (ii) the height–melting rate feedback (Huybrechts and de Wolde, 1999) and (iii) surface melting–bottom lubrication (Zwally et al., 2002) (Figure 28.3). The first feedback occurs when an ice flow creates friction at the interface between ice and the sub-ice basal material such as bedrock or sediments. Frictional heating causes melting and the generation of water that further lubricates ice movement. Higher ice velocity leads to additional frictional heating, and so forth. The height–melt rate feedback occurs as the height of the melting ice surface lowers, bringing it into altitudes of higher temperature and accelerated melting. The feedback between surface-melt and bottom lubrication occurs when water from meltwater ponds at the surface penetrates crevices and works its way to and lubricates the ice sheet base. The resulting accelerated ice flow would trigger the first two feedbacks, and so on. These feedbacks also operate in the Greenland ice sheet, where an abrupt disintegration has also been hypothesized recently (Oppenheimer and Alley, 2005).

28.2.3 What levels of anthropogenic climate change may trigger these threshold responses?

The location of these climate thresholds (in terms of CO_2, temperature, or other parameter spaces) is at this time uncertain. Consider, for example, the large range in current predictions about the MOC response to anthropogenic CO_2 emissions (Gregory et al., 2005; Schmittner et al., 2005). One class of sensitive models (e.g., Stouffer and Manabe, 1999) shows a considerable MOC weakening starting around the year 2000. The other class of insensitive models (e.g., Latif et al., 2000) shows a much lower MOC sensitivity to anthropogenic CO_2 emissions. The divergence in the MOC predictions is due to differences in the representation of key processes. Relevant examples include the approximations to describe the hydrological cycle, the vertical transport of heat in the ocean interior, and deepwater formation. Given the caveats introduced by these large uncertainties, the results from the current models' generation suggest that stabilizing atmospheric CO_2 somewhere below approximately 700 ppm may weaken the MOC, but typically avoid a MOC collapse (Manabe and Stouffer, 1994; Stocker and Schmittner, 1997; Gregory et al., 2005; Schmittner et al., 2005).

Models describing the future response of the WAIS face similar problems associated with model approximations. WAIS models incorporating some of the positive feedback mechanisms do not exhibit rapid WAIS threshold behavior but rather a slow, millennial-scale melting in a warmer world (Macayeal, 1992). The ice sheet dynamics are, however, represented incompletely in these models (Bindschadler, 1997; Oppenheimer and Alley, 2004). Given the limited predictive capabilities of the current WAIS models, one might be able to derive a rough estimate of the WAIS sensitivity to anthropogenic climate change from an analysis of sea level changes during the last interglacial period. At that time, temperatures were probably polar 3–5 °C warmer than today (Petit et al., 1997; Oppenheimer and Alley, 2004). Some (rather uncertain) evidence based on coral stands and erosion patterns suggests that the sea level was 2–6 m higher during this period, a change consistent with the hypothesis of a

WAIS (or Greenland ice sheet) disintegration (Neumann and Hearty, 1996; Carew, 1997; Kindler and Strasser, 1997). Hence, a precautionary interpretation of the paleo-record may suggest that avoiding an anthropogenic warming beyond 2.5 °C would be necessary to reduce the risk of WAIS disintegration to low levels.

The discussion so far illustrates that predictions about the future behavior of the MOC and the WAIS are highly uncertain. The evidence suggests the approximate location of the climate limits for the WAIS and the MOC thresholds may be located around 2.5 °C and 700 ppm CO_2, respectively. Whether these threshold responses would indeed be triggered as these climate limits are crossed is currently uncertain. One relevant decision problem is how to design near-term strategy in the face of these uncertain thresholds. We analyze this decision problem using a simple integrated assessment model of climate change.

28.3 A simple integrated assessment model of climate change

We adopt the Dynamic Integrated assessment model of Climate and the Economy (DICE) (Nordhaus and Boyer, 2000) as a starting point. The DICE model has been used in many previous studies to analyze the tradeoffs involved in designing a climate strategy in the face of potential climate thresholds (Nordhaus, 1994; Keller et al., 2004). The DICE model is an integrated assessment model of climate change developed by William Nordhaus and colleagues (Nordhaus, 1993; Nordhaus and Boyer, 2000). The DICE model relies, as any model does, on numerous simplifying assumptions (Tol, 1994; Schulz and Kasting, 1997), but has the advantage of linking the complex interactions between the social, economic, and natural systems in a consistent and transparent way.

Because the DICE model is described in great detail in previous publications (Nordhaus, 1994; Nordhaus and Boyer, 2000), we outline here only the key structural elements. The model is based on a globally aggregated optimal growth model (Ramsey, 1928). A Cobb–Douglas production function describes the flow of economic output as a function of the exogenous factors population and level of technology and the endogenous factor capital. The output is allocated among consumption, investment in capital stock, and CO_2 abatement. The objective is to maximize the discounted sum of the utility of per capita consumption weighted by population size. The two instruments are investments into capital and into reduction of CO_2 emissions ("CO_2 abatement"). Carbon dioxide abatement can improve the objective function as it weakens the negative impact of production on the utility of consumption through global climate change. This negative impact occurs because production activity generates CO_2 that – in the absence of CO_2 abatement – is vented into the atmosphere. Carbon dioxide emissions into the atmosphere act to increase atmospheric CO_2 concentrations that, in turn, act to increase globally averaged temperatures. Increases in globally averaged temperature are taken as a simple approximation for climate change (including, for example, precipitation changes or sea-level rise) and cause economic damages that reduce economic output.

We modify the DICE-99 model (Nordhaus and Boyer, 2000) to account for uncertainty, climate thresholds, and learning in three ways. First, we represent uncertainty by uncertain states of the world (SOW). We consider two SOW with respect to climate thresholds: a sensitive and an insensitive case. In the sensitive SOW, the climate threshold is triggered once the climate limit (e.g., 2.5 °C for the WAIS threshold) is crossed. In the insensitive SOW, the climate threshold is not triggered. The decisionmaker is initially uncertain about which state of the world is the correct one and assumes equal probabilities of the sensitive and the insensitive state. At some future point in time (the learning point), the decision-maker learns about the true SOW. Before the learning point, decisions are made in ignorance of the true SOW. After the learning point, decisions become state-dependent as the decisionmaker learns about the true SOW; i.e., the decision-maker acts with perfect information about the SOW. Note that the decision-maker does not know which SOW will be revealed at the learning point. The problem is hence to design a strategy that is optimal over both SOWs before the learning point and that can then be adjusted to the revealed SOW after learning occurred. In other words, the decisionmaker has to choose a single strategy before the learning point and one strategy for each considered SOW after the learning point.

As a second change to the DICE-99 (Nordhaus and Boyer, 2000) structure, we consider a decision criterion that is risk averse with respect to crossing climate thresholds. Specifically, we adopt an interpretation that crossing a WAIS or MOC threshold would violate the objective in Article 2 of the United Nations Framework Convention on Climate Change to "prevent dangerous anthropogenic interference with the climate system" (UNFCCC, 1992). This interpretation follows several previous studies (O'Neill and Oppenheimer, 2002; Mastrandrea and Schneider, 2004; Keller et al., 2005) but is certainly open to debate. We will return to the effects of different decision criteria below. The last, but arguably minor change to the DICE-99 model is to adjust the parameters related to the exogenous trends in productivity growth and emissions-output ratio to result in a business-as-usual CO_2 carbon emissions in the mid-range of the IPCC Special Report on Emissions Scenarios (IPCC, 2000). We use this integrated assessment model to analyze the potential economic value of reducing key scientific uncertainties surrounding climate thresholds.

28.4 Results and discussion

We first analyze constrained optimal strategies with and without learning about the sensitivity of the climate threshold. This analysis is illustrated in Figure 28.4 for the example of the WAIS threshold located at a 2.5 °C climate limit. Note that a 2.5 °C warming limit relative to pre-industrial conditions is

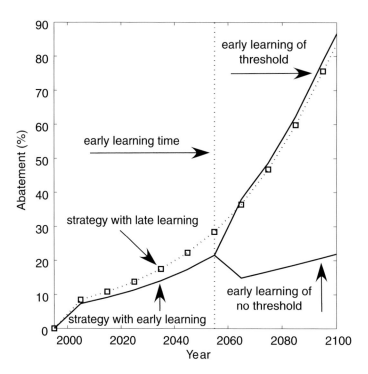

Figure 28.4 Constraint optimal abatement strategies in the modified DICE-99 model to avoid a climate threshold potentially located at 2.5 °C. Shown are the strategies for later learning (after 2200) and for early learning (at 2055).

roughly equivalent to a 2 °C limit relative to the current conditions. This is because globally averaged surface temperatures have increased by approximately 0.6 °C over the past 150 years (Folland et al., 2001). For the strategy without learning, abatement increases from roughly 10% in 2005 to approximately 90% within this century. The key to understanding this strategy lies in the objective to stay below the 2.5 °C limit as long as the uncertainty about the WAIS threshold is unresolved (Keller et al., 2005). A strategy to limit globally averaged temperature increase to 2.5 °C requires that the economy should be almost completely decarbonized on a century timescale (Nordhaus and Boyer, 2000; Keller et al., 2005).

The uncertainty about the WAIS might be reduced in the future as new observations become available or as the numerical models improve. In this case, the strategy can be designed with the projected learning date in mind. The effect of considering future learning on optimal strategy in the model is to decrease the near-term abatement (before the learning point). After the learning point, optimal abatement increases in the sensitive SOW, and decreases in the insensitive SOW. Before the learning point, optimal abatement under uncertainty and learning is below the optimal abatement without learning. After the learning point, the situation is reversed. This reversal compensates for the lower abatement levels under learning and is required to achieve the 2.5 °C temperature stabilization.

Reducing the scientific uncertainty can improve climate strategy and have an economic value (Manne and Richels, 1991; Nordhaus and Popp, 1997). The effect of early learning is shown in Figure 28.4. With later learning, the abatement is too high for both SOW before early learning occurs and is too high for the insensitive SOW after the early learning point. The expected value of information (EVI) quantifies how the availability of information improves strategy. We estimate the EVI by the net present value of consumption gains due to the availability of information. We analyze the EVI for the sensitivity of the MOC and the WAIS thresholds and as a function of the learning time (Figure 28.5). The EVI is higher for the more stringent climate limit (i.e., the 2.5 °C limit constrains the allowable CO_2 emissions more than the 700 ppm limit). The later learning occurs, the lower the EVI. Learning about the WAIS and MOC sensitivity within the next five decades has an EVI in this simple model of about 3 and 0.5% of gross world product (GWP), respectively.

A strategy to avoid a temperature threshold is very sensitive to the current uncertainty about the climate sensitivity (Keller et al., 2000; Caldeira et al., 2003). The climate sensitivity is the (hypothetical) equilibrium response of the globally averaged temperature to a doubling of the atmospheric CO_2 concentration. Current estimates of the climate sensitivity show a considerable uncertainty, predominantly due to uncertainties in the climate forcing, the climate observations, and the kinetics of oceanic heat uptake (Andronova and Schlesinger, 2001). Considering climate thresholds dramatically increases the economic value of information about the climate sensitivity (Figure 28.6). We represent the uncertainty

The value of information about climate thresholds 349

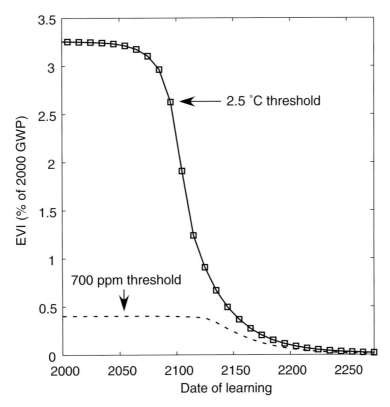

Figure 28.5 The economic value of early learning about potential climate thresholds located at a carbon dioxide concentration of 700 ppm (dashed line) and at 2.5 °C (solid line). See text for details.

about the climate sensitivity by three states of the world based on stratified Latin Hypercube samples from the estimated probability density function (McKay *et al.*, 1979; Andronova and Schlesinger, 2001). Learning about the climate sensitivity in 2015 in the case of no climate threshold (i.e., the original structure of the DICE-99 model) has an economic value of information of roughly 10 billion US$. (Figure 28.6). This is the same order of magnitude as the previous estimate of Nordhaus and Popp (1997) using the DICE-94 model structure. Considering temperature limits increases the EVI about the climate sensitivity considerably. For a learning time in 2045, the EVI about the climate sensitivity increase by more than an order of magnitude as the strategy objective changes from no temperature limit to a temperature limit of 2.5 °C (Figure 28.6). The EVI increases as the temperature limit is reduced even further (illustrated by the EVI for a temperature limit of 1.5 °C).

Our economic analysis suggests that information about the sensitivity of climate thresholds can have considerable economic value. This poses the question of how one might gather information about the threshold sensitivity and whether investments into such an information gathering process would pass a benefit–cost test. In the following section, we briefly outline key steps towards such an analysis for the example of the potential MOC threshold response.

28.5 An outline for an economic analysis of MOC observation systems

Numerous studies address the question of when one might detect a potential MOC change (Santer *et al.*, 1995; Kleinen *et al.*, 2003; Vellinga and Wood, 2004; Keller *et al.*, 2007; Baehr *et al.*, 2007). These studies use model simulations of the future MOC behavior, sample possible observations, and use detection methods to derive statistical statements regarding the detection probability over time. Santer *et al.* (1995) analyzed the task of detecting a MOC trend using annual and perfect observations in a stochastically forced ocean model. They conclude that MOC changes would not be detected within the first 100 years, predominantly because of the high internal MOC variability in their model simulations and the adopted detection methodology. Baehr *et al.* (2007) analyzed another coupled ocean–atmosphere model and considered the effects of observation errors. They conclude that observations at high (i.e., annual) frequency and a low observation error might well detect a MOC trend within a few decades in the ECHAM5/MPI-OM model (Marsland *et al.*, 2003). Vellinga and Wood (2004) come to similar conclusions to Baehr *et al.* (2007) in yet another model, but with the assumption of perfect observations. These studies illustrate the potentially large effects of differences in the analyzed physical model, the detection

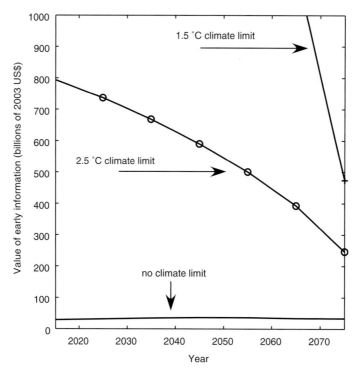

Figure 28.6 Sensitivity of the economic value of early learning about the climate sensitivity to the adopted climate objective.

methodology, and the observation system on the detection probability of MOC changes.

One key question often neglected in MOC detection studies is whether the assumed observations are logistically feasible. Studies assuming perfect observations at very high frequencies (e.g., Vellinga and Wood, 2004) may yield important theoretical insights, but are perhaps only of limited use for the decision problem in the real world. The MOC observation systems that have been implemented so far yield infrequent and uncertain MOC estimates. These current observation systems fall into two broad categories. The first category of MOC estimations relies on the inversion of hydrographic observations from ship-based transoceanic sections (Ganachaud and Wunsch, 2000; Peacock et al., 2000). This observation method has a likely observation error between 2 and 5 Sverdrups (Sv) (Ganachaud and Wunsch, 2000; Peacock et al., 2000; 1 Sv is equal to one million cubic meters of water per second). The observation frequency for this method is very low, since these observations are costly and personnel-intensive. The second method is based on mooring arrays and yields higher observation frequencies. One example of such a high frequency observation system is the recently installed array at 26 °N (Marotzke et al., 2002; Schiermeier, 2004). The setup of the array has been extensively tested in numerical models (Hirschi et al., 2003; Baehr et al., 2004, 2007). These model-based studies suggest that this observation system has the potential for lower observation errors than the ship-based method. Estimating the observation error for this array is an area of ongoing research. Here we explore the implications of observation errors of 1 and 3 Sv (Figure 28.7).

Detecting changes occurs – in general – faster for more frequent and less uncertain observations (Figure 28.7). One measure of how fast detection would occur is the detection time, the time period needed to be able to reject the null-hypothesis of no trend at $p < 0.05$ (Santer et al., 1995). We explore the range of MOC detection times in the model runs of Manabe and Stouffer (1994), using a simple frequentist detection method that accounts for autocorrelation and observation errors (Baehr et al., 2007). In this simple example, detection would occur on a decadal timescale (Figure 28.7). An observation system with observation errors of between 1 and 3 Sv and annual observations would yield median detection times between approximately 10 to 50 years for the considered model. Our economic analysis summarized in Figure 28.5 suggests that resolving the current uncertainty about the MOC response on a decadal timescale would have considerable economic value in the framework of the adopted economic model. This opens up the possibility that the economic benefits associated with MOC observation systems may well exceed the necessary investments. It is important to stress that detection of MOC changes is not equivalent to the prediction of an MOC threshold response (the information analyzed in the economic analysis, above).

28.6 Open research questions

Our analysis is adorned by numerous caveats that point to future research questions. Here we briefly discuss two issues

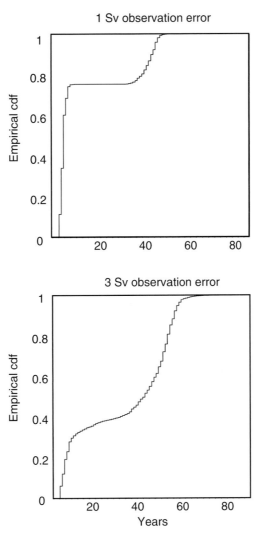

Figure 28.7 Model-based detection analysis of MOC changes in the model simulations of Manabe and Stouffer (1994). Shown are the cumulative density functions of the detection times (when the null-hypothesis of no change compared with the unforced control run can be rejected at $p < 0.05$) for the $4 \times CO_2$ scenario in Manabe and Stouffer (1994). The analysis uses the detection method described in Baehr et al. (2006). Compared are two observation systems with observation errors of 1 Sv and 3 Sv (upper and lower panel, respectively). See text as well as Baehr et al. (2007) for details.

relevant to the design of sound climate strategies in the face of uncertain climate thresholds.

First, similar to Baehr et al. (2007), the detection analysis shown in Figure 28.7 neglects the potential that additional tracers could be used in an optimal fingerprint analysis (Bell, 1982). MOC changes are associated with specific trends of tracers such as oxygen (Joos et al., 2003; Matear and Hirst, 2003) or chlorofluorocarbons (Schlosser et al., 1991). The predicted oxygen changes, for example, exceed the typical data-based estimation errors (Matear et al., 2000; Keller et al., 2002). This suggests that adding tracer observations to an MOC observation system might be a promising strategy to improve the detection and prediction capabilities.

Second, our current study expands on previous work (Keller et al., 2004, 2007) by considering a precautionary decision-making criterion. It can be argued that the decision criterion applied in this study is a more appropriate description of Article 2 of the United Nations Framework Convention on Climate Change (UNFCCC, 1992) than the expected utility maximization framework often applied to this problem (e.g., Nordhaus, 1992). To characterize the applied decision criterion of stakeholders and decisionmakers in this problem is important because different choices (e.g., expected utility maximization without a-priori constraints (Nordhaus and Boyer, 2000), reliability constraints (Keller et al., 2000), or robust decisionmaking (Lempert, 2002) would be likely to lead to different EVI estimates and different suggestions on how to design an economically efficient MOC observation system.

28.7 Conclusions

We draw from our analysis four main conclusions. First, crossing climate thresholds may be interpreted as "dangerous anthropogenic interference with the climate system." Second, strategies addressing climate thresholds have to be made in a situation of deep uncertainty. Third, investments into well-designed observation systems have the potential to reduce key uncertainties about climate thresholds. Finally, important issues (e.g., what framework best approximates the applied decision criteria; how this affects the design of "optimal" observation system; or how one might detect early warning signs of climate thresholds besides the MOC) require more detailed attention.

Acknowledgements

This paper is dedicated to the memory of David F. Bradford, who tragically died on February 22, 2005. We thank R. Alley, C. Brennan, K. Panchuk, E. Naevdal, D.-H. Min, and the participants of the EMF Integrated Assessment Workshop in Snowmass (2004) for helpful discussions. Financial support from the National Science Foundation (SES 0345925 to K.K.) is gratefully acknowledged. Any opinions, findings, and conclusions or recommendations expressed in this material are those of the authors and do not necessarily reflect the views of the funding agencies.

References

Alley, R. B. (2000). Ice-core evidence of abrupt climate changes. *Proceedings of the National Academy of Sciences of the United States of America* **97** (4), 1331–1334.

Alley, R. B., Marotzke, J., Nordhaus, W. *et al.* (2002). *Abrupt Climate Change: Inevitable Surprises*. National Research Council.

Alley, R. B., Marotzke, J., Nordhaus, W. D. *et al.* (2003). Abrupt climate change. *Science* **299**, 2005–2010.

Andronova, N. G. and Schlesinger, M. E. (2001). Objective estimation of the probability density function for climate sensitivity. *Journal of Geophysical Research-Atmospheres* **106** (D19), 22 605–22 611.

Baehr, J., Hirschi, J., Beismann, J. O. and Marotzke, J. (2004). Monitoring the meridional overturning circulation in the North Atlantic: a model-based array design study. *Journal of Marine Research* **62** (3), 283–312.

Baehr, J., Keller, K. and Marotzke, J. (2007). Detecting potential changes in the meridional overturning circulation at 26° N in the Atlantic, *Climatic Change*, published online 10 January 2007, doi: 2010.1007/S10584-10006-19153-Z.

Baumgartner, A. and Reichel, E. (1975). *The World Water Balance*. Amsterdam and New York: Elsevier.

Bell, T. L. (1982). Optimal weighting of data to detect climatic change: application to the carbon dioxide problem. *Journal of Geophysical Research* **87** (C13), 11 161–11 170.

Berger, A. and Loutre, M. F. (1991). Insolation values for the climate of the last 10 000 000 years. *Quaternary Science Reviews* **10** (4), 297–317.

Berger, W. H. (1990). The Younger Dryas cold spell – a quest for causes. *Global and Planetary Change* **89** (3), 219–237.

Bindschadler, R. (1997). West Antarctic ice sheet collapse?, *Science* **276**, 662–663.

Broecker, W. (1991). The great ocean conveyor. *Oceanography* **4** (2), 79–89.

Caldeira, K., Jain, A. K. and Hoffert, M. I. (2003). Climate sensitivity uncertainty and the need for energy without CO_2 emission. *Science* **299**, 2052–2054.

Carew, J. L. (1997). Rapid sea-level changes at the close of the last interglacial (substage 5e) recorded in Bahamian island geology: comments and reply. *Geology* **25** (6), 572–573.

Claussen, M., Kubatzki, C., Brovkin, V. *et al.* (1999). Simulation of an abrupt change in Saharan vegetation in the mid-Holocene. *Geophysical Research Letters* **26** (14), 2037–2040.

Clement, A. C., Cane, M. A. and Seager, R. (2001). An orbitally driven tropical source for abrupt climate change. *Journal of Climate* **14** (11), 2369–2375.

Crutzen, P. J. (2002). Geology of mankind. *Nature* **415**, 23–23.

Cubasch, U. and Meehl, G. A. (2001). Projections of future climate change, In *Climate Change 2001: The Scientific Basis. Contribution of Working Group I of the Third Assessment Report of the Intergovernmental Panel on Climate Change*, ed. J. T. Houghton, Y. Ding, D. J. Griggs *et al.* Cambridge: Cambridge University Press, pp. 526–582.

Cuffey, K. M., Clow, G. D. Alley, R. B. *et al.* (1995). Large Arctic temperature change at the Wisconsin–Holocene glacial transition. *Science* **270**, 455–458.

Denton, G. H. and Hendy, C. H. (1994). Younger Dryas age advance of Franz Josef Glacier in the Southern Alps of New Zealand. *Science* **264**, 1434–1437.

Dickson, R. R. and Brown, J. (1994). The production of North Atlantic deep water: Sources, rates, and pathways. *Journal of Geophysical Research* **99** (C6), 12 319–12 341.

Fedorov, A. V. and Philander, S. G. (2000). Is El Niño changing?. *Science* **288**, 1997–2002.

Folland, C. K., Rayner, N. A. Brown, S. J. *et al.* (2001). Global temperature change and its uncertainties since 1861. *Geophysical Research Letters* **28** (13), 2621–2624.

Ganachaud, A. and Wunsch, C. (2000). Improved estimates of global ocean circulation, heat transport and mixing from hydrographic data. *Nature* **408**, 453–457.

Gregory, J. M., Dixon, K. W. Stouffer, R. J. *et al.* (2005). A model intercomparison of changes in the Atlantic thermohaline circulation in response to increasing atmospheric CO_2 concentration. *Geophysical Research Letters* **32** L12703; doi: 10.1029/2005GL023209.

Hall, M. M. and Bryden, H. L. (1982). Direct estimates and mechanisms of ocean heat transport. *Deep-Sea Research* **29**, 339–359.

Hirschi, J., Baehr, J., Marotzke, J. *et al.* (2003). A monitoring design for the Atlantic meridional overturning circulation.

Geophysical Research Letters **38** (7) 1413; doi: 1410.1029/2002 GL016776.

Holland, D. M., Jacobs, S. S. and Jenkins, A. (2003). Modelling the ocean circulation beneath the Ross Ice Shelf. *Antarctic Science* **15** (1), 13–23.

Huybrechts, P. and de Wolde, J. (1999). The dynamic response of the Greenland and Antarctic ice sheets to multiple-century climatic warming. *Journal of Climate* **12** (8), 2169–2188.

IPCC (2000). *Emissions Scenarios: A Special Report of Working Group III of the Intergovernmental Panel on Climate Change*, ed. N. Nakicenovic and R. Swart. Cambridge: Cambridge University Press.

Joos, F., Plattner, G.-K., Stocker, T. F, Körtzinger, A. and Wallace, D. W. R. (2003). Trends in marine dissolved oxygen: implications for ocean circulation changes and the carbon budget. *EOS* **84**, 197–204.

Keller, K., Tan, K., Morel, F. M. M., and Bradford, D. F. (2000). Preserving the ocean circulation: implications for climate policy. *Climatic Change* **47**, 17–43.

Keller, K., Slater, R., Bender, M. and Key, R. M. (2002). Possible biological or physical explanations for decadal scale trends in North Pacific nutrient concentrations and oxygen utilization. *Deep-Sea-Research II* **49**, 345–362.

Keller, K., Bolker, B. M. and Bradford, D. F, (2004). Uncertain climate thresholds and optimal economic growth. *Journal of Environmental Economics and Management* **48**, 723–741.

Keller, K., Hall, M., Kim, S.-R., Bradford, D. F. and Oppenheimer, M. (2005). Avoiding dangerous anthropogenic interference with the climate system. *Climatic Change* **73**, 227–238.

Keller, K., Deutsch, C., Hall, M. G. and Bradford, D. F. (2007). Early detection of changes in the North Atlantic meridional overturning circulation: implications for the design of ocean observation systems. *Journal of Climate* **20**, 145–157.

Keller, K., Schlesinger, M. and Yohe, G. (2007). Managing the risks of climate thresholds: uncertainties and information needs. (An editorial essay). *Climatic Change*, in press. Published online: 23 January 2007, http://dx.doi.org/2010.1007/S10584-10006-19114-10586.

Kindler, P. and Strasser, A. (1997). Rapid sea-level changes at the close of the last interglacial (substage 5e) recorded in Bahamian island geology: Comment. *Geology* **25** (12), 1147.

Kleinen, T., Held, H. and Petschel-Held, G. (2003). The potential role of spectral properties in detecting thresholds in the Earth system: application to the thermohaline circulation. *Ocean Dynamics* **53**, 53–63.

Latif, M., Roeckner, E., Mikolajewski, U. and Voss, R. (2000). Tropical stabilization of the thermohaline circulation in a greenhouse warming simulation. *Journal of Climate* **13**, 1809–1813.

Lempert, R. J. (2002). A new decision sciences for complex systems. *Proceedings of the National Academy of Sciences of the United States of America* **99**, 7309–7313.

Macayeal, D. R. (1992). Irregular oscillations of the West Antarctic Ice-Sheet. *Nature* **359**, 29–32.

Macayeal, D. R. (1993). Binge/purge oscillations of the Laurentide ice-sheet as a cause of the North-Atlantics Heinrich events. *Paleoceanography* **8** (6), 775–784.

Manabe, S. and Stouffer, R. J. (1994). Multiple-century response of a coupled ocean-atmosphere model to an increase of atmospheric carbon dioxide. *Journal of Climate* **7** (1), 5–23.

Manabe, S. and Stouffer, R. J. (1995). Simulation of abrupt climate change induced by freshwater input to the North Atlantic ocean. *Nature* **378**, 165–167.

Manne, A. S. and Richels, R. (1991). Buying greenhouse insurance. *Energy Policy* **19**, 543–552.

Marotzke, J., Cunningham, S. A. and Bryden, H. L. (2002). *Monitoring the Atlantic Meridional Overturning Circulation at 26.5° N*. Proposal accepted by the Natural Environment Research Council (UK). Available at www.noc.soton.ac.uk/rapidmoc.

Marsland, S. J., Haak, H., Jungclaus, J. H., Latif, M. and Roske, F. (2003). The Max-Planck-Institute global ocean/sea ice model with orthogonal curvilinear coordinates. *Ocean Modelling* **5** (2), 91–127.

Mastrandrea, M. D. and Schneider, S. H. (2004). Probabilistic integrated assessment of "dangerous" climate change. *Science* **304**, 571–575.

Matear, R. J. and Hirst, A. C. (2003). Long-term changes in dissolved oxygen concentrations in the oceans caused by protracted global warming. *Global Biogeochemical Cycles* **17** (4), doi: 10.1029/2002GB001997.

Matear, R. J., Hirst, A. C. and McNeil, B. I. (2000). Changes in dissolved oxygen in the Southern ocean with climate change. *Geochemistry, Geophysics, Geosystems* **1**, Paper number 2000GC000086.

McKay, M. D., Beckman, R. J. and Conover, W. J. (1979). Comparison of 3 methods for selecting values of input variables in the analysis of output from a computer code. *Technometrics* **21** (2), 239–245.

McManus, J. F., Francois, R., Gherardi, J. M. Keigwin, L. D., and Brown-Leger, S. (2004). Collapse and rapid resumption of Atlantic meridional circulation linked to deglacial climate changes. *Nature* **428**, 834–837.

Meese, D. A., Alley, R. B., Fiacco, R. J. *et al.* (1994a). Preliminary depth-age scale of the GISP2 ice core. *Special CRREL Report, 94–1*.

Meese, D. A., Gow, A. J. Grootes, P. *et al.* (1994b). The accumulation record from the GISP2 core as an indicator of climate change throughout the Holocene. *Nature* **266**, 1680–1682.

Mercer, J. H. (1978). West Antarctic ice sheet and CO_2 greenhouse effect: a threat of disaster. *Nature* **271**, 321–325.

Neumann, A. C. and Hearty, P. J. (1996). Rapid sea-level changes at the close of the last interglacial (substage 5e) recorded in Bahamian island geology. *Geology* **24** (9), 775–778.

Nordhaus, W. D. (1991). To slow or not to slow – the economics of the greenhouse-effect. *Economic Journal* **101** (407), 920–937.

Nordhaus, W. D. (1992). An optimal transition path for controlling greenhouse gases. *Science* **258**, 1315–1319.

Nordhaus, W. D. (1993). Optimal greenhouse-gas reductions and tax policy in the DICE model. *American Economic Review* **83** (2), 313–317.

Nordhaus, W. D. (1994). *Managing the Global Commons: The Economics of Climate Change.* Cambridge, MA: MIT Press.

Nordhaus, W. D. and Boyer, J. (2000). *Warming the World: Economic Models of Global Warming.* Cambridge, MA: MIT Press.

Nordhaus, W. D. and Popp, D. (1997). What is the value of scientific knowledge? An application to global warming using the PRICE model. *Energy Journal* **18** (1), 1–45.

O'Neill, B. C. and Oppenheimer, M. (2002). Climate change – dangerous climate impacts and the Kyoto protocol. *Science* **296**, 1971–1972.

Oppenheimer, M. (1998). Global warming and the stability of the West Antarctic ice sheet. *Nature* **393**, 322–325.

Oppenheimer, M. and Alley, R. B. (2004). The West Antarctic ice sheet and long term climate policy: an editorial comment. *Climatic Change* **64** (1–2), 1–10.

Oppenheimer, M. and Alley, R. B. (2005). Ice sheets, global warming, and Article 2 of the UNFCCC. *Climatic Change* **68** (3), 257–267.

Peacock, S., Visbeck, M. and Broecker, W. (2000). Deep water formation rates inferred from global tracer distributions: an inverse approach. In *Inverse Methods in Global Biogeochemical Cycles*, ed. P. Kasibhatla, M. Heiman, P. Rayner *et al.* American Geophysical Union, pp. 185–195.

Petit, J. R., Basile, I., Leruyuet, A. *et al.* (1997). Four climate cycles in Vostok ice core. *Nature* **387**, 359.

Rahmstorf, S. (2000). The thermohaline ocean circulation: a system with a dangerous threshold? *Climatic Change* **46**, 247–256.

Ramsey, F. (1928). A mathematical theory of saving. *Economic Journal* **37**, 543–559.

Santer, B. D., Mikolajewicz, U., Bruggemann, W. *et al.* (1995). Ocean variability and its influence on the detectability of greenhouse warming signals. *Journal of Geophysical Research: Oceans* **100** (C6), 10 693–10 725.

Scheffer, M., Carpenter, S., Foley, J. A. Folke, C. and Walker, B. (2001). Catastrophic shifts in ecosystems. *Nature* **413**, 591–596.

Schiermeier, Q. (2004). Gulf Stream probed for early warnings of system failure. *Nature* **427**, 769.

Schlosser, P., Bönisch, G. Rhein, M. and Bayer, R. (1991). Reduction of deepwater formation in the Greenland Sea during the 1980's: evidence from tracer data. *Science* **251**, 1054–1056.

Schmittner, A. and Stocker, T. F. (1999). The stability of the thermohaline circulation in global warming experiments. *Journal of Climate* **12**, 1117–1133.

Schmittner, A., Latif, M., and Schneider, B. (2005). Model projections of the North Atlantic thermohaline circulation for the 21st century assessed by observations. *Geophysical Research Letters* **32** L23710, doi: 23710.21029/22005GL024368.

Schulz, P. A. and Kasting, J. F. (1997). Optimal reduction in CO_2 emissions. *Energy Policy* **25** (5), 491–500.

Stocker, T. F. and Schmittner, A. (1997). Influence of CO_2 emission rates on the stability of the thermohaline circulation. *Nature* **388**, 862–865.

Stocker, T. F., Clarke, G. K. C., Treut, H. L. *et al.* (2001). Physical climate processes and feedbacks. In *Climate Change 2001: The Scientific Basis. Contribution of Working Group I to the Third Assessment Report of the Intergovernmental Panel on Climate Change*, ed. J. T. Houghton, Y. Ding, D. J. Griggs *et al.* Cambridge: Cambridge University Press, pp. 417–470.

Stommel, H. (1961). Thermohaline convection with two stable regimes of flow. *Tellus* **13** (2), 224–230.

Stouffer, R. J. and Manabe, S. (1999). Response of a coupled ocean-atmosphere model to increasing atmospheric carbon dioxide: sensitivity to the rate of increase. *Journal of Climate* **12**, 2224–2237.

Stuiver, M., Grootes, P. M. and Braziunas, T. F. (1995). The GISP2 ^{18}O climate record of the past 16 500 years and the role of the sun, ocean and volcanoes. *Quaternary Research* **44**, 341–354.

Teller, J. T. Leverington, D. W. and Mann, J. D. (2002). Freshwater outbursts to the oceans from glacial lake Agassiz and their role in climate change during the last deglaciation. *Quaternary Science Reviews* **21** (8–9), 879–887.

Thomas, R., Rignot, E., Casassa, G. *et al.* (2004). Accelerated sea-level rise from West Antarctica. *Science* **306**, 255–258.

Timmermann, A. (2001). Changes of ENSO stability due to greenhouse warming. *Geophysical Research Letters* **28** (10), 2061–2064.

Tol, R. S. J. (1994). The damage costs of climate-change: a note on tangibles and intangibles, applied to DICE. *Energy Policy* **22** (5), 436–438.

Tol, R. S. J. (1997). On the optimal control of carbon dioxide emissions: an application of FUND. *Environmental Modeling and Assessment* **2**, 151–163.

UNFCCC (1992). *UN Framework Convention on Climate Change*, Geneva: Palais des Nations. www.unfccc.de/index.html.

Vellinga, M. and Wood, R. A. (2002). Global climatic impacts of a collapse of the Atlantic thermohaline circulation. *Climatic Change* **54** (3), 251–267.

Vellinga, M. and Wood, R. A. (2004). Timely detection of anthropogenic change in the Atlantic meridional overturning circulation. *Geophysical Research Letters* **31**, L14203, doi: 14210.11029/12004GL02036.

Yu, Z. C. and Wright, H. E. (2001). Response of interior North America to abrupt climate oscillations in the North Atlantic region during the last deglaciation. *Earth-Science Reviews* **52** (4), 333–369.

Zwally, H. J., Abdalati, W., Herring, T. *et al* (2002). Surface melt-induced acceleration of Greenland ice-sheet flow. *Science* **297**, 218–222.

29

Boiled frogs and path dependency in climate policy decisions

Mort Webster

29.1 Introduction

Formulating a policy response to the threat of global climate change is one of the most complex public policy challenges of our time. At its core a classic public-good problem, mitigating anthropogenic greenhouse gas emissions is likely to be very costly to any nation that undertakes it, while all would share the benefits. This dynamic creates a temptation to free-ride on others' efforts. This will require coordination among nations and the development of new institutional capacities. The heterogeneity across nations adds complexity; the costs of reducing emissions will not be the same, nor will the benefits of avoiding climate change. Another troubling characteristic is the enormous uncertainty involved, both in the magnitude of future climate change, and therefore the value of avoiding it, and in the costs of reducing emissions. The long timescales of the climate system, decades to centuries, add a final dimension to the policy dilemma. Given the stock nature of greenhouse gases, which build slowly over time, should we delay mitigation activities until some of the uncertainties are reduced? Or wait until technology improves to the point that mitigation is less costly?

We need not decide today on the amount of emissions reductions for all time. Given the degree of uncertainty, it would not make sense. Over time, we will revise the level of policy activities to respond to new information and changing conditions. The relevant question is how much greenhouse gas emissions abatement should be undertaken today. However, the choice of the "right" level of stringency does depend on our current expectations of what we will do later.

Given the policy question – how much effort to exert today when we can learn and revise in the future – and given some of the salient characteristics of the problem – uncertainty, sequential decision over time – a sensible choice for an analytical framework is that of decision analysis. Decision analytic tools have been developed to provide insight into precisely this kind of decision problem.

A number of studies have explicitly modeled the policy decision as a sequential choice under uncertainty, and allowed for learning and adaptation in policy over time (Hammitt *et al.*, 1992; Nordhaus, 1994a; Manne and Richels, 1995; Kolstad, 1996; Ulph and Ulph, 1997; Valverde *et al.*, 1999; Webster, 2002). The general result from all of these studies is that, given the ability to learn and adapt later, the optimal choice for the initial decision period is to undertake very little or no abatement activity. There are several reasons for this result. First, greenhouse gases are stock pollutants, which build up slowly in the atmosphere. This means that there is time to address the problem over the next century and that the impact of reductions in the next decade or two is relatively small. The second reason is that the uncertainty in future climate change is sufficiently large that if there is the ability to reduce this uncertainty and respond within a few decades, it is better to wait. Third, most of these models assume that technological options continue to be developed and improve over time, so that the cost of responding will fall over several decades. Finally, the use of a discount rate to reflect the opportunity cost of capital necessarily implies that policy costs in the near term will be weighed more heavily than either costs later or benefits later, which further biases the optimal choice to be one of waiting.

Human-induced Climate Change: An Interdisciplinary Assessment, ed. Michael Schlesinger, Haroon Kheshgi, Joel Smith, Francisco de la Chesnaye, John M. Reilly, Tom Wilson and Charles Kolstad. Published by Cambridge University Press. © Cambridge University Press 2007.

In this paper, I begin by exploring an apparent paradox. A common approach in modeling climate policy and other long-term problems is to simplify to a two-period decision, in which the first period represents "today" or the near term, and the second period represents "later" or further in the future, perhaps after some uncertainty has been reduced. As discussed below, typical results from such models suggest that very little or no reduction in greenhouse gas emissions in the near term is optimal. If the problem turns out to be serious, fairly stringent reductions are undertaken in period 2. Even without resolution of uncertainty, more stringent abatement is optimal in the latter period because of the other reasons outlined above. The dilemma is as follows. Suppose we apply this approach in 2005 and find that very little reduction is warranted for the next 10 years. Suppose we then follow this strategy, and then in 2015 we again apply a two-period model. Will that model not also recommend doing little or nothing, because of the same factors described above? Are we doomed to repeat this cycle, until it is "too late" in terms of avoiding climate impacts? I dub this problem the "Boiled Frog dilemma."[1]

In the next section, I describe the numerical modeling system, and explore the Boiled Frog dilemma using a variety of two-period and three-period decision trees. In the traditional application of decision techniques, I will demonstrate that there is no Boiled Frog. In Section 29.3, I will demonstrate an alternative formulation of the decision model that captures the intuition behind the Boiled Frog problem, namely path dependency in political systems. The final section discusses the implications both for climate policy and for research.

29.2. The modeling system and the Boiled Frog

There is a wide spectrum of models that can be used to project the impacts of greenhouse gas emissions and resulting climate change as well as the economic costs of constraining those emissions. These range from very simple approximations to very large sophisticated models that require weeks on a supercomputer for a single simulation. The advantage of the more complex models is that they represent many of the non-linearities and complexities that make climate change a cause for concern. On the other hand, solving the dynamic optimization problem under uncertainty requires some simplification to make the analysis feasible. The approach used here is to fit reduced-form models to a climate model of intermediate complexity and to use a relatively detailed computable general equilibrium model of the global economy. The reduced-form models are then embedded within a decision tree framework to solve for optimal decisions.

---

[1] This name derives from the apocryphal advice for cooking a live frog: if you drop a frog into boiling water, it will jump out. If you put a live frog in cool water and slowly heat over a stove, it will never jump out before boiling to death. Presumably the frog always believes it has more time, until it is too late. While this is apparently untrue, it provides a useful image for society's potential response to climate change.

29.2.1 The MIT Integrated Global System Model

The integrated assessment model used is the MIT Integrated Global System Model (IGSM) (Prinn et al., 1999), augmented with a damage function related to change in global mean temperature. The economic component of the model, the Emissions Projections and Policy Analysis (EPPA) model Version 3 (Babiker et al., 2001) is a recursive-dynamic computable general equilibrium model, consisting of 12 geopolitical regions linked by international trade, 10 production sectors in each region, and 4 consumption sectors. The climate component is a two-dimensional (zonal averaged) representation of the atmosphere and $\Delta T(t)$ (Sokolov and Stone, 1998). The climate model includes parameterizations of all the main physical atmospheric processes, and is capable of reproducing many of the non-linear interactions simulated by atmospheric general circulation models.

In order to choose one set of strategies as "optimal," a basis is required for comparing the costs of reducing emissions with the benefits of avoiding damages. I augment the EPPA mitigation cost model with the Nordhaus damage function (Nordhaus, 1994a). This damage function has been widely used (e.g., Peck and Teisberg, 1992; Kolstad, 1996; Lempert et al., 1996; Pizer, 1999), and facilitates the comparison of results here with other studies. The Nordhaus damage function estimates the percentage loss of gross world product as a function of the global mean temperature change,

$$d(t) = \eta[\Delta T(t)]^\pi \quad (29.1)$$

where $d(t)$ is the fraction of world product lost because of climate damages in year t, and $\Delta T(t)$ is the increase in global mean temperature from pre-industrial levels. Consistent with previous studies, I assume that $\pi = 2$ and vary η as an uncertain parameter.

29.2.2 Fitting reduced-form models

Solving for an optimal sequential decision under uncertainty requires a large number of simulations of the numerical economic–climate model. Used directly, the IGSM requires too much computation time for this application, so instead I estimate a reduced-form version using least squares regression on a data set of 1500 runs. I derive simple non-linear representations of temperature change as a function of uncertain climate parameters and emissions from EPPA. These simpler functional forms replicate the results of the original IGSM to within a 1% error of the mean. The reduced-form models are used in all calculations below.

29.2.3 The decision model

I use the results from EPPA and the reduced-form fits of the climate model to frame a two-period sequential decision under uncertainty. The decisionmaker represents an aggregate decisionmaker for the world. The decisionmaker seeks to

Table 29.1 Strategy choices in each period: reduction below unconstrained emission growth rate (% per 5 years).

Case	Number of periods	Period 1 choice set	Period 2 choice set	Period 3 choice set
A	2	2010–2029: {0%, 2%, 4%, 6%, 8%, 10%}	2030–2100: {0%, 1%, 2%, 3%, 4%, 5%}	none
B	3	2010–2029: {0%, 2%, 4%, 6%, 8%, 10%}	2030–2049: {0%, 2%, 4%, 6%, 8%, 10%}	2050–2100: {0%, 1%, 2%, 3%, 4%, 5%}
C	2	2010–2029 Fixed at 0%	2030–2049 {0%, 2%, 4%, 6%, 8%, 10%}	2050–2100: {0%, 1%, 2%, 3%, 4%, 5%}
D	2	2010–2049: Fixed at 0%	2050–2069 {0%, 2%, 4%, 6%, 8%, 10%}	2050–2100: {0%, 1%, 2%, 3%, 4%, 5%}

Table 29.2 Impacts of period 1 strategy choice in 2030 (median growth case).

Reduction rate (% per 5 yrs)	CO_2 (GtC)	% change CO_2	Carbon price ($/tonne C)	Consumption	%Chg cons.
0%	12.5		0.0	5094.2	
2%	11.8	−5%	19.3	5091.8	−0.05%
4%	11.1	−11%	45.1	5087.9	−0.12%
6%	10.5	−16%	79.7	5082.1	−0.24%
8%	9.9	−20%	125.1	5073.3	−0.41%
10%	9.4	−25%	181.6	5060.8	−0.66%

Chg. cons. = change in consumption relative to the no policy (0%) case.

minimize the net present value of the total consumption losses. These losses result both from constraints on carbon emissions and from impacts of climate change. The stream of costs over time is discounted at a rate of 5%. The possible strategies represent choice over levels of emissions abatement only; other possible complementary policies of research, adaptation, and geoengineering are not considered here. In order to illustrate the point here about sequential decision, I make the simplification of aggregating the globe; in reality, climate policy will be determined by negotiations among sovereign nations. Also, I wish to explore the implications for overall level of effort, and not get into orthogonal questions of relative burden sharing. For this reason, I assume global trading of emissions permits between countries, and only examine the total global losses. One adjustment to account for equity concerns for developing nations is that the emissions reductions described below are cut in half for all developing (non-Annex I) nations between 2010 and 2040. After 2040, all policies apply equally.

The strategies are defined as the reduction required in the rate of growth of carbon emissions, relative to the unconstrained case. Thus, these policies will have differential effects depending on the region's reference growth path and will also vary with the (uncertain) rate of economic growth. Thus, "0%" means no emissions constraints at all, and "5%" means a 5 percentage point reduction in the CO_2 growth rate over that 5-year period, relative to the reference rate of emissions growth.[2] Smaller rates of reduction will result in slowed growth of CO_2 emissions while larger rates will actually reduce global emission over time.

Table 29.1 shows four different multi-period decision models that I explore here. Case A is the basic two-period model where the first period represents the 20 years from 2010–2029. Case B is a three-period model for comparison. Case C does not reduce emissions during 2010–2029 and defines a new period 1 from 2030–2049. Case D imposes no constraints for 40 years, from 2010–2049, choosing reduction rates for 2050–2069 and 2070–2100. To provide context for the relative stringency of these strategies, the impacts on several variables in 2030 are given in Table 29.2 for the period 1 strategies. To put these policy choices in more familiar terms, Table 29.2 lists the impacts of each possible first-period strategy by 2030 for the median productivity growth case, and the initial carbon price in 2010.

Based on previous work (Webster et al., 2002, 2003), I consider five uncertain parameters that have the greatest impact on damage costs:

- Labor productivity growth rate (LPG): this parameter drives the overall rate of economic growth in EPPA.

---

[2] For example, if emissions grow 8% over the 5 years in the reference, then a policy of "5%" would limit emissions at the end of that 5-year period to be no more than 3% higher than the previous period.

Table 29.3 Distributions for uncertain quantities in decision model.

	Branch 1 ($P = 0.185$)	Branch 2 ($P = 0.63$)	Branch 3 ($P = 0.185$)
Labor productivity growth rate (relative to reference rates)	0.8	1.0	1.2
Temperature change (degrees C)	5th percentile	Median	95th percentile
Damage cost coefficient (%)	0.02	0.04	0.16

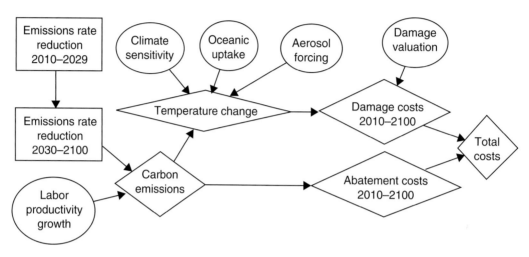

Figure 29.1 Influence diagram for standard two-period decision model.

Higher LPG results in higher carbon emissions and therefore higher temperature change and climate damages. (Webster et al., 2002).

- Climate sensitivity (CS): this parameter determines the change in global mean temperature at equilibrium that results from a doubling of CO_2 (Forest et al., 2002).
- Rate of ocean uptake (vertical diffusion coefficient, Kv): the 2D climate model parameterizes the mixing of both heat and carbon from the mixed-layer ocean into the deep ocean. A slower ocean will result in both higher carbon concentrations in the atmosphere and in more rapid warming (Forest et al., 2002).
- Strength of aerosol radiative forcing (Fa): this parameter represents the uncertainty in the magnitude of radiative forcing from sulfate aerosols, which are negative (cooling) (Forest et al., 2002).
- Damage valuation (DV): to reflect the large uncertainty in the valuation of climate change impacts, the damage coefficient η is uncertain (Nordhaus, 1994b).

The three uncertain climate parameters, CS, Kv, and Fa, are combined for each possible emissions path by performing a Monte Carlo simulation of 10 000 trials on the reduced-form climate models. The total resulting uncertainty in temperature change is then summarized by a three-point Tukey–Pearson approximation (Keefer and Bodily, 1983) using the 5th, 50th, and 95th percentiles. The climate parameter probability distributions are constrained by observations of twentieth century climate (Forest et al., 2002; Webster et al., 2003). When sampling from the climate parameter distributions, correlations are imposed of $\rho_{SK} = 0.004$, $\rho_{KA} = 0.093$, $\rho_{SA} = 0.243$, (where SK means correlation between sensitivity and Kv, SA between sensitivity and aerosol forcing, etc.) as consistent with twentieth-century observations.

The uncertainties in labor productivity and damage valuation are also represented in the decision tree with three-point discrete approximations (Table 29.3). The reference continuous distributions for these parameters are obtained from expert elicitation. The distribution for the damage valuation is taken from Roughgarden and Schneider (1999), based on the assessment by Nordhaus (1994b). The joint distribution of climate uncertainties, the labor productivity uncertainty, and the damage valuation uncertainty are assumed to be mutually probabilistically independent.

The basic two-period decision model is shown in Figure 29.1 as an influence diagram. Influence diagrams are a graphic representation of a sequential decision model. The rectangles represent the two decision points, 2010 and 2030, the circles represent the five uncertain parameters described above, and the diamonds represent the intermediate outcomes – carbon emissions, temperature change, damage costs and abatement costs, as well as the final outcome, total costs – of any decision path. Arrows pointing to decision nodes indicate the time ordering of decision and information. Arrows to outcome nodes indicate the

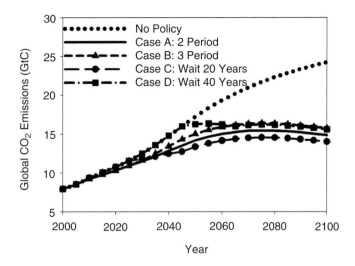

Figure 29.2 Optimal emissions paths for two- and three-period decision models.

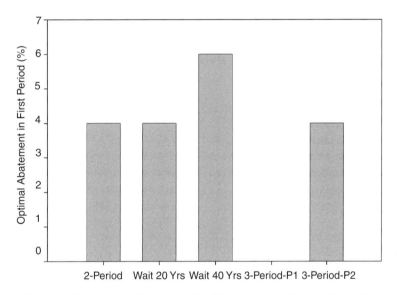

Figure 29.3 Optimal near-term abatement for two- and three-period decision models with resolution under uncertainty.

functional dependence; e.g., temperature change depends on the three uncertain climate parameters and on emissions. Figure 29.1 represents a situation with no resolution of uncertainty before the period 2 decision. The corresponding influence diagram for the case with learning before period 2 would add arrows from each uncertainty node to the 2030–2100 decision node.

I first show the results for each case where there is no resolution of uncertainty before the second-period decision. While this does not address the decision under uncertainty aspect, it does capture the other justifications for delaying abatement to the future: slow build-up of CO_2, technological change, and the discount rate.

Each of the decision models listed in Table 29.1 is calculated and solved for the optimal strategies in each period. The resulting global CO_2 emissions path under each optimal policy is shown in Figure 29.2. Note that the emissions from each two-period case closely approximate the three-period emissions. In other words, all cases are approximating the same continuous-time dynamic optimal path. The more flexibility in the model (i.e., more decision periods), the closer to the continuous-time optimal path the emissions will be. Waiting before imposing reductions does not cause the new period 1 choice to be less stringent; it is in fact more stringent to compensate for the delay.

Next, consider the situation in which uncertainty is revealed before the second-period decision in the two-period models or before the third-period decision in the three-period model. The optimal emissions paths are shown in Figure 29.2, and the optimal abatement strategy in the first period is shown in Figure 29.3. For the three-period model, the optimal decision in the first periods is 0% (no reduction) in period 1 and 4% in period 2. The two-period model, forced to choose one abatement level, has an optimal strategy of 4%. If the first decision period is delayed for 20 years with no abatement, the optimal strategy is 4%, and if delayed 40 years, the optimal strategy is 6%. The optimal abatement in the final period in all models is a probability distribution, since the optimal choice depends on what is learned.

If there were a Boiled Frog situation, it would appear in the form of first period optimal strategies that are equally or less stringent after delays. But in fact, if the first decision is delayed longer, with or without resolution of uncertainty, the optimal strategy becomes more stringent, not less. There is no Boiled Frog in this model!

This result should not be surprising to those familiar with dynamic optimization. Nevertheless, there is still something that may trouble some who consider the prospects for making dramatic emissions reductions in the future when little is undertaken in the near term. Why do we fear that each generation will continue to pass responsibility on to the next, never addressing climate change until impacts are already severe? There is a basis for this suspicion, but it is not represented in the decision models above. The problem, path dependency in political systems, is the topic of the next section.

29.3 Hysteresis and path dependency in climate policy decisions

29.3.1 The political context and path dependency

In applying decision analytic methods to thinking about appropriate levels of global greenhouse gas emission reductions over the next century, the "decisionmaker" is a fictional entity created for analytical convenience. The reality is that policy responses will emerge from extremely complex multi-level, multi-party negotiations that will occur continuously over the century. At one level, the negotiations are between nation states, such as the 188 parties to the Framework Convention on Climate Change (UNFCCC, 1992). However, the positions at this level are driven by the competing positions and interests in the domestic politics of each of those countries. The system as a whole has been compared to that of a "two-level game" (Putnam, 1988).

The competing domestic interests and their resolution in the form of an official national position cannot be represented as a static preference function, but a highly fluid stream of positions that evolve and change over time. This is particularly the case over the time horizon, decades to centuries, relevant to climate change policy.

The contrast between the decisionmaker as modeled and the actual decision process requires that we closely examine the assumptions of the analysis method. One particular aspect of the political decision process is relevant to the discussion here: its ability to make radical shifts over time. Political scientists have long noted the tendency of political systems to exhibit path dependency, and have used this feature to explain a number of political outcomes (Lipset and Rokkan, 1967; Sewell, 1996; Levi, 1997; Pierson, 2000). The idea of path dependency is that once a particular course of action has been chosen, it becomes increasingly difficult over time to reverse that course. Policies tend to exhibit "lock-in", and while a legislature might from time to time create a new bureaucratic agency, it is exceedingly difficult to eliminate one. Pierson (2000) suggests that the phenomenon can also be thought of as increasing returns to scale within the political system.

An examination of previous studies of climate policy as a sequential decision under uncertainty reveals that the characteristic of path dependency is largely absent. On the contrary, in keeping with the conventional approach, the range of emissions reductions from which to choose is the same or similar in each decision period, with no explicit constraints on future decisions depending on previous periods. One notable example is Hammitt (1999), which represented path dependency in the sense that reduced emissions in one period will result in lower emissions in all future periods at no cost.[3] This is one important type of path dependency, although this interaction is represented in CGE models such as EPPA. The concept of path dependency explored here is a stricter form: constraints on the choice set conditional on previous choices.

The idea of path dependency is part of the underlying intuition that delayed emissions reductions make more drastic future reductions more difficult or less likely. We found no Boiled Frog effect in the previous section because the model assumed that all period 2 strategy choices were available regardless of the period 1 decision made. If we believe that this effect is a salient characteristic of the political process, it must be included in the model.

29.3.2 Modeling path dependency

A challenge in exploring this issue is *how* to represent path dependency in dynamic optimization models of the type used here. Ideally, this feature would be represented somehow in the underlying representations of the costs and benefits of each decision path. Since it is not, however, the goal here is to add a simple adjustment to the decision model that has the desired effect and that makes sensitivity analysis straightforward.

There are a number of possibilities. One simple approach would be to assume that since the relevant decisionmakers are choosing for the present only and have no control over future decisions, one could model those future choices as uncertainties rather than decisions. The difficulty with this approach is that probability distributions are then needed for future political choices, where these distributions are conditional on the period 1 strategy that was chosen. It is not clear on what basis one could design these distributions or from whom one could elicit them. A second approach is to instead retain future policy choices as decisions, but allow for the possibility that the options may be limited. Thus there is an additional

---

[3] Hammitt (1999) shows that the impact of path dependency on near-term optimal abatement is in the same direction as the results shown in this study.

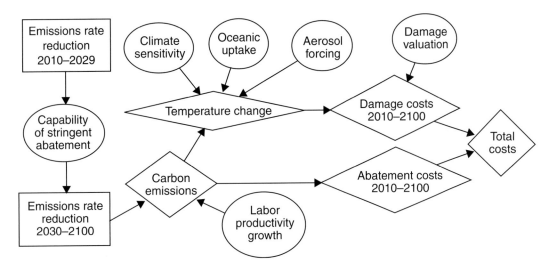

Figure 29.4 Influence diagram for decision model with hysteresis.

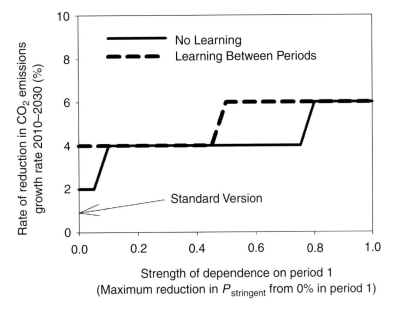

Figure 29.5 Optimal decision with path dependency.

uncertainty in period 1 over whether future decisions will have the full range of options to choose from. These constraints are then conditional on previous decisions. A third approach is to model the path dependency as an additional cost in the objective function to drastic changes in the level of stringency from previous periods, and which grows larger over time. This would have the same effect as the second approach, effectively ruling out some future options as non-optimal once the cost grew too large.

For analytical simplicity, this analysis uses the second approach of constraining future choice sets. Figure 29.4 shows how the influence diagram from Figure 29.1 changes for this variation (for the case without resolution of uncertainty). There is now an additional uncertainty node, "Capability of stringent abatement", which is resolved after period 1 but before period 2, and is influenced by the period 1 decision. This uncertainty determines the levels of emissions reductions that can be chosen from in period 2. There is some probability ($P_{stringent}$) that the full range of options from the original decision model is available to the period 2 decisionmaker. But now there is also the possibility, with probability $1 - P_{stringent}$, that only a limited range of emissions can be chosen from. In particular, the options no longer available are the most stringent reductions.

The critical aspect needed to capture the notion of path dependency, as described in the political science literature, requires that the range of available future actions *depends* on earlier choices. Unfortunately, the strength of this relationship is not known. Presumably, earlier actions have some influence on the set of future options, but there may be some random

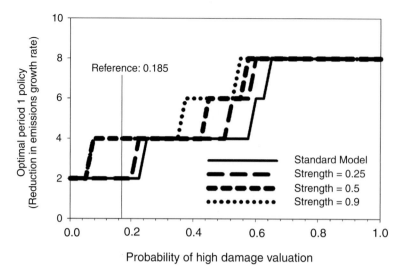

Figure 29.6 Sensitivity to damage uncertainty for varying degrees of path dependency (no learning between periods).

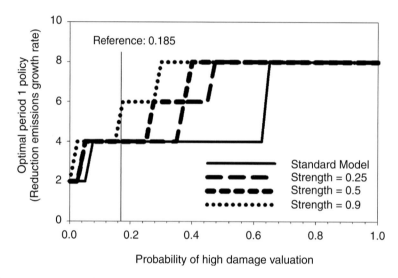

Figure 29.7 Sensitivity to damage uncertainty for different degrees of path dependency (complete resolution of uncertainty between periods).

stochastic element as well. I will use a simple formulation with extensive sensitivity analysis to explore the implications over a broad range of path dependency.

In this model of hysteresis, I define a second parameter $H_{strength}$ as the maximum reduction in probability that stringent reductions are possible. I then calibrate a linear function of the first period policy such that no reductions (0%) will result in the maximum decrease in $P_{stringent}$ and the most stringent policy (10%) will result in no change ($P_{stringent} = 1.0$). All other strategy choices result in proportional reductions in $P_{stringent}$. Thus, when $H_{strength}$ is zero, there is no hysteresis and the model is the same as in Section 29.2. When $H_{strength}$ is unity, the probability of stringent action in period 2 is entirely determined by the period 1 decision. If $H_{strength}$ is set to 0.2, this means that a period 1 strategy of 0% or no reductions will reduce the probability of stringent reductions in period 2 to 0.8, while a period 1 strategy of 10% (maximum reductions) would leave the probability of stringent reductions at 1.0. I explore the effect of this modification to the standard two-period model (Case A in Table 29.1).

29.3.3 Results with path dependency

In general, strengthening the hysteresis effect causes greater reductions in period 1 to be optimal, whether uncertainty is resolved between periods or not (Figure 29.5). If the period 1 strategy choice will alter the likelihood that dramatic reductions can be undertaken if necessary, this makes additional reductions desirable in the near term. A stronger hysteresis effect is needed to change the optimal decision when learning will occur, because in this model resolution of uncertainty already results in more stringent period 1 abatement than the no-learning case.

We can further illustrate the effect of hysteresis on the first period decision by varying the probability of the high damage valuation uncertainty (Figures 29.6 and 29.7). Increasing hysteresis leads to more stringent near-term abatement. The omission of any path dependency in a sequential model is likely to bias the optimal near-term abatement to be too little.

29.4 Discussion

The objective of this analysis is to draw attention to the implications for policy prescriptions and insights of methodological choices. We take a complex decision problem – climate policy; we observe some salient characteristics of that problem – uncertainty, ability to learn, ability to adapt and revise decisions over time; we choose a method suited to those characteristics – decision analysis. But another key feature of the new decision context of national governments, namely path dependency, is not typical of the individual decision-maker situations for which this tool was developed, and so is not normally accounted for in the way the decision models are constructed. But path dependency can lead to qualitatively different results and insights for near-term policy.

The formulation of path dependency presented here has been kept extremely simple to allow sensitivity analysis and to keep the focus on the basic concept. In fact, we do not know whether decisionmakers in future decades will be constrained or not in the range of emissions reductions that they can pursue, or how the likelihood of this constraint depends on today's policy choices. To get more robust insights from optimal sequential decision models, research into the magnitude of path dependency effects from institutional development and commitments is desirable. A related question where there is active research but more is needed is how technical change is influenced by regulation-driven price incentives.

In the interest of efficient policy, we often compare the costs of an action against its benefits. However, this static cost–benefit approach does not capture the whole picture, or even the most important part. On that basis alone, the optimal level of greenhouse gas mitigation in the near term is quite low. The true value of near-term mitigation policy is that of starting down the right path to maximize options for future policy adjustments. Recognizing this additional value justifies a greater level of abatement in the near term. This goal of creating and maximizing future options should be a primary focus in choosing the stringency and institutional design of climate policies.

References

Babiker, M., Reilly, J. M., Mayer, M. *et al.* (2001). The MIT emissions prediction and policy analysis (EPPA) model: revisions, sensitivities, and comparison of results. *MIT Joint Program on the Science and Policy of Global Change*. Report No. 71. Cambridge, MA: MIT.

Forest, C. E., Stone, P. H., Sokolov, A. P., Allen, M. R. and Webster, M. D. (2002). Quantifying uncertainties in climate system properties with the use of recent climate observations. *Science* **295**, 113–117.

Hammitt, J. K. (1999). Evaluation endpoints and climate policy: atmospheric stabilization, benefit–cost analysis, and near-term greenhouse-gas emissions. *Climatic Change* **41**, 447–468.

Hammitt, J. K., Lempert, R. A. and Schlesinger, M. E. (1992). A sequential-decision strategy for abating climate change. *Nature* **357**, 315–318.

Keefer, D. L. and Bodily, S. E. (1983). Three-point approximations for continuous random variables. *Management Science* **29**, 595–609.

Kolstad, C. D. (1996). Learning and stock effects in environmental regulation: the case of greenhouse gas emissions. *Journal of Environmental Economics and Management* **31**, 1–18.

Lempert, R. J., Schlesinger, M. E., Bankes, S. C. (1996). When we don't know the costs or the benefits: adaptive strategies for abating climate change. *Climatic Change* **33**, 235–274.

Levi, M. (1997). A model, a method, and a map: rational choice in comparative and historical analysis. In *Comparative Politics: Rationality, Culture, and Structure*, ed. M. I. Lichbach and A. S. Zuckerman. Cambridge: Cambridge University Press, pp. 19–41.

Lipset, S. M. and Rokkan, S. (1967). Cleavage structures, party systems and voter alignments: an introduction. In *Party Systems and Voter Alignments*, ed. S. M. Lipset and S. Rokkan. New York: Free Press, pp. 1–64.

Manne, A. S. and Richels, R. G. (1995). The greenhouse debate: economic efficiency, burden sharing and hedging strategies. *The Energy Journal* **16**(4), 1–37.

Nordhaus, W. D. (1994a). *Managing the Global Commons: The Economics of Climate Change*. Cambridge, MA: MIT Press.

Nordhaus, W. D. (1994b). Expert opinion on climatic change. *American Scientist* **82** (January), 45–51.

Peck, S. C. and Teisberg, T. J. (1992). CETA: a model for carbon emissions trajectory assessment. *The Energy Journal* **13**(1), 55–77.

Pierson, P. (2000). Increasing returns, path dependence, and the study of politics. *American Political Science Review* **94**, (2), 251.

Pizer, W. A. (1999). The optimal choice of climate change policy in the presence of uncertainty. *Resource and Energy Economics* **21** (3,4), 255–287.

Prinn, R., Jacoby, H., Sokolov, A. *et al.* (1999). Integrated global system model for climate policy assessment: feedbacks and sensitivity studies. *Climatic Change* **41** (3/4), 469–546.

Putnam, R. D. (1988). Diplomacy and domestic politics: the logic of two-level games. *International Organization* **42** (3), 427–460.

Roughgarden, T. and Schneider, S. H. (1999). Climate change policy: quantifying uncertainties for damages and optimal carbon taxes. *Energy Policy* **27**, 415–429.

Sewell, W. H. (1996). Three temporalities: toward an eventful sociology. In *The Historic Turn in the Human Sciences*, ed. T. J. McDonald. Ann Arbor: University of Michigan Press, pp. 245–263.

Sokolov, A. P. and Stone, P. H. (1998). A flexible climate model for use in integrated assessments. *Climate Dynamics* **14**, 291–303.

Ulph, A. and Ulph, D. (1997). Global warming, irreversibility and learning. *Economic Journal* **107** (442), 636–650.

UNFCCC (1992). United Nations Framework Convention on Climate Change. *International Legal Materials* **31**, 849–873.

Valverde, L. J. Jr, Jacoby, H. D. and Kaufman, G. (1999). Sequential climate decisions under uncertainty: an integrated framework. *Journal of Environmental Modeling and Assessment* **4**, 87–101.

Webster, M. D. (2002). The curious role of learning: should we wait for more data? *The Energy Journal* **23** (2), 97–119.

Webster, M. D., Babiker, M., Mayer, M. *et al.* (2002). Uncertainty in emissions projections for climate models. *Atmospheric Environment* **36** (22), 3659–3670.

Webster, M. D., Forest, C., Reilly, J. *et al.* (2003). Uncertainty analysis of climate change and policy response. *Climatic Change* **61** (3), 295–320.

30

Article 2 and long-term climate stabilization: methods and models for decisionmaking under uncertainty

Ferenc L. Toth

30.1 Introduction

The policy-oriented framing of the question about how much anthropogenic climate change the Earth's societies and ecosystems could endure is at least 20 years old. International conferences in Villach (Austria) in 1985 and 1987, and especially the one in Bellagio (Italy) in 1987, involved both scientists and policymakers in contemplating climate change and proposed that long-term environmental targets, such as the rates of global mean temperature increase or sea-level rise, should be used in policymaking (WCP, 1988). Referring to observed historical values, it was recommended to keep the rate of temperature increase below 0.1 °C per decade, based primarily on the estimated rate of ecological adaptation. This seemingly arbitrary proposition is rather specific compared with the riddle implied in the formulation of Article 2 about the ultimate objective of the United Nations Framework Convention on Climate Change (UNFCCC).

Article 2 of the UNFCCC frames the requirement for long-term climate policy in terms of an environmental objective "to prevent dangerous anthropogenic interference with the climate system." This calls for "inverse approaches" that provide information about possible emission strategies with respect to externally specified environmental targets. Early attempts by Working Group I of the Intergovernmental Panel on Climate Change (IPCC, 1994a) and by Wigley *et al.* (1996) depict emission paths with respect to given CO_2 concentration targets. Subsequent work takes *climate change* attributes (magnitude and rate of change in global mean temperature) or geophysical consequences (magnitude and rate of sea-level rise) as environmental targets (Alcamo and Kreileman, 1996; Toth *et al.*, 1997; Swart *et al.*, 1998) to guide long-term climate policies. While these analyses provide useful insights into the issue of concentration/climate stabilization, they are only remotely related to the ultimate concerns about climate change: its possible adverse effects on ecosystems, food production, and sustainable development.

In response to this challenge, the ICLIPS Integrated Assessment Model (IAM) (Toth, 2003) combines *impact analysis and cost estimates* and requires two types of normative inputs – social actors' willingness to accept a certain amount of climate change impact and their willingness to pay for climate change mitigation – to determine whether there exists a corresponding corridor of permitted emission paths. There are efforts to establish *impact-driven concentration pathways* by using arbitrary intermediate signposts (O'Neill and Oppenheimer, 2004). Other approaches adopt advanced modeling techniques to identify *robust policies* that perform well under a wide range of plausible futures (Lempert and Schlesinger, 2000) or use a cost–benefit framework to estimate probabilities of crossing "dangerous climate change" *thresholds* under deep uncertainties (Mastandrea and Schneider, 2004). Table 30.1 presents an overview of the approaches and the relevant literature sources.

Various authors have assessed the treatment of uncertainties in different types of integrated assessment models (Shackley and Wynne, 1995; van Asselt and Rotmans, 1996; Pizer, 1999; Kann and Weyant, 2000; Visser *et al.*, 2000). In this chapter, we explicitly address uncertainties from the perspectives of decisionmaking concerning Article 2.

This chapter provides a comparative appraisal of the diverse concepts and approaches to long-term climate stabilization. It

---

Human-induced Climate Change: An Interdisciplinary Assessment, ed. Michael Schlesinger, Haroon Kheshgi, Joel Smith, Francisco de la Chesnaye, John M. Reilly, Tom Wilson and Charles Kolstad. Published by Cambridge University Press. © Cambridge University Press 2007.

Table 30.1 Approaches to identifying climate policies to avoid "dangerous" climate change.

Method	Summary	Reference
Smooth concentration stabilization paths	Create emission paths to stabilize CO_2 concentrations at different levels	IPCC (1994a)
Cost-minimizing concentration stabilization	Identify emission pathways minimizing the total costs of achieving alternative concentration stabilization targets	Wigley et al. (1996); Models and results reviewed in IPCC (2001b)
Safe landing analysis	Derive ranges of emissions to keep climate change within predefined ranges (temperature change and sea-level rise and their rates)	Alcamo and Kreileman (1996); Swart et al. (1998); van Vuuren and de Vries (2000); van Vuuren et al. (2003)
Exploratory modeling	Find robust strategies (including mitigation pathways) under conditions of deep uncertainty	Lempert et al. (2000); Lempert and Schlesinger (2000)
Tolerable windows/Inverse/Guardrails approach	Derive emission corridors including all pathways that satisfy normative constraints on tolerable climate change impacts and mitigation costs	Toth et al. (1997); Toth et al. (2002); Toth (2003)
Concentration stabilization pathways	Obtain CO_2-equivalent stabilization paths by using standardized specification rules for emissions and approaching the target	O'Neill and Oppenheimer (2002); O'Neill and Oppenheimer (2004)
Probabilistic integrated assessment	Establish probabilities of "dangerous anthropogenic interference" from probability distributions of future climate change in a global cost–benefit framework	Mastandrea and Schneider (2004)

highlights benefits and shortcomings of existing techniques and compares their messages for near-term decisionmaking. Section 30.2 illustrates the diversity of ideas triggered by Article 2 concerning "dangerous" climate change over the past decade and introduces a simple typology of the associated uncertainties. Section 30.3 presents a sampler of analytical frameworks, their model implementations and results. The approaches and models are compared in Section 30.4 according to their vicinity to the "ultimate concerns" (to climate change impacts), their spatial level (global or regional), the model sophistication (loosely connected assembly of simple models or comprehensive integrated models), their consideration of vulnerability and adaptation, the decision analytical framework adopted, the treatment of uncertainties, and the delineation of normative judgements and technical analysis. The main conclusions are summarized in Section 30.5.

30.2 Interpretations of Article 2 and related uncertainties

Ever since the UNFCCC was adopted in 1992, scientists, modellers, national policymakers, and international negotiators have been considering the meaning of Article 2: what constitutes a "dangerous anthropogenic interference with the climate system" that should be prevented? Swart and Vellinga (1994) propose a six-step approach to identify the implications of Article 2. Parry et al. (1996) define thresholds of weather or climate events and critical levels of climate change that might signal danger. Schneider (2001) argues that subjective probabilities should be assigned to the scenarios presented in the Special Report on Emissions Scenarios (SRES) by the IPCC (2000) in order to help policy analysts "to assess the seriousness of the implied impacts" (p. 18) and assist policymakers to make proper judgements about how to avoid "dangerous anthropogenic interference." More recently, Desai et al. (2004) propose that in the climate change context both external definitions of danger (science-based risk analysis conducted by experts) and internal definitions (based on individual or collective experience or perception) need to be considered.

As part of its work on the Second Assessment Report (SAR), the IPCC devoted a special conference to this issue (IPCC, 1994b) and the synthesis document of the Second Assessment (IPCC, 1995) was framed to summarize information relevant to interpreting Article 2. The document presents an analytical approach to stabilizing concentration, the relevant technology and policy options for mitigation, as well as equity and sustainability considerations. The SAR synthesis document concludes that uncertainties remain over what constitutes a "dangerous anthropogenic interference," and emphasizes the importance of sequential decisionmaking.

After several years of additional research and deliberation, the problem of "dangerous interference" was one of the nine policy-relevant scientific questions along which main conclusions of the Third Assessment Report (TAR) were summarized. The Synthesis Report (IPCC, 2001a) reconfirms that while scientific information and evidence are crucial for deciding what constitutes dangerous interference, the decisions are primarily determined by value judgements and as such they are outcomes of socio-political processes.

The uncertainty of our knowledge about many components of the climate system and its interactions with human development has been a recurring issue in the climate change literature over the past few years and it is also one of the main topics in this volume. The presence of uncertainty has grave implications for making decisions about climate policy (sequential decisionmaking with learning and course correction). Uncertainty also affects the choice of the analytical frameworks that are intended to support climate stabilization decisions. Much effort has been devoted to improving our understanding of the implications of uncertainties and to representing them better in climate-economy models. Chapters in this volume by Pizer (Chapter 25) and Keller *et al.* (Chapter 28) are good examples.

Many integrated assessment modelers adopt some practical classification of uncertainties relevant to their own framing of the decision problem. There are also systematic efforts to develop general uncertainty typologies (e.g., Dowlatabadi and Morgan, 1993; van Asselt and Rotmans, 2002; Peterson, 2004). Considering lessons from earlier efforts, a simple typology is proposed here with a view to the comparative evaluation of the decision-analytical frameworks related to Article 2. From this perspective, it seems to be useful to distinguish four main types of uncertainties.

> Type 1: *Scientific uncertainties* concerning the dynamics of the various components of socio-economic and climate systems and their interactions: the fundamental drivers of development and greenhouse gas (GHG) emissions, the transfer of anthropogenic emissions across the climate system, their immediate and long-term effects, and many others.
> Type 2: *Modeling uncertainties*: the ways in which the actual best understanding of the socio-economic and climate systems is mapped into a set of equations (model structure) and the values chosen for characterizing the relationships (model parameters).

These two types of uncertainties have been widely discussed in the literature. There is a broad consensus that scientific uncertainties can be reduced by investing in research. Improved understanding of the Earth system and the processes shaping climate and weather will also foster their better representation in model structures and more accurate specification of key parameters.

Until recently, less attention has been devoted to two other uncertainty categories.

> Type 3: *Decision target uncertainties*: what is "dangerous" climate change depends on many factors, such as (i) the climate impact sector (those under more intensive human management have more opportunity to adapt, hence the threshold for danger is likely to be higher); (ii) for any given impact sector, the geographical location and pressures from other sources (pollution, harvesting); (iii) also depending on the impact sector, the socio-economic conditions, including adaptive capacity, risk perception, and risk management culture. As a result, there is a great deal of uncertainty about the acceptable degree of climate change under Article 2: is it a 20% change in water resource supplies in a given region, or is it a 3 °C increase in mean annual temperature in another region or, as suggested by the European Union, is it a 2 °C increase in global mean temperature?
> Type 4: *Implementation uncertainties*: even if the target related to Article 2 was clearly defined, and reliable scientific and modeling analyses charted the way to attain it, there would still be no guarantee of achieving the target, because of the imprecision of policy instruments, the limited and uncertain effectiveness of measures, and the overall operational uncertainty of the institutional arrangements (international agreements, government policies, private sector capacity to act). Moreover, since implementation of any target related to Article 2 inevitably involves several generations of decisionmakers, uncertainty over whether and how long future decisionmakers with follow the stabilization course is also an important factor.

The latter two types of uncertainties cannot be resolved by investing in research and learning. They require other techniques of uncertainty management. Target-related uncertainties can be moderated by risk mitigation processes, such as transfer payments to increase adaptive capacity, education to foster public understanding and risk evaluation, and the like. Uncertainties in implementation can be reduced by flexible and adaptive procedures that incorporate monitoring and course-correction elements.

Cutting across these uncertainty categories are the issues of conceptual and technical capacities of different decision-analytical frameworks to explore the implications of the above four types of uncertainties. They will also be examined in the next section.

30.3 Analytical frameworks and models: a sampler

This section presents a concise overview of selected integrated assessment efforts tackling the challenge of Article 2. The objective is to highlight the key features of the analytical frameworks, to illustrate the kinds of results they provide, and to prepare their comparison in the next section.

30.3.1 Smooth versus optimal stabilization paths

In an early attempt to assess stabilization options at different levels of atmospheric GHGs concentrations, the IPCC (1994a) discusses atmospheric processes of several gases. Carbon dioxide is given special attention because of its importance and the complexities of the processes underlying the relationships between emissions and concentrations. The IPCC (1994a) authors develop a set of concentration profiles that stabilize

CO_2 concentrations in the range between 350 and 750 ppmv and calculate the associated anthropogenic CO_2 emissions pathways. While the concentration profiles represent a smooth transition from the recent rate of CO_2 concentration increase towards stabilization, they imply near-term and fast departure from the projected CO_2 emission pathways (see IPCC, 1994a, p. 23). Given the inertia of the global energy system due to the long economic lifetime of the capital stock, following these paths towards any of the stabilization levels would imply considerable economic losses. Wigley et al. (1996) take the same concentration targets and apply two integrated assessment models to derive emission paths that involve much lower mitigation costs. They are known as the Wigley–Richels–Edmonds (WRE) paths (IPCC, 1996). Since the total amount of cumulative carbon emissions is relatively insensitive to the emission profile, the Wigley et al. emission paths portray higher emissions in the early decades and lower emissions later relative to the IPCC paths (see Wigley et al., 1996, p. 243).

Several factors explain the significantly lower costs of stabilization along the WRE paths. The most notable is that they largely avoid abandoning existing carbon-related fixed assets well ahead of their economic lifetime and they assume that technological development triggered by the anticipated increasing carbon constraint will make later reductions cheaper. The general insights are apparent: a slow and gradual diversion from the non-intervention emission path allows more time for all affected actors to prepare themselves (change investment strategies, invest in technological development) and is therefore less costly. It should be noted, however, that these least-cost paths ignore the possible differences in damage costs associated with alternative stabilization pathways.

The cost-efficiency framing of concentration stabilization has become a widespread exercise. A large number of studies are assessed by IPCC (2001b). The estimated costs of stabilization depend on the baseline emissions scenarios and differ widely across the models. It is important to note a common confusion concerning concentration targets. It is sometimes not clear whether CO_2 only or CO_2-equivalent GHG concentration is meant. The former ignores the radiative forcing of non-carbon GHGs and the actual climate change which might be higher, corresponding to about an additional 100 ppmv increase in CO_2 concentration. The latter raises the problem of GHG accounting in terms of CO_2-equivalence. Confusion prevails even in high-level policy pronouncements, like the 1996 European Union declaration that global average temperature should not exceed the pre-industrial level by more than 2 °C and the CO_2 concentration levels should not increase above 550 ppmv (EU, 1996).

30.3.2 Safe landing analysis

The main objective of the safe landing analysis (SLA) (Alcamo and Kreileman, 1996; Swart et al., 1998) is to establish ranges of near-term emissions called "safe emissions corridors" that keep long-term climate change (i.e., long-term stabilization objectives) within predefined limits. The safe emissions corridor determines a range of possible near-term (for the next decade or two) GHG emissions levels including at least one long-term (century-scale) emissions path that satisfies the externally defined long-term and intermediate climate protection goals. The numerical implementation of SLA is a software tool based upon regressions of many scenarios calculated with the IMAGE 2 model (Kreileman and Berk, 1997).

Swart et al. (1998) present an example of using the safe landing analysis. They take the 1996 declaration of the European Union according to which the global average temperature should not exceed the pre-industrial level by more than 2 °C. As noted above, the EU declaration also specifies a concentration limit for CO_2 at 550 ppmv, but considering the forcing by non-CO_2 GHGs, this limit would imply an increase in global mean temperature above 2.5 °C unless climate sensitivity is assumed to be in the very low segment of the IPCC range of 1.5–4.5 °C. Relating the 2 °C limit to the pre-industrial level implies a constraint of about 1.5 °C temperature increase relative to 1990. Another constraint shapes the mitigation side: the maximum rate of reduction of global annual emission is 2% per year. The target horizon is the year 2100.

Swart et al. (1998) find that up to 2010 none of the IPCC IS92 emissions scenarios leaves the safe corridor, i.e., no intervention is needed before 2010 if the 1.5 °C is the only climate constraint. However, the corridor becomes narrower if additional constraints are introduced. For example, if the rate of temperature increase is limited to 0.15 °C/decade (which can be violated in up to two decades during the twenty-first century) and sea-level rise should not exceed 30 cm, CO_2-equivalent emissions should not exceed 12.4 GtC/year any time between 1990 and 2010. If the constraints are even tighter and the rate of temperature change is not allowed to exceed 0.1 °C per decade (except for up to two decades), this logically narrows the safe corridor further, with an upper limit of less than 11 GtC/year around 2000 declining to a maximum of 9.6 GtC by 2010. If non-Annex I countries are allowed to increase their emissions according to the IS92a scenario, then under the same climate change constraints of 0.15 and 0.1°C/decade the corresponding emissions corridors for Annex I countries shrink by 2010 to about 6.1 and 3.3 GtC CO_2-equivalent emissions, respectively (see Swart et al., 1998, pp. 196–197).

More recent work with a new version of the IMAGE/TIMER model includes various stabilization exercises. Van Vuuren and de Vries (2000) take the SRES B1 scenario (IPCC, 2000) as baseline and study the implications of a 450 ppmv CO_2 concentration target. Van Vuuren et al. (2003) examine the environmental, economic, and technical aspects of different post-Kyoto architectures where the objectives are to stabilize atmospheric concentration of all Kyoto-constrained GHGs at 550 and 650 ppmv CO_2-equivalent levels, respectively. These efforts nonetheless resemble the concentration stabilization exercises rather than the classic safe landing/safe emissions corridor analyses.

30.3.3 Exploratory modeling

IAMs are conceived in the vein of different decision-analytical frameworks and they are formulated according to various modeling paradigms. Largely determined by these two factors, modelers take different routes to explore the implications of uncertainties on their results. The most typical techniques include simple sensitivity analysis, uncertainty propagation, and sequential decisionmaking. Lempert *et al.* (2000) and Lempert and Schlesinger (2000) take a different approach. The objective of their Exploratory Modeling technique is to find robust strategies, including mitigation and stabilization pathways, under the conditions of deep uncertainty.

Traditional approaches to climate change decisions define a set of alternative actions one might take, model the consequences of each action (including the associated costs and benefits), then adopt a metric (in most cases monetary units) and rank the actions according to the relative preferences based on the adopted metric. In contrast, the Exploratory Modeling technique summarizes all possible consequences of an action with average (expected) value. The optimal policy is the one that on average performs better than all the others.

The numerical implementation of the Exploratory Modeling technique is somewhat similar to the SLA software tool derived from regressions of many scenarios calculated with the IMAGE 2 model, although the objectives differ. In the SLA case the emphasis is on the existence of a long-term emission path departing from the short-term emissions corridor whereas the Exploratory Modeling focuses on the robustness of certain long-term emission paths selected from a large number of scenario runs.

30.3.4 Tolerable Windows/Inverse/Guardrails Approach

The initial approaches to Article 2 formulating climate stabilization in terms of GHG concentrations and temperature change relative to pre-industrial levels provide good measures of the perturbations caused by anthropogenic emissions. Yet most lay people and responsible policymakers would primarily worry about the impacts of climate change rather than about remote and abstract atmospheric indicators. They are inclined to raise the question: what magnitude (and rate) of climate change would imply unbearable impacts on their societies and environments? This requires deriving the answer to the "dangerous interference" question from the impacts of climate change and tracing the implications all the way back to climate change, concentrations, cumulative emissions limits, and emissions pathways. The Tolerable Windows Approach (TWA) as a decision-analytical framework and the Integrated Assessment of Climate Protection Strategies (ICLIPS) IAM do just that.

Early versions of the ICLIPS model start the inverse calculations from exogenously given constraints on the magnitude and rate of increase in the global mean temperature and sea level whereas the proxy for the maximum acceptable costs is the annual rate of change in GHG emissions (Toth *et al.*, 1997). The full version of the integrated model derives CO_2 emission corridors containing all pathways that satisfy user-specified normative constraints on tolerable climate change impacts and mitigation costs (Toth *et al.*, 2002). The underlying concept is that what is "dangerous" in the spirit of Article 2 is a social decision. It depends on the projected biophysical and socio-economic impacts, the risk perception, and the social attitudes to risks in a given society.

In order to help make informed judgments about what might constitute "dangerous" climate change impacts, biophysical climate impact response functions (CIRFs) have been developed for a series of impact sectors, such as natural ecosystems, agriculture, and water resources. CIRFs depict the evolution of impacts driven by incremental changes in climate (and, where relevant, atmospheric CO_2 concentrations) in climate-sensitive sectors (see Toth *et al.*, 2000; Füssel *et al.*, 2003). The second category of user inputs reflects another social decision concerning climate protection and relates to the specification of the willingness to pay in terms of income loss globally, in regions, and for generations. Based on these two normative, user-provided inputs, the model determines the range of long-term future emissions paths (corridors) that keep the climate systems within the tolerable impact limits at costs not exceeding the specified limits (see Toth *et al.*, 2003a, 2003b). If such a corridor does not exist, the user should consider increasing the level of climate change impacts that societies would be willing to accept, increasing the level of mitigation costs that societies would be willing to pay, or increasing the flexibility arrangements to reduce mitigation costs.

The ICLIPS model can also calculate cost-minimizing emission paths under given impact and cost constraints. Yet it is possibly more relevant that it produces long-term emissions corridors within which the actual paths can be chosen flexibly by considering additional criteria or preferences in the decisionmaking process.

30.3.5 Concentration stabilization pathways

Another impact-based stabilization assessment with an emphasis on linkages to near-term emissions targets is presented by O'Neill and Oppenheimer (2002). Without making use of a climate model, the authors speculate about climate change limits with a view to three impacts: sustained coral bleaching and the resulting loss of the reefs, the disintegration of the West Antarctic Ice Sheet, and the shutdown of the thermohaline circulation. They cite selected studies on the temperature constraints and IPCC summary results that link global mean temperature increases to atmospheric CO_2 concentrations. O'Neill and Oppenheimer settle the plausible targets relative to 1990 global temperatures at 1 °C for preventing severe damage to coral reefs, 2 °C for protecting the West Antarctic Ice Sheet, and 3 °C for averting the collapse of the thermohaline circulation, but then they focus on the prospects for achieving the 450 ppmv CO_2 concentration targets. They find that delaying the implementation of the emissions target specified by the Kyoto Protocol to 2020 would risk foreclosing the option of stabilizing concentrations at 450 ppmv.

In a recent paper, O'Neill and Oppenheimer (2004) derive CO_2-equivalent stabilization paths by using standardized specification rules for emissions and for approaching the ultimate target. They speculate about possible impacts on geophysical and ecological systems of a range of pathways involving different warming rates, including temporary overshoot of the target value. They consider coral reefs, spatially limited ecosystems, the Greenland and West Antarctic ice sheets, and the thermohaline circulation, but no models are used to link the pathways to the impact sectors.

30.3.6 Probabilistic integrated assessment

An explicit attempt to derive clues for what might be a dangerous anthropogenic interference (DAI) with the climate system from scientific information is made by Mastandrea and Schneider (2004). They establish probabilities of the dangerous interference from probability distributions of future climate change in a global cost–benefit framework developed by Nordhaus (1992). They take the impact synthesis chart prepared by Working Group II for the IPCC Third Assessment Report (see IPCC, 2001a, p. 11) as a starting point and establish a metric for DAI based on a cumulative density function of the threshold for dangerous climate change. This metric is applied to a range of results derived from uncertainty in climate sensitivity, climate change damage, and the discount rate. The model runs provide information on hedging against the risk of crossing particular thresholds by adopting some mitigation policies. For example, a carbon tax of the order of US$ 150–200 per tonne of carbon reduces the probability of crossing the 50 percentile DAI threshold from about 45% to virtually zero.

The authors recognize that the formulation of the DAI function is not a scientific question. The principle is similar to the case of CIRFs in the Tolerable Windows analysis: they display the relationships between incremental climate change and impacts on valued attributes of climate-sensitive systems, but it is a perception-driven social choice question (and as such, a normative question) what level of change is the acceptance limit beyond which societies find the impacts unacceptable and therefore "dangerous" for themselves. Since social perception of risks is complex and risk behavior is rather inconsistent not only across social groups and individuals but also across risks for the same individual, there is no science-based choice of what is dangerous. The IPCC chart simply shows levels of climate change beyond which the aggregated balance of benefits and damages in the given impact category becomes negative or overwhelmingly negative. The result is a somewhat arbitrary quasi-linear threshold function. The real value of the Mastandrea–Schneider (2004) framework is the approach to uncertainty analysis rather than the results concerning specific thresholds.

30.4 Comparative evaluation

The approaches and models presented in the previous section make up a diverse lot. The diversity is largely driven by the specific issues or aspects that modelers want to emphasize. This section compares the approaches according to a set of criteria that seem to be relevant for science-based policy advice on Article 2. The list of criteria includes:

- domain: for which component of the emissions–impacts chain stabilization is considered;
- spatial disaggregation: what geographical scale is considered in the local–global span;
- model sophistication: what is the complexity level of the adopted modules and the degree of their integration;
- vulnerability/adaptation: are these factors considered in specifying the "dangerous interference" limits;
- DAF: what decision-analytical framework is adopted;
- treatment of uncertainty: the degree and form of addressing uncertainties about the dangerous limits and the model results;
- science–normative disclosure: to what extent authors reveal the boundaries between normative choices of the stabilization constraints and the scientific assessment of their implications for policy interventions.

The evaluation also considers the extent to which the analytical frameworks deal with the four categories of uncertainties defined in Section 30.2 The main findings are summarized in Table 30.2.

Starting with the *domain* in which stabilization is analyzed, a clear chronological slant can be observed. Early assessments (IPCC WGI and WRE) focus on stabilizing atmospheric CO_2 concentrations. They are followed by many calculations of cost-minimizing stabilization paths, some of which include other GHGs to allow stabilization analyses of CO_2-equivalent GHG concentrations. The SLA and early version of the TWA expand the stabilization enquiry to the climate domain and use magnitudes and rates of global mean temperature change and sea-level rise to explore the implications of hypothetical danger values for near-term (Safe Landing) and long-term (Tolerable Windows) emissions corridors. By adding geographically explicit climate impact response functions to the Tolerable Windows analysis, it becomes possible to calculate global emission corridors on the basis of local or regional impact constraints. The Concentration Stabilization Pathways are based on global impact constraints, whereas the Probabilistic Integrated Assessment demarcates the danger zone on the basis of aggregated impacts synthesized from myriads of local, regional, national, and global impact studies assessed by IPCC (2001c).

The *spatial disaggregation* taken by the different approaches is partly determined by the domain and partly by the analytical framework. Obviously, studies focusing on concentration stabilization and global temperature or sea-level constraints are

Table 30.2 Main characteristics of the selected approaches to Article 2.

Approach	Domain	Spatial disaggregation	Model sophistication	Vulnerability/ adaptation	DAF	Treatment of uncertainty	Science–normative disclosure
Smooth concentration stabilization paths	CO_2 concentrations	Global	Simple C cycle	na	PEM	No	Implicit
Cost-minimizing concentration stabilization	CO_2, CO_2 equivalent or GHG-basket concentrations	Global	Integrated Simple C cycle and medium-complexity economy	na	CEA	No	Implicit
Safe Landing Analysis	Climate: T, dT, SLR, dSLR	Global	Medium-complexity climate	na	SLA/TWA: Proxy-cost constrained PEM	No	Yes
Exploratory Modeling	Flexible	Global	Medium-complexity climate	indirectly	PEM + post-processing	Core	Yes
TWA/Inverse/ Guardrails Approach	Impacts	Local to global	Medium-complexity integrated climate-economy	indirectly	SLA/TWA: impact and cost-constrained CEA	Parameter sensitivity; Target/cost decision sensitivity	Yes
Concentration stabilization pathways	Impacts	Global	Patches of C cycle and simple climate	no	Impact-constrained PEM	No	No
Probabilistic integrated assessment	Aggregate impacts	Global	Medium-complexity integrated climate-economy	no	Probabilistic CBA	Core	Yes

Notes:
DAF: decision analytical framework.
PEM: policy evaluation model.
CEA: cost-efficiency analysis.
CBA: cost-benefit analysis.
T and dT: temperate change and rate of temperature change.
SLR and dSLR: sea-level rise and rate of sea-level rise.

meaningful only at the global scale. Similarly, global studies are required if the dangerous interference is derived from large-scale or global impacts (coral reefs, thermohaline circulation) or from a globally aggregated impact function. The TWA/ICLIPS model is the only one in the sampler that is capable of taking regional or local impact constraints and performing the inverse analysis back to emissions corridors via global climate/concentration constraints.

The *sophistication of models* applied in the Article 2 analyses is rather diverse. At the one extreme, the IPCC exercise is based on a carbon-cycle model while at the other end of the spectrum the Tolerable Windows and the Probabilistic Integrated Assessment adopt models with fully integrated climate–economy modules. The models behind the cost-minimizing stabilization exercises also incorporate high degrees of integration but their modules are somewhat asymmetric: their economic components are often much more sophisticated than their carbon-cycle/atmosphere modules.

Considering *vulnerability/adaptation* is perhaps the weakest point in the "dangerous interference" exercises. This is remarkable because, at least in the impact sectors under human management and depending on the scope and costs, adaptation could shift the tolerable climate change upward owing to the possibility of offsetting at least a part of the negative impacts. For those approaches that look into the stabilization of CO_2/GHG concentrations or climate, adaptation has only indirect relevance. Similarly, the scope for adaptation within some global ecosystems or geophysical systems is unknown or assumed to be non-existent: for example, how could one help corals to adapt to changing water temperatures or fast-rising mean water levels; or could there be any geo-engineering option to sustain the thermohaline circulation despite higher temperature and freshwater inflow in the North Atlantic region? The climate impact response functions in the TWA incorporate only biophysical relationships of several human-managed sectors (agriculture, water supply) but they are void of information about adaptation.

The *decision-analytical frameworks* (DAFs) underlying the Article 2 assessments in our sampler are also diverse and they are not easy to cluster into traditional categories (see Toth, 2000). The first group includes policy evaluation models (see Weyant *et al.*, 1996). In the applications presented in the previous section, the original simulation models are supplied with additional constraints depicting the stabilization targets (smooth stabilization paths, Concentration Stabilization Pathways) or with some post-processing to select those model runs from a large ensemble that stay within the specified concentration, climate, or impact limits (Exploratory Modeling). Cost-minimizing stabilization is a classic cost-effectiveness decision framework while the Probabilistic Integrated Assessment is an innovative application of the cost–benefit framework. The Safe Landing and the Tolerable Windows approaches form a unique cluster because they incorporate elements of well-established techniques such as cost-effectiveness and cost–benefit, but they also add novel features,
such as the proxy-cost constraints in SLA and the relaxed cost–benefit framing in TWA.

In principle, most of the reviewed approaches and the associated models can be used to perform standard sensitivity analyses to explore the implications of any of the four main uncertainty categories defined in Section 30.2 Yet the *treatment of scientific and modeling uncertainties* (Types 1 and 2) leaves a lot to be desired in some of them, at least in their implementations to date. By definition, uncertainty is at the core of the Exploratory Modeling and the Probabilistic Integrated Assessment approaches to Article 2. Published results of the TWA/ICLIPS model present implications for the emissions corridors of uncertainties in some key model variables (e.g., climate sensitivity) and in the targets/costs decisions (acceptable impacts, tolerable mitigation costs). A recent extended version of the cost-minimizing stabilization approach moves from concentrations to temperature change limits, accounts for several GHGs, and adopts an innovative probabilistic framework (Richels *et al.*, 2007). The other approaches acknowledge uncertainties but do not detail their implications.

Concerning Type 3 uncertainties (about danger or target levels), all approaches can and most do undertake analyses to produce emissions pathways towards different stabilization or impact levels. For example, the most valuable insights from the cost-minimizing concentration stabilization exercises are the exponentially increasing mitigation costs as the target is gradually moved from 750 to 450 ppmv CO_2 concentration. Similarly, the Tolerable Windows analysis depicts drastically shrinking CO_2 emission corridors when the limits to the global share of nature reserves undergoing major ecosystem transformation (biome change) are gradually reduced from 50% to 30% (see Toth *et al.*, 2003b). In contrast, the implications of implementation-related uncertainties remain largely unexplored in all Article 2 assessments, although several frameworks could be extended in this direction, especially some cost-minimizing IAMs and the Probabilistic Integrated Assessment.

Embarking on an assessment of the implications of Article 2, one must walk a fine line between subjective judgements and objective analysis. The latter involves a number of subjective assessments of uncertain relationships, model formulations, and model parameters itself. Yet the really slippery ground of beliefs, perceptions, values, and claims is in the choices of what kind or magnitude of climate change or impacts are endurable and beyond which point they would become unbearable, for whom, where, when, and under what circumstances. The Probabilistic Integrated Assessment, the Exploratory Modeling, and the Safe Landing/Tolerable Windows approaches are plainly formulated to support the assessment of the implications of user-provided choices concerning dangerous climate change or intolerable impacts. They clearly disclose the boundary between choices and analysis. The smooth and cost-minimizing stabilization paths offer a range of options together with their implications and leave the choice implicitly to the users of the results. Finally, the Concentration Stabilization Pathways analysis takes some ecosystems and geophysical constraints and presupposes that valid social choices exist to avoid crossing them.

The nature of the recommendations for near-term policy (next 10–15 years) in the light of the four types of uncertainties is remarkably straightforward. For a given climate or impact constraint or probability threshold, most approaches prescribe one long-term emission path with distinctive features, most typically optimality in some sense (least-cost, robust, welfare-maximizing, etc.). There are two notable exceptions. The Safe Landing Analysis can check the feasibility of a near-term emission strategy with respect to a long-term climate stabilization constraint. The Tolerable Windows Approach provides a long-term emission corridor within which there is some flexibility to choose the near-term emission strategy albeit with implications for the range of permissible paths in subsequent decades.

The content of the near-term policy recommendations varies as well. For similar concentration or impact constraints, the cost-effectiveness frameworks allow the highest near-term emission paths while the impact-driven Concentration Stabilization Pathways involve the most stringent paths, because the former ones ignore the possibly larger damages associated with the "spiky" stabilization trajectories, especially if temporary overshooting of the ultimate target is allowed, whereas the latter is dominated by damage concerns. Since climate change is a global stock pollutant issue, some diversion from the calculated emission paths might be possible in both cases. For the same reason, near-term emissions are only modestly sensitive to all four types of uncertainties. The most fundamental exception is when targets for impact and/or for the associated temperature or concentration stabilization are low (about 500 ppmv CO_2-equivalent or below). In this case, there is little room for flexibility of any kind and rapid decline of emissions is prescribed by all models.

A substantial part of the variation in policy recommendations stems from the framing of the decision problem: based on prevailing social conditions (especially beliefs, values, preferences), the analysis intended to support decisionmaking can be formulated in various frameworks. As a general rule, the adopted frameworks need to reflect the social context and must be compatible with the decision criteria, public choice, etc., of the society and public policy they intend to serve. Given the diversity of views and values in any society, different DAFs (emphasizing environmental, ethical, or economic concerns) are used and they produce diverging results.

30.5 Summary and conclusions

Article 2 of the UNFCCC specifies the "dangerous anthropogenic interference" to be avoided in terms of GHG concentrations and mentions two impact sectors as examples of key impact concerns: food supply and natural ecosystems. Yet the most intensely scrutinized segment of the emissions-to-impacts chain has so far been the alternative emissions pathways stabilizing CO_2 or GHG concentrations at a given level. Following the WRE analysis, many other models adopted the cost-efficiency framework to calculate the costs of CO_2 stabilization. The central issue in the debate is technological development: its sources, processes, driving forces, timing, diffusion, and transfer. Its crucial importance is reinforced by the widely held view that a variety of technological innovations and improvements rather than profound value and lifestyle changes will provide the solution to the climate change problem.

Discussions and analyses of "dangerous" radiative forcing limits are rare and the studies conducted in the Energy Modeling Forum EMF-21 are probably the only notable exception. Political pronouncements regarding the maximum permissible increase in global mean temperature have been abundant for years but model-based climate change stabilization assessments are a relatively recent development. This is not surprising for two main reasons. The first reason is that the analysis requires prespecified relationships or probability density functions along the emissions–temperature sequence (see Richels et al., 2007). Alternatively, a dynamic GHG–climate module needs to be included into the integrated assessment model to perform the necessary accounting tasks (see Bruckner et al., 2003). Nonetheless, there is no way around the fact that uncertainties are prevalent. Moreover, looking at climate change (typically approximated by the increase in global mean temperature) is still far from impacts, the real reasons for concern about climate change. One of the major challenges for climate change science is to improve our knowledge about the dynamics of climate–impact relationships in different regions of the world. This will need to be complemented by integrating biophysical sensitivity and socio-economic vulnerability in human-managed impact sectors and by techniques that help to identify local or regional vulnerable spots.

The second reason for the difficulties in moving to the climate domain and beyond is that as uncertainties build up in moving from concentrations to radiative forcing to climate change, the stabilization in terms of temperature change becomes more precarious. Cascading uncertainties in the emissions-to-climate direction mean that a wide range of climate change may result from a single emission path. In the opposite direction they mean that a climate change limit may be attained by considerably different magnitudes and trajectories of GHG emissions. Analysts have come a long way in developing new concepts and techniques to handle uncertainties. Yet there is still a great need for creative improvements in measuring and understanding the implications of uncertainties for managing the climate risk in the near and the long term.

It has been repeatedly emphasized that the directive of Article 2 involves social choices based on ethical principles, value judgements, risk perceptions, and risk attitudes. In an early note on the "dangerous" interference, Moss (1995) indicates the need for regular interactions between the scientific and policy communities in order to resolve the challenge of Article 2. Several approaches in our sampler are suitable for supporting joint exploration of policy options by representatives of the two communities, such as the Safe Landing and Tolerable Windows approaches, the Exploratory Modeling and the Probabilistic Integrated Assessment, although different techniques and designs would be required to imbed them into a participatory exercise (see Toth and Hizsnyik, 2007). There are two basic strategies for scientists to follow. The first one is to work with policymakers in a participatory process either "online" by using models directly or with "canned" model runs to solicit choices, explore the implications, and modify the choices. The second option is the offline version in which analysts need to produce a large number of possible decisions, present the outcomes of all options, and sort them along different criteria so that policymakers can easily understand the implications of and the tradeoffs between different choices.

Some analytical frameworks emphasize economic efficiency (cost–benefit or cost-effectiveness analysis), others focus on environmental concerns (Concentration Stabilization Pathways), yet others combine various criteria (Safe Landing and Tolerable Windows Approach) or adopt one of these in a probabilistic structure (Robust Modeling and Probabilistic Integrated Assessment). The nature (specific optimal policies versus qualitative insights) and the content (fast emission reductions versus gradual departure from historical emissions trends) of messages from different frameworks diverge accordingly. Depending on the target, early and bold reductions emerge as the proposed strategy from ecology-oriented frameworks in contrast to the more delayed aggressive reductions recommended by economics-based frameworks, especially if technological development is explicitly represented in their model implementations.

The comparative analysis of the analytical frameworks and models adapted to provide solid scientific insights for policy decisions concerning Article 2 of the UNFCCC indicates that the conceptual and practical difficulties hidden in the notion of "dangerous anthropogenic interference" have triggered a great diversity of approaches. They emphasize different aspects of the stabilization dilemma and frame the problem according to difference decisionmaking paradigms. As a result, they produce useful insights into the tradeoffs, costs and risks of different strategies and contribute to solving the Article 2 puzzle. Yet this appraisal also reveals the need for more new approaches, innovative tools, and inspired modeling techniques that better represent the distinctive features of managing the risk of anthropogenic climate change.

Acknowledgements

The author is indebted to Charles Kolstad and the referees for their recommendations for improving the draft version. Parts of this material were presented at the IPCC Expert Meeting on the science related to UNFCCC Article 2 including key vulnerabilities, in Buenos Aires (May 18–20, 2004), and at the Energy Modeling Forum Summer Workshop on Climate Change Impacts and Integrated Assessments in Snowmass, Colorado (July 27– August 5, 2004). The author is grateful to participants in these events for their comments but remains solely responsible for all remaining errors. The views expressed in this paper are those of the author and do not necessarily represent the views of the IAEA or its Member States.

References

Alcamo, J. and Kreileman, E. (1996). Emission scenarios and global climate protection. *Global Environmental Change* **6**, 305–334.

Bruckner, T., Hoos, G., Füssel., H.-M. and Hasselmann, K. (2003). Climate system modeling in the framework of the tolerable windows approach: the ICLIPS Climate model. *Climatic Change* **56**, 119–137.

Desai, S., Adger, W. N., Hulme, M. *et al.* (2004). Defining and experiencing dangerous climate change. *Climatic Change* **64**, 11–25.

Dowlatabadi, H. and Morgan, M. G. (1993). A model framework for integrated studies of the climate problem. *Energy Policy* **21**, 209–221.

EU (European Union) (1996). *Communication on Community Strategy on Climate Change. Council Conclusions*. Brussels, Belgium: European Commission.

Füssel, H.-M., Toth, F. L., van Minnen, J. G. and Kaspar, F. (2003). Climate impact response functions as impact tools in the tolerable windows approach. *Climatic Change* **56**, 91–117.

IPCC (1994a). *Radiative Forcing of Climate Change and an Evaluation of the IPCC IS92 Emission Scenarios. Special Report of the Intergovernmental Panel on Climate Change*, ed. J. T. Houghton, L. G. Meira Filho, J. Bruce *et al.* Cambridge: Cambridge University Press.

IPCC (1994b). *Intergovernmental Panel on Climate Change Special Workshop: Article 2 of the United Nations Framework Convention on Climate Change*. Geneva, Switzerland: IPCC.

IPCC (1995). *Intergovernmental Panel on Climate Change Second Assessment Synthesis of Scientific-Technical Information Relevant to Interpreting Article 2 of the UN Framework Convention on Climate Change*. Geneva, Switzerland: IPCC.

IPCC (1996). *Climate Change 1995: The Science of Climate Change. Contribution of Working Group I to the Second Assessment Report of the Intergovernmental Panel on Climate Change*, ed. J. T. Houghton, L. G. Meira Filho, B. A. Callender *et al.* Cambridge, UK: Cambridge University Press.

IPCC (2000). *Emissions Scenarios: A Special Report of Working Group III of the Intergovernmental Panel on Climate Change*, ed. N. Nakicenovic and R. Swart. Cambridge, UK: Cambridge University Press.

IPCC (2001a). *Synthesis Report: Third Assessment Report of the Intergovernmental Panel on Climate Change*, ed. R. T. Watson and the core Writing Team. Cambridge, UK: Cambridge University Press.

IPCC (2001b). *Climate Change 2001: Mitigation. Contribution of Working Group III to the Third Assessment Report of*

Intergovernmental Panel on Climate Change, ed. B. Metz, O. Davidson, R. Swart and J. Pan. Cambridge: Cambridge University Press.

IPCC (2001c). *Climate Change 2001: Impacts, Adaptation, and Vulnerability. Contribution of Working Group II to the Third Assessment Report of the Intergovernmental Panel on Climate Change*, ed. J. J. McCarthy, O. F. Canziani, N. A. Leary, D. J. Dokken and K. S. White. Cambridge: Cambridge University Press.

Kann, A. and Weyant, J. P. (2000). Approaches for performing uncertainty analysis in large-scale energy/economic policy models. *Environmental Modeling & Assessment* **5**, 29–46.

Kreileman, G. J. J. and Berk, M. M. (1997). *The Safe Landing Analysis: Users Manual.* RIVM Report No 481508003. Bilthoven, The Netherlands: RIVM.

Lempert, R. J. and Schlesinger, M. E. (2000). Robust strategies for abating climate change. *Climatic Change* **45**, 387–401.

Lempert, R. J., Schlesinger, M. E., Bankes, S. C. and Andronova, N. G. (2000). The impacts of variability on near-term policy choices and the value of information. *Climatic Change* **45**, 129–161.

Mastandrea, M. D. and Schneider, S. H. (2004). Probabilistic integrated assessment of "dangerous" climate change. *Science* **304**, 571–575.

Moss, R. H. (1995). Avoiding 'dangerous' interference in the climate system. The roles of values, science and policy. *Global Environmental Change* **5**, 3–6.

Nordhaus, W. D. (1992). An optimal transition path for controlling greenhouse gases. *Science* **258**, 1315–1319.

O'Neill, B. C. and Oppenheimer, M. (2002). Dangerous climate impacts and the Kyoto Protocol. *Science* **296**, 1971–1972.

O'Neill, B. C. and Oppenheimer, M. (2004). Climate change impacts sensitive to path stabilization. *Proceedings of the National Academy of Sciences* **47**, 16411–16416.

Parry, M. L., Carter, T. R. and Hulme, M. (1996). What is a dangerous climate change? *Global Environmental Change* **6**, 1–6.

Peterson, S. (2004). *The Contribution of Economics to the Analysis of Climate Change and Uncertainty: A Survey of Approaches and Findings.* Kiel Working Paper No. 1212. Kiel, Germany: Kiel Institute for World Economics.

Pizer, W. (1999). The optimal choice of climate change policy in the presence of uncertainty. *Resource and Energy Economics* **21**, 255–287.

Richels, R. G., Manne, A. S. and Wigley, T. M. L. (2007). Moving beyond concentrations: the challenge of limiting temperature change. *Climatic Change* (forthcoming).

Schneider, S. H. (2001). What is 'dangerous' climate change? *Nature* **411**, 17–19.

Shackley, S. and Wynne, B. (1995). Integrating knowledges for climate change: pyramids, nets and uncertainties. *Global Environmental Change* **5**, 113–126.

Swart, R. M. and Vellinga, P. (1994). The ultimate objective of the Framework Convention on Climate Change requires a new approach in climate change research. *Climatic Change* **30**, 1–7.

Swart, R., Berk, M., Janssen, M., Kreileman, E. and Leemans, R. (1998). The safe landing approach: risks and trade-offs in climate change. In *Global Change Scenarios of the 21st Century*, ed. J. Alcamo, R. Leemans, and E. Kreileman. Oxford: Elsevier, pp. 193–218.

Toth, F. L. (2000). Decision Analysis Frameworks in the Third Assessment Report. In *Guidance Papers on the Cross Cutting Issues of the Third Assessment Report of the Intergovernmental Panel on Climate Change*, ed. R. Pachauri, T. Taniguchi and K. Tanaka. Geneva, Switzerland: IPCC, pp. 53–68.

Toth, F. L. (2003). Climate policy in light of climate science: the ICLIPS project. *Climatic Change* **56**, 7–36.

Toth, F. L. and Hizsnyik, E. (2007). Managing the inconceivable: participatory assessments of impacts and responses to extreme climate change. *Climatic Change*, forthcoming.

Toth, F. L., Bruckner, T., Füssel, H.-M. *et al.* (1997). The tolerable windows approach to integrated assessments. In *Climate Change and Integrated Assessment Models (IAMs): Bridging the Gaps*, ed. O. K. Cameron, K. Fukuwatari and T. Morita. Tsukuba, Japan: Center for Global Environmental Research, National Institute for Environmental Studies, pp. 401–430.

Toth, F. L., Cramer, W. and Hizsnyik, E. (2000). Climate impact response functions: an introduction. *Climatic Change* **46**, 225–246.

Toth, F. L., Bruckner, T., Füssel, H.-M. *et al.* (2002). Exploring options for global climate policy in an inverse integrated assessment framework. *Environment* **44** (No. 5, June), 23–34.

Toth, F. L., Bruckner, T., Füssel, H.-M., Leimbach, M. and Petschel-Held, G. (2003a). Integrated assessment of long-term climate policies: Part 1. Model presentation. *Climatic Change* **56**, 37–56.

Toth, F. L., Bruckner, T., Füssel, H.-M., Leimbach, M. and Petschel-Held, G. (2003b). Integrated assessment of long-term climate policies: Part 2. Model results and uncertainty analysis. *Climatic Change* **56**, 57–72.

van Asselt, M. and Rotmans, J. (1996). Uncertainty in perspective. *Global Environmental Change* **6**, 121–157.

van Asselt, M. and Rotmans, J. (2002). Uncertainty in integrated assessment modelling: from positivism to pluralism. *Climatic Change* **54**, 75–105.

van Vuuren, D. P. and de Vries, H. J. M. (2000). *Mitigation Scenarios in a World Oriented at Sustainable Development: The Role of Technology, Efficiency, and Timing.* RIVM report 490200 001. Bilthoven, The Netherlands: RIVM.

van Vuuren, D. P., den Elzen, M. G. J., Berk, M. M. *et al.* (2003). *Regional Costs and Benefits of Alternative Post-Kyoto Climate Regimes.* RIVM Report 728001025/2003. Bilthoven, The Netherlands: RIVM.

Visser, K., Folkert, R. J. M., Hoekstra, J. and De Wolff, J. J. (2000). Identifying key sources of uncertainty in climate change projections. *Climatic Change* **45**, 421–457.

WCP (World Climate Programme) (1988). *Developing Policies for Responding to Climatic Change. A Summary of the Discussions and Recommendations of the Workshops held in Villach,*

Austria, September–October 1987, and Bellagio, November 1987. Stockholm, Sweden: WMO, UNEP, Beijer Institute.

Weyant, J. (lead author), Davidson, O., Dowlatabadi, H. *et al.* (1996). Integrated assessment of climate change: an overview and comparison of approaches and results. In *Climate Change 1995: Economic and Social Dimensions of Climate Change. Scientific-Technical Analysis*, ed. J.P. Bruce, H. Lee and E.F. Haites. Cambridge, UK: Cambridge University Press, pp. 367–396.

Wigley, T.M.L., Richels, R. and Edmonds, J.A. (1996). Economic and environmental choices in the stabilization of atmospheric CO_2 concentrations. *Nature* **379**, 240–243.

31

Whither integrated assessment? Reflections from the leading edge

Hugh M. Pitcher, Gerald M. Stokes and Elizabeth L. Malone

31.1 Introduction

After 10 years of Climate Change Impacts and Integrated Assessment (CCI & IA) workshops under the aegis of the Energy Modeling Forum (EMF), it is appropriate to consider what progress has been made and what additional tasks confront us. The breadth and scope of the papers in this volume provide ample evidence of the progress. In this paper, we consider the additional tasks before us as a community interested in applying integrated assessment to a wide range of research and policy issues.

Integrated assessment (IA) has arisen and continues to grow in significance because it provides insights and understanding not available from research and analysis conducted from the perspective of individual disciplines. Increasingly, the questions arising from such integrated consideration of issues pose questions new to the respective disciplines, thus driving disciplinary research also.

The particular approach to integrated analysis of historic interest to EMF emphasizes solutions to the problems that have implications for both program direction and major resource allocation. Because of this, individuals who have, or represent, major political and economic interests in the outcome subject the results and components of the models to an unprecedented level of scrutiny. This scrutiny will affect not only model components but the structure of feedbacks and the tools for measuring impacts as well.

Increased scrutiny comes at a time when the demand is growing for the kind of analyses IA can provide. Both major components of the United States government's climate program, the Climate Change Science Program (CCSP) and the Climate Change Technology Program (CCTP), emphasize the need for improved decision support systems to assist in managing the risks associated with anthropogenic climate change. The completed Millennium Ecosystem Assessment used IA tools to create the base scenarios, assess the potential impacts of various scenarios on climate, and understand the impact of the associated climatic changes on ecosystems. The Intergovernmental Panel on Climate Change (IPCC) is likely to make increasing use of IA tools.

The types of uses for IA tools are also expanding. They may be used in an attempt to shorten the five-year-plus time-frame that is now entailed in running a series of scenarios through large-scale climate models and then using the climate models in impacts research. Moreover, IA may offer the only way to allow the full set of detailed results to be used to establish the level of control at which benefits and costs of mitigation policies might be roughly in balance. Finally, IA offers one of the few practical approaches to many more immediate issues concerning local air, water, and solid waste pollution that successful economic development will force many countries to confront over and over as the century progresses.

We begin with a description of the basic structure and function of modern IA models. Then in Section 31.3, we take a critical view of the state of the art of IA modeling systems and their suitability for the emerging problems that IA will be asked to address. In Section 31.4 we suggest steps to improve IA models to meet the new demands for their application. Section 31.5 calls for the establishment of a program that will develop IA community standards, including base scenario data; serve as a repository of model elements and results;

Human-induced Climate Change: An Interdisciplinary Assessment, ed. Michael Schlesinger, Haroon Kheshgi, Joel Smith, Francisco de la Chesnaye, John M. Reilly, Tom Wilson and Charles Kolstad. Published by Cambridge University Press. © Cambridge University Press 2007.

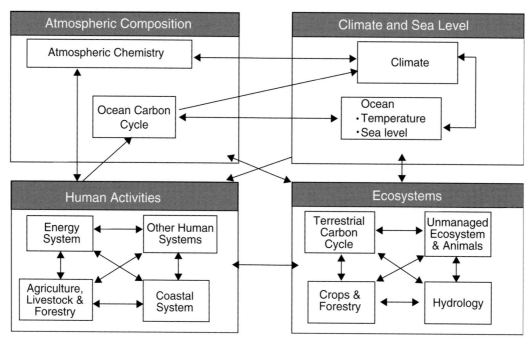

Figure 31.1 A schematic diagram of the major components of a climate-oriented integrated assessment model.

support international processes such as those of the IPCC; and facilitate modeling skill and career development.

31.2 The IA model for climate: a stylized description

Even consideration of the climate problem without the oft-mentioned issues of development, local pollution and disease reveals an extremely complex set of concerns. On the one hand, complexity grows as we achieve greater understanding of the climate system, and the human systems that are both implicated in changing the climate and affected by it. On the other hand, the various relationships among systems force us to consider things at deeper levels of complexity. Over the past decade, IA models, as a new class of tools, have been applied to climate change research and policy to address this complexity. These tools are attempts to provide a comprehensive view of the climate problem to support policy and decisionmaking.

An IA model potentially involves many complex systems and their interrelationships. Conceptually, integrated assessment modeling for climate change attempts to bring together a representation of all relevant components of the climate system into a single modeling framework, schematically shown in Figure 31.1 (colour plates section). The details of this figure are less important than the inclusion of the broad sweep of systems from the various components of the carbon cycle to the global energy system.

There are several major points to be made about Figure 31.1.

1. The system boundaries are unclear. While most IA models include the broad functions depicted in Figure 31.1, not all integrated assessment models explicitly incorporate all of the systems and connections shown. Neither are the items shown in the figure a complete set of possible systems and connections.

2. The appropriate level of complexity of the components is unknown. IA models differ in the complexity of their treatment of the various components of their modeling framework. Some have very complicated atmospheric and carbon cycle components; others emphasize features such as the technology in the global energy system or various components of the "other human systems" like health.

3. There are multiple ways to treat the uncertainty within the system. IA models can be analytic systems used in schemes such as the one depicted in Figure 31.2 (color plates) where key elements of the future are exogenously specified through scenarios. Here the analytic process is one of trying to understand the consequence of different policy pre-scriptions or options. It is this framework that makes IA models "decision support" systems. Alternatively, IA models can undergo full-scale Monte-Carlo-based uncertainty analysis, giving a sense of system limits and behavior. The likelihood of different scenarios and potential ways in which the approaches might be combined remain major topics of debate within the IA community.

4. IA systems must be able to capture the tradeoffs that real policymakers have to make and the finiteness of resources they control, making economics a critical component of an IA system. An important part of this is the representation and clearing of markets such as energy demand, energy supply, and land. The global nature of the climate system requires the IA system results to have a global

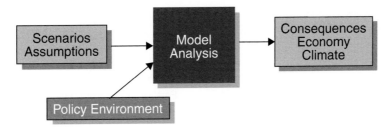

Figure 31.2 A schematic diagram of the elements of an IA model-based policy process.

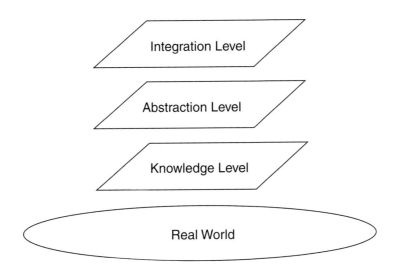

Figure 31.3 Representation of the relationship among the various levels at which relevant systems can be modeled.

representation of emissions and economic activity, while the spatial character of economic decisions requires a country or regional representation of resource allocations.

One of the most striking things about IA modeling is its relatively low computational intensity. Climate models run on the largest available supercomputers; IA models generally run on desktop computers. Considering that climate is just one component of the IA modeling framework shown in Figure 31.1, this may seem incongruous. But the boxes and arrows in Figure 31.1 are themselves abstractions of the very complex processes. Figure 31.3 shows the processes that make the framework in Figure 31.1 possible.

IA modeling involves a process of abstracting "reduced form" or simple structural model representations of complex processes. As shown in Figure 31.3, there are four levels in the picture decreasing in complexity as one moves up from the "real world" at the bottom to what could be called the "integration level" at the top. The integration level corresponds to Figure 31.1 and is explicitly the level at which the individual components of the system are connected. The first level above the real world is the "knowledge level". This is the domain of disciplinary experts who collect data about and construct detailed models of the real world. This knowledge is too extensive and the models too detailed to be used at the integration level, so the information at the knowledge level goes through a further abstraction process that connects it to the integration level. In reality, there may be several knowledge and abstraction levels between the real world and the eventual IA representation at the integration level.

The abstraction process needs to accomplish several important tasks. First, it must facilitate the connection among the various components by ensuring that the inputs and outputs to various top-level processes are both appropriate and matched. Next, it needs to select and model the qualitatively important determinants of the important processes, particularly as they relate to the problems being analyzed at the integration level. An important characteristic of the abstraction process illustrated by the layers in Figure 31.3 is that one such column of layers underpins each of the sub-models (boxes) in Figure 31.1.

It is important to understand behavior in the processes below the levels at which connection and integration occur. This is one of the areas where the "art" of model building becomes crucial. As noted, the level of detail in many process or "bottom-up" models is both conceptually and computationally too extensive to be fully incorporated into the integration tool, necessitating a careful process of reducing complex processes and data so that only important behavioral processes and data are represented in the IA tool. Because adding detail in the IA tools means a much greater increase in computational intensity – the "curse of dimensionality" – processes and feedbacks must be limited.

Important approaches to reduce the dimensionality of detailed process level tools are:

- Simplification of the model structure: in this approach, only crucial behavioral elements are included within the IA tool.
- Parameterizations: a parameterization is an attempt to capture the essence of a real-world process in a simple form, often a simple number or relationship, without having to model it in detail. For example, the statistical details of how individuals might react to the price of a good or service can be summarized by the elasticity of demand, which describes how much aggregate demand changes in response to a change in price.
- Reduced form models: in this case the knowledge in a complicated model is abstracted by creating a representation of the results of many modeling runs and using statistical methods to summarize the relationships among key variables.
- Calibration: here a simple model of the relationship among key variables is postulated for use at the integration level. Adjustable parameters, which capture the strength of the key relationships, are determined statistically from past data and assumed to be constant as a function of time.

None of the abstraction processes is either simple or suitable for automation. For example, reduced form models are problematic because the time frames are long enough and the potential changes in key variables are large enough that the implicit assumptions embodied in "reduced form" models may not continue to hold. Such models need to be examined with care to ensure that major constraints implicit in historical data, such as physical limits on the efficiency of energy conversion, continue to hold in projections.

The need for each element in the IA structure depicted in Figure 31.1 to have a supporting analytic and empirical structure illustrates the key point that an IA model is not just a collection of models but represents a significant intellectual reworking of the underlying subject matter.

Part of the complexity of IA tools is that they must, by the nature of the problem, incorporate multiple flows: not only the economic flows of value, but also the associated physical flows and the impacts that these flows have on the natural systems within which human activity takes place. Physical flows are what matter to the atmosphere and climate – not dollars or euros or rupees or yen – and they provide very real and important constraints on policy options.

IA models also face a variety of scales. By this we mean that there are significant physical and economic processes that take place on wide ranges of spatial and temporal scales – global, regional, and local; and hours, days, years, and decades. Significant connections among processes and associated feedbacks can exist at very different scales.

IA models track physical changes in the environment arising from human activity (and even some natural processes). For example, the conversion of emissions into global concentrations of greenhouse gases in the atmosphere is a complex product of atmospheric chemistry and flows of gases into the atmosphere and thence into and through the biosphere and the ocean. The changing concentration of greenhouse gases in the atmosphere creates feedbacks through the climate system on water vapor. And while still not completely understood, these feedbacks create the largest (and still poorly understood) components of the resulting change in global, regional, and local temperature. Although it is widely assumed that in a warmer world the hydrological cycle will be faster, it is not yet understood whether increases in precipitation will outpace increases in evapotranspiration, leaving us with highly uncertain estimates of the potential net effect of climate change on stream flow. The combination of human activity and changes in temperature and precipitation regimes has the potential to create large changes in land use and land cover, further complicating the complex set of relationships between human activity and climatic conditions.

Adaptation to these changed conditions may be a surmountable challenge for developed economies, but may severely test developing economies who cannot effectively manage even the current problems posed by climate.

Uncertainty is a critical factor that affects both the structure of the IA tool and the way in which it is used. Structural uncertainty arises from an increasingly narrow set of considerations, beginning with issues of system boundaries: what processes should be included within the IA tool, what should be treated as exogenous inputs and what should be ignored? The past decade has seen a continual expansion of the processes included within the boundary, but this is still a critical source of uncertainty. Another level of uncertainty emerges in considering the relationships among the various modules. Future treatment of demographics, now an external input into the IA tools and a likely area where the system will expand, illustrates both points. Including first-order demographic effects within the modeling framework is not difficult, as the age cohort system is computationally straightforward. What *is* difficult is creating a demographic module in which the attributes of a population can respond to economic, environmental, and policy changes. This is in fact a significant research problem. Further, understanding how to accommodate the obvious pressures to change policy as a result of the global decline in birth rates and deciding how socioeconomic forces might modify total completed fertility raise even more difficult questions.

The second set of more conventional measures of uncertainty focuses more on quantitative uncertainty arising from parameter variance and uncertainty, as well as data uncertainty associated, for example, with calibration. Other discussions of uncertainty have focused on these issues, such as the range of climate sensitivity (Andronova and Schlesinger, 2001) and the range of estimates for base period emissions in the Special Report on Emissions Scenarios (SRES) (IPCC, 2000).

The third aspect of uncertainty relates to the treatment in models of the evolution of human activity and physical systems over time. The usual approach has been to develop a scenario or a set of scenarios. This approach rarely tries to use some systematic approach to potential sources of variation and as a result experiences considerable criticism, normally captured in a demand for scenario likelihoods or probability (Schneider, 2003). An alternative approach has been to use Monte Carlo techniques to assess the range of the emissions (Webster *et al.*, 2003). What appears to be developing at the moment is a hybrid approach, which embeds scenarios, with their level of detail providing some basis for "traction" with decisionmakers, within a range developed using Monte Carlo analysis. The climate system is equally subject to this set of concerns, since such issues as the likelihood of changes in major droughts still lie beyond the skill of these systems to predict. Paleoclimatic data make it abundantly clear that such changes occur, but our ability to forecast them is highly limited.

These problems all become more focused as one considers the difficulties in providing support to decisionmakers. How does one provide a systematic array of supportable evidence for decisions about the direction of technical change or the level of fossil fuel prices given the long list of modeling, estimation, and data uncertainties described above? IA models, given their computational efficiency, make the best tools for evaluating the implications of uncertainty across the domain of issues they consider, which can provide useful support both for policy and research planning.

31.3 Where we are now: a critical view

Integrated assessment has made great progress during the decade of CCI&IA workshops. The range of economic and physical processes included in the models and especially in mitigation options has increased substantially. Models now include a full range of greenhouse gases and are beginning to explore aerosol emissions and controls. Options for controlling non-CO_2 gases now include marginal abatement cost curves for all major sources of the Kyoto-cited gases, and work is beginning on control options for aerosols. Strategies to manage carbon have been widely extended and now routinely include the ability to include carbon capture and geologic storage as well as land management and afforestation options. Work is under way to improve the land-based component of the carbon cycle, and the representation of the oceanic carbon cycle is also being examined. The climate representation has been systematically improved. Simple climate models are now routinely available to the community (such as COSMIC and MAGICC), and several models include more elaborate representations of the climate system.

The number of IA models has increased dramatically, as has the range of regions actively involved in creating and using models. Several dozen models now participate in EMF exercises, and the list now includes all of the Annex 1 countries. Representatives of key developing economies also participate in a range of modeling and policy activities. A major area where we might anticipate improvement over the next decade is in the involvement of modelers from non-Annex 1 countries, who now participate largely through joint projects with developed countries. The skill and talent is clearly present to develop and exercise appropriate models within these countries, but resources and demands of other pressing problems remain significant limitations.

Some models now extend the climate results to include impacts. The Asia–Pacific Integrated Model (AIM) group has a module to explore the impact of climate change on malaria, and some explore the implications of climate change for natural systems.

However, many areas remain where IA models need substantial improvement. Challenges to the IA community include adaptation, a fuller representation of cost, better representation of policies, endogenous technical change, land use, water, and ecosystems. To capture the relevant behavioral detail, the models are likely to need finer spatial and timescales than the scales used in the other major components of typical IA systems. This scaling problem exists, for example, for almost all of the major parts of the ecosystem box of Figure 31.1. The issue is relevant both for creating estimates of the economic, social, atmospheric, and climate conditions under which, for example, an analysis of agricultural production occurs, and for the related process of aggregating the results of an agricultural analysis back up to the level at which economic and climate systems operate.

This process of relating different scales is complicated because the systems, while linear under small perturbations, may well not be linear under larger perturbations. For example, above a certain temperature, crops stop creating biomass through the photosynthetic pathway and switch to a photo-respiration pathway to preserve chlorophyll. Soil moisture, arguably the major determinant of net physical productivity, is determined by a complex mix of precipitation, relative humidity, cloud cover, and wind speed at scales far below those at which climate models run. At a somewhat larger scale, stream flow is a non-linear result of precipitation and evapo-transpiration. This implies that even if the extremely uncertain changes in precipitation attributable to climate change could be pinned down, we still could not be sure of what would happen to available water supplies. On top of this, uncertainties in land-use patterns and climate projections balloon at the scales used in process-based crop productivity models.

At a still larger scale, climate models are thought to have little skill at predicting multi-year changes in weather patterns that result in major droughts. For the oceans, the potential for and characteristics of changes in the thermohaline circulation are debated intensely. All of these factors argue that a full integration of ecosystems into IA tools will require a major research effort aimed at developing tools that incorporate scale effects across orders-of-magnitude differences in time and physical scales.

Books such as Jared Diamond's *Guns, Germs and Steel* or *Collapse* make it clear that even a full understanding of ecosystem effects is not sufficient. The next step is to articulate how ecosystem changes affect health status, agricultural crops, and water availability, and then how these effects impact society's ability to create human, social, and physical capital. This capacity gives a society its ability to adapt to climatic variation and its resilience to natural disasters. The complex set of relationships among climatic, geographic, and natural resources and the economic capacity of a society is not yet completely understood. The answer to the critical question of how much climate change societies can tolerate must be embedded in this larger understanding.

Clearly, significant changes must be made in the treatment of demographics and institutions in IA models. The economic basis of development can no longer simply be the number of people. Their educational and health status matters, as do the institutions necessary to provide the services that improve both these critical human characteristics.

Further, the recent debate over the appropriate measure of economic output has raised the issue of relative prices and how they change as countries develop. The pattern of development is also crucial and affects the ability of regions and countries to focus on climate adaptation and mitigation, in addition to other pressing concerns more immediately related to development. This entangles the economic models in the ongoing discussion about the major determinants of economic growth, such as the debate about whether institutions or human capital are the primary driver of development.

And finally, there is the common assumption that there is some high-level decisionmaker who can implement policies as if by magic, whereas in the real world we are faced with multiple decisionmakers with multiple criteria, leaving us to wonder how our policies will be implemented and with what effectiveness. IA models have clear relevance in assessing alternative policies both as to stringency and composition. These tools can also help prioritize the allocation of research and development on the broad range of issues, including science, vulnerability, and adaptation as well as mitigation options. To fully realize this potential, however, improvements in the IA tools are clearly warranted.

Decisions about the size and allocation of research and development expenditures cannot wait for the ideal model to be created. What is needed is a set of practical steps that can improve the tools on relatively short timescales (US CCSP, 2003). It would be nice to be able to predict how efforts to improve various components of the models would turn out. However, setting general goals such as reducing the overall uncertainty of the system (which, given the history of the climate problem, is likely to be slow) does not appear to be the appropriate way to approach the problem. It is probably better to systematically identify critical issues and focus research on the associated questions and model improvements. Given resource constraints and the already heavy demands on research teams from national and international programs, care in choosing new activities is required, suggesting that a deeper assessment of the IA tools is required than can be attained in this paper.

One motivation for such an assessment is the major role of IA models in creating the current generation of IPCC scenarios as reported in SRES. There are some areas where better IA tools could substantially improve the scenario process. Whether this would change the estimated range of emissions or the conditions under which adaptation, mitigation, and impacts would occur is a question whose answer must await those improvements. A persistent problem faced in scenario development is the assumed independence of the demographic and socio-economic assumptions alluded to previously. Since socio-economic conditions clearly affect births, deaths, and migration, the assumption can be problematic. For example, the projected path of total completed fertility in the scenarios used in SRES is substantially higher for all three population scenarios than the estimated levels were. This discrepancy was evident even before the end of the SRES process. Although undertaking a projection of total completed fertility is not for the faint of heart, an examination of the underlying trends for at least a couple of the structural models should help to explain why there should be a noticeable break in the trend at the point where projections start or an explanation for why the trends should be expected to continue. The same can be expected to be the case for life expectancy. Perhaps an even harder question is what will happen to life expectancy in the long run. Currently there seems to be substantial reason to believe that it will take a major medical breakthrough to move life expectancy beyond the current upper bound of 85. Migration is a much harder problem again, as it clearly depends on political decisions as well as underlying trends in socio-economic conditions. But it could have major impacts on population levels in at least some developed countries.

31.4 What do we need to do to move ahead?

In this part of the paper we begin the discussion of what is needed in order to create an improved set of IA tools. One goal must be to increase confidence in their results. This requires good documentation and a clear set of runs whose full results are available for review and assessment. A second goal is to critically review model capacity in the light of the questions that form important topics for analysis by IA models. A third is to develop a process for creating access to model documentation and results of experiments. The idea is to build on the EMF process to enhance transparency as well as model capacity.

One way to enhance model capacity, and perceived reliability and relevance, is to extend the current EMF process. This essentially consists of a number of parallel processes located in major centers around the world. These models share little in the way of components and produce results that are difficult to compare because of differences in baselines, in regional composition, in definitions of units, in projected non-intervention scenarios, and so on.

One major step forward would be the *development of a consistent set of base period data*. This would involve working with the major sources of primary data, such as the Global Trade Analysis Project (GTAP); the United Nations statistical groups, including economic and population data; the International Energy Agency (IEA) energy data; the Carbon Dioxide Information Analysis Center (CDIAC) at Oak Ridge National Laboratory; the inventory data on greenhouse gas emissions, and certainly other sources. It is essential to understand not only the data but also their accuracy and definitional limitations. Providing a common data set at a country level would help not only the IA modeling teams but also those working to understand adaptation, mitigation, and impacts. Getting the data right and keeping it up to date will be a really large-scale task, probably best carried out under the aegis of the United Nations or other major statistical agency. Once the data is collected, quality controlled, and documented it then needs to be made available in electronic form. The terms under which this is done also need to be addressed, as the cost of acquiring data from multiple sources can be a major component of the expense of developing and maintaining a model.

A second task required is to *document the models and make this documentation readily available*. Documentation needs to include the theoretical underpinnings of the model as well as the implementation. Doing this for a full-scale IA model is a daunting task but will be increasingly important if the IA models are to fulfill their potential, and help provide insight into the potential solutions to the broad list of economic and environmental programs that society will face in the twenty-first century. As with the input data, the model documentation needs to be made available electronically to a wide audience.

A third task is to *assess the performance of the models*. This work should include diagnostic intercomparisons and other tools designed to aid the development of trust in the model results and the process of improving the models. Both are critical to fulfilling the need for tools to help decisionmakers understand the complex interplay of decisions and results necessary to the creation of effective policies that achieve intended targets and avoid unanticipated results.

A fourth task is to *expand the range of the models*, both in the detail of the processes that are now represented, and in new areas of capability that emerge during the course of analysis work and monitoring of the real world. A list of the needed additional capability in IA models is already enough to be well beyond the capacity of any single individual and any short-term project.

At the same time, there is a continued need to explore the impact of different approaches to such problems as land use, economic growth and mitigation strategies. This suggests that a fifth task would be to *develop a common set of interfaces among model components*. This would serve the community well by aiding in the design of new modules and in the creation of tools that build on the community efforts in efficient ways. Such a set of interfaces would allow a mix of parallel and common efforts that could serve the community well.

A sixth task would be to *create community modules that meet accepted documentation and validation standards*, thus easing the burden of completing full models, while still allowing experimentation with new algorithms and extended model capacity. It would economize on the resources needed to create a set of tools that can be a significant resource for understanding and decisionmaking in what is sure to be a rather turbulent century.

The issue is then how to create the resources that are necessary to support what is a major set of data, modeling, and analysis tasks. Indefinite expansion of a series of fully parallel efforts does not appear efficient given the common elements of data and perhaps some model elements.

In the climate modeling community, the provision of data, design and monitoring of experiments, and the provision of software reporting tools has been assisted by PCMDI (Program for Climate Model Diagnosis and Intercomparison). This group has facilitated the organization of a number of intercomparisons, providing software to allow standardized reporting, tools for doing the intercomparison, and storage facilities for the model results. PCMDI also participates in the process of model development, providing model parameterization testbeds and tools for detection and attribution of climate change. Currently, PCMDI is also providing data storage and access for a wide variety of coupled ocean–atmospheric runs being made in support of the IPCC's Fourth Assessment.

A comparable organization, building on the PCMDI experience as well as the EMF capacity, would substantially enhance the ability of the IA community to build better tools and create the credibility, through model experiments and assessments, that is essential if the tools are to fulfill their potential to help solve the difficult set of issues confronting humankind during the twenty-first century.

It is, however, not a given that the desire to create such an organization will translate into an effective organization. Participation in PCMDI experiments is resource-intensive, as it involves setting up experiments, significant machine time to run the experiments, and further staff and machine time to verify the results and transform them into a format acceptable to PCMDI. Reading the output and format requirements for the IPCC experiments should readily convince any skeptic of the level of effort involved. So it is essential that an IAMDI process be seen as providing real value to the participants. Today, after 15 years, participation in PCMDI experiments is seen as a critical aspect of being a serious climate model, and all major climate model groups participate. This outcome was not a sure thing and required substantial resources, care in the design of experiments, and a community committed to the testing and improvement of their tools.

The impact of the work of PCMDI can be seen on its influence on the International Geosphere–Biosphere Program (IGBP) and its Global Analysis, Integration and Modeling (GAIM) component. GAIM has built on the original sets of intercomparisons at PCMDI and extended them to paleo-climatological models and more recently to those models that now integrate the carbon cycle with the atmosphere and ocean in a single modeling framework.

Some critical questions confront the IA community if there is a decision to create an IAMDI process. First, it will be essential to establish access to a substantial resource base. This will undoubtedly have to be an iterative process, building on some initial successes funded in a temporary manner. Critical to the success of such an initial step will be the design of some good experiments supported by a transparent, community-supported protocol. These experiments must be seen to illuminate the behavior of the models as well as provide data that would allow some verification or comparison with real-world, historical data. In order to allow such comparisons, modelers would need to start the models at a point in time early enough to allow real validation of various key concepts, including emissions estimates as a function of population, income per capita, energy consumption per unit of income, and carbon per unit of energy – in other words, the elements of the Kaya identity.

Thus, one task of an IAMDI process would be to change the current practice of revising the input data sets to reflect the most recent available data. Instead, IAMDI would encourage modelers to create historical outcomes, including time series estimates of key concepts that can be compared to actual data, and supply historical data in a format that would facilitate such comparisons. Since few of the models run on annual time steps or include any representation of the business cycle, figuring out how best to make comparisons between model results and historical outcomes will be difficult. Pursuing this will require substantial resources, persuasion, and the development of a community spirit that accepts feedback in the spirit of seeking to do better, rather than judgements that might affect funding or a perception of merit. The current atmosphere of intense scrutiny by climate skeptics only makes the task that much more difficult. Yet the task seems essential if the models are to achieve a level of acceptance that will support their use in decisions with real resource implications.

A second set of model outputs concerns the consequences of emissions. Here issues of the behavior of the carbon cycle lead to conclusions about the emissions trajectory that will achieve stabilization in the atmosphere at different levels of concentration. This involves the behavior of natural systems, changes in land use resulting from human activities, and feedbacks, as well as interactions arising from the combination of climate change and human activities. Substantial uncertainties in both the natural and human systems are involved in the carbon cycle, so historical data provide only loose constraints on model behavior. One possible experiment here would be to see where the models convert land to human use and compare this to what is known about historical patterns of land-use conversion. This will be a difficult pattern to model well, as institutional decisions have major impacts on where and what kind of land-use conversion takes place. But land use has been identified as a major area needing improvement (Nakicenovic et al., 2000), so attempting such a check is worthwhile.

Moving from concentrations to climatic change involves still more uncertainty, but would seem to allow for at least some historical validation if the IA models can be started as early as 1970. Again, as with the carbon cycle, the implied constraints are only loose, but still allow for some significant information about level and pattern to be created. Again, as with economic and land-use data, considerable skill and thought will have to go into the comparisons.

What tests could be made for impacts? We might begin by considering land use, already mentioned in the context of the carbon cycle. But we can also consider projections of changes in boundaries for species, of changes in the length of growing season, and of glacial melting. Many of these systems are not included in current IA models, but observed changes in these systems would provide a natural set of experiments and might serve to constrain parameters earlier in the system. It would also be possible to conceive of experiments aimed at understanding whether crop cultivars should change or whether stream flow predictions match records. Finally, one could consider model results for sea-level rise as compared to tide gauge measurements.

The upshot is that many experiments could be used to provide feedback on model skill and accuracy, but getting the models to the point where they could usefully be tested in this way will involve pushing their starting points much earlier than the current norm. The resource requirements here are substantial and may well exceed what makes sense for an initial experiment. Many of the models have 1990 as a beginning year; we might be able to construct an experiment looking initially at the 1990 to 2000 time-frame and then to the 2005 time-frame.

A second set of tasks involves understanding the structure of the models and then seeing if there are areas where a better representation of the science (an improved structural model) would strengthen the entire IA community, such as:

- Adding links between economic growth and population dynamics;
- Improving the representation of economic growth, especially the transition between subsistence and modern economic sectors;
- Understanding the drivers for land-use change more fully, particularly as this relates to the carbon cycle;
- Improving the ability of the models to downscale broad economic and demographic trends to a scale appropriate for impacts and adaptation;
- Developing better schemes to link the outcome of a variety of process-level tools, such as technology and crop models, to their corresponding aggregate concepts within the IA models.

Here the development of a community modeling framework or set of interfaces, as well as modules to fill in the various elements could result in a rapid improvement in quality and expansion of capability. It seems reasonable that this should proceed on a slower time-frame than the initial set of experiments.

The discourse now shifts from strategy or long-term goals, to tactics – how can we get started? The obvious starting point is to use the current EMF exercise, EMF-22. Several topics in EMF-22 could be the basis for a set of simple experiments. One would be the construction of at least a minimal common data set for the base period. We could meet and determine what year this should be, and discuss the variables that should be included in this data set. The discussion earlier in the paper would support using 1990 as a base period with additional data elements for 1995 and 2000. Population, and associated demographic statistics, as well as GDP, energy use, and emissions could be provided. This would begin to meet the need for a common data set, so that models could at least start from a common point of view. A second stage in the development of this common data set would be to assess the range of errors in the data. This would require working closely with the various statistical data agencies and might prove illuminating.

An additional data task would evaluate the current status and completeness of the purchasing power parity (PPP) data, including coverage, timing, completeness and potential error levels. Key for many of the models will be an understanding if the potential exists to express key data sets, such as input/output tables in a PPP framework. Until this is achieved, construction of CGE models on a PPP framework will not be possible.

With a time series database in hand, it will be possible to assess how well the models reproduce key variables, such as population levels, per capita income and energy use, and carbon as well as other emissions. To do this will require running the models in a mode where they do not tune the output to match current results. Constructing valid experiments to run at this scale may prove difficult, but is crucial to begin to create a sense of the skill levels of the models.

31.5 Conclusion

The assets in the community are considerable. A substantial and growing group of energetic, talented individuals and groups participate in IA modeling. Ongoing programs in a variety of areas engage these groups in extending and strengthening their systems and grasp of the issues and problems, including such activities as the CCI & IA workshops at Snowmass, EMF, Ensembles, the European technology exercise, and the various AIM team activities. The community clearly has the capacity to take on board major critiques and engage significant outside professionals in the process, as is evidenced by the PPP/MER discussion.

However, considerable efforts are needed in order to realize the full potential of the IA tools:

1. A systematic effort to develop an understanding of why results are different;
2. Better representation, on multiple scales, of ecosystem behavior;
3. Incorporation of a select list of additional behaviors, such as demographics and water, into a wider range of IA models;
4. A major effort to incorporate appropriate impacts and feedbacks into the economic system at a level beyond a temperature-driven aggregate damage function.

The challenge in all of these efforts is to extend the system in such a way that the underlying behavior is credible. To do all this requires a systematic and well-resourced program of ongoing model development and intercomparison to create a better-understood, better-documented set of tools. Within the program, IA modelers can develop common standards and model elements, establish a central location for model results, articulate research needs and associated resource needs, and provide for the development of IA skills and tools. IA practice will then become an IA discipline, which should lead to stronger role for IA modeling in decisions support and resource allocation within the climate management arenas and across a variety of environmental concerns.

Implementing such a program will be a fitting capstone to the decade of successful summer workshops on IA. Immediate steps to be taken include the following:

- Use the current EMF-22 exercise to begin development of baseline data for IA models.
- Establish an ad hoc committee to begin the process of initiating an IAMDI.
- Explore what aspects of the Atmospheric Model Implementation Project (AMIP) and companion comparisons within PCMDI can be used to advantage for IA.
- Determine what services and functions would be needed to support IAMDI.

An IAMDI program will require additional resources and the willingness of the community to participate actively and constructively – but the ability of IA tools to address societal needs along the increasing range of issues involving human and natural systems warrants both the resources and community-building efforts.

References

Andronova, N. G. and Schlesinger, M. E. (2001). Objective estimation of the probability distribution for climate sensitivity. *Journal of Geophysical Research* **106** (D19) 22 605–22 612.

Diamond, J. (1997). *Guns, Germs and Steel: The Fates of Human Societies*. New York: W. W. Norton.

Diamond, J. (2005). *Collapse: How Societies Choose to Fail Or Succeed*. New York: Viking Press, Penguin Group.

IPCC (2000). *Emissions Scenarios: Special Report of Working Group III of the Intergovernmental Panel on Climate Change*, ed. N. Nakicenovic and R. Swart. Cambridge: Cambridge University Press.

Schneider, S. (2003). Editorial comment. *Climatic Change* **52**, 441–451.

US Climate Change Science Program (2003). *Strategic Plan for the US Climate Change Science Program*, www.climatescience.gov/Library/stratplan2003/final/default.htm

Webster, M. D., Forest, C., Reilly, J. *et al.* (2003). Uncertainty analysis of climate change and policy response. *Climatic Change* **61** (3), 295–320.

32

Moving beyond concentrations: the challenge of limiting temperature change

Richard G. Richels, Alan S. Manne and Tom M. L. Wigley

32.1 Introduction

The UN Framework Convention on Climate Change (UNFCCC) shifted the attention of the policy community from stabilizing greenhouse gas emissions to stabilizing atmospheric greenhouse gas concentrations. While this represents a step forward, it does not go far enough. We find that, given the uncertainty in the climate system, focusing on atmospheric concentrations is likely to convey a false sense of precision. The causal chain between human activity and impacts is laden with uncertainty. From a benefit–cost perspective, it would be desirable to minimize the sum of mitigation costs and damages. Unfortunately, our ability to quantify and value impacts is limited. For the time being, we must rely on a surrogate. Focusing on temperature rather than on concentrations provides much more information on what constitutes an ample margin of safety. Concentrations mask too many uncertainties that are crucial for policymaking.

The climate debate is fraught with uncertainty. In order to better understand the link between human activities and impacts, we must first understand the causal chain between the two, i.e., the relationship between human activities, emissions, concentrations, radiative forcing, temperature, climate, and impacts. The focus of the UNFCCC is on atmospheric concentrations of greenhouse gases. Although this represents a major step forward by advancing the debate beyond emissions, it does not go far enough. In this paper, we carry the analysis beyond atmospheric concentrations to temperature change. Although closely linked to concentrations, we believe that temperature is a more meaningful metric in that it incorporates several additional considerations critical for informed policymaking. In particular, the uncertainty related to climate sensitivity can dramatically alter the effectiveness of a prescribed concentration ceiling when trying to control temperature change. If the focus is on limiting atmospheric concentrations, policymakers may be given a false impression of the impact of their actions.

Reducing greenhouse gas emissions will require fundamental changes in the way in which we produce, transform, and use energy (Nakićenović et al., 1998; Edmonds and Wise, 1999; Hoffert et al., 2002;). How we go about making these changes will, in large part, determine the price tag for dealing with the threat of climate change. We can either make the necessary investments today to ensure ample supplies of low-cost alternatives in the future or we can continue the current decline in energy research, development, and demonstration (RD&D) and make do with high-cost substitutes that are already on the shelf. The present analysis examines the implications of choosing one path over the other.

Providing a range of possible outcomes with no indication of their likelihood can be misleading (Moss et al., 2000). Yet, the "curse of dimensionality" makes it extremely difficult for any one analysis to provide a formal treatment of all relevant uncertainties. At the very least, analysts can explore those uncertainties that may be most critical to the issue at hand. Here we attach *subjective* probabilities to: (a) income growth, (b) climate sensitivity, and (c) the rate of heat uptake by the deep ocean. We also deal with the uncertainty surrounding a temperature cap, but only through sensitivity analysis.

Three caveats are in order. First, we note that specifying an absolute limit, whether it be on concentrations or temperature, implies that damages are infinite beyond that limit. It would be

---

Human-induced Climate Change: An Interdisciplinary Assessment, ed. Michael Schlesinger, Haroon Kheshgi, Joel Smith, Francisco de la Chesnaye, John M. Reilly, Tom Wilson and Charles Kolstad. Published by Cambridge University Press. © Cambridge University Press 2007.

preferable to employ a cost–benefit approach, balancing the costs of climate policy with what such a policy buys in terms of reduced damages (Reilly *et al.*, 1999). Unfortunately, our understanding of the nature of future damages and how to value them is so rudimentary that a formal cost–benefit analysis would be questionable. For the current analysis, we use temperature as a surrogate, assuming that an absolute limit reflects a political decision as to what constitutes an "ample margin of safety."

Second, a critical assumption underlying the present analysis is that of complete "where" and "when" flexibility. That is, through trade in emissions rights, reductions will be made wherever it is cheapest to do so, regardless of the geographical location. Similarly, we allow banking and borrowing over time so that reductions will take place whenever it is cheapest to do so. This approach does not imply that reductions can be delayed indefinitely. Eventually, any temperature ceiling will become a binding constraint. To the extent that these two tenets of economic efficiency (where and when flexibility) are violated, the costs of a particular ceiling will be higher (Manne and Richels, 1997).

Finally, we need to recognize the "act, then learn, then act again" nature of the decision problem. A key issue facing today's decisionmakers is to specify the rate and magnitude of greenhouse gas reductions. This is not a once-and-for-all decision, but one that will be revisited over time. There will be ample opportunity for learning and mid-course corrections. The challenge is to identify a sensible set of near-term decisions in the face of the many long-term uncertainties. This paper provides useful information for the decisionmaking process, but stops short of analyzing the question of what to do now. For an example of how such information can be used to determine a rational hedging strategy, see Manne and Richels (1995). We believe that this is a crucial area for future research.

32.2 The model

The analysis is based on MERGE (a model for evaluating the regional and global effects of greenhouse gas reduction policies). This section provides a brief overview of MERGE. For details on the model's structure, data, and key information sources, the reader is encouraged to visit our website: www.stanford.edu/group/MERGE.

MERGE is an intertemporal general equilibrium model. There is international trade in oil, gas, and energy-intensive goods. Each of the model's nine regions maximizes the discounted utility of its consumption subject to an intertemporal budget constraint. Each region's wealth includes capital, labor, and exhaustible resources.

Like its predecessors, the current version (MERGE 5.0) is designed to be sufficiently transparent that one can explore the implications of alternative viewpoints in the greenhouse debate. The model integrates sub-modules that provide reduced-form descriptions of the energy sector, the economy, emissions, concentrations, temperature change, and damage assessment.[1]

MERGE combines a bottom-up representation of the energy supply sector together with a top-down perspective on the remainder of the economy. For a particular scenario, a choice is made among specific activities for the generation of electricity and for the production of non-electric energy. Oil, gas, and coal are viewed as exhaustible resources. There are introduction constraints on new technologies and decline constraints on existing technologies.

Outside the energy sector, the economy is modeled through nested, constant-elasticity production functions. The production functions determine how aggregate economic output depends upon the inputs of capital, labor, and electric and non-electric energy. In this way, the model allows for both price-induced and autonomous (non-price) energy conservation and for interfuel substitution. It also allows for macroeconomic feedbacks. Higher energy and/or environmental costs will lead to fewer resources available for current consumption and for investment in the accumulation of capital stocks. Economic values are reported in US dollars of constant 2000 purchasing power.

A number of gases have been identified as having a positive effect on radiative forcing (IPCC, 1994). In addition to CO_2, methane (CH_4) and nitrous oxide (N_2O), MERGE 5.0 has been extended to incorporate the so-called "second basket" of greenhouse gases included in the Kyoto Protocol. These are the hydrofluorocarbons (HFCs), the perfluorocarbons (PFCs), and sulfur hexafluoride (SF_6).

For CO_2, we relate emissions to concentrations using a convolution, ocean carbon-cycle model and assuming a neutral biosphere. The other gases are modeled with one-box models with constant lifetimes. In spite of these simple representations, projected gas concentrations agree well with those given in the Intergovernmental Panel on Climate Change's Third Assessment Report (TAR) (IPCC, 2001) for the SRES illustrative scenarios (IPCC, 2000).

We also consider the cooling effect of sulfate aerosols assuming that SO_2 emissions follow the SRES B2 scenario in all cases (i.e., we assume that there is no "feedback" effect of greenhouse gas mitigation policies on the emissions of SO_2).

For radiative forcing we use relationships consistent with the TAR for greenhouse gases, and the median aerosol forcing from Wigley and Raper (2001). As shown in the latter, temperature projections are relatively insensitive to aerosol forcing uncertainties.

Projections for the non-CO_2 greenhouse gases are based largely on the guidelines provided by EMF 21: Multi-Gas Mitigation and Climate Change (Stanford Energy Modeling Forum EMF-21, 2005). Reductions from the reference path are determined by a set of time-dependent marginal abatement-cost

---

[1] The current analysis, which focuses on cost-effectiveness analysis, does not use the damage assessment sub-module.

curves. For details, the reader is directed to the MERGE website.

When dealing with multiple gases, we need some way to establish equivalence among gases. The problem arises because the gases are not comparable. Each gas has its own lifetime and specific radiative forcing. The IPCC has suggested the use of global warming potentials (GWPs) to represent the relative contributions of different greenhouse gases to the radiative forcing of the atmosphere (IPCC, 1996). However, a number of studies have pointed out the limitations of this approach (Schmalensee, 1993). In MERGE 5.0, we adopt an alternative approach. We make an endogenous calculation of the incremental value of emission rights for the non-CO_2 greenhouse gases relative to CO_2 in each time period. The marginal abatement costs then provide the necessary basis for the tradeoffs among gases (Manne and Richels, 2001).

The Kyoto Protocol states that Annex B commitments may be met by "the net changes in greenhouse gas emissions from sources and removal by sinks resulting from direct human-induced land use change and forestry activities limited to afforestation, reforestation, and deforestation since 1990, measured as verifiable changes in stocks in each commitment period" (Conference of the Parties, 1997). MERGE incorporates this option for offsets. We suppose that marginal sink enhancement costs rise with the quantity of enhancement. We assume that the potential for sink enhancement increases over time, but is eventually limited by the cumulative capacity for carbon absorption in forests.

32.3 Treatment of technologies

For the present analysis, we construct two technology scenarios. In the pessimistic ("business as usual") case, we assume that the current downward spiral in energy RD&D continues unabated,[2] and that the transition to a less greenhouse-gas-intensive economy is achieved with technologies that are currently on the shelf or in the marketplace. In the optimistic scenario, we are much more sanguine. A reversal of current investment trends in energy RD&D leads to a much brighter technological future.

Clearly, technology investment will be influenced by the price that society places on greenhouse gases. If a high price is deemed warranted, the current downward trend in energy RD&D is more likely to be reversed. Although there are differences in RD&D costs between the two scenarios, we ignore these differences and quantify only the differences in payoffs. This is justifiable because, in general, we find that RD&D costs are measured in billions of dollars, but that the payoffs could run into the trillions. The payoff is determined by the costs of meeting a particular climate goal with and without the advanced supply and demand-side technologies described below.

The detail in which technology is described in a particular analysis depends on the focus. For the present analysis, we attempt to examine the differences between being in a technology-rich and technology-poor world. We are particularly interested in the impacts of such differences on the timing and costs of emission reductions. The level of detail for such an analysis requires assumptions about the availability, costs, performance characteristics, and greenhouse gas emissions from various categories of technologies and how these parameters change across space and time. The level of specificity would be much greater if we were to model the competition between different approaches within a particular category of technology. Whereas such detail is necessary to address certain questions, it is not called for here.

In MERGE, a distinction is made between electric and non-electric energy. Table 32.1 identifies the alternative sources of electricity supply. The first five technologies represent sources serving electricity demand in the base year (2000). The second group of technologies includes candidates for serving electricity needs in 2010 and beyond. The composition of the latter group differs in our two scenarios.

In the pessimistic scenario, we assume that future electricity demand will be met primarily with new, state-of-the-art gas and coal plants. In addition, there is a technology which we refer to as ADV-HC (advanced high-cost carbon-free electricity generation). Its distinguishing characteristic is that, once introduced, it is available at a high but constant marginal cost. Any of a number of technologies could be included in this category: solar (in several forms), advanced nuclear, biomass, and others. For a discussion of possible candidates see Hoffert et al. (2002). Given the enormous disagreement as to which of these technologies or combination of technologies will eventually win out, in terms of economic attractiveness and public acceptability, we refer to them generically rather than attempt to identify one or two winners.

Because knowledge is not fully appropriable, private markets are likely to under-invest in RD&D (Schneider and Goulder, 1997). For our optimistic scenario, we assume that this market imperfection is overcome through a sustained commitment on the part of the public sector to direct investment, the subsidy of private sector RD&D, or both. As a result, fuel cells, and integrated gasification combined cycle with carbon capture and sequestration are added to our list of technologies. We also add a category similar to ADV-HC, but whose learning costs decline by 20% for every doubling of capacity. This is called LBDE (learning-by-doing, electric). LBDE provides a learning component to those technologies grouped under ADV-HC.[3]

Table 32.2 identifies alternative sources of non-electric energy within the model. Notice that oil and gas supplies for

---

[2] According to the International Energy Agency, investment in energy RD&D declined by approximately 50% worldwide between 1980 and 1999 (see Runci et al., 2005).

[3] For a discussion of learning-by-doing, see Clark and Weyant, 2002.

Table 32.1 Electricity generation technologies available to US[a] (shared rows represent technologies only available in the optimistic scenario).

Technology name	Identification/examples	Earliest possible introduction date	Costs in 2000[b] (Mills/kWh)	Potential cost reduction due to learning-by-doing (LBD) (Mills/kWh)	Carbon emission coefficients (billion tonnes per TWH)
HYDRO	Hydroelectric, geothermal and other renewables	Existing in base year	40.0		0.0000
NUC	Remaining initial nuclear	Existing in base year	50.0		0.0000
GAS-R	Remaining initial gas fired	Existing in base year	35.7		0.1443
OIL-R	Remaining initial oil fired	Existing in base year	37.8		0.2094
COAL-R	Remaining initial coal fired	Existing in base year	20.3		0.2533
GAS-N	Advanced combined cycle	2010	30.3		0.0935
GAS-A	Fuel cells with capture and sequestration	2030	47.7		0.0000
COAL-N	Pulverized coal without CO_2 recovery	2010	45.0		0.1955
COAL-A	Fuel cells with capture and sequestration	2040	55.9		0.0068
IGCC	Integrated gasification and combined cycle with capture and sequestration	2020[c]	52.0		0.0240
ADV-HC	Carbon-free technologies; costs do not decline with LBD	2010	100.0		0.0000
LBDE[d]	Carbon-free technologies; costs decline with LBD	2010	100.0	70.0	0.0000

[a] Introduction dates and costs may vary by region.
[b] Except for oil and gas costs and the learning-by-doing component, we assume that the cost of all technologies decline at a rate of 0.5% per year beginning in 2000. Note that this column is used to calculate the autonomous learning component. The earliest possible introduction date is specified in the previous column.
[c] IGCC is currently available, but *without* capture and sequestration.
[d] For these technologies, it is necessary to specify an initial quantity. We assume that the cumulative global experience prior to 2000 is only 0.2 TkWh.

each region are divided into 10 cost categories. The higher cost categories reflect the potential use of non-conventional sources. With regard to carbon-free alternatives, the choices have been grouped into two broad categories: RNEW (low-cost renewables such as ethanol from biomass) and NE-BAK (high cost backstops such as hydrogen produced through photovoltaics and electrolysis). The key distinction is that RNEW is in limited supply, but NE-BAK is assumed available in unlimited quantities at a constant but considerably higher marginal cost. As in the case of electric energy, we have added a new category of technologies for our optimistic scenario. This is called LBDN (learning-by-doing, non-electric). As with its counterpart in the electric sector, costs are a declining function of cumulative experience.

Except for the learning-by-doing component, we assume that the costs of all technologies decline at a rate of 0.5% per year beginning in 2000. This is the case for both the pessimistic and optimistic scenarios.

The energy-producing capital stock is typically long-lived. In MERGE, introduction and decline constraints are placed on all technologies. For new technologies, we assume that production in each region is constrained to 1% of total production in the year in which it is initially introduced and can increase by a factor of three for each decade thereafter. The decline rate is limited to 2% per year. The decline rate, however, does not apply to existing technologies. This is to allow for the possibility that some climate constraints may be sufficiently tight to force premature retirement of existing capital stock.

Turning to the demand-side, to allow for greater progress at the point of end-use, we assume that the long-run price elasticities are 25% higher in the optimistic technology scenario. Here we assume that we succeed in removing the barriers to

Table 32.2 Non-electric energy supplies available to United States[a] (bottom row represents technology only available in the optimistic scenario).

Technology name	Description	Cost in 2000 ($/GJ)	Potential cost reduction due to learning-by-doing ($/GJ)	Carbon emission coefficients (tonnes of carbon per GJ)
CLDU	Coal-direct uses	2.50		0.0241
OIL-1–10	Oil	3.00–5.25		0.0199
GAS-1–10	Gas	2.00–4.25		0.0137
SYNF	Coal-based synthetic fuels	8.33		0.0400
RNEW	Renewables	6.00		0.0000
NEB-HC	Non-electric backstop	14.00		0.0000
LBDN[b]	Carbon-free technologies; costs decline with learning-by-doing	14.00	6.00	0.0000

[a] Costs may vary by region. Except for the learning-by-doing component, we assume that the costs of all technologies decline at a rate of 0.5% per year beginning in 2000.
[b] We assume that cumulative experience prior to 2000 is only 1 GJ.

increased efficiency and that the costs of doing so do not outweigh the benefits.

32.4 Treatment of uncertainty

In this paper, we attempt to compare the costs of stabilizing global-mean temperature in a technology-rich and a technology-poor world. Mitigation costs will depend not only on the characteristics of the energy system, but also on a number of socio-economic and scientific considerations each of which is highly uncertain. These include factors influencing future greenhouse gas emissions, the carbon cycle, radiative forcing, climate sensitivity, and the efficiency with which heat is transferred from the surface into the deeper ocean. It would be virtually impossible to include all of these factors in a rigorous probabilistic analysis. Rather, we have chosen to focus on three areas of uncertainty that we feel are particularly relevant to the present analysis. The dominant importance of future emissions and climate sensitivity to global-mean temperature is well documented (Wigley and Raper, 2001; Caldeira et al., 2003). In addition, because of the importance of the lag between potential and realized temperature change to the present analysis, we add response time to our list of critical uncertainties.[4]

Future greenhouse gas emissions are particularly difficult to project. In a previous study, we examined the sensitivity of the emissions of carbon dioxide to five factors: potential economic growth; the elasticity of price-induced substitution between energy, capital, and labor; the rate of non-price-induced energy efficiency improvements; the availability of economically competitive carbon-free alternatives to coal-fired electricity; and the costs of the non-electric backstop alternative to liquid fuels (Manne and Richels, 1994). The analysis showed that economic growth was by far the most important determinant of future emissions. Figure 32.1 shows five projections of growth over the twenty-first century and the authors' subjective probability for each.[5]

Figure 32.2 shows the resulting projections of future carbon emissions. The solid lines and the dashed lines correspond to the pessimistic and optimistic technology scenarios, respectively.

Climate sensitivity is defined as the equilibrium global-mean surface temperature change in response to a doubling of the CO_2 concentration. The cumulative distribution function (CDF) in Figure 32.3 corresponds to the log-normal probability distribution adopted by Wigley and Raper (Wigley and Raper, 2001). For the purposes of the present analysis, we focus on the tails of the distribution. This yields discrete probabilities of 5%, 50%, and 95% for climate sensitivities of 1.3, 2.6, and 5.0 °C, respectively.

As noted in Wigley and Raper (2001) and Forest et al. (2002), the two properties that control the climate system's decadal-to-century response to radiative forcing are the climate sensitivity and the rate of heat uptake by the ocean. The rate at which heat is transferred from the surface into the deeper ocean is determined by the climate sensitivity, the ocean's effective vertical diffusivity, and changes in the ocean's thermohaline circulation. In the MERGE climate model, the rate of ocean heat uptake is characterized by a single (response time) parameter. The values used for this timescale are based on a calibration of the model against the upwelling-diffusion, energy-balance model (MAGICC) used in the IPCC TAR (IPCC, 2001) and in Wigley and Raper (2001) (see Appendix A32.1). Figure 32.4 shows the response time for alternative

---

[4] The response time is defined as the time it takes for the temperature to reach $(1 - 1/e)$ of the equilibrium response – see Appendix A32.1.

[5] These projections coincide remarkably well with the full range of SRES (IPCC, 2000) scenarios for the year 2100.

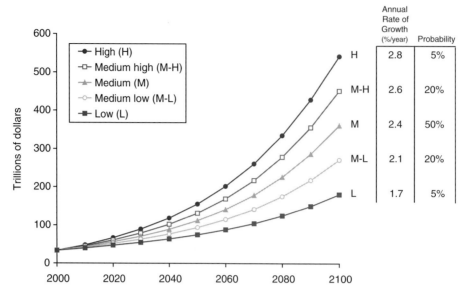

Figure 32.1 Potential gross world product.

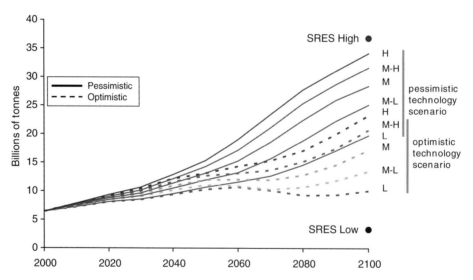

Figure 32.2 Carbon emissions baselines. High (H); Medium high (M-H); Medium (M); Medium low (M-L); Low (L).

climate sensitivities, accounting for thermohaline circulation changes and uncertainties in vertical diffusivity.

For the purposes of the analysis that follows, we calculate discrete conditional probabilities based on each of the three CDFs in Figure 32.4, again focusing on the tails of the distribution: see Table 32.3.

32.5 Why is temperature a more meaningful metric than atmospheric concentrations?

Having discussed our numerical inputs, we now turn to the analysis. First, we address the issue of a stabilization metric. The UN Framework Convention on Climate Change shifted the attention of the policy community from stabilizing greenhouse gas emissions to stabilizing atmospheric greenhouse gas concentrations. While this represented a step forward, it did not go far enough. Concentrations are not the end of the line, but only one more link in the causal chain between human activities and impacts. In the present analysis, we go beyond atmospheric concentrations to temperature. Not only does the focus on temperature avoid the problems associated with the use of GWPs, but it also provides a more meaningful metric for policymaking purposes.

The following experiment illustrates why we believe temperature to be a better choice of metric. Suppose that there are (1) two alternative temperature ceilings: 2 °C and 3 °C, (2) two

Moving beyond concentrations

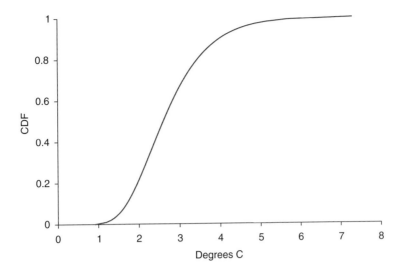

Figure 32.3 Climate sensitivity.

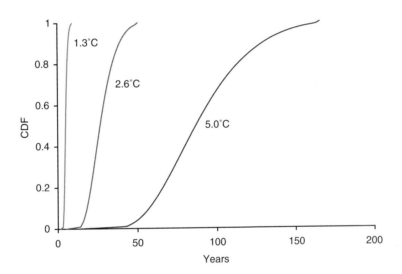

Figure 32.4 Response time for alternative climate sensitivities.

Table 32.3 MERGE climate model response times (years).

	Climate sensitivity		
Percentile	**1.3 °C**	**2.6 °C**	**5.0 °C**
5%	4	15	57
50%	5	25	96
95%	8	39	158

possible technology futures as defined earlier: pessimistic and optimistic, and (3) economic growth follows our median projections. The calculations shown in Table 32.4 are based on the MERGE model. Note that for a particular temperature cap and technological future, the associated CO_2 concentrations can vary widely. Interestingly, although ocean heat uptake has some influence, we find that climate sensitivity turns out to be far more important in determining allowable concentrations.

Suppose that the ceiling is 2 °C and the pessimistic technology scenario materializes. In this case, the corresponding CO_2 concentration target spans a range approaching 300 ppmv. There is one chance in 20 that the concentration ceiling required to stabilize temperature at the prescribed level will be 452 ppmv or less. This corresponds to high climate sensitivity. There is also one chance in 20 that the limit in 2100 exceeds 743 ppmv. Here we have low climate sensitivity. The median value turns out to be 482 ppmv and corresponds to the median value for climate sensitivity. The range is somewhat narrower for the 2 °C ceiling under the optimistic technology scenario. This is because the cost of carbon-free substitutes is such that more emphasis is placed on reducing CO_2 emissions and less emphasis on the other alternative

Table 32.4 Peak atmospheric CO_2 concentrations (ppmv) under alternative temperature caps.

		5th percentile	50th percentile	95th percentile
2 °C	Pessimistic	452	482	743*
	Optimistic	445	465	654*
3 °C	Pessimistic	499	587	788†
	Optimistic	485	580	658†

* Concentrations to rise after 2100. Emissions reduced below baseline.
† Concentrations to rise after 2100. Emissions unconstrained.

Table 32.5 Likelihood of temperature change over the twenty-first century in the absence of climate policy.

	5th percentile	50th percentile	95th percentile
Pessimistic technology scenario	2.1 °C	3.1 °C	3.6 °C
Optimistic technology scenario	1.9 °C	2.7 °C	3.3 °C

gases. The findings are qualitatively similar for a 3 °C cap. However, here there is much less pressure to constrain emissions. Indeed, for the 95th percentile, emissions track the baseline. Again, the latter scenarios correspond to low climate sensitivity.

Given the current uncertainties in our understanding of the climate system, it is impossible to project with any degree of confidence the effect of a given concentration ceiling on temperature. Or conversely, for a particular temperature cap, the required concentration ceiling is highly uncertain. This calls into question the current focus on atmospheric concentrations. At the very least, we may want to shift the focus to temperature and then identify the implications for concentrations. This would give policymakers a more realistic understanding of the potential consequences of their actions.

32.6 Temperature change in the absence of climate policy

We next turn to the issue of the impacts of a temperature constraint on mitigation costs. The first step is to examine the non-policy baseline. Using MERGE, we estimate cumulative distribution functions (CDFs) for our two technology scenarios in the absence of policy to mitigate climate change. From Figure 32.5, we see that the range is consistent with the recent IPCC projections (IPCC, 2000, 2001). Whereas the IPCC presents a range of 1.4 to 5.8 °C between 1990 and 2100, we project a range of 1.2 to 5.5 °C between 2000 and 2100.[6] Allowing for the warming from 1990 to 2000, these two sets of projections are virtually identical.

However, unlike the IPCC, we assign probabilities to various outcomes. The analysis suggests that the extreme ends of the range are highly unlikely. Table 32.5 shows the 5th, 50th and 95th percentiles for each of the two technology scenarios. For the pessimistic scenario, there is only one chance in 20 that the temperature increase will be less than 2.1 °C or more than 3.6 °C.

Notice that shifting from the pessimistic to the optimistic scenario results in only a modest shift in the CDF. The explanation is straightforward. Suppose that all parameters were to take on their median values. The difference in emissions between the two technology scenarios can be attributed to two factors: (1) learning-by-doing which drives down the price of the electric backstop technologies to the point where they are economically competitive with conventional gas and coal, and (2) the long-run price elasticities. However, from Figure 32.2, note that the emission baselines representing the median values do not diverge substantially until the second half of the century. With a climate sensitivity of 2.6 °C, the response time is of the order of 25 years. Hence, we should not be surprised to see so little difference in the median values for temperature change in 2100. With the higher climate sensitivity, the response time is such that there is insufficient time for substantial divergence in temperature by 2100. With the lower climate sensitivity, the rate of temperature change is so small that the faster response times have little influence.

Because we have focused on the tails of the distributions for climate sensitivity and response time, the distributions tend to rise sharply for the middle fractiles. We would see a less rapid rise in the CDFs if we were to use more points in characterizing the individual distributions, but there would be little change in the tails of the distributions and in their median values.

32.7 A ceiling on temperature increase

We now turn to the issue of temperature ceilings. We begin with a 2 °C cap on temperature increase from 2000. This may seem ambitious given that approximately 95% of the outcomes in Figure 32.5 exceed this level by 2100. One measure of the difficulty of meeting a temperature ceiling is how fast we must depart from the emissions baseline. Although we are focusing on temperature as a constraint rather than atmospheric concentrations, the issue remains one of cumulative CO_2 emissions. For any given climate sensitivity, a global-warming ceiling defines a carbon budget. The challenge is to determine how the budget should be allocated over time to meet the climate goal at minimum cost.

Figure 32.6 shows the cost-effective rate of departure for the pessimistic and optimistic scenarios. Notice that the rate of departure begins slowly and increases over time. This is

---
[6] Note that all estimates of warming in this paper are measured from the year 2000.

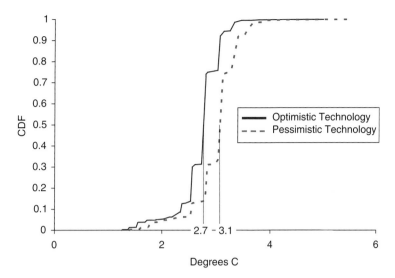

Figure 32.5 Temperature increase during twenty-first century in the absence of mitigation policy (50th percentile values highlighted).

consistent with "when" flexibility (Wigley et al., 1996). A gradual rate of departure reduces the pressure to retire existing long-lived capital stock prematurely (e.g., power plants, buildings, and transport) and provides more time to develop and introduce new, economically competitive carbon-free technologies into the energy system.

Interestingly, the rate of the departure from the baseline through 2030 is virtually insensitive to the technology scenario. The explanation has to do with the timing and costs of the new technologies and the size of the carbon budget. The near-term options for reducing CO_2 emissions are limited to fuel switching from coal and oil to natural gas, and to price-induced conservation. This is because the payoff from technology investment is "back loaded." The development and deployment of new technologies does not happen overnight. The payoff is initially modest, but increases over time. Fortunately, with a 2 °C cap there is still some leeway to emit CO_2. It makes sense to use what remains of the carbon budget in the early years when the alternatives are expensive and to transition gradually to a less carbon-intensive economy.

Even if we are pessimistic about the technological future, using the remainder of the carbon budget early on reduces the need for a precipitous reduction in the existing carbon-producing and carbon-using capital stock. Hence, regardless of one's views on technology, this makes little difference in the initial rate of departure from the baseline. This is the case whether we are focusing on the median or the tails of the CDFs; see Table 32.6.

A second measure of the difficulty of meeting a constraint is the implicit price that would have to be placed on carbon to meet the particular goal. That is, how high would we have to raise the price of carbon-intensive technologies to make them less desirable than the non-carbon-venting alternatives?

Figure 32.7 shows the magnitude of the carbon tax that would be required to limit the temperature increase to 2 °C. The tax is computed for 2010, 2020, and 2030 for each of the technology scenarios. Note that the results appear sensitive to our technological perspective. That is, the less sanguine we are about the prospects for low-cost alternatives, the higher the carbon tax in the early years. This is because, when accounting for future developments, the economically efficient tax will rise at a rate approximating the return on capital. The long-term price of energy will govern the initial level of the tax. The more pessimistic we are about the long term, the higher the tax in the near term.

From Table 32.7, we see that the distributions are skewed to the right. This reflects the difficulty of maintaining a temperature cap of 2 °C when climate sensitivity is high and/or we have a rapid response time.

But what if we were to have a higher cap, say 3 °C? From Figure 32.5, approximately 95% of the outcomes exceed 2 °C. Two-thirds of the outcomes exceed 3 °C for the pessimistic technology scenario and only one-quarter of the outcomes exceed 3 °C for the optimistic technology scenario. Also, from Figure 32.5, note that for the majority of the outcomes that exceed 3 °C, the amount by which this threshold is exceeded tends to be minor.

Figure 32.8 compares the carbon taxes in 2010 required to maintain 2 and 3 °C caps. In each case, the tax is computed for the pessimistic and optimistic technology scenarios. As we would expect, the difficulty of maintaining a 3 °C cap is considerably less than that for a 2 °C cap. With a 3 °C ceiling, we would also expect a lower carbon tax trajectory and a slower rate at which emissions depart from the baseline.

32.8 The role of technology in containing the costs of climate policy

There has been a 50% decline in energy RD&D worldwide since 1980. For the present analysis, we suppose that the continuation of this trend will result in the pessimistic technology

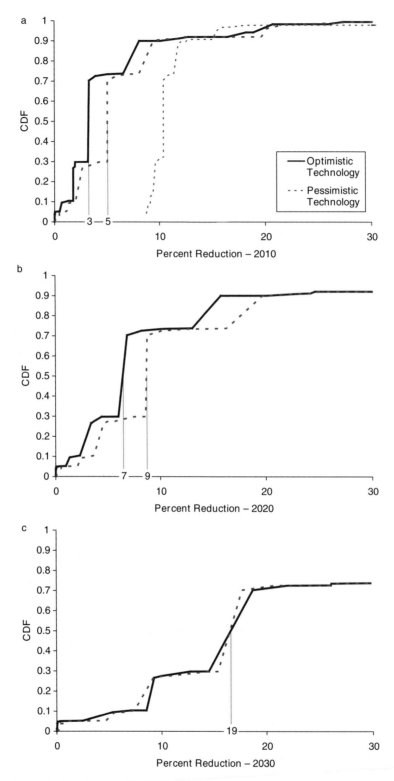

Figure 32.6 Reductions in carbon emissions from the baseline with 2 °C temperature cap during twenty-first century (50th percentile values highlighted).

Table 32.6 Percentage reduction from the baseline for a 2 °C cap on temperature.

	Pessimistic technology scenario			Optimistic technology scenario		
	5th percentile	50th percentile	95th percentile	5th percentile	50th percentile	95th percentile
2010	0	5	20	0	3	19
2020	1	9	37	0	7	40
2030	1	19	53	0	19	60

scenario. Conversely, the optimistic scenario is designed to reflect a reversal of current trends. In this section, we explore the benefits from an RD&D effort sufficient to bring about the more optimistic of our two technological futures.

Care must be taken to define what is meant by benefits. Losses are incurred when the imposition of a temperature constraint leads to a reallocation of resources from the patterns that would be preferred in the absence of the constraint. A temperature constraint will lead to fuel switching and to more expensive price-induced conservation activities. There are also changes in domestic and international prices. In most cases, these forced adjustments result in a reduction in economic performance. Low-cost, carbon-free substitutes can reduce this loss in economic performance. It is this reduction in losses that is referred to as the benefits of RD&D.

In calculating benefits, we do not subtract the costs of the additional RD&D. That is, we deal with gross, not net, benefits. Nor do we account for the reduced environmental damages resulting from the temperature constraint. In the case of the latter, we assume that climate goals will be met with whatever technologies are available. Hence, environmental benefits will be the same in both the pessimistic and optimistic scenarios.

We begin with a 2 °C temperature cap. Figure 32.9 compares discounted consumption losses for each of the two technology scenarios. Over the period of a century the losses can be of the order of trillions of dollars. However, the figure suggests that the losses can be reduced substantially if we are successful in achieving the more ambitious technology objectives.

Table 32.8 shows the discounted present value of consumption losses under the two technology scenarios for the 5th, 50th, and 95th percentiles. The table also shows the differences in consumption losses between the two scenarios. That is, these are the benefits from moving from the pessimistic to the optimistic technology scenario.

Figure 32.10 compares the benefits for the 2 and 3 °C temperature caps. As we would expect, the payoff declines as the stringency of the constraint is weakened. Nevertheless, the payoff is still likely to be substantial even with the higher temperature cap.

32.9 The relative contribution of the various greenhouse gases to radiative forcing

As noted earlier, the analysis encompasses the six categories of greenhouse gases identified in the Kyoto Protocol. With our focus on the energy sector and the costs of meeting a particular temperature ceiling in a technology-rich and technology-poor world, our attention has been on CO_2. It is interesting, however, to note the contribution of the other gases in meeting the temperature constraints. Figure 32.11 shows our projections for the globally and annually averaged anthropogenic radiative forcing due to changes in the concentrations of greenhouse gases over the twenty-first century.[7] Here the second basket of gases (the HFCs, PFCs, and SF_6) are combined under just two categories: short-lived fluorinated gases (SLF) and long-lived fluorinated gases (LLF).[8]

Among the non-CO_2 greenhouse gases, CH_4 with its relatively short lifetime (12 years) makes the greatest contribution to meeting the temperature ceilings. The impact of the relative cost of abatement can be seen when comparing the 3 °C cases. Under the optimistic technology scenario, less pressure is placed on reducing CH_4. This is because CO_2 abatement is relatively inexpensive when compared with the pessimistic scenario. The short-lived fluorinated gases also play a role, but there are insufficient quantities to offset the need for large CO_2 reductions.

32.10 Some concluding remarks

The analysis has yielded some policy-relevant results. We find that, given the uncertainty in the climate system, focusing on atmospheric concentrations is likely to convey a false sense of precision. The causal chain between human activity and impacts is fraught with uncertainty. From a benefit–cost perspective, it would be desirable to minimize the sum of mitigation costs and damages. Unfortunately, our ability to quantify and value impacts is limited. For the time being, we must rely on a surrogate. Focusing on temperature

---

[7] The temperature cap is imposed in all periods. Because MERGE is an intertemporal optimization model, there may be some minor differences between scenarios with regard to the year that the cap becomes binding. For this particular example, we adopt the median values for income growth, climate sensitivity, and response time.

[8] There are a large number of second-basket gases, which are modeled in MERGE using a representative short-lived fluorinated gas (HCF134a) and a representative long-lived fluorinated gas (SF_6). An equivalent concentration of HCF134a is used to represent all gases with short lifetimes (less than 65 years), while an equivalent concentration of SF_6 is used to represent all gases with longer lifetimes. Total radiative forcing changes for all second-basket gases can be modeled quite accurately by this simple representation.

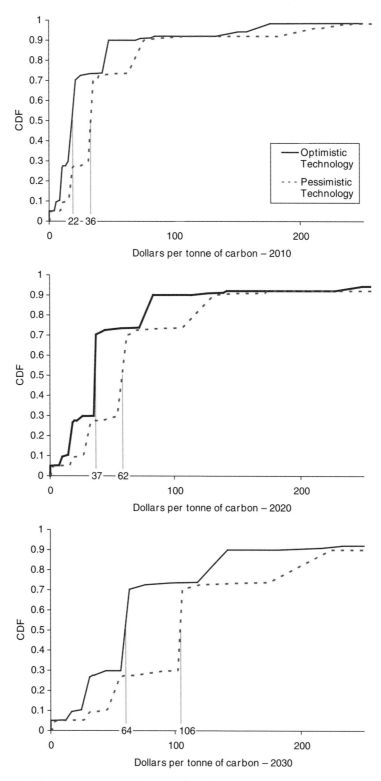

Figure 32.7 Carbon prices with 2 °C temperature cap (50th percentile values highlighted).

rather than on concentrations provides much more information on what constitutes an ample margin of safety. Concentrations mask too many uncertainties that are crucial for policymaking.

For the "no policy" case, the analysis produces a temperature range for 2100 that is similar to that of the IPCC. However, unlike the IPCC, we attempt to determine the likelihood of various temperature outcomes. This is done by

Table 32.7 Price of carbon ($/tonne) with a 2 °C cap on temperature increase from 2000.

	Pessimistic technology scenario			Optimistic technology scenario		
	5th percentile	50th percentile	95th percentile	5th percentile	50th percentile	95th percentile
2010	2	36	212	0	22	176
2020	4	62	355	0	37	255
2030	6	106	637	0	64	409

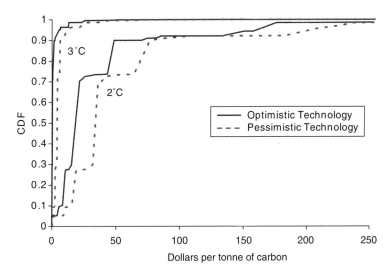

Figure 32.8 Carbon prices in 2010 with 2 and 3 °C temperature caps.

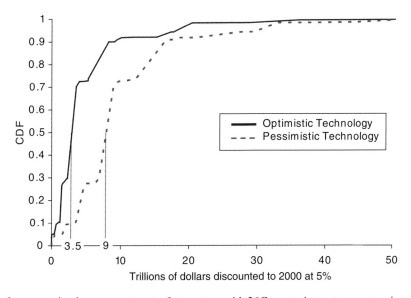

Figure 32.9 Present value of consumption losses over twenty-first century with 2 °C constraint on temperature increase (50th percentile values highlighted).

Table 32.8 Difference in consumption losses for two technology scenarios under a 2 °C temperature cap – discounted to 2000 at 5% in trillions of dollars.

	5th percentile	50th percentile	95th percentile
Consumption losses under pessimistic scenario	1.0	9.0	30.0
Consumption losses under optimistic scenario	0.0	3.5	17.7
Benefits of optimistic scenario	1.0	5.5	12.3

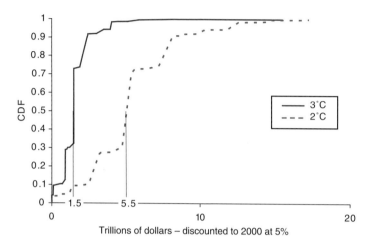

Figure 32.10 Gross benefits from R&D program under alternative temperature constraints (50th percentile values highlighted).

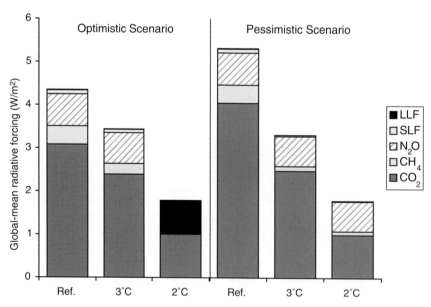

Figure 32.11 Contributions to radiative forcing, 2000–2100.

assigning probabilities to three critical areas of uncertainty: those relating to future economic activity, climate sensitivity, and how quickly the temperature system responds to changes in radiative forcing. The results suggest that the temperature projections at the tails of the range are far less likely than those in the middle.

Focusing on the energy sector and CO_2, the analysis confirms previous findings suggesting that, for a given constraint,

Table A32.1 Response times (in years) from MAGICC climate model.

	$\Delta T_{2x}=1.0\,°C$	$\Delta T_{2x}=2.0\,°C$	$\Delta T_{2x}=3.0\,°C$	$\Delta T_{2x}=4.0\,°C$	$\Delta T_{2x}=5.0\,°C$
$K_z=1.0\,cm^2/s$	2.1 (2.0)	7.6 (7.5)	16.6 (16.7)	29.4 (29.4)	46.3 (45.7)
$K_z=2.0\,cm^2/s$	3.3 (2.8)	12.6 (12.6)	29.0 (29.4)	53.4 (53.4)	86.0 (84.4)
$K_z=3.0\,cm^2/s$	4.3 (3.5)	17.5 (17.5)	41.5 (42.0)	77.3 (77.2)	125.1 (123.0)

a gradual departure from the emissions baseline is preferable to a more rapid departure. This result appears to be insensitive to one's expectations about the long-term price of greenhouse gas abatement. However, such expectations do have a substantial effect on the near-term price of abatement. Specifically, the more optimistic one's views about the future availability of low-cost carbon-free substitutes, the lower the near-term carbon tax.

The analysis also suggests that investment in energy RD&D is no "magic bullet," but it can substantially reduce the economic losses arising from mitigation associated with a temperature constraint. Stabilizing temperature is likely to require a fundamental restructuring of the global energy system. It is hard to imagine that the costs will not be substantial. But investments in the broad portfolio of energy technologies required to meet the emerging needs of both developed and developing countries can dramatically reduce the size of the price tag.

Appendix A32.1 Climate model

The climate model in MERGE is a simple one-box model where the box represents the ocean and its size defines the thermal inertia of the climate system. This in turn determines the lag between externally imposed forcing and global-mean temperature response. While this is clearly an oversimplification of the climate system, such a model can still be used to characterize the response to external forcing in a quantitatively realistic way by calibrating the model against more realistic models.

In a one-box model, the response is determined by two parameters: a climate sensitivity that defines the equilibrium response and a timescale or "response time" (equivalent to the box size) that defines how rapidly the system approaches equilibrium. Defining a suitable single timescale for global-mean temperature response is difficult because the ocean, a primary determinant of the response time, operates on multiple timescales. There is, therefore, no unique way to define a response time – and different ways to define a response time will lead to different values. The way the response time is defined here is to consider the response to a step forcing change of $5\,W/m^2$, and define the response time as if the response were exponential. The response time (τ) is then how long it takes for the temperature to reach $(1 - 1/e)$ of the equilibrium response.

Note that, if the response were exponential, and characterized by a single timescale, then one would reach $(1 - 1/e^2)$ of the equilibrium response after a time equal to 2τ. In fact, it takes much longer than this to reach this point – a consequence of the fact that, as time goes by, the influence of deeper layers in the ocean becomes increasingly more important. This effectively causes the thermal inertia of the system to increase with time, so the system's characteristic response timescale also increases with time. Equally, the initial response is much more rapid than the exponential decay model would lead one to expect – representing the response of the oceanic mixed layer with its relatively small thermal inertia. In spite of these deficiencies, a one-box model still captures the essential features of the system's response, provided an appropriate response time is used. Here we choose appropriate response times using the upwelling–diffusion, energy-balance climate model MAGICC (Wigley and Raper, 1992, 2002), the same model that was used for global-mean temperature projections in the IPCC TAR. MAGICC, in turn, has been calibrated against a number of state-of-the-art coupled atmosphere–ocean GCMs.

The main determinants of the response time, τ, are the climate sensitivity (ΔT_{2x}; °C), and the effective vertical diffusivity of the ocean (K_z; cm^2/s). Table A32.1 gives τ results (in years) from MAGICC, using TAR best-estimate results for all other parameters. In parentheses are approximate results obtained using the following best-fit formula:

$$\tau = [a + b(\Delta T_{2x})]^2$$
$$a = 0.04233(K_z)^2 - 0.4261(K_z) + 0.466 \qquad (32.1)$$
$$b = -0.06071(K_z)^2 + 0.7277(K_z) + 0.668$$

Note that the dependence of τ on ΔT_{2x} is crucial.

As a test, the best-estimate (median) values for ΔT_{2x} and K_z are 2.6 °C and 2.3 cm^2/s (see Wigley and Raper, 2001). The MAGICC value for τ is 24.3 years. The above formula gives a value of 24.6 years. Extrapolation outside the above parameter ranges will lead to errors, but the probability of being outside the above ranges is small.

Acknowledgements

This paper was motivated by our participation in the Energy Modeling Forum (EMF) 21 Study. We have benefited from discussions with Francisco de la Chesnaye, Jae Edmonds, Marty Hoffert, Haroon

Kheshgi, John Reilly, Steve Smith, Richard Tol, John Weyant, Jonathan Wiener, Larry Williams, and Tom Wilson. We are indebted to Christopher Gerlach and Charles Ng for research assistance. Funding was provided by the Electric Power Research Institute (EPRI). The views presented here are solely those of the authors. Comments should be sent to rrichels@epri.com.

References

Caldeira, K., Jain, A. K. and Hoffert, M. I. (2003). Climate sensitivity uncertainty and the need for energy without CO_2 emission. *Science*, **299**, 2052–2054.

Clark, L. and Weyant, J. (2002). *Modeling Induced Technical Change: An Overview*. Working Paper. Stanford University.

Conference of the Parties (1997). Kyoto Protocol to the United Nations Framework Convention on Climate Change. *Report of the Conference of the Parties, Third Session*, Kyoto, 1–10 December, FCCC/CP/1997/L.7/Add.1.

Edmonds, J. and Wise, M. (1999). Exploring a technology strategy for stabilizing atmospheric CO_2. In *International Environmental Agreements on Climate Change*. Dordrecht: Kluwer Academic Publishers.

Forest, C. E., Stone, P. H., Sokolov, A. P., Allen, M. R. and Webster, M. (2002). Quantifying uncertainties in climate system properties with the use of recent climate observations. *Science* **295**, 113–117.

Hoffert, M. I., Caldeira, K., Benford, G. *et al.* (2002). Advanced technology paths to global climate stability: energy for a greenhouse planet. *Science*, **298**, 981–987.

IPCC (1994). *Radiative Forcing of Climate Change and an Evaluation of the IPCC IS92 Emissions Scenarios. Special Report of the Intergovernmental Panel on Climate Change*, ed. J. T. Houghton, L. G. Meira Filho, J. Bruce *et al.* Cambridge: Cambridge University Press.

IPCC (1996). *Climate Change 1995: Economic and Social Dimensions of Climate Change. Contribution of Working Group III to the Second Assessment Report of the Intergovernmental Panel on Climate Change*, ed. J. P. Bruce, H. Lee, E. F. Haites. *et al.* Cambridge:Cambridge University Press.

IPCC (2000). *Emissions Scenarios: A Special Report of Working Group III of the Intergovernmental Panel on Climate Change*, ed. N. Nakicenovic and R. Swart. Cambridge: Cambridge University Press.

IPCC (2001). *Climate Change 2001: The Scientific Basis. Contribution of Working Group I to the Third Assesment Report of the Intergovernmental Panel on Climate Change*, ed. J. T. Houghton, Y. Ding, D. J. Griggs, *et al.* Cambridge:Cambridge University Press.

Manne, A. S. and Richels, R. G. (1994). The costs of stabilizing global CO_2 emissions: a probabilistic analysis based on expert judgments. *Energy Journal* **15**(1), 31–56.

Manne, A. S. and Richels, R. G. (1995). The greenhouse debate: Economic efficiency, burden sharing and hedging strategies. *The Energy Journal*, **16**(4), 1–38.

Manne, A. S. and Richels, R. G. (1997). On stabilizing CO_2 concentrations: cost-effective emission reductions strategies. *Environmental Modeling and Assessment*, **2**(4), 251–265.

Manne, A. S. and Richels, R. G. (2001). An alternative approach to establishing trade-offs among greenhouse gases. *Nature* **410**, 675–677.

Moss, R. H. and Schneider, S. H. (2000). Uncertainties in the IPCC Third Assessment Report: Recommendations to lead authors for more consistent assessment and reporting. In *Guidance Papers on the Cross Cutting Issues of the Third Assessment Report of the IPCC*, ed. R. Pachauri, K. Tanaka and T. Taniguchi. Geneva: IPCC, pp. 33–51.

Nakićenović, N., Grubler, A. and McDonald, A. (1998). *Global Energy Perspectives*. New York: Cambridge University Press.

Reilly, J., Prinn, R., Harnisch, J. *et al.* (1999). Multigas assessment of the Kyoto Protocol. *Nature* **401**, 549–555.

Runci, P., Clarke, L. and Dooley, J. (2005). *Energy R&D Investment in the Industrialized World: Historic and Future Directions*. Working Paper PNNL-SA-47701. Pacific Northwest National Laboratory.

Schmalensee, R. (1993). Comparing greenhouse gases for policy purposes. *The Energy Journal*, **14**(1), 245–255.

Schneider, S. H. and Goulder, L. (1997). Achieving low-cost emission targets. *Nature* **389**, 13–14.

Stanford Energy Modeling Forum (EMF-21) (2005). *Multi-gas Mitigation and Climate Control*, January 31, 2005, Washington, DC. www.stanford.edu/group/EMF/

Wigley, T. M. L. and Raper, S. C. B. (1992). Implications for climate and sea level of revised IPCC emissions scenarios. *Nature* **357**, 293–300.

Wigley, T. M. L. and Raper, S. C. B. (2001). Interpretation of high projections for global-mean warming. *Science* **293**, 451–454.

Wigley, T. M. L. and Raper, S. C. B. (2002). Reasons for larger warming projections in the IPCC Third Assessment Report. *Journal of Climate* **15**, 2945–2952.

Wigley, T. M. L., Richels, R. and Edmonds, J. A. (1996). Economic and environmental choices in the stabilization of atmospheric CO_2 concentrations. *Nature* **379**, 240–243.

33

International climate policy: approaches to policies and measures, and international coordination and cooperation

Brian S. Fisher, A. L. Matysek, M. A. Ford and K. Woffenden

33.1 Introduction

Since the industrial revolution, anthropogenic activities such as the burning of fossil fuels, and agricultural and industrial production, have led to growing emissions of greenhouse gases. It is projected that a sustained increase in the demand for energy, particularly in developing countries, and the continued reliance on fossil fuels as an energy source, will cause anthropogenic emissions to rise substantially into the future. The continued growth in emissions is expected to significantly increase atmospheric concentrations of greenhouse gases beyond current levels. Evidence now suggests that these human-induced increases in atmospheric greenhouse gas concentrations have the potential to alter the Earth's climate dramatically, leading to possible environmental and economic damage (Schneider and Sarukhan, 2001).

Human-induced climate change is a global phenomenon that is expected to have impacts in all regions of the world, although with differing specific effects. As a result, responding to the climate problem poses several challenges. One key challenge is to design policies that balance the cost of any damage from climate change with the cost of actions to reduce that damage. However, the considerable uncertainties surrounding the causes, nature, and impacts of possible climate change magnify this challenge. A second challenge will be for countries to manage adaptation, which could involve major investments and managing economic and social change. Of course, significant adaptation will occur as a result of economic agents responding to changes in both their physical and economic environments. Determining both the optimal level and type of policy intervention to facilitate adaptation without hindering the autonomous process will not be straightforward. A third challenge will be to engage all major emitters in meaningful efforts to reduce emissions. Emission abatement may involve reduced or more expensive energy use, with potential to hamper the development prospects of many countries.

The Kyoto Protocol, negotiated under the United Nations Framework Convention on Climate Change, is an initial attempt to address the climate problem. However, the Protocol is flawed because it fails to meet the three principles for an effective response. A new path forward must be found without the pitfalls inherent in the processes that led to the Kyoto Protocol.

33.2 Principles for effective policy

Policy to deal with the potential threats of human-induced climate change without compromising countries' capacity for development must adhere to three fundamental principles: environmental effectiveness; economic efficiency; and equity.

33.2.1 Environmental effectiveness

Two elements are important for environmental effectiveness as follows.

Human-induced Climate Change: An Interdisciplinary Assessment, ed. Michael Schlesinger, Haroon Kheshgi, Joel Smith, Francisco de la Chesnaye, John M. Reilly, Tom Wilson and Charles Kolstad. Published by Cambridge University Press. © Cambridge University Press 2007.

Focus on the right environmental objective
Atmospheric concentrations of greenhouse gases are understood to drive changes in global climate. Therefore the focus of the policy response must ultimately be on concentrations or some measure of climate change itself and not simply on the emissions of a select group of developed countries over a relatively short time-frame.

Involve all large emitters
A particularly important determinant of environmental effectiveness is that all major emitters need to be included to make significant reductions in greenhouse gas emissions and atmospheric concentrations in the long term. Excluding any major emitters would undermine the environmental effectiveness of abatement action in two ways. First, unless all major emitters are included, the total extent of action undertaken is reduced or the share to be done by others to achieve a given environmental goal is increased. Ultimately, controlling growth in atmospheric concentrations of greenhouse gases will not be possible without the involvement of all large emitters.

Second, emitters that do not take part in abatement action may gain a competitive advantage in production, inducing movement of emission-intensive industries to these countries from countries where emission constraints apply. This emission "leakage" partly offsets abatement undertaken elsewhere, and increases the economic costs of participating in any emission reduction process. An analysis of the Kyoto Protocol found that the carbon equivalent leakage rate was 12.4% in 2015 for non-participating countries (Jakeman et al., 2002). That is, carbon equivalent emissions from the United States and non-Annex B regions were projected to increase by 12.4 tonnes for every 100 tonne reduction in carbon equivalent emissions in participating Annex B countries. For a detailed discussion of carbon leakage see Hourcade et al. (2001), pp. 542–3.

In the short to medium term, however, it may be possible to achieve meaningful reductions in greenhouse gas emissions if only some of the major emitters agree to constraints on greenhouse gas emissions. For example, in 2002, China and the United States accounted for approximately 14 and 22% of global greenhouse gas emissions respectively. By 2050, it is projected that strong economic and population growth in China will see China's share of global greenhouse gas emissions rise to approximately 20%, while the United States' share will fall to about 16%. A bilateral agreement between these countries encouraging improvements in energy efficiency and energy intensity, and the transfer of new energy technologies could constrain emissions growth in both countries and thereby substantially reduce the growth in global greenhouse gas emissions simply as a consequence of the absolute sizes of these two economies.

In the long term, however, the involvement of all major emitters will be required to achieve an environmentally effective and equitable outcome. Bilateral agreements between major emitters can be seen as a way to make short-term cuts in emissions, while working toward achieving multilateral agreements in the longer term that will enable greater emissions abatement to be achieved.

33.2.2 Economic efficiency

An economically efficient policy to respond to climate change minimizes the cost or overall welfare impact of meeting an environmental objective. There are six important components to an economically efficient policy framework.

Embrace all opportunities for mitigation
Emission abatement opportunities exist in many sectors of economies. There is also potential to sequester carbon dioxide in soils and vegetation, as well as in geological formations. Increasing the range of potential abatement activities (and including all low-cost options) can reduce the cost of abatement since the lower the marginal cost of abatement, the more abatement can be achieved for a given level of economic effort.

Facilitate market-based solutions
Although most governments will adopt a portfolio of policies and measures to reduce greenhouse gas emissions it has been shown that market-based solutions are generally less costly than command and control approaches to abatement (Tietenberg, 1985; Hahn and Stavins 1992; Fisher et al., 1996).

Recognize the role of technology
The underlying demand for energy is rising strongly over time, both globally and for most individual countries (IEA, 2002). There are various ways to mitigate the growth in net greenhouse gas emissions under such circumstances: reduce economic growth and therefore energy use; discourage the use of emission-intensive technologies by increasing their cost relative to less emission-intensive technologies; and encourage development of cost-effective and less emission-intensive technologies. This can be done, for example, through investments in new technology to increase the efficiency of fossil-fuel-based energy production, by capturing and geologically sequestering carbon dioxide, or by developing renewable sources of energy. In practice, governments might choose a combination of approaches, but, given the importance of energy in driving economic growth, a technology solution is required to make a significant impact on greenhouse gas emissions without hampering development prospects. Development of new alternative sources of energy may also be desirable as a means to improve domestic energy security in some countries, by reducing reliance on imported fuels. Whether such a strategy is economically efficient will depend on both the nature and the consequences of possible energy supply disruptions.

Act over the appropriate time-frame
There are an infinite number of emission pathways that can achieve a given level of concentrations of greenhouse gases in the atmosphere. Hourcade *et al.* (2001) reviewed a number of studies that compared the cost of stabilizing atmospheric concentrations of carbon dioxide at a given level. It was found that a gradual transition away from carbon-intensive fuels in the short term, followed by a sharp reduction in emissions at some point in the future, was less costly for achieving a given atmospheric concentration than sharp reductions in the near term followed by less rapid declines in long-term emissions.

This cost effect occurs for several reasons. First, capital stock in energy-intensive industries is typically long-lived. Therefore, emission reductions may be achieved more cost-effectively at the point of capital stock turnover than by the costly early retirement of capital required for large reductions in near-term emissions (Hourcade *et al.*, 2001).

Second, projected improvements in the costs and efficiencies of a range of energy technologies suggest that a given level of emissions reduction will become cheaper in the future. Positive returns to capital mean that a given level of reduction in the future can be achieved using a smaller amount of today's resources, provided that today's resource rents are invested and earn a positive return (Hourcade *et al.*, 2001).

While delaying some climate mitigation actions to a point in the future may achieve a given level of atmospheric concentrations at a lower cost than the same actions today, other considerations may lead to early outcomes. For example, a decision to lower the target atmospheric concentration would bring forward the point in time at which substantial reductions in emissions would need to be achieved (Wigley *et al.*, 1996).

Be flexible in the light of new knowledge
Making projections about greenhouse gas emissions based on expected rates of economic growth, the cost and availability of technologies and the pattern of consumer demands is complicated and uncertain. Current understanding of global warming, and of associated climate changes and potential impacts, is quite limited. However, it is likely that understanding will improve over time and given the long time horizon over which action will be required, it is essential to build flexibility into any response so that strategies can be varied as knowledge improves.

Include adaptation strategies
Even immediate and severe emission abatement will not avert some degree of global warming. It follows that strategies for adapting to change will be required. A cost-effective approach to the climate problem, and therefore one that is sustainable and invites participation, would combine mitigation activities with adaptation.

A fundamental aim in developing a policy response to global warming must be to achieve the maximum benefit from mitigating adverse impacts of climate change while at the same time striving to minimize the total cost of actions required and the adaptation that will inevitably need to take place. At the point where the marginal benefit of avoiding the adverse impacts of global warming is equal to the marginal cost of the action required, which includes adaptation responses and emission abatement, global welfare will be maximized.

33.2.3 Equity

Because the problem of climate change transcends national boundaries, it requires an international response and a framework that is perceived as fair. An equitable framework would have three elements.

Consistency with sustainable economic development
The Convention recognizes that environmental objectives need to be met in a way that facilitates countries' expectations of future economic growth and development: "all countries ... need access to resources required to achieve sustainable social and economic development and that, in order for developing countries to progress toward that goal, their energy consumption will need to grow ..." (UNFCCC, 1999, p. 4).

While there may often be complex tradeoffs between economic and environmental objectives, sustainable development strategies seek those actions that are beneficial to both economic growth and the environment. Without recognition of the need to achieve environmental goals within economic and social realities, countries will not participate in an international regime designed to address global warming.

No coercion
Forcing countries to agree to mitigation activities is unlikely to prove successful in the long run. A legally binding framework may reduce the incentive for global participation and the threat of punishment for failure provides an incentive to withdraw altogether if meeting tightly defined targets becomes impossible. In the long run, international agreements need a strong element of cooperation to remain successful. A framework designed to promote self-interested participation is likely to be far more successful than one brought about through coercion and punitive threats (Barrett, 2002).

Technology transfer
Technology exists today that can put developing countries on a lower emissions trajectory than their developed country counterparts were at the same stage of economic development. Despite the apparent win–win associated with technology transfer from developed to developing countries, there are a number of barriers that impede large-scale transfer. It is

essential to work toward reducing these barriers to allow access to existing and new technologies. These issues are discussed further below.

33.3 The current policy framework scorecard

Reflecting widespread concerns about potential climate change, the great majority of national governments are parties to the UNFCCC. The Convention commits parties to taking action aimed at achieving the objective of stabilizing "greenhouse gas concentrations in the atmosphere at a level that would prevent dangerous anthropogenic interference with the climate system."

The Kyoto Protocol to the Convention, adopted in 1997, is the most significant outcome of the international negotiations on climate change response policy so far. Given Russia's decision to ratify the Protocol it entered into force in February 2005 despite its repudiation by the United States.

Some 12 years after 155 countries originally signed the Convention, we have, in the Protocol, an agreement that will do little to curb global greenhouse gas emissions or to move toward stabilizing atmospheric concentrations of greenhouse gases. Given the extent of activity that has taken place under the Convention, this outcome is surely not for lack of resources or international negotiating effort. Why has the process failed the environment so far, and can things improve?

Repudiation of the Protocol in March 2001 by the single largest emitter significantly reduced the potential emission reduction the Protocol could achieve. The opportunity that the US renunciation provided for reflection on the direction of global climate policy was unfortunately lost as a result of the prevailing attachment by many to Kyoto as "the only game in town" and as a "first step." But does it represent a first step along a pathway toward an effective global regime? Before considering that question it is worth reflecting on what the Convention, and under it, the Protocol, have achieved.

International awareness of global climate change has certainly been greatly heightened over the past decade. The Convention has provided a forum for exchange of information and ideas. A range of non-binding actions have been promoted and techniques for measuring and reporting emissions developed. However, negotiations have often been conducted in a divisive atmosphere that has strained international relations, something that is hardly conducive to progress when so much depends on collaboration.

Clearly, the policy framework has several significant shortcomings. Some of these are inherent in the nature of the UN negotiations; others have been built into the detail of the Convention and the Protocol. Bodansky (2001) enumerates some of the lessons to be learnt from the Kyoto process.

Bodansky first points to the complexity of the Protocol and its elaborate architecture. The institutional support required to maintain the many aspects of ensuring that parties meet their targets is enormous. In addition, there are a very large number of players in any UN negotiation, and they have a diverse range of interests and agendas. Progress is therefore cumbersome and slow as a result of politics and difficult logistics.

Second, contrary to typical practice in treaty making, the Kyoto Protocol does not involve a mutual exchange of promises among parties. Rather, commitments have been made by a specific group of parties according to rules negotiated by all the parties, including those to whom group commitments do not apply. Developing countries participated fully in the negotiation process yet were specifically absolved from taking on commitments to emission reduction (Bodansky, 2001).

The origin of this problem is inextricably embedded in the Convention itself. The Convention contains a number of clauses that have been used by the developing country bloc to justify their unwillingness to accept mitigation commitments. Article 3.1 of the Convention states that "Parties should protect the climate system ... in accordance with their common but differentiated responsibilities and respective capabilities ... developed country Parties should take the lead ... "

Developing countries have interpreted this to mean that developed countries must reduce their emissions before developing countries will take on any emission reduction commitments of their own, and moreover, that developed countries must first have made demonstrable progress in reducing emissions.

Article 4.7 states that developing country commitments "will depend on the effective implementation by developed country Parties of their commitment under the Convention related to financial resources and transfer of technology ... "

Coupled with the reference in Article 3.1 to "respective capabilities" this has been used to underpin the developing countries' argument that they cannot take on mitigation unless and until developed countries provide the resources to increase their capacity to do so. Article 4.7 also contains the phrase: " ... economic and social development and poverty eradication are the first and over-riding priorities of the developing country Parties."

Developing countries argue that they cannot afford to divert resources to mitigation activities as development will always be the priority. This is of course a reasonable argument, but enshrining this and related principles in the Convention has inevitably led to many unproductive hours of north–south debate in the negotiations. Thus, under the current framework, not only is there no emission abatement required by developing countries, but there is also little prospect of any way forward for their future engagement.

These problems are exacerbated by the fact that inclusion in Annex I to the Convention or Annex B to the Protocol has not been determined strictly on the basis of a country's per person income. A significant number of countries not included in Annex B currently have per person incomes well in excess of some countries that have taken on targets under the Protocol.

33.4 A new framework?

The framework associated with the Kyoto Protocol is unlikely to lead to effective environmental outcomes. What, if any, framework could lead to policy outcomes that meet the principles of economic efficiency, environmental effectiveness, and equity, as outlined earlier? The answer depends on finding a framework that more effectively aligns the national interests of major emitters with globally optimal environmental outcomes than is the case under the Kyoto Protocol.

33.4.1 Existing drivers

There are several existing drivers that could move countries toward reducing greenhouse gas emissions without resorting to an over-arching multilaterally negotiated framework such as the Kyoto Protocol. These drivers typically act in ways that address climate change indirectly, owing to the occurrence of ancillary benefits associated with other developmental, social, environmental or economic policies.

Economic – There are a number of options for energy efficiency improvements within economies that can reduce costs over time. Often energy savings emerge as a result of broader economic reforms. This is evidenced by the reforms in China that have led to energy efficiency improvements in coal-fired power generation and industry, as well as a decline in direct household use of coal (and a shift to electricity).

Reforms may often involve removal of energy subsidies, which not only distort the choice between competing fuels, but may also fail to achieve their stated goals. IEA and OECD studies have concluded that removing subsidies from the energy sector alone across a sample of eight countries – India, China, Russia, Iran, South Africa, Indonesia, Kazakhstan, and Venezuela – could result in a reduction in global emissions of almost 5% (IEA, 1999, 2000). This reduction is greater than that achievable under the Kyoto Protocol during the first commitment period.

International trade and investment – Some forms of trade and investment can lead to mutual economic and environmental benefits. For example, increased investment in the transfer of more energy-efficient technologies to developing countries offers potentially valuable long-term commercial relationships and profits for the investor and also more energy-efficient production processes in the developing country. Significant reductions in greenhouse gas emissions in developing countries could be achieved with the transfer and diffusion of more energy-efficient technologies that are currently in use or being developed for commercialization in developed regions.

A number of voluntary agreements on climate already exist. For example, the Australia–China bilateral agreement on climate, outlined in the Memorandum of Understanding between the Government of Australia and the Government of the People's Republic of China on Climate Change Activities, aims to encourage joint activities in a range of areas including climate change research; climate change policies and measures; climate change impacts and adaptation; greenhouse gas inventories and projections; technology cooperation; capacity building; public awareness; and renewable energy and energy efficiency.

It may also be in the interests of multinational companies to transfer efficient energy technologies to all foreign subsidiaries in response to emphasis on triple bottom line reporting. Companies wishing to demonstrate corporate social responsibility may initiate the transfer and diffusion of efficient technologies into developing regions at a quickened pace if the investment environment is suitable.

Energy efficiency standards in sectors such as transport and manufacturing could also "pull" newer technologies into trade-affected markets. The greater the number of countries that apply a given environmental standard, the greater the incentive for other countries to implement similar standards, since a larger number of trading partners are affected. Adopting uniform standards under such circumstances would be important in remaining competitive in the global market and this could promote adoption of similar standards in domestic markets, since it would become less viable to maintain production of both types of commodities (Buchner and Carraro, 2004). These effects may be particularly strong in sectors where there are a limited number of global manufacturers. For example, in the automobile industry there are fewer than a dozen major manufacturers, so it is possible that fuel efficiency standards adopted in major country markets could have flow-on effects to the global industry. The European Union already has a voluntary agreement in place with Japanese and Korean car manufacturers to limit carbon dioxide emissions. Such agreements could potentially be extended to other importing regions given that existing production standards are in place elsewhere.

Domestic pollution – Increasing urbanization and industrialization have led to severe local and regional air pollution problems in some countries, particularly in the urban areas of developing nations. Major sources of pollution include the combustion of fossil fuels in electricity production, transport, households, and agriculture and forest fires. Declining air quality can have a number of impacts including widespread health effects, adverse effects on infrastructure and agricultural production losses (Aunan *et al.*, 2004).

The level of total suspended particles in the atmosphere in some of the Association of Southeast Asian Nations (ASEAN) is well above World Health Organization standards and polices have been implemented to try to reduce the level of local air pollutants and mitigate the impacts of declining air quality. For example, most ASEAN member countries have phased out leaded fuel and are adopting higher air quality standards. Improvements in energy production, transportation emissions, and diversification to less carbon-intensive fuels are also being encouraged (Karki *et al.*, 2005).

China is experiencing adverse health impacts from degraded air quality in many of its major cities. The Bureau of East Asian and Pacific Affairs (2004) estimates the cost of pollution at between 7 and 10% of China's GDP each year. There is therefore a significant direct economic incentive for China, and other countries with similar air quality problems, to reduce local air pollution. Since many of the local pollutants are associated with burning fossil fuels, efforts to improve local air quality may also provide secondary climate benefits, although there are complicated atmospheric chemistry effects that make these secondary climate effects difficult to assess (Wang and Prinn, 1997; Reilly et al., 2003).

India also suffers from poor air quality and has adopted a range of legislation designed to improve air quality standards, including emission standards for vehicles and national air quality standards for sulfur dioxide and other pollutants. But enforcement of these standards is quite lax, and there has been recent judicial action to try to enforce previously ignored legislation dealing with air pollution (EIA, 2003; GAO, 2003).

The environmental, social, and economic impacts of cross-boundary air pollution are also recognized in the Convention on Long-range Transboundary Air Pollution. This Convention was adopted in 1979 and establishes a broad framework throughout European and North American regions for co-operative action on air pollution by promoting research, monitoring and the development of emission reduction strategies on a regional scale. Seven legally binding protocols establishing controls on specific pollutants have also entered into force and provide an example of agreements that may have important secondary climate benefits (UNECE, 2004).

Domestic desire to deal with the climate problem – Ahmad et al. (2001) report that there are likely to be regionally differentiated environmental and economic consequences of climate change in the long term. Regional capacities to adapt to climate change will also differ because of variations in the climate impact, economic resources, agricultural systems, and human health facilities. There is, therefore, an underlying economic incentive for some countries to mitigate the impacts of climate change to the extent that it is cost-effective to do so in conjunction with adaptation activities.

A number of developed countries already have domestic policies specifically designed to reduce greenhouse gas emissions. The United States, for example, has a target of reducing the emission intensity of its economy by 18% between 2002 and 2012, in order to reduce forecast emissions by 4.5% over that period. The United States is also investing heavily in technology development and climate research (White House Council on Environmental Policy, 2004). A number of European countries also implemented a range of abatement actions prior to EU ratification of the Kyoto Protocol.

Although climate change remains a more marginal issue in developing countries compared with issues such as poverty alleviation, food security, and economic and social development, developing countries are also initiating developmental policies that may have significant climate benefits. For example, agro-forestry projects designed to improve food production and prevent soil erosion can maintain carbon levels in the soil, transforming some agricultural soils from carbon sources to carbon sinks (Davidson et al., 2003).

The expansion of the Brazilian ethanol program has demonstrated the technical feasibility of large-scale ethanol production and has provided ethanol and ethanol–gasoline mixes for the Brazilian car fleet and industry. The widespread use of ethanol has reduced city air pollution and is reported to have led to the avoidance of 6–10 million tonnes of carbon emissions per year since 1980 (Huq et al., 2003). Brazilian grain producers have also commonly adopted minimum tillage practices to reduce fuel and labor costs. It is estimated that in 2000, the minimum tillage cultivation of 12 million hectares led to the sequestration of 20–35 million tonnes of carbon (Huq et al., 2003).

Energy security – Many countries are becoming increasingly concerned about domestic energy security issues. Energy security broadly refers to the provision of a reliable and adequate supply of energy at "reasonable" prices in order to sustain economic growth. Energy security issues may arise from concerns about the source and type of fuel supplies and the maintenance and security of energy infrastructure. The projected expansion of many economies and populations throughout the world, particularly in developing and transition countries, and the continued heavy reliance on oil in the transport sector is expected to lead to ongoing concerns about energy security. Some policy actions designed to address energy security concerns may also concurrently address the climate change problem.

Projected increases in demand for energy throughout the world are expected to place increasing pressure on domestic energy reserves and increase the import dependence of many countries such as the United States, China, and Japan. For example, China's total energy consumption is projected to increase by approximately 250% by 2030, with much of this growth in energy consumption expected to be supported by energy imports. China's net oil imports are projected to rise on average by 8.3% a year between 2002 and 2030, which would increase China's oil import dependence from about 30% in 2004 to around 85% by 2030 (Hogan et al., 2005).

Achieving continued and sustainable growth is heavily dependent on the reliable supply of energy. Continued geopolitical tensions and volatility in the world oil market have led to increased concern in many nations about rising dependence on energy imports. Policies that could be adopted to address energy security concerns may also have secondary climate benefits. Such policies include the improvement of supply- and demand-side energy efficiency and the diversification of fuel supplies to less carbon-intensive sources such as natural gas and biofuels. Renewable energy is also seen as a means to greatly reduce the risk of external shocks to fuel types, sources, and prices while simultaneously providing significant greenhouse gas abatement.

The existence of domestic drivers that work toward improved climate outcomes is critical for an effective global policy framework. These drivers can provide a basis for actions that address the climate problem while being consistent with national interests, including economic and development interests. Each country will have different domestic drivers, and place different emphasis on the importance of various drivers, reflecting the diverse structures and circumstances of economies.

33.4.2 Leveraging existing domestic drivers for further international cooperation

International trade and cooperation have the potential to deliver mutual economic benefits to all involved parties. It may therefore be possible to develop an international framework under which countries cooperate to achieve more cost-effectively goals relating to their domestic abatement drivers, but with ancillary benefits for climate change. Under such a framework, existing domestic drivers would be used to leverage trade and international cooperation to achieve significant emission reductions at a reduced cost.

One way forward is through bilateral agreements that aid and promote cooperation between concerned countries in achieving national interest goals that are also consistent with positive climate change outcomes. These agreements could cover a number of areas. For example: facilitating foreign direct investment in alternative or more energy-efficient technology; facilitating investment flows that assist in dealing with adaptation to climate change; facilitating investment flows that generate capital structures more consistent with meeting domestic pollution reduction objectives; providing assistance in adopting economic reforms that result in reduced greenhouse gas emissions; providing capacity building assistance to strengthen legal and regulatory environments, and facilitate technology transfer and development, through, for example, protection of intellectual property rights; liberalizing trade flows to ensure production is taking place in regions employing resources most efficiently; and sharing of scientific and economic data and exchanges of relevant climate and technological expertise.

A key element of a climate change forward strategy that leverages the aforementioned policy drivers is technology. Several authors have advocated the potential benefits of technology agreements. Buchner and Carraro (2004) note that a self-enforcing agreement is more likely to emerge when countries cooperate on environmental technological innovation than when they attempt to cooperate on emission abatement, in part because such agreements are not plagued by the free-rider problem.

Since many countries, including those among the 12 largest emitters defined in this paper (Figure 33.1), already have in place agreements to deal with technology cooperation, using cooperative mechanisms to drive technology research, deployment, and diffusion is a logical pathway forward. Many of the obstacles inherent in a Kyoto-type process have already been overcome where countries have voluntarily agreed to cooperate in their self-interest. This self-interest has been underpinned by the adoption of policies and measures that improve access to advanced technologies and enhance growth prospects while simultaneously reducing emissions.

However, it is not merely the existence of bilateral or multilateral technology agreements between countries that might drive a technology solution to the climate problem. It is equally feasible that increasing energy security concerns, domestic pollution issues, and trade and investment could catalyze a more rapid introduction of low and zero emission technologies throughout various sectors of the global economy.

33.5 Technology: a key element for success

The increasing global demand for energy is a key indicator that technology must play a crucial role in any future climate framework. The following section provides some examples of the possibilities afforded by new and existing technology options for achieving significant, cost-effective emission reductions.

33.5.1 Examples of potential technology opportunities

Advances in technology provide opportunities for emissions abatement throughout the economy while still allowing economic growth and development.

Significant technological advances designed to reduce greenhouse gas emissions are occurring in the industrial sector, which currently accounts for approximately 41% of global carbon dioxide emissions (Worrell and Levine, 1999). For example, technological advancements in the iron ore and steel industry that reduce carbon dioxide emissions include improvements in process energy efficiency and the recovery of carbon dioxide from blast furnace gas.

The aluminum industry is also experimenting with inert, non-carbon anodes that could potentially reduce carbon dioxide emissions by around 10%. Drained/wettable cathodes are another new technology with the potential to reduce electricity use in aluminium smelting by between 10 and 20% from around 2010 onwards. It is expected that a combination of the two types of technology will not only give greater energy savings, but also eliminate anode emissions. The combined savings would result in around a 35% reduction in carbon dioxide emissions per tonne of aluminum produced for producers relying on electricity generated using fossil fuels.

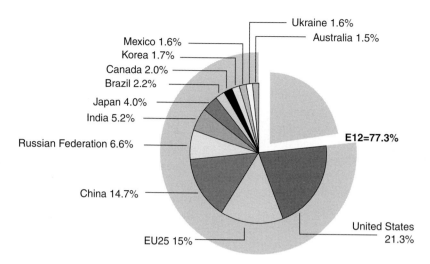

Figure 33.1 Share of global greenhouse gas emissions for selected regions (2002). Data shown are CO_2 equivalent calculated using 100-year GWPs from the IPCC Second Assessment Report. IPCC (1996).

Technological improvements that could have beneficial impacts in the transport sector include hybrid gasoline–electric cars, which are expected to improve fuel efficiency by between 50 and 65% compared with conventional vehicles, and fuel cell vehicles powered by hydrogen, which could potentially reduce carbon dioxide emissions by 45% compared with conventional engines, depending on the source used to generate the hydrogen (IEA, 2000).

Moomaw and Moreira (2001) refer to numerous technology options for reducing emissions in the buildings, manufacturing and energy sectors as well as in the agriculture and waste industries. Examples of technologies that could reduce emissions in the buildings sector, which contributes to approximately 31% of global greenhouse gas emissions (Millhone, 1999), include installation of more efficient heating, cooling, and lighting equipment, and advanced window and insulation retrofits.

One of the major opportunities for global emissions abatement is from technological improvements in electricity generation. The large, stationary source nature of the electricity sector makes it a prime candidate for targeting significant reductions in greenhouse gas emissions.

There are three main technological options for reducing greenhouse gas emissions from electricity generation: improving the efficiency of electricity generation by reducing the amount of fuel input per unit of energy output; fuel switching from more carbon-intensive to less carbon-intensive fuels or zero carbon-emitting sources of energy; and carbon capture and storage to allow for near zero emissions from power generation.

The potential for low and zero emission technologies such as carbon capture and storage, renewables, and hydrogen to transform the power and transport systems and thereby decouple economic growth from emissions is the fundamental tenet of a technology solution. While the domestic abatement drivers referred to above may provide enough incentive along with inherent cost reductions associated with learning effects to push advanced technologies into the system, policy interventions that promote the uptake of such technologies and reduce impediments to their uptake may provide a fast track for an environmentally effective outcome.

Policy interventions to speed the uptake of new, more efficient technologies may be necessary to achieve an environmentally effective outcome. This is because if market forces are left to act in isolation, the uptake of new technology will be limited to that which occurs at the natural turnover rate, or once capital has been fully amortized. To achieve the abatement necessary to avoid dangerous interference with the climate system, more significant near-term abatement may be required than that which would be achieved from new technology uptake associated with the natural turnover of capital stock.

33.5.2 Impediments to the diffusion of technology

There are a number of factors that influence the rapid diffusion of technology. First, macroeconomic conditions greatly influence the potential for success of technical transfer. For example, high inflation, exchange rate fluctuations, and trade policies that disallow free movement of capital all act as impediments to technology transfer by significantly increasing the risk associated with investment and reducing credit availability. Risk also increases the discount rate, thereby affecting the attractiveness of investments.

Second, since technical innovation is linked to R&D expenditure, having insufficient resources invested in research and development could also act as an impediment to innovation and subsequent technology diffusion. As such, declining government expenditures on energy R&D across all types of energy research, including fossil fuels, nuclear, and renewables may represent a policy issue to be addressed.

Third, concern about protection of intellectual property rights is a key issue. Ogonowski *et al.* (2004) found that countries with stronger intellectual property rights receive more foreign direct investment and imports. Strengthening intellectual property rights in developing countries would increase the transfer of technology.

Fourth, inadequate institutional and human capital may also hamper diffusion of new technology. Institutional capacity is important not only for providing adequate intellectual property rights protection, but also for providing necessary linkages between technology providers, users, and developers. Inadequate human capacity, or a lack of the knowledge, skills, and practical experience required to implement, operate, and maintain new technology reduces productivity and impedes rapid technology transfer (UNEP, 2003).

Fifth, inadequate infrastructure can impede investment in new technologies, as projects can be dependent on external infrastructure, such as gas pipelines or electricity grids. If the infrastructure is unreliable or of poor quality, the project will be less likely to go ahead.

Sixth, policy environments may act to deter investment if issues such as double taxation are not adequately dealt with. Market interventions such as tax and subsidy distortions have the potential to alter the relative prices of energy and may distort incentives for fuel switching to low-carbon fuels.

Many such barriers to technology transfer could potentially be significantly reduced if governments, acknowledging the economic and environmental benefits, worked together to overcome them. In particular, recognizing success stories and engaging developed countries to assist developing countries in implementing strategies to overcome developmental and capacity issues would provide a tangible step forward.

Technology research, deployment and diffusion are not "one-size-fits-all" processes: solutions must be tailored to suit individual circumstances. There are many factors that influence the rate of technological change, potentially making it less than optimal. Clearly there is a role for governments to play in addressing many of these influences. This makes the international framework proposed here a particularly suitable one for focusing on technology while at the same time helping to address the climate change problem.

33.6 Advantages of the proposed framework

The voluntary framework outlined in this paper has a number of advantages over the established approach under the UNFCCC. Under a set of cooperative bilateral partnerships the extent of commitment would be a question for each partner, rather than jointly determined by a larger group. This is in contrast to the UNFCCC consensus-based approach to negotiations, which is not conducive to outcomes tailored to particular countries' circumstances.

In 2002, twelve countries (if the European Union is counted as a single unit) gave rise to almost 80% of global greenhouse gas emissions (Figure 33.1). Together, China and the United States are likely to account for 40–50% of global carbon dioxide emissions by 2050. Significant progress could be made in abatement if agreements could be struck between these countries. This is not an argument for excluding smaller emitters – which should be free to engage in discussions at any time – but an argument for focusing on agreements between key players that meet the conditions of environmental effectiveness, economic efficiency, and equity.

A bilateral framework such as this avoids the demanding global negotiating and legal framework since a large international bureaucracy is not required to enforce actions. Negotiating effort can be focused where it brings results rather than being dissipated by side issues and obstructionism. The network of bilateral relationships that form could be multilateralized at some later date and this might be a natural evolution if there were found to be mutual benefits; for example, an Australian agreement with the United States might link into agreements between the United States and China, and between Australia and China.

Multilateralizing such agreements should not prove a complex process. Unlike trade agreements, which can be difficult to multilateralize on account of rules of origin, climate agreements would suffer no such impediment. For example, not only are environmental standards agreements achieved under bilateral partnerships legally enforceable under the WTO but multilateralizing such arrangements may actually augment economic efficiency. This is because production could be made more uniform in a market operating under a harmonized standard than would be the case in a differentiated regulatory environment.

Under the framework presented here, the individual circumstances of each country can be addressed and leveraged to maximize the overall climate change response. The motivation for participation will differ between countries.

The importance of technology in finding a solution to the climate problem cannot be overemphasized. While there are already many bilateral agreements that focus on funding and capacity building, this is often not sufficient for technology transfer. Full technical diffusion will inevitably be integrated with trade, globalization, and development. For this reason, policies that promote economic reforms that are also consistent with reduced greenhouse gas emissions are an essential component of any international arrangement focused on technology diffusion. An approach such as the one outlined in this paper could potentially address some of the key impediments to transfer of technology. Importantly, because the nature of any technology transferred is initially the subject of bilateral arrangements only, parties are free to choose the investment that best meets their needs.

This approach relies on market forces and domestic policy drivers rather than an enforced multilateral regime. A set of bilateral partnerships or multilateral arrangements between major developed and developing emitters to reduce greenhouse gas emissions is not inconsistent, however, with the continued existence of the UNFCCC. The Convention could continue to offer a useful forum for exchange of scientific knowledge, and sharing data and technical information.

33.7 Conclusions

The approach to the climate problem to date – based on mandatory targets and culminating in the Kyoto Protocol – has an enticing appearance of environmental certainty. The targets, processes, and disciplinary focus create the illusion of achievement. But the reality is that such an approach is unlikely to work. Notwithstanding its entry into force in 2005, the withdrawal of the single largest emitter, the United States, from the Protocol was a clear signal that another approach needs to be considered.

Although all countries have a long-term interest in finding a solution, not all have the same means at their disposal to take action, not all are motivated by the same drivers, and not all can have the same effect on the outcome. An approach such as the one outlined here acknowledges these facts, and builds on actions that relate to national interests, while at the same time recognizing that it is essential to have developing country participation if the climate problem is to be effectively addressed. Most importantly, because the actions are the subject of voluntary arrangements without externally imposed regulation, parties can choose the best investments for their needs and there is no risk of generating unenforceable arrangements.

The approach postulated here moves away from the punitive approach brought by some parties to the Kyoto negotiations and seeks to rely on positive drivers for change. Everybody can be better off, which encourages participation and allows the action to develop its own momentum, without expensive negotiations that end in international stand-offs. Coercion is not required if there are demonstrable and shared benefits for developed countries, developing countries, and the environment.

References

Ahmad, Q. K., Anisimov, O., Arnell, N. *et al.* (2001). 'Summary for Policymakers'. In *Climate Change 2001: Impacts, Adaptation and Vulnerability. Contribution of Working Group II to the Third Assessment Report of the Intergovernmental Panel on Climate Change*, ed. J. J. McCarthy, O. F. Canziani, N. A. Leary, D. J. Dokken and K. S. White. Cambridge University Press, Cambridge.

Aunan, K. Fang, J. Vennemo, H. Oye, K. and Seip, H. M. (2004). Co-benefits of climate policy: lessons learned from a study in Shanxi, China. *Energy Policy* **32**, 567–571.

Barrett, S. (2002). *Towards a Better Climate Treaty*. Fondazione Eni Enrico Mattei Nota di Lavoro 54.2002.

Bodansky, D. (2001). Bonn voyage: Kyoto's uncertain revival. *The National Interest*, Fall.

Buchner, B. and Carraro, C. (2004). *Economic and Environmental Effectiveness of a Technology-based Climate Protocol*. Fondazione Eni Enrico Mattei Nota di Lavoro 61.2004.

Bureau of East Asian and Pacific Affairs (2004). Background Note: China, www.state.gov/r/pa/ei/bgn/18902.htm (accessed 15 September 2004).

Davidson, O., Halsnaes, K., Huq, S. *et al.* (2003). The development and climate nexus: the case of sub-Saharan Africa. *Climate Policy* **3S1** S97–S113.

Energy Information Administration (EIA) (2003). *International Energy Outlook 2003*. Washington: EIA.

Fisher, B. S., Barrett, S., Bohm, P. *et al.* (1996). An economic assessment of policy instruments for combating climate change. In *Climate Change 1995: Economic and Social Dimensions of Climate Change*. Contribution of Working Group III to the Second Assessment Report of the Intergovernmental Panel on Climate Change, ed. J. P. Bruce, H. Lee and E. F. Haites. Cambridge University Press, Cambridge, pp. 397–439.

General Accounting Office (GAO) (2003). *Climate Change. Information on Three Air Pollutants' Climate Effects and Emissions Trends*. United States General Accounting Office report to Congressional Requesters, House of Representatives www.gao.gov/new.items/d0325.pdf. (accessed 26 August 2004).

Hahn, R. W. and Stavins, R. (1992). Economic incentives for environmental protection: integrating theory and practice. *American Economic Review* **82**, 464–468.

Hogan, L., Fairhead, L., Gurney, A. and Pritchard, R. (2005). *Energy Security in APEC: Assessing the Costs of Energy Supply Disruptions and the Impacts of Alternative Energy Security Strategies*. ABARE Research Report 05.2. Canberra: ABARE.

Hourcade, J. C., Shukla, P. Kverndokk, S. *et al.* (2001). Global, regional, and national costs and ancillary benefits of mitigation. In *Climate Change 2001: Mitigation. Contribution of Working Group III to the Third Assessment Report of the Intergovernmental Panel on Climate Change*, ed. B. Metz, O. Davidson, R. Swart and J. Pan. Cambridge University Press, Cambridge.

Huq, S., Reid, H. and Kok, M. (2003). *Development and Climate. Bridging the Gap between National Development Policies and Dealing with Climate Change*. Results of Phase 1. Bilthoven: The Netherlands Environmental Assessment Agency.

International Energy Agency (IEA) (1999). *World Energy Outlook: Looking at Energy Subsidies. Getting the Prices Right*. Paris: OECD/IEA.

IEA (2000). *Energy Technology and Climate Change: A Call to Action*. Paris: OECD/IEA.

IEA (2002). *World Energy Outlook 2002*. Paris: IEA/OECD.

Jakeman, G., Hester, S., Woffenden, K. and Fisher, B. (2002). Kyoto Protocol. The first commitment and beyond. In *Australian Commodities March 2002, Forecasts and Issues*, **9**(1), ed. A. Wright. Canberra: ABARE.

Karki, S. K., Mann, M. D. and Salehfar, H. (2005). Energy and environment in the ASEAN: challenges and opportunities. *Energy Policy* **33**, 499–509.

Millhone, J. (1999). Residential, commercial, and institutional buildings sector. In *Methodological and Technological Issues in Technology Transfer. Special Report of the Intergovernmental Panel on Climate Change*, ed. B. Metz, D. Davidson, J. Martens, S. van Rooijen and L. Van Wie McGrory. United Nations Environment Programme/World Meteorological Organization. Cambridge: Cambridge University Press.

Moomaw, W. R. and Moreira, J. R. (2001). Technological and economic potential of greenhouse gas emissions reduction. In *Climate Change 2001: Mitigation. Contribution of Working Group III to the Third Assessment Report of the Intergovernmental Panel on Climate Change*, ed. B. Metz, O. Davidson, R. Swart and J. Pan. Cambridge University Press, Cambridge.

Ogonowski, M., Schmidt, J. and Williams, E. (2004). *Advancing Climate Change Technologies through Shared Intellectual Property Rights*. Working paper prepared for the Dialogue on Future International Actions to Address Global Climate Change, Stockholm, 16–19 May.

Reilly, J. M., Jacoby, H. D. and Prinn, R. G. (2003). *Multi-Gas Contributors to Global Climate Change: Climate Impacts and Mitigation Costs of Non-CO_2 Gases*. Arlington, VA: Pew Center on Global Climate Change.

Schneider, S. and Sarukhan, J. (2001). Overview of impacts, adaptation, and vulnerability to climate change. In *Climate Change 2001: Impacts, Adaptation and Vulnerability. Contribution of Working Group II to the Third Assessment Report of the Intergovernmental Panel on Climate Change*, ed. J. J. McCarthy, O. F. Canziani, N. A. Leary, D. J. Dokken and K. S. White. Cambridge University Press, Cambridge.

Tietenberg, T. H. (1985). *Emissions Trading: An Exercise in Reforming Pollution Policy*. Washington, DC: Resources for the Future.

United Nations Economics Commission for Europe (UNECE) (2004). *Convention on Long-range Transboundary Air Pollution*, www.unece.org/env/lrtap/lrtap_h1.htm. (accessed 20 August 2004).

United Nations Environment Programme (UNEP) (2003). *Technology Transfer: The Seven C's for the Successful Transfer and Uptake of Environmentally Sound Technologies*. Osaka: International Environmental Technology Centre.

United Nations Framework Convention on Climate Change (UNFCCC) (1999). *Convention on Climate Change*. Bonn: UNEP Information Unit for Conventions.

Wang, C. and Prinn, R. (1997). *Interactions among Emissions, Atmospheric Chemistry and Climate Change*. MIT Report No. 25.

White House Council on Environmental Policy (2004). *US Climate Change Policy*. www.state.gov/g/oes/rls/fs/2004/36425.htm (accessed 20 August 2004).

Wigley, T. M. L., Richels, R. and Edmonds, J. A. (1996). Economic and environmental choices in the stabilization of atmospheric CO_2 concentrations. *Nature* **379**, 240–243.

Worrell, E. and Levine, M. (1999). Industry. In *Methodological and Technological Issues in Technology Transfer. Special Report of the Intergovernmental Panel on Climate Change*, ed. B. Metz, D. Davidson, J. Martens, S. van Rooijen and L. Van Wie McGrory Edited by United Nations Environment Programme/World Meteorological Organization. Cambridge University Press, Cambridge.

Index

Page numbers in *italic* denote figures. Page numbers in **bold** denote tables.

abstraction, model 379–80
accounting models, energy 172–4
acid production 271–2
acid rain 94, 99
adaptation 156–9, 403, 405
   analysis, global-mean sea-level rise 127–8
   dynamic 110
   global mean sea-level rise 127–8
adaptive capacity 157–8
adipic acid production 271–2
adjoint method 11, 15
advection, low-latitude 345
AEROCOM project 35
aerosol-cloud interaction 34–46
aerosols
   biomass burning 18, 23–31
   black carbon
      effect on climate 18–31, 55, 93, 94; regional 41, 43–6
      effect on clouds 37–8, 96
   carbonaceous
      and climate 34–46; regional impact 41, 43–6
   fossil fuel 23–31
   and photochemistry 95
   in radiative forcing 54–5, 57
   sea-salt 34–6
   sulfate 23–31
   and temperature trends 23–31
   volcanic 20, 25
afforestation 231
   offsets 239–41, *242*, 243, *244*, 245
Africa
   agriculture 105, 116
   carbon dioxide storage **182**, *191–2*
   climate sensitivity, and economic growth 139
   economic development 163–4, **165**, 166
   impact of climate change 161–6
   malaria, and economic growth 139
aggregation 178
agriculture
   and climate change
      Africa 105, 116, 161–6
      Asia 116
   fertilizers 273
   global land-use 253–5
   greenhouse gas emissions 238–47, 252, 253, 259
      ASMGHG model 248–50
      mitigation 239–47
   impact forecasting 110–12, 116
   impact of sea-level rise 121, 122
   land
      abandonment *73*, 79; biomass and carbon plantations 74, 75
      albedo 78
   methane emission 272–5
   models 228
   nitrous oxide emissions 252, 259, 272–5
agro-ecological zones
   crop production 254–5, 256–7, **259**, *260*, *261*
   FAO data 256, 262–4
   global 255–6
   and mitigation cost 257–9
air pollution 407–8
   caps 96–101
   effect on human health 138
   urban
      control 93–101; effect on economy 99–100; effect on sea-level 100; effect on temperature 100
      effect on climate 93–101
      photochemistry 94–5
albedo
   effect of aerosols 35
   effect on global-mean surface-air temperature 79–80
   and land-use change 72, 78, 79–81
   snow 76–7
Amazonia, effect of biomass aerosols 41, 44, 45–6
AMIGA model 174
Annex B countries **284**
   and climate change policy 282–92
   Kyoto Protocol 282–92
Anopheles, malaria vector 136
Antarctica
   halocarbon concentration 20
   stratospheric ozone 20
   see also West Antarctic Ice Sheet
Arctic haze 35
Armington Formulation 218
Arrhenius, Svante August (1859–1927) 6

Index

Asia
  agriculture 116
  carbon dioxide storage **182**, *191–2*
Asia-Pacific Integrated Model (AIM) 314–27, *315*, 381
Asia-Pacific region
  climate change, crop productivity 315–19
  climate change policy 314–27
  greenhouse gas emissions 314
  greenhouse gas stabilization 318, 319–23
Asia-Pacific Vulnerability Analysis 122
ASMGHG model 248–50
Atlantic circulation *see* North Atlantic meridional overturning circulation
Atlantic conveyor *see* North Atlantic meridional overturning circulation
atmosphere general circulation model 6–7, 10–11
atmosphere–ocean general circulation model 6–8, 10–11, 50–3, 108–9, 315
Australia
  carbon dioxide storage **182**, 190, *191–2*
  global mean sea-level rise, effect on agriculture 127
  wheat productivity and climate change 317–18
Australia–China bilateral agreement on climate 407, 411
Autonomous Energy Efficiency Improvement 168

backstop technology 329, 332, 336–8
Bahamas, sea-level rise 127
Bangladesh, sea-level rise 126, 127
Barents Sea Oscillation 35
barley, production 257, *258*, **259**, *260*, *261*, **262**
beaches, erosion, impact on tourism 149
Belize, sea-level rise 127
Benin, sea-level rise 127
biodiversity loss 104
biofuels
  and greenhouse gas emissions 74–5, 77
  mitigation 72
  offsets 239–41, *242*, 243, *244*, 245
biomass
  aerosols, and cloud properties 34, 38
  burning 18, 23–31, 38
  hydrogen production technology 201, 202, 203, 209, 213, *214*
  plantations 72–81
    and climate change 78–80
'boiled frog dilemma' 356
bottom-up modeling 171
Brainard model 295–6
  and climate policy 296–8
Brazil
  ethanol program 408
  rice productivity and climate change 317–18, *319*
burden sharing 320, 322–3
Bush Administration
  greenhouse gas policy 285, *286*, 294, 308
    Clear Skies Initiative 309

California, peak power demand, accounting model 173
Cameroon, sea-level rise 127
Canada
  carbon dioxide storage **182**, 190, *191–2*
  carbon price 286–9, **287**, 291, 292
Canadian Climate Centre Model 108, 110–16
  climate change Africa 162–6
  impact forecasts 108, 110–16
cap and trade programs 282, 285, 306, 328–32, 333
caps
  air pollution 96–101
  emission 219–21, 305, 320, 322–3
  temperature 395, *396*, *397*, *398*
capture, carbon dioxide 183, 184, 187
  see also carbon dioxide, capture and storage
carbon
  black
    aerosol effects on clouds 37–8, 96
    effect on temperature 18–31
    emission 26
  capture and storage *see* carbon dioxide, capture and storage
  global value 186, 193
  land-use emissions 67, 68–9, 252
  marginal value 186–7
  organic, soil 63, 64
  plantations 72–81, *230*
    and climate change 78–80, 229–32
    fast-growing timber 229
    tree types *76*
  rental rage 228, 237
  sequestration 72–81, 252–3, 404
    forests 227–35
      sensitivity analysis 229–35, *231*, *233*
    leakage 228, 404
    technology 340–1
  storage
    marginal cost, regional 187, *188–9*, 190, 193, *194*
    regional pattern, technology choice 193
  vegetation 63–4
    see also forests
  see also carbon dioxide
carbon cycle
  feedback 85
  oceanic 54
carbon dioxide 93
  atmospheric
    change 62, 64
    land-use experiments 77–8
  capture and storage 181–96, 206, 208–9
    capture 184, 185, **185**
    cost 183, 184, **184**, 187
    economic value 195
    models 183–5
    storage 184–5; geologic reservoirs 181–2, 183, 184, 208–9, 212

carbon (cont.)
  regional marginal cost 187, *188–9*, 190; regional pattern 190, *191–2*, 193, *194;* value-added 186, 190
  technology choice 193, 195
  vegetation 63–4
 concentration, targets 365, 368
 concentration stabilization 84–91, *87*, *88*, 183, 368
  carbon capture and storage 181–96, 206, 208–9
  cost **195**
  effect of plantations 77–8, 80
  overshoot pathways 84–91
  Wigley–Richels–Edmonds pathways 84–90, 367–8
 emissions
  global *321*
  hydrogen production technology 199, 200–1, *202*
  light duty vehicles 204–6, *207*
  reduction, Japan 323–6
  reference scenario 185–6, *186*
  United States of America 198, *199*
 fertilization 111
  feedback 65–6
 land–atmosphere flux 65, 69–70, 72, 75
 and net primary productivity 63–4
carbon flux
 land–atmosphere 65, 69–70, 72, 75
 land-use experiments 77–8
 simulation 62–3
carbon monoxide 93, 94
 emission control 97–100
carbon price 183, 219, *220*, 230, 283, **287**, 289, 335–8, **397**, *398*, *399*
carbon tax 222, 223–4, *224*, 324, *325*, **326**, 334–8, 395, *398*
cassava, production 257, *258*, **259**, *260*, **262**
CCSR climate model, impact forecasts 108, *111*, *112*, *113*, 114, *115*, *116*
CCSR/NIES model 315
Center for Sustainability and the Global Environment (SAGE), global land-use data 253–5, 259, **261**
cereals, production 257, *258*, *260*, **262**, *295*
CGCM2 model 315
CH_4 *see* methane
Charney report 1979 5, 8
China
 air pollution 408
 Australia-China bilateral agreement on climate 407, 411
 carbon storage **182**, 190, *191–2*
 effect of aerosols on precipitation 43–4
 energy consumption 408
 energy efficiency improvement 407
 greenhouse gas emissions 404, 411
CIMS model 174
circulation
 general, models 50–3
 thermohaline 345
  change 90
 see also North Atlantic meridional overturning circulation

Clean Development Mechanism 328
Clear Skies Initiative 309
climate
 tourism and recreation 147–54
  attitudinal studies 148
  behavioral studies 148–9
 variability
  modeling 55
  proxy data 55
climate change
 Africa 161–6
 agriculture 105
 and government 105
 human health impact 104, 135–8
 impact forecasts 107–17
 impact measurement 104
  cross-sectional studies 109
  experimental studies 109
 impact response functions 369
 integrated assessment models 104, 384
 policies
  cap and trade programs 328–32
  costs 282–92; national differences 287–92
  design under uncertainty 305–12
  drivers 407–9
  expectations: cost modeling 216–26; uncertainty 222–3
  framework 403–12
  horizon 216
  international 403–12; time-frame 405
  lessons from monetary policy 294–302
  path dependency 360–3
  taxes 282–3, 290–1
  uncertainty 334
 probabilistic estimation 49–59
  models 49, *50*, 51–9; parametric sensitivity 51; state-of-the-art 50–1
  observation-based 50, 55–6
  statistical 56–8
 and society 105
 tourism and recreation 105, 149–54
 uncertainty 49, 50, 52–3, **53**, 56, 58, 59, 355
 welfare-maximizing 309–10
climate forcing *see* forcing, climate
climate impact response functions 369
climate sensitivity 5–15, 391–4, *393*
 concept 5–15, 391–4, *393*
  future perspectives 14–15
  history 5–10
  recent development 10–14
 effective 6
 equilibrium 6, 10, 51
 estimation 8, *9*, 10, 52
  inverse 7, 11
 modeling 5–15, 49, 50
  feedback 7, 8, *9*, 10, 51

paleodata 7
probability density function 7, 8, 11
transient 6
uncertainty 10, 14, 298–301, 348–9
climate stabilization 365–74
Climate Stewardship Act 2003 219–21
see also McCain-Lieberman bill
climate system, stabilization 84
climate thresholds 343–52, 365
uncertainty 347–8
CLIO general circulation model 75–7
clouds
black carbon aerosols 37–8
convective
and carbonaceous aerosols 34, 38–41; regional impact 41
droplet number concentration 35–6, 38, *45*, 46
and photochemistry 95
CO_2 *see* carbon dioxide
co-operation, international 405–6, 409, 411
coal
in hydrogen production 200, *202*, 203, *207*, 213, *214*
industry, carbon sequestration technology 340–1
mining, methane emissions 271, 277, **279**
coasts
beach erosion, impact on tourism 149
flooding, direct cost estimate 124
impact forecasting 113–14, *115*
management, national level 128–9
population 122–3, *124*
relative sea-level rise 120
flooding 122–3, *124*, 129–30
wetlands
global-mean sea-level rise 123–4, *125*, 126–7, *126*;
direct cost estimate 124
computable general equilibrium model 127
Multi-Region National model 217–21
Social Accounting Matrix 138, 217, 218–19
concentration stabilization pathways, impact-driven 365, **366**, 367–73
Congo, sea-level rise 127
consortia, research and development 340
convection
and carbonaceous aerosols 34, 38–41
radiative-convective model 6
Convention on Long-range Transboundary Air Pollution 408
coral bleaching 369, 370
corridors, safe emissions 368, 369
cotton, production 257, *258*, **259**, *260*, *261*, **262**
cropland
abandonement *73*
biomass and carbon plantations 75
see also agriculture, land, abandonment
carbon flux 64, 65, 68, 69
global data 253–4

nitrous oxide emission 272–3
crops
distribution 254–5, *258*, 259, **259**, *260*, 261
harvested area 259, 261
management, fossil fuel conservation 239, *240*, *242*, *244*
production 255–7, *258*, **259**
productivity, Asia-Pacific region 315–19
yields, FAO data 262–4
cross-sectional studies, impact measurement 109
CSIRO global climate model 18, **22–3**, 23, 108, 110–16, 315
impact forecasts 108, *111*, *112*, *113*, 114, *115*, *116*
cumulus cloud 34

D-FARM model 228
damage valuation 358
decision making
adaptation 156–9
sequential 355, 356, 366
uncertainty 358–60
climate change policy 222
decision models 356–63
decomposition analysis 178
deforestation, carbon emissions 230, 234
Delft Hydraulics, dike building costs 124
deltas, global mean sea-level rise 123, 126
development
economic 158
Africa 163–4, **165**, 166
socio-economic, and malaria 136–7, 139
DICE model 230, 295, 298–9, **300**, **301**, 347, *348*, 349
dikes, coastal protection
direct cost estimate 124, 128
standards 122, 123
DINAS-COAST project 132
direct cost estimates, sea-level rise 124, 126–7
Disability Adjusted Life Years lost (DALYs) 137
disease
climate-sensitive 135
models 135–6
Global Burden of Disease study 137–8
health transitions 139–40, 142
and life expectancy 142
lifestyle-related 140, *141*, 142
drought, Africa 161

ECBilt-CLIO atmosphere-ocean-cryosphere model 75–7, 78
ECHAM4 model 315
economic analysis
global-mean sea-level rise 124, 126–8
adaptation analysis 127–8
direct cost estimates 124, 126–7
economy-wide impact estimate 127
economic efficiency 404–5, 407
economic potential, mitigation technologies 176–7, *176*
economy, effect of air pollution 99–100

ecosystems
    effect of air pollution control 99
    flooding risk 122
effectiveness, environmental 403–4
efficiency, economic 404–5, 407
EIA model 174
El Chichon volcanic eruption 22
El Niño Southern Oscillation 20, 22, 343
electricity
    competition for hydrogen technology 215
    as energy carrier 198–9
    generation 389–90
        new technology 410
    utility sector, partial equilibrium model 223
electrolysis, hydrogen production technology 201, 202, 203, 206, 209
emissions *see* greenhouse gas, emissions
Emissions Prediction and Policy Analysis model 138, 283–4, 356
    submodel 95
energy
    accounting models 172–4
    balance model 6, 49, 51, 52–4
    carbon-free 390, **390**, **391**
    carriers
        electricity 198–9
        molecular hydrogen 198
    demand and supply, bottom-up modeling 171–9
    efficiency 171
        barriers 171, 175–7; empirical evidence 177–9
    electric 389, **390**
    impact forecasting 112
    non-electric 389–90, **391**
    production, new technology 330–2
    renewable, hydrogen technology 201
    security issues 408–9
    services, technology 172, 181, 184, **185**
    subsidies, and reduced emissions 407
Energy Modeling Forum
    integrated assessment models 135, 238, 377
    NCGG mitigation 270–1
engineering optimization models 174
enhanced coal bed methane recovery 186
enhanced oil recovery 186
ensembles 51–2, 77
enteric fermentation, methane emission 273–4
environment, impact of carbon sequestration 235, 245–6
environmental effectiveness 403–4
equity, international 405–6
erosion, beach, impact on tourism 149
ethanol program, Brazil 408
Europe, carbon dioxide storage **182**, *191–2*
European Union
    carbon price 286–92, **287**, *289*, 291, 292
    Emissions Trading Scheme 305, 308
    Strategy on Climate Change 368
expectation, policy 219–22

changes in legislation 225
expected value of information 348–9, *350*
expected value problem 223, 224
experimental studies, impact measurement 109
exploratory modeling 369, **371**, 372
extra-tropics, albedo and land-use change 75, 81

Falciparum malaria *see Plasmodium falciparum* malaria
FAO, agro-ecological zones 256, 262–4
FARM computable general equilibrium model 127
FASOMGHG model 250–1
FCCC *see* United Nations Framework Convention on Climate Change
feedback
    albedo-snow 76–7
    carbon cycle 85
    climate and carbon dioxide fertilization 65–6
    climate sensitivity models 7, 8, *9*, 10, 51
fermentation, enteric, methane emission 273–4
fertilizer, nitrogen-based 273
Fiji, coastal protection 128
flood risk analysis
    national scale 129–30, 132
    sub-national scale 130, 132
flooding, coastal 122–3, *124*
floodplain
    coastal 129–30
    fluvial 129
    indicative 129
fluorinated gases, emission 266, *267*, 268–9, **270**, 275–7, 279, 397
forcing
    anthropogenic 343–52
    biomass burning 23–31
    black carbon 23–31
    climate 5, 54
    greenhouse gas 23–31, 54
    ozone 23
    radiative 5, 6, 7, 10, *13*, 14, 52–5, 57, 358, 391, 397, *400*
        anthropogenic aerosol 7, 8, 11, 23
    relaxed 25
    solar 12, 23, 55
    sulfate aerosol 23–31
    volcanic 23, 25
foresight, perfect 217–19, 221
forest management, offsets 239, *240*, *242*, *244*
forestry
    dynamic models 228
    greenhouse gas emissions 238–47
        FASOMGHG model 250–1
    mitigation 239–47
    impact forecasting 113
forests
    albedo 78
    boreal 230, 231, 232
    carbon sequestration 227–35

Index

sensitivity analysis 229–35, *231*, *233*
  global model 236–7
Former Soviet Union, carbon storage **182**, 190, *191–2*
fossil fuel
  air pollution 407–8
  black carbon 18–31
  emissions 26–7, 77, 181
  hydrogen production technology 200, 202, 203
Framework for Uncertainty, Negotiation and Distribution (FUND) model 138–9
free radicals 93, 94–5, 98
fuel-cells, hydrogen, vehicle technology 204, 207–8

gas *see* natural gas
gasoline
  competition for hydrogen technology 215
  dual fueling 211
  tax 209–10, *210*, *211*
GCOMAP model 228
general circulation models 50–3
geographic information system 136
Germany, wheat productivity and climate change 317–18
GFDL-R30 model 315
GIEPIE model 174
Global Analysis, Integration and Modeling 384
Global Burden of Disease study 137–8
global forestry model 234, 236–7
Global Impact Model 162
global mean sea-level rise 119–32
  assessment
    national 128–30, 132
    regional to global 121–8
    sub-national 130, 132
  methane and nitrous oxide reduction 88–90, *89*
global mean temperature 7
  land-use experiments 78, *79*, *80*
  methane and nitrous oxide reduction 88–90, *89*
Global Trade Analysis Project (GTAP) 253–64, 283
Global Vulnerability Assessment 121, 122, 124, 127
global warming potential 267
Goddard Institute for Space Studies ModelE 35
government, and climate change 105
grants, research and development 339
greenhouse effect, work of Arrhenius 6
greenhouse gas 12, 94
  concentration stabilization 84–91, 181–96, 387, 392, 397–8
    Asia-Pacific Region 318, 319–23
    Time-frame 405
  emissions 108
    agriculture and forestry studies 238–47; mitigation 239–47; environmental co-effects 245–6; models 248–51; offsets 239–45
    Asia-Pacific region 314, 318, 319–23
    and biofuels 74–81
    bottom-up models 171–9

mitigation potential 175–7
  caps 219–21, 305, 320, 322–3, 333–4
    cap and trade programs 328–32, 333
  control, health benefits 137, 138
  integrated assessment models 384
  international policy 404–12
  mitigation
    benefits 308, 309
    GTAP 253
  price-based approach 309
  quotas, trading 282, 283, 285, 289–90, 291–2, 305–9
  reduction target, Japan 323–6
  research and development 330, 331, 332–41
  safe emissions corridors 368, 369
  safe landing analysis 368
  safety-valve mechanism 309–10, 334
  see also Intergovernmental Panel on Climate Control, Special Report on Emission Scenarios
forcing 23, 54, 397
non-carbon dioxide, mitigation
  analysis 266–79
  early work 269–70
  new directions 277–9
  recent work 270–7
Greenland Ice Sheet 119, 346, 370
groundnut production 257, *258*, **259**, *260*, **262**
growing period, length 255, *256*, **256**, 257, *258*, *260*
growth, economic
  effect of malaria 139
  and public health 140

Hadley Meteorological Centre model 108, 110–16, *117*
  climate change Africa 162–6
  crop productivity Asia-Pacific region 315
  impact forecasts 108, 110–16, *117*
  temperature change and black carbon 38
halocarbon concentration, Antarctica 20
Hamburg Tourism Model 151–2
haze, Arctic 35
health
  and climate change 104
    economic cost 138–9
    Global Burden of Disease study 137–8
    models 135–44
  greenhouse gas emission control 138
  public
    and economic growth 140
    twentieth century 140
  transitions 139–40, 142
HIV/AIDS, Africa 140
'hot air', Russian 282, 283, 285
hybrid models, bottom-up 174
hybrid vehicles 204
  competition for hydrogen technology 215
hydrofluorocarbons, emissions 268–9, **270**, 275–7, 279

hydrogen
    as energy carrier 198
    light duty vehicles 198–215
        barriers 207–15; competitive technologies 215; fueling costs 209–11; resource limitation 211–13; safety 213–14; vehicle technology 207–8
        carbon dioxide emissions 204–6, *207*
        dual fueling 211
        fuel cost 208
        fuel use 204, *206*, 208
        gasoline tax 209–10, *210*, *211*
        sales 204, *205*
        scenarios 204–6
        subsidization 209–10, *210*, 211, *211*
    production technology
        carbon dioxide release 200–1, *202*
        cost 201–3, *202*, 209–10
    supply technology, barriers 208–9
hydroxyl free radicals
    photochemistry 93, 94–5
    and pollution control 98
hysteresis 361–3, *361*

ice *see* Greenland Ice Sheet; West Antarctic Ice Sheet
IMAGE integrated assessment model 73, 74–5, 77, 79, 80
    carbon sequestration 228
    climate stabilization 368
impact, market 109–10, 114–16, *117*
impact distribution 104
impact forecasting 107–17
    agriculture 110–12, 116
    coasts 113–14, *115*
    energy 112
    forestry 113
    model 109–10
    regional variables 110, 111, 116
    sea-level 113–14
    timber 113
    water 113, *114*
impact measurement 104
    cross-sectional studies 109
    experimental studies 109
IMPLAN SAM 217
incentive
    cap and trade programs 329–30, 331, 333
    carbon tax 335–8
    tax 339
inconsistency, dynamic 330, 335–8
India
    air pollution 408
    carbon storage **182**, 190, *191–2*
    crop productivity, and climate change 316–19
    effect of aerosols on precipitation 43–4
    life expectancy 143

Indian Ocean Experiment 43
induced technological change 330–2, 333, 335–8
innovation 329–33, 335–8
integrated Assessment of Climate Protection Strategies (ICLIPS) 365, 369, 372
integrated assessment models 104–5, 135, 168, 238, 347
    Article 2, UNFCCC 365, 367–74
        evaluation 370–3, **371**
    diagnosis and intercomparison 384
    Energy Modeling Forum 135, 238, 377
    future development 383–6
    land-use 252–64
    probabilistic 370, **371**, 372
    state of the art 381–3
    structure and function 378–81
    uncertainty 380–1, 384
Integrated Global System Model (IGSM) 95–6, *96*, 356
Integrated Science Assessment model 62, *63*, 65, *66*, **67**
intellectual property 411
intensity targets 310–11
interest rate 110
Intergovernmental Panel on Climate Control (IPCC)
    Assessment Reports 5
        Fourth 156
        Second 10, 366
        Third 84, 85, 107, 366, 370;
            adaptation 156, 157, 159;
            climate sensitivity 6, 10
    Special Report on Emission Scenarios 73, 80, 85, **86**, 366
        crop productivity 316
        forests 227–8
        integrated assessment models 382
        life expectancy 143
        population 136
        socio-economic 121, 123
    Working Group I 365
interhemispheric temperature difference 7, 11, *13*, *14*, *22*, 55
International Geosphere-Biosphere Programme 254, 384
Intertropical Convergence Zone 38
investment, international 407, 410–11
IPCC *see* Intergovernmental Panel on Climate Control
islands, global mean sea-level rise 123, 126–7, 128

Japan
    carbon dioxide reduction 323–6
    carbon price 286–92, **287**, *289*
    carbon storage **182**, 190, *191–2*, 193
        cost 193, *194*
    Kyoto Protocol costs 283, 285
    New Climate Change Policy Programme 2002 323

Kaya identity 172
Kiribati, coastal protection 128
Korea, carbon storage **182**, 190, *191–2*, 193

Kuwait, oil fire cloud formation 44
Kyoto Protocol
   carbon sequestration 252
   cost 282–92
      Annex B countries 282–92, 389
   criticism 403, 406–7, 411, 412
   greenhouse gas emission targets 174, 216
      Japan 285, 323–6
   non-carbon dioxide greenhouse gas 282
   policy framework 406
   United States of America 198, 282, 285, *286*, 404

land endowment 254–5
land rent 254
land supply elasticity 229, 230, 231, 232–3, 234
land-use
   carbon emissions **67**, 68–9, 252
      GTAP 253
   change
      and climate 72–81, 253; experiments 75
      forestry and agriculture 228, 229, 231, 234
   global data 253–5
   integrated assessment modeling 252–64
landfill, methane emissions 271, 277, *278*, **279**
last glacial maximum 58
Latin America, carbon dioxide storage **182**, *191–2*
leakage, carbon sequestration 228, 234, 404
'learning by doing' 332, 339, 389, 390, **390**, **391**
legislation, changes, and policy expectation 225
life
   loss of 104
   quality of 104
life expectancy, as indicator of population health 142–3
lifestyle, health risks 140, *141*, 142
livestock, methane emissions 273–4
Lucas critique 216

Macau, sea-level rise 126
McCain-Lieberman bill 219–21
macroeconomics 174
MAGICC model 86, 391, 401
maize, production 257, *258*, **259**, *260*, *261*, **262**
malaria
   and economic growth 139
   mortality 139
   risk estimates 137–8
   and socioeconomic development 136–7
   transmission, models 136
Malaysia, sea-level rise 127
Maldives, sea-level rise 126, 127
mangroves, global-mean sea-level rise 124
manure, methane emission 274
MARA/ARMA model 136, 137
MARKAL model 174, 178
market barriers

   energy efficiency 171, 175–7
      empirical evidence 177–9
market impact 109–10, 114–16, *117*
market potential, mitigation technologies 175–7, 176
Marshall Islands, coastal protection 128
mercury, emissions reduction 309
MERGE model 84, 85, 388–9, 390, 391, **393**, 394, 401
meridional overturning circulation *see* North Atlantic meridional overturning circulation
methane
   emissions 267, *268*, **268**, **269**, *272*
      agriculture 252, 259, 272–5; enteric fermentation 273–4; management 239, *240*, *242*, *244*; manure 274; rice production 259, 274
      energy and waste sectors 271, 277–8, **279**
      mitigation 271, 272, 273–5
      oil sector 271–2
      reduction 84–91, 100; agricultural management 239, *240*, *242*, *244*
   photochemistry 94–5
   and pollution control 98–9
   temperature ceilings 397
metrics, impact measurement 104
Mexico, methane emissions, landfill 277, *278*
Micronesia
   coastal protection 128
   sea-level rise 126
Microwave Sounding Unit, satellite temperature data 18–20, 22, **22–3**
Middle East, carbon dioxide storage **182**, *191–2*
millet, production 257, *258*, **259**, *260*, *261*, **262**
MiniCAM model 181, 182, 183
ministries, decision-intensive 159
mitigation 119
   agriculture and forestry 239–47, 248–51
      environmental co-effects 235, 245–6
      models 248–51
   analysis
      non-carbon dioxide greenhouse gas 266–79; marginal abatement 271, 275–8, 288–91, *289*
   benefits 308, 309
   biofuels 72–81
   carbon plantations 72–81
   cost, and agro-ecological zones 257–9
   lessons from monetary policy 294–302
   potential 175–7, *176*
   sea-level rise 122–3, 124, 132
   time-frame 405
   uncertainty 294, 299, 308
modeling
   bottom-up 171–9
   climate change cost
      policy expectations 216–26; perfect foresight 217–19, 221
   exploratory 369, **371**
   top-down 171, 172

models 108–9
　accounting, energy 172–4
　adaptation 127–8
　adjoint method 11, 15
　agriculture and forestry 248–51
　atmosphere-ocean general circulation 6–8, 10–11, 50–3, 108–9, 315
　carbon capture and storage 183–5
　climate policy decision 355–63
　climate sensitivity
　　feedback 7, 8, *9*, 10, 51
　　history 5–10
　　recent developments 10–14
　computable general equilibrium 127
　　Social Accounting Matrix 138, 217, 218–19
　disease 136, 143–4
　energy accounting 172–4
　energy balance 6
　engineering optimization 174
　ensemble 51–2, 77
　forestry 228, 234, 236–7
　global, state-of-the-art 50–1, 53
　health 136, 143–4
　hybrid 174
　impact 109–10
　　cross-sectional 109
　　experimental 109
　integrated assessment 104–5, 135, 168, 238, 347, 377–86
　　diagnosis and intercomparison 384
　　future development 383–6
　　state of the art 381–3
　　structure and function 378–81
　　uncertainty 380–1
　　UNFCCC Article 2 365, 367–74; evaluation 370–3, **371**
　neural network method 15
　ocean general circulation 6
　parametric sensitivity 51
　partial equilibrium 174
　radiative-convective 6
　reduced form 379, 380
　simplified climate 7
　temperature trends, and black carbon aerosols 20, 23–31
monetary loss 104
monetary policy, United States of America, lessons for mitigation 294–302
morbidity, Global Burden of Disease study 137–8
morality
　childhood 139
　Global Burden of Disease study 137–8
　twentieth century 139
Multi-Region National model 217–21
Multi-Sector Multi-Region Trade model 218

National Flood and Coastal Defence Database 129
Natural Emissions Model 95–6

natural gas
　in hydrogen production 200, *202*, 203, *207*, 212
　methane emission 271, 278, **279**
　supply vulnerability 212
NEMS model 174
net ecosystem production *98*, 99
net primary productivity
　and carbon dioxide 63–4
　and ozone concentration *98*, 99
neural network method 15
New Caledonia, coastal protection 128
New Zealand, coastal protection 128
nitric acid production 271–2
nitrogen oxides, emission
　control 97–101
　OTC program *307*, 308, 309
　and ozone concentration 93, 94
　RECLAIM trading program 305, 306, 307–8, *307*, 309
nitrous oxide, emissions 267, *268*, **268**, **269**
　acid production 271–2
　agriculture 252, 259, 272–5
　cap and trade programs 329
　mitigation 272, 273–5
　reduction 84–91
　　agricultural management 239, *240*, *242*, *244*, 273
NO$_x$ *see* nitrogen oxides
N$_2$O *see* nitrous oxide
non-carbon dioxide greenhouse gas
　Kyoto Protocol 282
　mitigation
　　analysis 266–79; marginal abatement 271, 275–8, 288–91, *289*
　　early work 269–70
　　new directions 277–9
　　recent work 270–7
　see also fluorinated gases; methane; nitrous oxide
North Atlantic meridional overturning
　circulation 343, 344–7, 369, 370
　climate threshold 348–52
　observation systems 349–52
northern hemisphere, ozone 98
nuclear power, hydrogen production technology 201, 203

obesity 140, *141*, 142
oceans
　atmosphere-ocean model 6–8, 10–11
　carbon cycle 54
　general circulation model 6
　heat uptake 54, 358, 391
　temperature change 8
offsets, greenhouse gas emissions, agriculture and forestry studies 239–45
oil
　demand, developing countries, accounting model 173–4
　supply vulnerability 198, 212
oilpalm, production 257, *258*, **259**, *260*, **262**

Index

optimization, stochastic 222–3, 224
organic compounds, volatile
   emissions control 97–100
   photochemistry 93, 94–5
overshoot pathways, carbon dioxide 84–91
ozone
   dissociation 94
   forcing 23, 54
   and net primary production *98*, 99
   and pollution control 98
   stratospheric 19, 20, 25, 93, 94
   tropospheric 12, 23, 25, 54, 93, 95
   and VOC emission 93
Ozone Transport Commission *307*, 308

Pakistan, sea-level rise 127
Palau, coastal protection 128
paleoclimate, reconstruction, proxy data 7, 51
Papua New Guinea, sea-level rise 127
Parallel Climate Model 18, 20, **22–3**, 23, 108–9, 110–16, *117*
   climate change Africa 162–6
   impact forecasts 108–9, 110–16, *117*
parameterization, model 51, 380
partial equilibrium models 174
   electricity utility sector 223
pastureland
   abandonment *73*,
   see also agriculture, land, abandonment
   carbon flux 64, 65, 69
patent rights 339
path dependency 360–3
perfluorocarbons, emissions 268–9, **270**, 275–7
Perturbed Physics Ensemble Method 11
photochemistry
   hydroxyl free radical 93, *94*
   troposphere, and air pollution 94–5
photosynthesis 63, 64
photovoltaics, hydrogen technology 201, 203, 209
Physiological Equivalent Temperature 148
Pinatubo volcanic aerosols 20
plantations
   biomass 72–81
      and climate change 78–80
      impact on carbon dioxide emissions 77–8
   carbon 72–81
      and climate change 78–80
      impact on carbon dioxide emissions 77–8
   fast-growing timber 229
Plasmodium falciparum malaria, models 136
policy
   analysis 222–3
   climate *see* climate change, policies
   hedging 298–9, 300–1
   international, greenhouse gas emissions 404–12
   monetary, United States of America, lessons for mitigation 294–302
pollution
   air 407–8
      control 93–101; caps 96–101; effect on economy 99–100; effect on ecosystem 99; effect on sea-level 100; effect on temperature 100
   effect on climate 93–101
   effect on human health 138
   trading programs 305
Pooled Travel Cost Model 149
population
   coastal flooding risk 122–3, *124*
   growth, and malaria risk 136
   health model 142–3
potatoes, production 257, *258*, **259**, *260*, *261*, **262**
potential, mitigation 175–7, *176*
precipitation
   Africa 161, 162
   climate change models *108*
   effect of aerosols 34, *42*, 43, *43*, 44
productivity, primary 63–4, *98*, 99
   labor 357–8
Program for Climate Model Diagnosis and Intercomparison 383–4
protection, coastal 124, 126–8
proxy data, paleoclimate 7
pulses, production 257, *258*, **259**, *260*, *261*, **262**

quasi-rent 330, 333, 334, 335–6

radiation
   solar 10–11
   top atmosphere 10
radiative forcing 5, 7, 10, 11–12, *13*, 14, 52–5, 57, 358, 391, 397, *400*
   anthropogenic aerosol 7, 8, 11, 23
   solar 7, 12, 23
   volcanic 7, 23
radiative-convective model 6
rapeseed, production 257, *258*, **259**, *260*, *261*, **262**
recreation 147–54
   impact of climate change 149–54
   importance of weather and climate 147–9
rent
   carbon 228, 237
   land 254
   see also quasi-rent
research and development 330, 331, 332–41, 389, 395, 397, 410
   consortia 340
reservoirs, geologic
   carbon dioxide storage 181–2, 183, 184, 208–9, 212
   availability *188–9*, 190, 193
   regional marginal cost 187, *188–9*, 190
   value-added 186, 190

rice
    production 257, *258*, **259**, *260*, *261*, **262**
        methane emission 259, 274
    productivity, Asia-Pacific region 316–19, *318*
risk management 294–5, 299
rivers, flooding 129
Russian 'hot air' 282, 283, 285
rye, production 257, *258*, **259**, *260*, *261*, **262**

safe emissions corridors 368, 369
safe landing analysis 368, 370–3, **371**, 372, 373
safety-valve mechanism, greenhouse gas emissions 309–10, 334
SAGE, global land-use data 253–5, 259, **261**
Sahara, agriculture and climate change 163
saltmarshes, global-mean sea-level rise 124
scenario analysis 222–3, 224
sea-level
    climate change models 108–9
    effect of air pollution control 100
    global mean rise 119–32
        coastal flood risk 122–3, *124*
        coastal wetlands 123–4, *125*, 126–7, *126*
        economic analysis 124, 126–8
        methane and nitrous oxide reduction 88–90, *89*
        impact forecasting 113–14
        relative rise 120–1
            socio-economic impacts 120–1, **121**
        stabilization 90
semi-direct effect, and aerosols 44–5
Senegal, sea-level rise 127
sensitivity
    climate *see* climate sensitivity
    parametric 51
sensitivity analysis, forest carbon sequestration 229–35
    carbon 72–81, 252–3, 404
        forests 227–35
        technology 340–1
    soil 239–41, *242*, 243, *244*
severe acute respiratory syndrome (SARS) 140
simplified climate model 7
ski resorts, impact of climate change 149, 152
smoke, biomass burning 26
SO_2 *see* sulfur dioxide
Social Accounting Matrix 138, 217, 218–19
society, and climate change 105
socio-cultural potential, mitigation technologies 176–7, *176*
soil, organic carbon 63, 64
soil sequestration 239–41, *242*, 243, *244*
sorghum, production 257, *258*, **259**, *260*, *261*, **262**
southern hemisphere, ozone 98
soybean, production 257, *258*, **259**, *260*, *261*, **262**
species loss 104
stabilization paths 367–8, 369–70

storage
    carbon dioxide 181–2, 183, 184–5, 208–9, 212
        marginal cost 187, *188–9*, 190
        regional pattern 190, *191–2*, 193
            technology choice 193
        reservoirs 181–2, 212
stratocumulus cloud 34
stratosphere, ozone 19, 20, 25, 93, 94
subsidies, energy sector, and reduced emissions 407
substitution, global, carbon sequestration 228
sugarbeet, production 257, *258*, **259**, *260*, **262**
sugarcane, production 257, *258*, **259**, *260*, *261*, **262**
sulfate
    aerosol, and pollution control 98
    anthropogenic 7, 12, 23–31
    radiative forcing 57
sulfur dioxide 93, *94*
    emissions
        control 97–9
        trading programs 305, 306–7, *306*, 308, 309, 329
sulfur hexafluoride, emissions 268–9, **270**, 275–7
sunflower seed, production 257, *258*, **259**, *260*, *261*, **262**
Sweden, coastal protection 128

Tangent and Adjoint Model Compiler 11, 15
tax incentives 339
taxes
    carbon 222, 223–4, *224*, 324, *325*, **326**, 334–8, 395, *398*
    and climate change policy 282–3, 290–1
    gasoline 209–10, *210*, *211*
    Multi-Region National model 219
technical potential, mitigation technologies 175–7, *176*
technology
    backstop 329, 332, 336–8
    change
        cost 177
        electricity generation 410
        induced 330–2, *333*, 335–8
        'learning by doing' 332, 339, 389, 390, **390**
        patent rights 339
        prize competition 339–40
        research and development 330, 331, 332–41, 389, 395, 397, 410;
            consortia 340
    energy and greenhouse gas emission 171–9
        mitigation scenarios 172
    energy services 172, 181, 184, **185**
    new
        and emission reduction 404, 409–10; impediments 410–11
    regional carbon dioxide storage 193
temperature
    2LT 20
    Africa 162
    caps 395, *396*, **397**, *398*
    ceilings 394–5, 397

climate change models *108*
  in climate sensitivity estimates 7
  effect of air pollution 100
  effect of black carbon 18–31
  global mean 7, 55
    change *321*
    and land-use change 78–80
    methane and nitrous oxide reduction 88–90, *89*
    stabilization 391–401
  impact 109, 111–13
    tourism and recreation 148–54
  interhemispheric difference 7, 11, *13*, *14*, 22, 55
  near-surface 5, 6, 10
  paleodata 7
  stabilization 387
  surface 20–3, 56
  tourism and recreation 148–54
  troposphere 56
    mid, effect of black carbon 18–31
Terrestrial Ecosystems Model 95
Thailand, rice productivity and climate change 317–18, *319*
thermohaline circulation 90, 345
  see also North Atlantic meridional overturning circulation
timber
  global market 228, 229
    model 236–7
      sensitivity analysis 229–35
  impact forecasting 113
TIMER model 74
  climate stabilization 368
Tolerable Windows Approach 369, 370–3, **371**
top-down modeling 171, 172
tourism 147–54
  attractiveness of destination 149, 151, 152
  and climate change 105
  impact of climate change 149–54
    demand 151
    global flow 151–2, *153*
  importance of weather and climate 147–9
trade
  international 407
  MS-MRT model 218
trading
  emission quotas 282, 283, 285, 289–90, 291–2, 305–9, 333–4
    cap and trade programs 282, 285, 306, 328–32, 333
transaction costs 177
tropics, land-use change 81
troposphere
  mid, effect of black carbon 18–31
  photochemistry 94–5
  temperature 56
Tuvalu, coastal protection 128

ultraviolet light, photochemistry 94

uncertainty 158
  climate change 49, 50, 355
    mitigation 294, 299, 308
    policy design 305–12
    policy expectations 222–3, 334, 367
    probabilistic estimation 49, 52–3, **53**, 56, 58, 59
  climate sensitivity 10, 14, 298–301, 348–9
  climate thresholds 347–8
  decision making 358–60
  decision target 367, 372
  implementation 367, 372
  integrated assessment models 380–1, 384
  modeling 367, 372
  scientific 367, 372
  temperature stabilization 391
  UNFCCC Article 2 365, 366–7
United Kingdom
  coastal management, and sea-level rise 129
  flood risk analysis 129–30, **131**, 132
United Nations Framework Convention on Climate Change
  Article 2 84, 365–74
    uncertainty 365, 366–7
  Conference of the Parties 156, 159
  'dangerous anthropogenic interference' 308, 347, 365, 366, 369, 370, 372, 373
  integrated assessment models 367–74
    evaluation 370–3, **371**
  see also Kyoto Protocol
United States of America
  air pollution, economic and health impacts 138
  bilateral agreement, China 411
  Bush Administration
    greenhouse gas policy 285, *286*, 294, 308; Clear Skies Initiative 309
  carbon dioxide emissions 198, *199*
  carbon price 286–92, **287**, 291, 292
  carbon storage **182**, 190, 191–2, 193; cost 193
  Clean Air Act 308
  greenhouse gas emissions 404, 408, 411
  Kyoto Protocol target 285, *286*
  molecular hydrogen, light duty vehicles 198–215
  monetary policy, lessons for mitigation 294–302
Urban Air Pollution submodel 95

vegetation
  carbon storage 63–4
  natural 64
vehicles, light-duty, hydrogen powered 204–6, *205*
Vietnam, sea-level rise 126
volatile organic compounds
  emissions, control 97–100
  photochemistry 93, 94–5
  and pollution control 98

water, impact forecasting 113, *114*
weather
   tourism and recreation 147–9
      attitudinal studies 148
      behavioral studies 148–9
welfare, Multi-Region National model 219–21
welfare-maximizing 309–10
West Antarctic Ice Sheet 119, 343, 345–7, 369, 370
   climate threshold 347–8
West Nile virus 140

wetlands, global-mean sea-level rise 123–4, *125*, 126–7, *126*
wheat
   production 257, *258*, **259**, *260*, *261*, **262**
   productivity, Asia-Pacific region 316–19, *318*
Wigley–Richels–Edmonds pathways 84–90, 368, 370
wind patterns, and aerosols 35
wind turbines, hydrogen technology 201, 203, 209
winter sports, impact of climate change 149, 152

Younger Dryas, climate change 344